# Ciliate Atlas: Species Found in the South China Sea

Xiaozhong Hu • Xiaofeng Lin • Weibo Song
Editors

# Ciliate Atlas: Species Found in the South China Sea

Science Press
Beijing

Springer

*Editors*
Xiaozhong Hu
Fisheries College
Ocean University of China
Qingdao, Shandong, China

Xiaofeng Lin
College of Life Science
South China Normal University
Guangzhou, Guangdong, China

Weibo Song
Institute of Evolution & Marine Biodiversity
Ocean University of China
Qingdao, Shandong, China

ISBN 978-981-13-5900-2 ISBN 978-981-13-5901-9 (eBook)
https://doi.org/10.1007/978-981-13-5901-9



This Springer imprint is published by the registered company Springer Nature Singapore Pte Ltd.
The registered company address is: 152 Beach Road, #21-01/04 Gateway East, Singapore 189721, Singapore

*We dedicate this book to our close friend and colleague,*
***Dr. Alan Warren***
*The Natural History Museum, London, UK,*
*who has contributed tremendously to the ciliatology and*
*given great help and kindness to the present authors*

# Preface

The South China Sea, located in the western Pacific Ocean, is one of China's three marginal seas (Fig. 1). It extends from the Tropic of Cancer at its northernmost point (23°26′N), crosses the equator, and reaches a latitude of 3°58′N at its southernmost point. It interlinks the Pacific and Indian oceans. Its geographical area is about 3.5 million square kilometers with an average water depth of 1212 m.

The South China Sea has a remarkable tropical marine climate with a long summer and short spring and autumn and is without snow and ice in winter. The air is moist and rainfall is abundant (the average annual precipitation is over 1200 mm). The average surface water temperature is 25-28°C. The annual salinity of surface seawater fluctuates between 32‰ and 34.5‰, but that of coastal water is slightly lower owing to land runoff.

The South China Sea possesses rich biological resources. According to the *Report on the State of the Environment in China 2016*, over 600 species of zooplankton and >1200 species of benthic organisms had been recorded in the South China Sea. Basic information on the diversity of microorganisms, especially protozoans in coastal environments, however, is extremely scarce.

Ciliated protozoa are one of important components in marine microbial food webs and play vital roles in the flux of materials and energy of the aquatic ecosystem. In a first attempt to determine the ciliate diversity of the South China Sea, the authors made an extensive survey of various biotopes along its northern coast. This work was carried out during a series of projects funded by the National Natural Science Foundation of China. Water and sediment samples were collected from multiple habitats including natural open waters, marine aquaculture waters, estuaries, intertidal zones, and mangrove wetlands. Based on the application of modern ciliate taxonomic methods (silver staining, rDNA sequencing), a total of 312 species have been identified belonging to 11 classes, 30 orders, 67 families, and 136 genera (Fig. 2). This includes 105 new species, 18 new genera, 4 new families, and a large number of new records for China. In addition to morphological taxonomy, these studies included investigations of temporal and spatial patterns of distribution, the evolution of community structure and its relationships with environmental factors, genomic DNA bank construction, molecular systematics, and morphogenesis. This work results in 124 articles in the international peer-reviewed journals (please find details in relevant references of this book).

The biodiversity of ciliates is extremely high due to their specialized cell structures which enable them to occupy a wide range of ecological niches. Studies of ciliate diversity are, however, constrained by a lack of modern monographs and guides for the identification of tropical marine ciliates. Therefore, an image-rich guide or atlas will undoubtedly be of great benefit for those who are not trained in ciliate taxonomy but who encounter ciliates in disciplines such as ecology, environmental biology, aquaculture, and other related fields. This atlas aims to fulfil this need by providing a "one-stop" reference source of basic data on tropical marine ciliates that should be used for researchers worldwide. It also provides maps marking the localities and images showing the habitats of each species. Included in the atlas are detailed illustrations and habitat data for every species. A list of key references is also provided for those who require more detailed information on the morphology, ecology, and systematics of any given species.

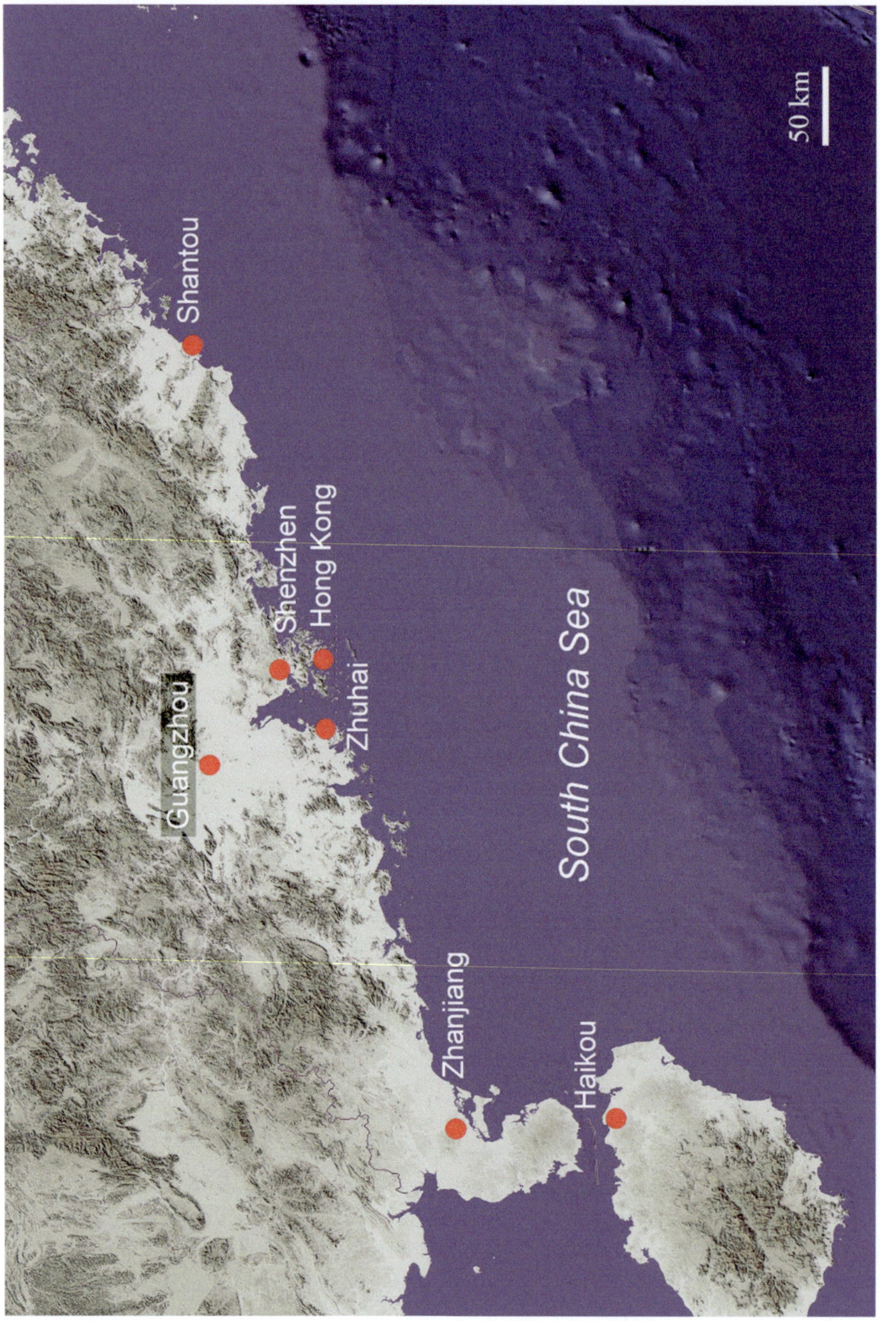


**Fig. 1** Sampling map showing several cities (red dots) along the coast of the South China Sea

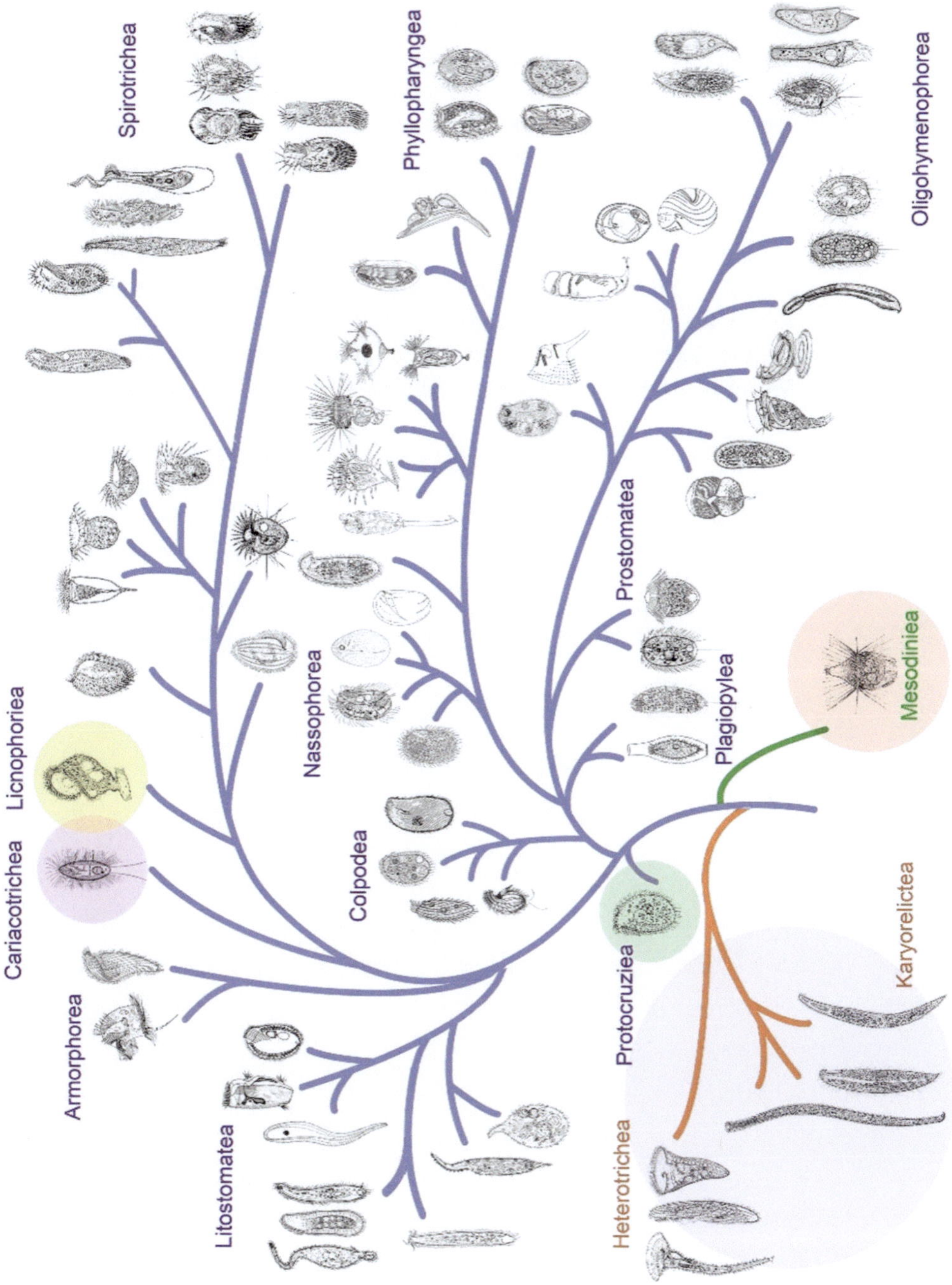


**Fig. 2** Hypothetical evolution of ciliated protozoa. (Modified from Gao et al. 2016, Scientific Reports, 6: 24874, https://www.nature.com/articles/srep24874)

The majority of data in the present atlas comes from original research carried out by graduates and ex-graduates from authors' groups. The names of contributors are listed as follows (the number of species illustrations supplied by each is in parentheses): Ms. Wenping Chen (27), Dr. Xiangrui Chen (5), Dr. Xumiao Chen (13), Xinpeng Fan (10), Dr. Yangbo Fan (3), Ms. Tiantian Gong (24), Ms. Xueying Han (24), Mr. Chunxu Hu (11), Dr. Jiamei Jiang (15), Dr. Mingjian Liu (8), Dr. Weiwei Liu (52), Dr. Xiaotian Luo (32), Dr. Hongbo Pan (20), Dr. Xuming Pan (10), Dr. Zhishuai Qu (33), Dr. Zhuo Shen (22), Dr. Wen Song (19), Dr. Lei Wu (45), Dr. Yuan Xu (5), and Dr. Ying Yan (2). We here acknowledge their hard work and professional dedication that enabled this volume to be completed with high quality and in a timely fashion.

The findings presented here are a result of a series of projects supported by the National Natural Science Foundation of China, Qingdao National Laboratory for Marine Science and Technology "Aoshan Science and Technology Innovation Program," and the Fundamental Research Funds for the Central Universities and several international collaboration projects including the Darwin Initiative CHN-UK (14-015), CHN-USA (31111120437), and CHN-AT (31761133001). Special thanks are due to our longtime partner, Dr. Alan Warren of the Natural History Museum, London, UK, for his participation in several collaborative studies which form the basis for much of the material included in this volume and his selfless help in improving the English text. We also thank our colleagues, Khaled A. S. Al-Rasheid and Saleh A. Al-Farraj (Saudi Arabia).

Qingdao, Shandong, China — Xiaozhong Hu
Guangzhou, Guangdong, China — Xiaofeng Lin
Qingdao, Shandong, China — Weibo Song
May 10, 2018

# Acknowledgments

Funded by Qingdao National Laboratory for Marine Science and Technology "Aoshan Science and Technology Innovation Program," the National Natural Science Foundation of China (project numbers: 31430077, 41576148, 41376141, 31222050, 41176119, 30870280, 40506033), and the Fundamental Research Funds for the Central Universities (201762017).

# Brief Introduction

This atlas will assemble diagrams and photomicrographs of 312 free-living species of ciliate that have been found in the coastal area of the South China Sea. To be reader-friendly, habitats, ecological features, and key references as well as systematics will be provided for every species.

This book fits for researchers, educators, students, environmental officers, conservationists, and amateurs with interests in ciliate biodiversity and/or biogeography.

# Contents

# Contributors

**Wenping Chen** South China Normal University, Guangzhou

**Xiangrui Chen** Ningbo University, Ningbo

**Xumiao Chen** Institute of Oceanology, Chinese Academy of Sciences, Qingdao

**Xinpeng Fan** East China Normal University, Shanghai

**Tiantian Gong** South China Normal University, Guangzhou

**Xueying Han** South China Normal University, Guangzhou

**Chunxu Hu** South China Normal University, Guangzhou

**Jiamei Jiang** Shanghai Ocean University, Shanghai

**Mingjian Liu** Ocean University of China, Qingdao

**Weiwei Liu** South China Sea Institute of Oceanology, Guangzhou

**Xiaotian Luo** Ocean University of China, Qingdao

**Hongbo Pan** Shanghai Ocean University, Shanghai

**Xuming Pan** Harbin Normal University, Harbin

**Zhishuai Qu** Ocean University of China, Qingdao

**Zhuo Shen** Sun Yat-sen University, Guangzhou

**Wen Song** Ocean University of China, Qingdao

**Lei Wu** South China Normal University, Guangzhou

**Yuan Xu** East China Normal University, Shanghai

**Ying Yan** Ocean University of China, Qingdao

**Yangbo Fan** Harbin Institute of Technology Shenzhen Graduate School, Shenzhen

# Class Mesodiniea

Xiaozhong Hu, Xiaofeng Lin, and Weibo Song

Class Mesodiniea Chen et al., 2015
Order Mesodiniida Grain, 1994
Family Mesodiniidae Jankowski, 1980
*Askenasia* cf. *stellaris*

**Ecological feature:** mangrove
**Water temperature:** 19.7 °C
**pH:** 7.8
**Salinity:** 23.9‰

X. Hu (✉)
Fisheries College, Ocean University of China,
Qingdao, Shandong, China
e-mail: xiaozhonghu@ouc.edu.cn

X. Lin
College of Life Science, South China Normal University,
Guangzhou, Guangdong, China
e-mail: xlin@oucscnu.edu.cn

W. Song
Institute of Evolution & Marine Biodiversity,
Ocean University of China, Qingdao, Shandong, China
e-mail: wsong@ouc.edu.cn


X. Hu et al. (eds.), *Ciliate Atlas: Species Found in the South China Sea*,
https://doi.org/10.1007/978-981-13-5901-9_1

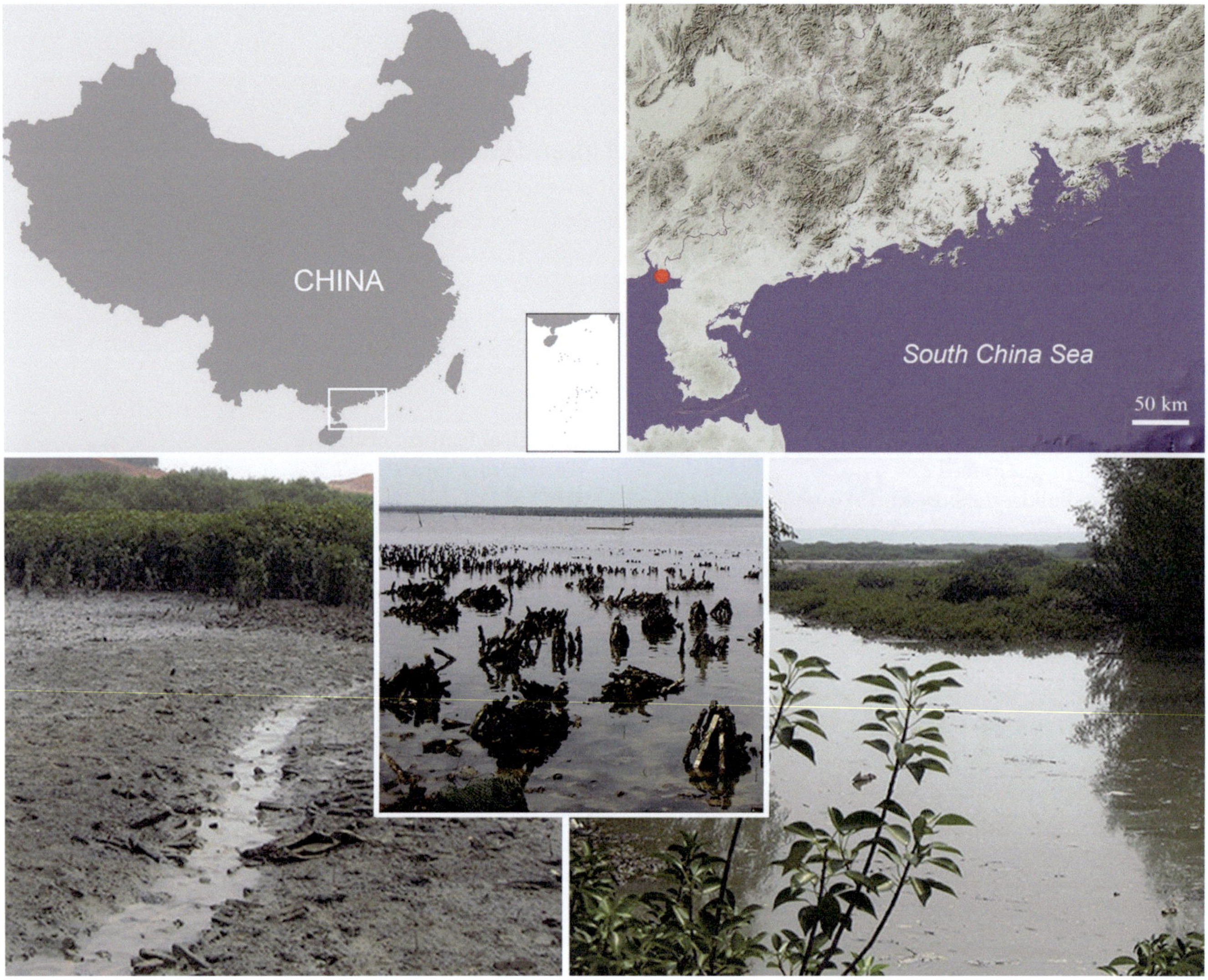


**Fig. 1A** Sample site

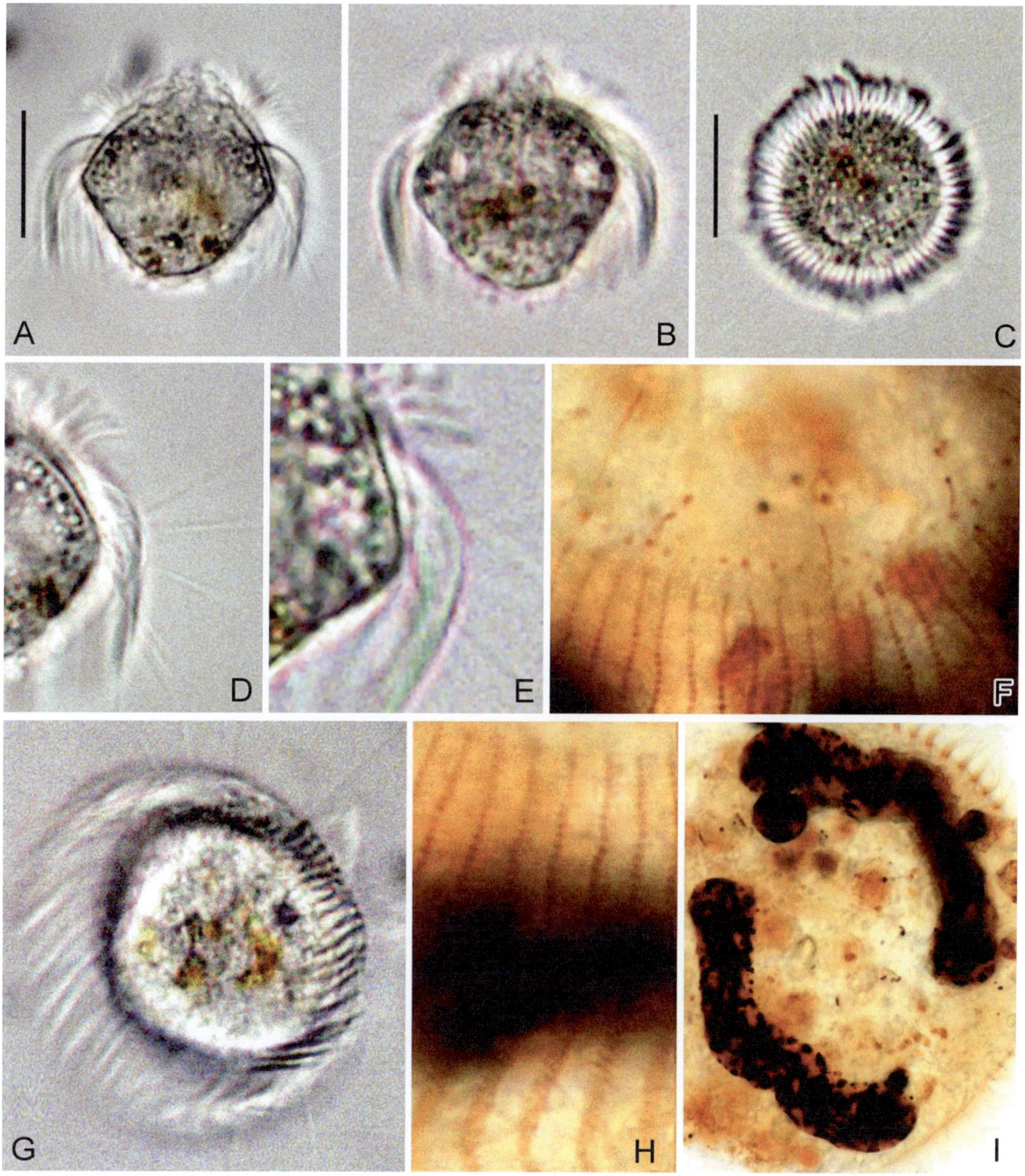


**Fig. 1B** *Askenasia* cf. *stellaris*
(**a**, **b**) Lateral views of live individuals; (**c**) apical view of live cell; (**d**, **e**) lateral views to show the bristle complexes and somatic cilia; (**f**) apical-lateral view, indicating somatic ciliature and the circumoral wreath of granules kineties; (**g**) poster-lateral view, showing the somatic cilia; (**h**) lateral view to show somatic kineties; (i) to show macronuclei. Scale bars: 30 μm

## Reference

Krainer K, Foissner W (1990) Revision of the genus *Askenasia* Blochmann, 1895, with proposal of two new species, and description of *Rhabdoaskenasia minima* n. g., n. sp. (Ciliophora, Cyclotrichida). J Protozool 37:414–427

Class Mesodiniea Chen et al., 2015
Order Mesodiniida Grain, 1994
Family Mesodiniidae Jankowski, 1980
*Mesodinium pulex* Claparède & Lachmann, 1858

**Ecological feature:** mangrove
**Water temperature:** 23.1 °C
**pH:** 6.3
**Salinity:** 10.0‰

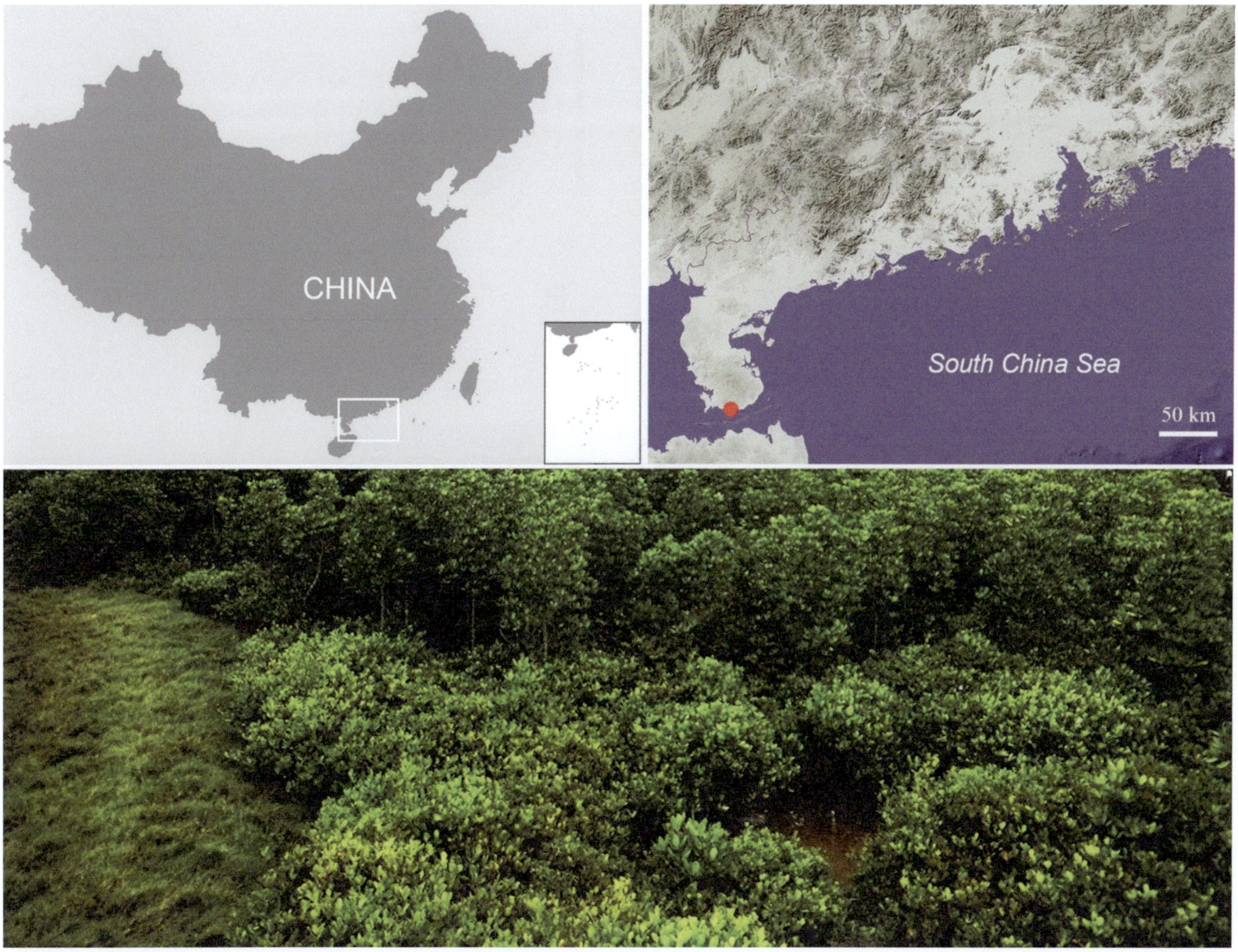


**Fig. 2A** Sample site

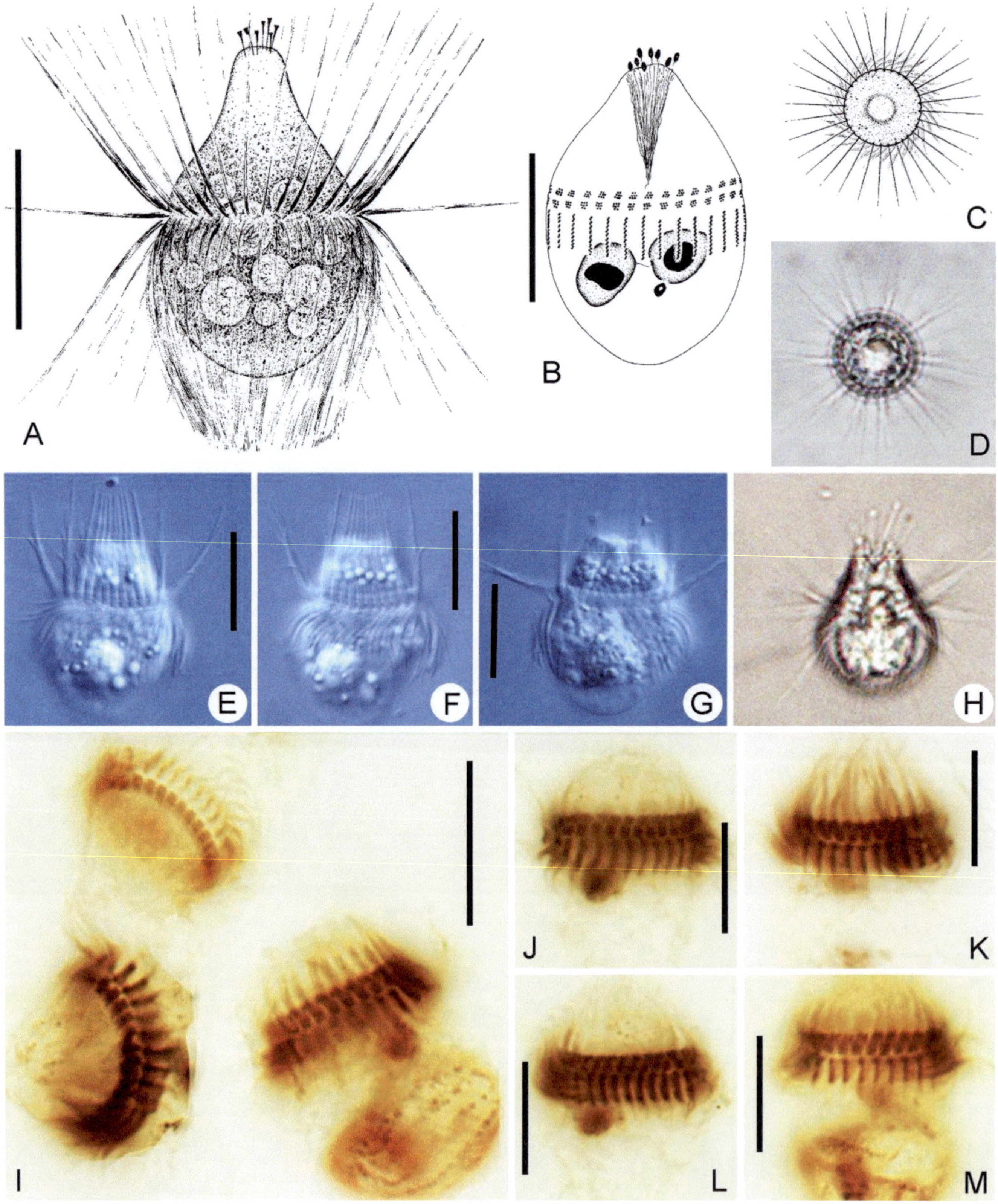


**Fig. 2B** Mesodinium pulex

(**a, e–h**) Lateral views of live cells, showing body shapes; (**b, i–m**) ciliature; (**c, d**) apical views of live cells. Scale bars: 10 μm

## References

Chen X, Ma H, Al-Rasheid KAS, Miao M (2015) Molecular data suggests the ciliate *Mesodinium* (Protista: Ciliophora) might represent an undescribed taxon at class level. Zool Syst 40:31–40

Song WB, Hu X, Wei J (1997) Morphological and comparative studies on *Mesodinium* red tidal ciliates in the marine biotope (Protozoa, Ciliophora). Chin J Appl Environ Biol 3:163–167

# Class Karyorelictea

Xiaozhong Hu, Xiaofeng Lin, and Weibo Song

Class Karyorelictea Corliss, 1974
Order Loxodida Jankowski, 1980
Family Loxodidae Bütschli, 1889
*Remanella achroma* Xu et al., 2013

**Ecological feature:** estuary
**Water temperature:** 25.0 °C
**pH:** no data available
**Salinity:** 15.0‰

X. Hu (✉)
Fisheries College, Ocean University of China,
Qingdao, Shandong, China
e-mail: xiaozhonghu@ouc.edu.cn

X. Lin
College of Life Science, South China Normal University,
Guangzhou, Guangdong, China
e-mail: xlin@oucscnu.edu.cn

W. Song
Institute of Evolution & Marine Biodiversity,
Ocean University of China, Qingdao, Shandong, China
e-mail: wsong@ouc.edu.cn


X. Hu et al. (eds.), *Ciliate Atlas: Species Found in the South China Sea*,
https://doi.org/10.1007/978-981-13-5901-9_2

**Fig. 1A** Sample site

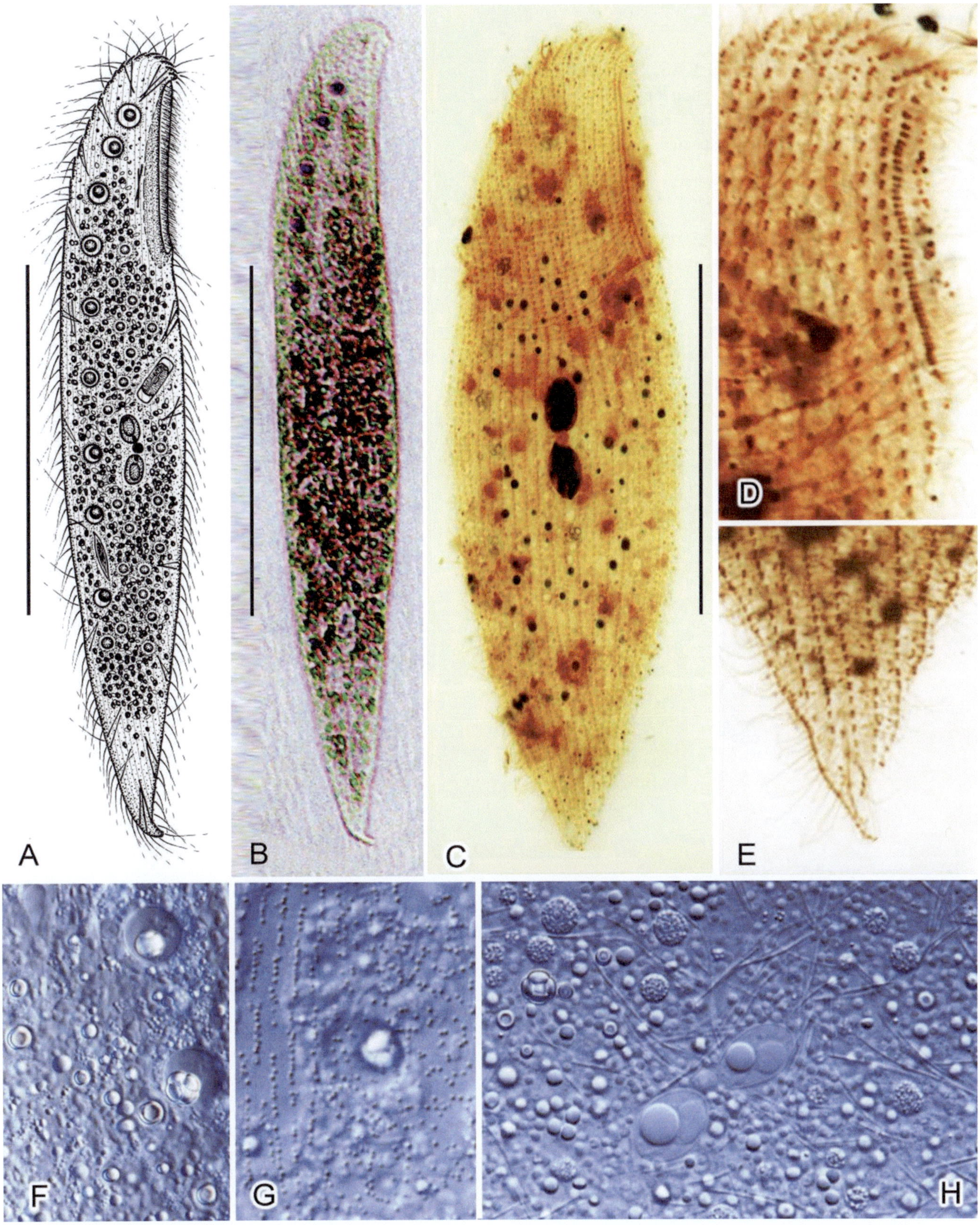


**Fig. 1B** *Remanella achroma*
(**a**, **b**) Lateral views of live cells; (**c**) general ciliature; (**d**) buccal ciliature, indicating right buccal kinety; (**e**) ciliature of posterior end, showing dorsolateral kinety; (**f**) Müller vesicles; (**g**) distribution of cortical granules; (**h**) cytoplasmic spicules and macronuclei. Scale bars: 120 μm

## Reference

Xu Y, Gao S, Hu X, Al-Rasheid KAS, Song WB (2013) Phylogeny and systematic revision of the karyorelictid genus *Remanella* (Ciliophora, Karyorelictea) with descriptions of two new species. Eur J Protistol 49:438–452

Class Karyorelictea Corliss, 1974
Order Protostomatida Small & Lynn, 1985
Family Trachelocercidae Kent, 1881
*Trachelocerca chinensis* Xu et al., 2015

**Ecological feature:** mangrove
**Water temperature:** no data available
**pH:** no data available
**Salinity:** 14.5‰

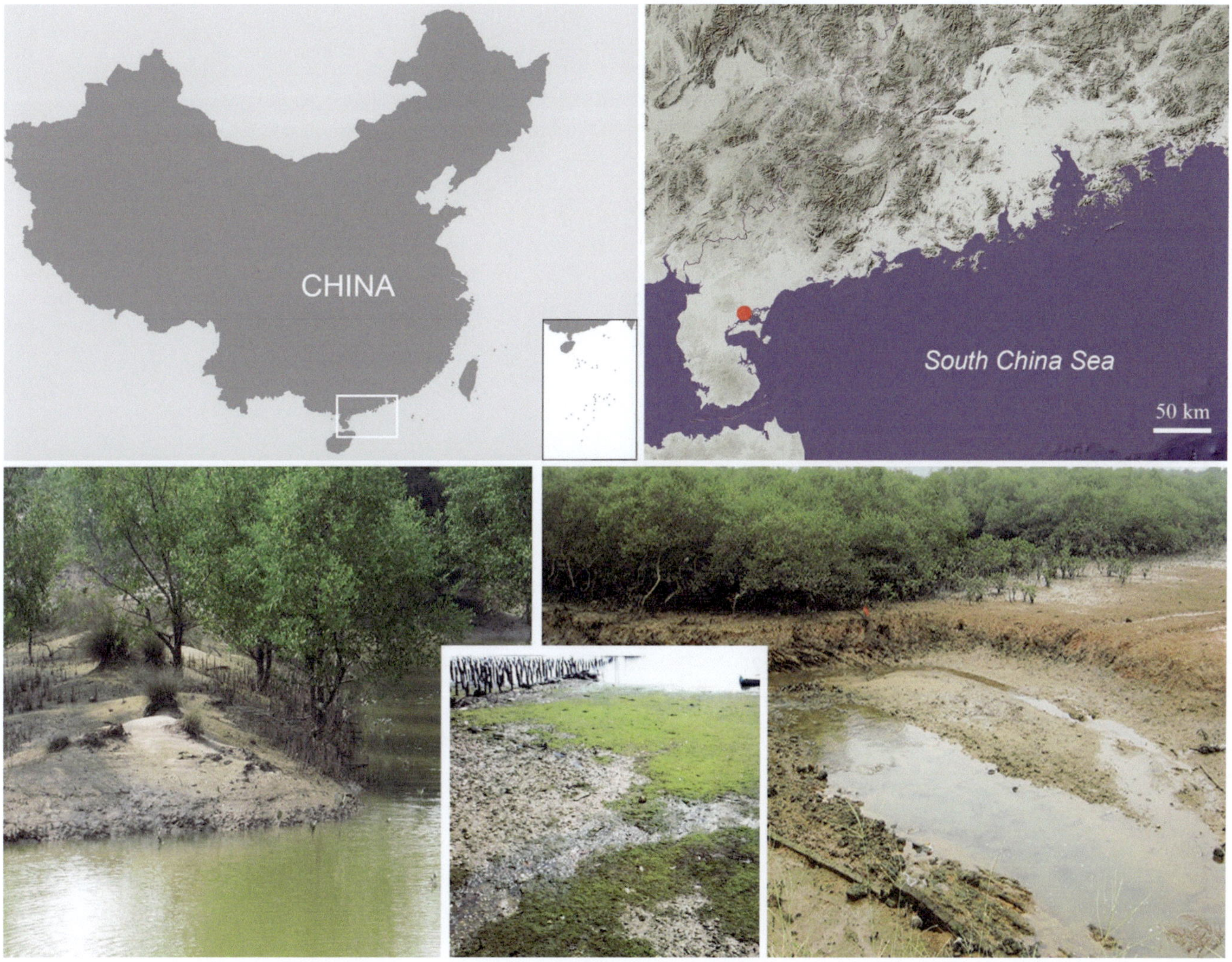


**Fig. 2A** Sample site

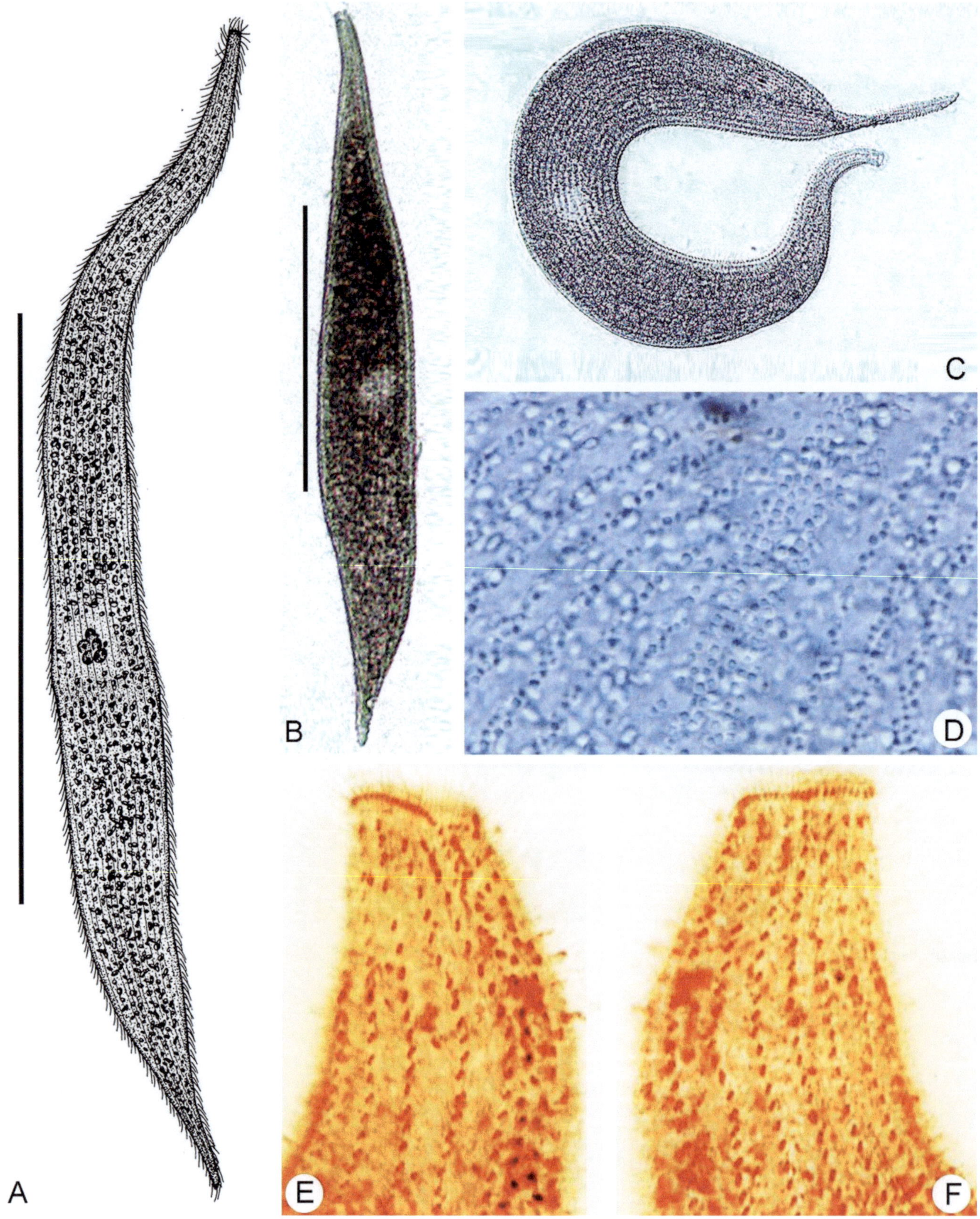


**Fig. 2B** *Trachelocerca chinensis*
(**a**, **b**) Lateral views of live individuals; (**c**) contorted cell; (**d**) distribution of cortical granules between ciliary rows and in glabrous stripe; (**e**, **f**) ciliature of anterior part of body, showing circumoral kinety, bristle kinety, and somatic kineties. Scale bars: 350 μm

## Reference

Xu Y, Yan Y, Li L, Al-Rasheid KAS, Al-Farraj SA, Song WB (2014) Morphology and phylogeny of three karyorelictean ciliates (Protista, Ciliophora), including two new species, *Trachelocerca chinensis* sp. n. and *Tracheloraphis dragescoi* sp. n. Int J Syst Evol Microbiol 64:4084–4097

Class Karyorelictea Corliss, 1974
Order Protostomatida Small & Lynn, 1985
Family Trachelocercidae Kent, 1881
*Tracheloraphis similis* Raikov & Kovaleva, 1968

**Ecological feature:** intertidal zone
**Water temperature:** 22.0 °C
**pH:** no data available
**Salinity:** 32.0‰

**Fig. 3A** Sample site

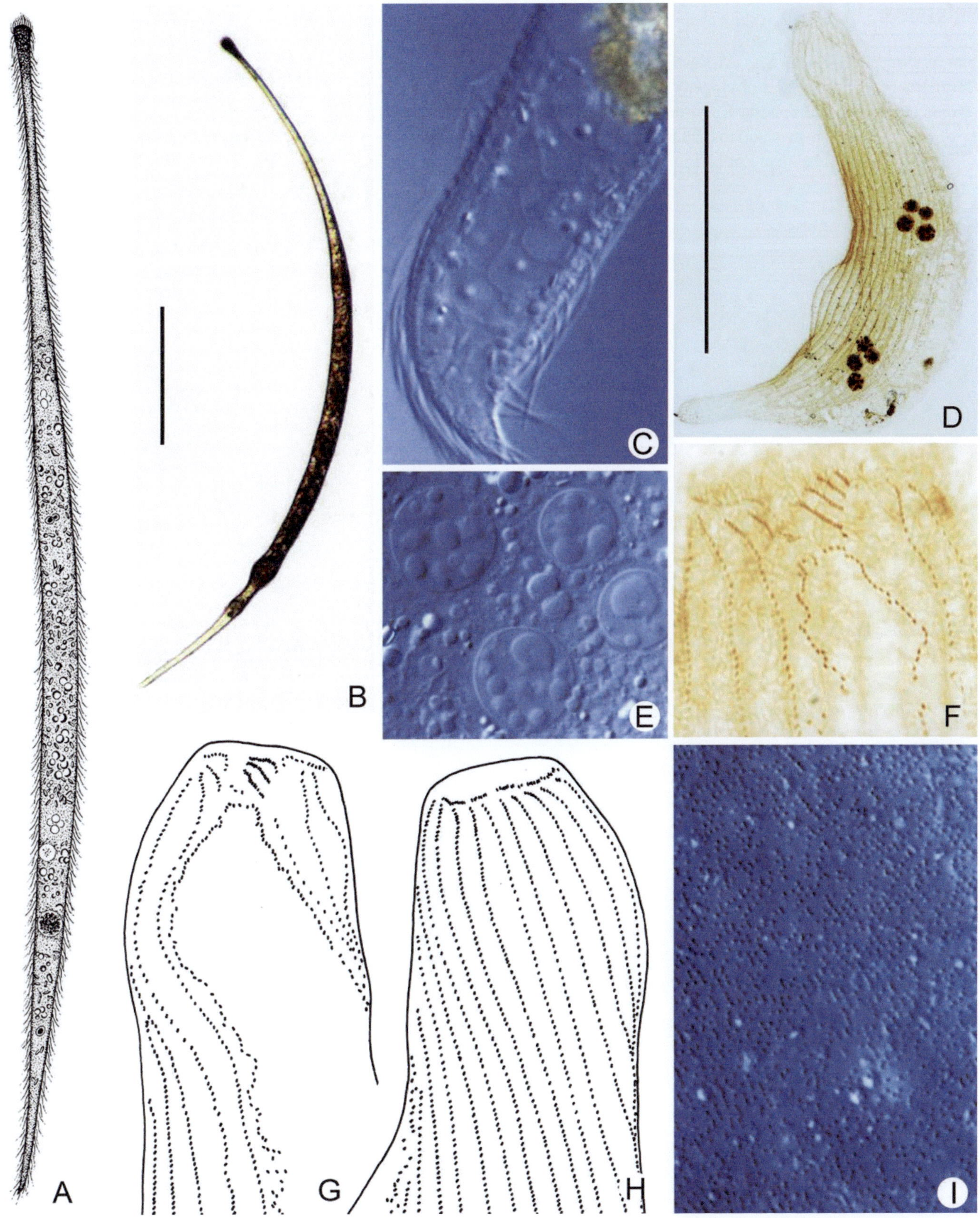


**Fig. 3B** *Tracheloraphis similis* (**a**, **b**) Lateral views of live individuals; (**c**) wedge-shaped tail; (**d**) general ciliature; (**e**) nuclear group formed by four macronuclei and two micronuclei; (**f–h**) ciliature of anterior portion of body; (**i**) distribution of cortical granules. Scale bars: 150 μm

## Reference

Yan Y, Xu Y, Al-Farraj SA, Al-Rasheid KAS, Song WB (2016) Morphology and phylogeny of three trachelocercids (Protozoa, Ciliophora, Karyorelictea), with description of two new species and insight to the evolution of the family Trachelocercidae. Zool J Linn Soc 177:306–319

# Class Heterotrichea

Xiaozhong Hu, Xiaofeng Lin, and Weibo Song

Class Heterotrichea Stein, 1859
Order Heterotrichida Stein, 1859
Family Condylostomatidae Kahl in Doflein & Reichenow, 1929
*Condylostoma* cf. *magnum*

**Ecological feature:** estuary area
**Water temperature:** 24.5 °C
**pH:** 7.5
**Salinity:** 27.4‰

X. Hu (✉)
Fisheries College, Ocean University of China,
Qingdao, Shandong, China
e-mail: xiaozhonghu@ouc.edu.cn

X. Lin
College of Life Science, South China Normal University,
Guangzhou, Guangdong, China
e-mail: xlin@oucscnu.edu.cn

W. Song
Institute of Evolution & Marine Biodiversity,
Ocean University of China, Qingdao, Shandong, China
e-mail: wsong@ouc.edu.cn


X. Hu et al. (eds.), *Ciliate Atlas: Species Found in the South China Sea*,
https://doi.org/10.1007/978-981-13-5901-9_3

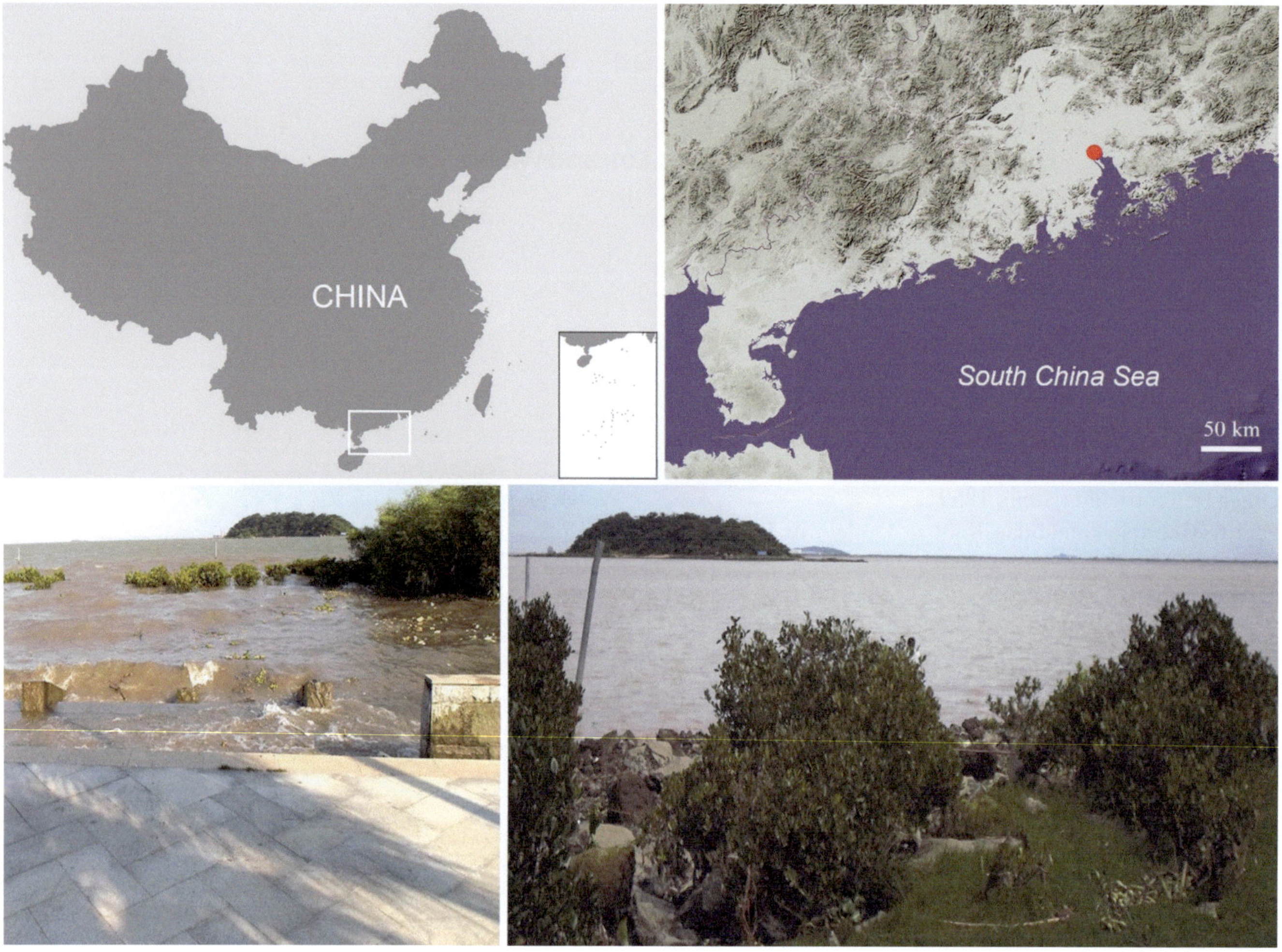


**Fig. 1A** Sample site

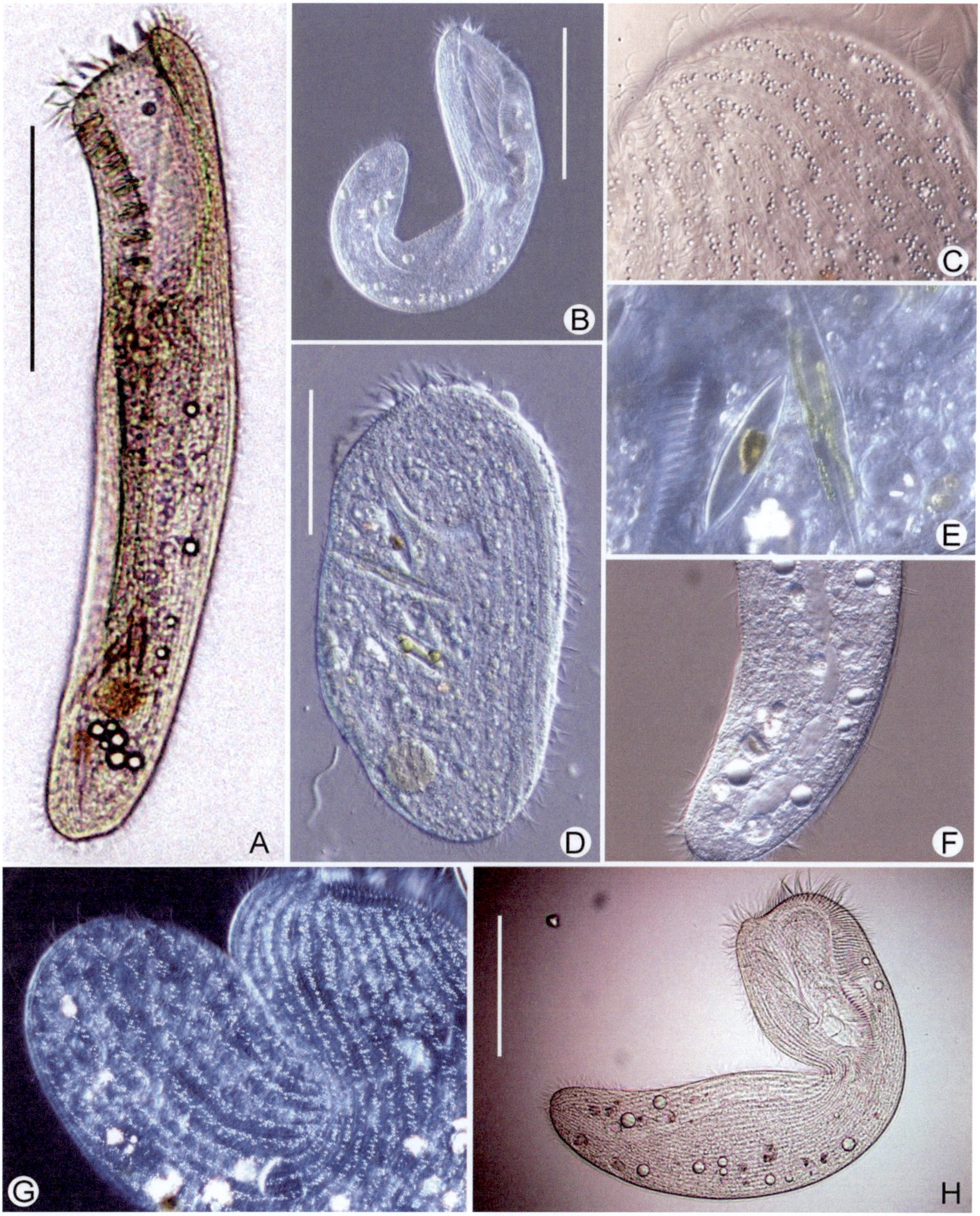


**Fig. 1B** *Condylostoma* cf. *magnum*
(**a**, **b**, **h**) Ventral views of live individuals; (**c**, **g**) distribution of cortical granules; (**d**) contracted body, showing food vacuoles; (**e**) ingested diatoms; (**f**) posterior part of cell, showing macronuclear nodules. Scale bars: 200 μm

## References

Bohatier J (1979) Morphogenèse de régénération chez le cilié *Condylostoma magnum* (Spiegel): étude ultrastructurale. J Protozool 26:404–414

Song WB, Wilbert N (1997) Morphological investigations on some free living ciliates (Protozoa, Ciliophora) from China Sea with description of a new hypotrichous genus, *Hemigastrostyla* nov. gen. Arch Protistenk 148:413–444

Class Heterotrichea Stein, 1859
Order Heterotrichida Stein, 1859
Family Condylostomatidae Kahl in Doflein & Reichenow, 1929
*Condylostoma curvum* Burkovsky, 1970

**Ecological feature:** estuary (Guangzhou); mangrove (Zhanjiang, Shenzhen)
**Water temperature:** 25.2 °C; 24.0 °C; no data available
**pH:** 8.5; 6.8; 6.8
**Salinity:** 6.3‰; 15.0‰; 10.0‰

**Fig. 2A** Sample site

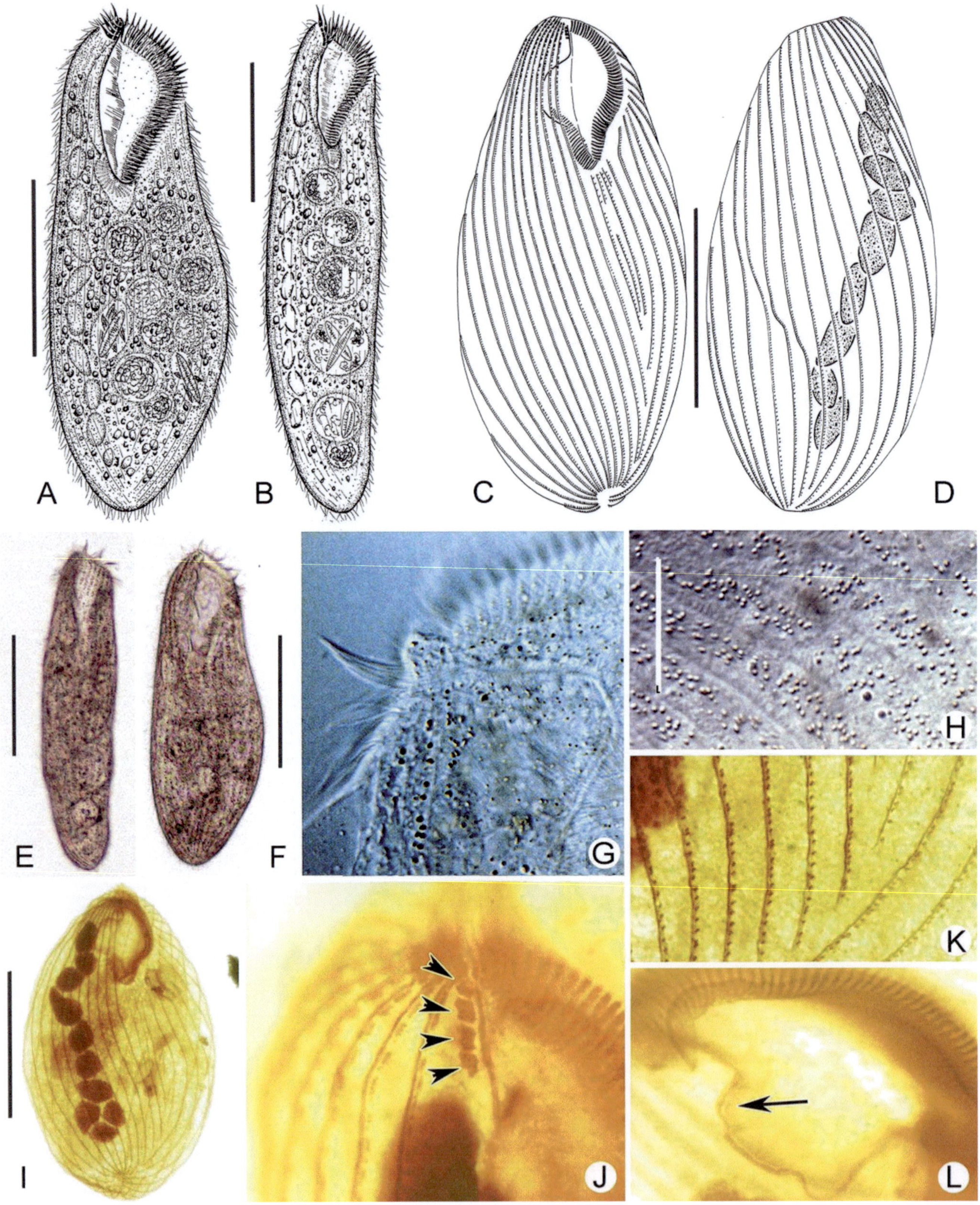


**Fig. 2B** *Condylostoma curvum*
(**a**, **b**, **e**, **f**) Ventral views of live individuals; (**c**, **d**) ventral (**c**) and dorsal (**d**) views of ciliature; (**g**, **h**) distribution of cortical granules; (**i–l**) ventral views of ciliature, arrowheads and arrow show frontal cirri and paroral membrane, respectively. Scale bars: 100 μm (**a–f**, **i**); 30 μm (**h**)

## References

Song WB, Warren A, Ji D, Wang M, Al-Rasheid KAS (2003) New contributions to two heterotrichous Ciliates, *Folliculina simplex* (Dons, 1917), *Condylostoma curva* Burkovsky, 1970 and one licnophorid, *Licnophora lyngbycola* Fauré-Fremiet, 1937 (Protozoa, Ciliophora): descriptions of morphology and infraciliature. J Eukaryot Microbiol 50:449–462

Yan Y, Chen X, Chen X, Gao F, Al-Farraj SA, Al-Rasheid KAS (2015) Morphology and molecular phylogeny of three marine *Condylostoma* species from China, including two new ones (Ciliophora, Heterotrichea). Eur J Protistol 51:66–78

Class Heterotrichea Stein, 1859
Order Heterotrichida Stein, 1859
Family Condylostomatidae Kahl in Doflein & Reichenow, 1929
*Condylostoma elongatum* Yan et al., 2015

**Ecological feature:** sandy sediment
**Water temperature:** 26.0 °C
**pH:** no data available
**Salinity:** 24.5‰

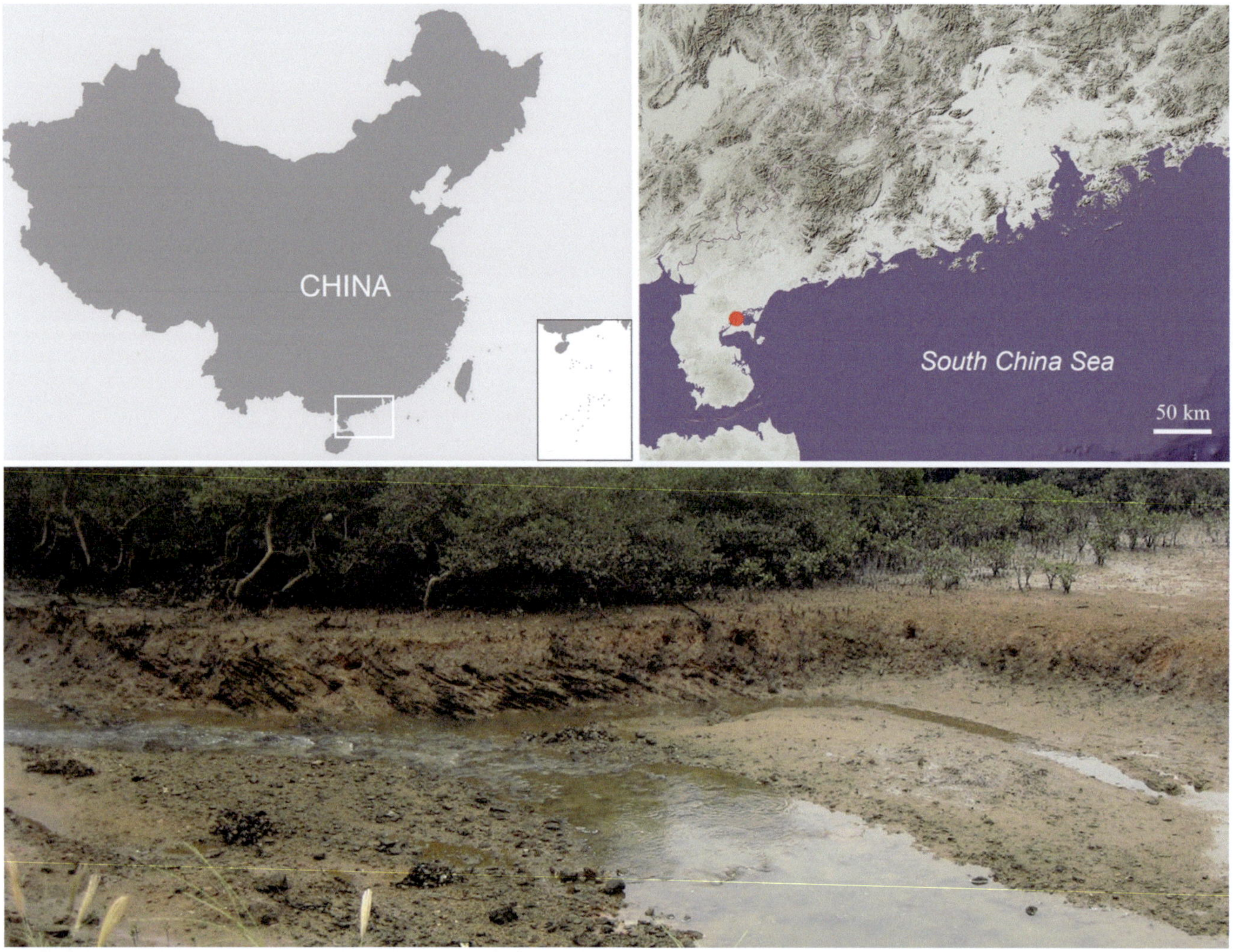


**Fig. 3A** Sample site

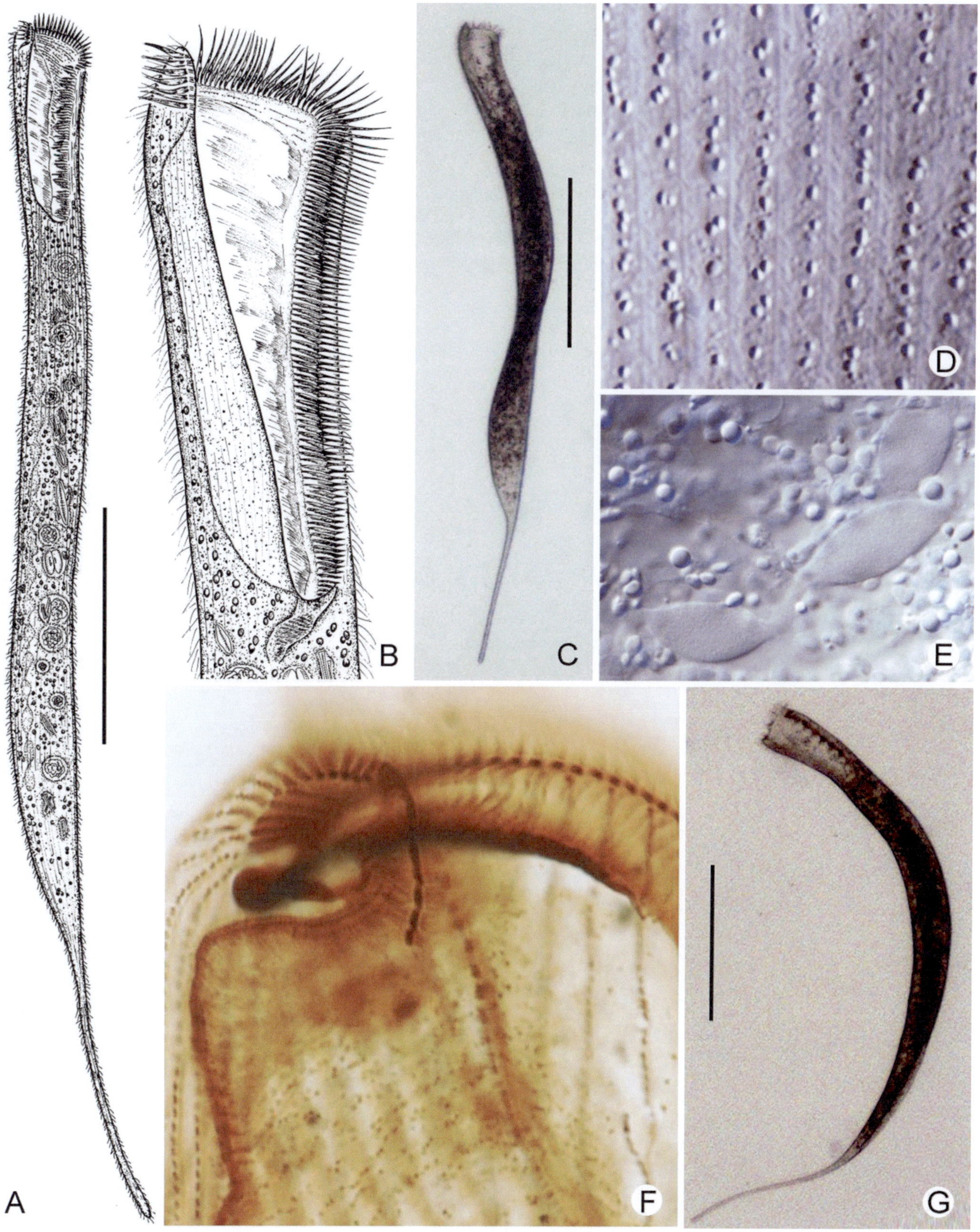


**Fig. 3B** *Condylostoma elongatum*
(**a**, **c**, **g**) Ventral views of live cells; (**b**, **f**) ventral view of anterior portion of cell to show the prominent oral cavity; (**d**) distribution of cortical granules; (**e**) moniliform macronucleus. Scale bars: 200 μm

## Reference

Yan Y, Chen XM, Chen XR, Gao F, Al-Farraj SA, Al-Rasheid KAS (2015) Morphology and molecular phylogeny of three marine *Condylostoma* species from China, including two new ones (Ciliophora, Heterotrichea). Eur J Protistol 51:66–78

Class Heterotrichea Stein, 1859
Order Heterotrichida Stein, 1859
Family Condylostomatidae Kahl in Doflein & Reichenow, 1929
*Condylostoma minutum* Bullington, 1940

**Ecological feature:** mangrove
**Water temperature:** 27.0 °C
**pH:** 6.0
**Salinity:** 10.0‰

**Fig. 4A** Sample site

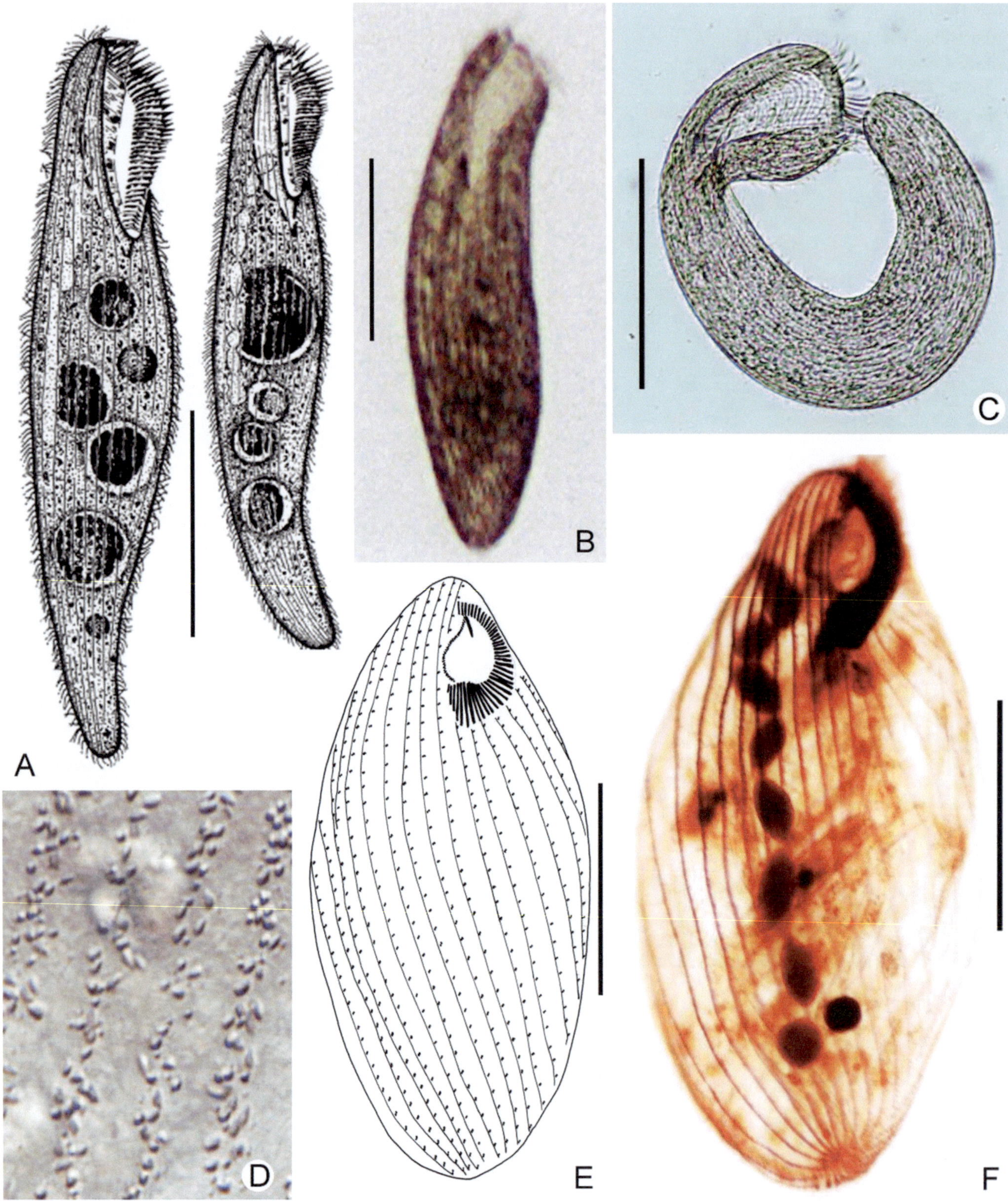


**Fig. 4B** *Condylostoma minutum*
(**a**, **b**) Ventral views of live individuals; (**c**) twisted individual; (**d**) arrangement of cortical granules; (**e**, **f**) ventral views of ciliature. Scale bars: 50 μm

## Reference

Chen X, Song WB, Al-Rasheid KAS, Warren A, Long H, Shao C, Al-Farraj SA, Gong J, Hu X (2007) The morphology of three marine heterotrichous ciliates, *Condylostentor auriculatus* (Kahl, 1932) Jankowski, 1978, *Condylostoma minutum* Bullington, 1940 and *C. spatiosum* Ozaki & Yagiu in Yagiu, 1944 (Ciliophora: Heterotrichida). Acta Protozool 46:289–309

Class Heterotrichea Stein, 1859

Order Heterotrichida Stein, 1859

Family Condylostomatidae Kahl in Doflein & Reichenow, 1929

*Condylostoma spatiosum* Ozaki & Yagiu in Yagiu, 1944

**Ecological feature:** mangrove (Shenzhen, left; Zhanjiang, right)

**Water temperature:** no data available; 24.0 °C

**pH:** 7.0; 6.4

**Salinity:** 15.0‰; 12.0‰

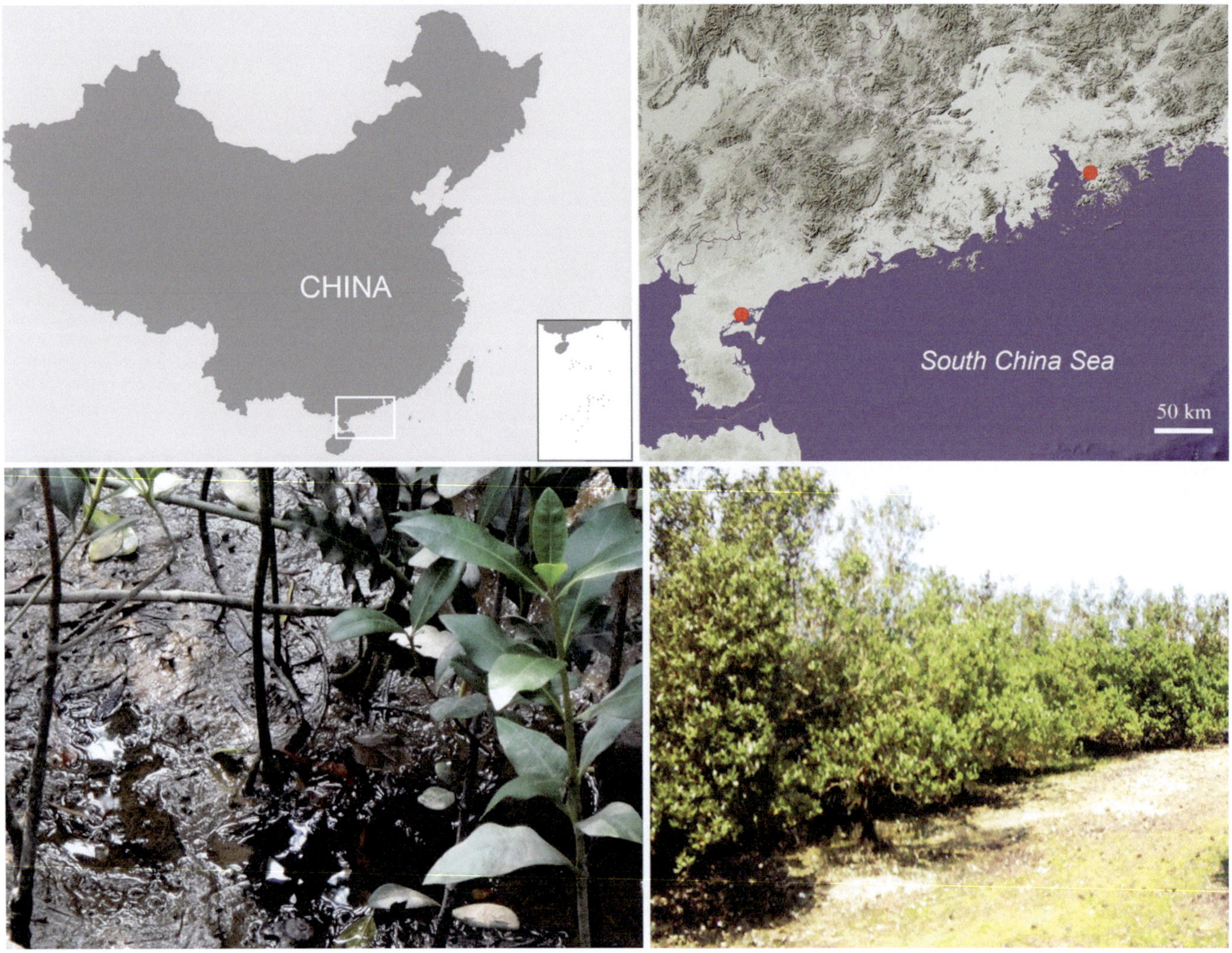


**Fig. 5A** Sample site

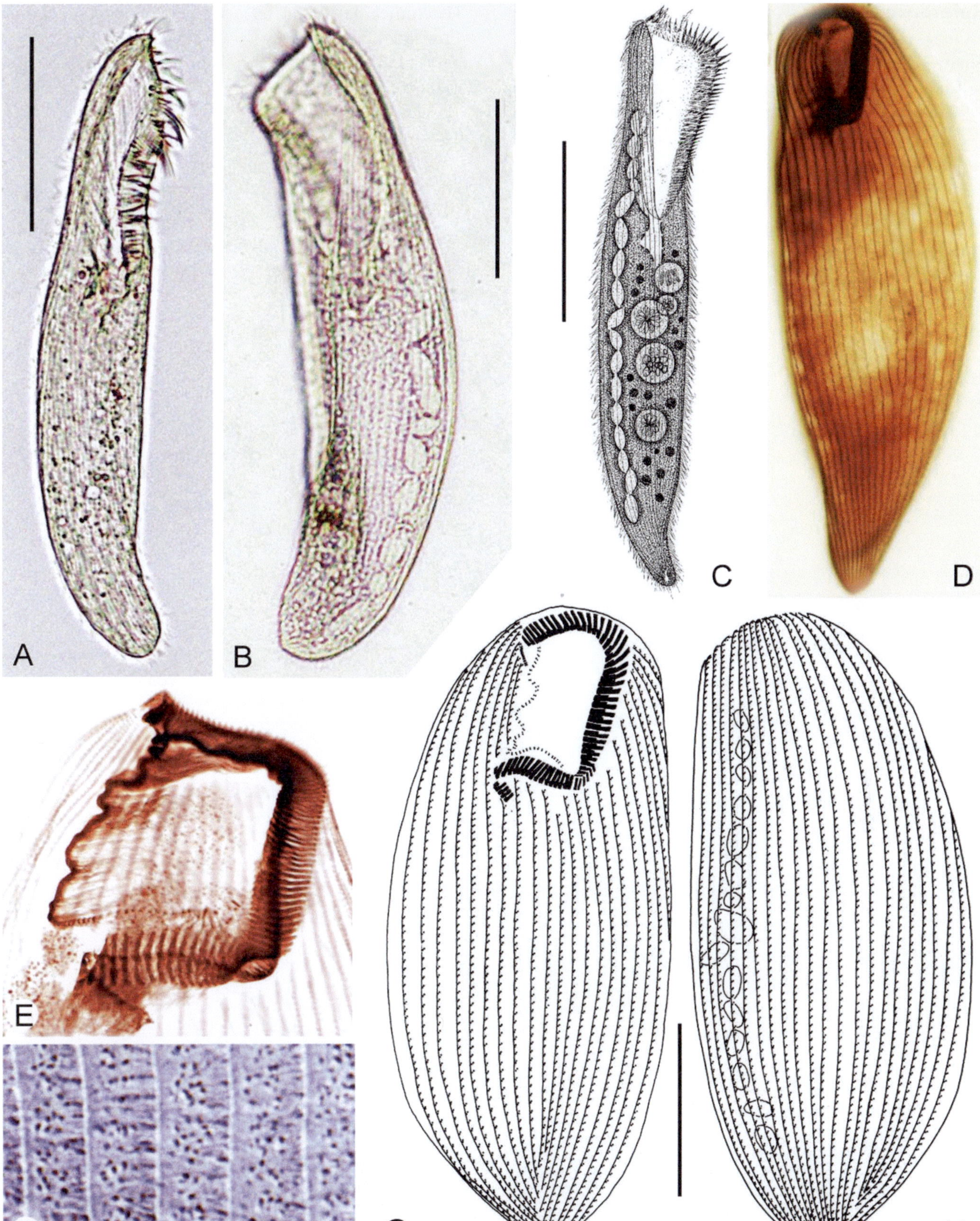


**Fig. 5B** *Condylostoma spatiosum*
(**a–c**) Ventral views in vivo; (**d**, **e**, **g**, **h**) ventral (**d**, **e**, **g**) and dorsal (**h**) views of ciliature; (**f**) distribution of cortical granules. Scale bars: 100 μm

## References

Shao C, Song WB, Hu X, Ma H, Zhu M, Wang M (2006) Cell division and morphology of the marine ciliate, *Condylostoma spatiosum* Ozaki and Yagiu (Ciliophora, Heterotrichida) based on a Chinese population. Eur J Protistol 42:9–19

Yagiu R (1944) Studies on the ciliates *Condylostoma* I: taxonomy. J Sci Hiroshima Univ (Ser B, Div 1) 10:161–184

Class Heterotrichea Stein, 1859
Order Heterotrichida Stein, 1859
Family Condylostomatidae Kahl in Doflein & Reichenow, 1929
*Condylostoma tropicum* Yan et al., 2015

**Ecological feature:** estuary
**Water temperature:** 25.2 °C
**pH:** 8.5
**Salinity:** 6.3‰

**Fig. 6A** Sample site

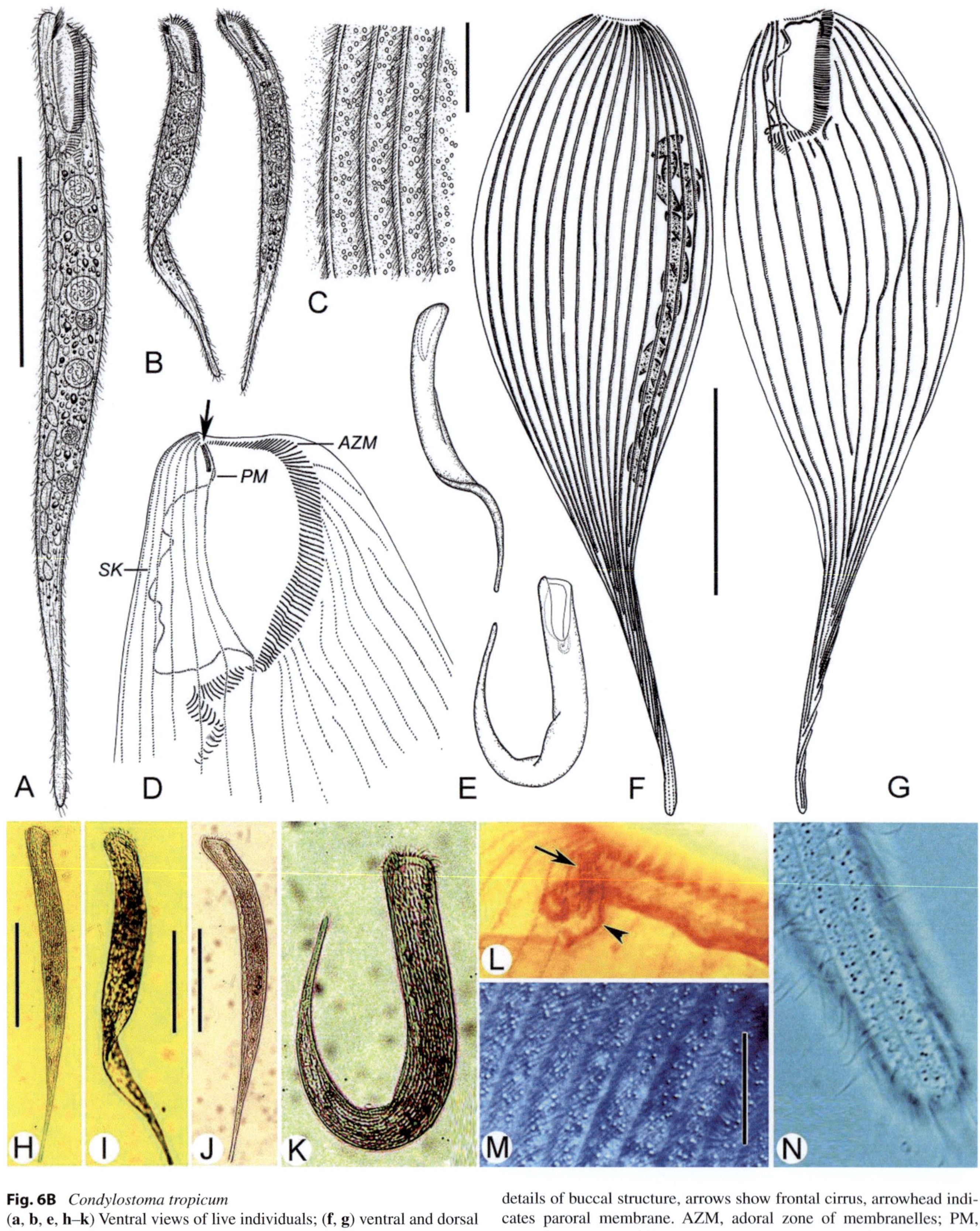


**Fig. 6B** *Condylostoma tropicum*
(**a**, **b**, **e**, **h**–**k**) Ventral views of live individuals; (**f**, **g**) ventral and dorsal views of ciliature; (**c**, **m**, **n**) distribution of cortical granules; (**d**, **l**) details of buccal structure, arrows show frontal cirrus, arrowhead indicates paroral membrane. AZM, adoral zone of membranelles; PM, paroral membrane; SK, somatic kineties. Scale bars: 200 μm (**a**, **f**–**j**); 30 μm (**c**, **m**)

## Reference

Yan Y, Chen X, Chen X, Gao F, Al-Farraj SA, Al-Rasheid KAS (2015) Morphology and molecular phylogeny of three marine *Condylostoma* species from China, including two new ones (Ciliophora, Heterotrichea). Eur J Protistol 51:66–78

Class Heterotrichea Stein, 1859
Order Heterotrichida Stein, 1859
Family Condylostomatidae Kahl in Doflein & Reichenow, 1929
*Linostomella vorticella* (Ehrenberg, 1833) Aescht in Foissner et al., 1999

**Ecological feature:** mangrove
**Water temperature:** no data available
**pH:** 7.0
**Salinity:** 3.0‰

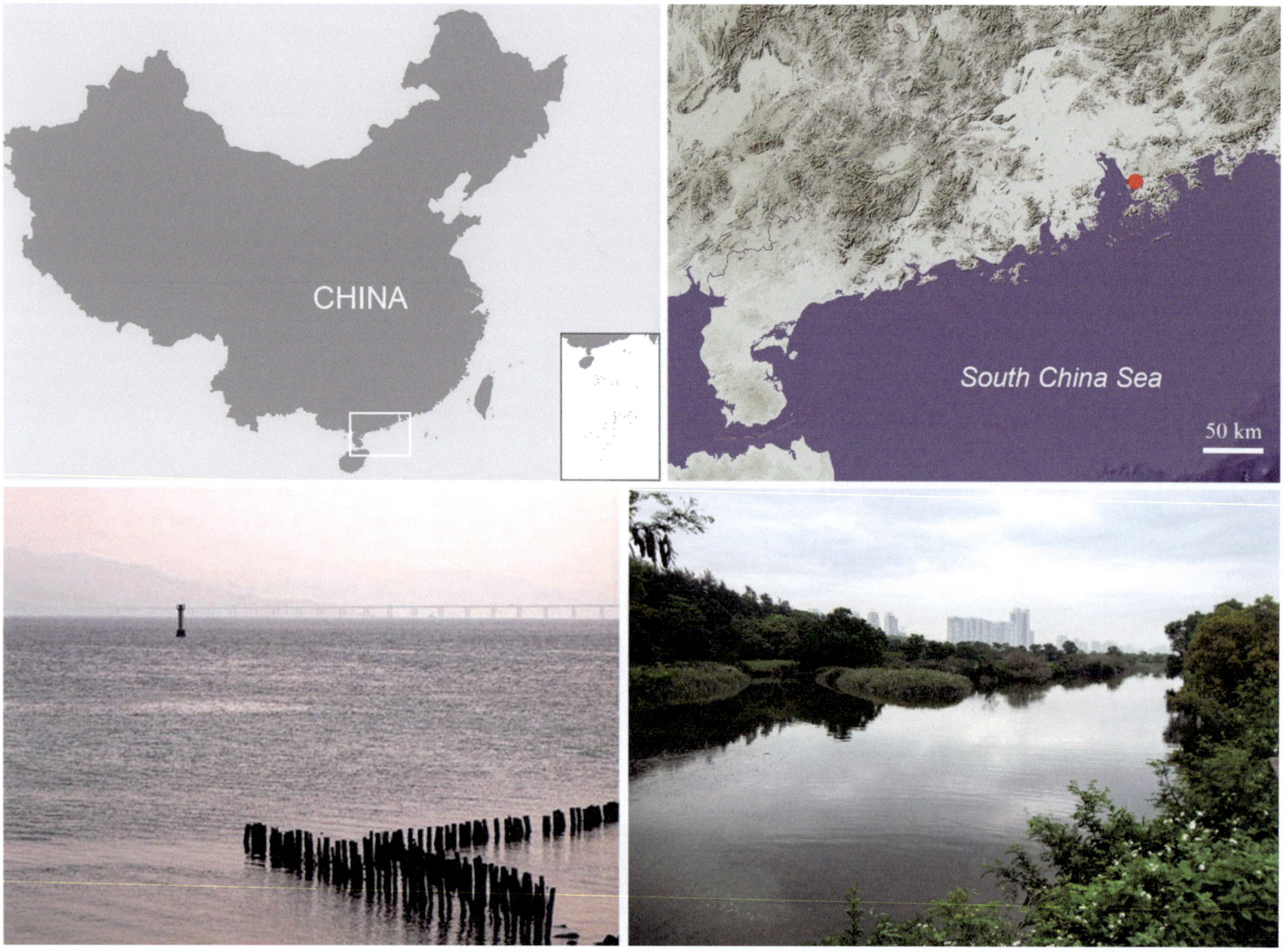


**Fig. 7A** Sample site

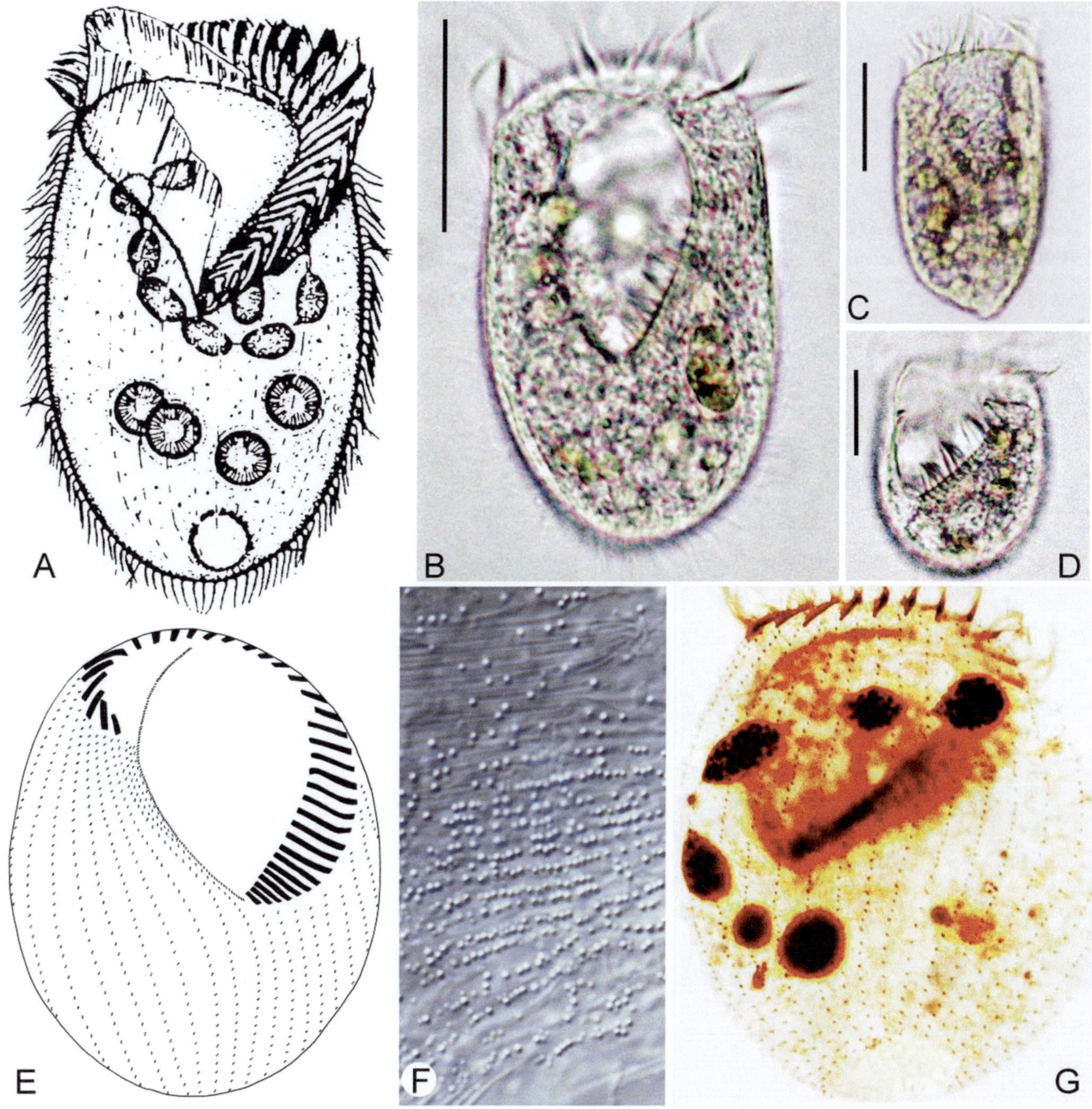


**Fig. 7B** *Linostomella vorticella*
(**a**) Ventral view of a live individual, from Foissner et al. (1999); (**b–d**) different sizes of cells in vivo; (**e**) ventral view of ciliature; (**f**) distribution of cortical granules; (**g**) dorsal view of ciliature. Scale bars: 50 μm

## Reference

Foissner W, Berger H, Schaumburg J (1999) Identification and ecology of limnetic plankton ciliates. Informationsberichte des Bayer. Landesamtes für Wasserwirtschaft 3/99:1–793

Class Heterotrichea Stein, 1859
Order Heterotrichida Stein, 1859
Family Folliculinidae Dons, 1914
*Folliculina simplex* (Dons, 1917) Song et al., 2003

**Ecological feature:** offshore water
**Water temperature:** 19.0 °C
**pH:** 8.0
**Salinity:** 29.0‰

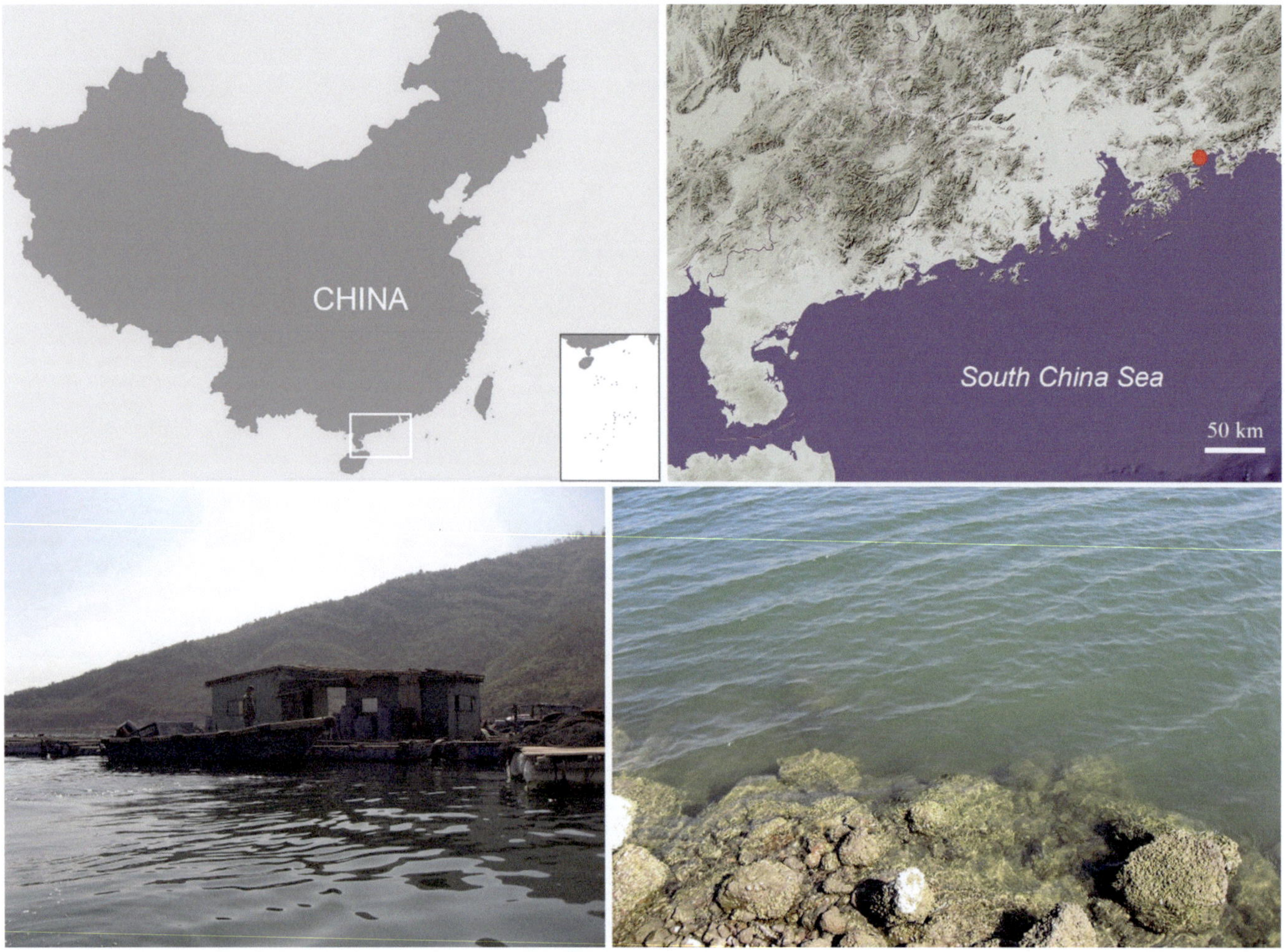


**Fig. 8A** Sample site

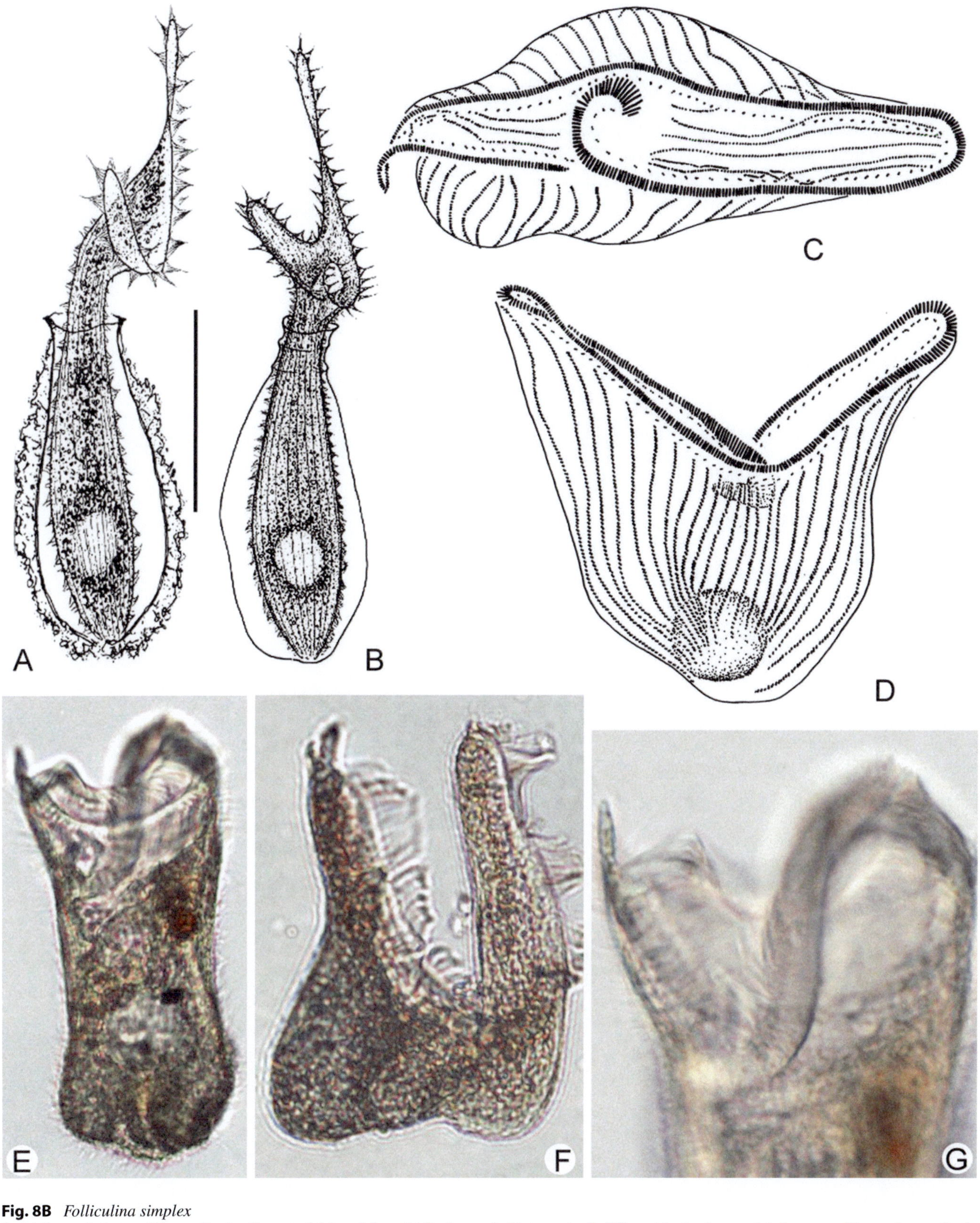


**Fig. 8B** *Folliculina simplex*
(**a**, **b**) Lateral views of live cells; (**c**, **d**) ventral (**c**) and dorsal (**d**) views of ciliature; (**e**, **f**) different body shapes; (**g**) the arrow-like oral area. Scale bar: 100 μm

## Reference

Song WB, Warren A, Ji D, Wang M, Al-Rasheid KAS (2003) New contribution to two heterotrichous ciliates, *Folliculina simplex* (Dons, 1917), *Condylostoma curva* Burkovsky, 1970 and one licnophorid, *Licnophora lyngbycola* Faure-Fremiet, 1937 (Protozoa, Ciliophora): description of morphology and infraciliature. J Eukaryot Microbiol 50:449–462

Class Heterotrichea Stein, 1859
Order Heterotrichida Stein, 1859
Family Folliculinidae Dons, 1914
*Folliculinopsis producta* (Wright, 1859) Frauré-Fremiet, 1936

**Ecological feature:** offshore water
**Water temperature:** 22.0 °C
**pH:** 8.0
**Salinity:** 29.0‰

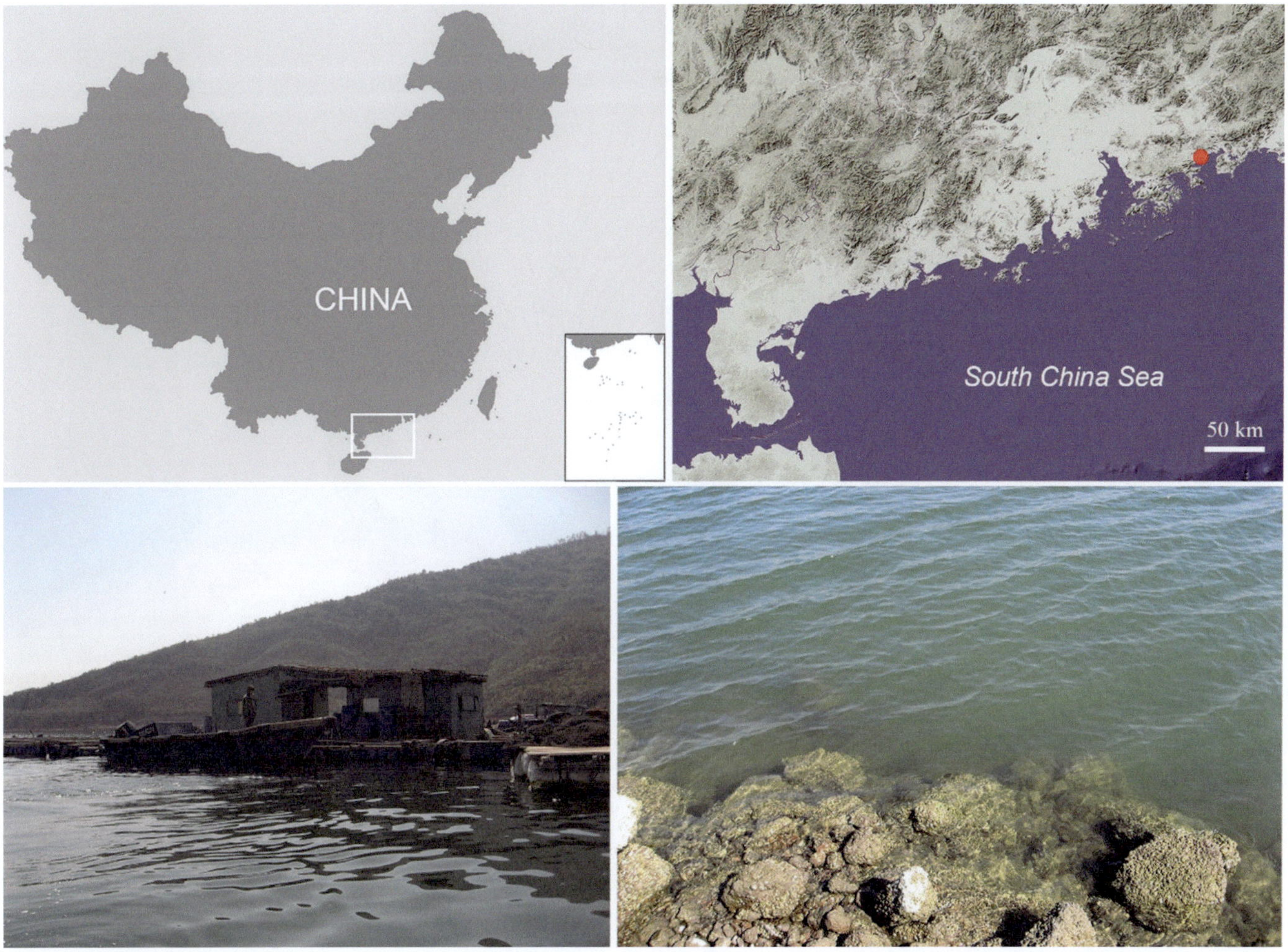


**Fig. 9A** Sample site

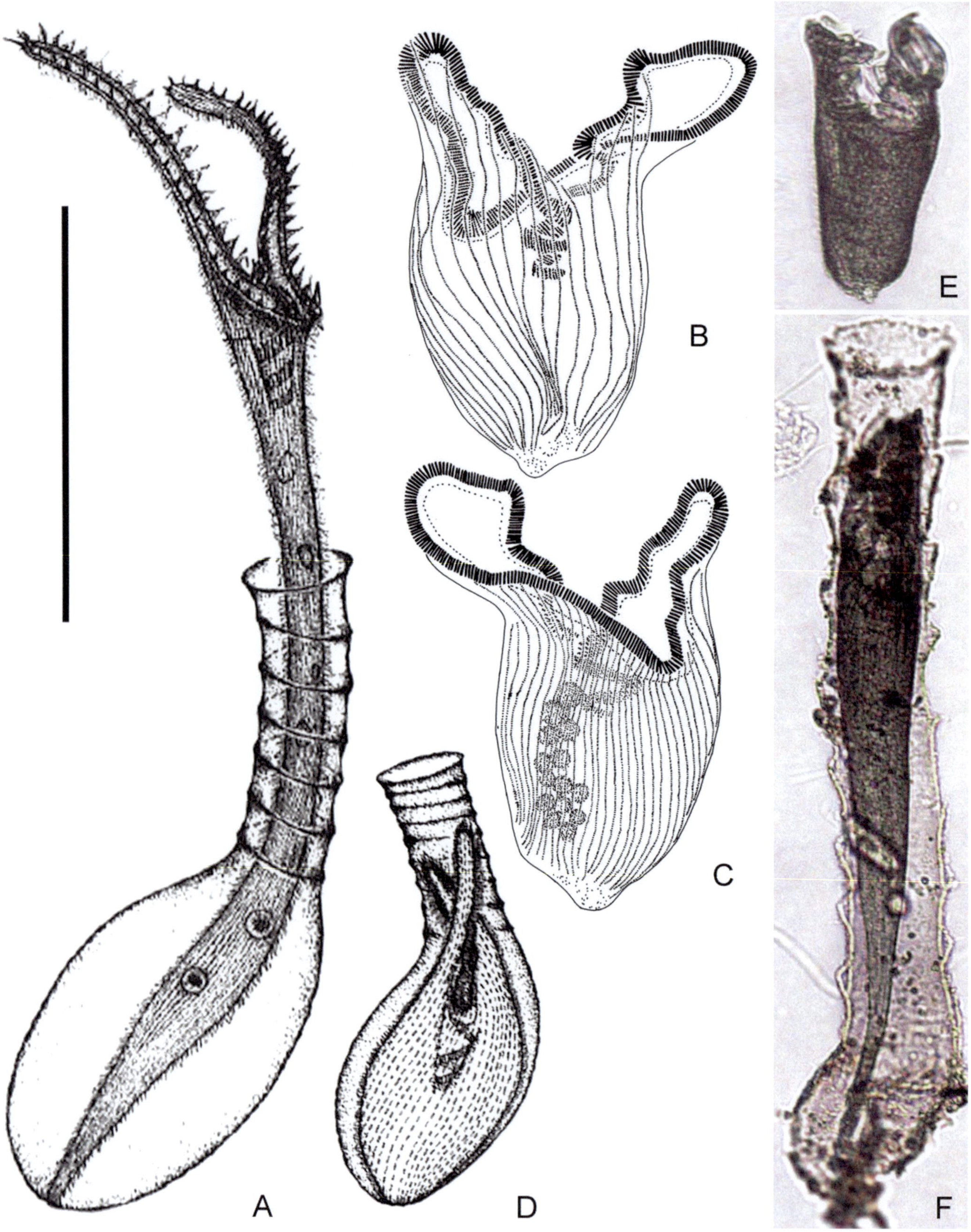


**Fig. 9B** *Folliculinopsis producta*
(**a**) General view of a representative cell; (**b**, **c**) dorsal (**b**) and ventral (**c**) views of ciliature; (**d**, **f**) live cells in lorica; (**e**) a live individual away from lorica. Scale bar: 300 μm

## Reference

Ji D, Lin X, Song WB (2004). Complementary notes on a 'well-known' marine heterotrichous ciliates, *Folliculinopsis producta* (Wright, 1895) Frauré-Fremiet, 1936 (Protozoa, Ciliophora). J Ocean Univ China 3:65–69

Class Heterotrichea Stein, 1859
Order Heterotrichida Stein, 1859
Family Peritromidae Stein, 1867
*Peritromus* cf. *faurei*

**Ecological feature:** intertidal zone
**Water temperature:** 25.0 °C
**pH:** 7.5
**Salinity:** 31.0‰

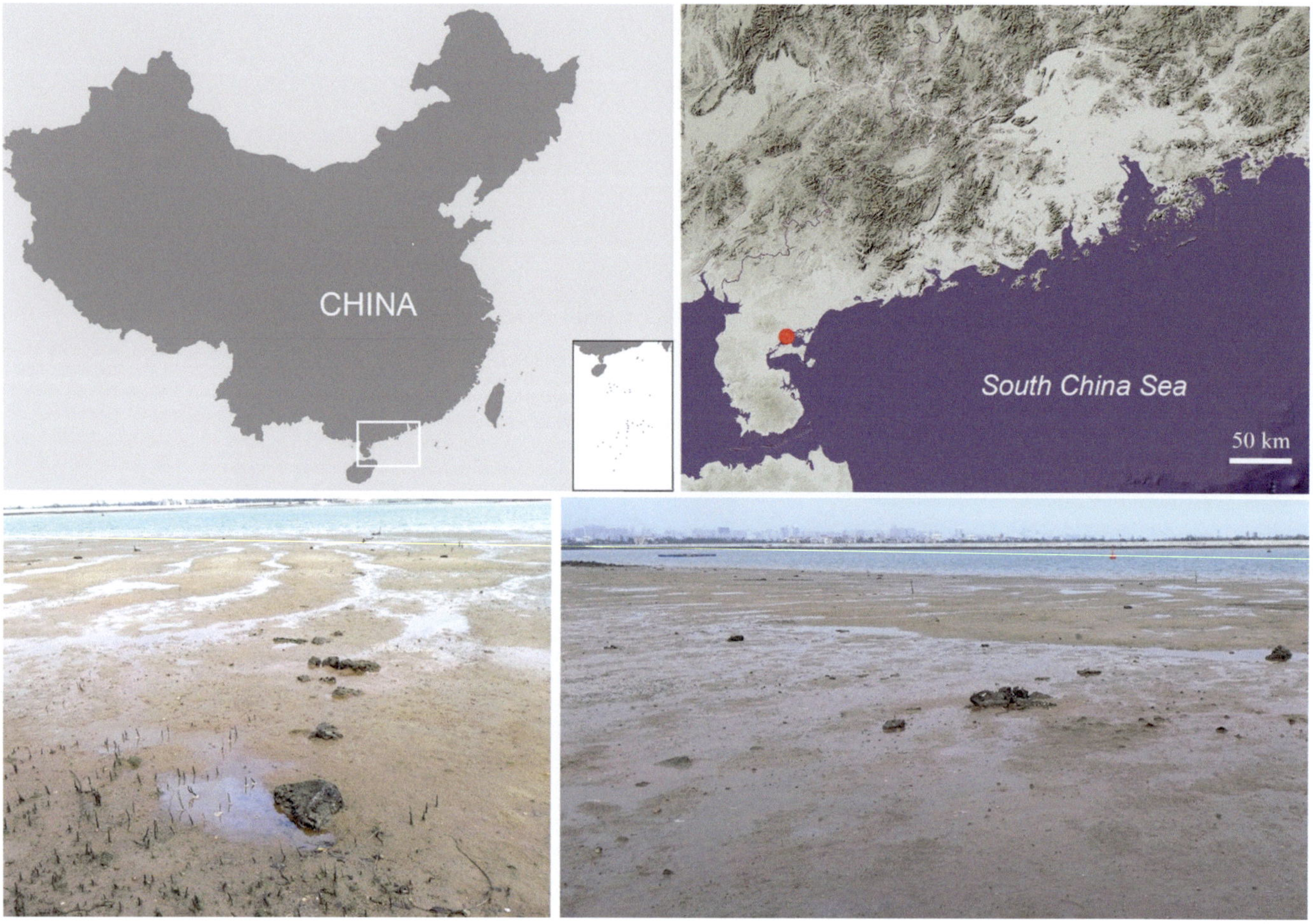


**Fig. 10A** Sample site

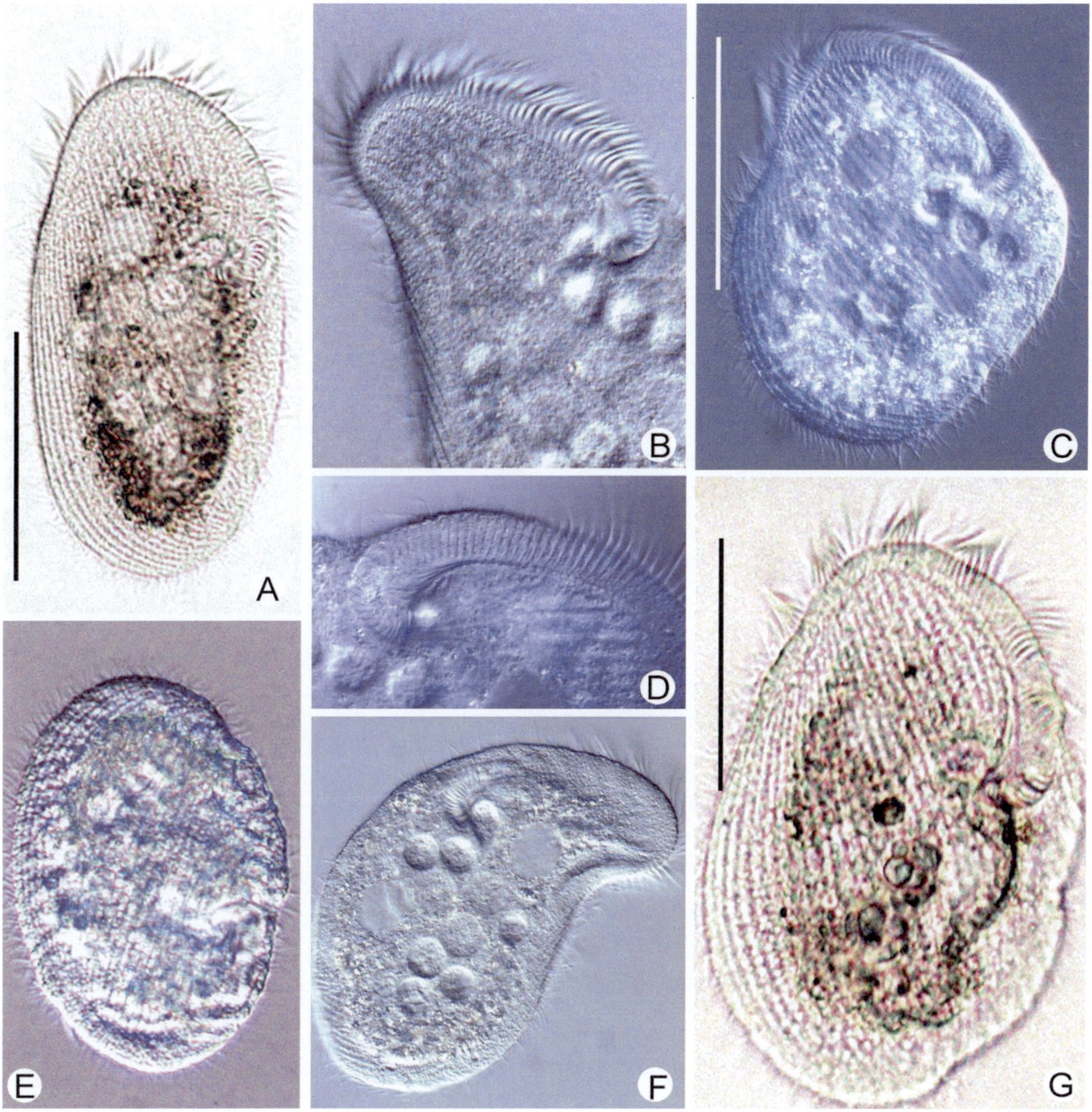


**Fig. 10B** *Peritromus* cf. *faurei*
(**a**) Ventral view of a representative cell; (**b**, **d**) buccal area showing adoral membranelles; (**c**, **e**–**c**) variation of body shapes. Scale bars: 50 μm

## References

Rosati G, Modeo L, Melai M, Petroni G, Verni F (2004) A multidisciplinary approach to describe protists: a morphological, ultrastructural, and molecular study on *Peritromus kahli* (Ciliophora, Heterotrichea). J Eukaryot Microbiol 51:49–59

Song WB, Wilbert N (1997) Morphological studies on some free living ciliates (Ciliophora: Heterotrichida, Hypotrichida) from marine biotopes in Qingdao, China, with descriptions of three new species: *Holosticha warreni* nov. spec, *Tachysoma ovata* nov. spec. and *T. dragescoi* nov. spec. Eur J Protistol 33:48–62

Class Heterotrichea Stein, 1859
Order Heterotrichida Stein, 1859
Family Peritromidae Stein, 1867
*Peritromus faurei* Kahl, 1932

**Ecological feature:** mangrove
**Water temperature:** 23.0 °C
**pH:** 6.8
**Salinity:** 15.0‰

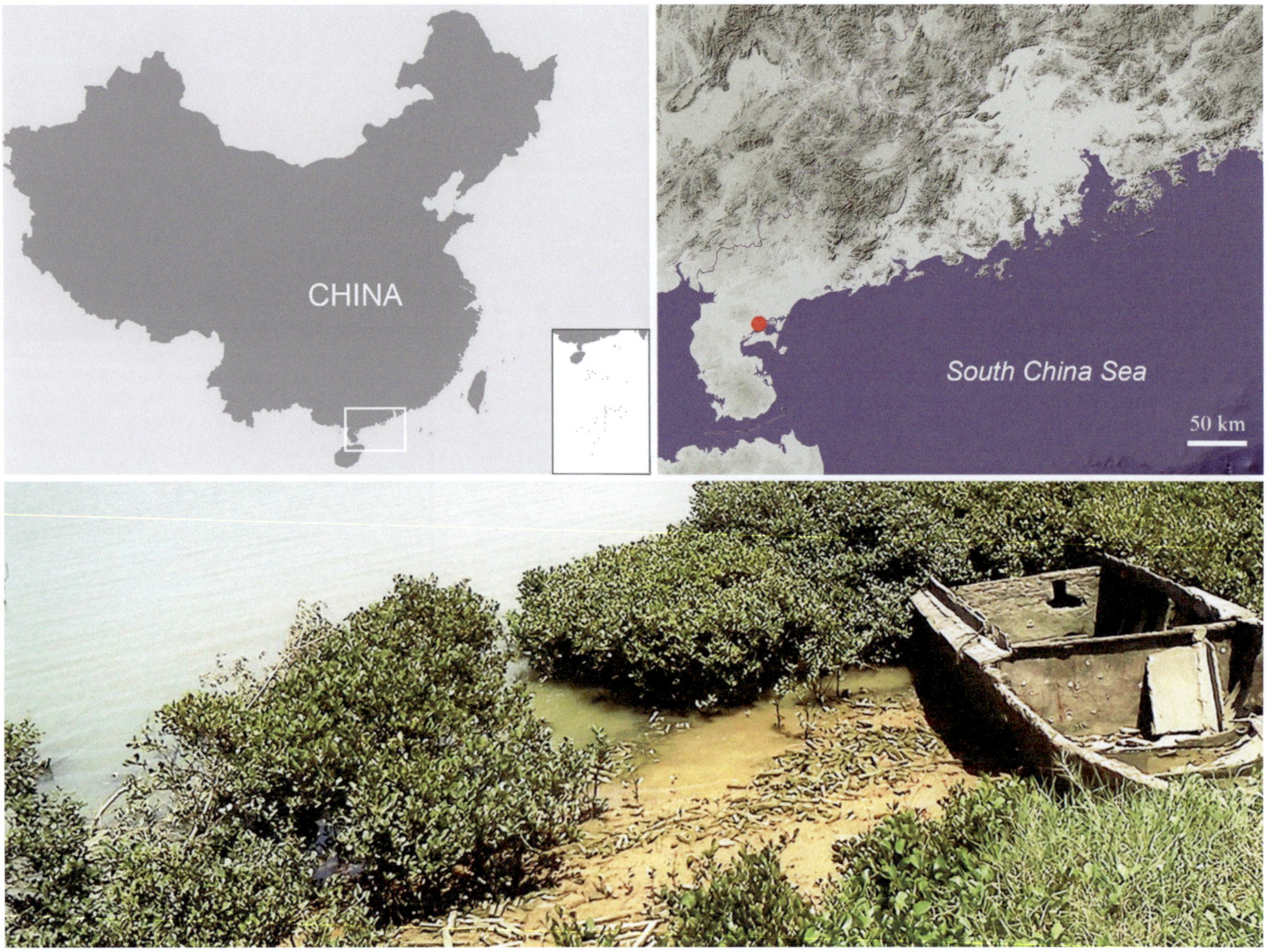


**Fig. 11A** Sample site

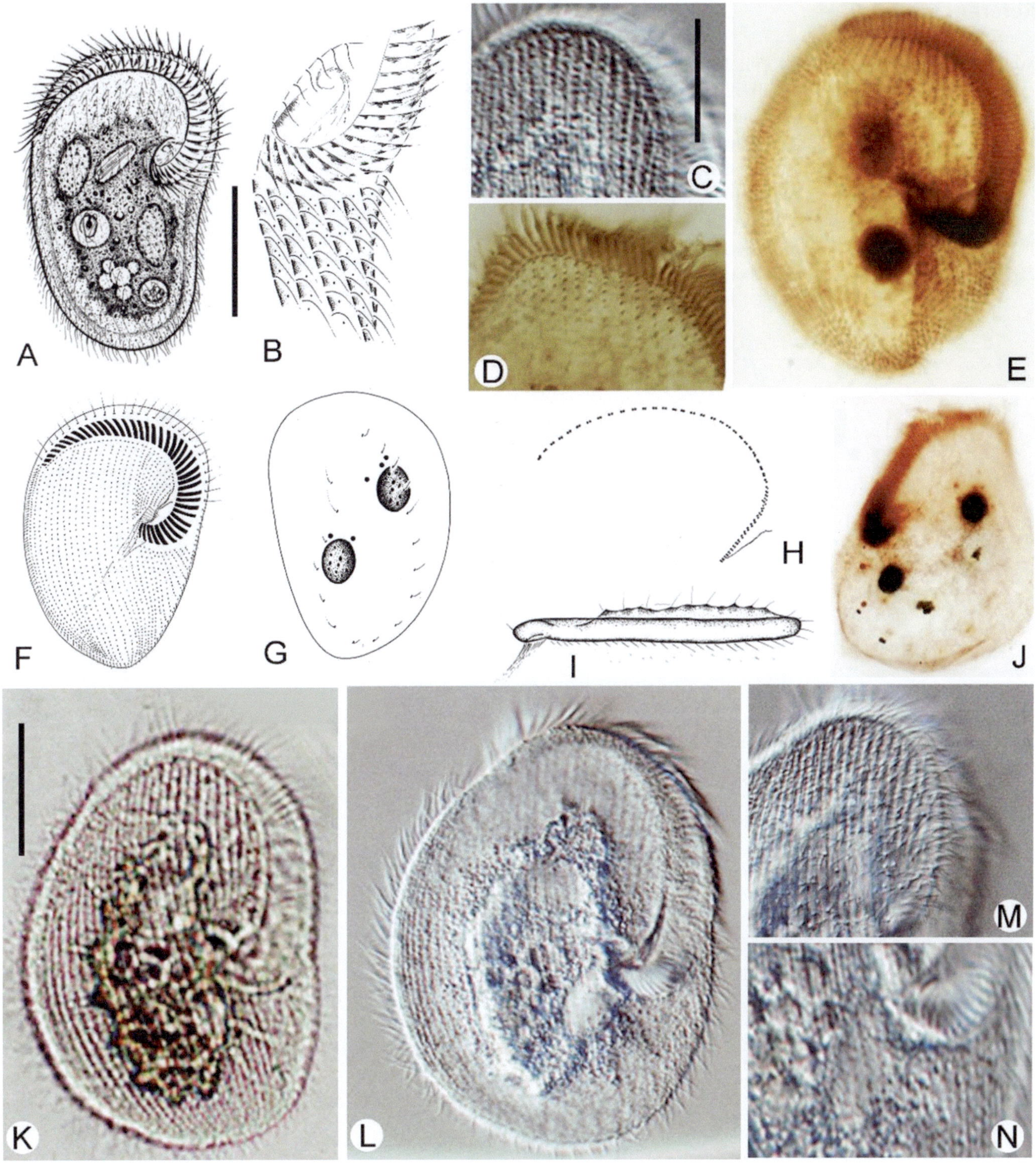


**Fig. 11B** *Peritromus faurei*
(**a**, **k**, **l**) Ventral views of live individuals; (**b**) detail of the buccal field and membranelles; (**c**) paroral kinety; (**d**) the anterior bristles; (**e**) ventral ciliature; (**f**, **g**) ventral and dorsal views of ciliature; (**h**) detail of the paroral kinety and paroral membrane; (**i**) lateral view of a live cell; (**j**) dorsal kinety and nuclei; (**m**) distribution of cortical granules; (**n**) posterior kinety. Scale bars: 40 μm

## Reference

Song WB, Wilbert N (1997) Morphological studies on some free living ciliates (Ciliophora: Heterotrichida, Hypotrichida) from marine biotopes in Qingdao, China, with descriptions of three new species: *Holosticha warreni* nov. spec, *Tachysoma ovata* nov. spec. and *T. dragescoi* nov. spec. Eur J Protistol 33:48–62

Class Heterotrichea Stein, 1859
Order Heterotrichida Stein, 1859
Family Spirostomidae Stein, 1867
*Spirostomum intermedium* Kahl, 1932

**Ecological feature:** mangrove
**Water temperature:** 26.0 °C
**pH:** 7.0
**Salinity:** 2.7‰

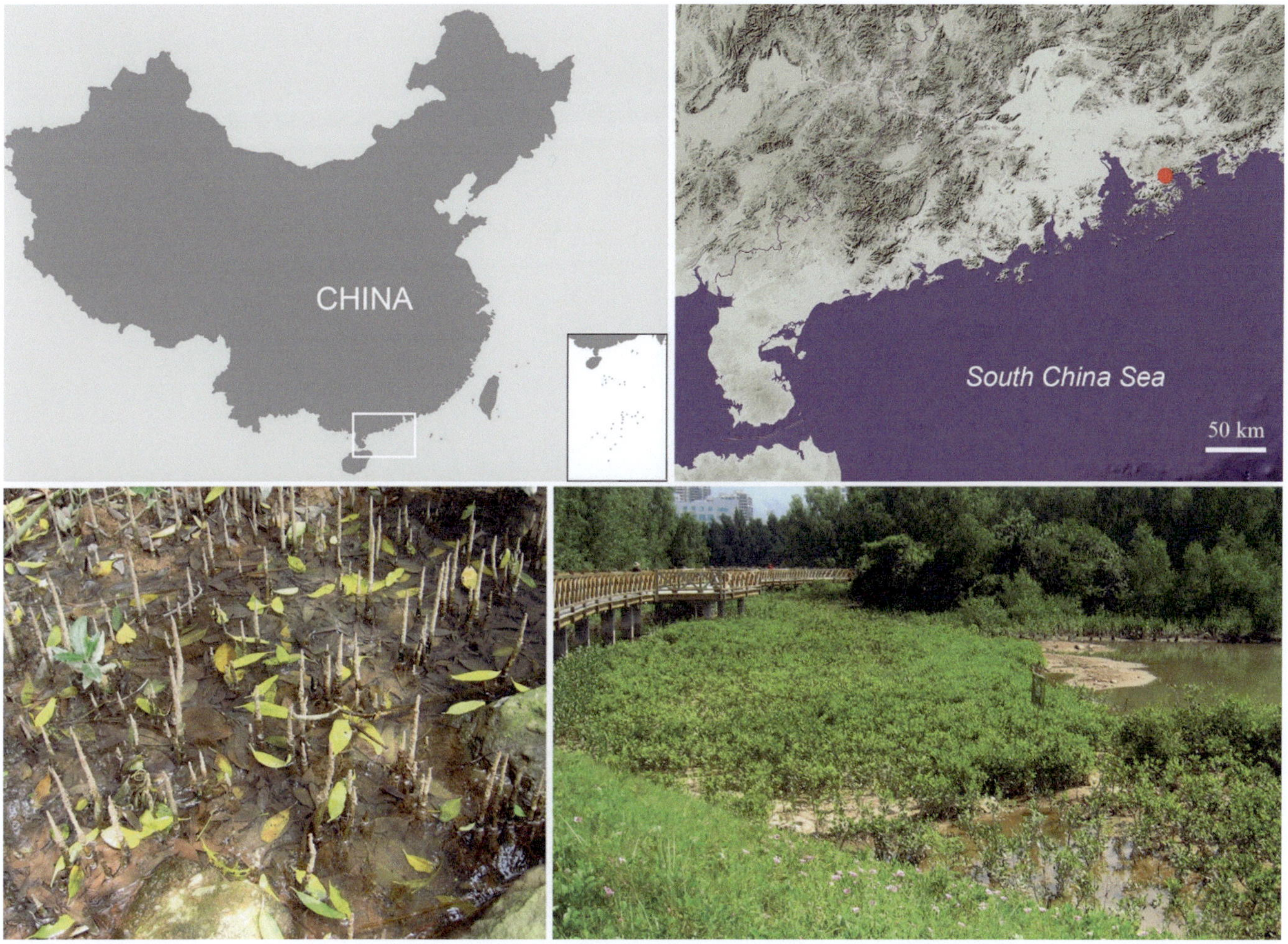


**Fig. 12A** Sample site

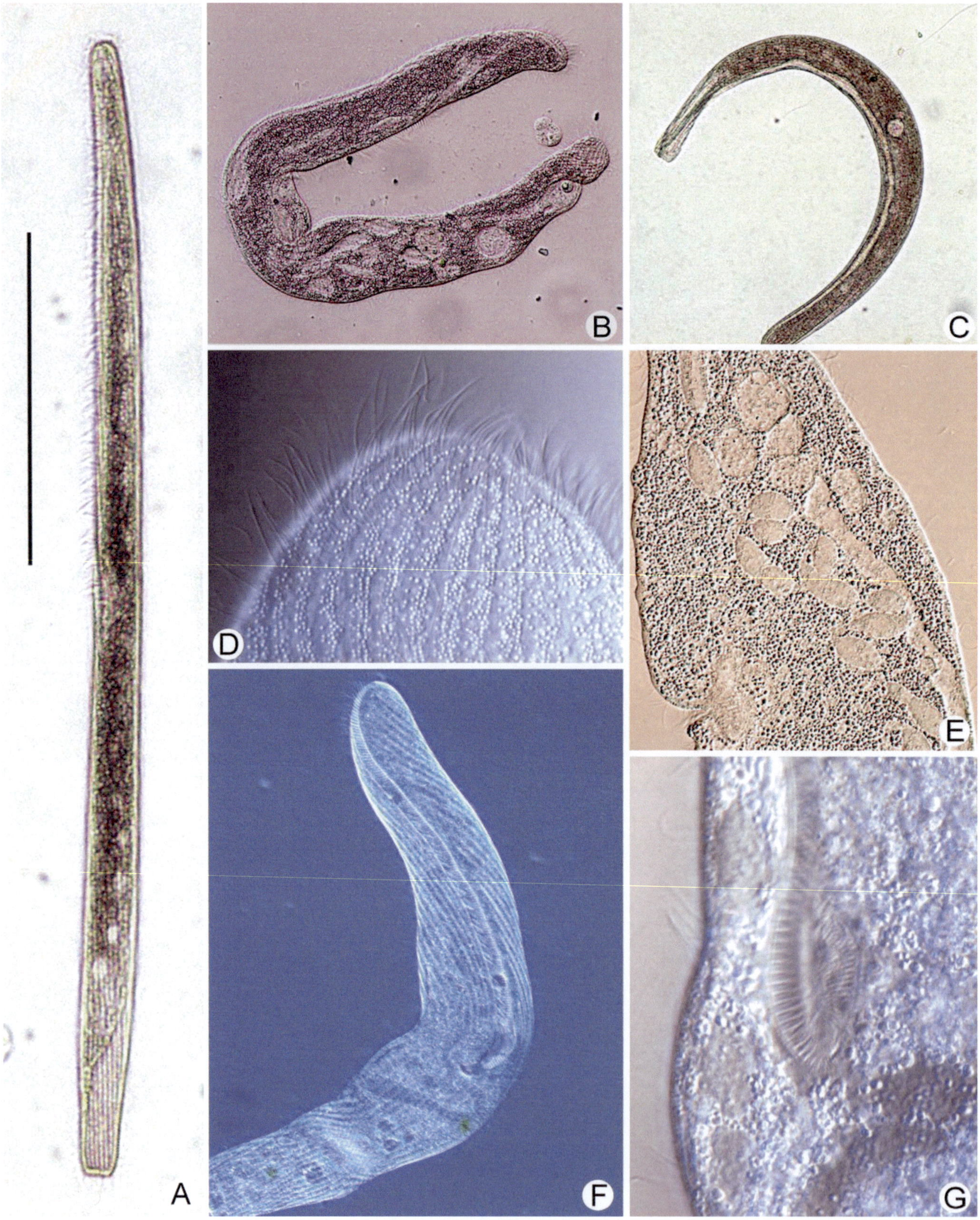


**Fig. 12B** *Spirostomum intermedium*
(**a**) Lateral view of a live cell; (**b**) contorted cell; (**c**) a live cell showing contractile vacuole with collecting canal; (**d**) anterior part of cell, to show the cortical granules; (**e**) macronuclei; (**f**) anterior part of body, to show the oral slit; (**g**) ventral side view, to show the oral base. Scale bar: 250 μm

## References

Boscaro V, Carducci D, Barbieri G, Senra MV, Andreoli I, Erra F, Petroni G, Verni F, Fokin SI (2014) Focusing on genera to improve species identification: revised systematics of the ciliate *Spirostomum*. Protist 165:527–541

Repak A, Isquith I (1974) The systematics of the genus *Spirostomum* Ehrenberg, 1838. Acta Protozool 12:325–333

Class Heterotrichea Stein, 1859
Order Heterotrichida Stein, 1859
Family Spirostomidae Stein, 1867
*Spirostomum teres* Claparède & Lachmann, 1859

**Ecological feature:** mangrove
**Water temperature:** 26.3 °C
**pH:** 8.0
**Salinity:** 0.5‰

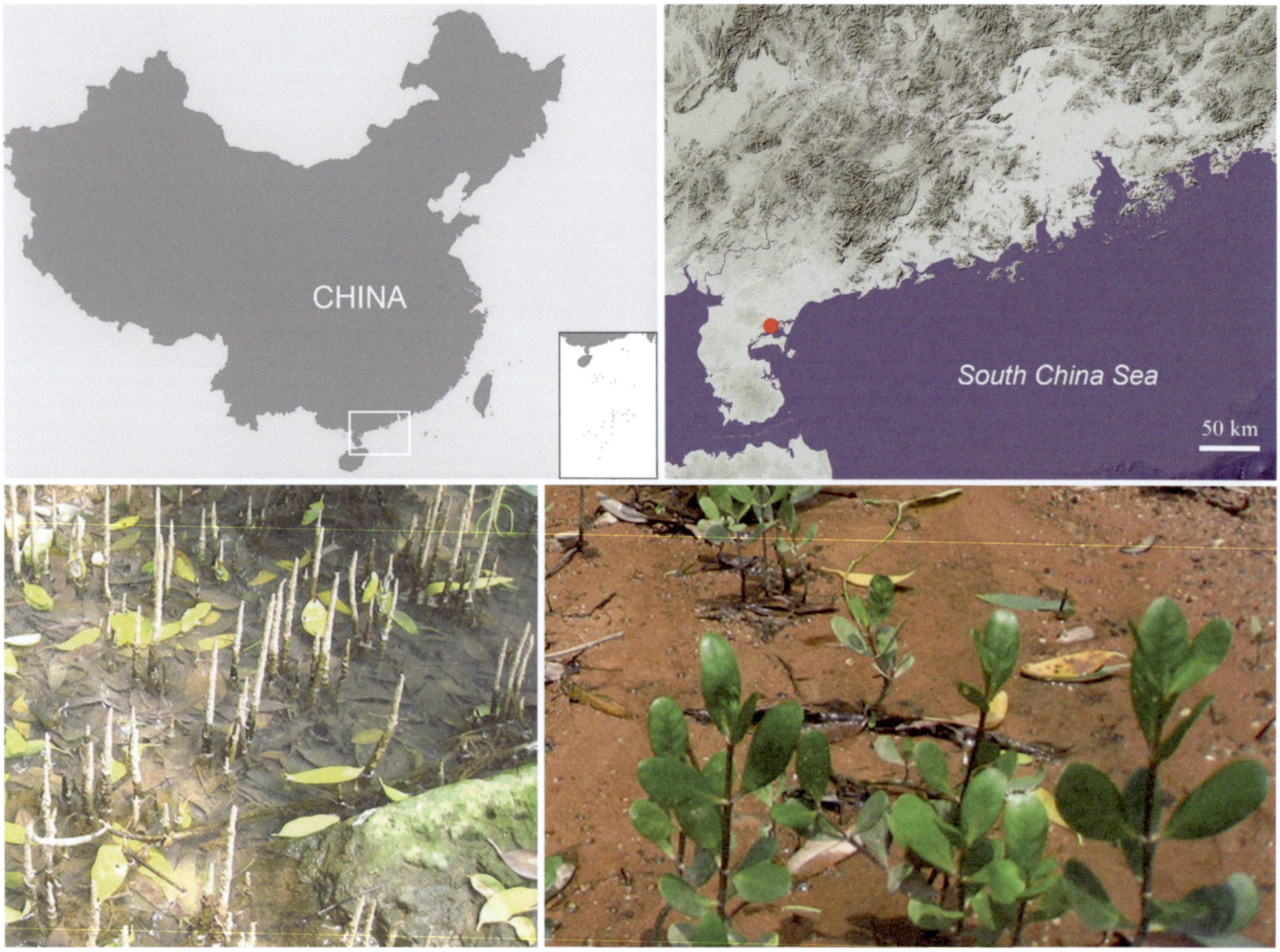


**Fig. 13A** Sample site

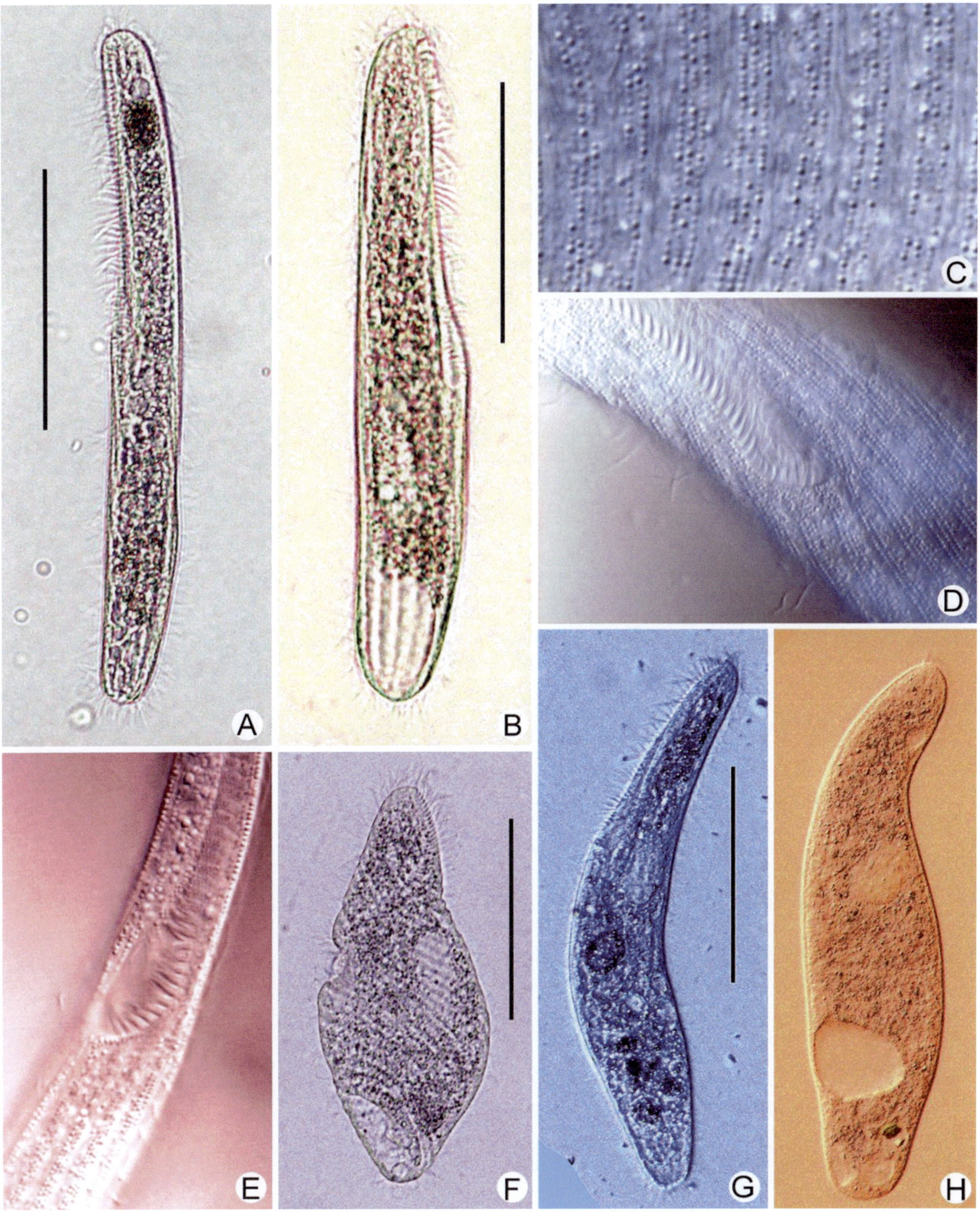


**Fig. 13B** *Spirostomum teres*
(**a**, **b**, **g**) Lateral views of live cells; (**c**) distribution of cortical granules; (**d**, **e**) ventral side view, to show the slit oral base; (**f**) contorted specimen; (**h**) a contracted cell with a subterminal vacuole. Scale bars: 250 μm

## References

Boscaro V, Carducci D, Barbieri G, Senra MV, Andreoli I, Erra F, Petroni G, Verni F, Fokin SI (2014) Focusing on genera to improve species identification: revised systematics of the ciliate *Spirostomum*. Protist 165:527–541

Fernandes NM, Neto IDDS (2013) Morphology and 18S rDNA gene sequence of *Spirostomum minus* and *Spirostomum teres* (Ciliophora: Heterotrichea) from Rio de Janeiro, Brazil. Zoologia (Curitiba) 30:72–79

# Class Armophorea

Xiaozhong Hu, Xiaofeng Lin, and Weibo Song

Class Armophorea Lynn, 2004
Order Armophorida Jankowski, 1964
Family Metopidae Kahl, 1927
*Metopus contortus* (Quennerstedt, 1867) Kahl, 1932

**Ecological feature:** mangrove
**Water temperature:** no data available
**pH:** 6.5
**Salinity:** 5.0‰

X. Hu (✉)
Fisheries College, Ocean University of China,
Qingdao, Shandong, China
e-mail: xiaozhonghu@ouc.edu.cn

X. Lin
College of Life Science, South China Normal University,
Guangzhou, Guangdong, China
e-mail: xlin@oucscnu.edu.cn

W. Song
Institute of Evolution & Marine Biodiversity,
Ocean University of China, Qingdao, Shandong, China
e-mail: wsong@ouc.edu.cn


X. Hu et al. (eds.), *Ciliate Atlas: Species Found in the South China Sea*,
https://doi.org/10.1007/978-981-13-5901-9_4

**Fig. 1A** Sample site

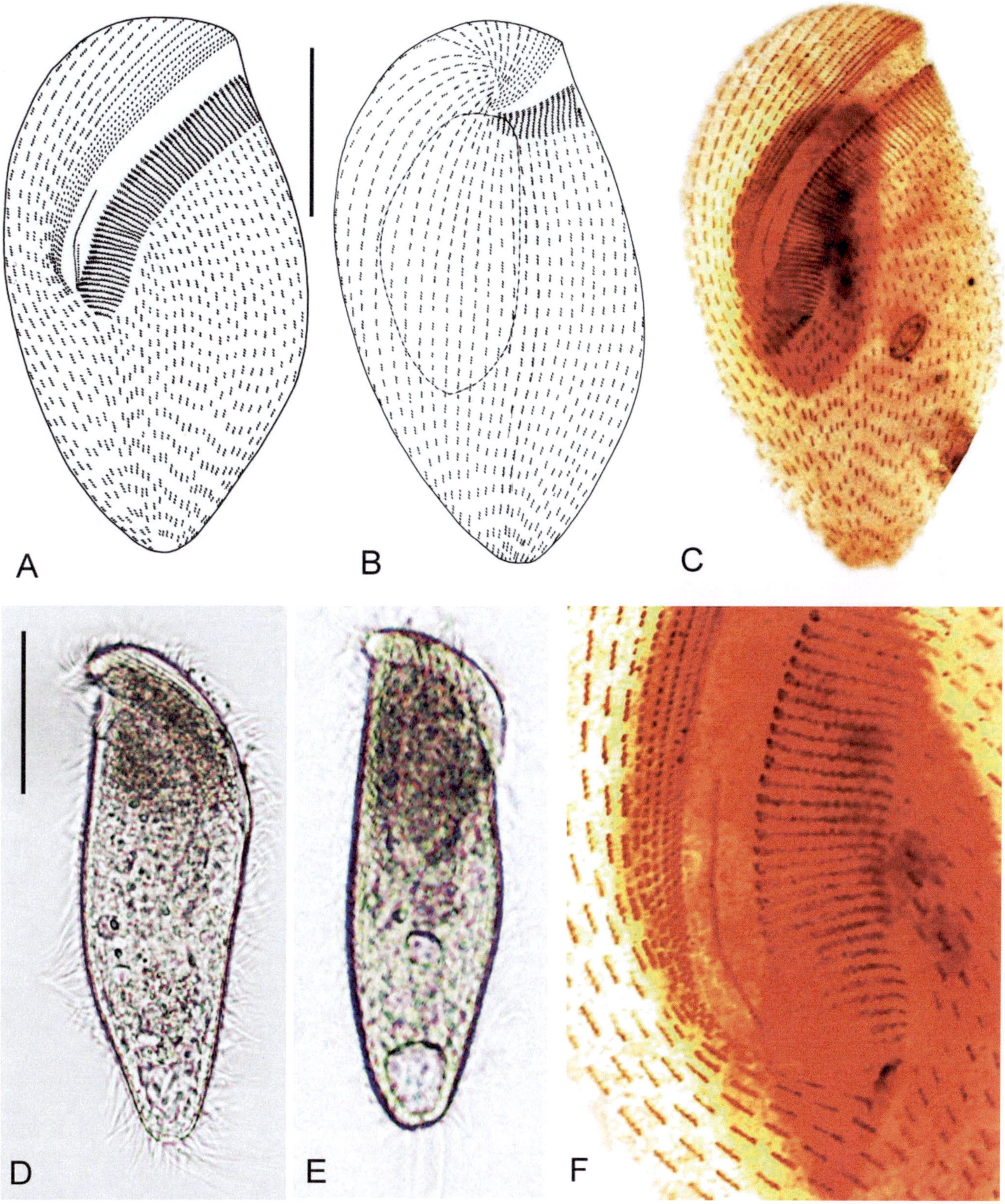


**Fig. 1B** *Metopus contortus*
(**a**–**c**) Ventral views of ciliature; (**d**, **e**) different cells showing contractile vacuole; (**f**) adoral membranelles and paroral membrane. Scale bars: 50 μm

## References

Dragesco J (1996) Infraciliature et morphométrie de cinq espèces de ciliés mésopsammiques méditerranéens. Cah Boil Mar 37:261–293

Esteban G, Fenchel T, Finlay B (1995) Diversity of free-living morphospecies in the ciliate genus Metopus Arch Protistenk 146:137–164

Kahl A (1932) Urtiere oder Protozoa I: Wimpertiere oder Ciliata (Infusoria) 3. Spirotricha. Tierwelt Dtl 25:399–650

# Class Litostomatea

Xiaozhong Hu, Xiaofeng Lin, and Weibo Song

Class Litostomatea Small & Lynn, 1981
Order Haptorida Corliss, 1974
Family Acropisthiidae Foissner & Foissner, 1988
*Chaenea* cf. *vorax*

**Ecological feature:** mollusc-farming pond
**Water temperature:** 19.4 °C
**pH:** 8.2
**Salinity:** 32.6‰

X. Hu (✉)
Fisheries College, Ocean University of China,
Qingdao, Shandong, China
e-mail: xiaozhonghu@ouc.edu.cn

X. Lin
College of Life Science, South China Normal University,
Guangzhou, Guangdong, China
e-mail: xlin@oucscnu.edu.cn

W. Song
Institute of Evolution & Marine Biodiversity,
Ocean University of China, Qingdao, Shandong, China
e-mail: wsong@ouc.edu.cn


X. Hu et al. (eds.), *Ciliate Atlas: Species Found in the South China Sea*,
https://doi.org/10.1007/978-981-13-5901-9_5

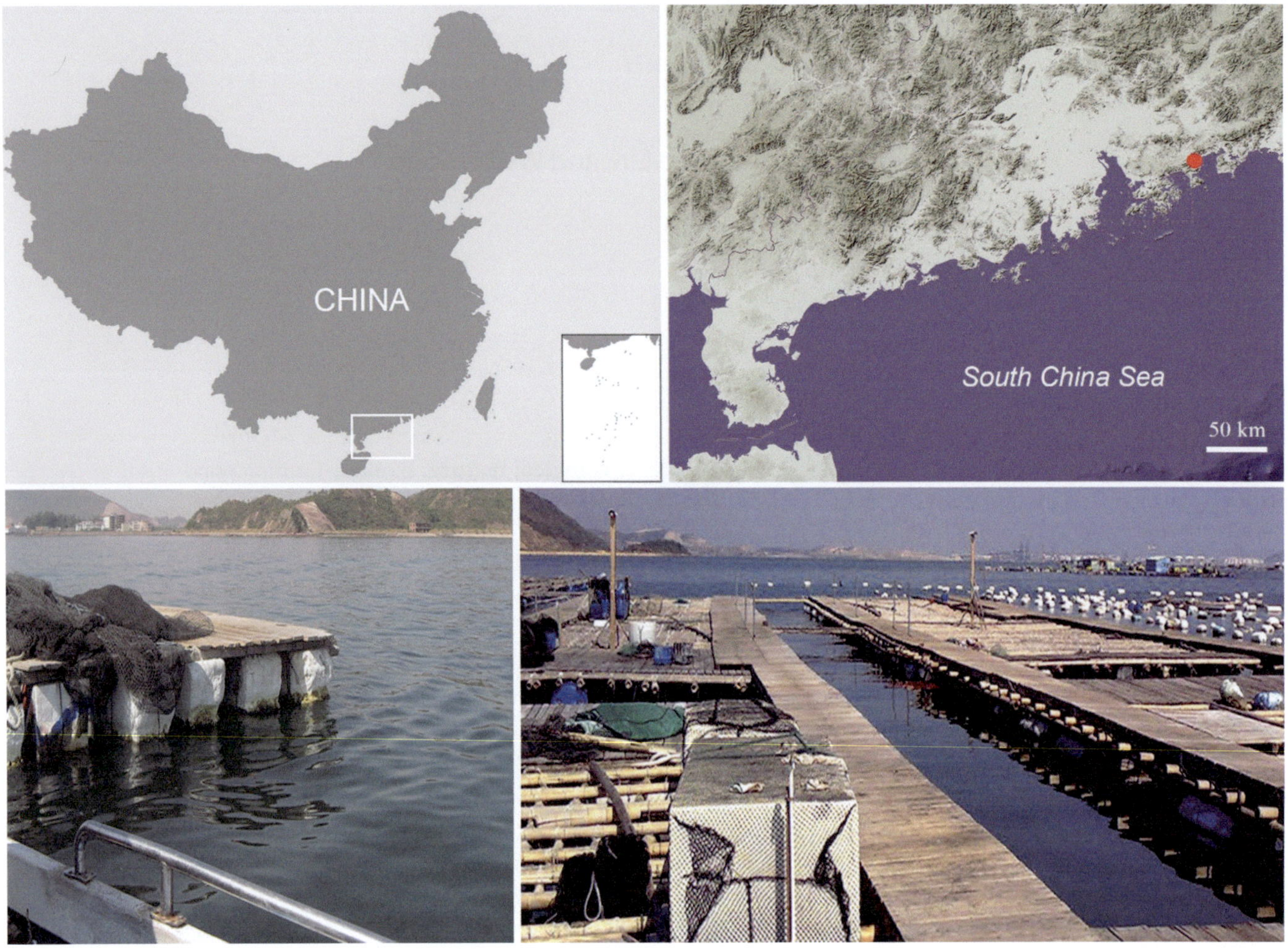


**Fig. 1A** Sample site

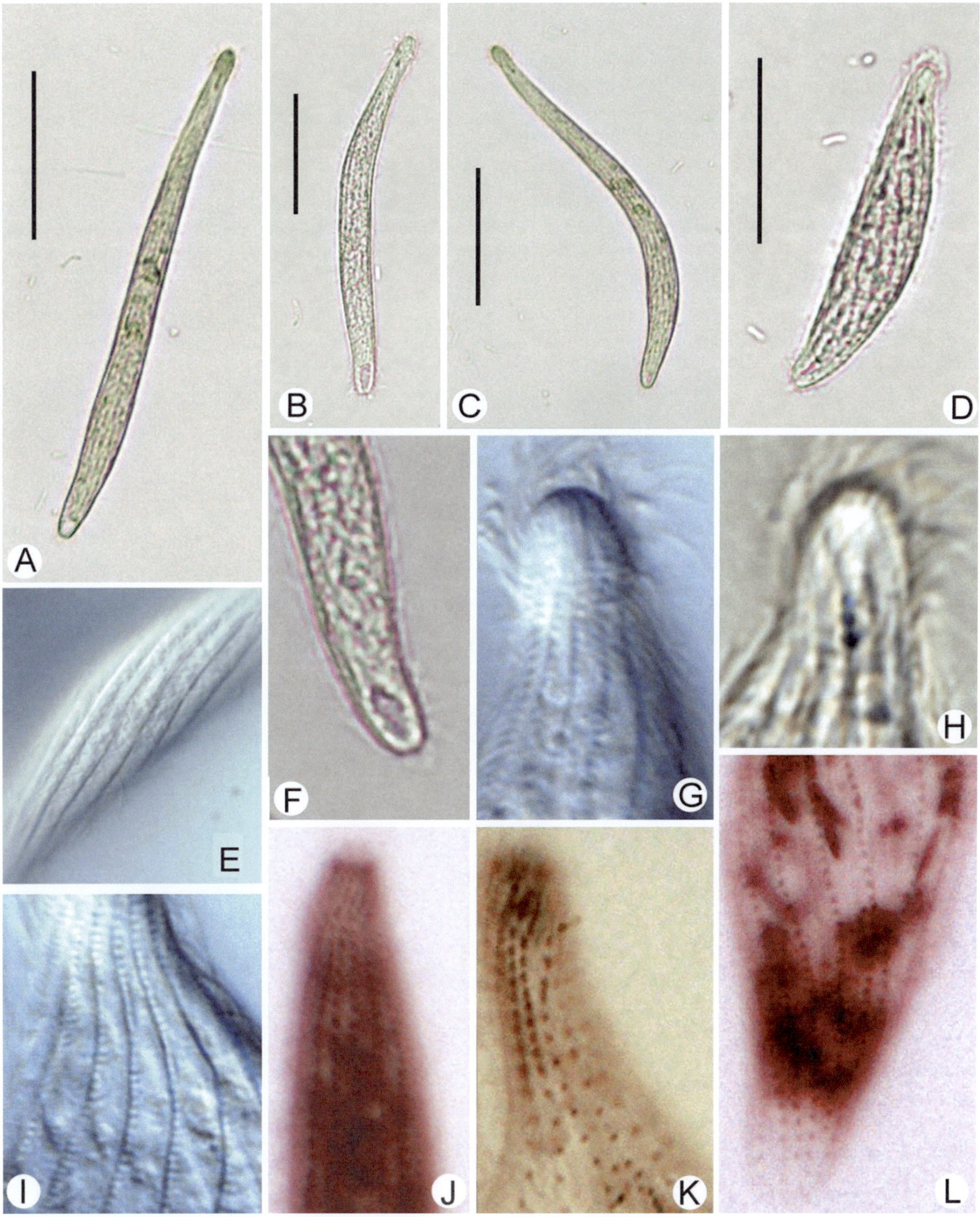


**Fig. 1B** *Chaenea* cf. *vorax*
(**a–d**) Lateral views of live cells; (**e**) spiraled ciliary rows; (**f**) posterior portion of cell showing contractile vacuole; (**g, h**) anterior portion of cell showing brush bristles; (**i**) somatic cilia; (**j**, **l**) anterior (**j**) and posterior (**l**) portion of cells showing somatic kineties; (**k**) showing brush kineties. Scale bars: 50 μm

## Reference

Song WB, Packroff G (1997) Taxonomische Untersuchungen an marinen Ciliaten aus China mit Beschreibungen von zwei neuen Arten, *Strombidium globosaneum* nov. spec. und *S. platum* nov. spec. (Protozoa, Ciliophora). Arch Protistenk 147:331–360

Class Litostomatea Small & Lynn, 1981
Order Haptorida Corliss, 1974
Family Acropisthiidae Foissner & Foissner, 1988
*Chaenea paucistriata* Fan et al., 2015

**Ecological feature:** coastal water
**Water temperature:** 21.0 °C
**pH:** 7.5
**Salinity:** 30.0‰

**Fig. 2A** Sample site

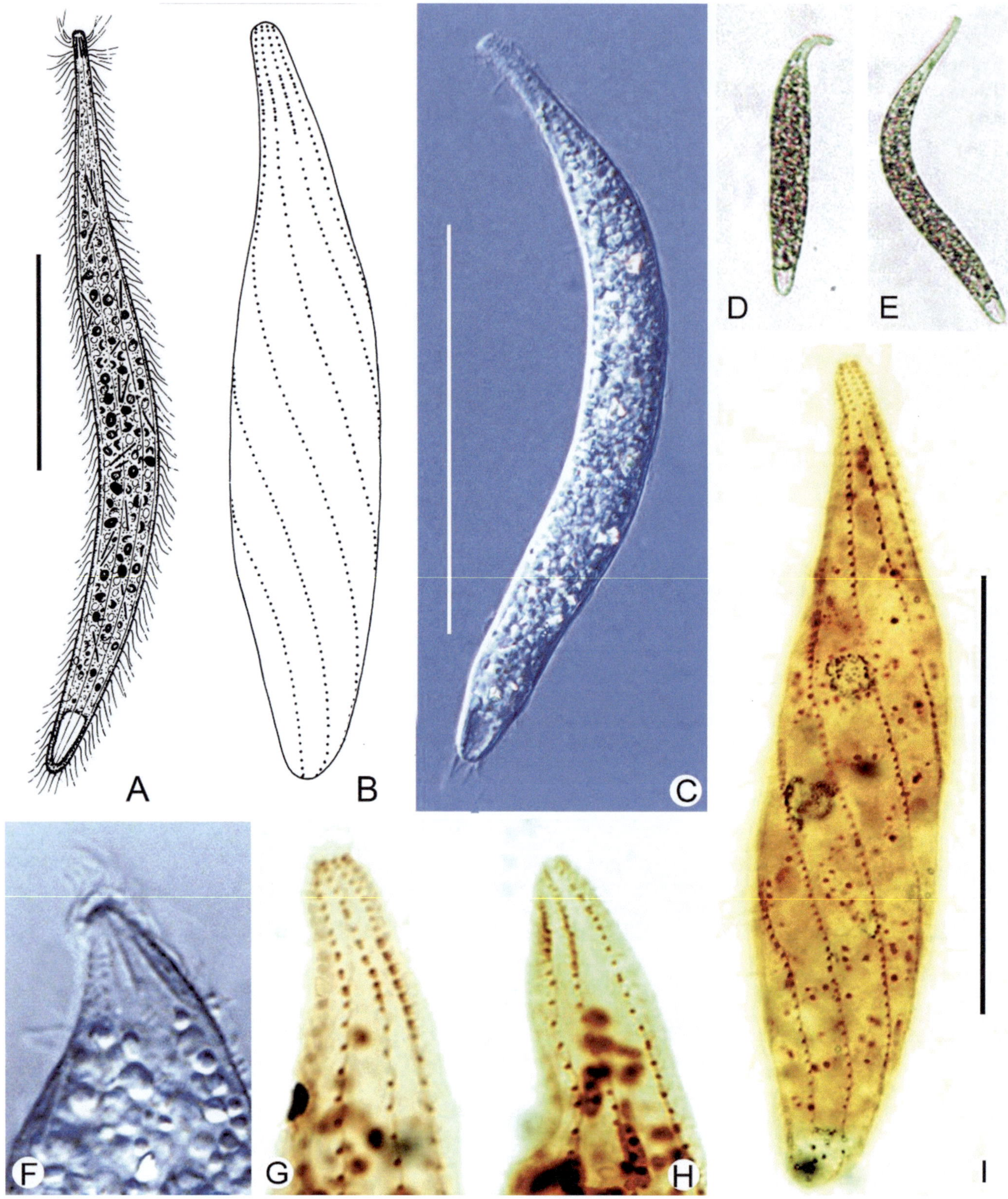


**Fig. 2B** *Chaenea paucistriata*
(**a**, **c**) General views of live cells; (**b**, **i**) general views of ciliature; (**d**, **e**) shape variance of the same body; (**f**) anterior portion of cell to show rod-shaped extrusomes; (**g**, **h**) anterior portion of cell showing the dorsal brush kineties 1–4. Scale bars: 50 μm (**a**); 90 μm (**c**); 70 μm (**i**)

## Reference

Fan X, Xu Y, Gu F, Li J, Al-Farraj SA, Al-Rasheid KAS, Hu X (2015) Morphology of two novel species of *Chaenea* (Ciliophora, Litostomatea): *Chaenea paucistriata* spec. nov. and *C. sinica* spec. nov. Acta Protozool 54:97–106

Class Litostomatea Small & Lynn, 1981
Order Haptorida Corliss, 1974
Family Acropisthiidae Foissner & Foissner, 1988
*Chaenea* sp.

**Ecological feature:** off-shore water
**Water temperature:** 23.5 °C
**pH:** 7.6
**Salinity:** 34.5‰

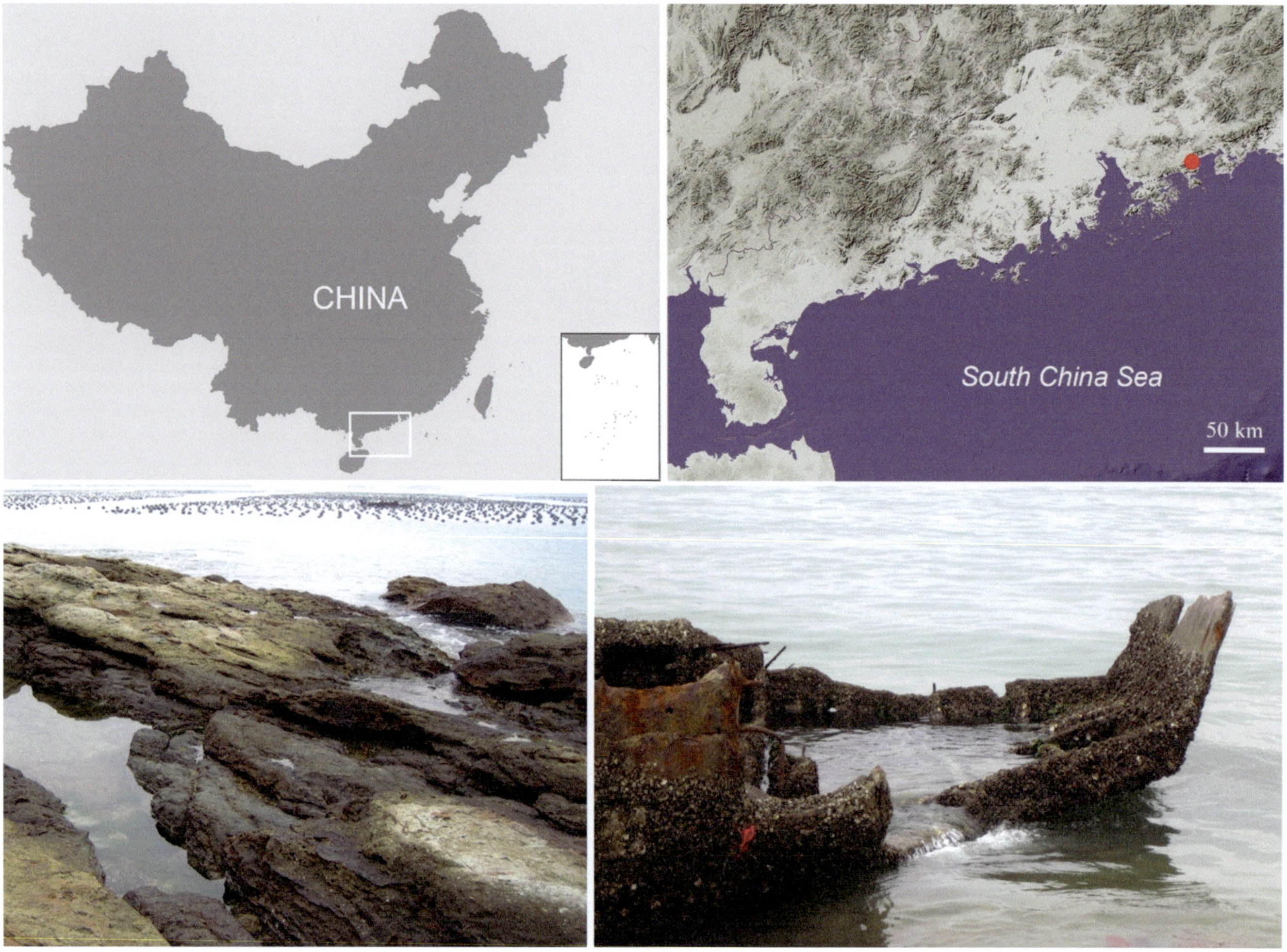


**Fig. 3A** Sample site

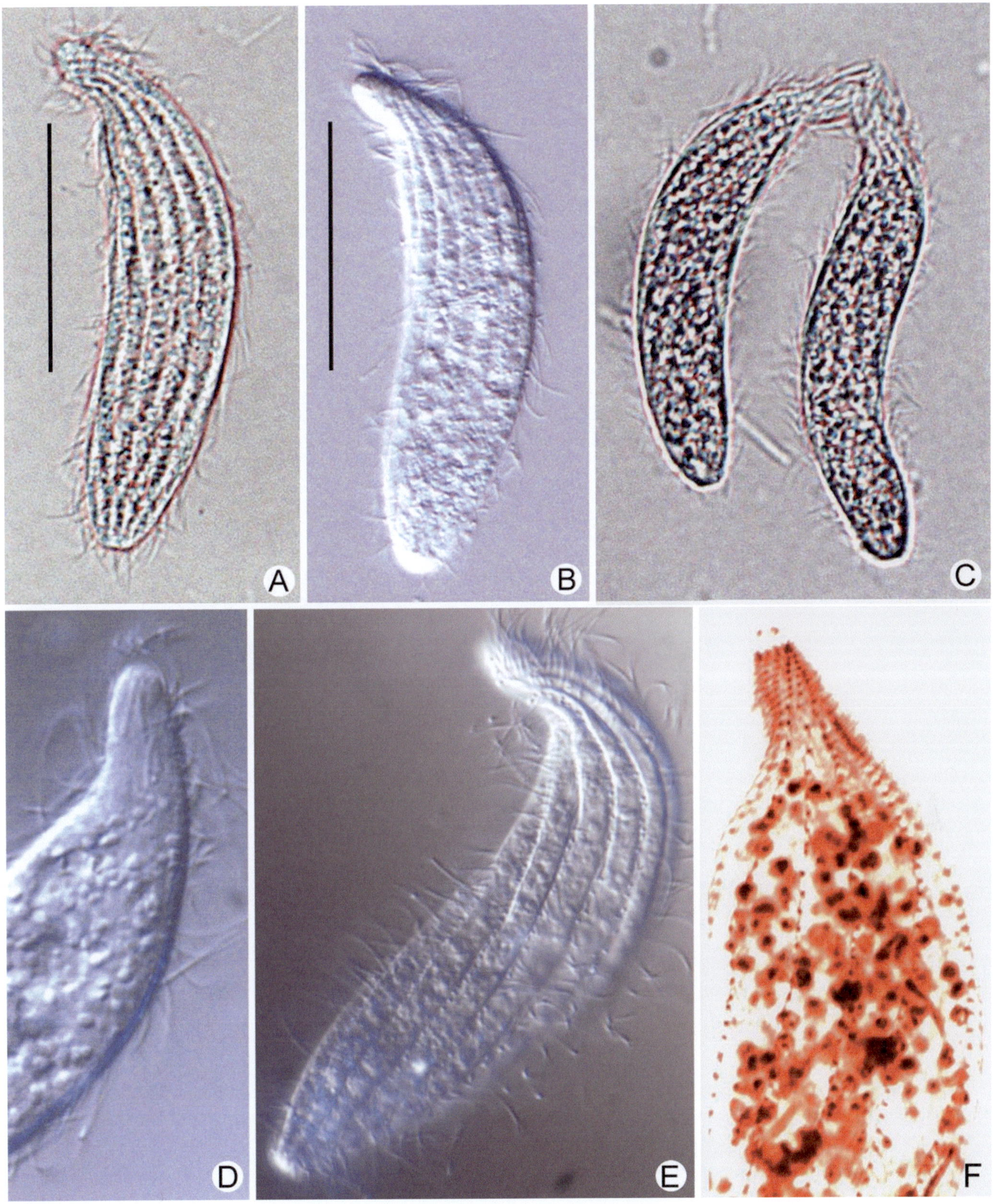


**Fig. 3B** *Chaenea* sp.
(**a**, **b**, **e**) Lateral views of live individuals; (**c**) conjugants; (**d**) anterior part of body, showing extrusomes; (**f**) lateral view of ciliature. Scale bars: 50 μm

## References

Carey PG (1992) Marine interstitial ciliates: an illustrated key. Chapman & Hall, London

Kwon CB, Vďačný P, Shazib SUA, Shin MK (2014) Morphology and molecular phylogeny of a new haptorian ciliate, *Chaenea mirabilis* sp. n., with implications for the evolution of the dorsal brush in haptorians (Ciliophora, Litostomatea). J Eukaryot Microbiol 61:278–292

Class Litostomatea Small & Lynn, 1981
Order Haptorida Corliss, 1974
Family Acropisthiidae Foissner & Foissner, 1988
*Chaenea teres* (Dujardin, 1841) Kent, 1881

**Ecological feature:** mangrove
**Water temperature:** 23.9 °C
**pH:** 6.6
**Salinity:** 13.0‰

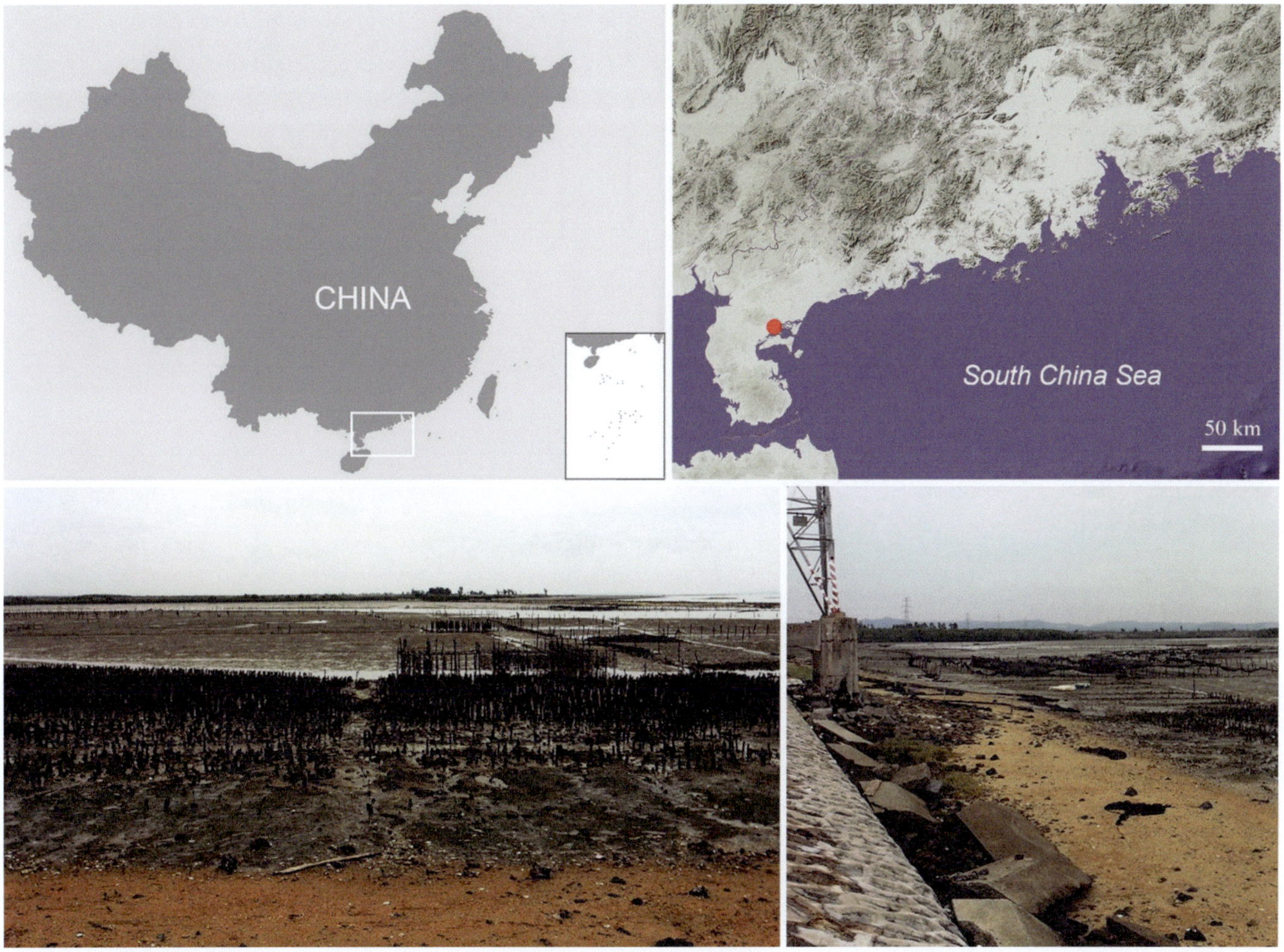


**Fig. 4A** Sample site

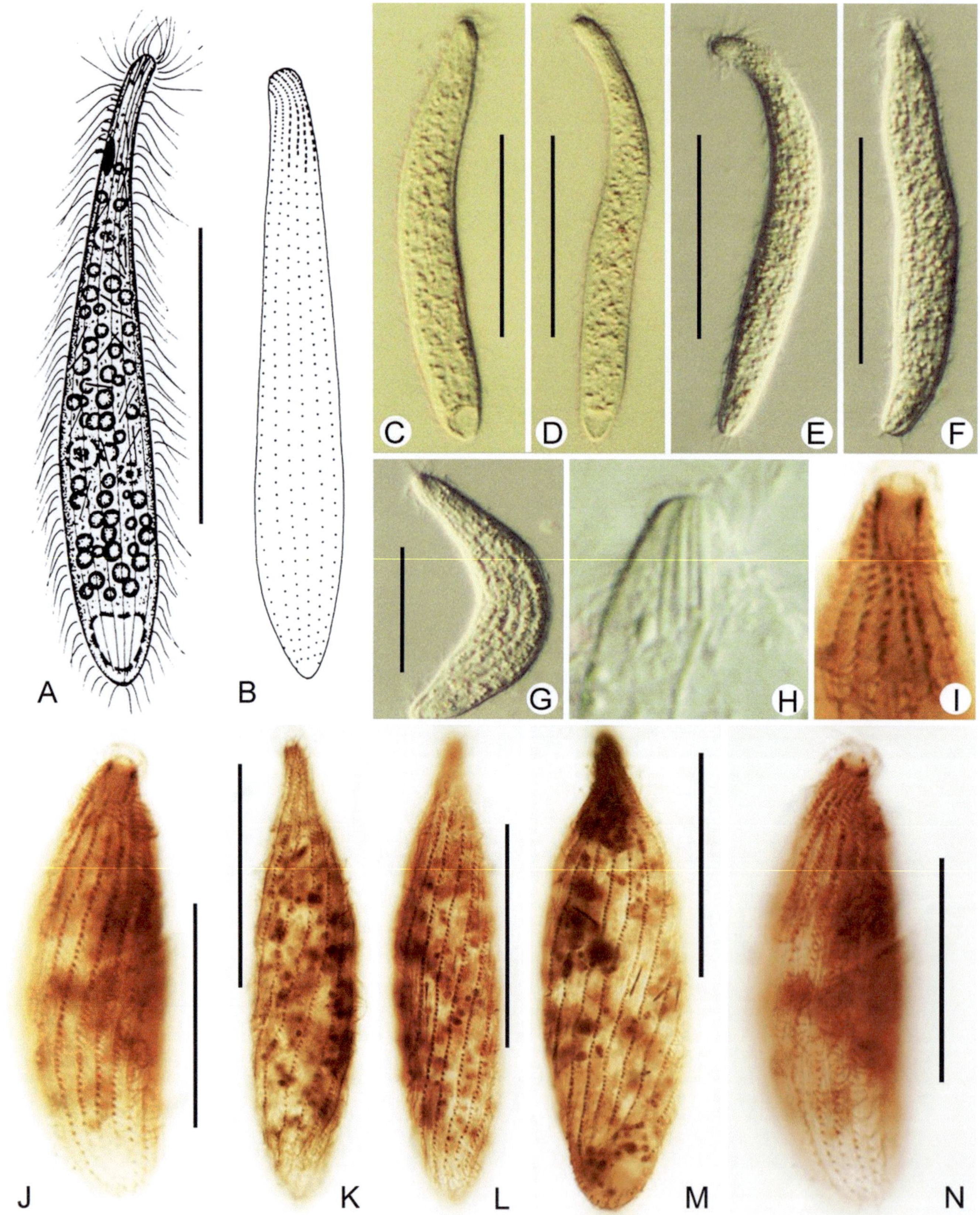


**Fig. 4B** *Chaenea teres*
(**a**, **c**–**g**) Live individuals in variable shapes; (**b**, **i**–**n**) lateral views of ciliature; (**h**) extrusomes in oral area. Scale bars: 45 μm

## Reference

Song W B, Warren A, Hu X (2009) Free-living ciliates in the Bohai and Yellow Seas, China. Science Press, Beijing

Class Litostomatea Small & Lynn, 1981
Order Haptorida Corliss, 1974
Family Lacrymariidae de Fromentel, 1876
*Lacrymaria coronata* Claparède & Lachmann, 1858

**Ecological feature:** mangrove
**Water temperature:** no data available
**pH: 7.0**
**Salinity:** 10.0‰

**Fig. 5A** Sample site

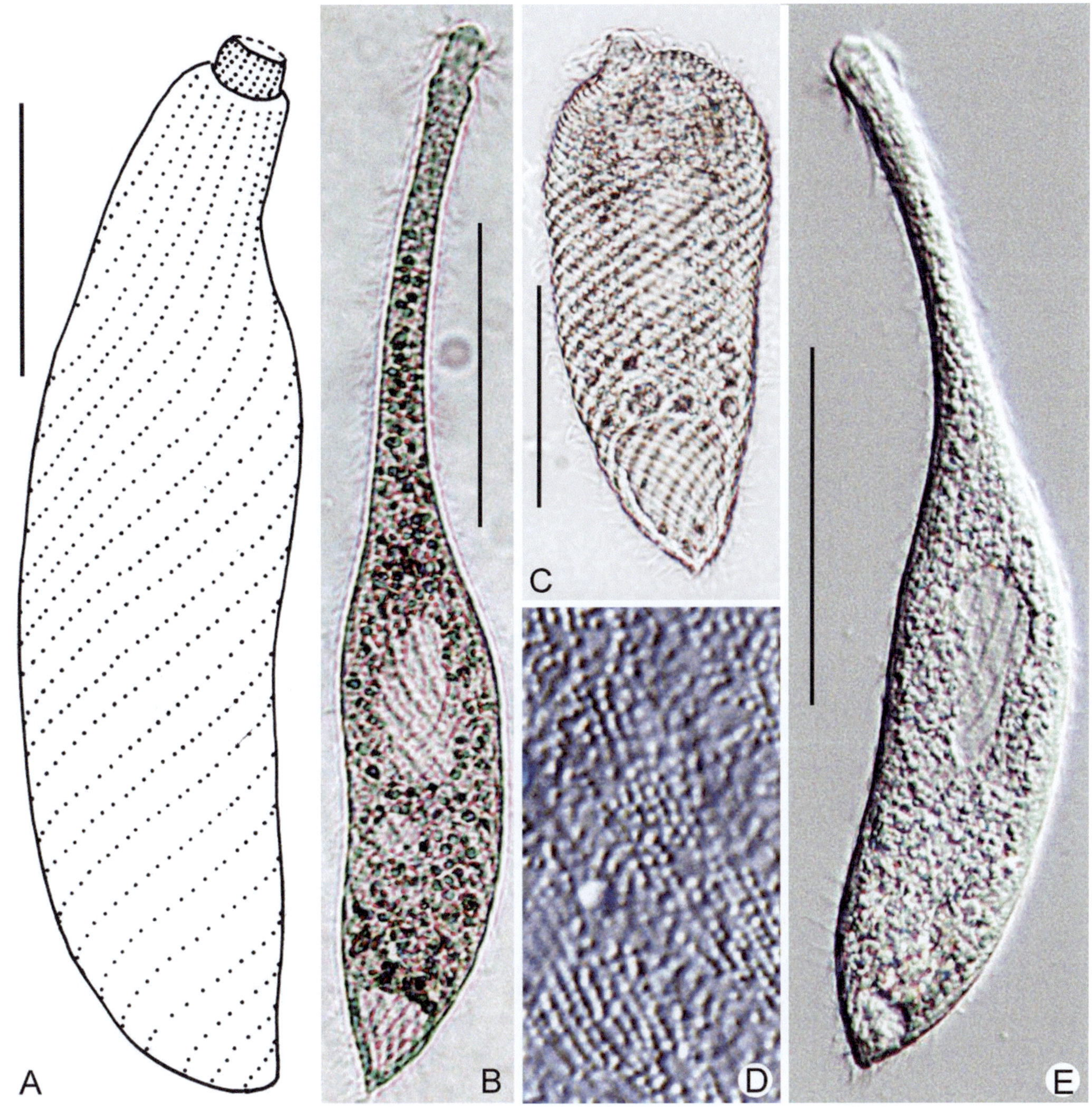


**Fig. 5B** *Lacrymaria coronata*
(**a**) Lateral view of ciliature; (**b**, **e**) extended cells; (**c**) contracted body; (**d**) distribution of cortical granules. Scale bars: 45 μm (**a**, **c**); 95 μm (**b**, **e**)

## References

Azovsky A, Mazei Y (2005) Distribution and community structure of benthic ciliates in the northeastern part of the Black Sea. Protistology 4:83–90

Bock K (1952) Zur Ökologie der Ciliaten des marinen Sandgrundes der Kieler Bucht I. Kieler Meeresforsch 9:77–89

Carey P (1992) Marine interstitial ciliates: an illustrated key. Chapman & Hall, London

Class Litostomatea Small & Lynn, 1981
Order Haptorida Corliss, 1974
Family Lacrymariidae de Fromentel, 1876
*Lacrymaria marina* Kahl, 1933

**Ecological feature:** mangrove
**Water temperature:** 24.8 °C
**pH:** 6.7
**Salinity:** 17.0‰

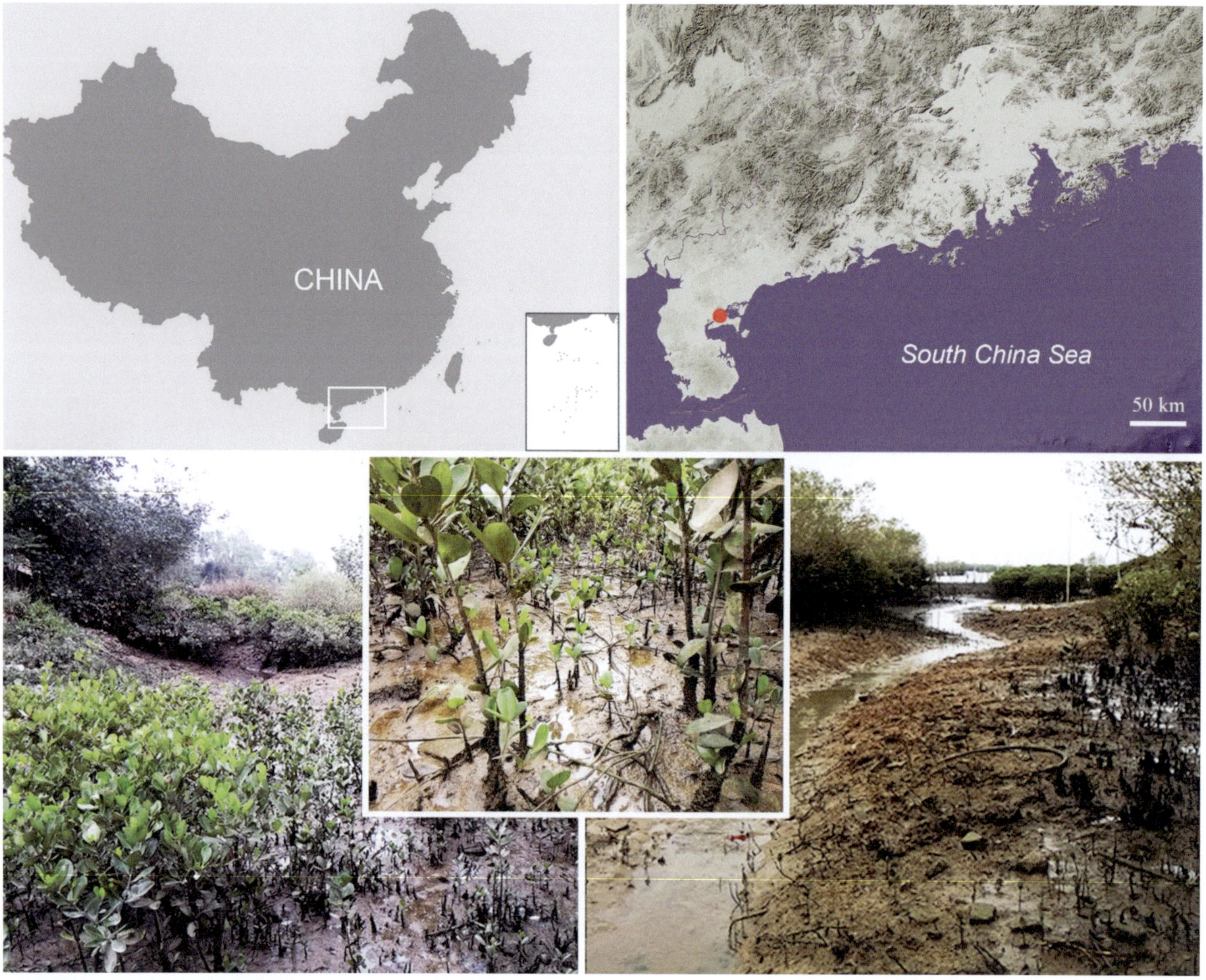


**Fig. 6A** Sample site

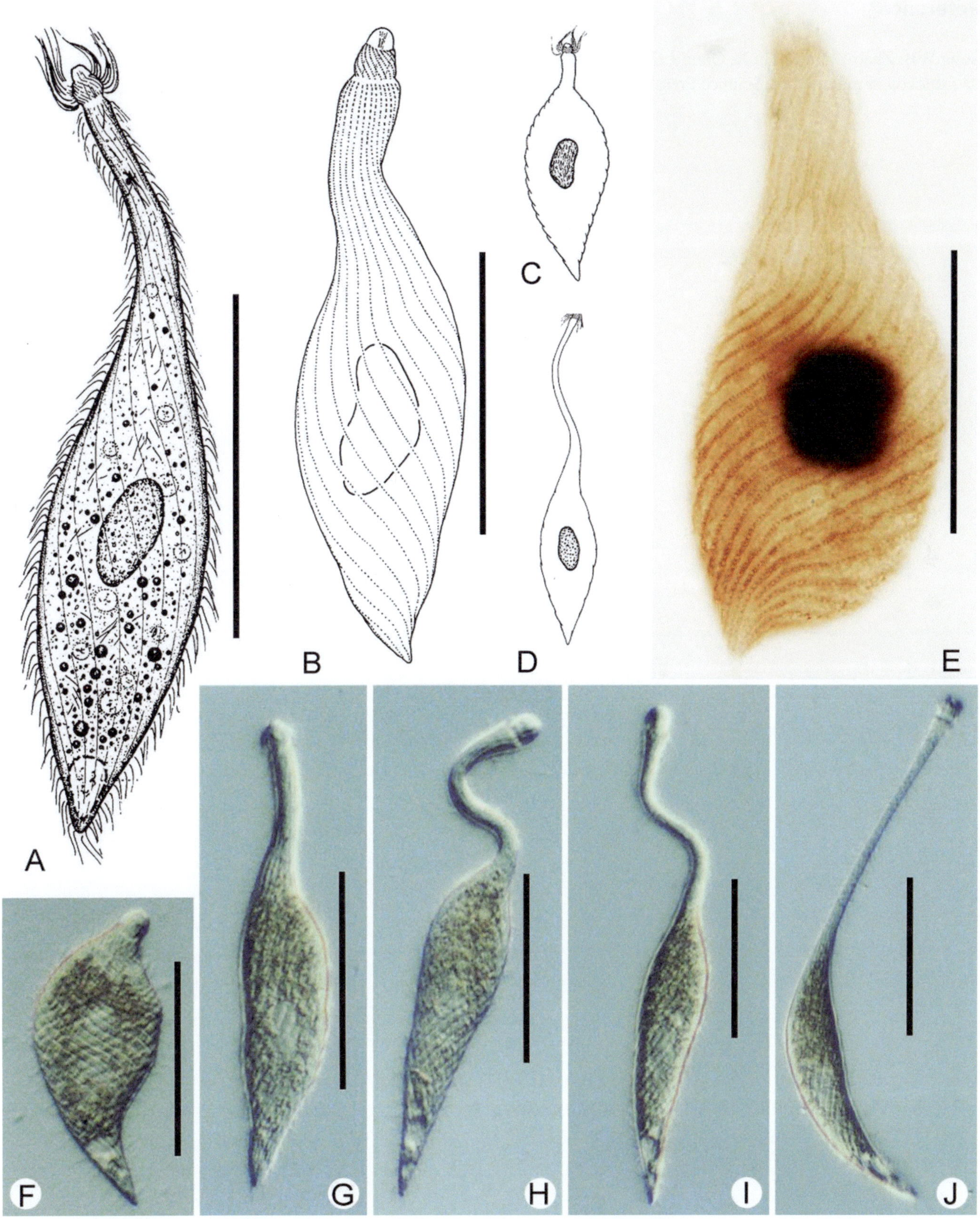


**Fig. 6B** *Lacrymaria marina*
(**a**, **c**, **d**, **f**–**j**) Live individuals in variable shapes; (**b**, **e**) ciliature and macronuclei. Scale bars: 30 μm

## Reference

Song WB, Zhao Y, Xu K, Hu X, Gong J (2003) Pathogenetic protozoa in mariculture. Science Press, Beijing

Class Litostomatea Small & Lynn, 1981
Order Haptorida Corliss, 1974
Family Lacrymariidae de Fromentel, 1876
*Lacrymaria maurea* Dragesco, 1966

**Ecological feature:** mangrove
**Water temperature:** no data available
**pH:** 7.0
**Salinity:** 10.0‰

**Fig. 7A** Sample site

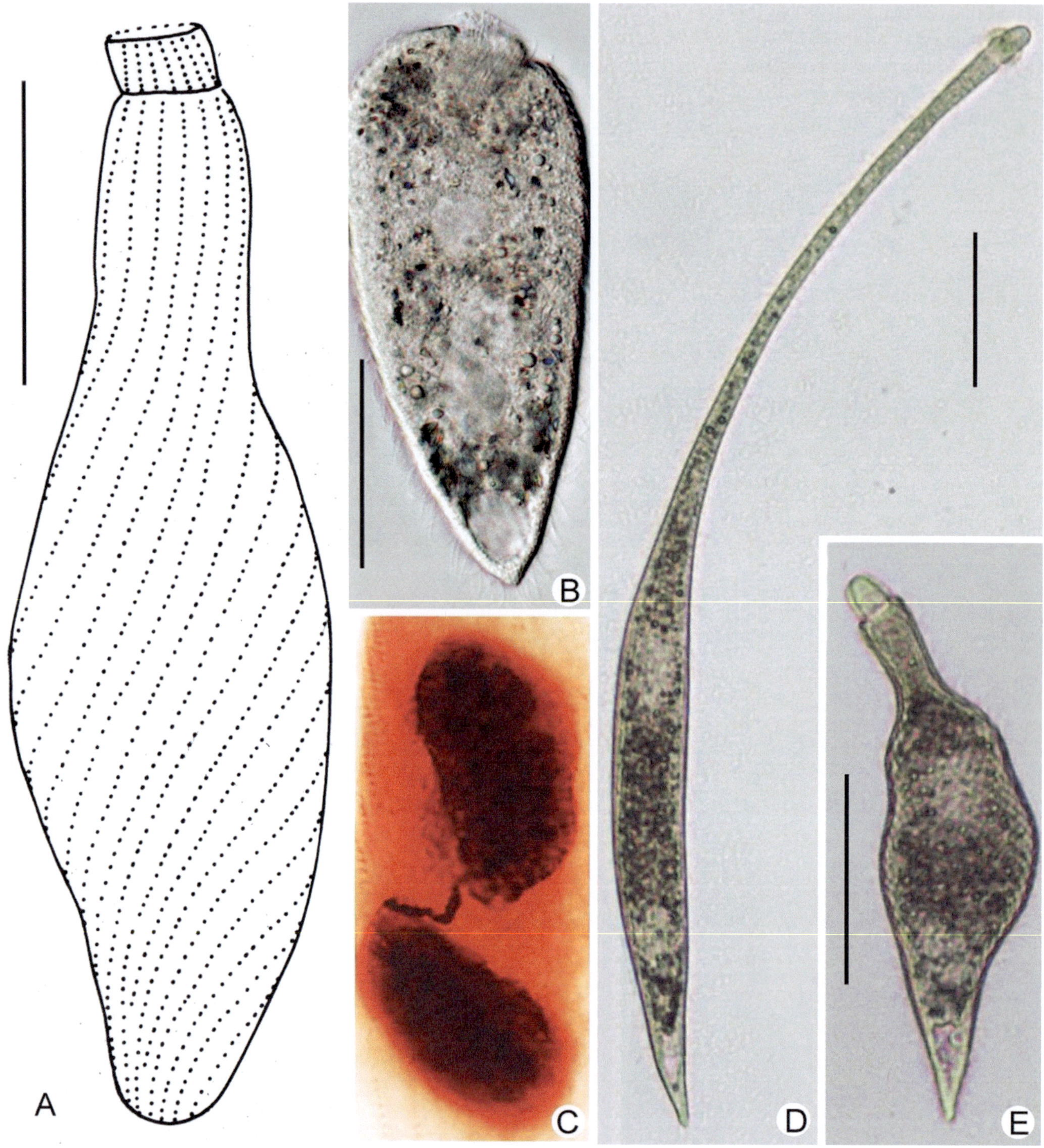


**Fig. 7B** *Lacrymaria maurea*
(**a**) Lateral view of ciliature; (**b**) contracted cell in vivo; (**c**) macronuclei after protargol staining; (**d**, **e**) different body shapes. Scale bars: 100 μm (**d**); 50 μm (**a**, **b**, **e**)

## Reference

Dragesco J (1965) Compléments à la connaissance de *Swedmarkia arenicola* et *Discocephalus ehrenbergi* ciliés hypotriches. Ann biol 4:187–204

Class Litostomatea Small & Lynn, 1981
Order Haptorida Corliss, 1974
Family Lacrymariidae de Fromentel, 1876
*Phialina salinarum* (Kahl, 1928) Song et al., 2003

**Ecological feature:** mangrove
**Water temperature:** 23.1 °C; 21.0 °C
**pH:** 6.3; 7.0
**Salinity:** 10.0‰; 10.0‰

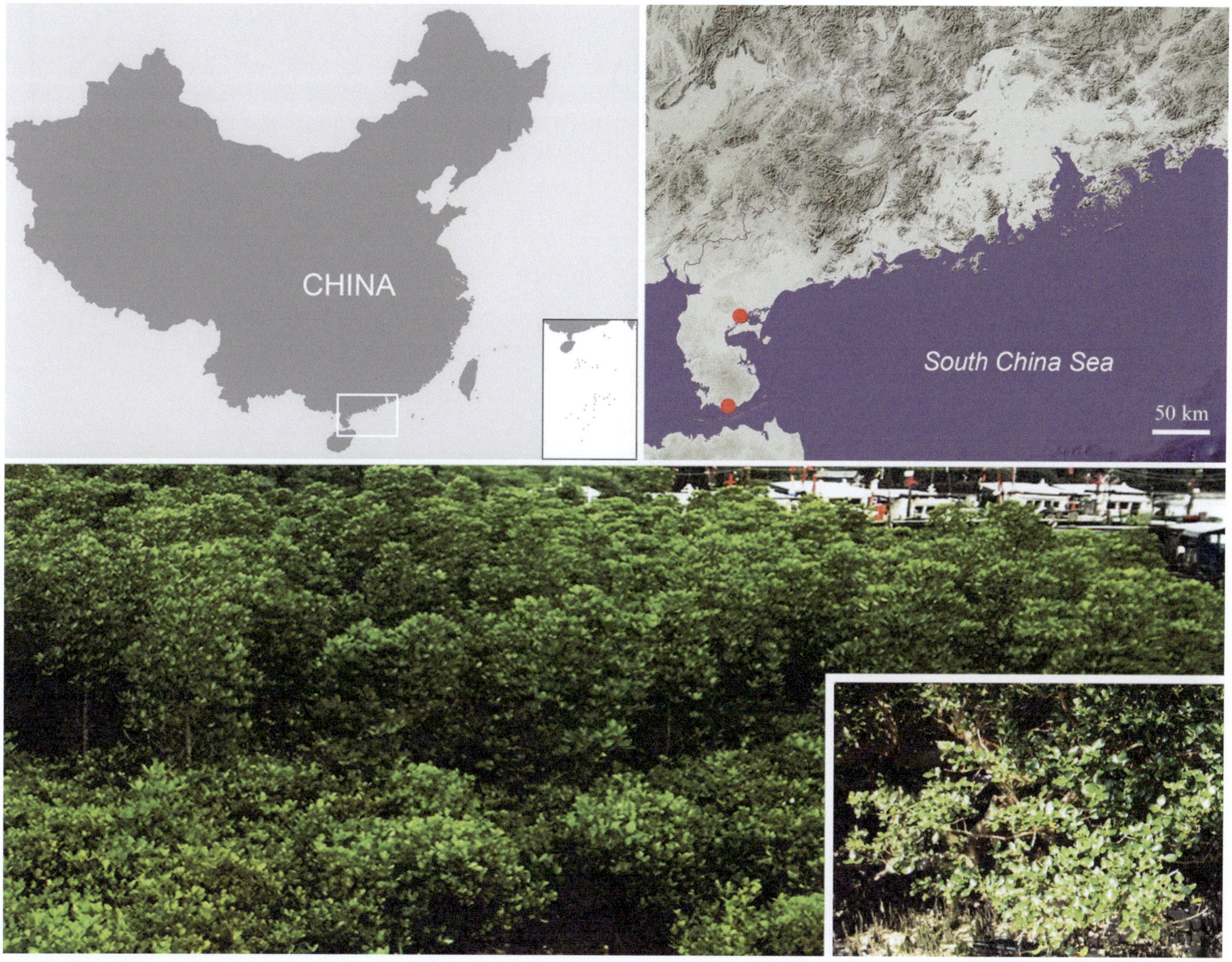


**Fig. 8A** Sample site

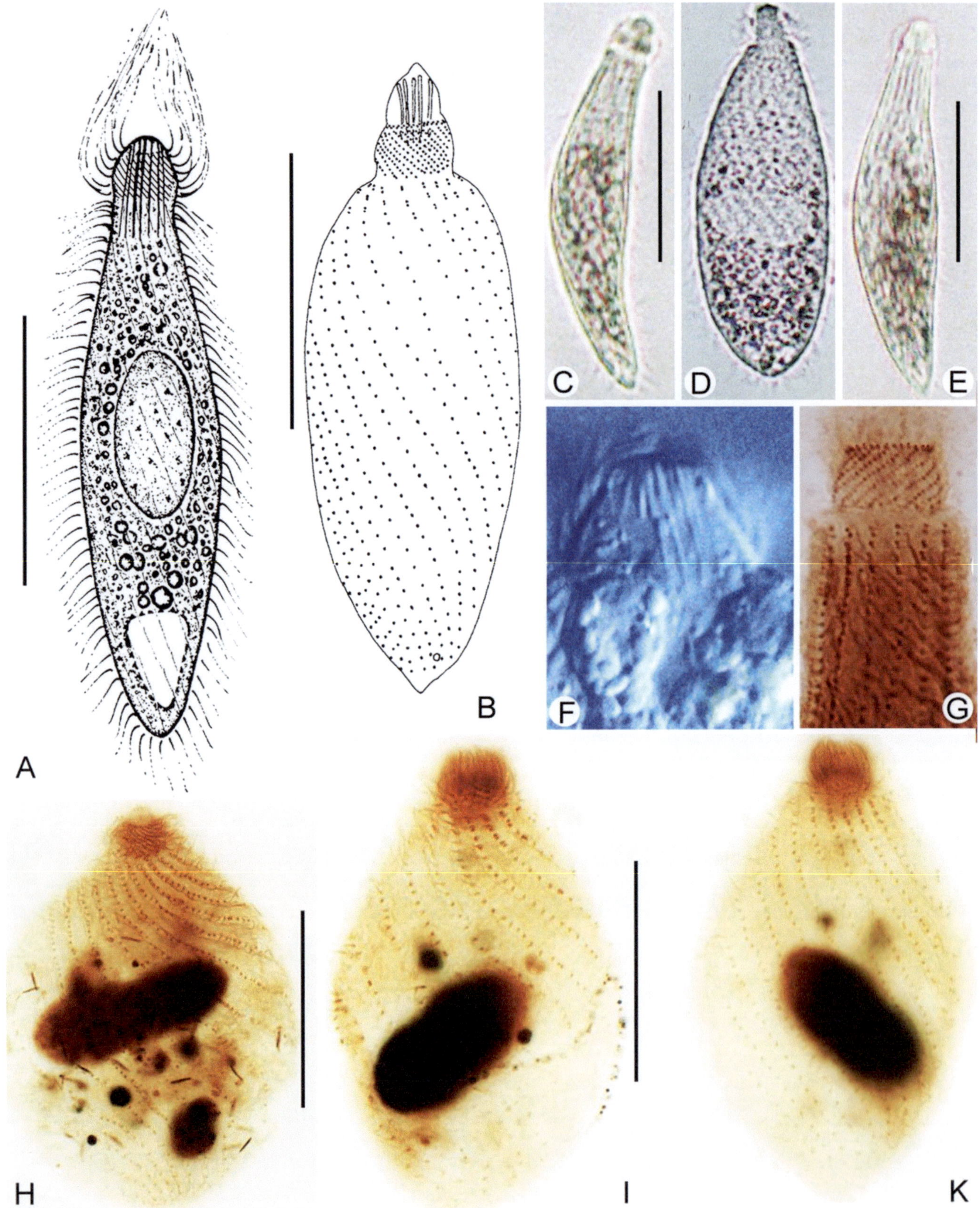


**Fig. 8B** *Phialina salinarum*
(**a**, **c–e**) Lateral views of live individuals, showing variation of body shapes; (**b**, **g–k**) specimens after protargol staining, showing ciliature and nuclear apparatus; (**f**) extrusomes in oral area. Scale bars: 40 μm

## References

Dong J, Liu W, Li J, Lin X (2010) Morphological studies on six Haptorida ciliates (Protozoa, Ciliophora) from the Daya Bay, South China. Acta Hydrobiol Sin 34(2):236–241

Song WB, Zhao Y, Xu K, Hu X, Gong J (2003) Pathogenetic protozoa in mariculture. Science Press, Beijing

Class Litostomatea Small & Lynn, 1981
Order Haptorida Corliss, 1974
Family Pseudotrachelocercidae Song, 1990
*Pseudotrachelocerca trepida* (Kahl, 1928) Song, 1990

**Ecological feature:** mangrove (Beihai); intertidal zone (Zhanjiang)
**Water temperature:** 24.2 °C; 21.0 °C; 30.2 °C
**pH:** 7.0; 7.0; 8.3
**Salinity:** 17.0‰; 10.0‰; 31.8‰

**Fig. 9A** Sample site

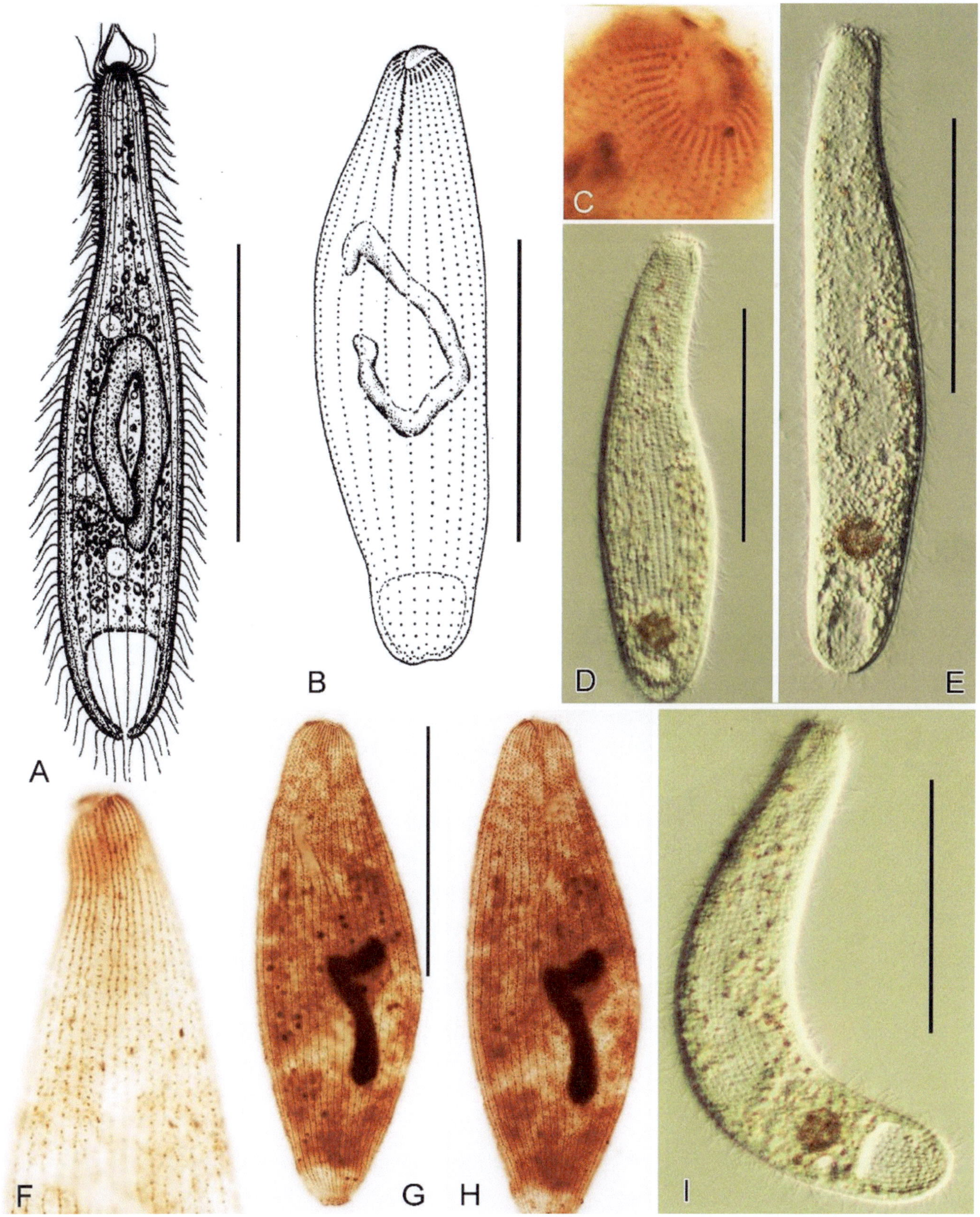


**Fig. 9B** *Pseudotrachelocerca trepida*
(**a**, **d**, **e**) Lateral views of live cells; (**c**) oral apparatus; (**b**, **f–h**) ciliature; (**i**) a bent live cell showing contractile vacuole at posterior end of cell. Scale bars: 100 µm

## Reference

Song WB, Zhao Y, Xu K, Hu X, Gong J (2003) Pathogenetic protozoa in mariculture. Science Press, Beijing

Class Litostomatea Small & Lynn, 1981
Order Haptorida Corliss, 1974
Family Spathidiidae Kahl in Doflein & Reichenow, 1929
*Epispathidium polynucleatum* Foissner et al., 2002

**Ecological feature:** mangrove
**Water temperature:** 21.0 °C
**pH:** 7.3
**Salinity:** 22.5‰

**Fig. 10A** Sample site

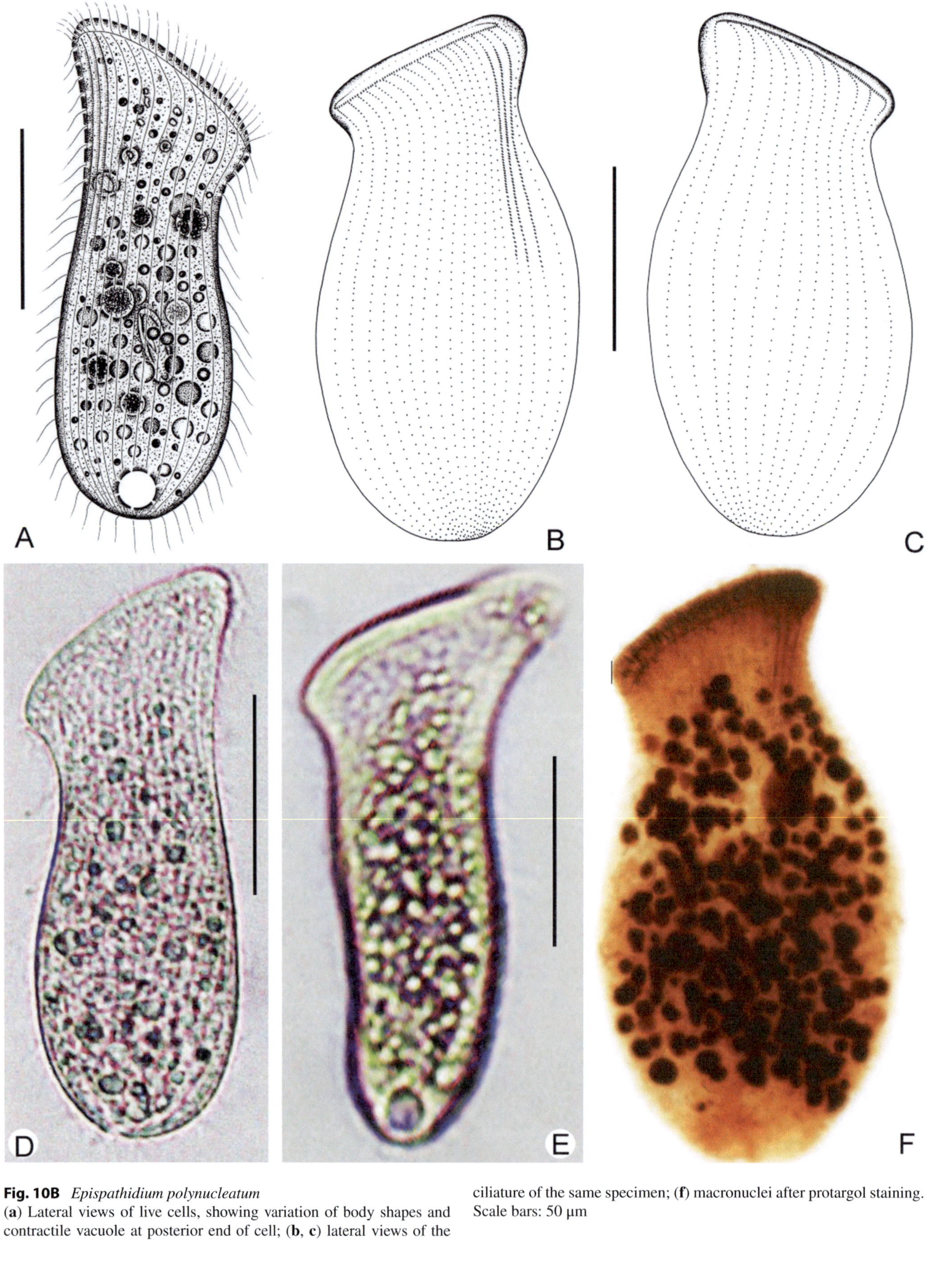


**Fig. 10B** *Epispathidium polynucleatum*
(**a**) Lateral views of live cells, showing variation of body shapes and contractile vacuole at posterior end of cell; (**b**, **c**) lateral views of the ciliature of the same specimen; (**f**) macronuclei after protargol staining. Scale bars: 50 μm

## Reference

Foissner W, Agatha S, Berger H (2002) Soil ciliates (Protozoa, Ciliophora) from Namibia (Southwest Africa), with emphasis on two contrasting environments, the Etosha region and the Namib desert. Denisia 5:1–1063

Class Litostomatea Small & Lynn, 1981
Order Haptorida Corliss, 1974
Family Trachelidae Ehrenberg, 1838
*Dileptus* sp.

**Ecological feature:** mangrove
**Water temperature:** 21.3 °C
**pH:** 8.0
**Salinity:** 6.0‰

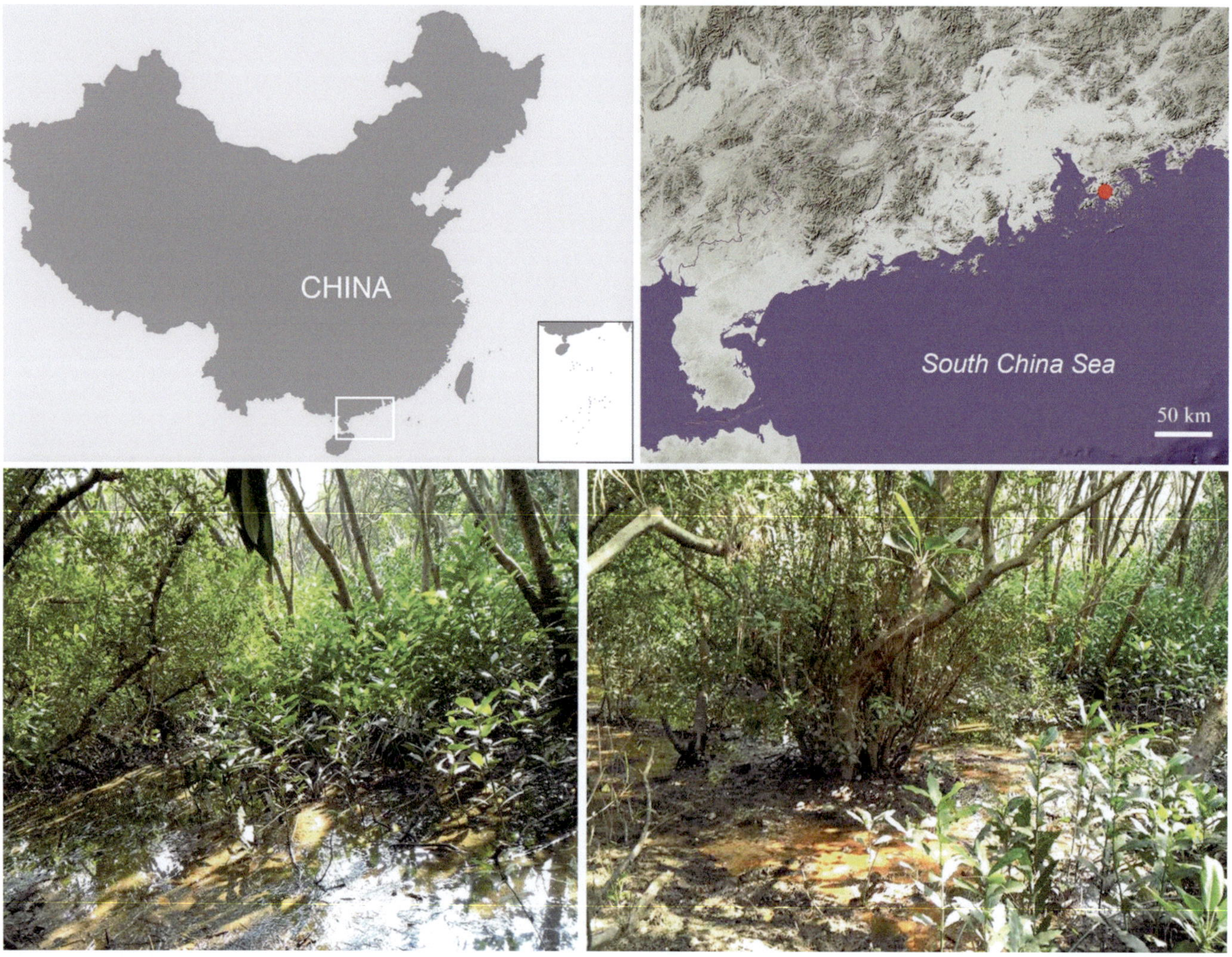


**Fig. 11A** Sample site

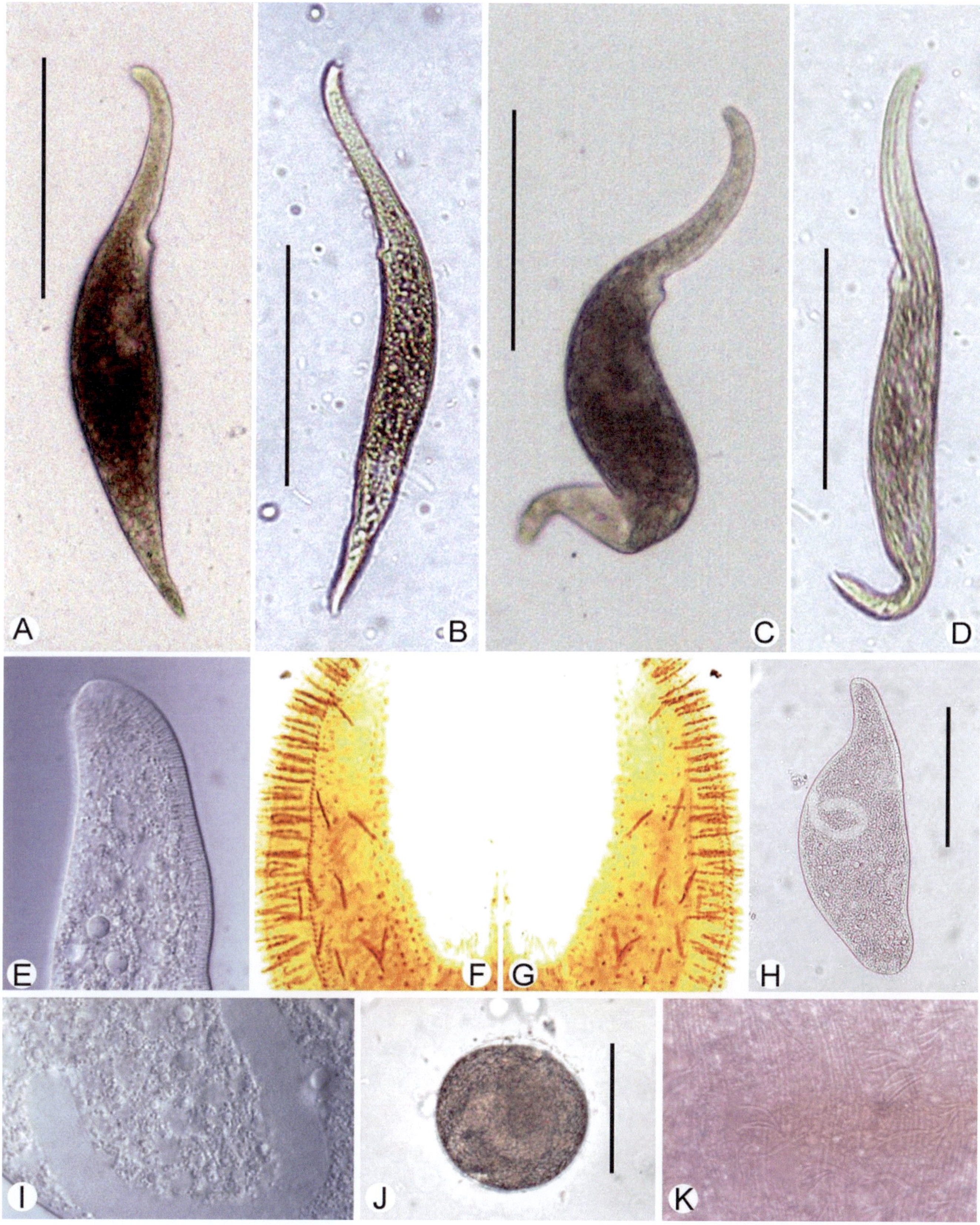


**Fig. 11B** *Dileptus* sp.
(**a–d**) Lateral views of live individuals, showing shape variants; (**e**) extrusomes in anterior part of cell; (**f, g**) ciliary pattern of anterior part of left (**f**) and right (**g**) sides; (**h**) right side view of a compressed cell; (**i**) macronucleus; (**j**) cyst; (**k**) cortical granules. Scale bars: 150 μm (**a–d, h**); 15 μm (**j**)

## References

Berger H, Foissner W (2014) Typification of the genus *Dileptus* Dujardin, 1841 (Ciliophora, Rhynchostomatia). Eur J Protistol 50:314–317

Golinska K (1982) Regulation of ciliary pattern in *Dileptus* (Ciliata). Develop 68:99–114

Vd'ačný P, Foissner W (2008) Morphology, conjugation, and postconjugational reorganization of *Dileptus tirjakovae* n. sp. (Ciliophora, Haptoria). J Eukaryot Microbiol 55:436–447

Class Litostomatea Small & Lynn, 1981
Order Pleurostomatida Schewiakoff, 1896
Family Amphileptidae Bütschli, 1889
*Amphileptus bellus* Wu et al., 2015

**Ecological feature:** mangrove
**Water temperature:** 21.8 °C
**pH:** 8.0
**Salinity:** 12.5‰

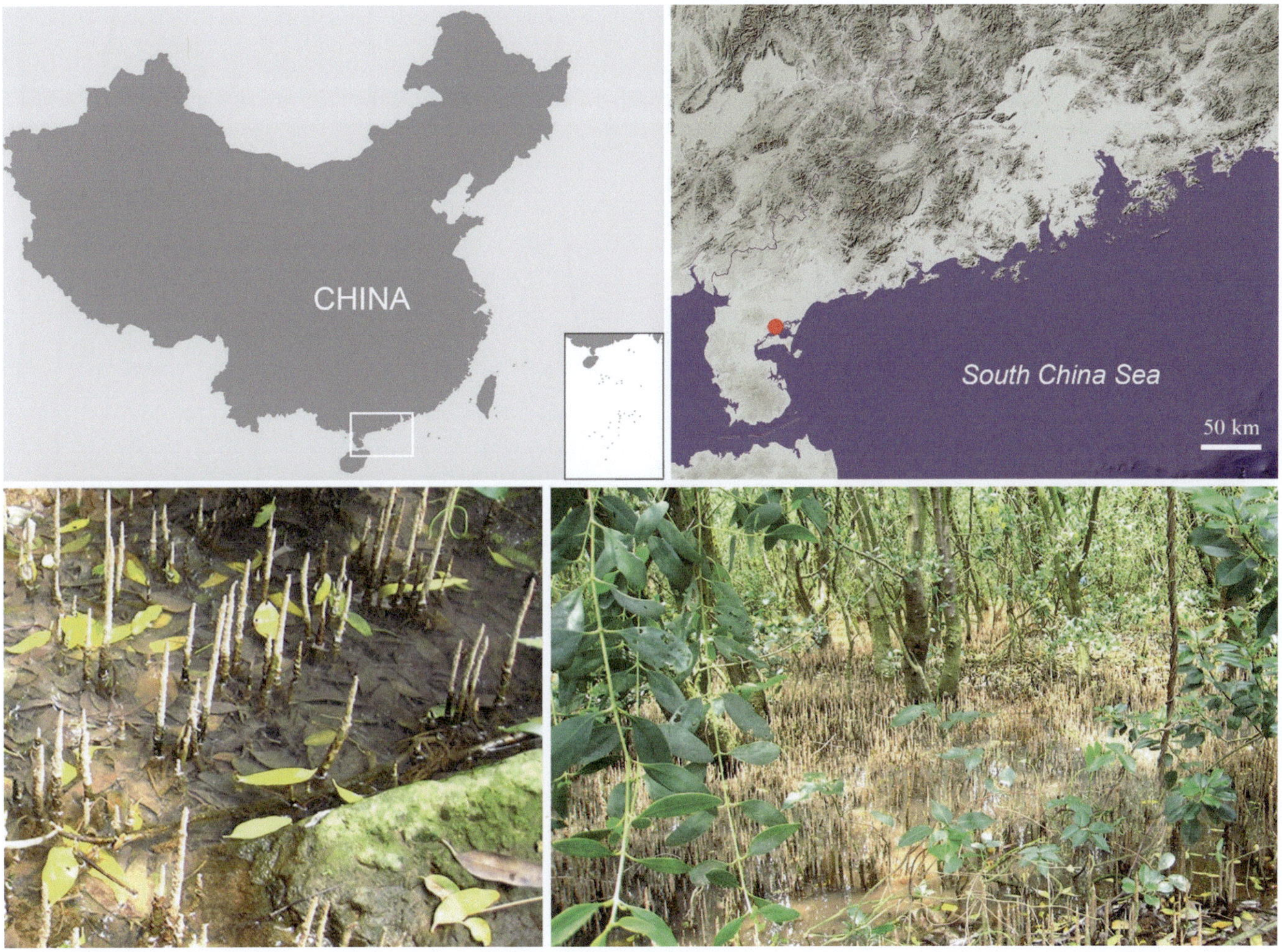


**Fig. 12A** Sample site

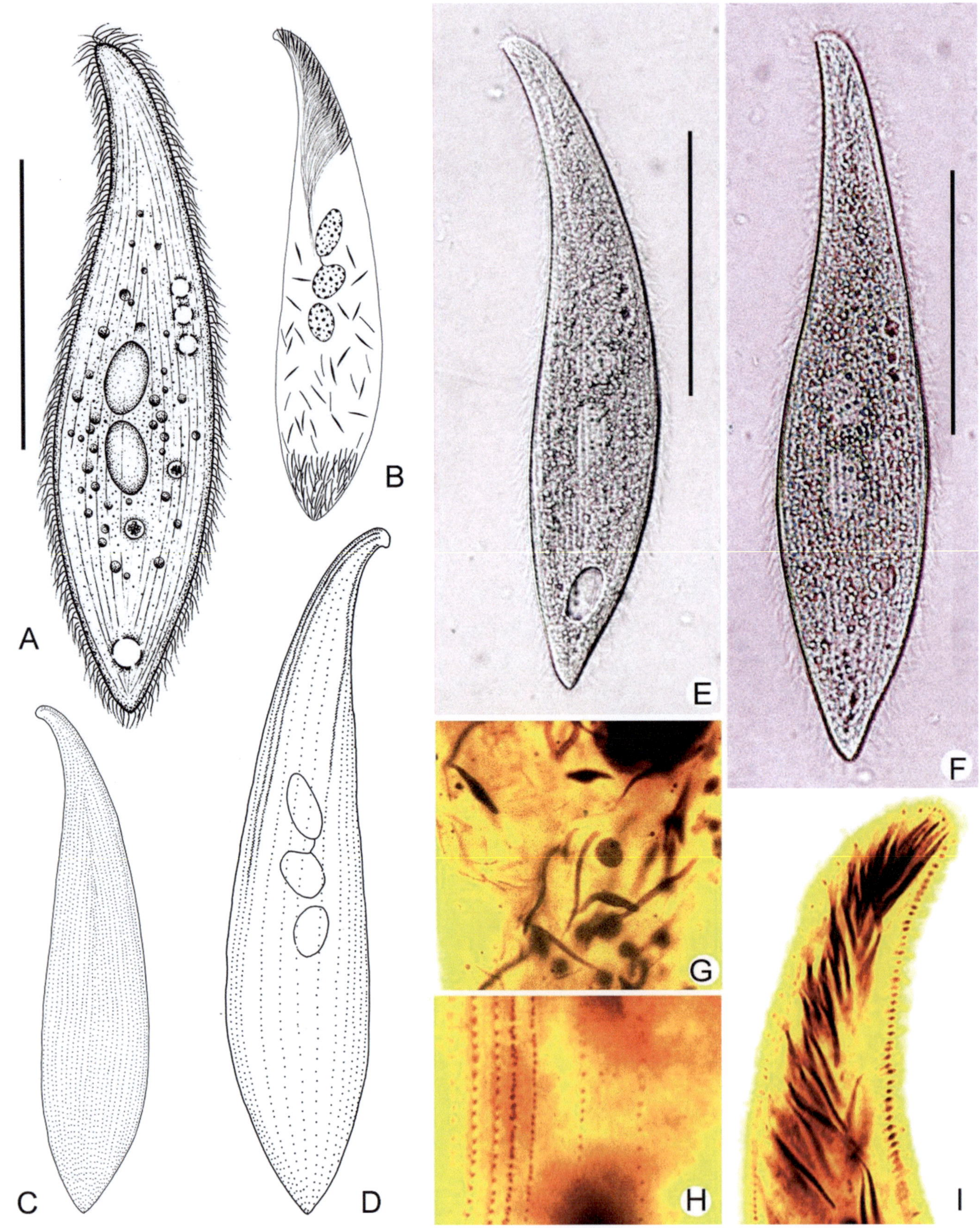


**Fig. 12B** *Amphileptus bellus*
(**a**, **e**, **f**) Right-lateral views of live cells; (**b**) extrusome pattern and nuclear apparatus; (**c**, **d**) ciliary pattern on right (**c**) and left (**d**) sides; (**g**) spindle-shaped extrusomes; (**h**) ventral view of middle portion of cell showing kineties; (**i**) anterior part of left side showing extrusomes. Scale bars: 150 μm

## Reference

Wu L, Yi Z, Li J, Warren A, Xu H, Lin X (2015) Two new brackish ciliates, *Amphileptus spiculatus* sp. n. and *A. bellus* sp. n. from mangrove wetlands in southern China, with notes on the molecular phylogeny of the family Amphileptidae (Protozoa, Ciliophora, Pleurostomatida). J Eukaryot Microbiol 62:662–669

Class Litostomatea Small & Lynn, 1981
Order Pleurostomatida Schewiakoff, 1896
Family Amphileptidae Bütschli, 1889
*Amphileptus dragescoi* Pan et al., 2014

**Ecological feature:** intertidal zone
**Water temperature:** 23.8 °C
**pH:** 7.4
**Salinity:** 23.9‰

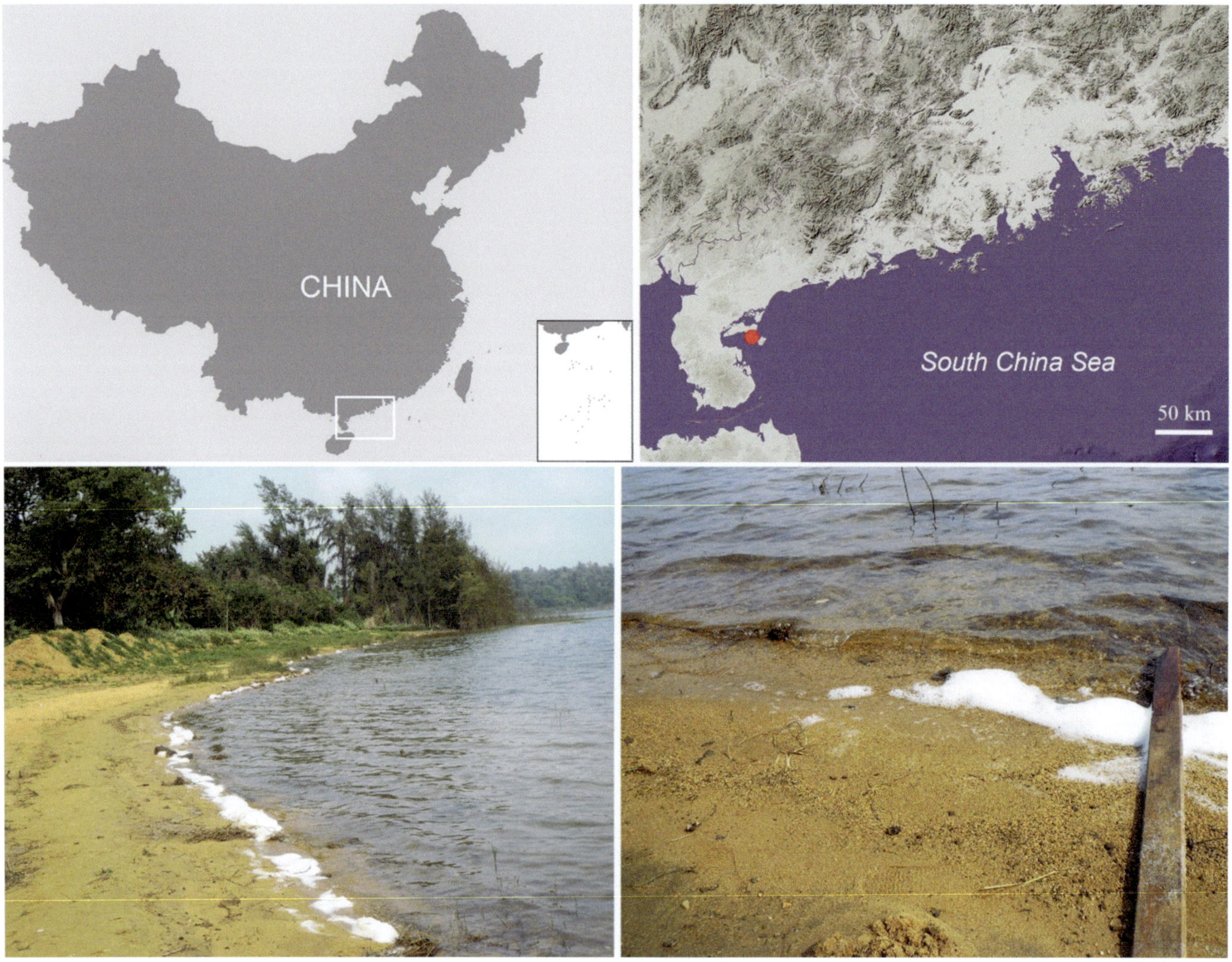


**Fig. 13A** Sample site

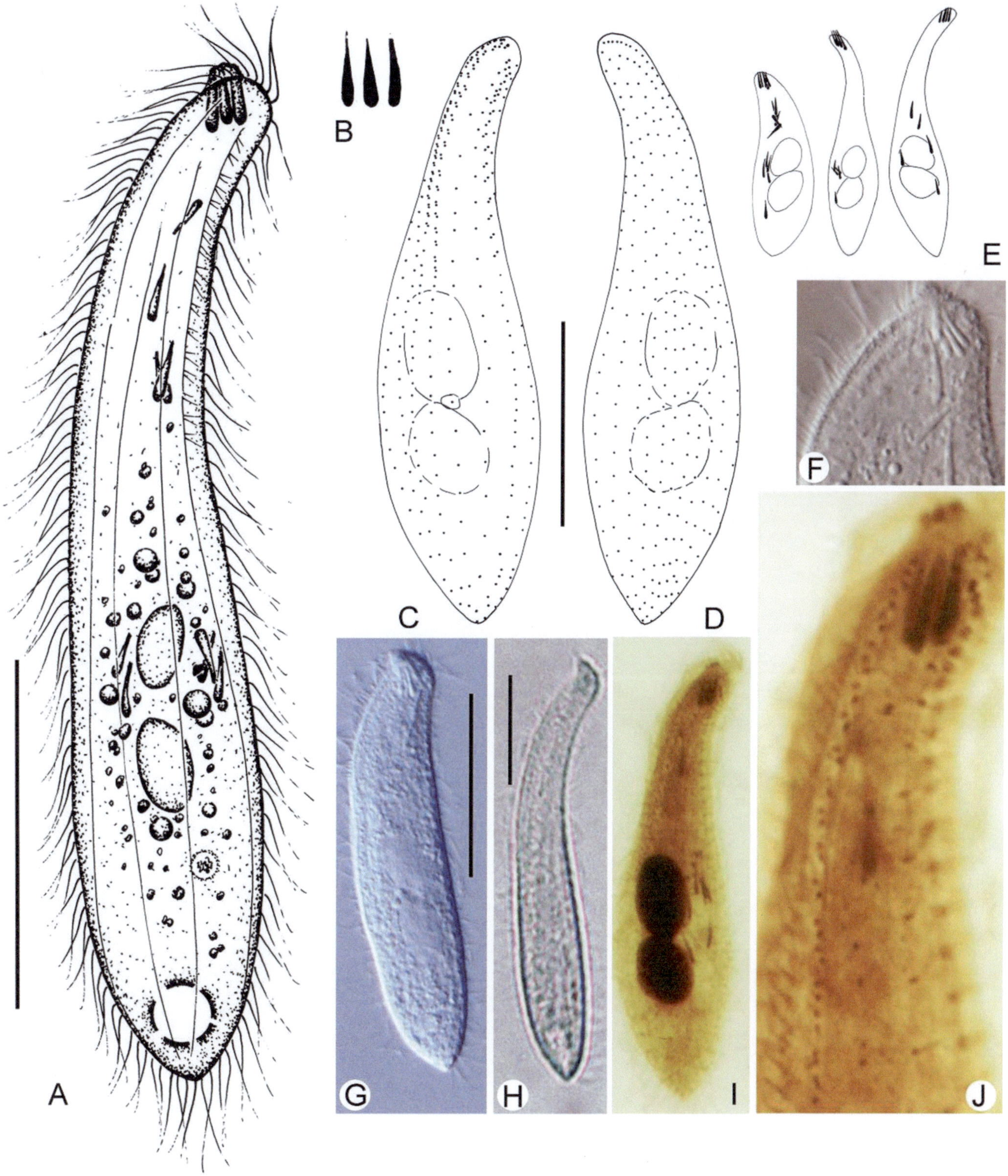


**Fig. 13B** *Amphileptus dragescoi*
(**a**, **g**, **h**) Left views of live individuals; (**b**, **f**) extrusomes; (**c**, **d**, **i**) left (**c**, **i**) and right (**d**) lateral views of cell to show ciliary pattern; (**e**) shape variants; (**j**) left-lateral view of ciliature of anterior body portion. Scale bars: 60 μm

## Reference

Pan H, Li L, Lin X, Li J, Al-Farraj SA, Al-Rasheid KAS (2014) Morphology of three species of *Amphileptus* (Protozoa, Ciliophora, Pleurostomatida) from the South China Sea, with note on phylogeny of *A. dragescoi* sp. n. J Eukaryot Microbiol 61:644–654

Class Litostomatea Small & Lynn, 1981
Order Pleurostomatida Schewiakoff, 1896
Family Amphileptidae Bütschli, 1889
*Amphileptus marinus* (Kahl, 1931) Song et al., 2004

**Ecological feature:** intertidal zone
**Water temperature:** 23.8 °C
**pH:** 7.4
**Salinity:** 23.9‰

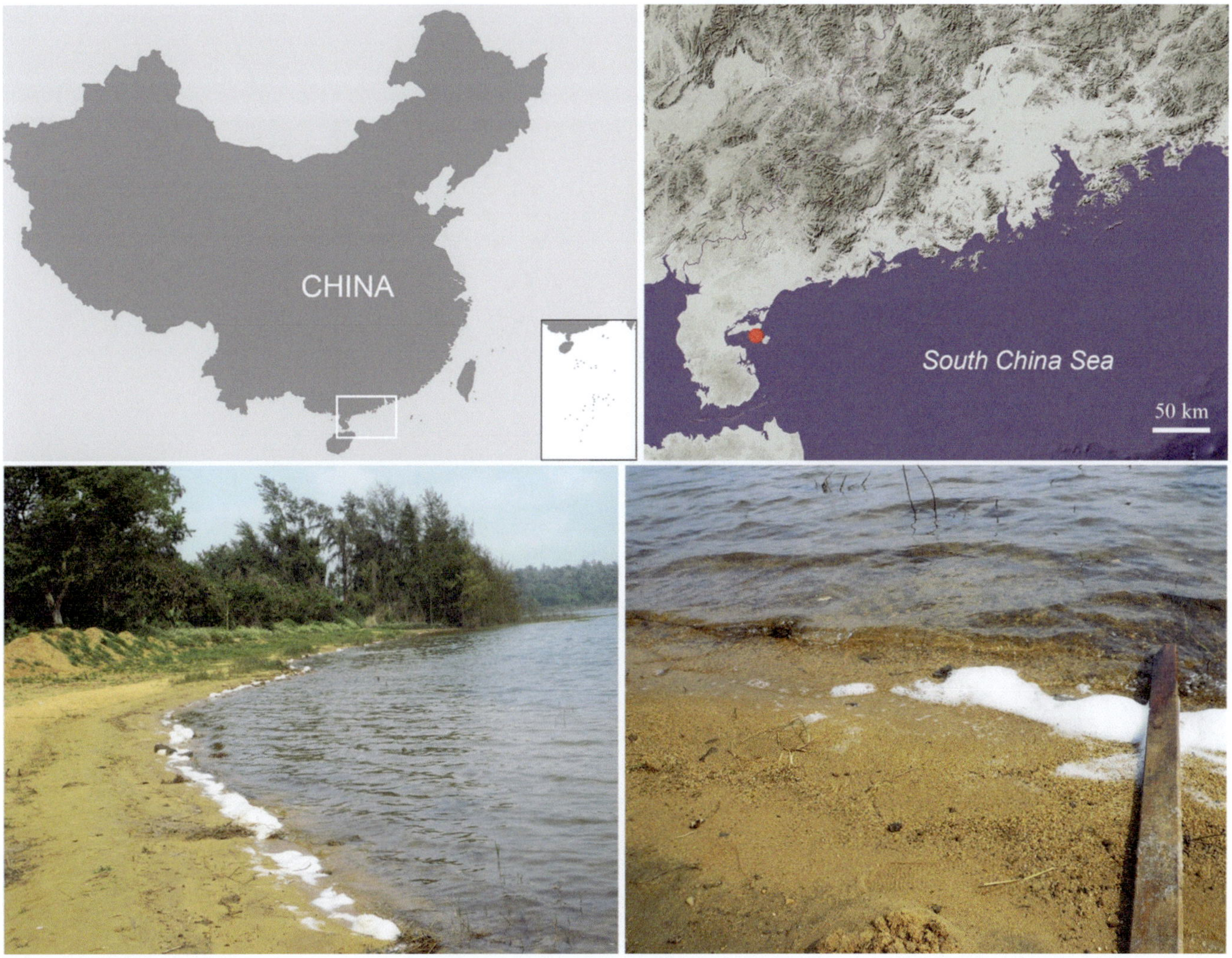


**Fig. 14A** Sample site

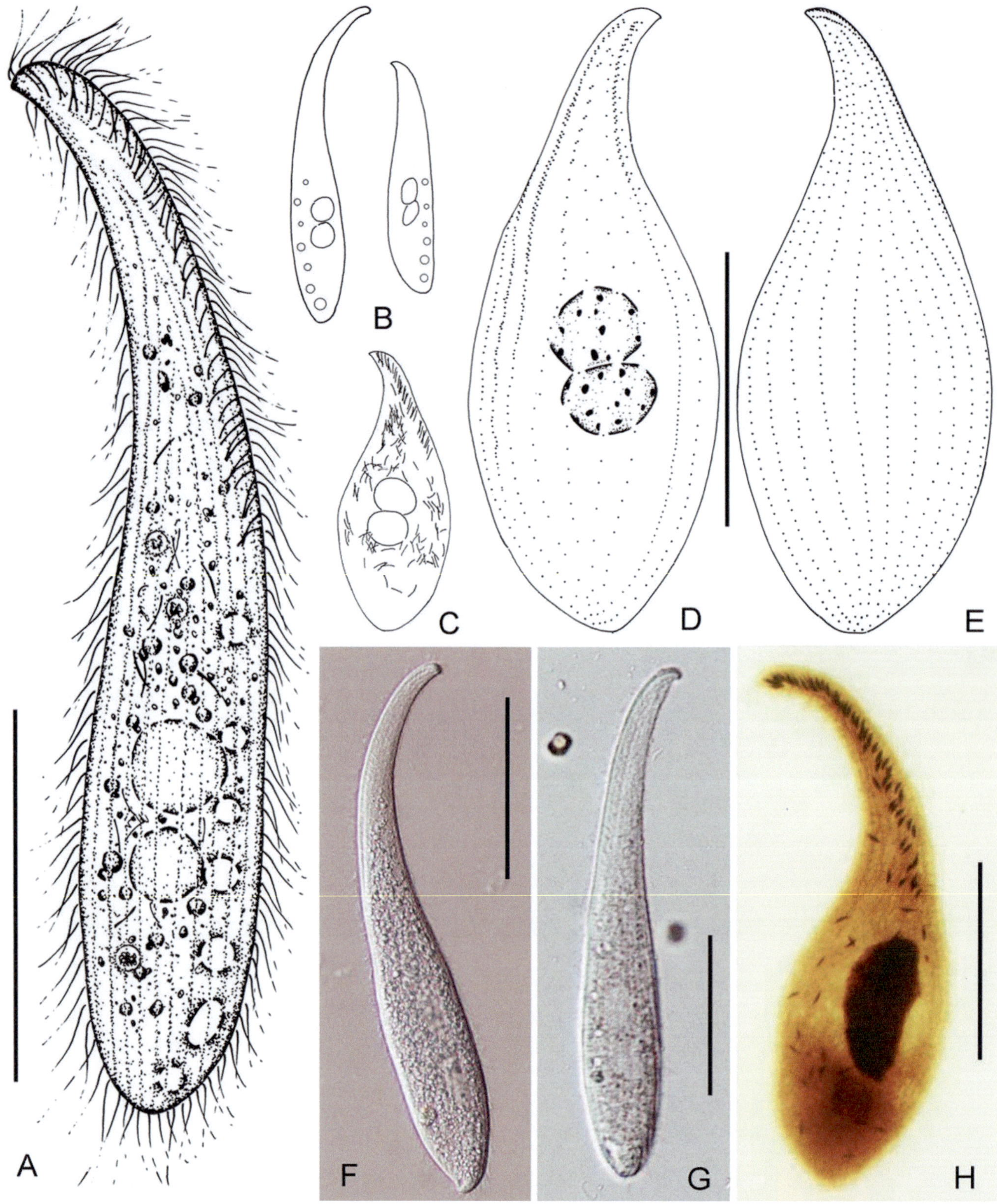


**Fig. 14B** *Amphileptus marinus*
(**a**) Right-lateral view of live cell; (**b**) different body outlines showing contractile vacuole and macronuclei; (**c**) distribution of extrusomes; (**d**, **e**) left (**d**) and right (**e**) lateral views to show ciliary pattern; (**f**, **g**) left-lateral view of live individuals; (**h**) right-lateral view of an individual with single macronucleus. Scale bars: 100 μm

## References

Pan H, Li L, Lin X, Li J, Al-Farraj SA, Al-Rasheid KAS (2014) Morphology of three species of *Amphileptus* (Protozoa, Ciliophora, Pleurostomatida) from the South China Sea, with note on phylogeny of *A. dragescoi* sp. n. J Eukaryot Microbiol 61:644–654

Song WB, Wilbert N, Hu X (2004) Redescription and neotypification of *Amphileptus marinus* (Kahl, 1931) nov. comb. (Ciliophora, Pleurostomatida), and reactivation of *A. parafusidens* Song & Wilbert, 1989. Eur J Protistol 40:1–11

Class Litostomatea Small & Lynn, 1981
Order Pleurostomatida Schewiakoff, 1896
Family Amphileptidae Bütschli, 1889
*Amphileptus salignus* Chen et al., 2011

**Ecological feature:** mangrove
**Water temperature:** 20.0–26.5 °C
**pH:** 7.0–8.0
**Salinity:** 9.0‰–14.0‰

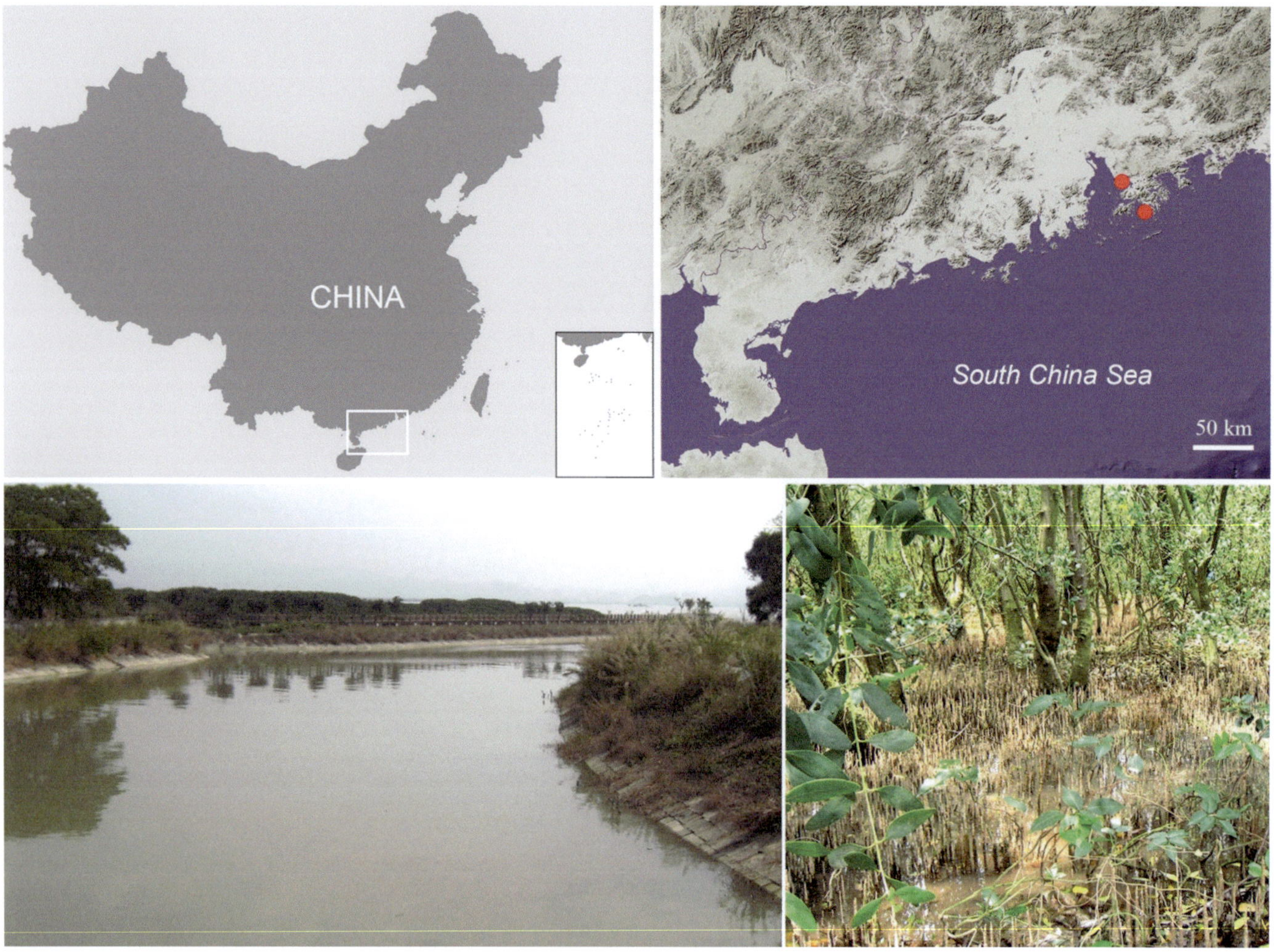


**Fig. 15A** Sample site

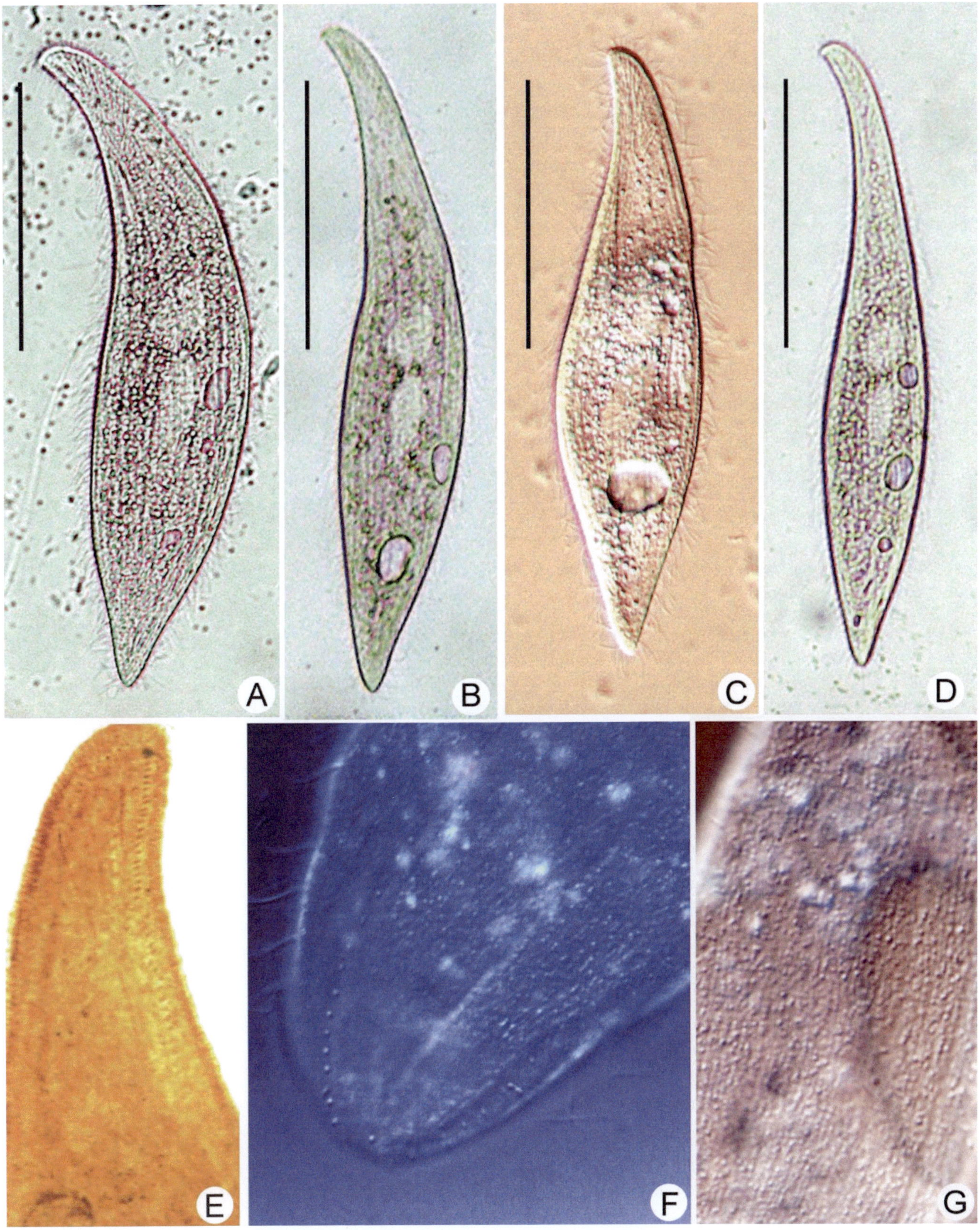


**Fig. 15B** *Amphileptus salignus*
(**a**) Right-lateral view of live cell; (**b–d**) different cells, showing body shape variants; (**e**) ciliary pattern of anterior left side; (**f**) posterior part of left side of body; (**g**) cortical granules. Scale bars: 100 μm

## Reference

Chen R, Lin X, Warren A (2011) A new pleurostomatid ciliate, *Amphileptus salignus* n. sp. (Protozoa, Ciliophora), from mangrove wetlands in southern China. Zootaxa 3048:62–68

Class Litostomatea Small & Lynn, 1981
Order Pleurostomatida Schewiakoff, 1896
Family Amphileptidae Bütschli, 1889
*Amphileptus spiculatus* Wu et al., 2015

**Ecological feature:** mangrove
**Water temperature:** 25.0 °C
**pH:** 7.0
**Salinity:** 15.0‰

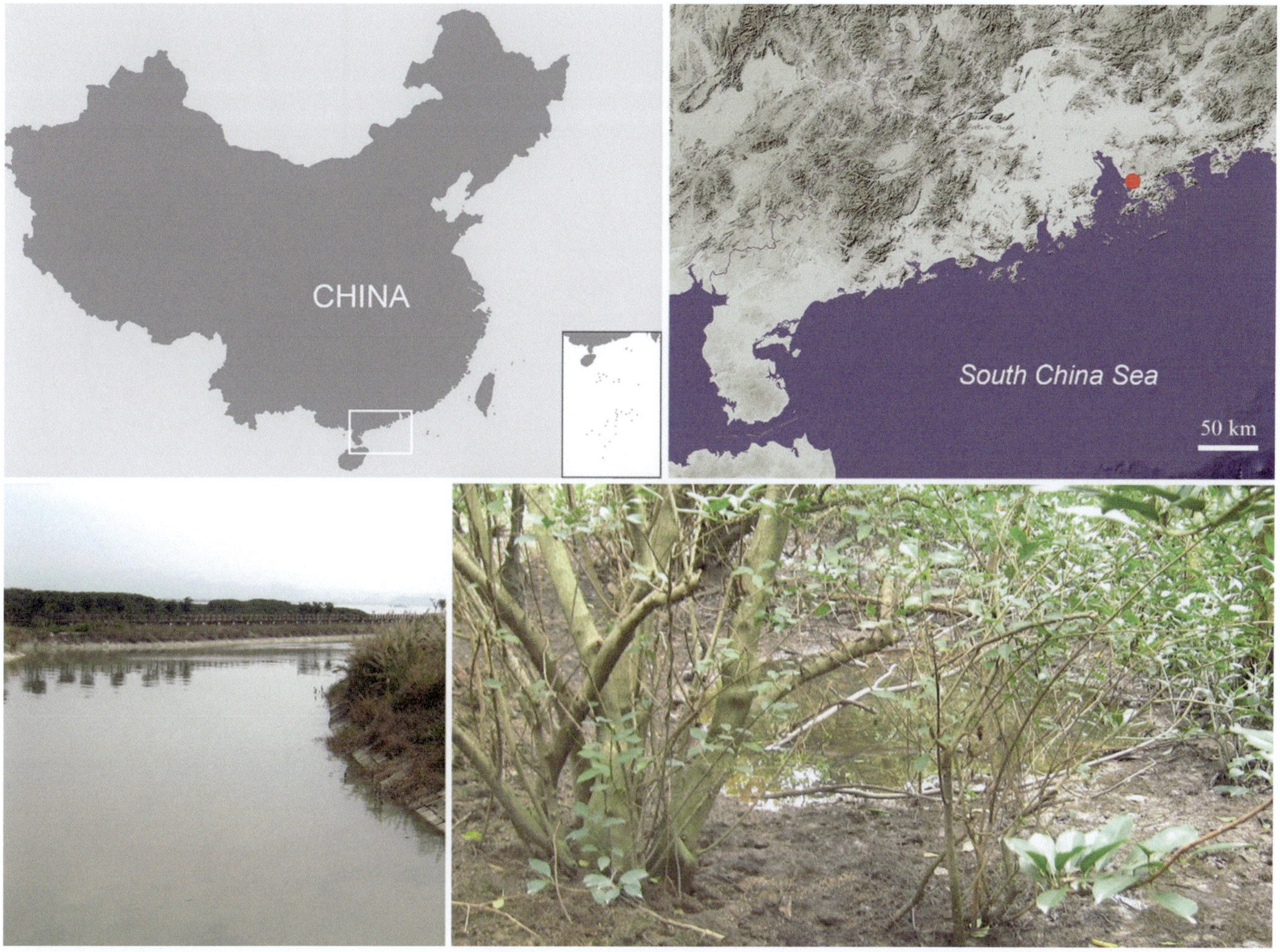


**Fig. 16A** Sample site

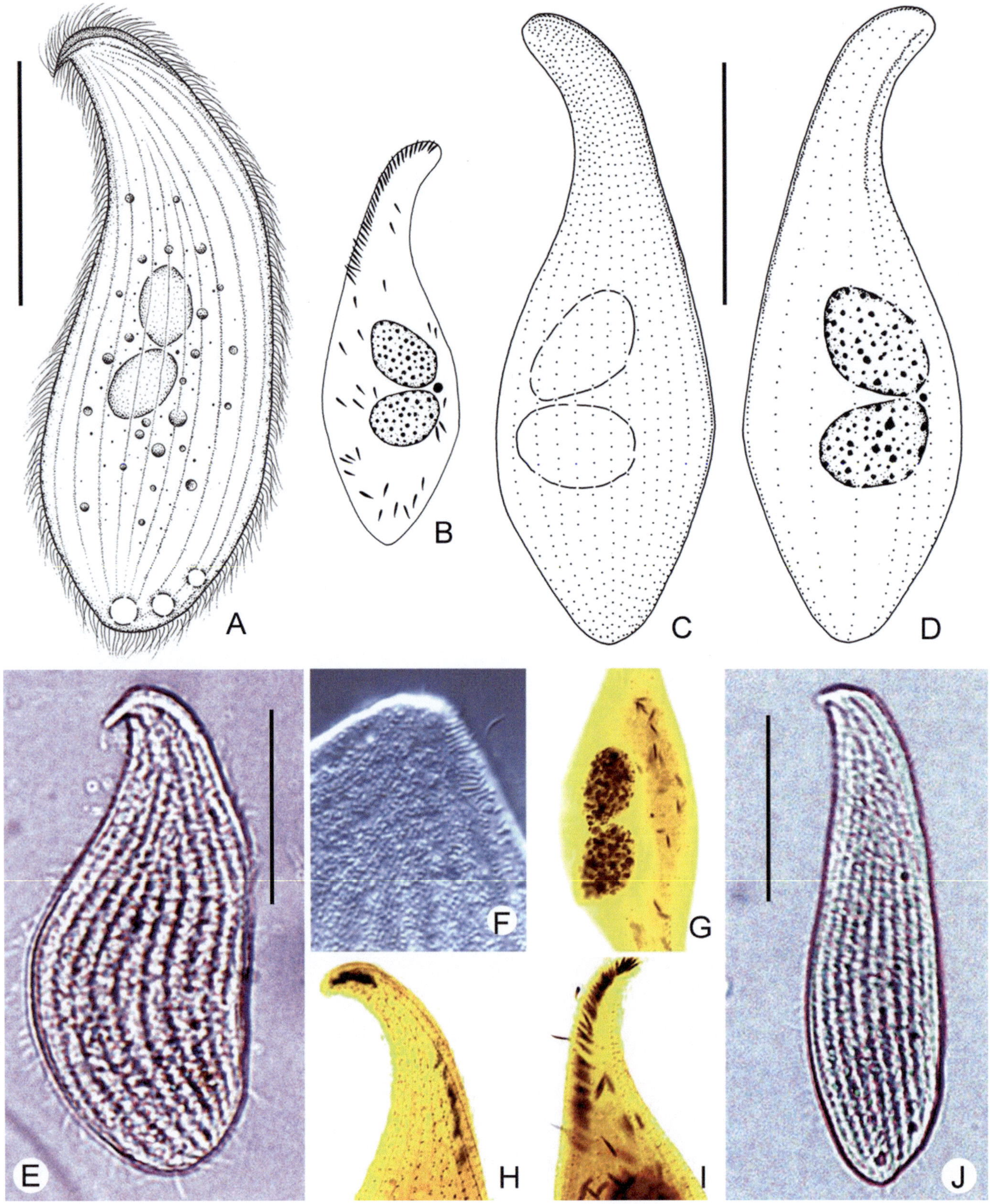


**Fig. 16B** *Amphileptus spiculatus*
(**a**, **e**, **j**) Right-lateral views of live individuals; (**b**, **g**) extrusomes pattern and nuclear apparatus; (**c**, **d**, **h**, **i**) ciliary pattern of right (**c**, **h**) and left (**d**, **i**) sides; (**f**) anterior right-lateral view of live cell; (**g**) Macronuclei and extrusomes. Scale bars: 50 μm

## Reference

Wu L, Yi Z, Li J, Warren A, Xu H, Lin X (2015) Two new brackish ciliates, *Amphileptus spiculatus* sp. n. and *A. bellus* sp. n. from mangrove wetlands in southern China, with notes on the molecular phylogeny of the family Amphileptidae (Protozoa, Ciliophora, Pleurostomatida). J Eukaryot Microbiol 62:662–669

Class Litostomatea Small & Lynn, 1981
Order Pleurostomatida Schewiakoff, 1896
Family Amphileptidae Bütschli, 1889
*Amphileptus wilberti* Pan et al., 2014

**Ecological feature:** mangrove
**Water temperature:** 24.6 °C
**pH:** 12.5
**Salinity:** 13.6‰

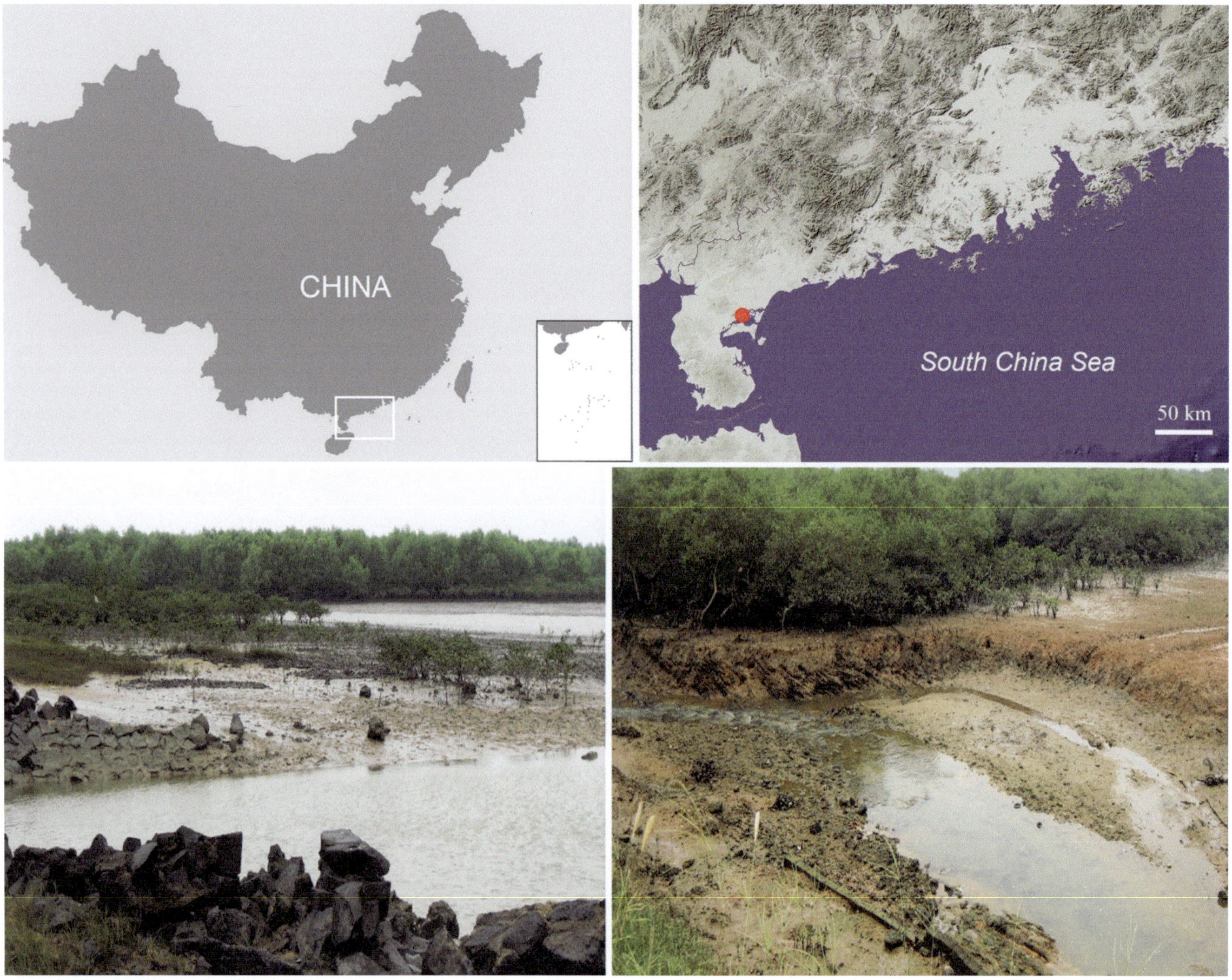


**Fig. 17A** Sample site

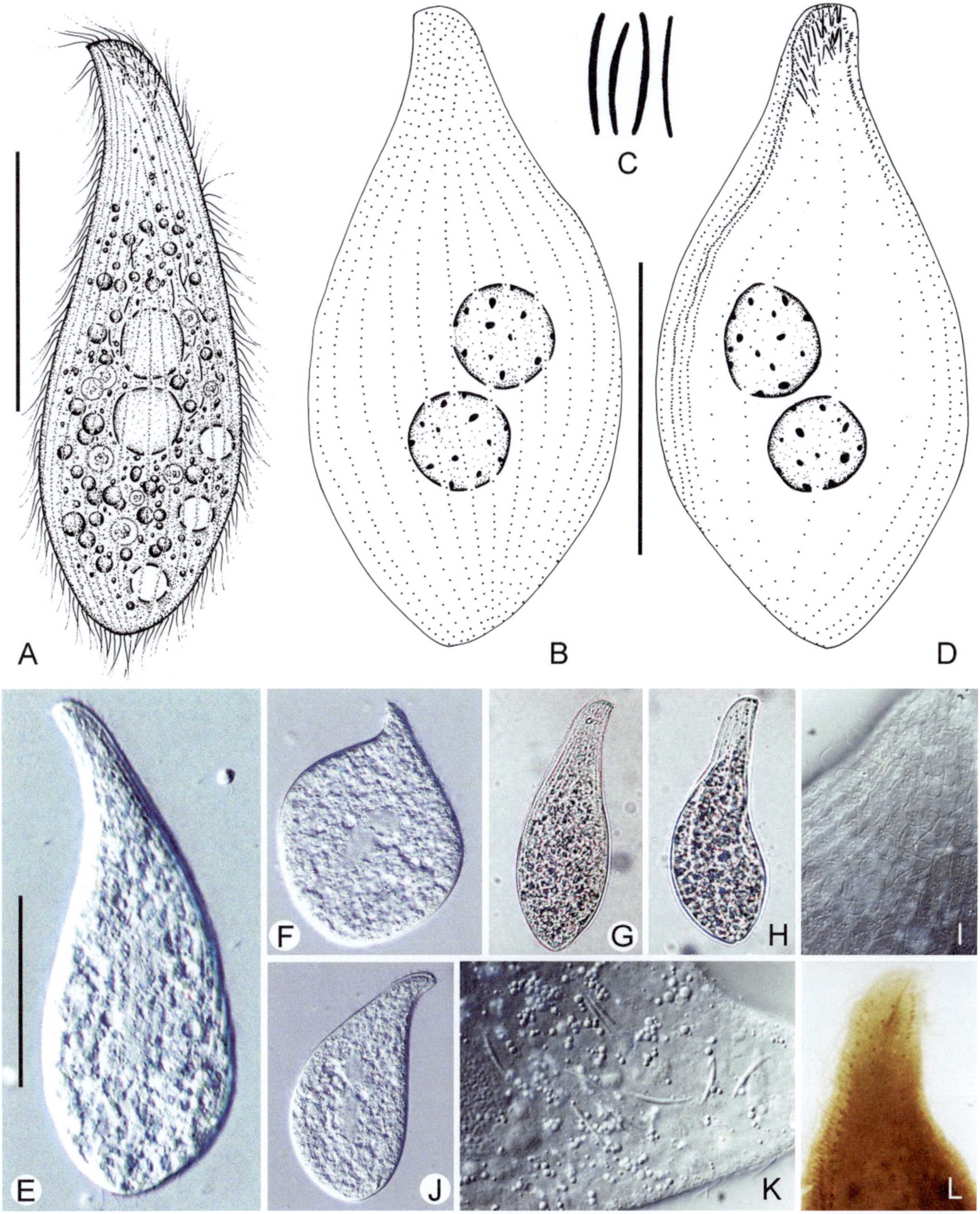


**Fig. 17B** *Amphileptus wilberti*
(**a**, **e**) Right-lateral views of live cells; (**b**, **d**) right (**b**) and left (**d**) views to show ciliary pattern; (**c**) extrusomes; (**f–h**, **j**) body shape variants; (**i**) cilia on right side of cell; (**k**) cytoplasmic extrusomes; (**l**) ciliature of anterior portion of left side. Scale bars: 90 μm

## Reference

Pan H, Li L, Lin X, Li J, Al-Farraj SA, Al-Rasheid KAS (2014) Morphology of three species of *Amphileptus* (Protozoa, Ciliophora, Pleurostomatida) from the South China Sea, with note on phylogeny of *A. dragescoi* sp. n. J Eukaryot Microbiol 61:644–654

Class Litostomatea Small & Lynn, 1981
Order Pleurostomatida Schewiakoff, 1896
Family Amphileptidae Bütschli, 1889
*Kentrophyllum bispinum* Wu et al., 2015

**Ecological feature:** intertidal zone
**Water temperature:** 26.0 °C
**pH:** 7.8
**Salinity:** 25.0‰

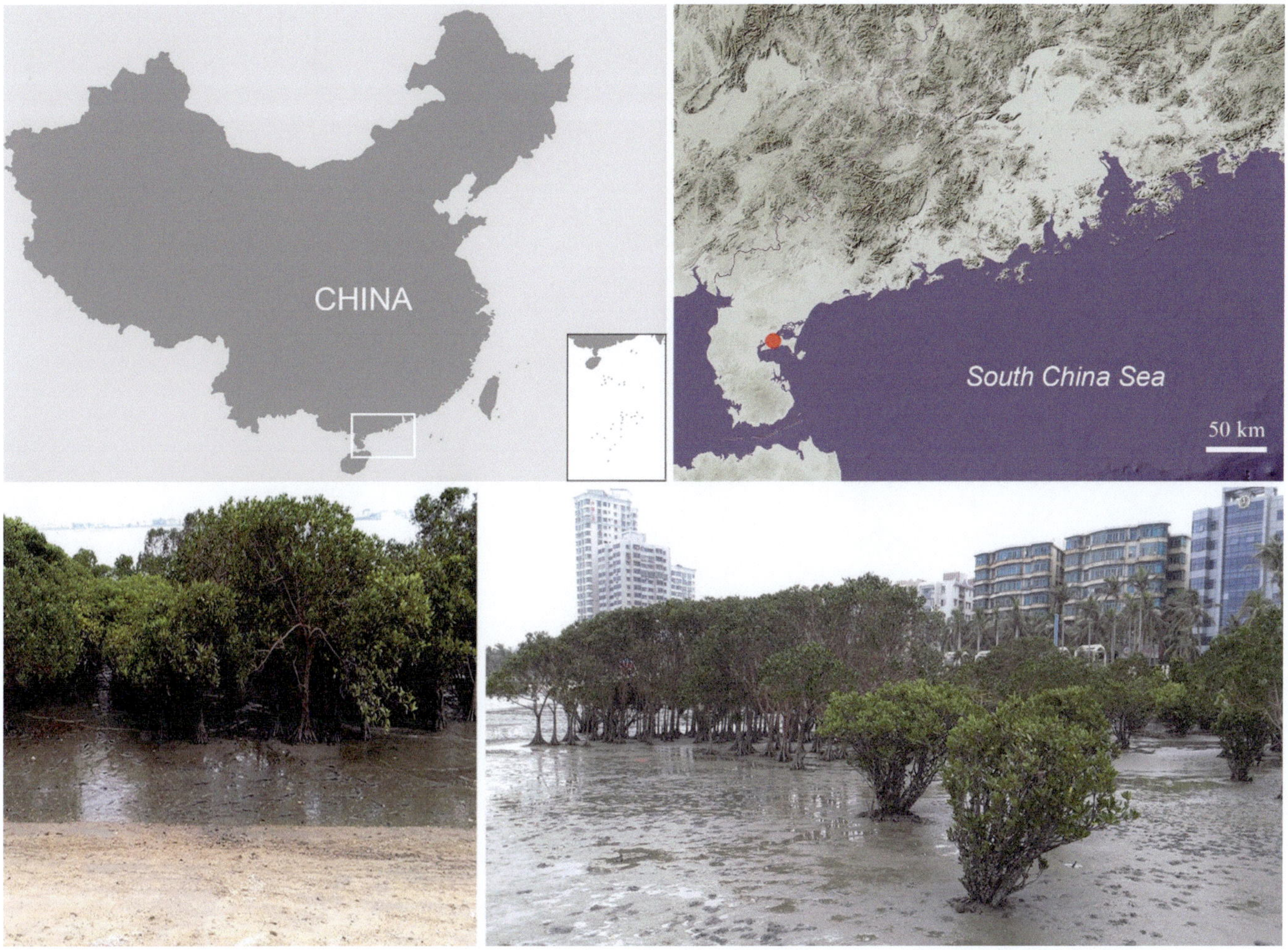


**Fig. 18A** Sample site

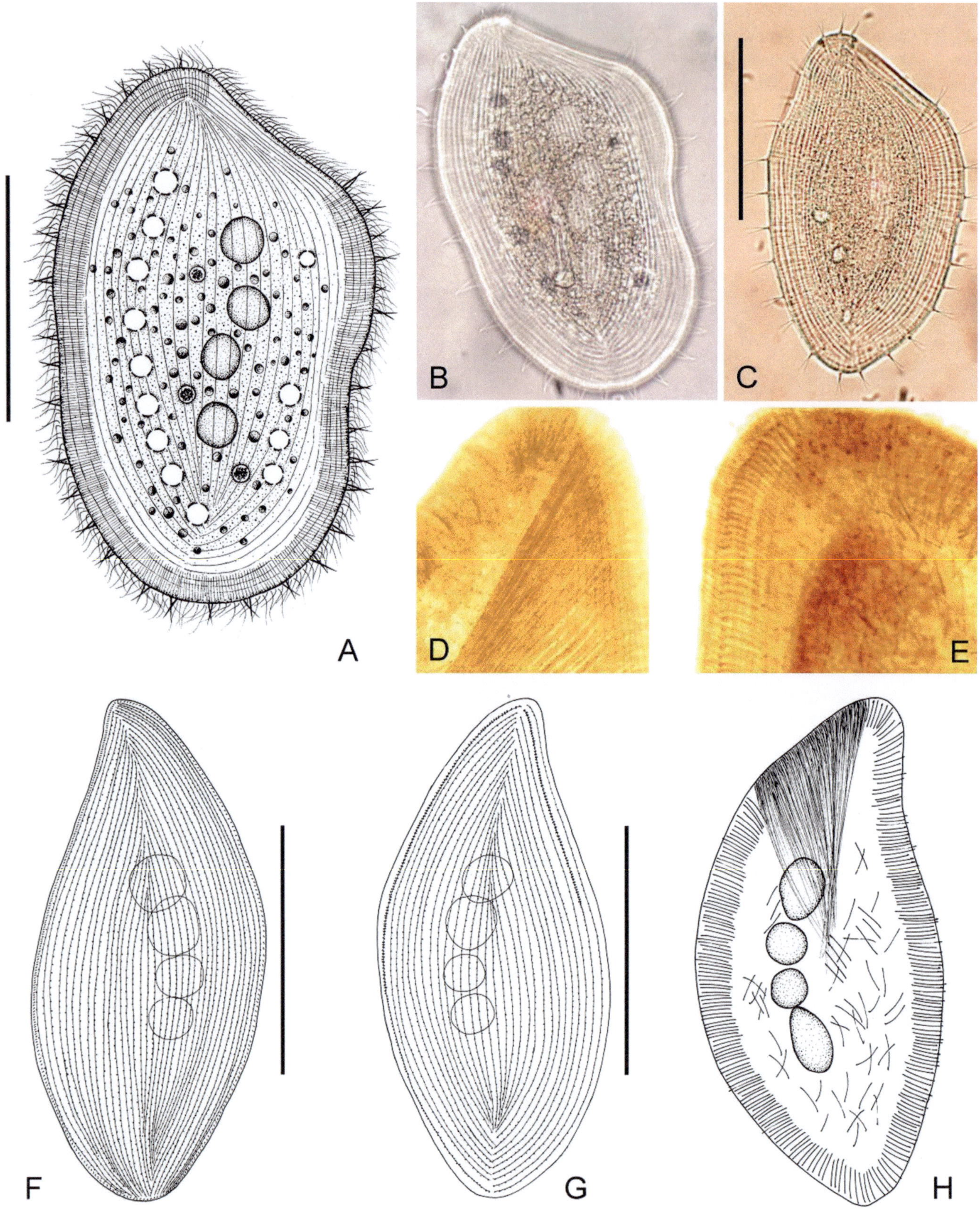


**Fig. 18B** *Kentrophyllum bispinum*
(**a–c**) Lateral views of live cells; (**d**, **e**) anterior part of right (**d**) and left (**e**) sides after silver staining; (**f**, **g**) ciliature on right (**f**) and left (**g**) sides; (**h**) extrusome pattern and nuclear apparatus. Scale bars: 80 μm

## Reference

Wu L, Clamp JC, Yi Z, Li J, Lin X (2015) Phylogenetic position and taxonomic revision of an enigmatic group of haptorian ciliates, with description of the new family, Kentrophyllidae fam. n. (Ciliophora, Litostomatea, Haptoria, Pleurostomatida). PLoS ONE 10(5):e0123720

Class Litostomatea Small & Lynn, 1981
Order Pleurostomatida Schewiakoff, 1896
Family Amphileptidae Bütschli, 1889
*Kentrophyllum shenzhenense* (Pan et al., 2010) Wu et al., 2015

**Ecological feature:** mangrove
**Water temperature:** 20.7 °C
**pH:** 7.5
**Salinity:** 13.8‰

**Fig. 19A** Sample site

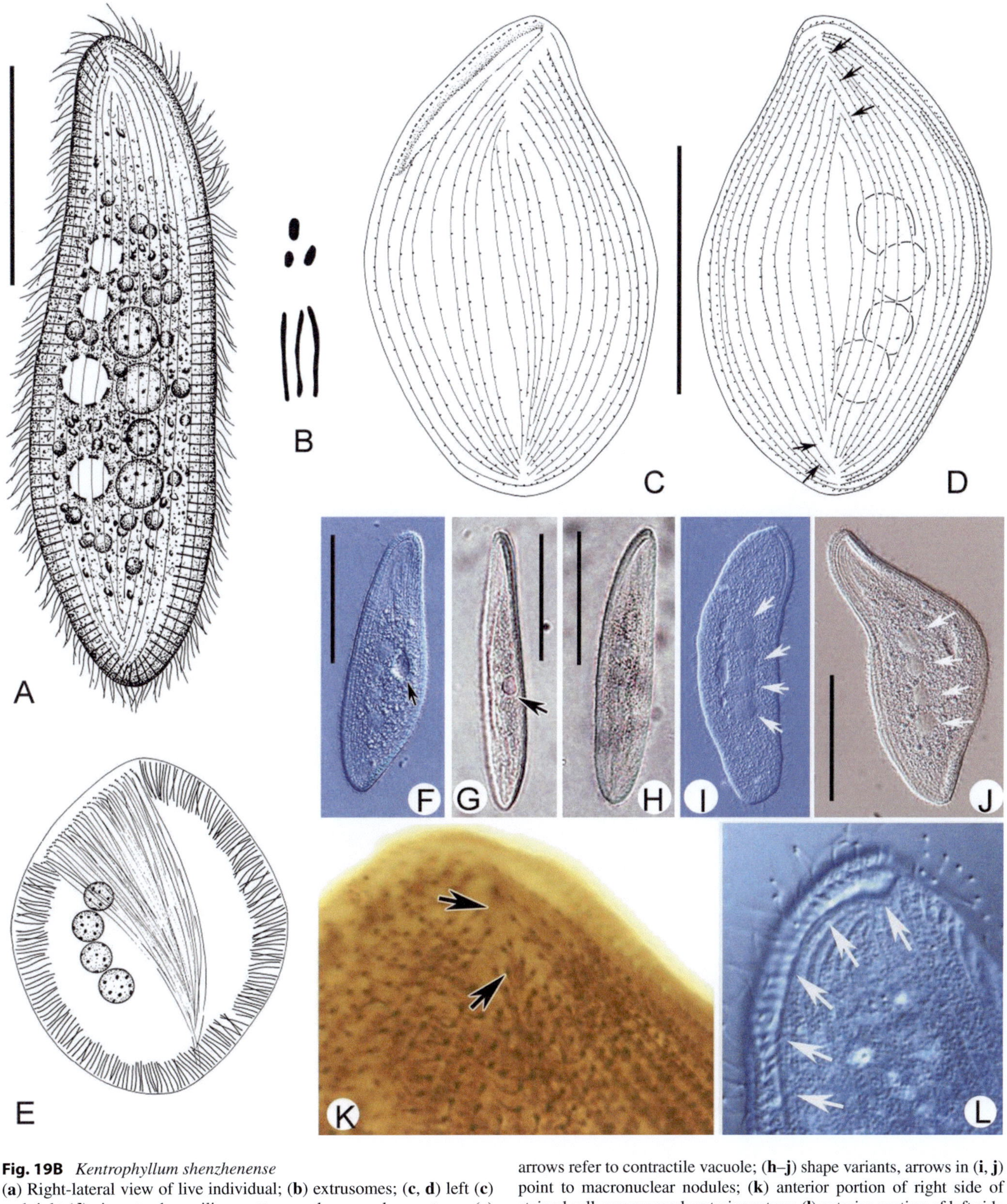


**Fig. 19B** *Kentrophyllum shenzhenense*
(**a**) Right-lateral view of live individual; (**b**) extrusomes; (**c**, **d**) left (**c**) and right (**d**) views to show ciliary pattern, and arrows show sutures; (**e**) distribution of extrusomes; (**f**, **g**) left-lateral views of live cells, and arrows refer to contractile vacuole; (**h**–**j**) shape variants, arrows in (**i**, **j**) point to macronuclear nodules; (**k**) anterior portion of right side of stained cell, arrows mark anterior suture; (**l**) anterior portion of left side of live cell, arrows point to oral slit. Scale bars: 70 μm

## References

Pan H, Gao F, Li J, Lin X, Al-Farraj SA, Al-Rasheid KAS (2010) Morphology and phylogeny of two new pleurostomatid ciliates, *Epiphyllum shenzhenense* n. sp. and *Loxophyllum spirellum* n. sp. (Protozoa, Ciliophora) from a mangrove wetland, South China. J Eukaryot Microbiol 57:421–428

Wu L, Clamp JC, Yi Z, Li J, Lin X (2015) Phylogenetic position and taxonomic revision of an enigmatic group of haptorian ciliates, with description of the new family, Kentrophyllidae fam. n. (Ciliophora, Litostomatea, Haptoria, Pleurostomatida). PLoS ONE 10(5):e0123720

Class Litostomatea Small & Lynn, 1981
Order Pleurostomatida Schewiakoff, 1896
Family Amphileptidae Bütschli, 1889
*Kentrophyllum strumosum* Wu et al., 2015

**Ecological feature:** mangrove
**Water temperature:** 23.5 °C
**pH:** 7.4
**Salinity:** 11.0‰

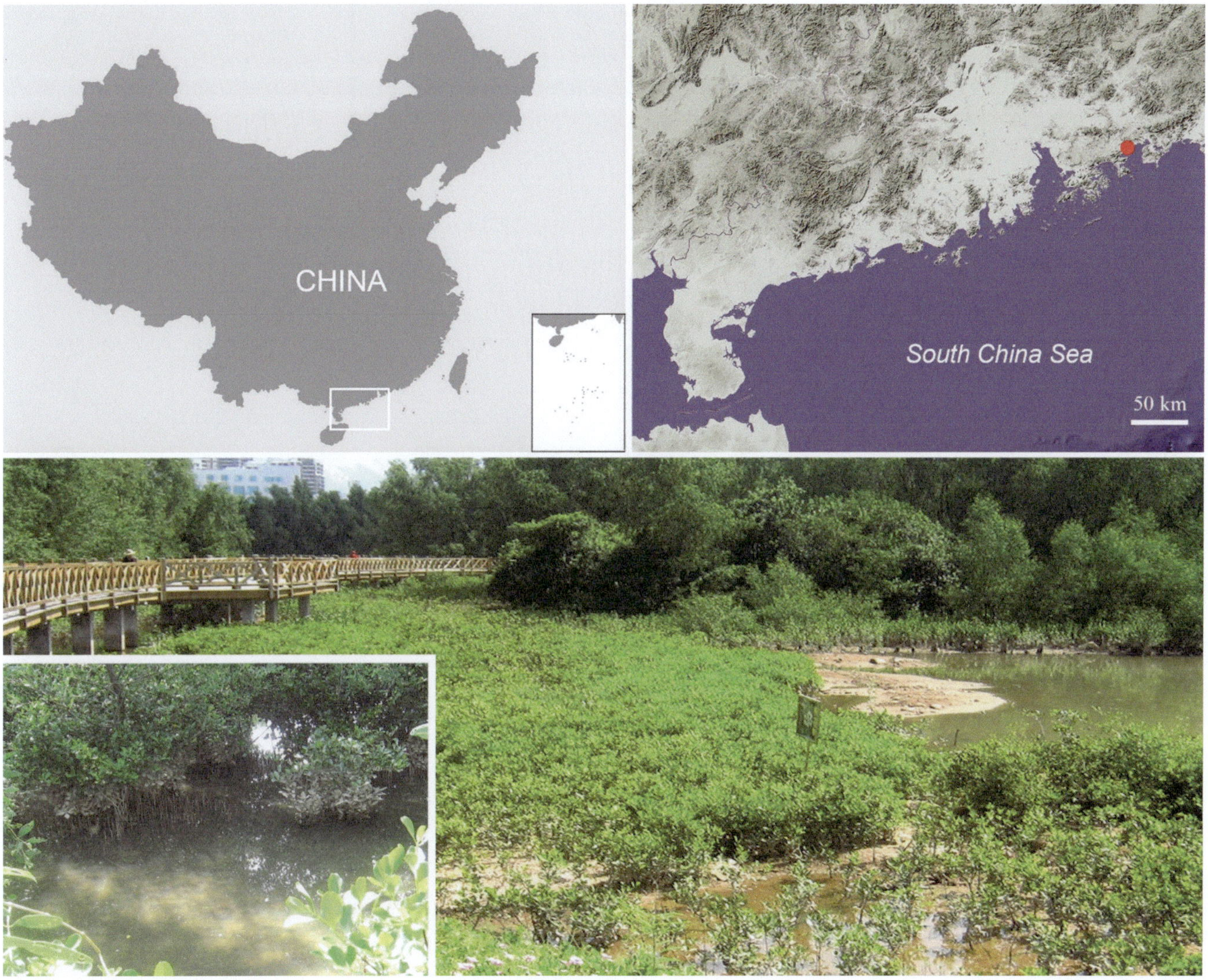


**Fig. 20A** Sample site

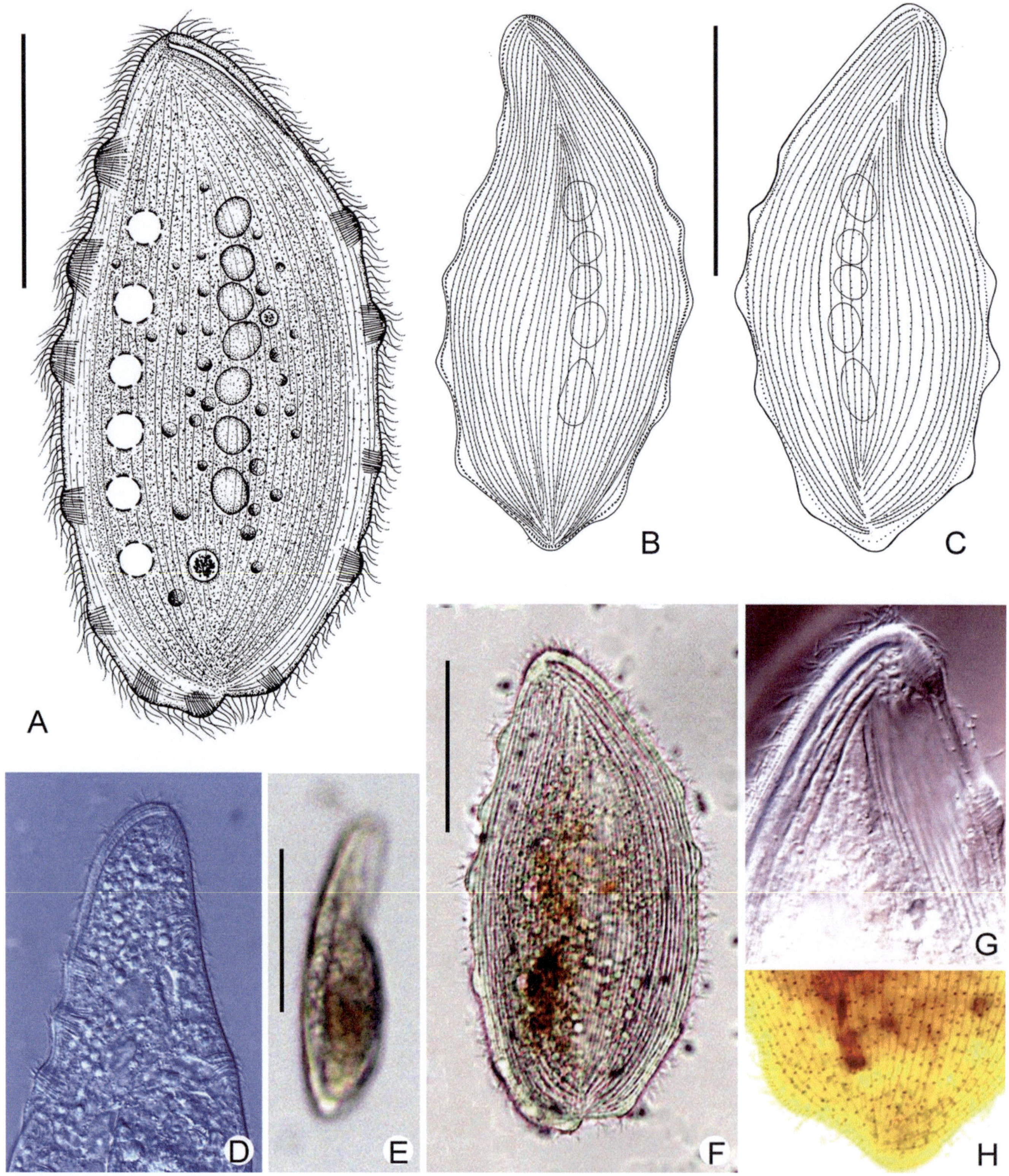


**Fig. 20B** *Kentrophyllum strumosum*
(**a**, **f**) Right-lateral views of live cells; (**b**, **c**) ciliary pattern on right (**b**) and left (**c**) sides of cells; (**d**, **g**) anterior part of lateral side of live cells; (**e**) dorsal side view of live cell, showing the arched left side; (**h**) posterior part of right-lateral view of stained cell. Scale bars: 80 μm

## Reference

Wu L, Clamp JC, Yi Z, Li J, Lin X (2015) Phylogenetic position and taxonomic revision of an enigmatic group of haptorian ciliates, with description of the new family, Kentrophyllidae fam. n. (Ciliophora, Litostomatea, Haptoria, Pleurostomatida). PLoS ONE 10(5):e0123720

Class Litostomatea Small & Lynn, 1981
Order Pleurostomatida Schewiakoff, 1896
Family Amphileptidae Bütschli, 1889
*Kentrophyllum verrucosum* (Stokes, 1893) Petz et al., 1995

**Ecological feature:** mangrove
**Water temperature:** 22.4 °C; 17.5 °C
**pH:** 7.8; 8.0
**Salinity:** 25.0‰; 27.0‰

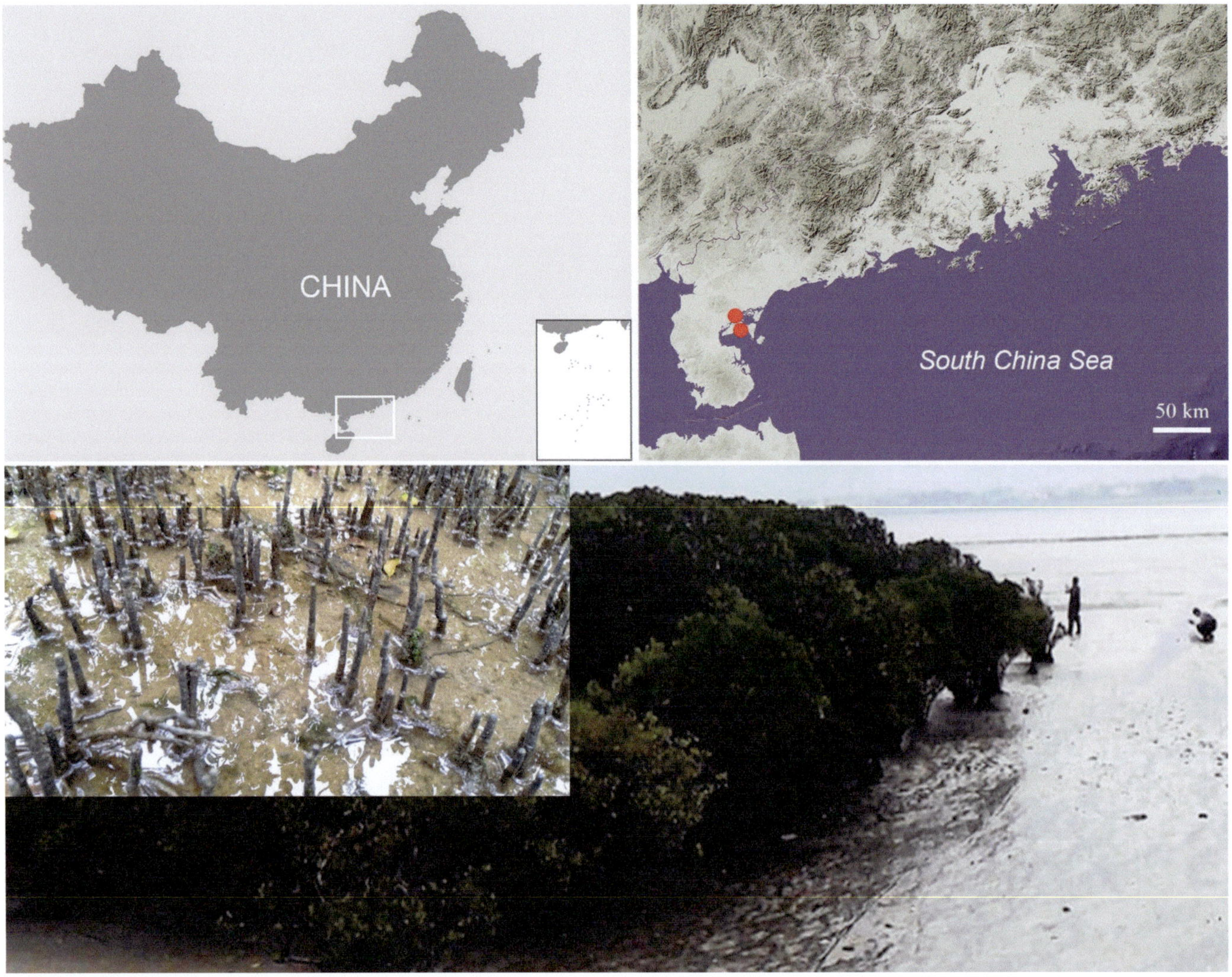


**Fig. 21A** Sample site

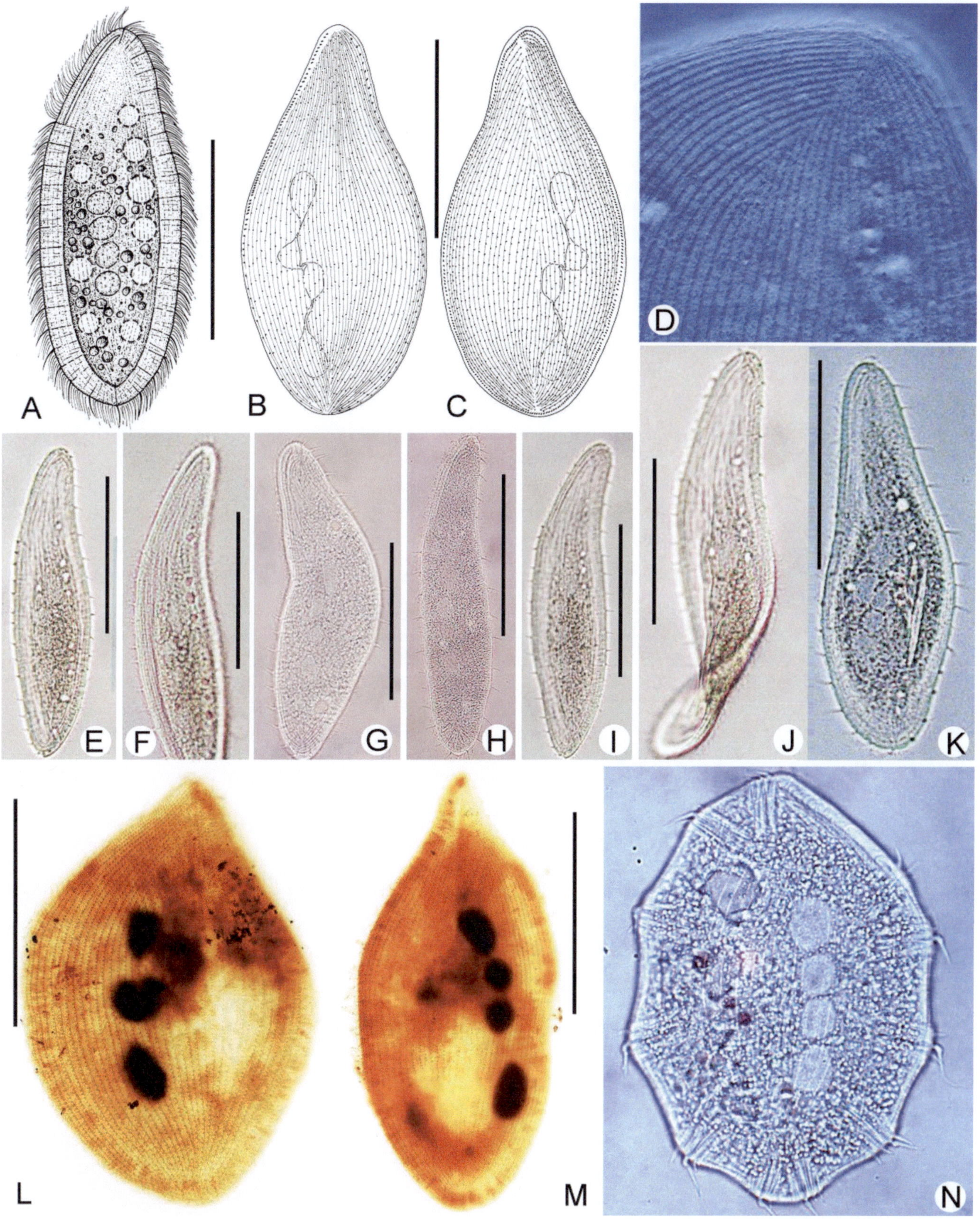


**Fig. 21B** *Kentrophyllum verrucosum*
(**a**, **e**, **f**, **h–j**) Left-lateral views of live cells; (**b**, **c**, **l**, **m**) left (**b**, **l**) and right (**c**, **m**) side views of ciliature; (**d**) anterior portion of lateral side of live cell, showing suture; (**b**, **k**, **n**) right-lateral views of live cells. Scale bars: 100 μm

## Reference

Lin X, Song WB, Warren A (2005) Taxonomic studies on three marine pleurostomatid ciliates: *Kentrophyllum verrucosum* (Stokes, 1893) Petz, Song et Wilbert, 1995, *Epiphyllum soliforme* (Faure-Fremiet, 1908) gen. n., comb. N. and *Amphileptus sikorai* sp. n. with the establishment of a new genus *Epiphyllum* (Ciliophora, Pleurostomatida). Acta Protozool 44:129–145

Class Litostomatea Small & Lynn, 1981
Order Pleurostomatida Schewiakoff, 1896
Family Litonotidae Kent, 1881
*Acineria incurvata* Dujardin, 1841

**Ecological feature:** mangrove
**Water temperature:** 25.3 °C
**pH:** 8.0
**Salinity:** 22.3‰

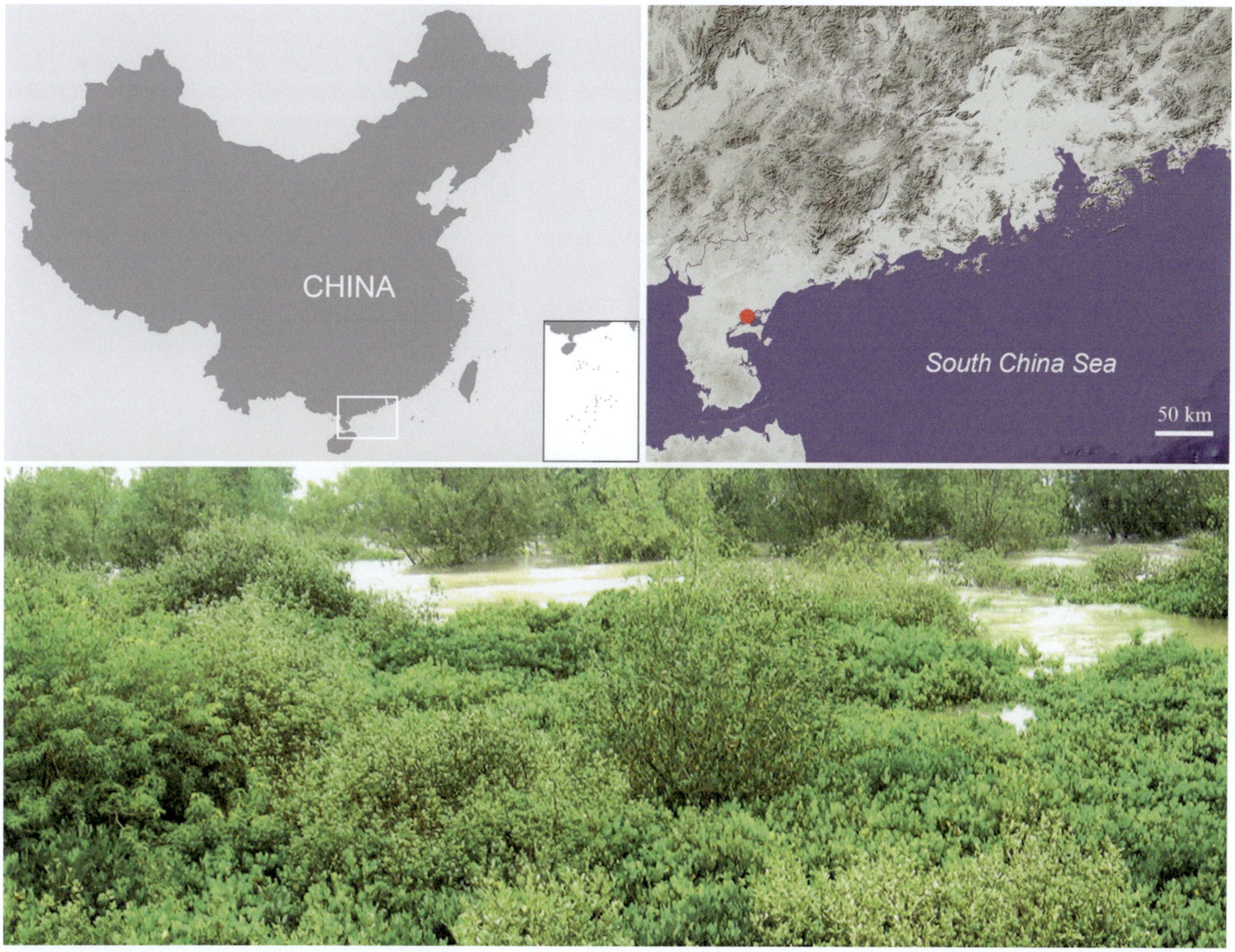


**Fig. 22A** Sample site

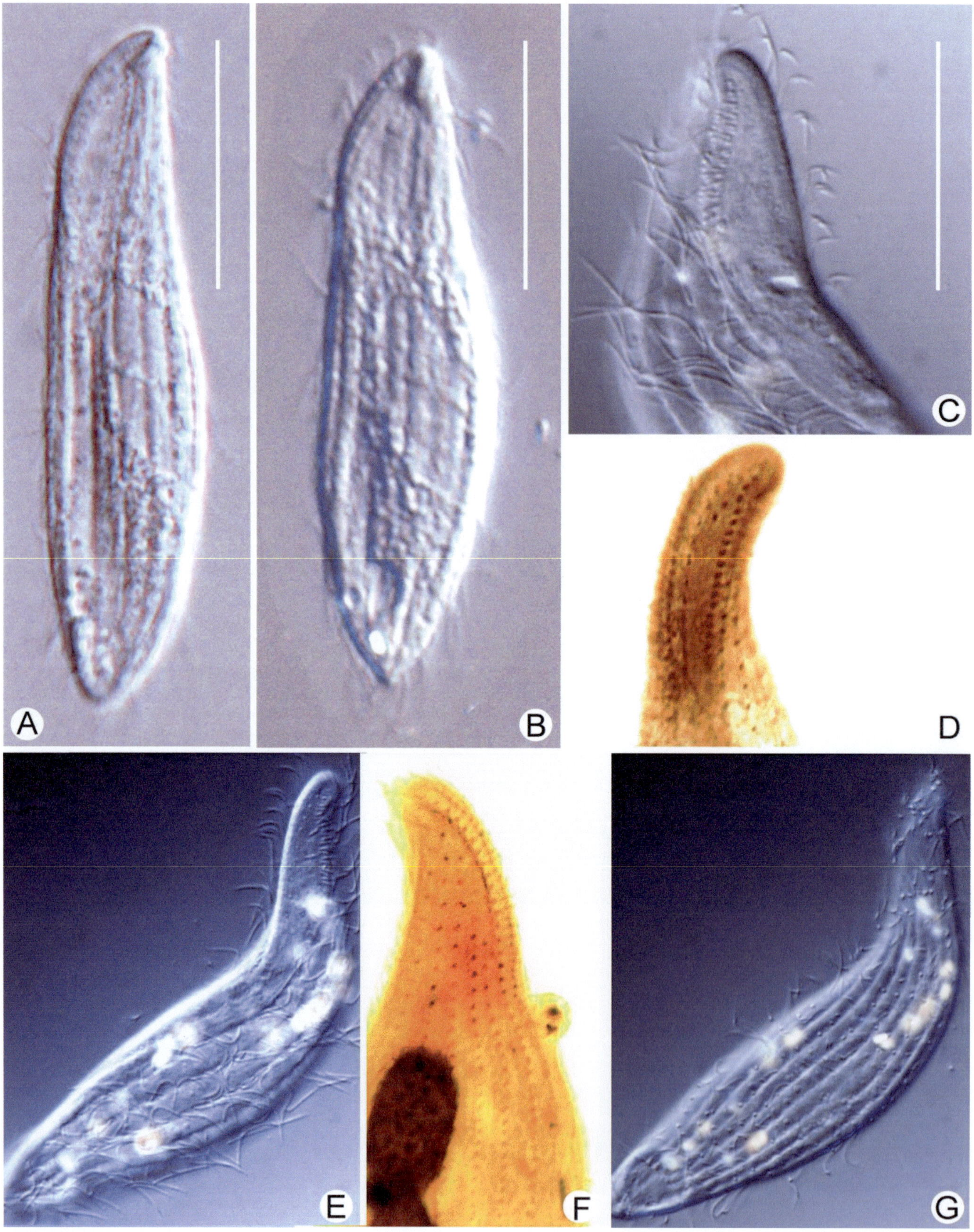


**Fig. 22B** *Acineria incurvata*
(**a**, **b**, **e**) Left-lateral views of live cells; (**c**, **d**) anterior part of cell, showing dorsal brush; (**f**) ciliature of anterior right side of cell; (**g**) right-lateral view of live cell. Scale bars: 30 μm (**a**, **b**); 20 μm (**c**)

## References

Foissner W (1977) Taxonomische Studien über die Ciliaten des Großglocknergebietes (Hohe Tauern, Österreich), 2. Familie Amphileptidae. Ber Haus Natur Salzburg 8:87–93

Kahl A (1933) Ciliata libera et ectocommensalia. *Die Tierwelt der Nord- und Ostsee*. Lieferung 23 (Teil II. $c_3$): 29–146.

Class Litostomatea Small & Lynn, 1981
Order Pleurostomatida Schewiakoff, 1896
Family Litonotidae Kent, 1881
*Litonotus bergeri* Lin et al., 2008

**Ecological feature:** mangrove
**Water temperature:** 24.0 °C
**pH:** 7.8
**Salinity:** 8.0‰

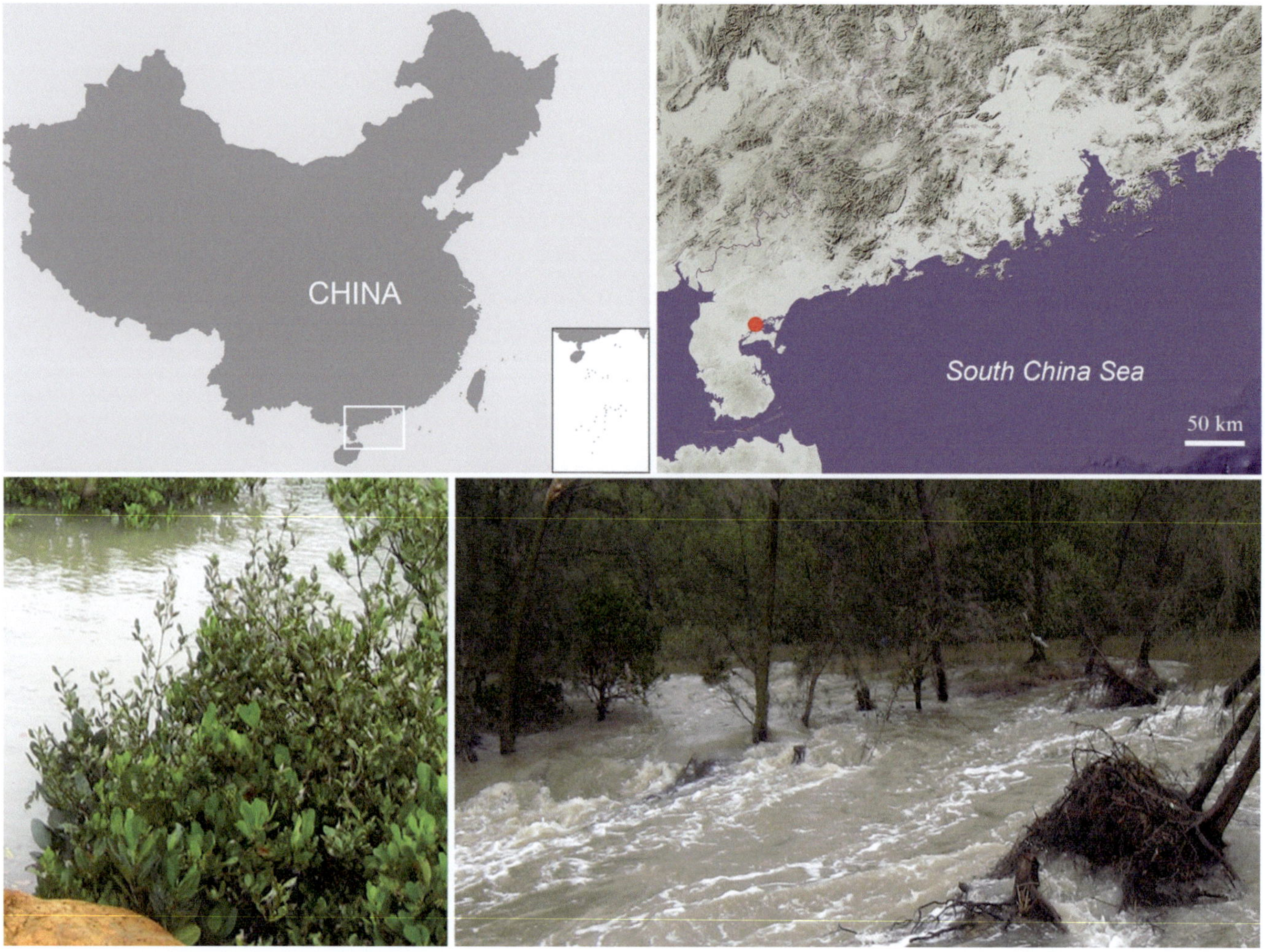


**Fig. 23A** Sample site

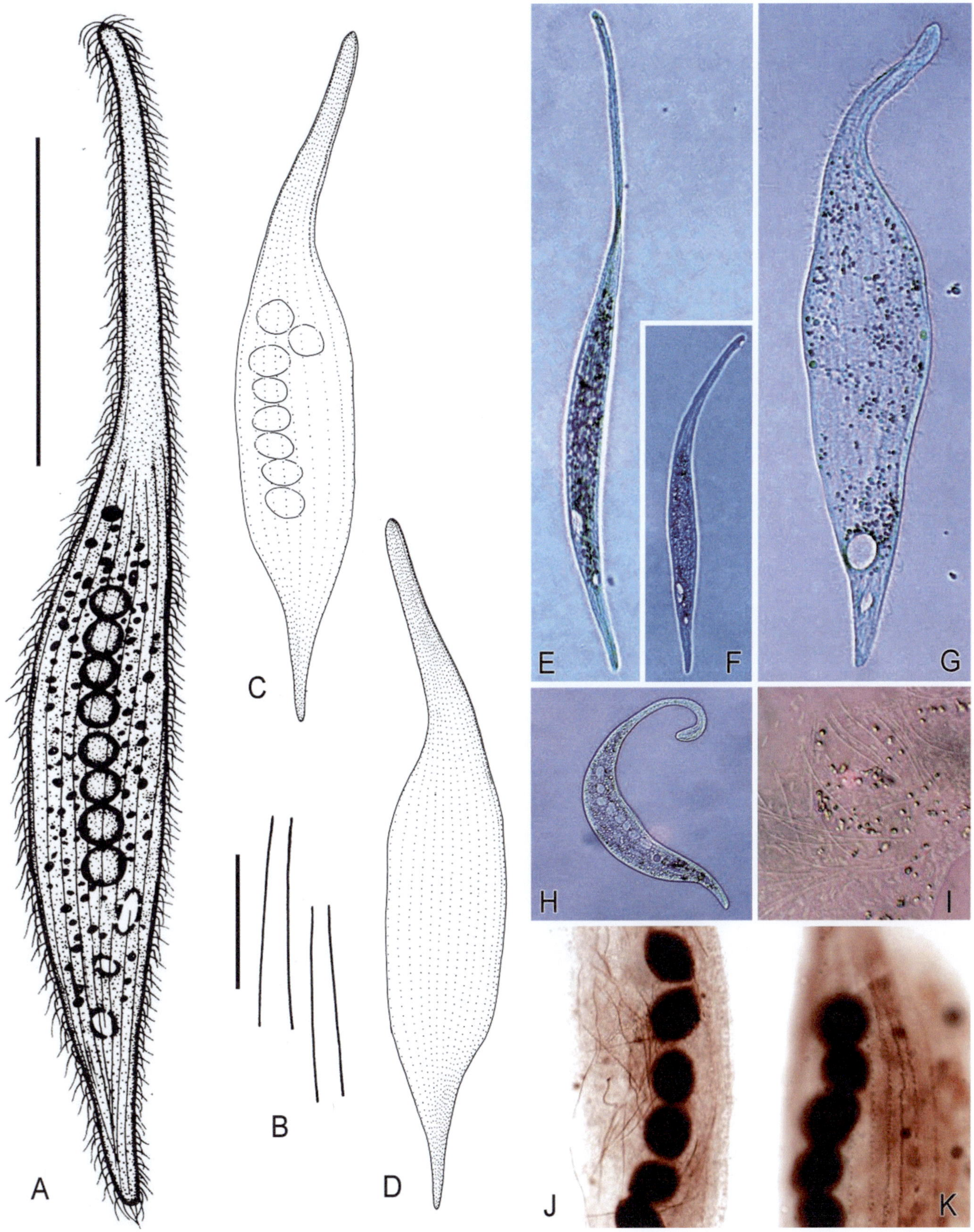


**Fig. 23B** *Litonotus bergeri*
(**a**, **e**) Right-lateral views of live cells; (**b**, **i**) extrusomes; (**c**, **d**) ciliature of left (**c**) and right (**d**) sides; (**f**–**h**) left-lateral views of live cells; (**j**) macronuclear nodules; (**k**) middle portion of cell showing perioral kinety. Scale bars: 150 um (**a**); 5 um (**b**)

## Reference

Lin X, Li J, Gong J, Warren A, Song WB (2008) Taxonomic studies on three marine pleurostomatid ciliates, *Litonotus bergeri* nov. spec., *L. blattereri* nov. spec. and *L. petzi* nov. spec. (Ciliophora, Pleurostomatida). Eur J Protistol 51:494–506

Class Litostomatea Small & Lynn, 1981
Order Pleurostomatida Schewiakoff, 1896
Family Litonotidae Kent, 1881
*Litonotus* cf. *triqueter*

**Ecological feature:** intertidal zone
**Water temperature:** 25.8 °C
**pH:** 8.9
**Salinity:** 12.2‰

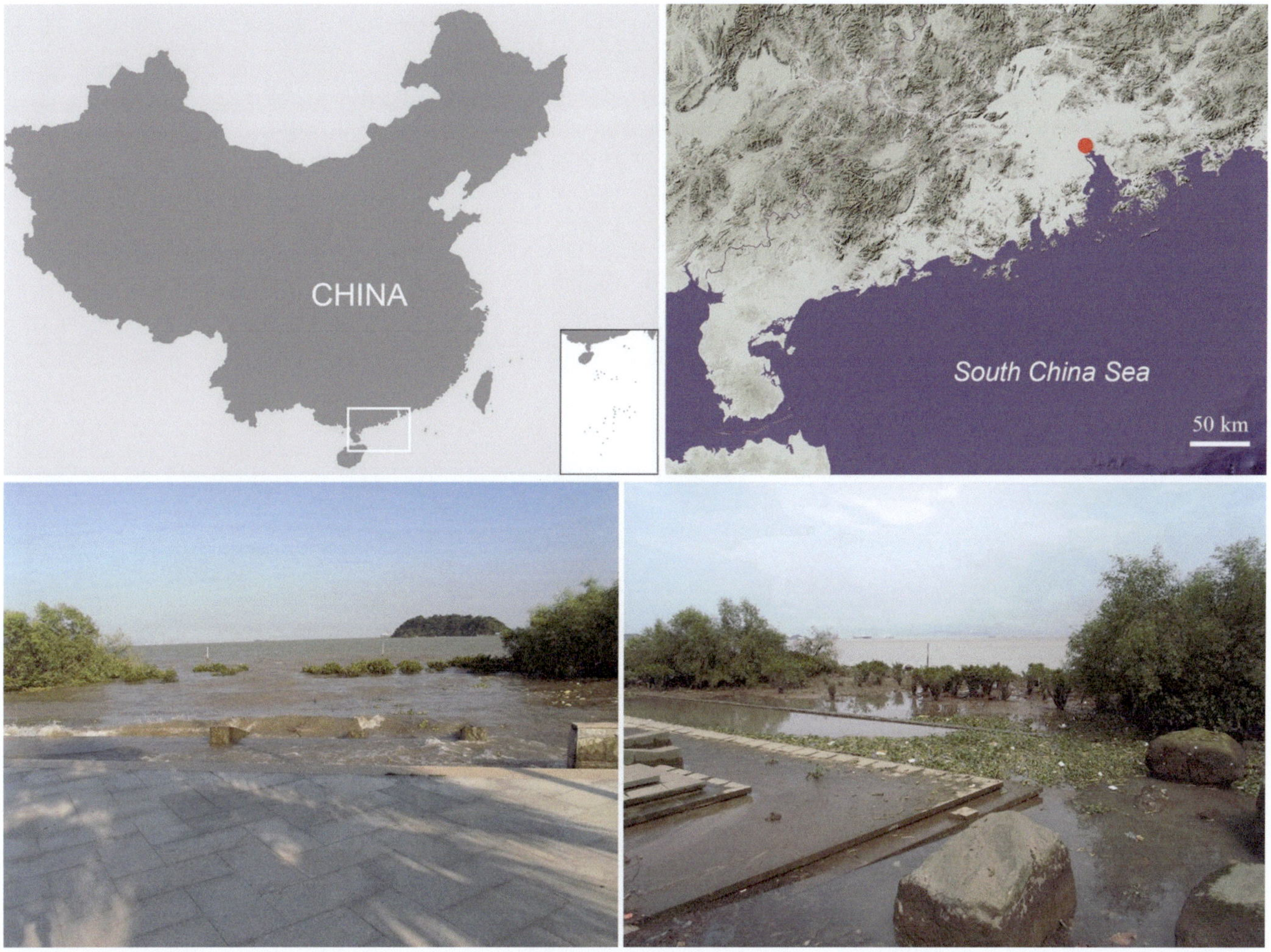


**Fig. 24A** Sample site

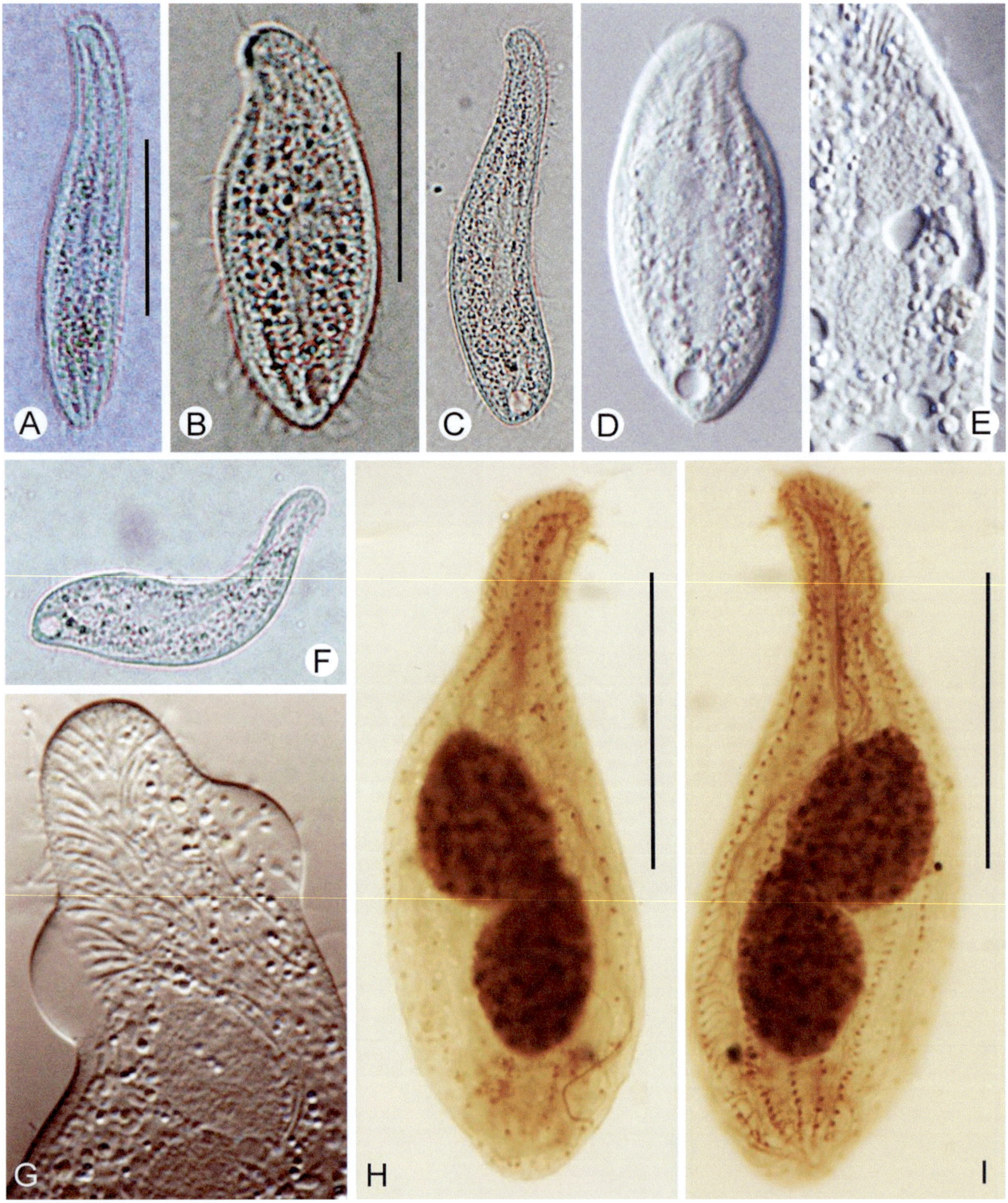


**Fig. 24B** *Litonotus* cf. *triqueter*
(**a**) Right view of live individual; (**b–d**, **f**) body shape variants showing contractile vacuole at posterior end of cell; (**e**) macronuclei and micronucleus; (**g**) extrusomes along oral slit; (**h**, **i**) left (**h**) and right (**i**) views to show ciliary pattern. Scale bars: 60 μm

## Reference

Song WB, Wilbert N (1989) Taxonomische Untersuchungen an Aufwuchsciliaten (Protozoa, Ciliophora) im Poppelsdorfer Weiher, Bonn. Lauterbornia 3:2–221

Class Litostomatea Small & Lynn, 1981
Order Pleurostomatida Schewiakoff, 1896
Family Litonotidae Kent, 181
*Litonotus gracilis* Pan et al., 2015

**Ecological feature:** mangrove
**Water temperature:** 19.7 °C
**pH:** no data available
**Salinity:** 23.9‰

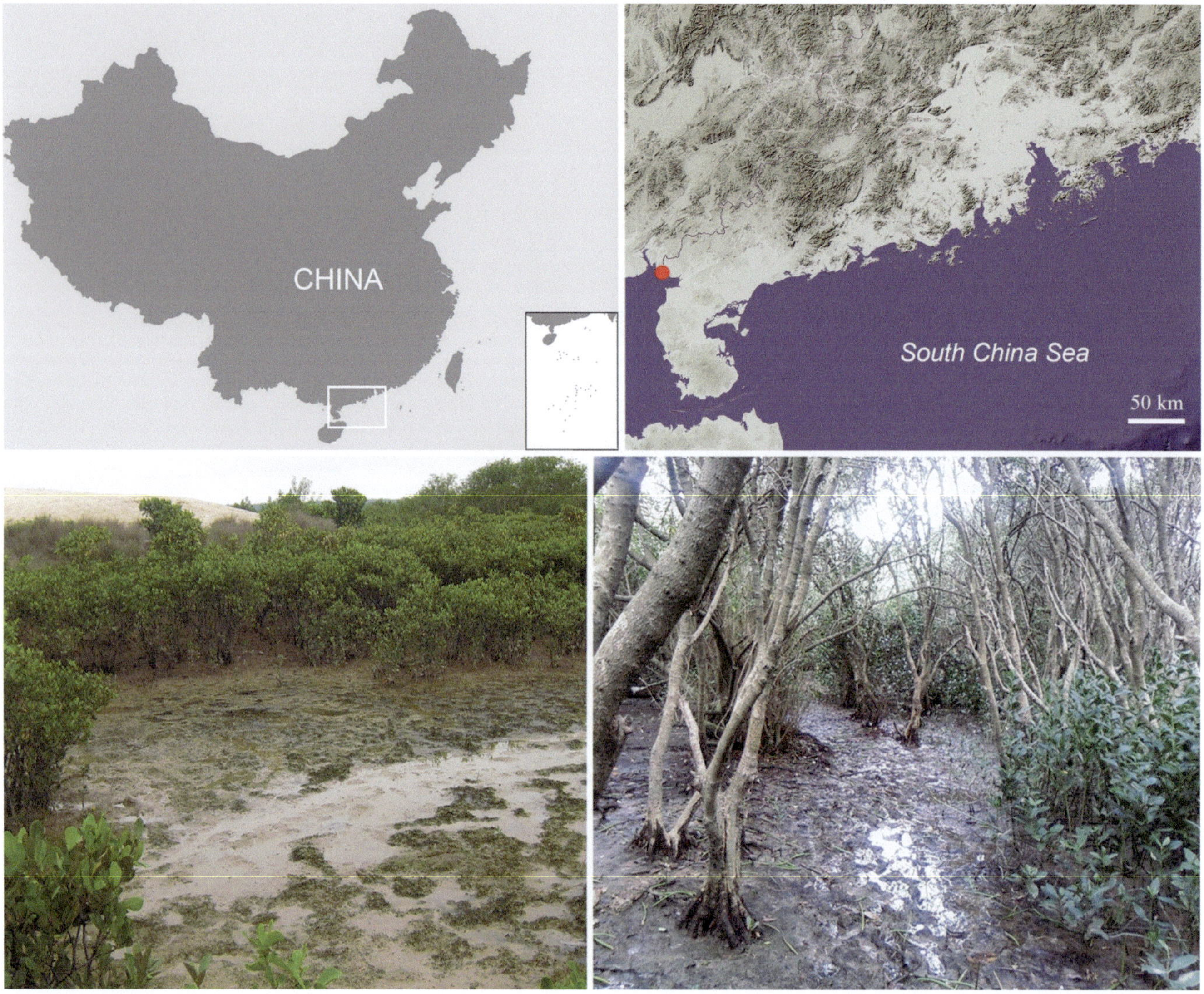


**Fig. 25A** Sample site

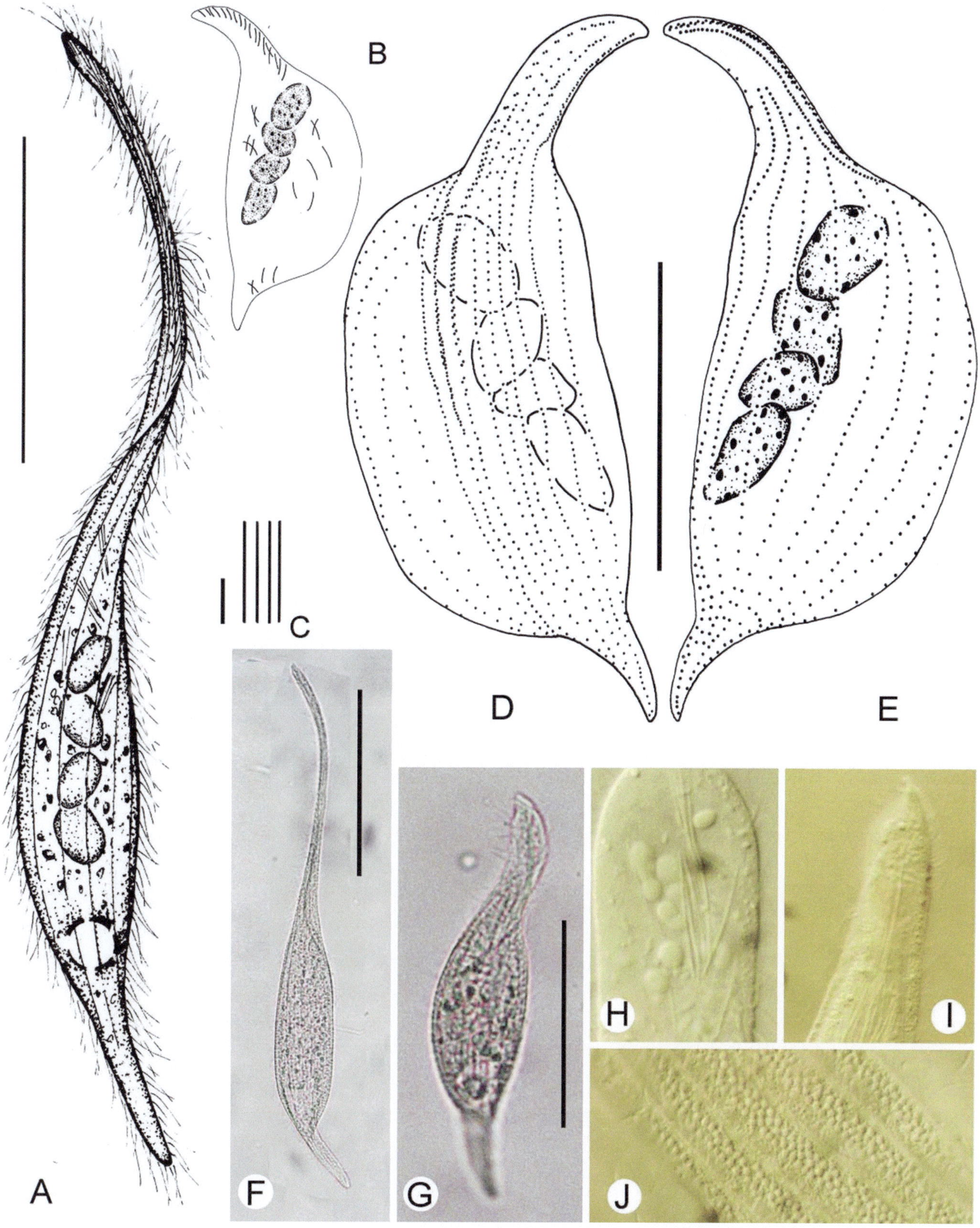


**Fig. 25B** *Litonotus gracilis*
(**a**, **f**) Lateral views of representative cells; (**b**) macronuclei and extrusomes; (**c**) rod-like extrusomes; (**d**, **e**) ciliature of left (**d**) and right (**e**) sides of cell; (**g**) a contracted cell; (**h**, **i**) anterior portion of live cells showing extrusomes; (**j**) cortical granules. Scale bars: 100 μm (**a**, **d**, **e**, **f**, **g**); 8 μm (**c**)

## Reference

Pan H, Li L, Wu L, Miao M, Al-Rasheid K, Song WB (2015) Morphology of three *Litonotus* species (Ciliophora: Pleurostomatida) from China seas, with brief notes on their SSU rDNA-based phylogeny. Eur J Protistol 51:494–506

Class Litostomatea Small & Lynn, 1981
Order Pleurostomatida Schewiakoff, 1896
Family Litonotidae Kent, 1881
*Litonotus* sp.

**Ecological feature:** mangrove
**Water temperature:** 24.0 °C
**pH:** 7.7
**Salinity:** 21.3‰

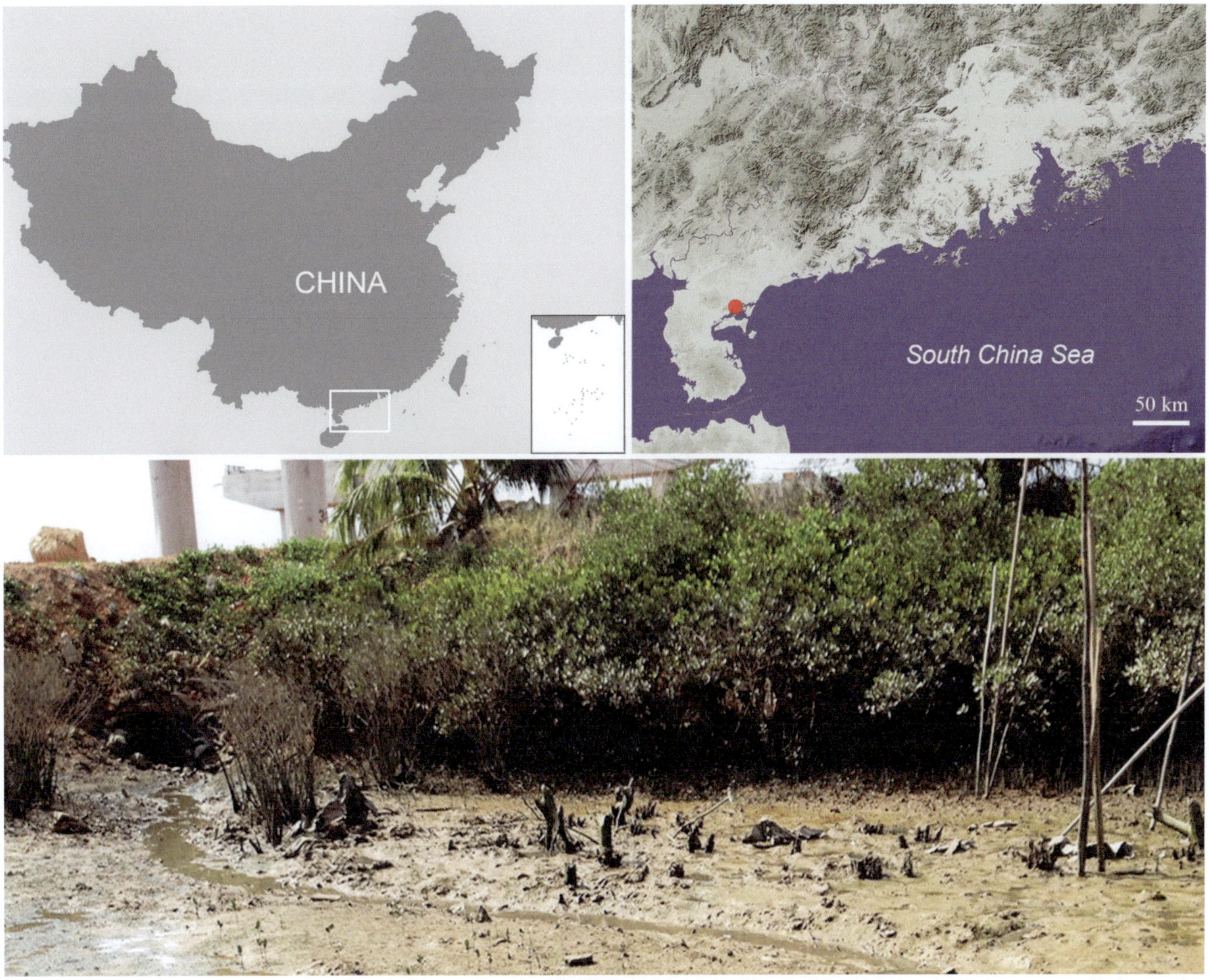


**Fig. 26A** Sample site

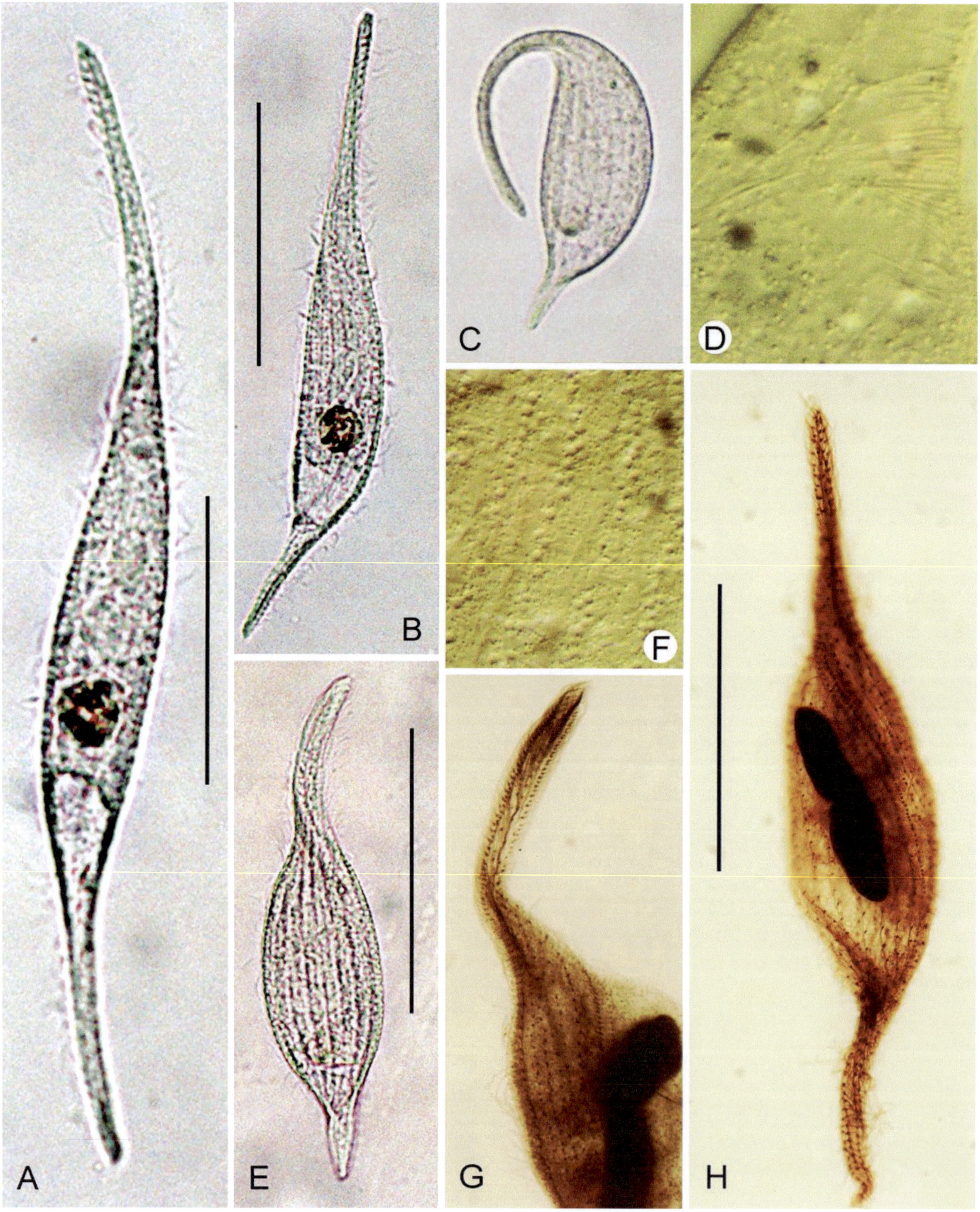


**Fig. 26B** *Litonotus* sp.
(**a**, **b**, **c**, **e**) Lateral views of live individuals, showing body shape variants; (**d**) extrusomes; (**f**) cortical granules; (**g**) anterior portion of stained cell, to show ciliature of left side; (**h**) ciliature of right side of cell. Scale bars: 100 μm

## Reference

Zhao L (2013) Morphological studies on fourteen marine cyrtophorid and pleurostomatid ciliates. Ocean University of China, Master thesis.

Class Litostomatea Small & Lynn, 1981
Order Pleurostomatida Schewiakoff, 1896
Family Litonotidae Kent, 181
*Litonotus tropicus* Pan et al., 2015

**Ecological feature:** intertidal zone
**Water temperature:** 24.2 °C
**pH:** no data available
**Salinity:** 13.9‰

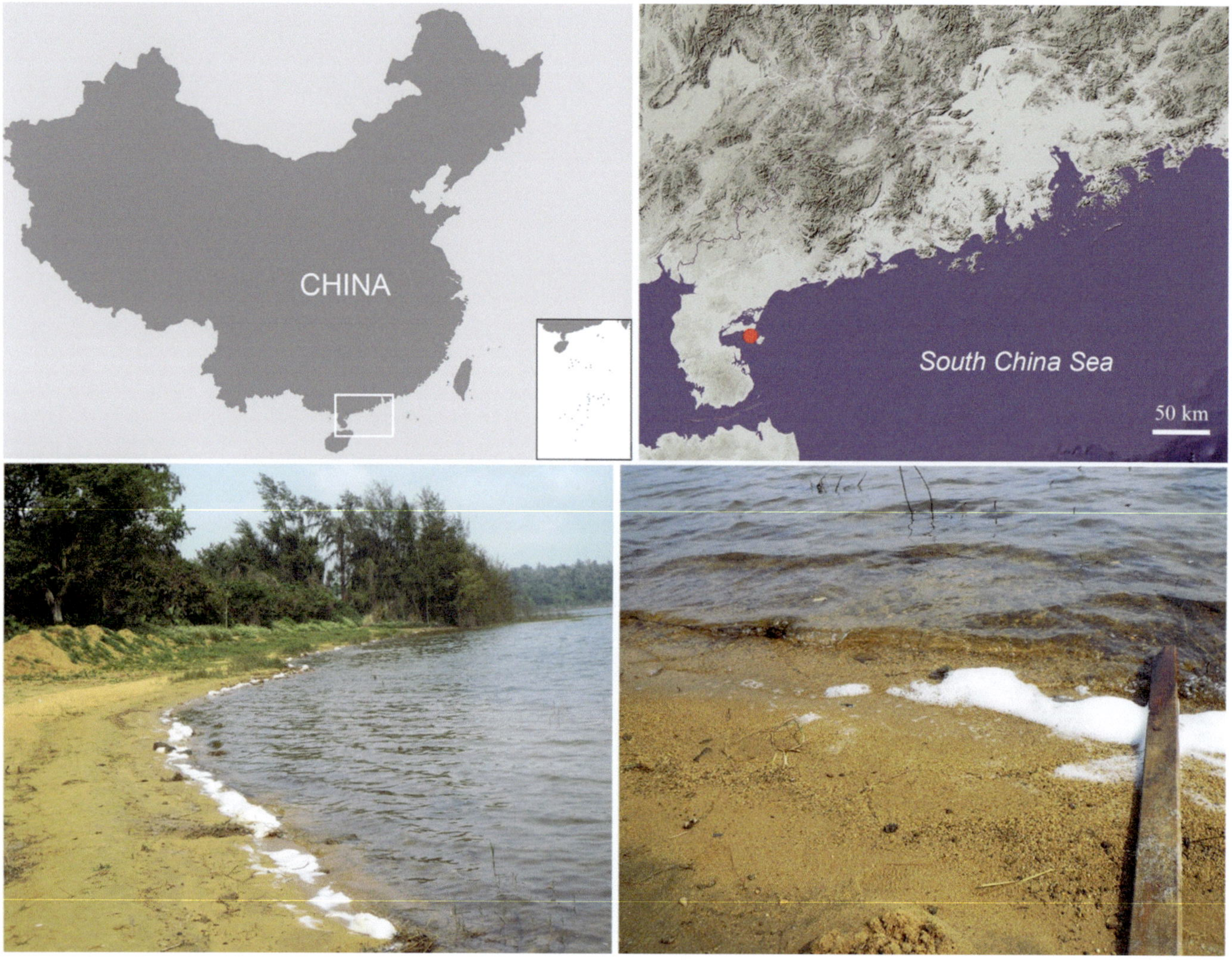


**Fig. 27A** Sample site

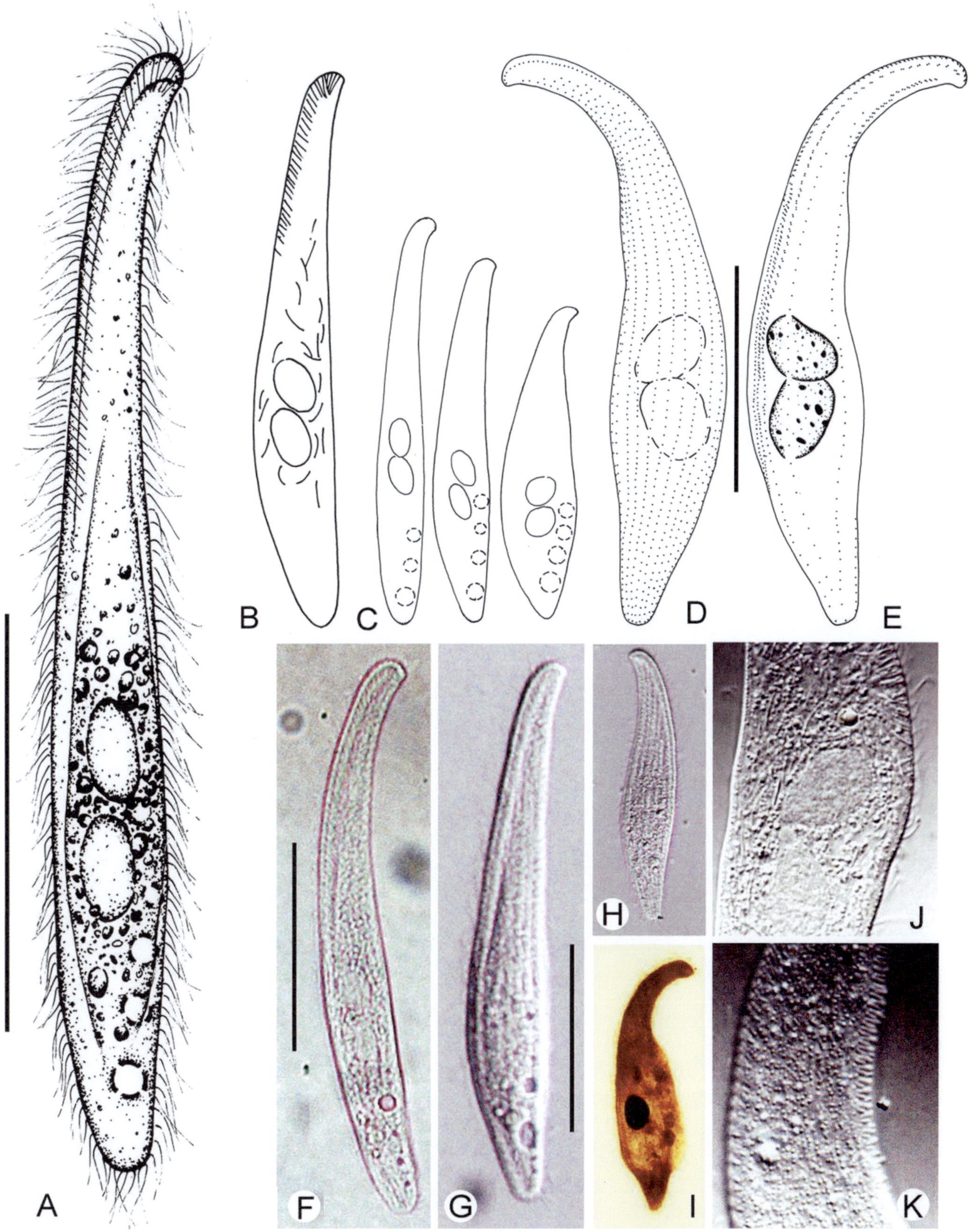


**Fig. 27B** *Litonotus tropicus*
(**a**, **f**, **g**, **h**) Lateral views of live cells; (**b**) extrusomes; (**c**) body shape variants; (**d**, **e**) ciliature of right (**d**) and left (**e**) sides; (**i**) a stained cell with single macronucleus; (**j**) details of cytoplasm showing nuclei; (**k**) part of cell surface on left side, showing cortical granules. Scale bars: 50 μm

## Reference

Pan H, Li L, Wu L, Miao M, Al-Rasheid KAS, Song WB (2015) Morphology of three *Litonotus* species (Ciliophora, Pleurostomatida) from China seas, with brief notes on their SSU rDNA-based phylogeny. Eur J Protistol 51:494–506

Class Litostomatea Small & Lynn, 1981
Order Pleurostomatida Schewiakoff, 1896
Family Litonotidae Kent, 181
*Loxophyllum caudatum* Pan et al., 2013

**Ecological feature:** mangrove
**Water temperature:** 23.8 °C
**pH:** 5.5
**Salinity:** 11.6‰

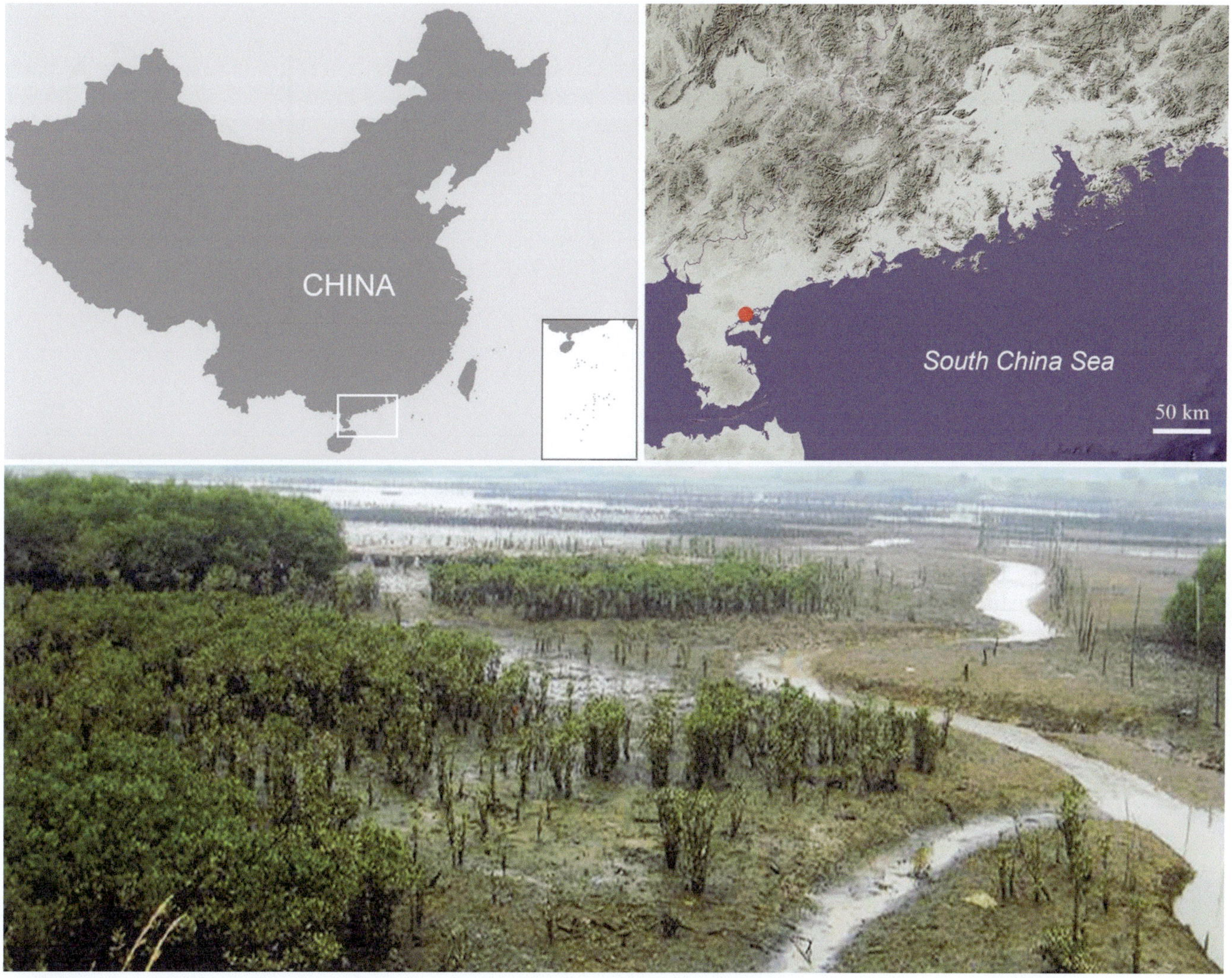


**Fig. 28A** Sample site

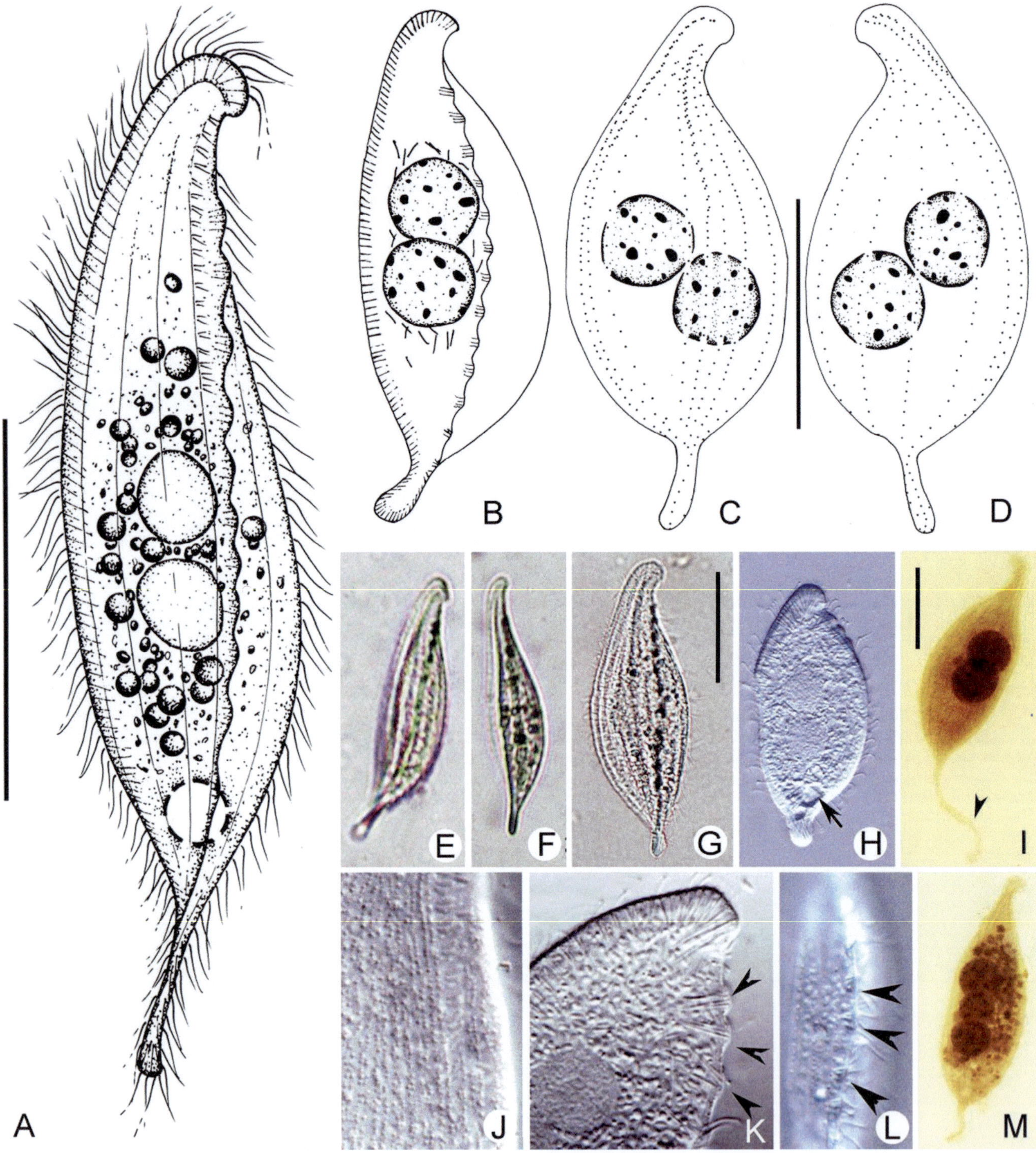


**Fig. 28B** *Loxophyllum caudatum*
(**a**, **e**, **g**) Left views of live cells; (**b**) distribution of extrusomes; (**c**, **d**) left (**c**) and right (**d**) views of ciliature; (**f**) dorsal view of live individual; (**h**) left view of live cell, arrow refers to contractile vacuole; (**i**) a stained cell, indicating nuclei and tail (arrowhead); (**j**) cortical granules; (**k**) anterior portion of live cell, arrowheads point to warts; (**l**) part of dorsal margin of live cell, arrowheads refer to warts; (**m**) left view of a stained specimen with three macronuclear nodules. Scale bars: 50 μm

## Reference

Pan H, Gao F, Lin X, Warren A, Song WB (2013) Three new *Loxophyllum* species (Ciliophora: Pleurostomatida) from China with a brief review of the marine and brackish *Loxophyllum* species. J Eukaryot Microbiol 60:44–56

Class Litostomatea Small & Lynn, 1981
Order Pleurostomatida Schewiakoff, 1896
Family Litonotidae Kent, 181
*Loxophyllum* cf. *caudatum*

**Ecological feature:** mangrove
**Water temperature:** 26.0 °C
**pH:** 7.8
**Salinity:** 7.0‰–14.8‰

**Fig. 29A** Sample site

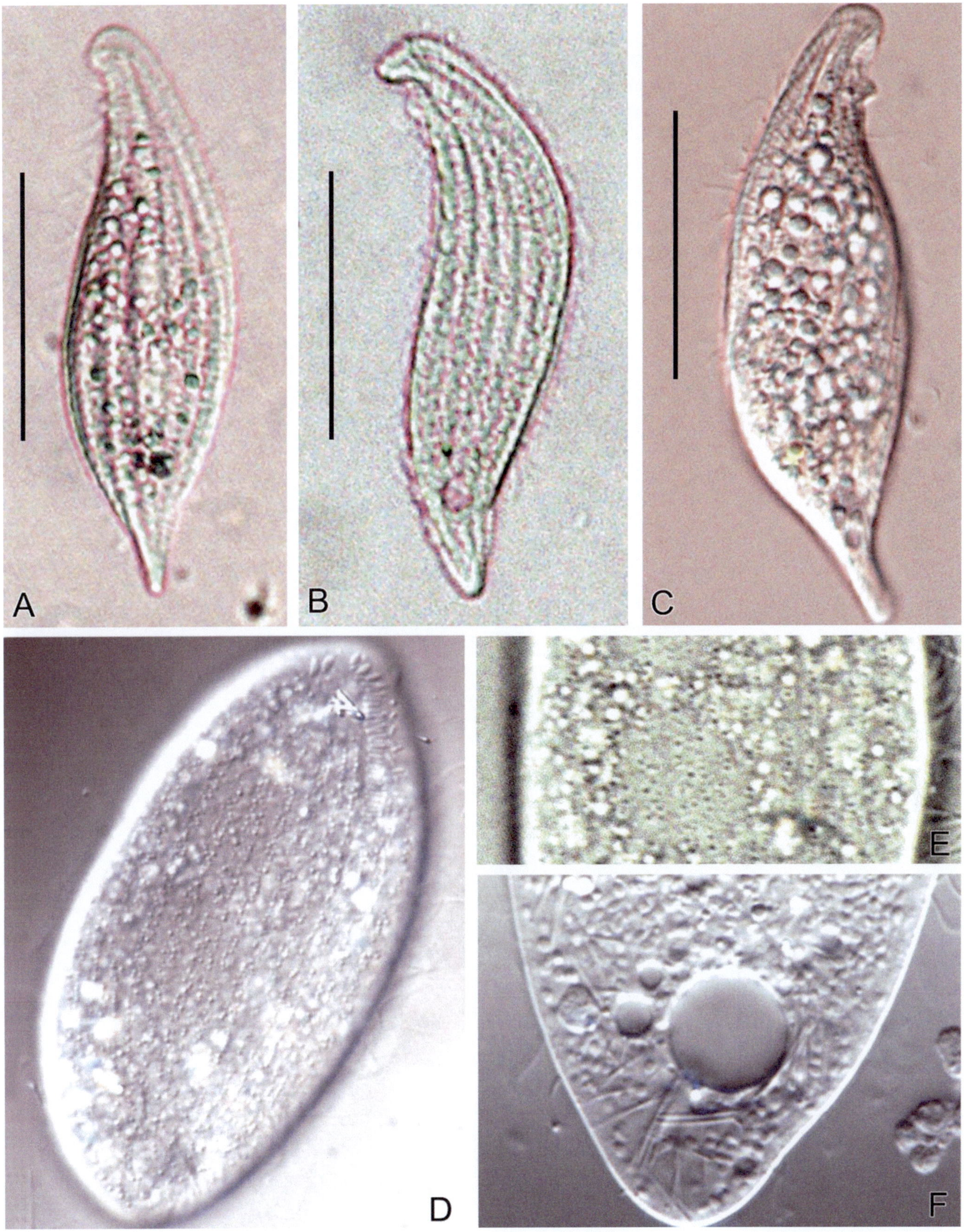


**Fig. 29B** *Loxophyllum* cf. *caudatum*
(**a–c**) Lateral views of live cells; (**d**) left side view of live cell, shows the cortical granules; (**e**) distribution of cortical granules; (**f**) extrusomes and contractile vacuole in posterior end of cell. Scale bars: 50 μm

## Reference

Pan H, Gao F, Lin X, Warren A, Song WB (2013) Three new *Loxophyllum* species (Ciliophora: Pleurostomatida) from China with a brief review of the marine and brackish *Loxophyllum* species. J Eukaryot Microbiol 60:44–56

Class Litostomatea Small & Lynn, 1981
Order Pleurostomatida Schewiakoff, 1896
Family Litonotidae Kent, 181
*Loxophyllum* cf. *chinense*

**Ecological feature:** estuary
**Water temperature:** 26.0 °C
**pH:** 7.5
**Salinity:** 16.0‰

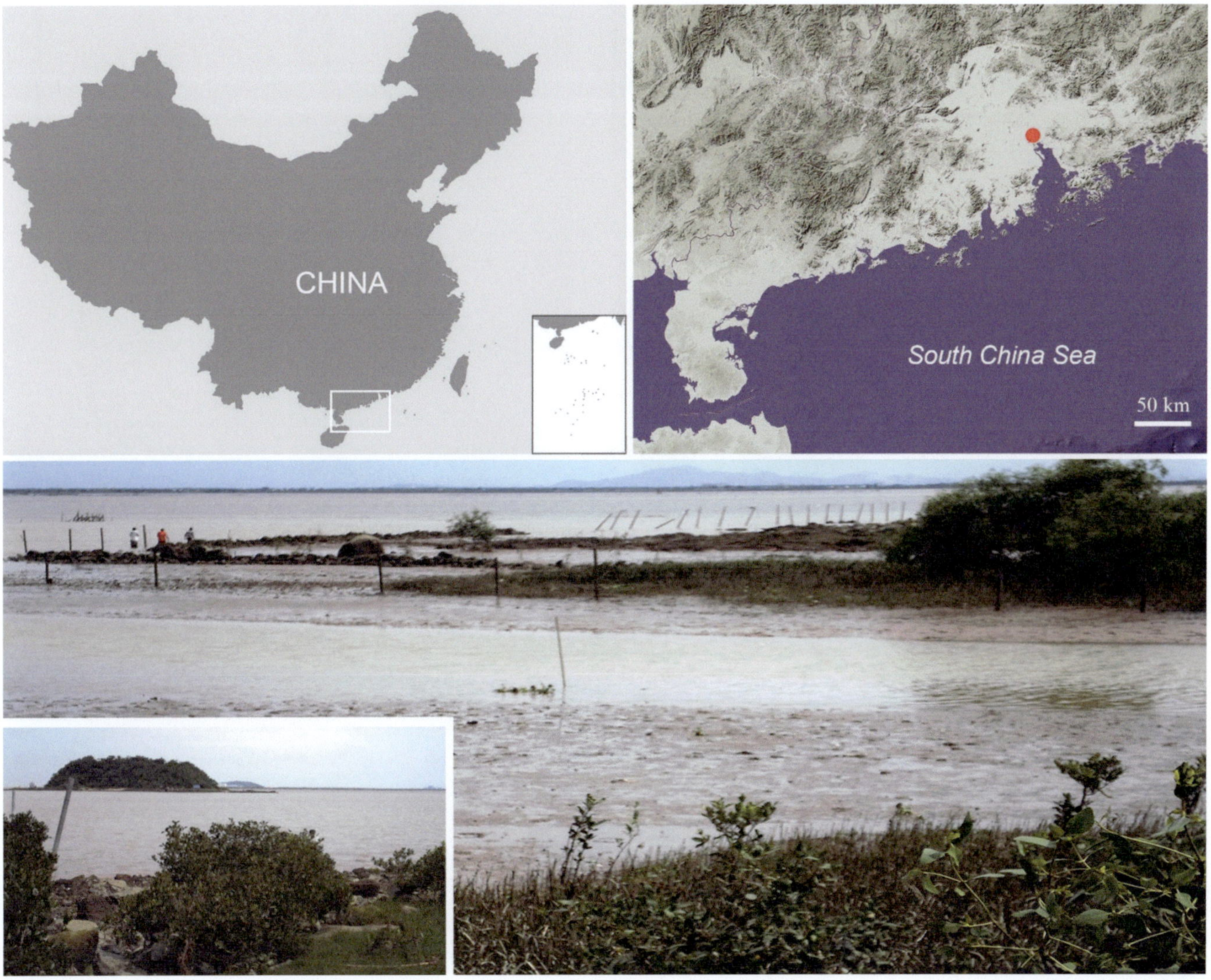


**Fig. 30A** Sample site

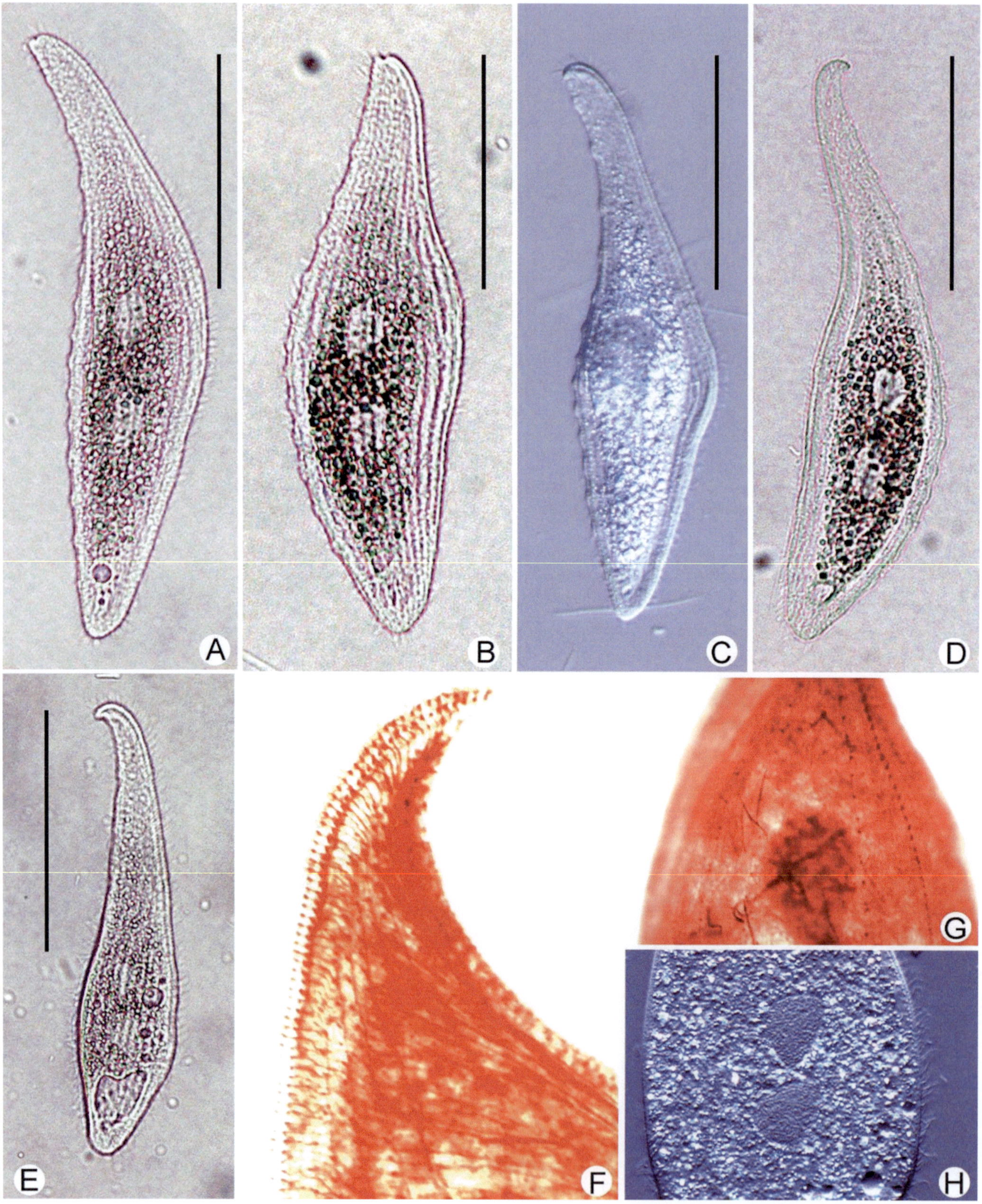


**Fig. 30B** *Loxophyllum* cf. *chinense*
(**a**–**c**, **e**) Right-lateral views of live individuals, showing body shape variants; (**d**) left side view of live cell, to show macronuclear nodules; (**f**) oral apparatus; (**g**) dorsal bristles and extrusomes; (**h**) middle portion of live cell, to show the macronuclear nodules. Scale bars: 50 μm

## Reference

Pan H, Gao F, Lin X, Warren A, Song WB (2013). Three new *Loxophyllum* species (Ciliophora: Pleurostomatida) from China with a brief review of the marine and brackish *Loxophyllum* species. J Eukaryot Microbiol 60:44–56

Class Litostomatea Small & Lynn, 1981
Order Pleurostomatida Schewiakoff, 1896
Family Litonotidae Kent, 181
*Loxophyllum lembum* Wu et al., 2014

**Ecological feature:** off-shore water/mangrove
**Water temperature:** 26.0 °C
**pH:** 7.8
**Salinity:** 27.2‰

**Fig. 31A** Sample site

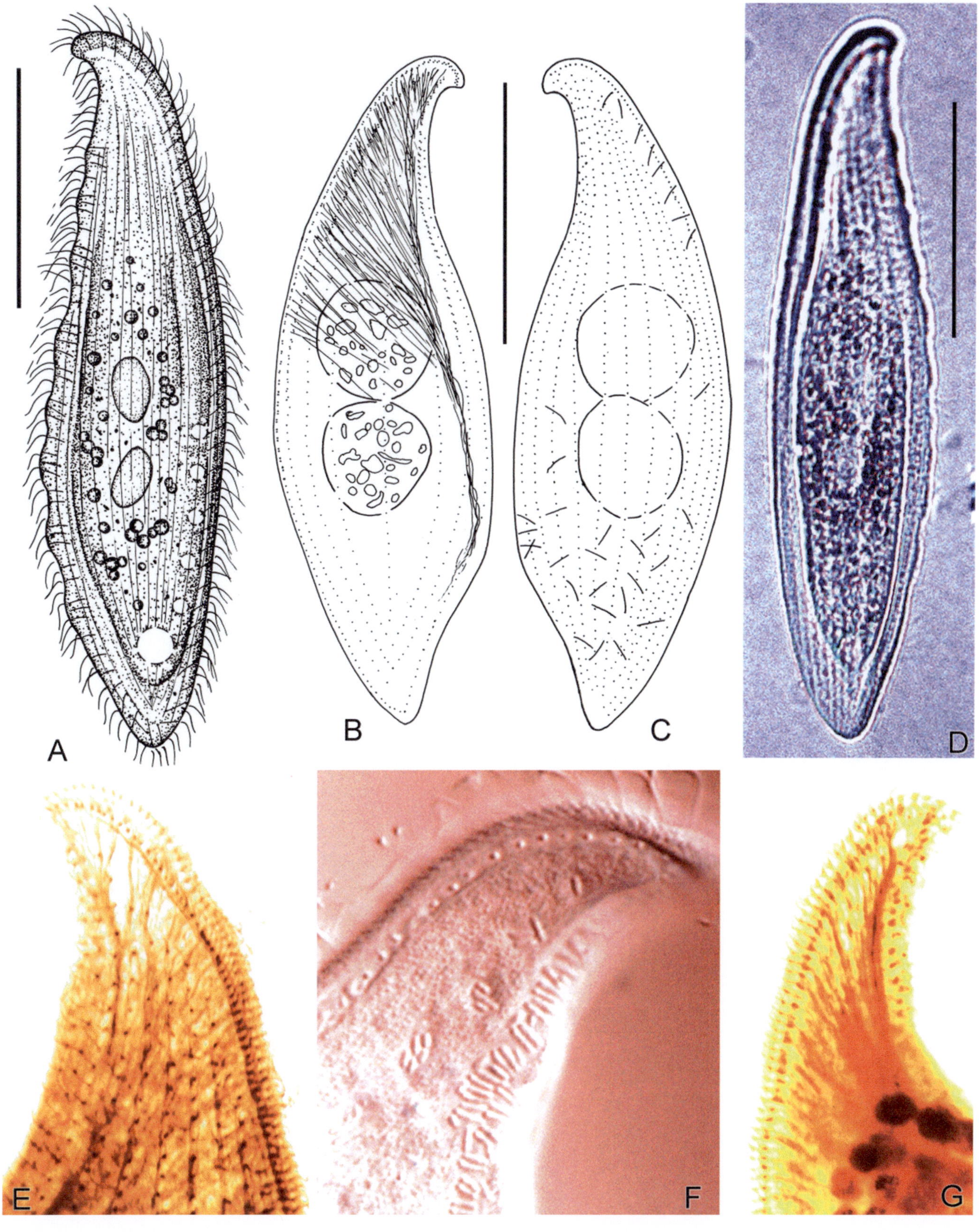


**Fig. 31B** *Loxophyllum lembum*
(**a**, **d**) Right-lateral (**a**) and left-lateral (**d**) views of live cells; (**b**, **c**) ciliature of left (**b**) and right (**c**) sides; (**e**, **g**) ciliature of anterior right (**e**) and anterior left (**g**) sides of cell; (**f**, **g**) perioral kinety 1 and dorsal brush. Scale bars: 80 μm

## Reference

Wu L, Chen R, Yi Z, Li J, Warren A, Lin X (2014) The morphology of three *Loxophyllum* species (Ciliophora, Pleurostomatida) from southern China, *L. lembum* sp. n., *L. vesiculosum* sp. n. and *L. perihoplophorum* Buddenbrock, 1920, with notes on the molecular phylogeny of *Loxophyllum*. J Eukaryot Microbiol 61:115–125

Class Litostomatea Small & Lynn, 1981
Order Pleurostomatida Schewiakoff, 1896
Family Litonotidae Kent, 181
*Loxophyllum meridionale* Wu et al., 2013

**Ecological feature:** mangrove/intertidal zone
**Temperature:** 19.7 °C
**pH:** 7.8
**Salinity:** 23.9‰

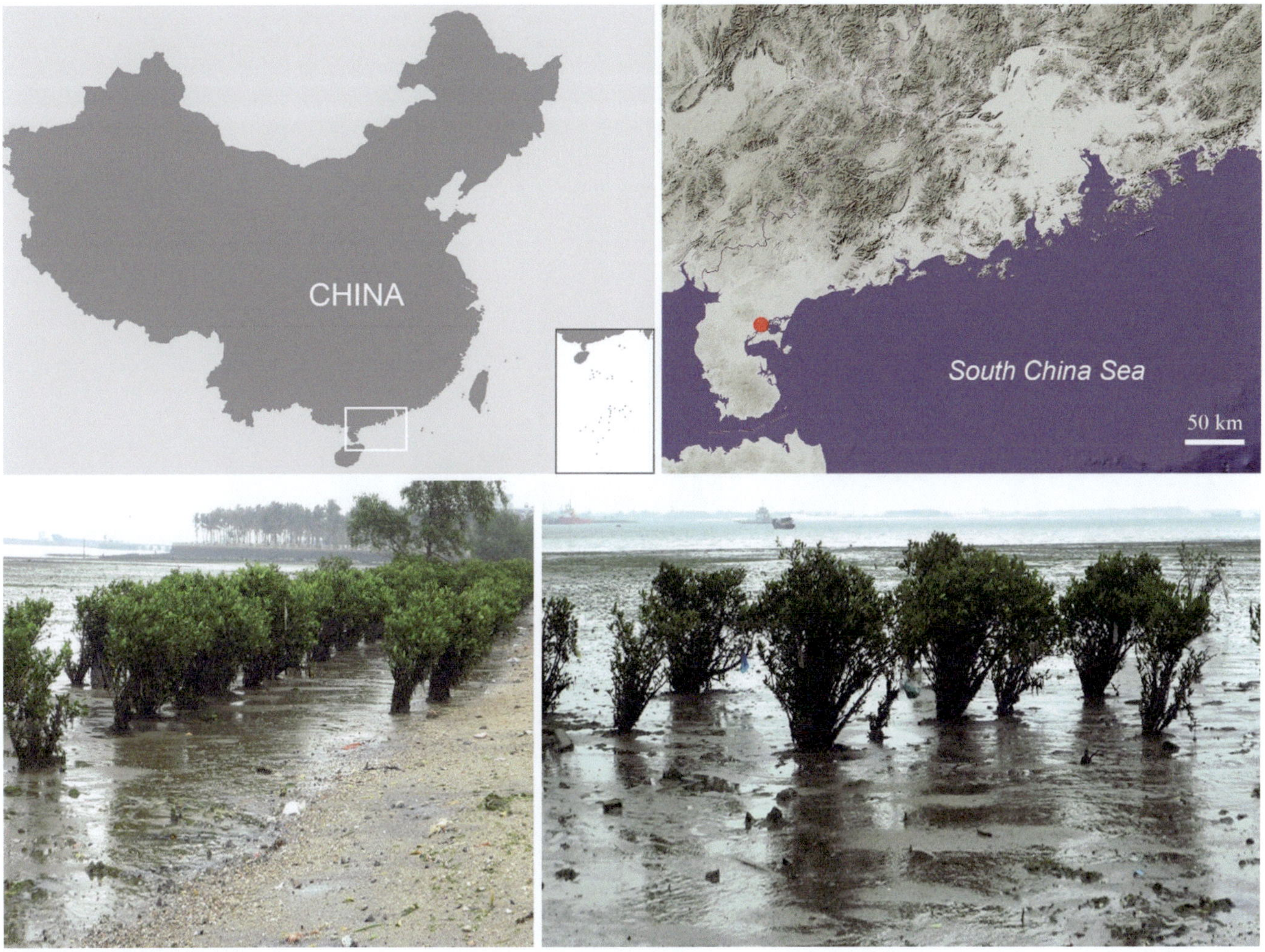


**Fig. 32A** Sample site

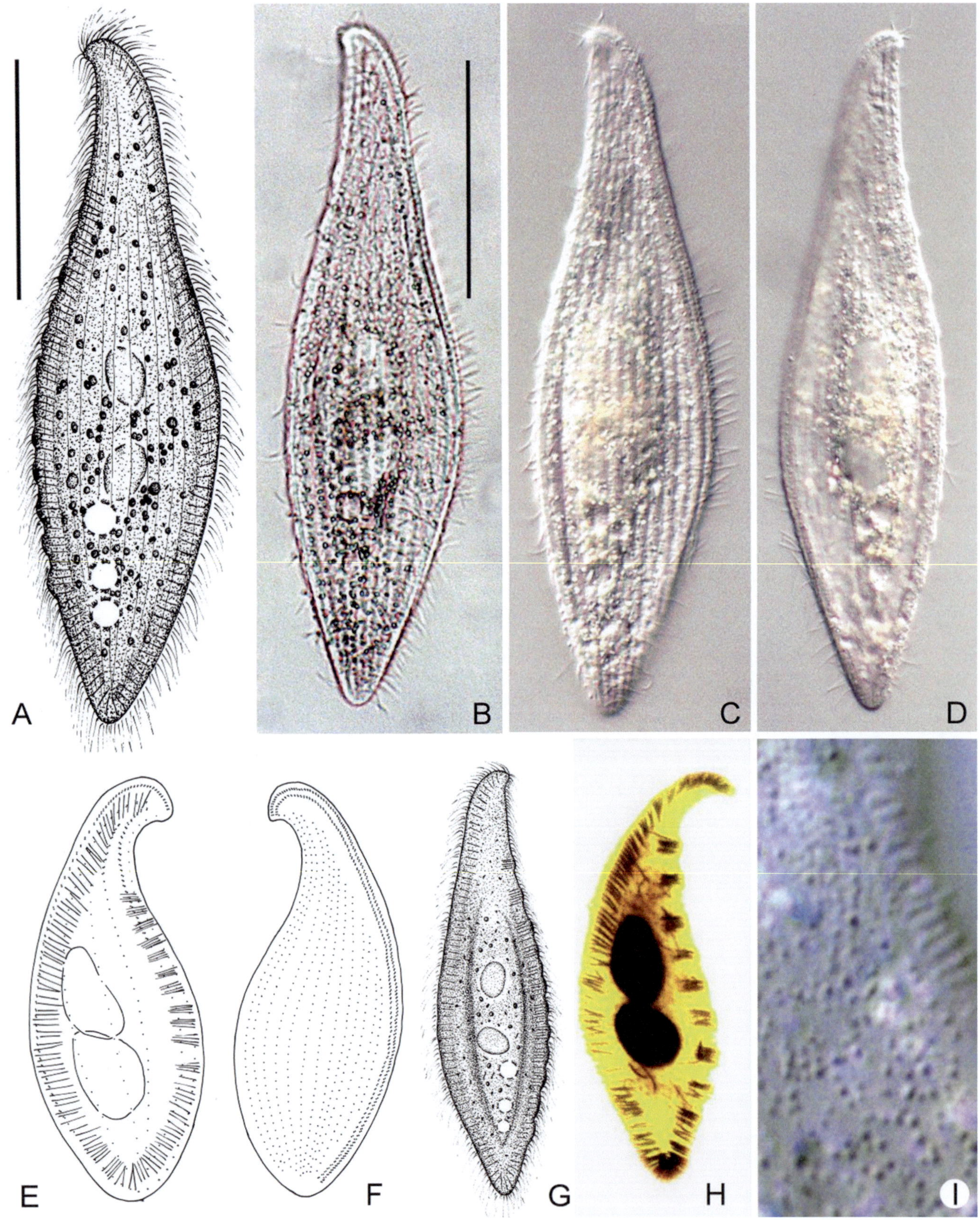


**Fig. 32B** *Loxophyllum meridionale*
(**a–d**, **g**) Right-lateral (**a–c**) and left-lateral (**d**, **g**) views of live cells; (**e**, **f**) ciliature of right (**e**) and left (**f**) sides; (**h**) extrusome pattern and nuclear apparatus; (**i**) cortical granules and dorsal brush. Scale bars: 80 μm

## Reference

Wu L, Chen R, Yi Z, Li J, Warren A, Lin X (2013) Morphology and phylogeny of three new *Loxophyllum* species (Ciliophora, Pleurostomatida) from mangrove wetlands of southern China. J Eukaryot Microbiol 60:267–281

Class Litostomatea Small & Lynn, 1981
Order Pleurostomatida Schewiakoff, 1896
Family Litonotidae Kent, 181
*Loxophyllum perihoplophorum* Buddenbrock, 1920

**Ecological feature:** mangrove
**Water temperature:** 22.1 °C
**pH:** 7.6
**Salinity:** 21.8‰

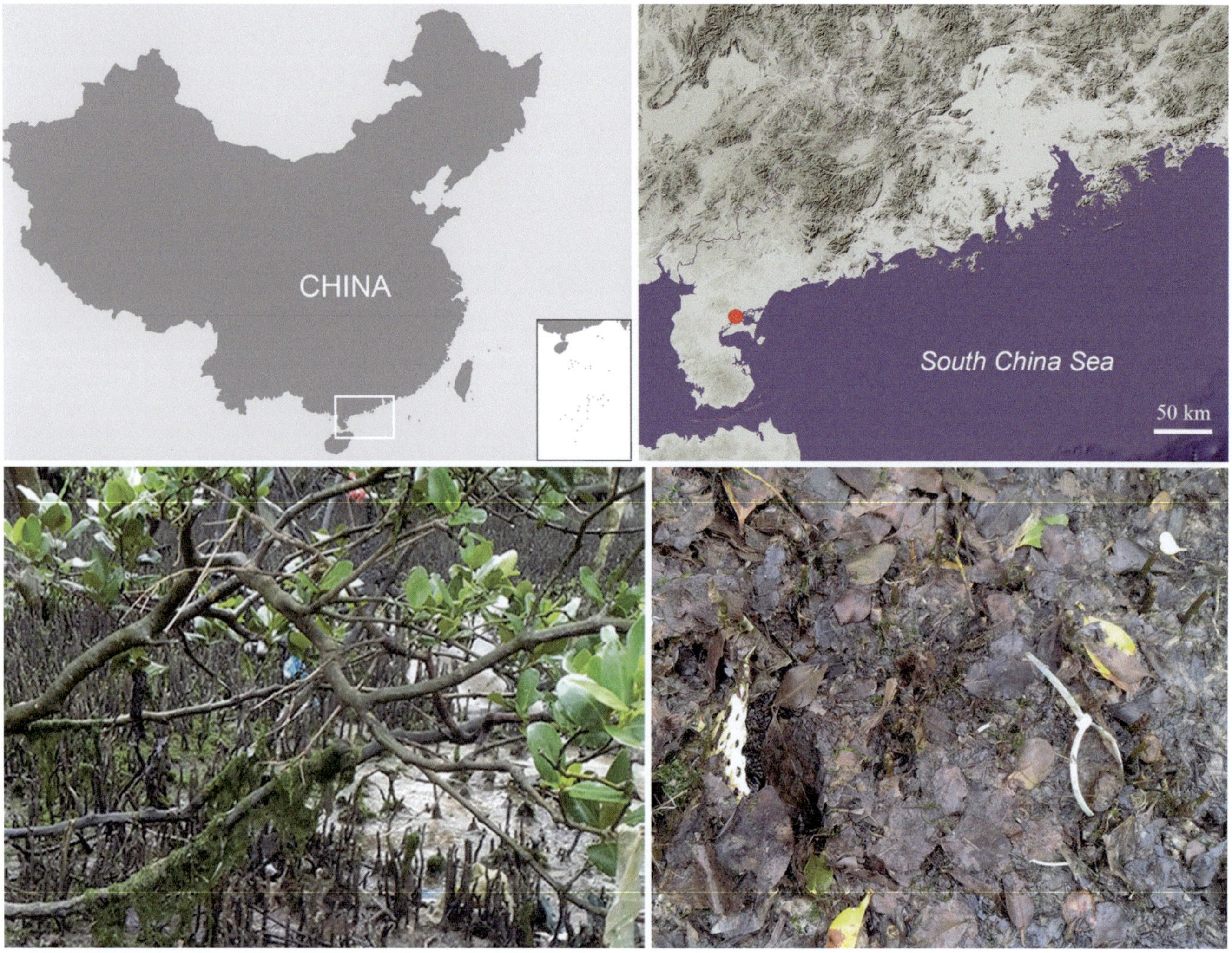


**Fig. 33A** Sample site

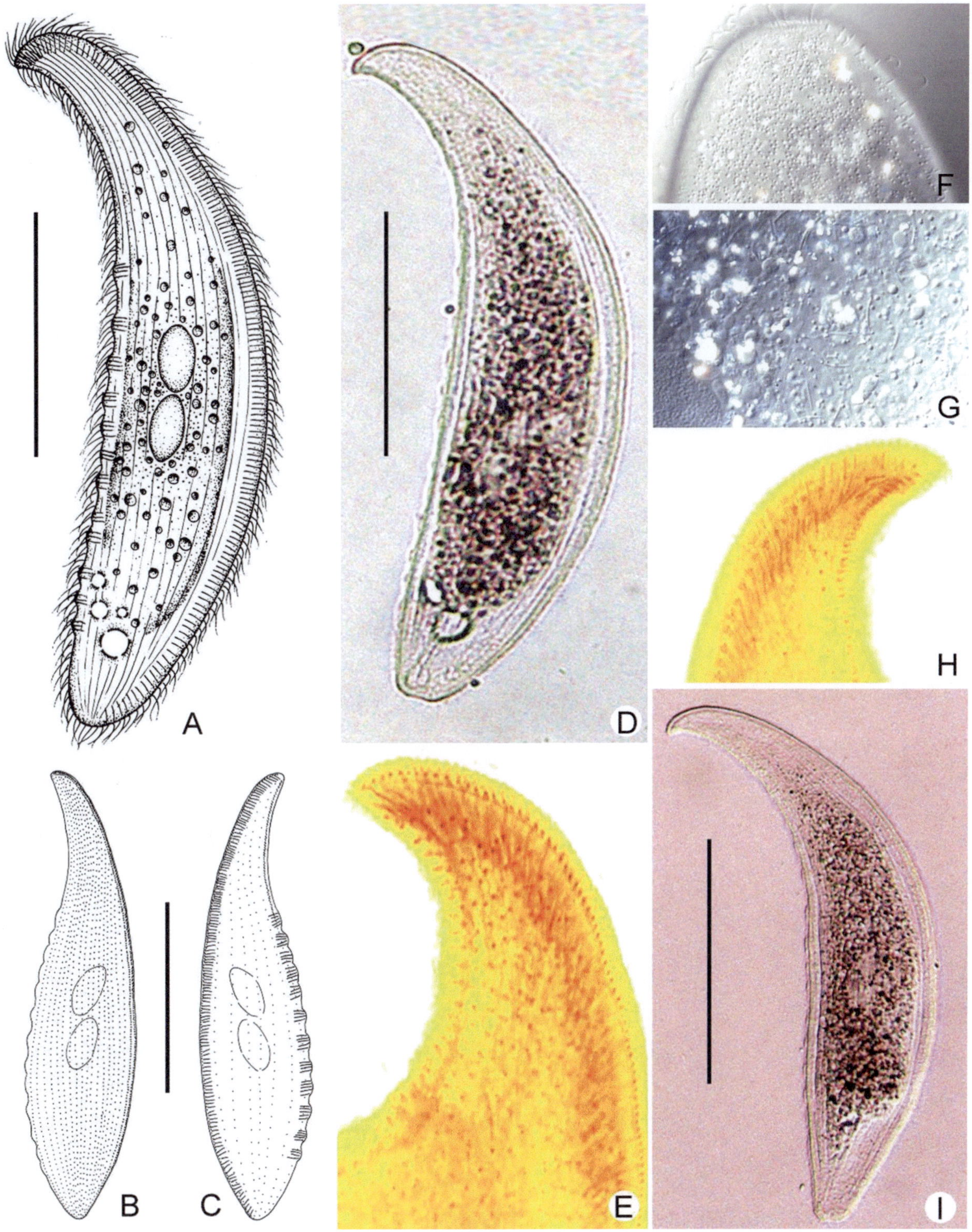


**Fig. 33B** *Loxophyllum perihoplophorum*
(**a**, **d**, **i**) Right-lateral views of live cells; (**b**, **c**, **e**, **h**) ciliature of right (**b**, **e**) and left (**c**, **h**) sides; (**f**) anterior portion of live cell, to show cortical granules; (**g**) middle part of live cell, to show the extrusomes. Scale bars: 100 μm

## Reference

Wu L, Chen R, Yi Z, Li J, Warren A, Lin X (2014) The morphology of three *Loxophyllum* species (Ciliophora, Pleurostomatida) from southern China, *L. lembum* sp. n., *L. vesiculosum* sp. n. and *L. perihoplophorum* Buddenbrock, 1920, with notes on the molecular phylogeny of *Loxophyllum*. J Eukaryot Microbiol 61:115–125

Class Litostomatea Small & Lynn, 1981
Order Pleurostomatida Schewiakoff, 1896
Family Litonotidae Kent, 181
*Loxophyllum planum* Wu et al., 2013

**Ecological feature:** mangrove
**Temperature:** 20.0 °C
**pH:** 7.8
**Salinity:** 15.7‰

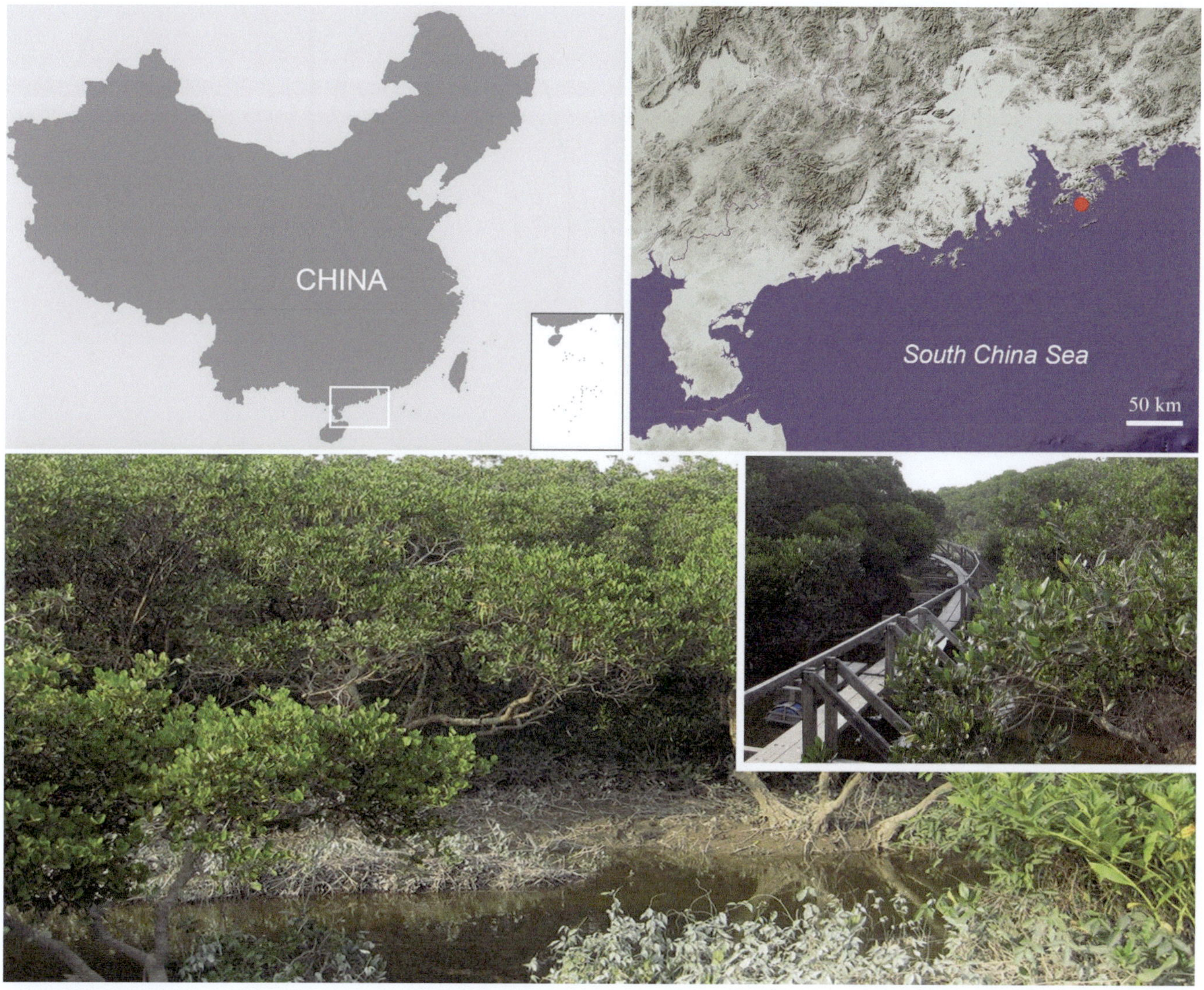


**Fig. 34A** Sample site

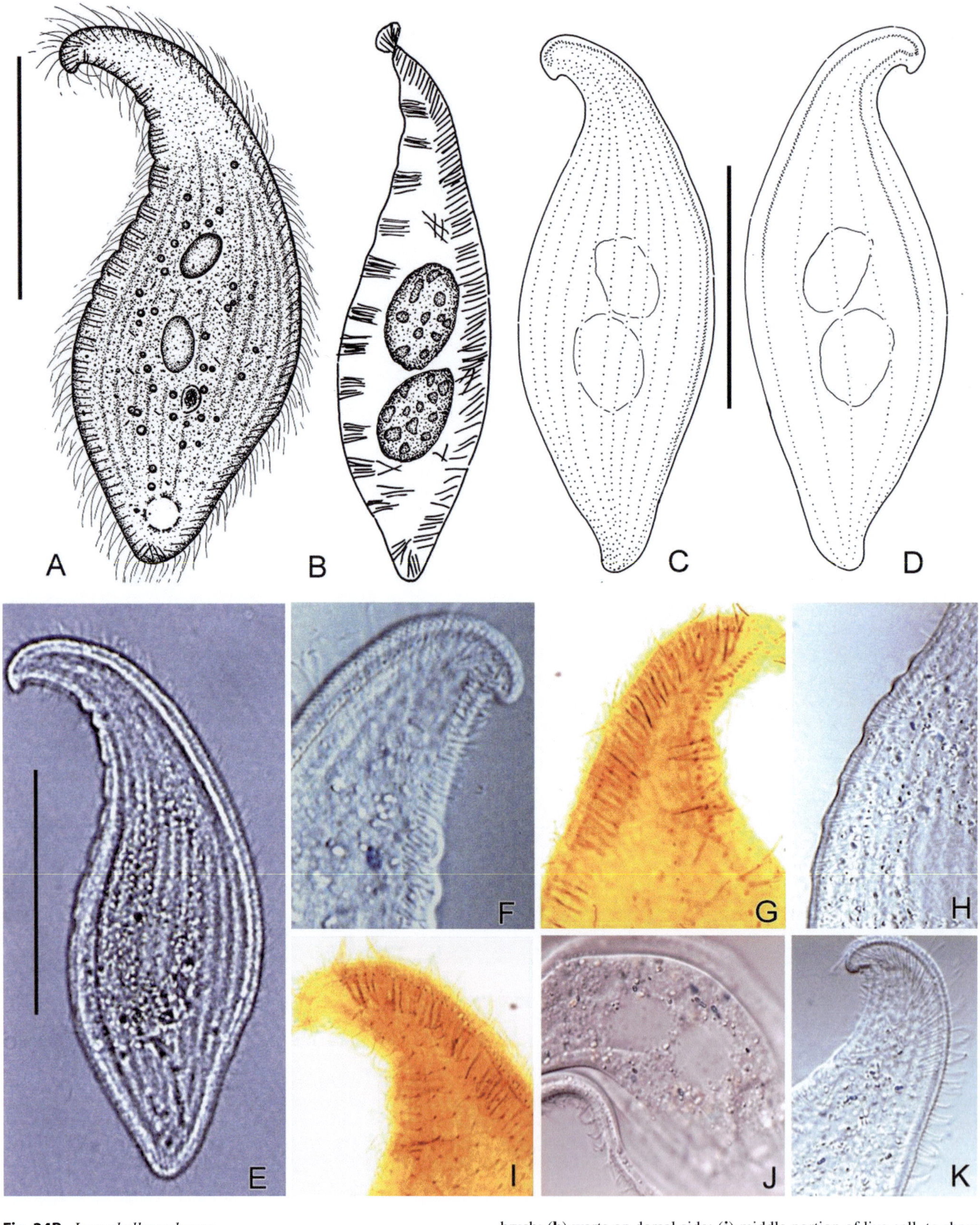


**Fig. 34B** *Loxophyllum planum*
(**a**, **e**) Right-lateral views of live cells; (**b**) extrusome pattern and macronuclei; (**c**, **d**, **g**, **i**) ciliature of right (**c**, **i**) and left (**d**, **g**) sides; (**f**) dorsal brush; (**h**) warts on dorsal side; (**j**) middle portion of live cell, to show macronuclei; (**k**) extrusomes along oral slit. Scale bars: 100 μm

## Reference

Wu L, Chen R, Yi Z, Li J, Warren A, Lin X (2013) Morphology and phylogeny of three new *Loxophyllum* species (Ciliophora, Pleurostomatida) from mangrove wetlands of southern China. J Eukaryot Microbiol 60:267–281

Class Litostomatea Small & Lynn, 1981
Order Pleurostomatida Schewiakoff, 1896
Family Litonotidae Kent, 181
*Loxophyllum rostratum* Cohn, 1866

**Ecological feature:** mangrove
**Water temperature:** 23.0 °C
**pH:** 6.3
**Salinity:** 16.0‰

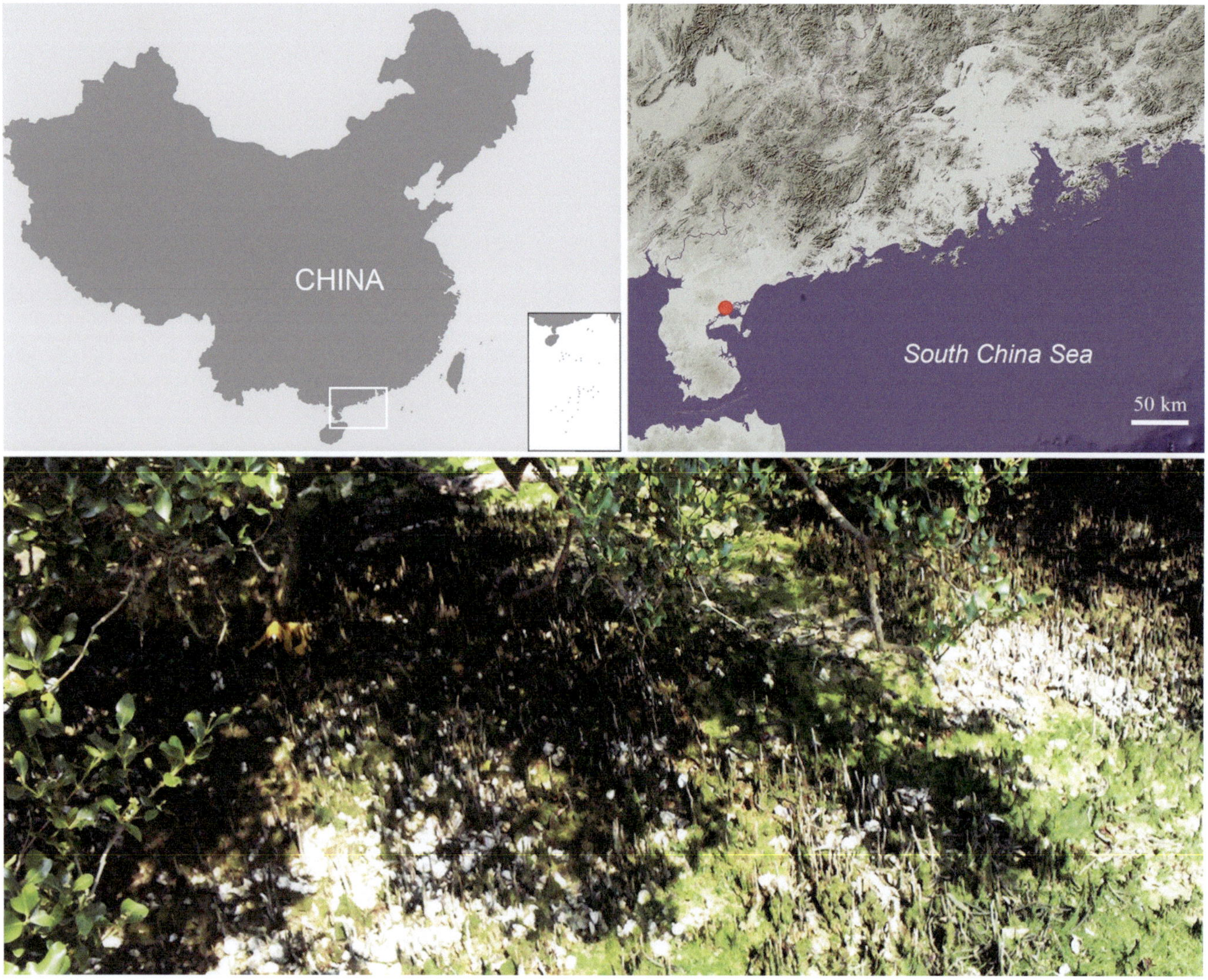


**Fig. 35A** Sample site

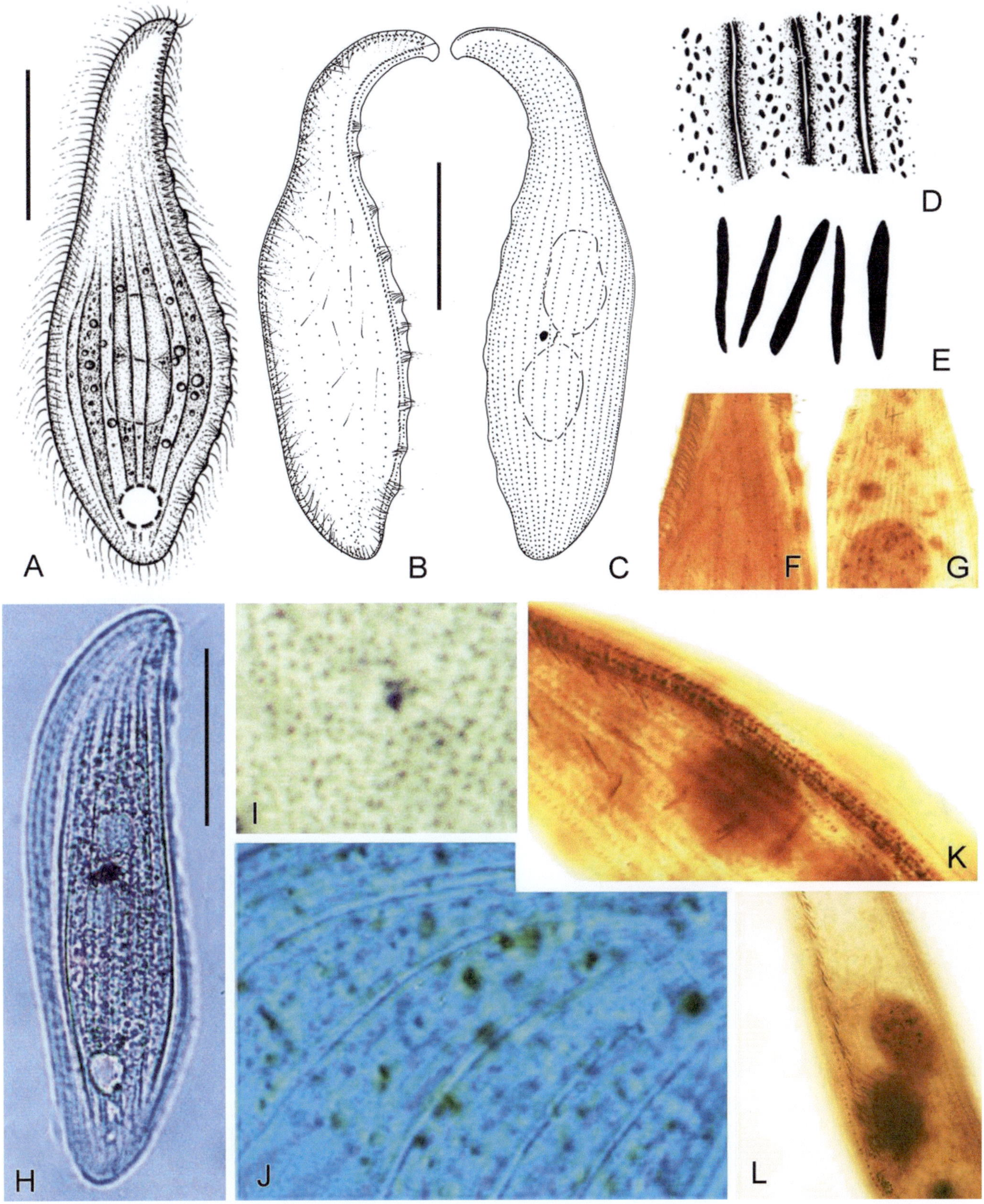


**Fig. 35B** *Loxophyllum rostratum*
(**a**, **h**) Left-lateral views of live cells; (**b**, **c**, **f**, **g**) ciliature of left (**b**, **f**) and right (**c**, **g**) sides; (**d**, **j**) the shallow grooves on left side; (**e**, **l**) extrusomes; (**i**) cortical granules; (**k**) top view of oral slit. Scale bars: 90 μm

## Reference

Lin X, Al-Rasheid KAS, Al-Quraishy SA, Al-Farraj SA, Song WB (2008) Identification of three highly confused marine *Loxophyllum* (Ciliophora: Pleurostomatida) with a key to seven congeners from the China Sea. J Eukaryot Microbiol 55:331–442

Class Litostomatea Small & Lynn, 1981
Order Pleurostomatida Schewiakoff, 1896
Family Litonotidae Kent, 181
*Loxophyllum salinum* Wu et al., 2013

**Ecological feature:** mangrove
**Temperature:** 20.0 °C
**pH:** 7.8
**Salinity:** 15.7‰

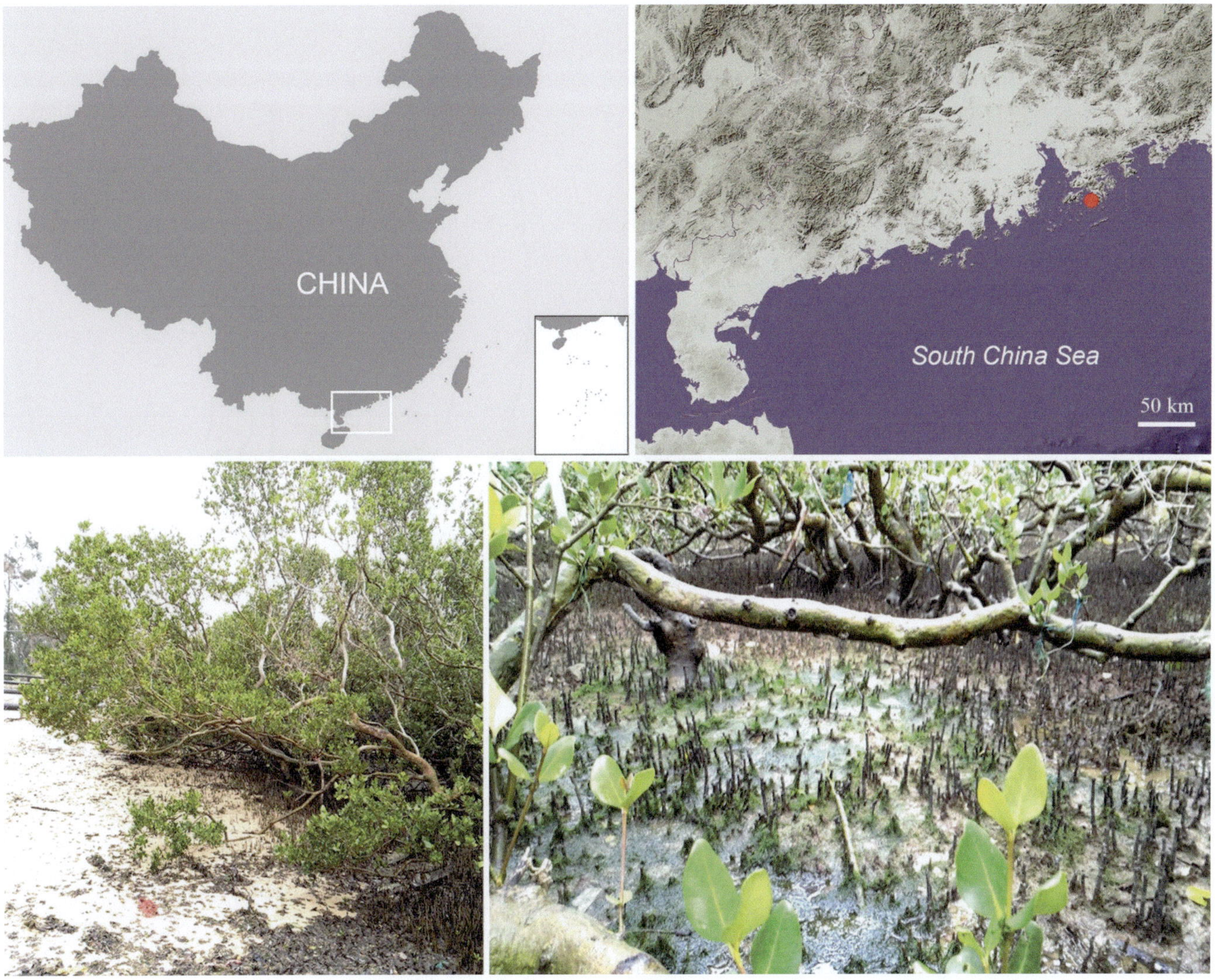


**Fig. 36A** Sample site

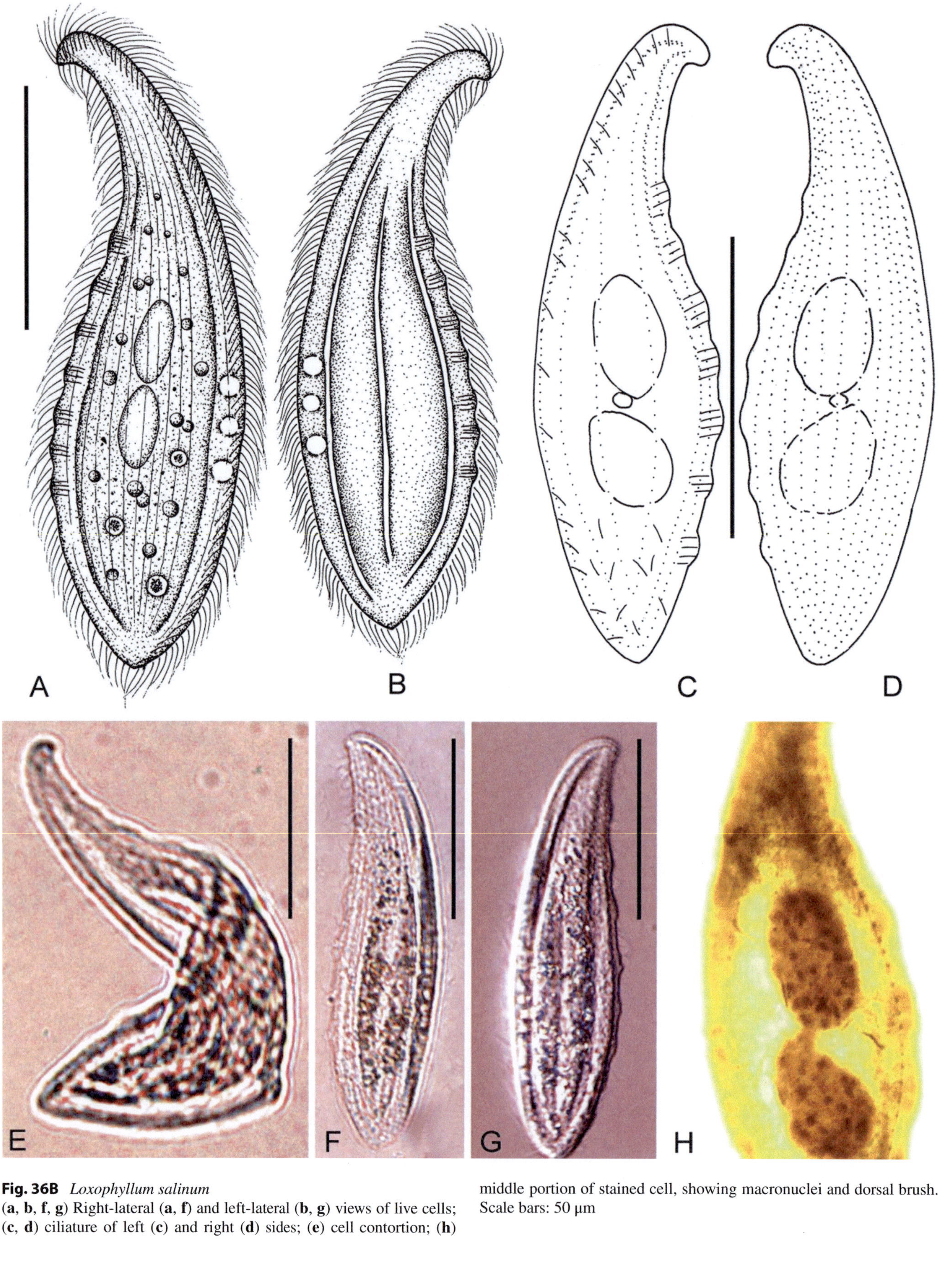


**Fig. 36B** *Loxophyllum salinum*
(**a**, **b**, **f**, **g**) Right-lateral (**a**, **f**) and left-lateral (**b**, **g**) views of live cells; (**c**, **d**) ciliature of left (**c**) and right (**d**) sides; (**e**) cell contortion; (**h**) middle portion of stained cell, showing macronuclei and dorsal brush. Scale bars: 50 μm

## Reference

Wu L, Chen R, Yi Z, Li J, Warren A, Lin X (2013) Morphology and phylogeny of three new *Loxophyllum* species (Ciliophora, Pleurostomatida) from mangrove wetlands of southern China. J Eukaryot Microbiol 60:267–281

Class Litostomatea Small & Lynn, 1981
Order Pleurostomatida Schewiakoff, 1896
Family Litonotidae Kent, 181
*Loxophyllum shini* Lin et al., 2007

**Ecological feature:** mangrove
**Water temperature:** 21.0 °C
**pH:** 7.0
**Salinity:** 17.0‰

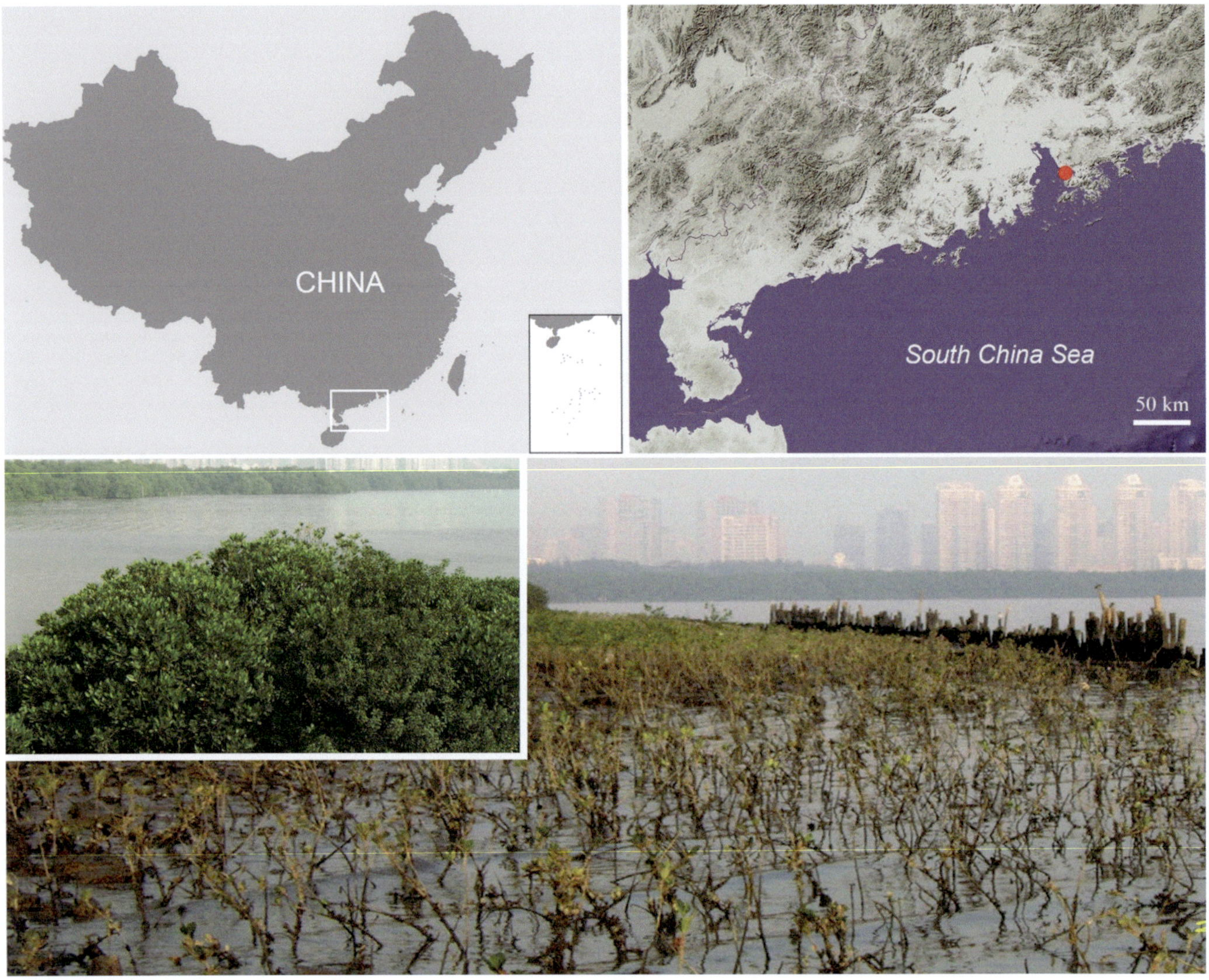


**Fig. 37A** Sample site

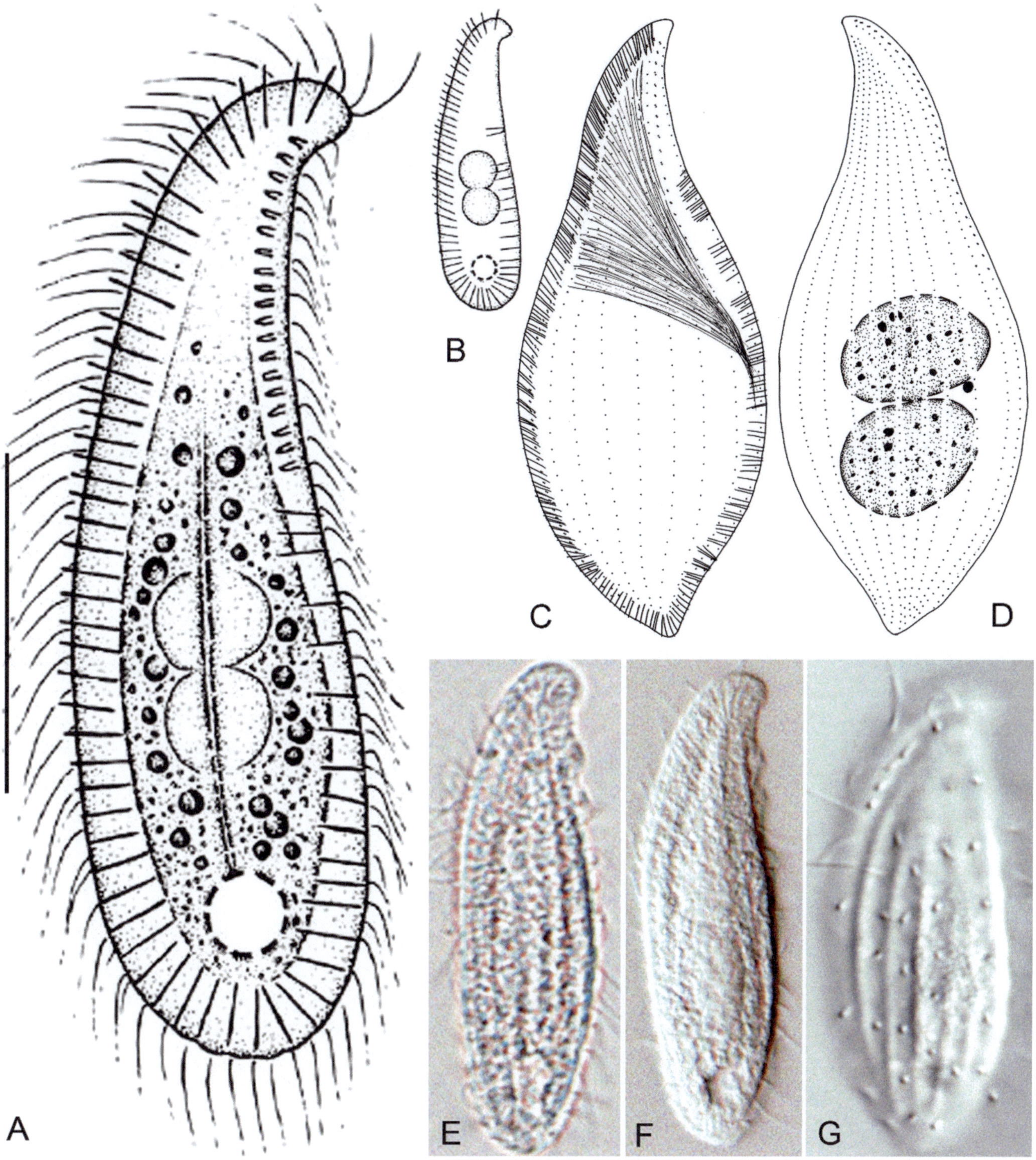


**Fig. 37B** *Loxophyllum shini*
(**a**, **e**) Left-lateral view of live individuals; (**b**) extrusomes and macronuclei; (**c**, **d**) ciliature of left (**c**) and right (**d**) sides; (**f**) left-lateral view of live cell, showing contractile vacuole and extrusomes; (**g**) ribs on left side of live cell. Scale bar: 20 μm

## Reference

Lin X, Song WB, Li J (2007) Description of two new marine pleurostomatid ciliates, *Loxophyllum choii* nov. spec. and *L. shini* nov. spec. (Ciliophora, Pleurostomatida) from China. Eur J Protistol 43:131–139

Class Litostomatea Small & Lynn, 1981
Order Pleurostomatida Schewiakoff, 1896
Family Litonotidae Kent, 181
*Loxophyllum spirellum* Pan et al., 2010

**Ecological feature:** mangrove
**Water temperature:** 23.4 °C; 24.5 °C; 26.0 °C
**pH:** 6.3; 8.2; 7.8
**Salinity:** 10.0‰; 13.8‰; 7.0‰

**Fig. 38A** Sample site

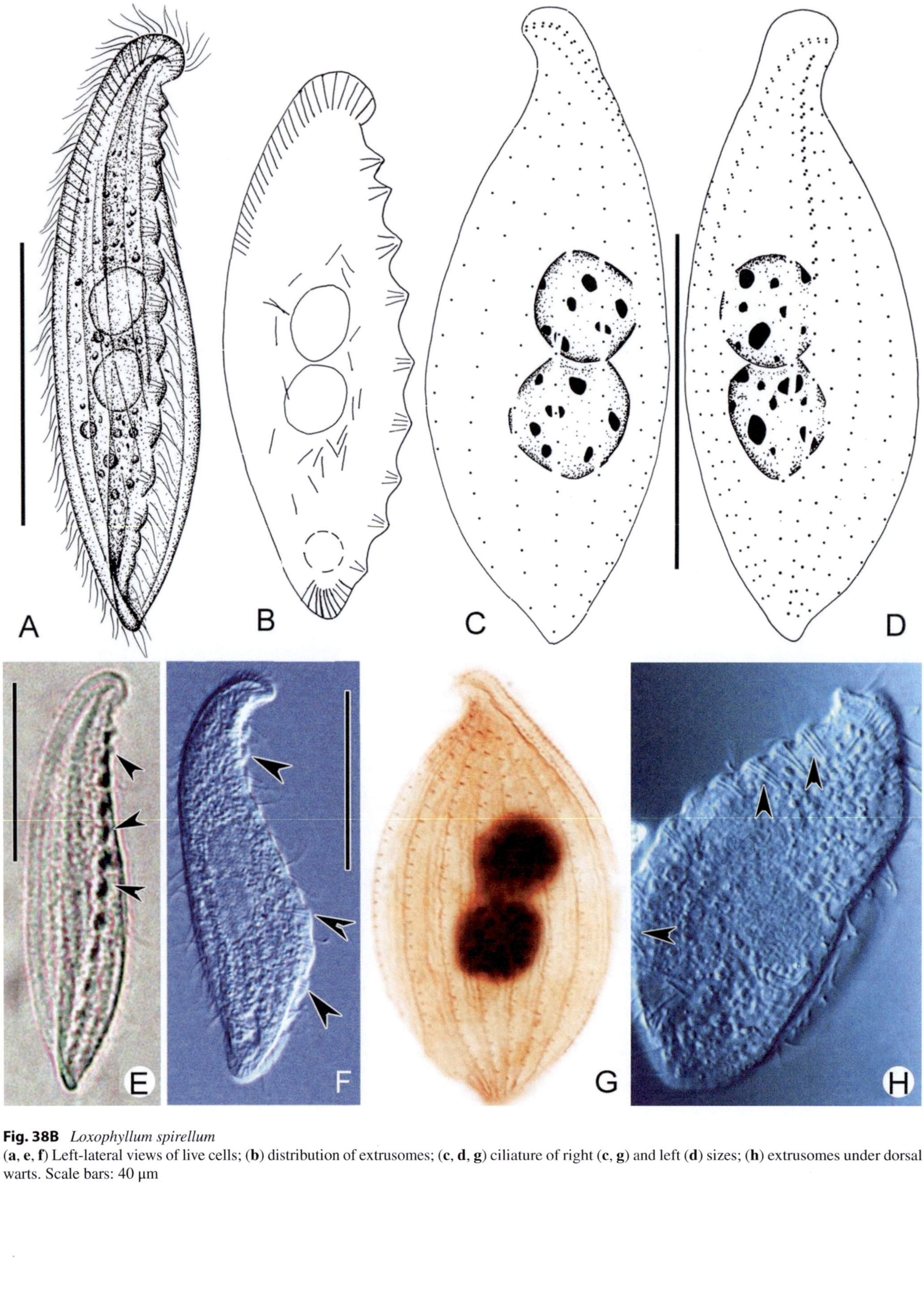


**Fig. 38B** *Loxophyllum spirellum*
(**a**, **e**, **f**) Left-lateral views of live cells; (**b**) distribution of extrusomes; (**c**, **d**, **g**) ciliature of right (**c**, **g**) and left (**d**) sizes; (**h**) extrusomes under dorsal warts. Scale bars: 40 μm

## Reference

Pan H, Gao F, Li J, Lin X, Al-Farraj SA, Al-Rasheid KAS (2010) Morphology and phylogeny of two new pleurostomatid ciliates, *Epiphyllum shenzhenense* n. sp. and *Loxophyllum spirellum* n. sp. (Protozoa, Ciliophora) from a mangrove wetland, South China. J Eukaryot Microbiol 57:421–428

Class Litostomatea Small & Lynn, 1981
Order Pleurostomatida Schewiakoff, 1896
Family Litonotidae Kent, 181
*Loxophyllum vesiculosum* Wu et al., 2014

**Ecological feature:** shrimp-farming pond (Zhanjiang); mangrove (Daya Bay)
**Water temperature:** 23.8–25.6 °C
**pH:** 7.4
**Salinity:** 22.3‰–24.8‰

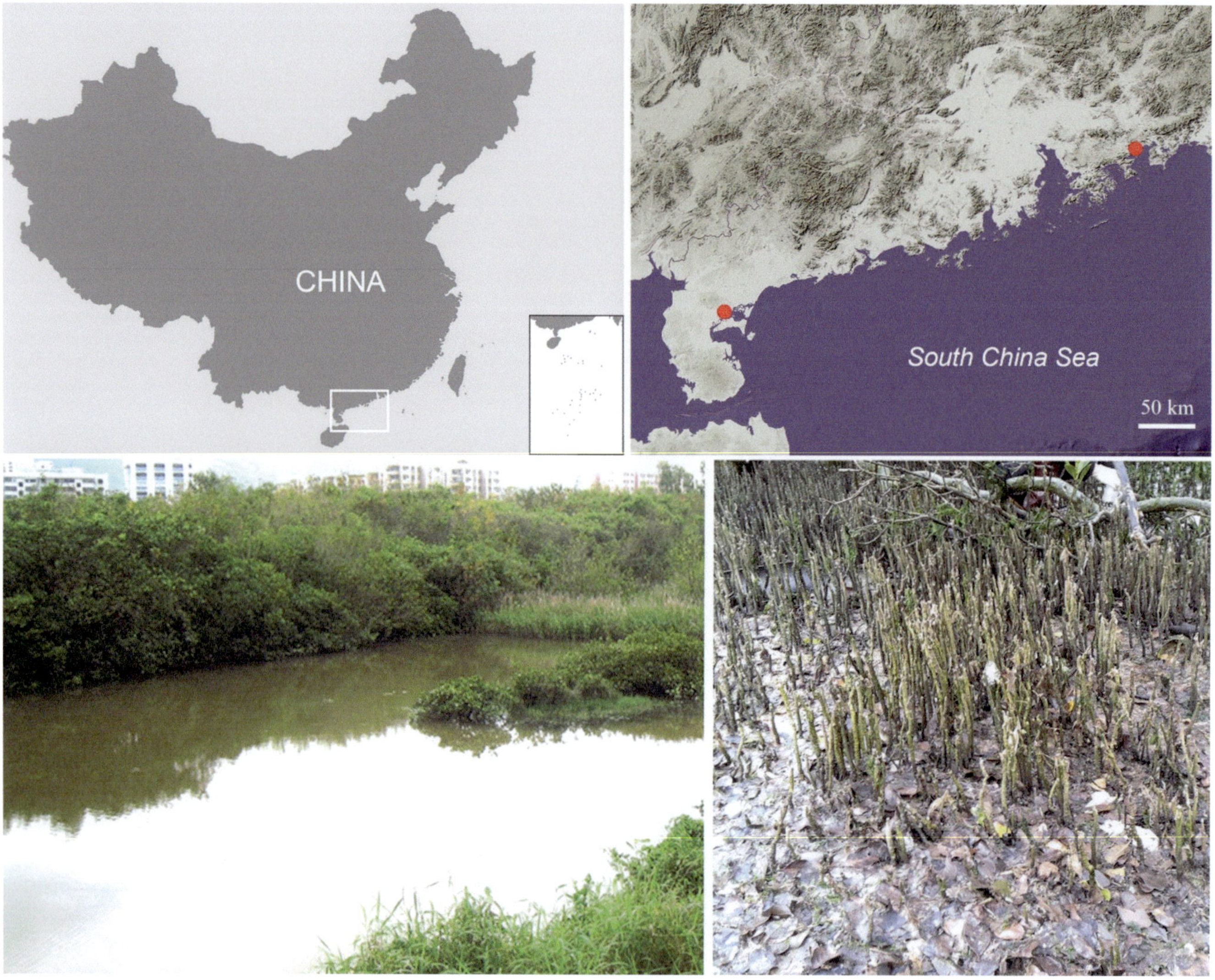


**Fig. 39A** Sample site

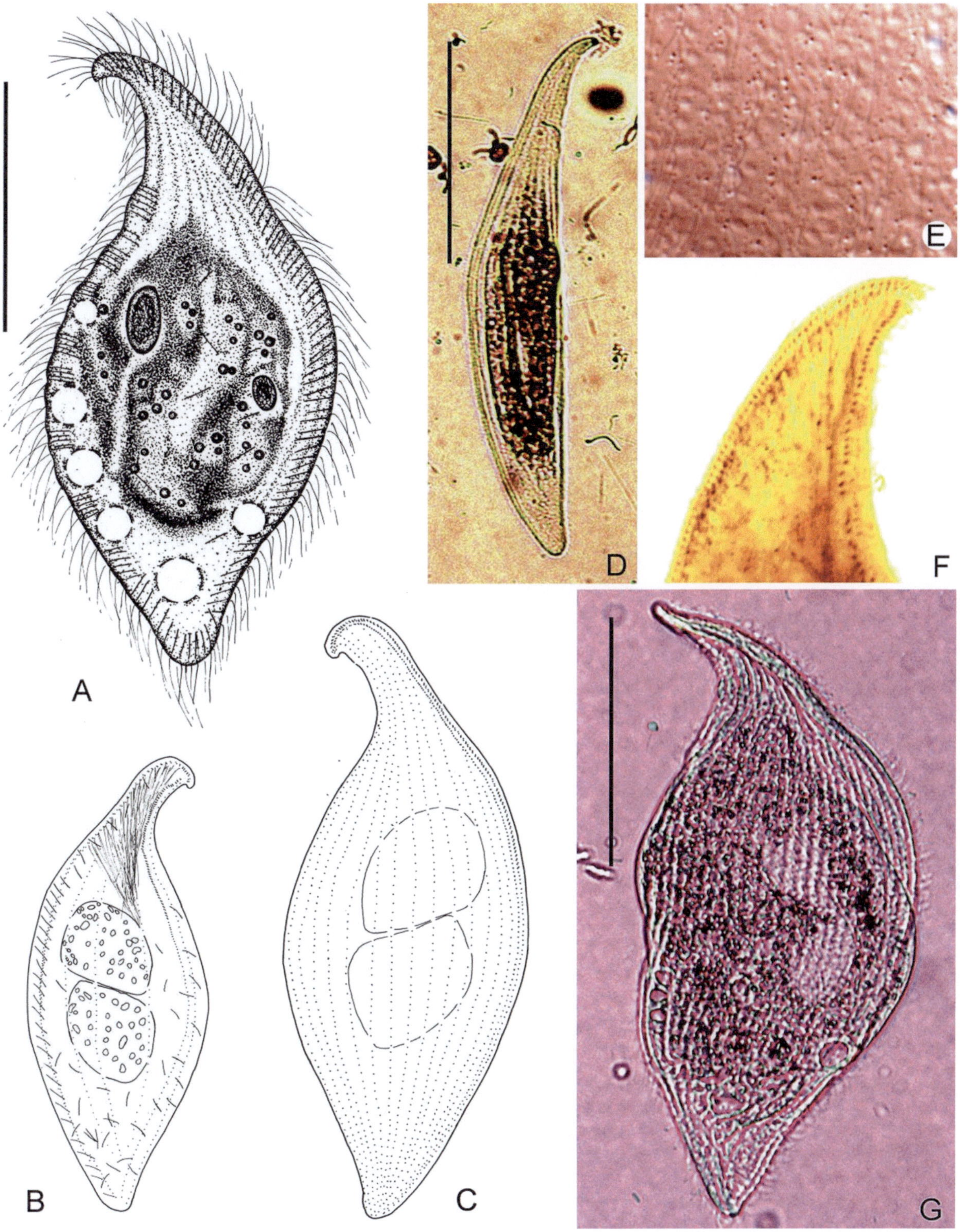


**Fig. 39B** *Loxophyllum lembum*
(**a**, **g**) Right-lateral views of live cells; (**b**, **c**) ciliature of left (**b**) and right (**c**) sides; (**d**) left-lateral view of live cell; (**e**) cortical granules; (**f**) perioral kineties and dorsal brush. Scale bars: 150 μm

## Reference

Wu L, Chen R, Yi Z, Li J, Warren A, Lin X (2014) The morphology of three *Loxophyllum* species (Ciliophora, Pleurostomatida) from southern China, *L. lembum* sp. n., *L. vesiculosum* sp. n. and *L. perihoplophorum* Buddenbrock, 1920, with notes on the molecular phylogeny of *Loxophyllum*. J Eukaryot Microbiol 61:115–125

Class Litostomatea Small & Lynn, 1981
Order Pleurostomatida Schewiakoff, 1896
Family Litonotidae Kent, 181
*Protolitonotus longus* Wu et al., 2017

**Ecological feature:** oyster-farming pond
**Water temperature:** 23.0 °C
**pH:** 8.0
**Salinity:** 12.0‰

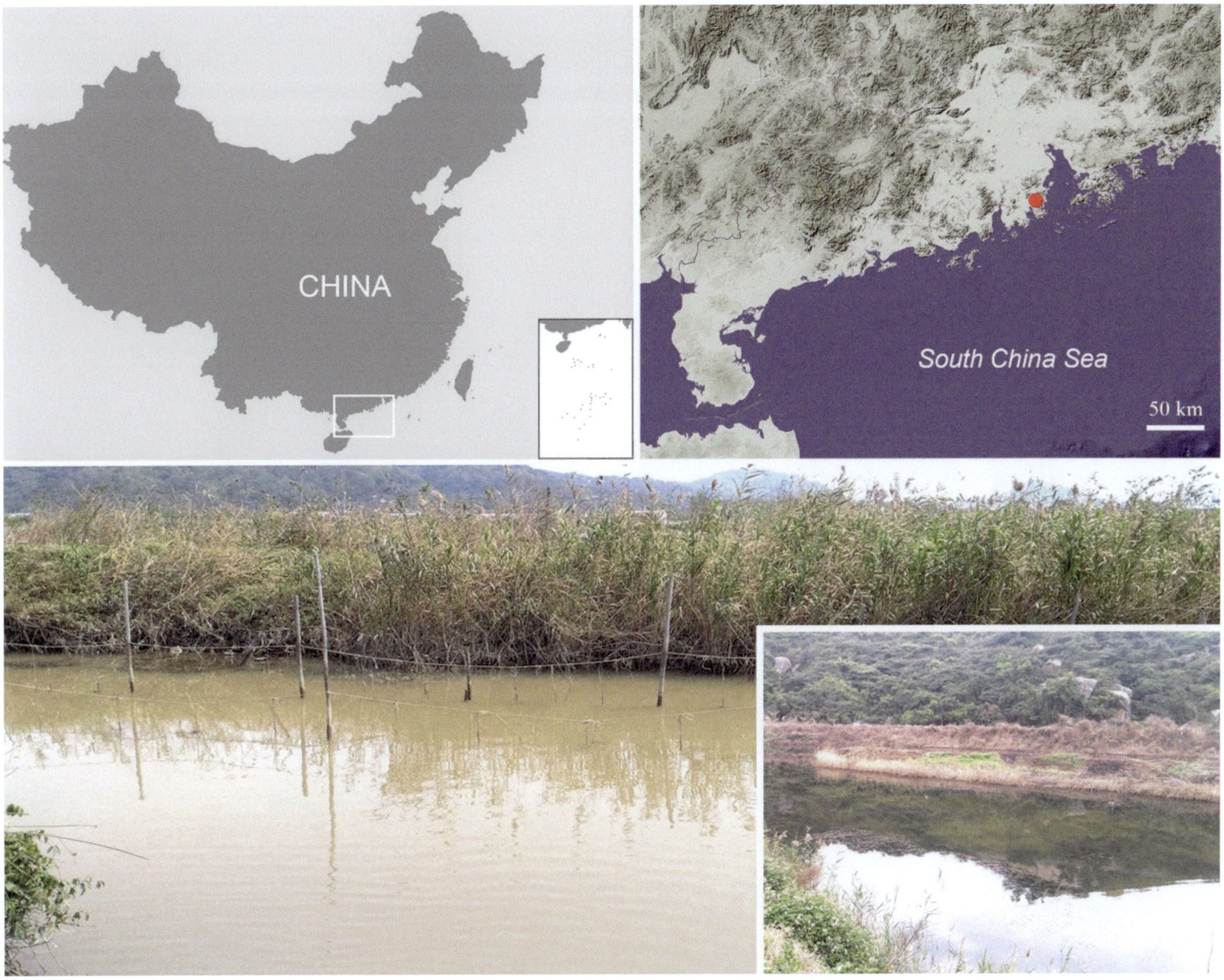


**Fig. 40A** Sample site

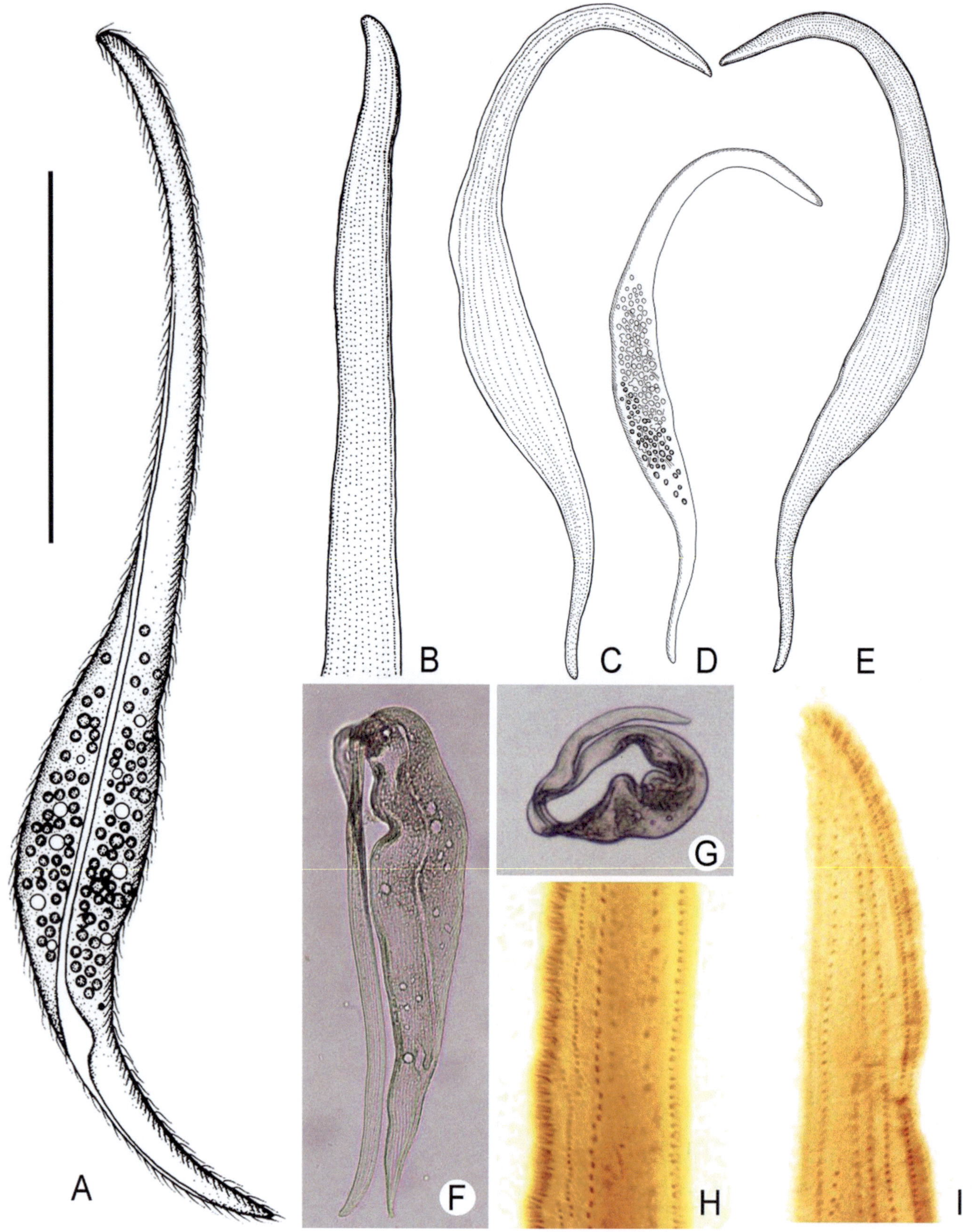


**Fig. 40B** *Protolitonotus longus*
(**a**) Right-lateral view of a live cell; (**b**) anterior part of right side indicating semi-suture; (**c**, **e**) Left (**c**) and right (**e**) views of ciliature; (**d**) macronuclear nodules and extrusomes; (**f**, **g**) cell contortions; (**h**) perioral kinety 1 and dorsal brush; (**i**) anterior right side view showing semi-suture. Scale bar: 250 μm

## Reference

Wu L, Jiao X, Shen Z, Yi Z, Li J, Warren A, Lin X (2017) New taxa refresh the phylogeny and classification of pleurostomatida ciliates (Protozoa, Ciliophora, Litostomatea). Zool Scr 46:245–253

Class Litostomatea Small & Lynn, 1981
Order Pleurostomatida Schewiakoff, 1896
Family Litonotidae Kent, 181
*Protolitonotus magnus* Wu et al., 2017

**Ecological feature:** mangrove (Zhanjing); intertidal zone (Daya Bay)
**Water temperature:** 27.0 °C; 30.0 °C
**pH:** 8.0; 8.0
**Salinity:** 28.0‰; 30.0‰

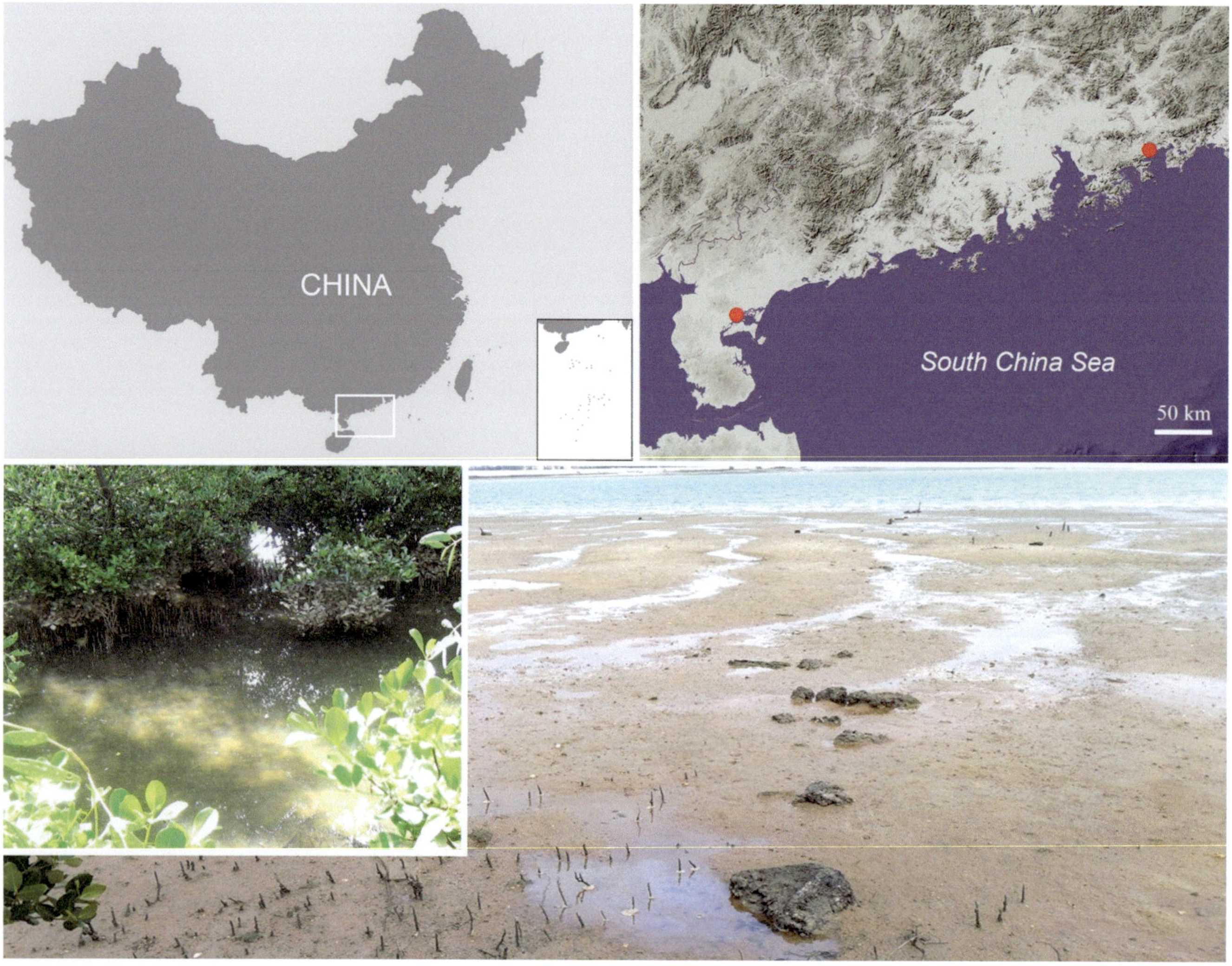


**Fig. 41A** Sample site

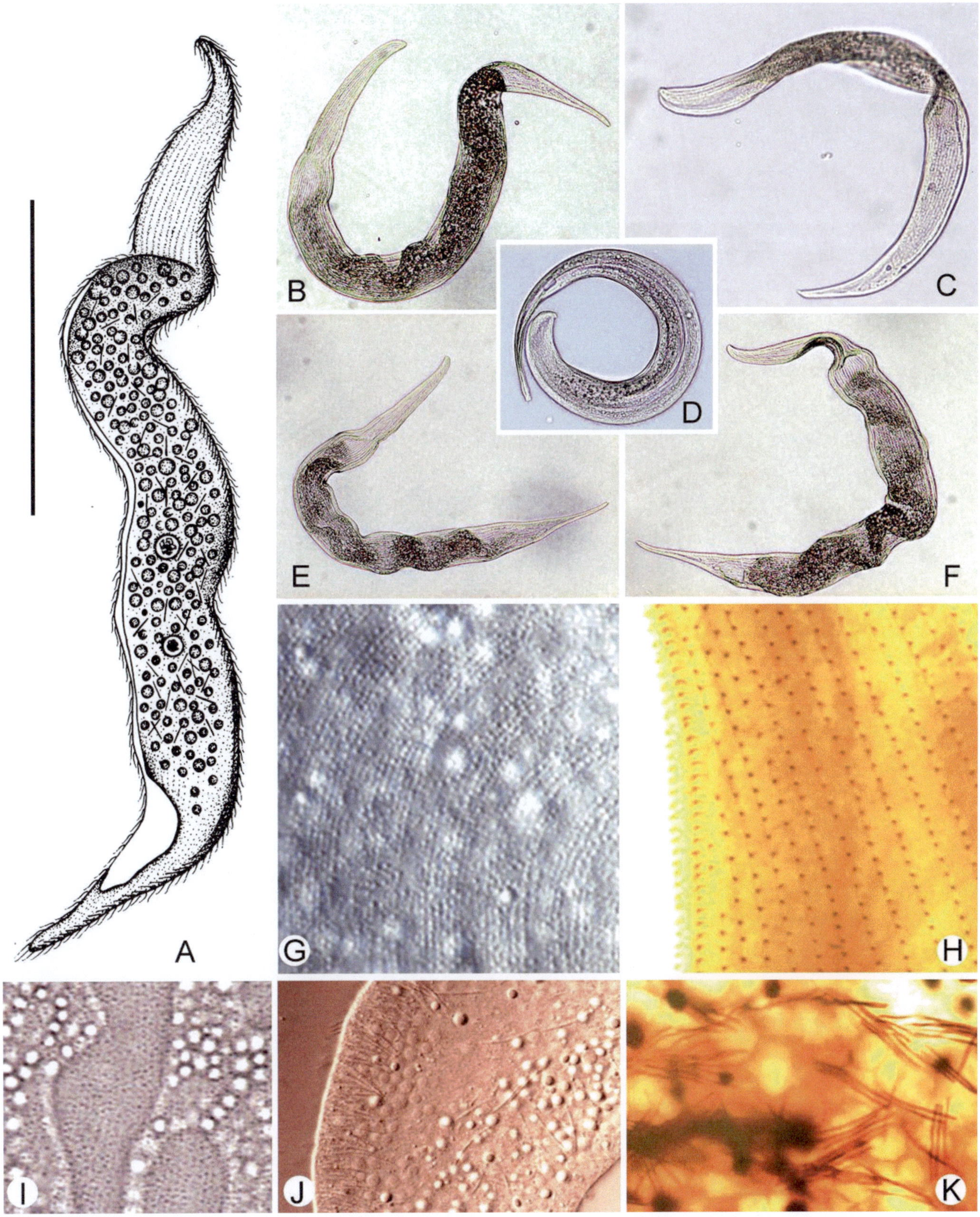


**Fig. 41B** *Protolitonotus magnus*
(**a**) Right-lateral view of a representative individual; (**b–f**) cell contortions; (**g**, **i**) cortical granules; (**h**) anterior part of right side showing the semi-suture; (**j**, **k**) extrusomes. Scale bar: 250 µm

## Reference

Wu L, Jiao X, Shen Z, Yi Z, Li J, Warren A, Lin X (2017) New taxa refresh the phylogeny and classification of pleurostomatida ciliates (Protozoa, Ciliophora, Litostomatea). Zool Scr 46:245–253

# Class Nassophorea

Xiaozhong Hu, Xiaofeng Lin, and Weibo Song

Class Nassophorea Small & Lynn, 1981
Order Microthoracida Jankowski, 1967
Family Microthoracidae Wrzesniowski, 1870
*Trochiliopsis australis* Foissner et al., 1988

**Ecological feature:** mangrove
**Water temperature:** 20.0 °C
**pH:** no data available
**Salinity:** 6.0‰

X. Hu (✉)
Fisheries College, Ocean University of China,
Qingdao, Shandong, China
e-mail: xiaozhonghu@ouc.edu.cn

X. Lin
College of Life Science, South China Normal University,
Guangzhou, Guangdong, China
e-mail: xlin@oucscnu.edu.cn

W. Song
Institute of Evolution & Marine Biodiversity,
Ocean University of China, Qingdao, Shandong, China
e-mail: wsong@ouc.edu.cn


X. Hu et al. (eds.), *Ciliate Atlas: Species Found in the South China Sea*,
https://doi.org/10.1007/978-981-13-5901-9_6

**Fig. 1A** Sample site

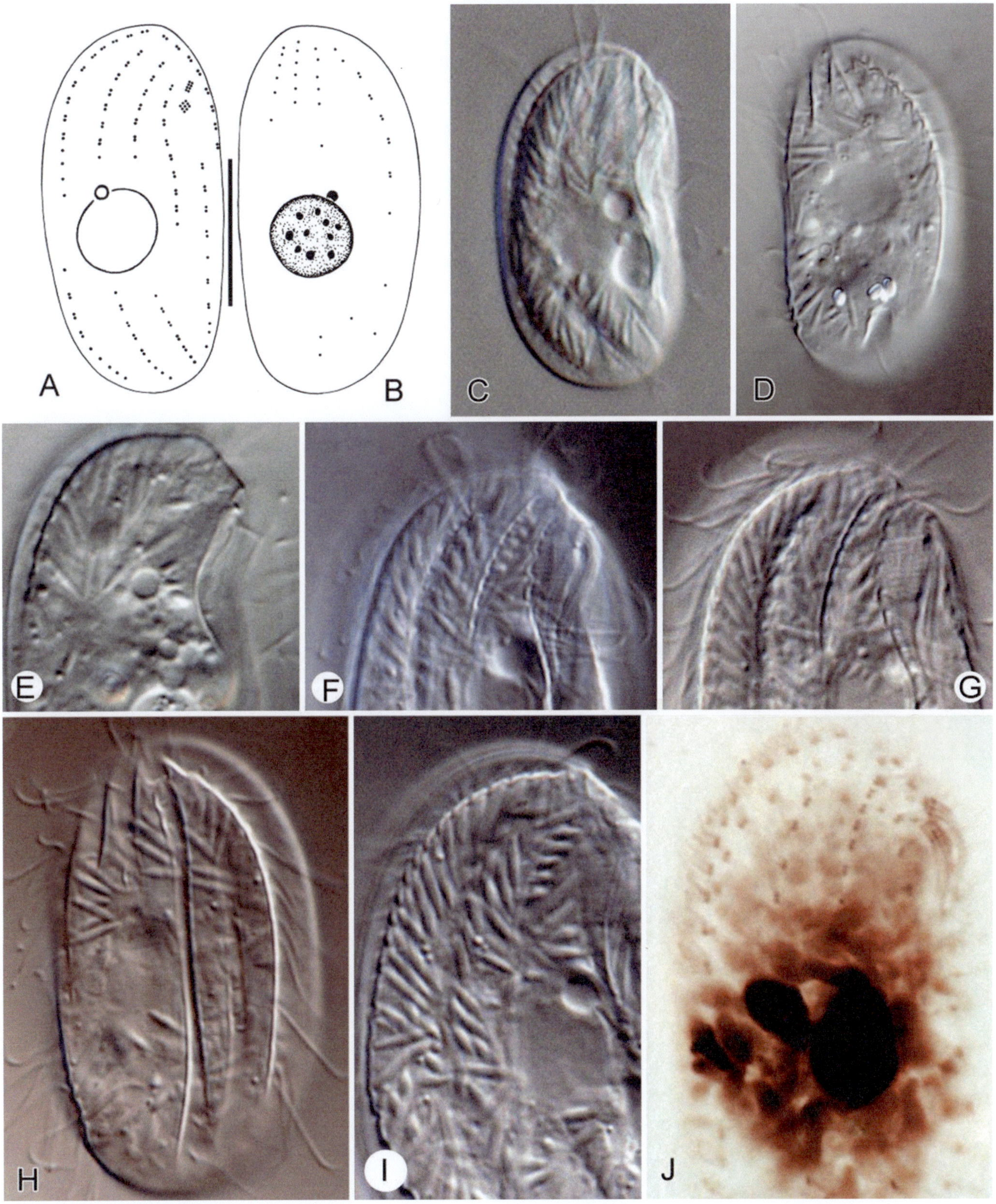


**Fig. 1B** *Trochiliopsis australis*
(**a**, **b**) Ventral (**a**) and dorsal (**b**) views of ciliature; (**c**, **d**, **i**) ventral (**c**, **i**) and dorsal (**d**) views of living cells; (**e**) showing the location of buccal opening; (**f**) oral membranelles; (**g**) part of cytopharyngeal basket; (**h**) dorsal view of cell, showing the concaves; (**j**) ventral view of ciliature. Scale bar: 20 μm

## References

Fan X, Pan H, Li L, Jiang J, Al-Rasheid K AS, Gu F (2014) Phylogeny of the poorly known ciliates, Microthoracida, a systematically confused taxon (Ciliophora), with morphological reports of Three Species. J Eukaryot Microbiol 61:227–237

Augustin H, Foissner W, Adam H (1987) Revision of the genera *Acineria*, *Trimyema* and *Trochiliopsis* (Protozoa, CIliophora). Bull Br Mus (Nat Hist) Zool 52:197–224

Class Nassophorea Small & Lynn, 1981
Order Nassulida Jankowski, 1967
Family Nassulidae de Fromentel, 1874
*Nassula* cf. *labiata*

**Ecological feature:** tidal pool
**Water temperature:** 22.0 °C
**pH:** 8.4
**Salinity:** 30.0‰

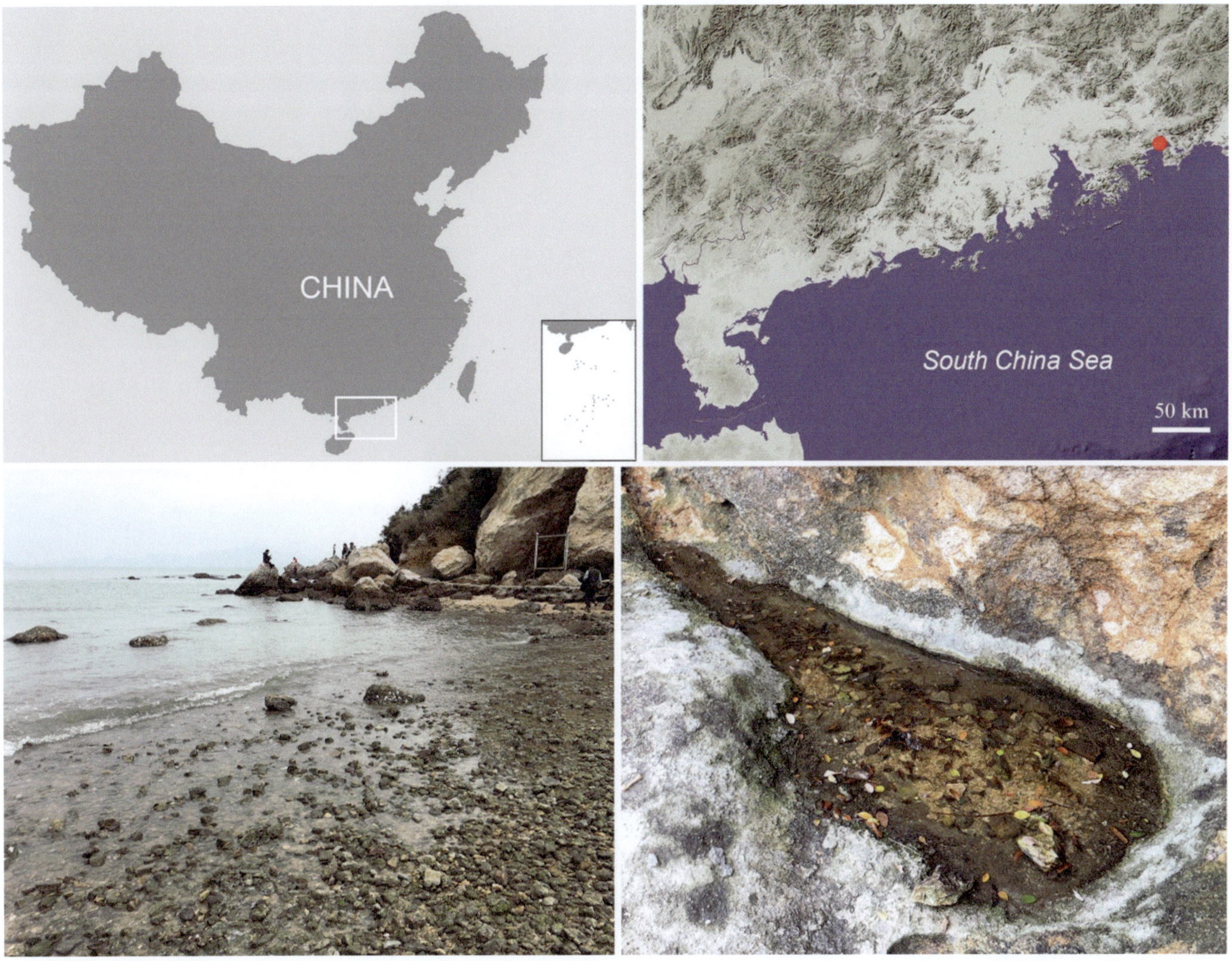


**Fig. 2A** Sample site

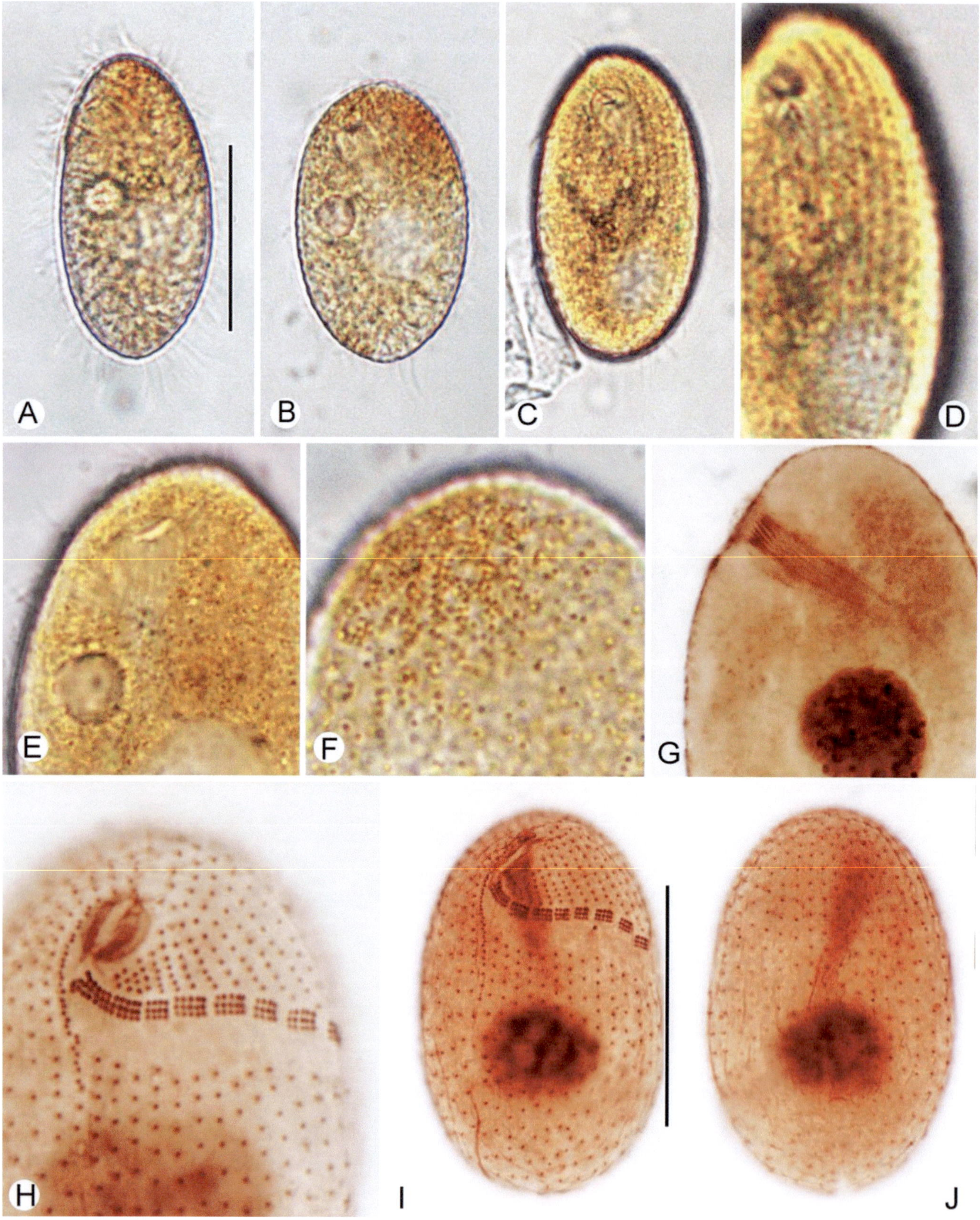


**Fig. 2B** *Nassula* cf. *labiata*
(**a–c**) Left ventrolateral views of live cells; (**d**) notched pellicle; (**e**) cytostome and contractile vacuole; (**f**) yellow and brown pigments on anterior end of cell; (**g**) cytopharyngeal rods and macronucleus; (**h–j**) ciliature of left ventrolateral (**h**, **i**) and right dorsolateral (**j**) sides. Scale bars: 30 μm

## References

Foissner W, Berger H, Kohmann F (1994) Taxonomische und ökologische Revision der Ciliaten des Saprobiensystems – Band III: Hymenostomata, Prostomatida, Nassulida. Munich: Informationsberichte des Bayer Landesamtes für Wasserwirtschaft, Heft 1/94

Song W B, Wei J (1998) Morphological studies on three marine pathogenetic ciliates. Acta Hydrobiol Sin 4:361–366 (In Chinese with English abstract)

Class Nassophorea Small & Lynn, 1981
Order Synhymeniida de Puytorac et al. in Deroux, 1978
Family Orthodonellidae Jankowski, 1968
*Orthodonella sinica* Long et al., 2008

**Ecological feature:** intertidal zone
**Water temperature:** 24.0 °C
**pH:** 8.1
**Salinity:** 33.5‰

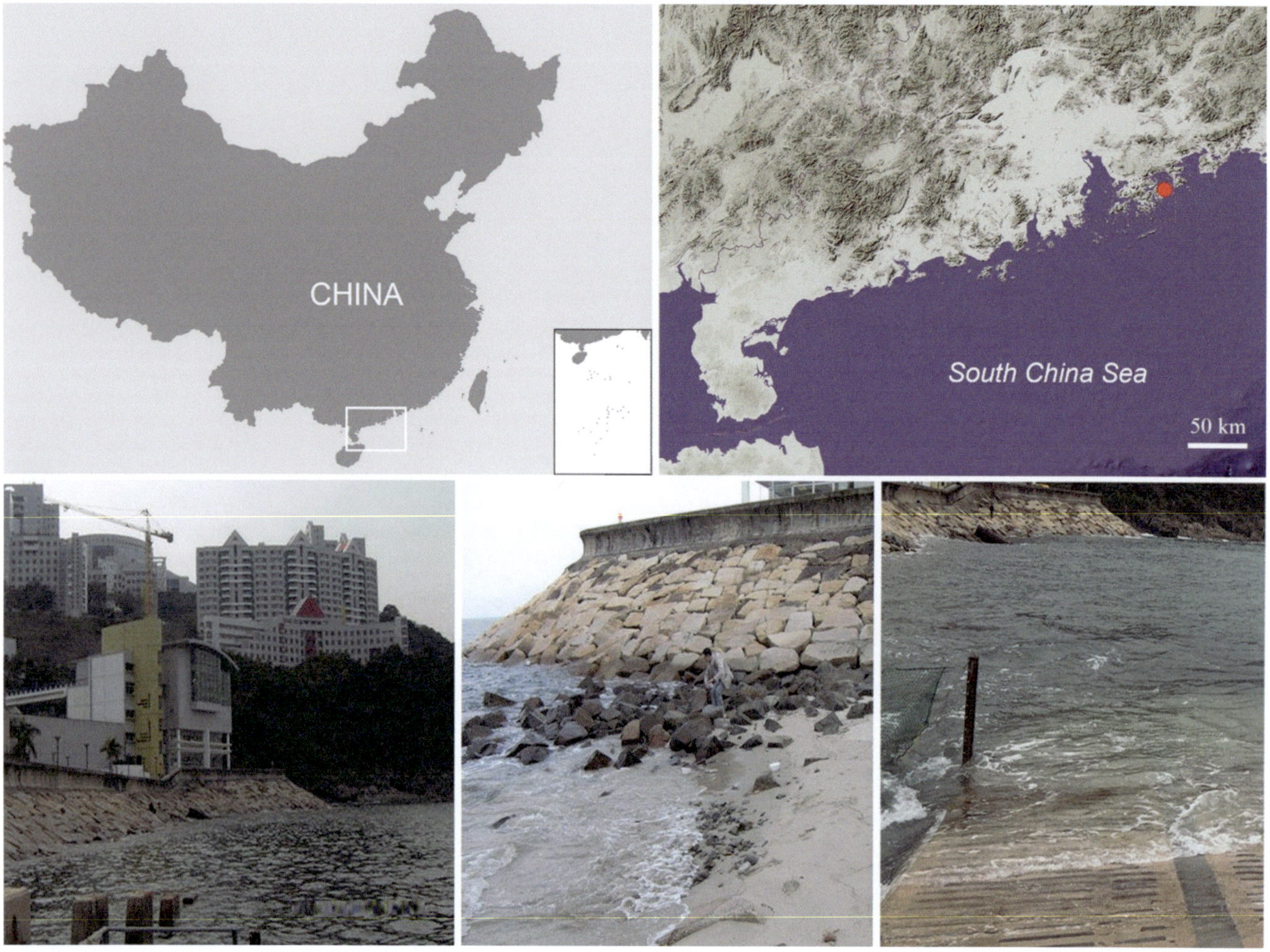


**Fig. 3A** Sample site

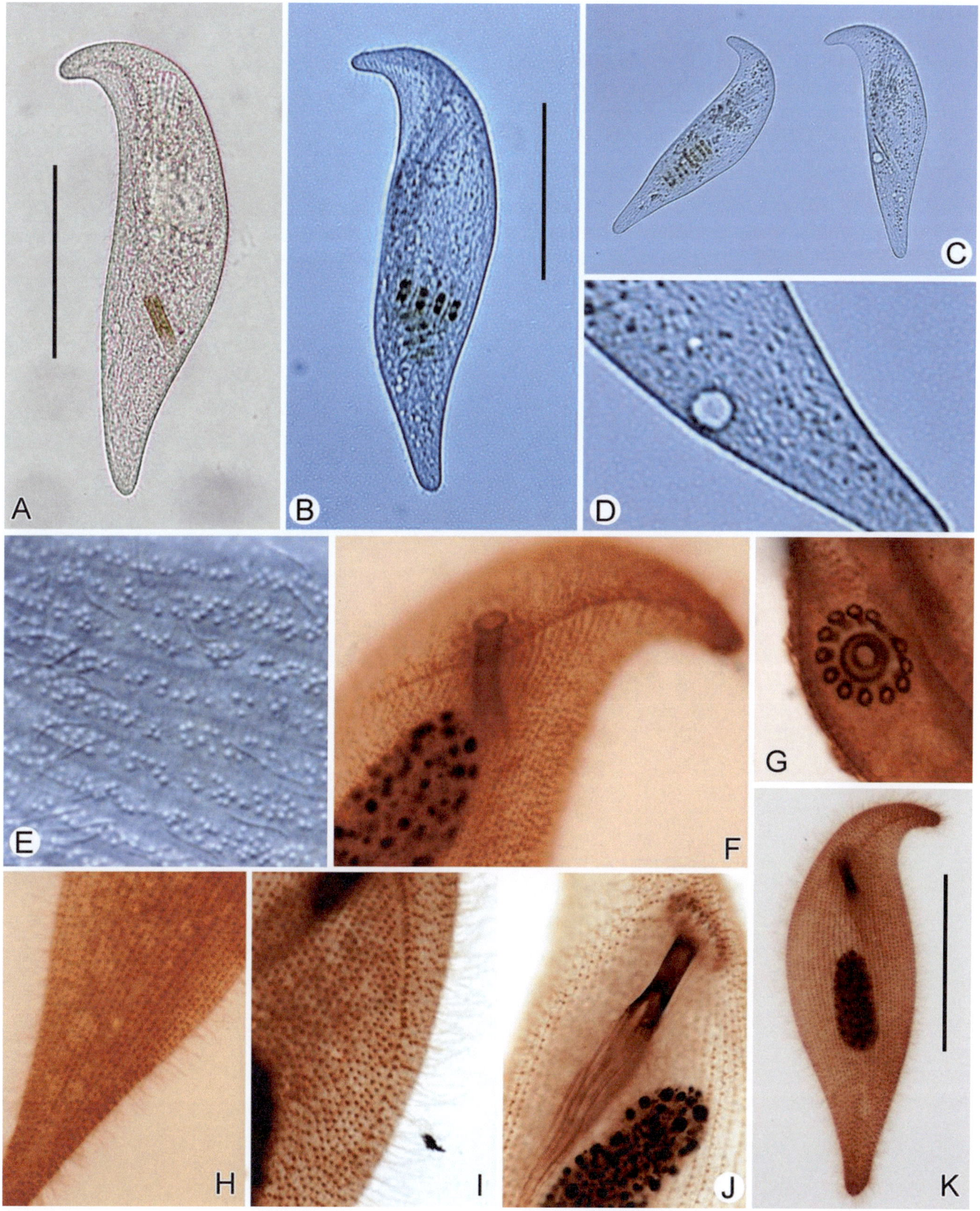


**Fig. 3B** *Orthodonella sinica*
(**a–c**) Dorsal views of live cells; (**d**) contractile vacuole in posterior third of cell; (**e**) cortical granules; (**f**, **g**, **j**) pharyngeal basket and nematodesmal rods; (**g**) part of synhymenium; (**h**) posterior somatic kineties; (**k**) ventral view of ciliature. Scale bars: 80 μm

## Reference

Long H, Liu H, Liu W, Miao M, Hu X, Lin X, Song WB (2008) Two new ciliates from Hong Kong coastal water: *Orthodonella sinica* n. sp. and *Apokeronopsis wrighti* n. sp. (Protozoa: Ciliophora). J Eukaryot Microbiol 55:321–330

Class Nassophorea Small & Lynn, 1981
Order Nassulida Jankowski, 1967
Family Orthodonellidae Jankowski, 1968
*Orthodonella* sp.

**Ecological feature:** mangrove
**Water temperature:** 26.8 °C
**pH:** 7.9
**Salinity:** 25.3‰

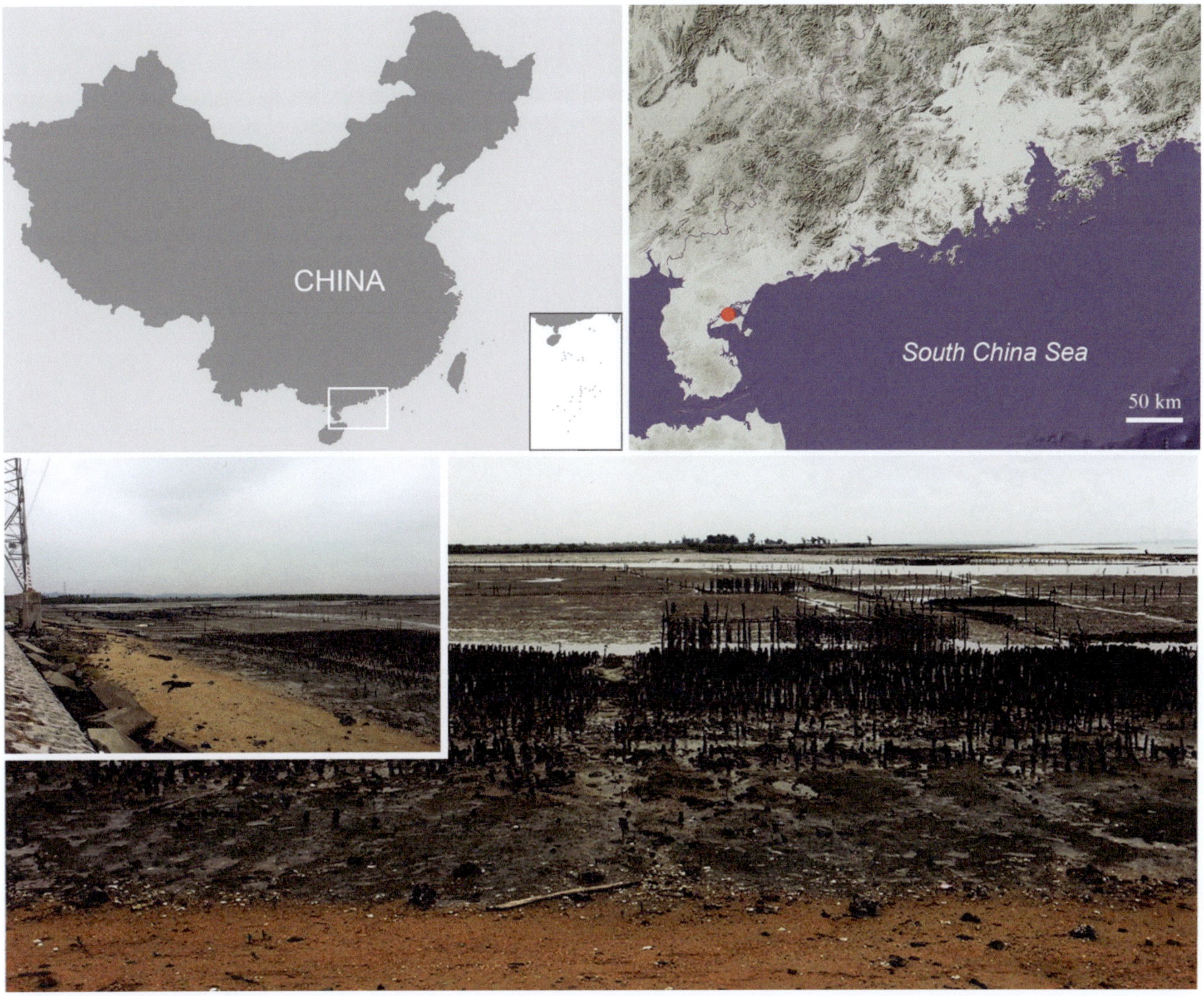


**Fig. 4A** Sample site

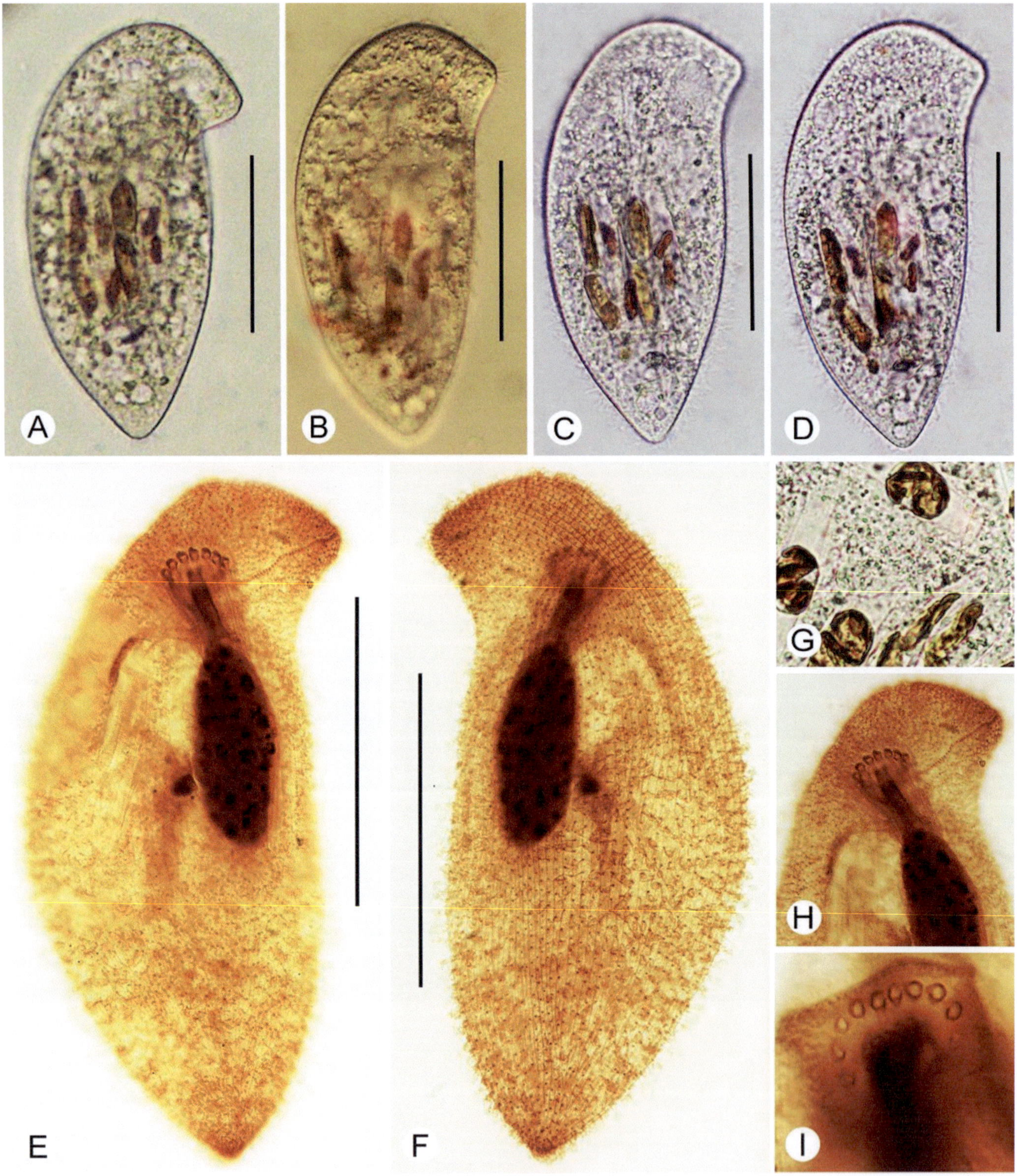


**Fig. 4B** *Orthodonella* sp
(**a–d**) Ventral views of live cells, showing variation of body shapes; (**e**, **h**) ventral view of ciliature, showing pharyngeal basket; (**f**) dorsal view of ciliature; (**g**) cortical granules and ingested diatoms; (**i**) nematodesmal rods. Scale bars: 70 μm

## Reference

Lin X, Gong J, Song WB (2004) Morphological studies on a new species of *Orthodonella*, with redescription of *O. gutta* (Cohn, 1866) Kahl, 1931 (Protozoa: Ciliophora: Synhemeniida) from coastal water off Qingdao, China. J Nat Hist 38:2001–2011

Class Nassophorea Small & Lynn, 1981
Order Synhymeniida de Puytorac et al. in Deroux, 1978
Family Orthodonellidae Jankowski, 1968
*Zosterodasys agamalievi* Deroux, 1978

**Ecological feature:** mangrove
**Water temperature:** 24.8 °C
**pH:** 6.7
**Salinity:** 17.0‰

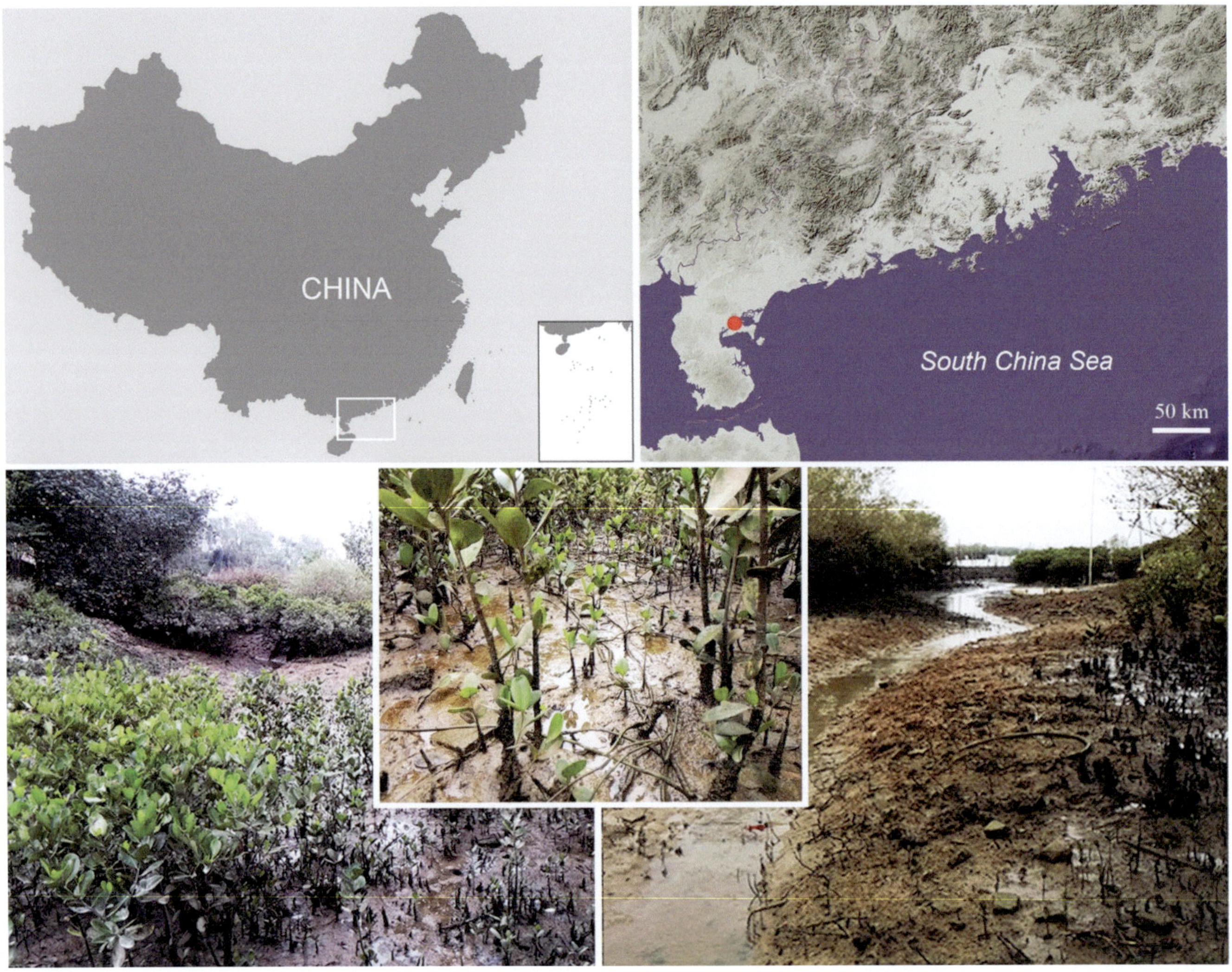


**Fig. 5A** Sample site

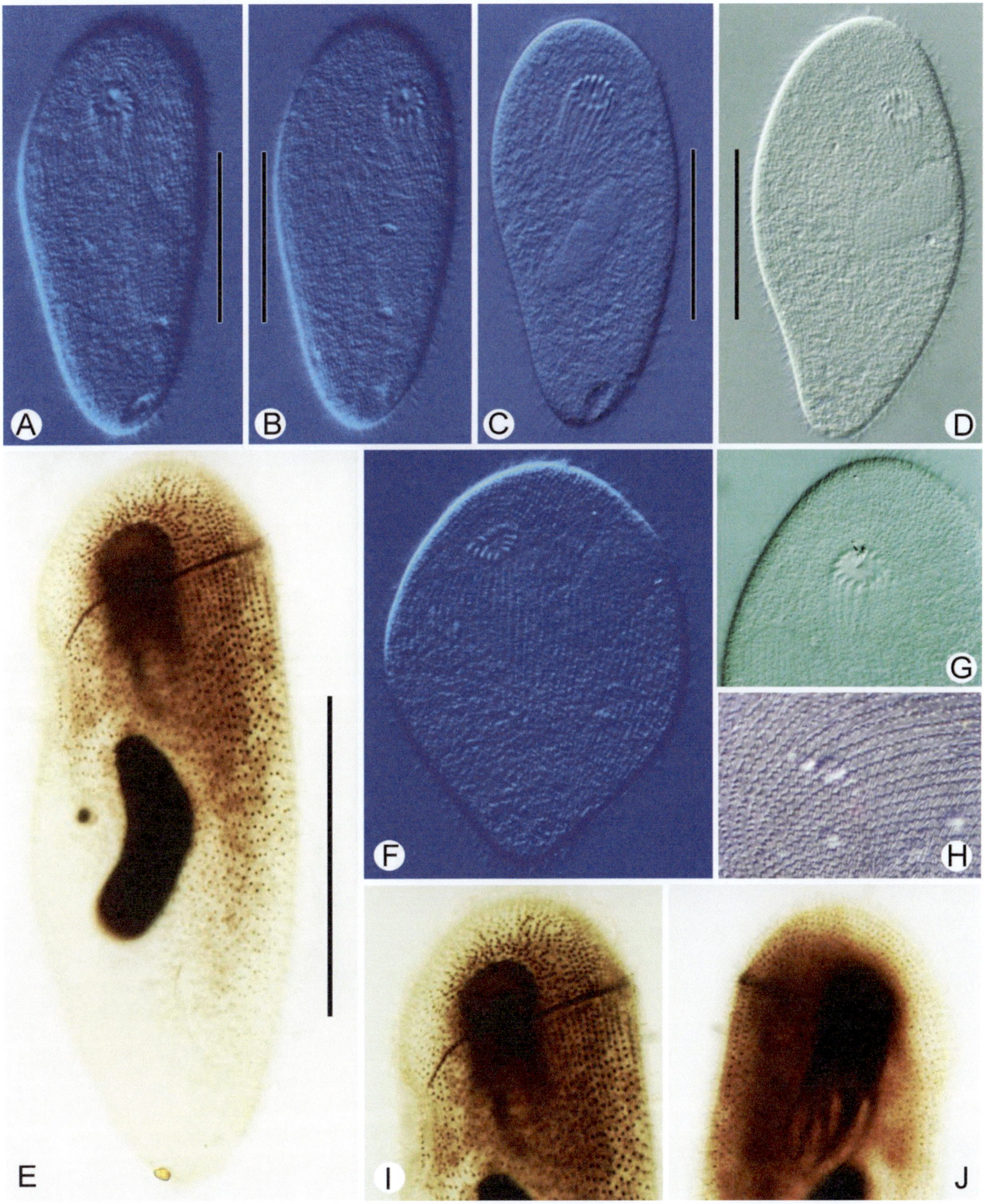


**Fig. 5B** *Zosterodasys agamalievi*
(**a–d**) Ventral views of live cells; (**e**, **i**, **j**) ciliature of ventral (**e**, **i**) and dorsal (**j**) sides; (**f**, **g**) oral area, to show nematodesmal rods; (**h**) detail of somatic kineties. Scale bars: 30 μm

## Reference

Petz W, Song WB, Wilbert N (1995) Taxonomy and ecology of the ciliate fauna (Protozoa, Ciliophora) in the endopagial and pelagial of the Weddell Sea, Antarctica. Stapfia 40:1–223

Class Nassophorea Small & Lynn, 1981
Order Synhymeniida de Puytorac et al. in Deroux, 1978
Family Orthodonellidae Jankowski, 1968
*Zosterodasys* cf. *cantabrica*

**Ecological feature:** mangrove
**Water temperature:** 17.5 °C
**pH:** 8.0
**Salinity:** 27.0‰

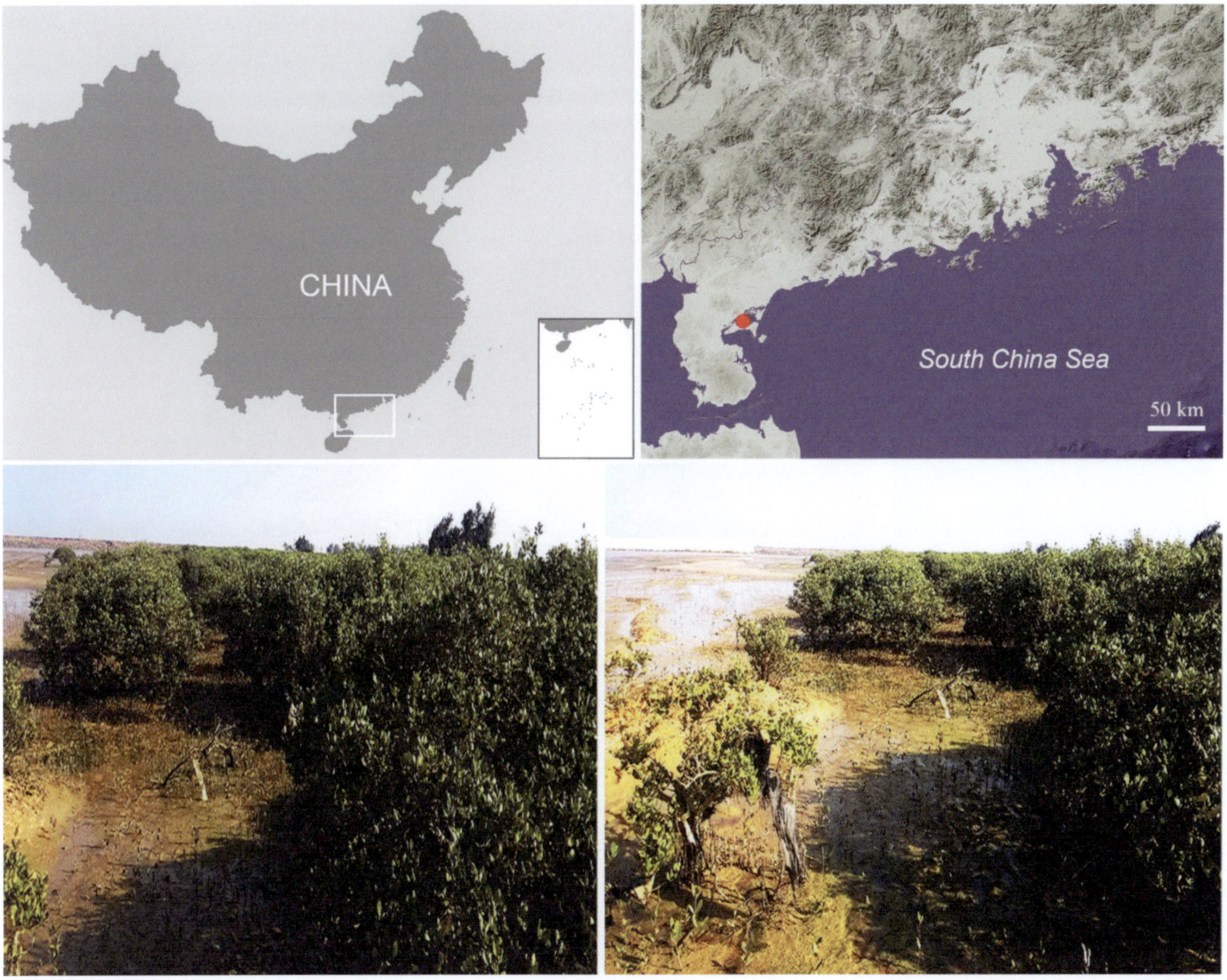


**Fig. 6A** Sample site

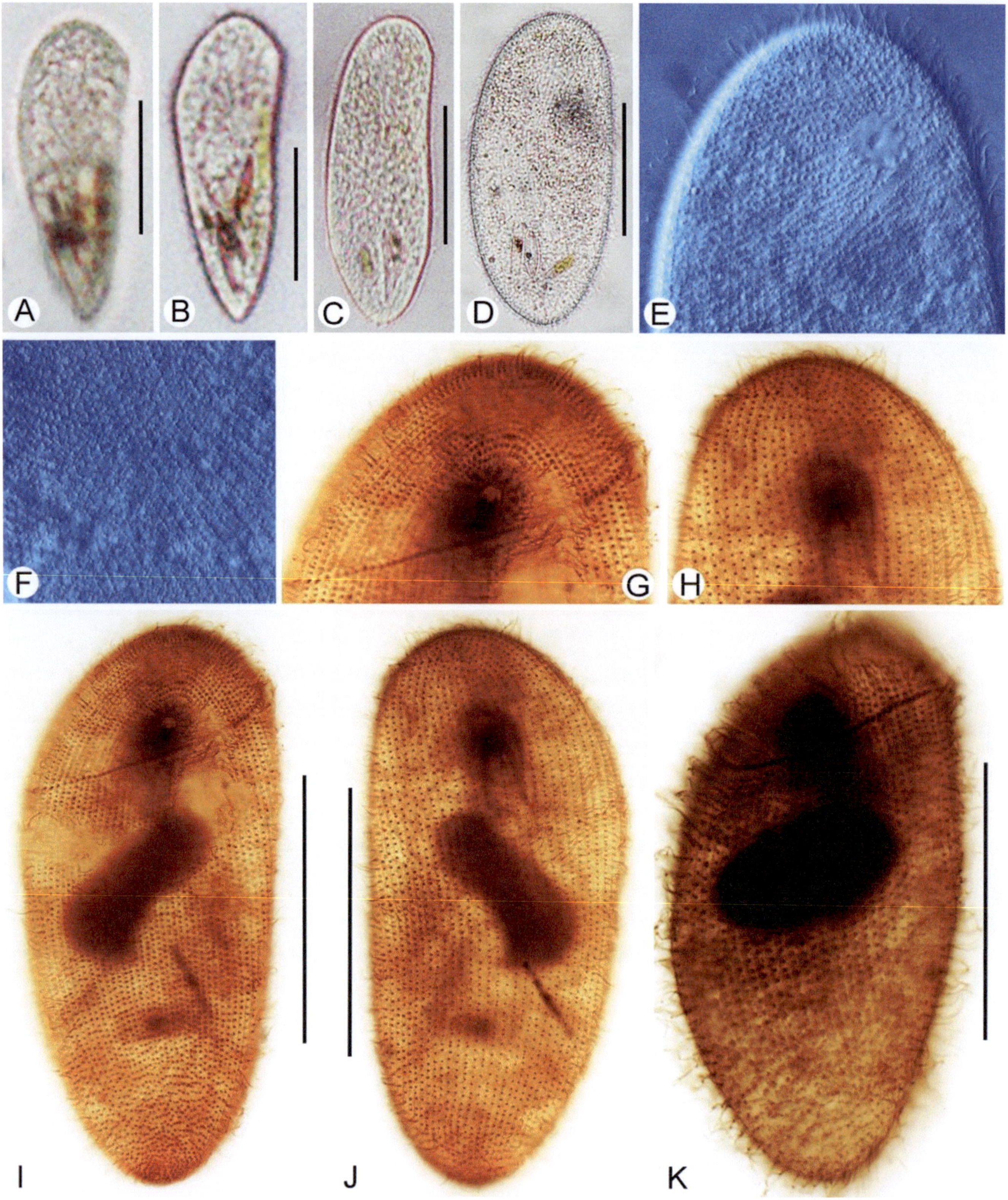


**Fig. 6B** *Zosterodasys* cf. *cantabrica*
(**a–d**) Ventral views of live cells; (**e**) ventral view of anterior part of cell, showing oral area; (**f**) cortical granules; (**g–k**) ciliature of ventral (**g**, **i**, **k**) and dorsal (**h**, **j**) sides. Scale bars: 40 μm

## References

Aliev AR (1990) New species of infusoria (Synhymeniida) of the genus *Zosterodasys* from natural-water bodies of Azerbaijan. Zool Zh 69:13–23

Fernandez-Leborans G, Alekperov IK (1996) Revision of the genus *Zosterodasys* with description of new species (Protozoa, Ciliophora). Zoosyst Rossica 4:3–16

Class Nassophorea Small & Lynn, 1981
Order Synhymeniida de Puytorac et al. in Deroux, 1978
Family Orthodonellidae Jankowski, 1968
*Zosterodasys* cf. *transversa*

**Ecological feature:** brackish water
**Water temperature:** no data available
**pH:** no data available
**Salinity:** 10.0‰

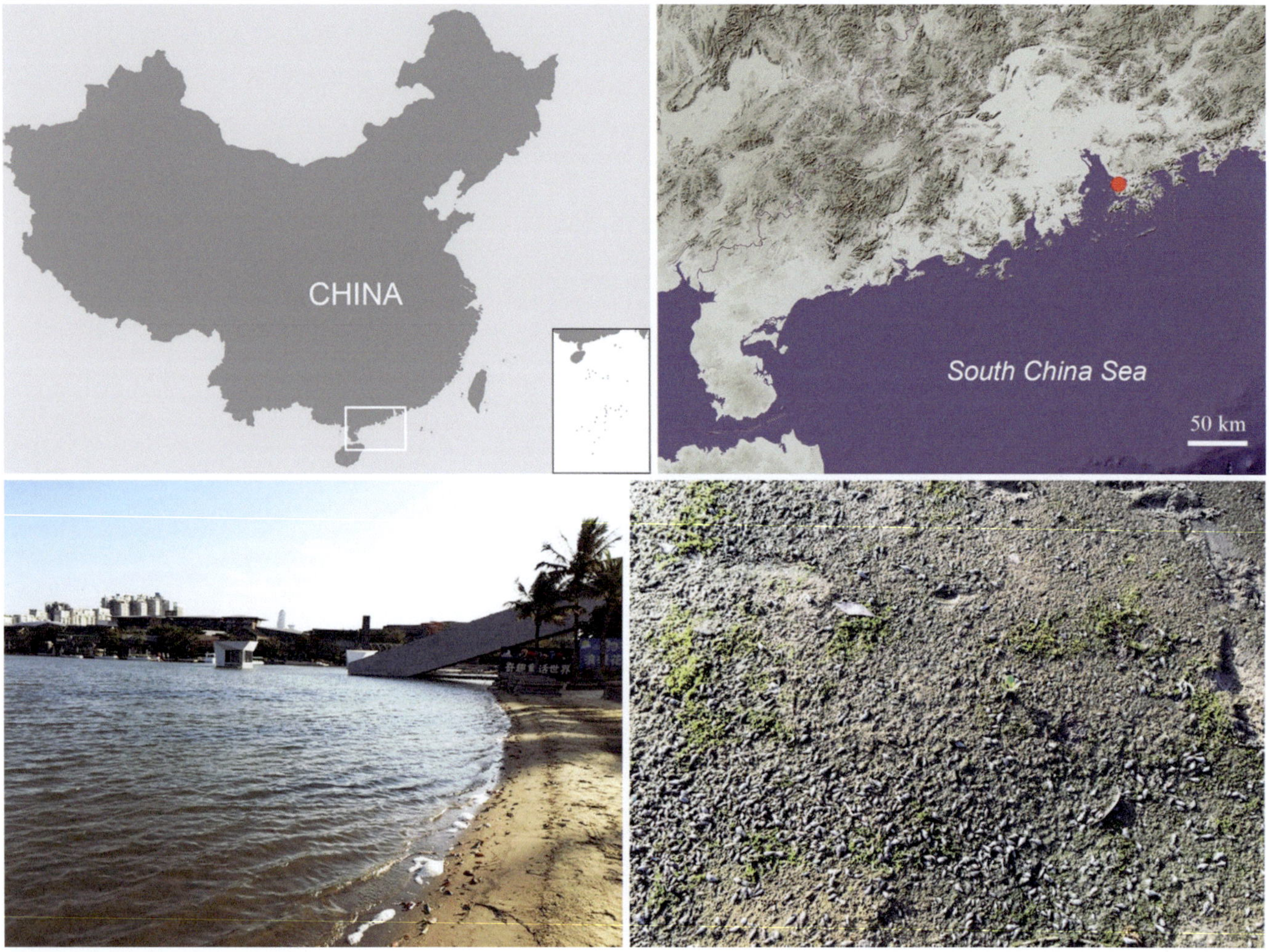


**Fig. 7A** Sample site

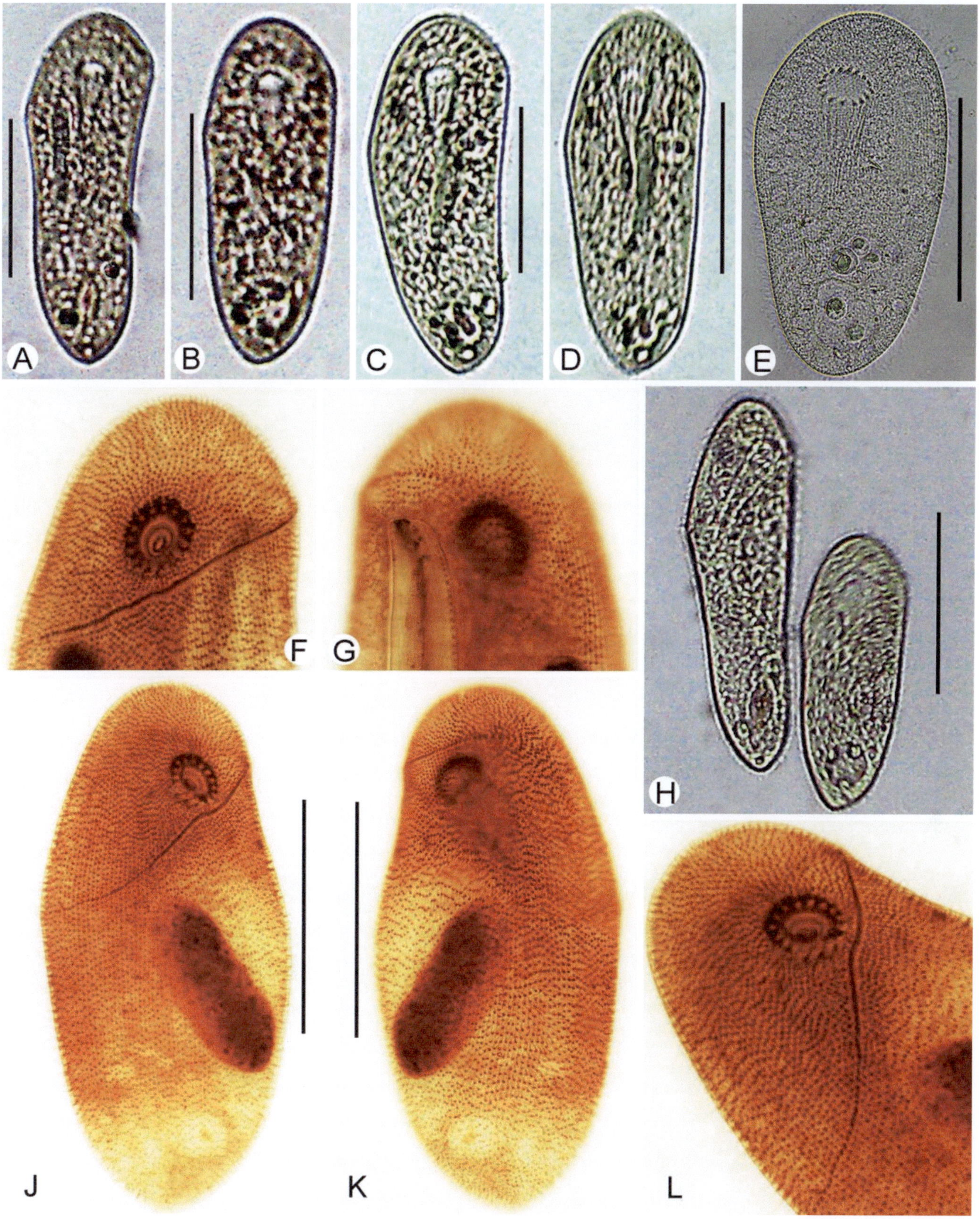


**Fig. 7B** *Zosterodasys* cf. *transversa*
(**a–e**, **h**) Ventral views of live cells; (**f**, **g**, **j–l**) ciliature of ventral (**f**, **j**, **l**) and dorsal (**g**, **k**) sides. Scale bars: 80 μm

## References

Petz W, Song WB, Wilbert N (1995) Taxonomy and ecology of the ciliate fauna (Protozoa, Ciliophora) in the endopagial and pelagial of the Weddell Sea, Antarctica. Stapfia 40:1–223

Wang L, Zhao Y, Gong J (2012) Redescription of two synhymeniid ciliates, *Chilodontopsis simplex* Ozaki & Yagiu, 1941 and *Zosterodasys transverses* (Kahl, 1928) Foissner et al., 1994 (Alveolata, Ciliophora, Phyllopharyngea). Zootaxa 3167:45–52

Class Nassophorea Small & Lynn, 1981
Order Synhymeniida de Puytorac et al. in Deroux, 1978
Family Orthodonellidae Jankowski, 1968
*Zosterodasys jankowskii* Aliev, 1990

**Ecological feature:** mangrove
**Water temperature:** 23.1 °C
**pH:** 6.3
**Salinity:** 10.0‰

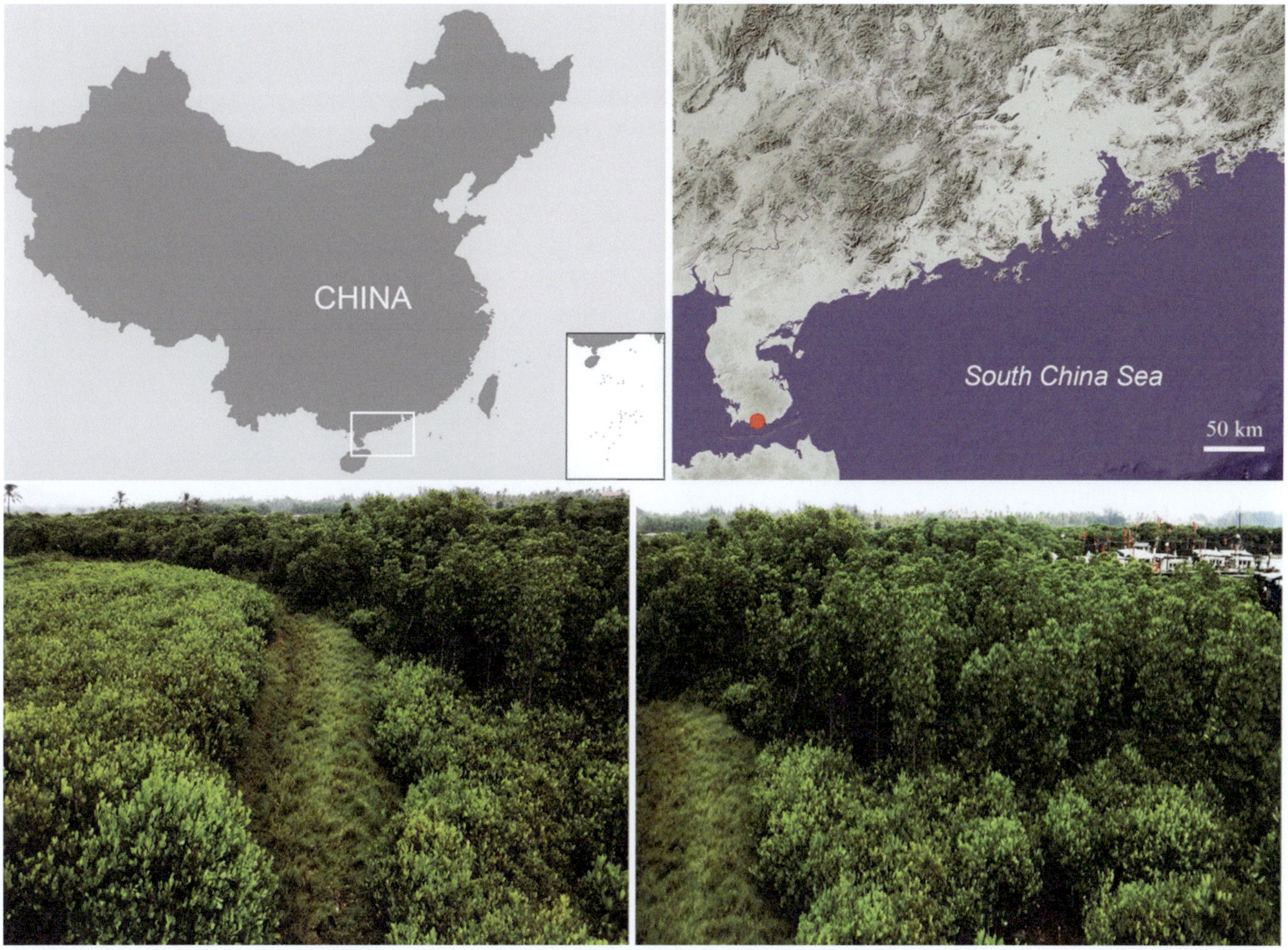


**Fig. 8A** Sample site

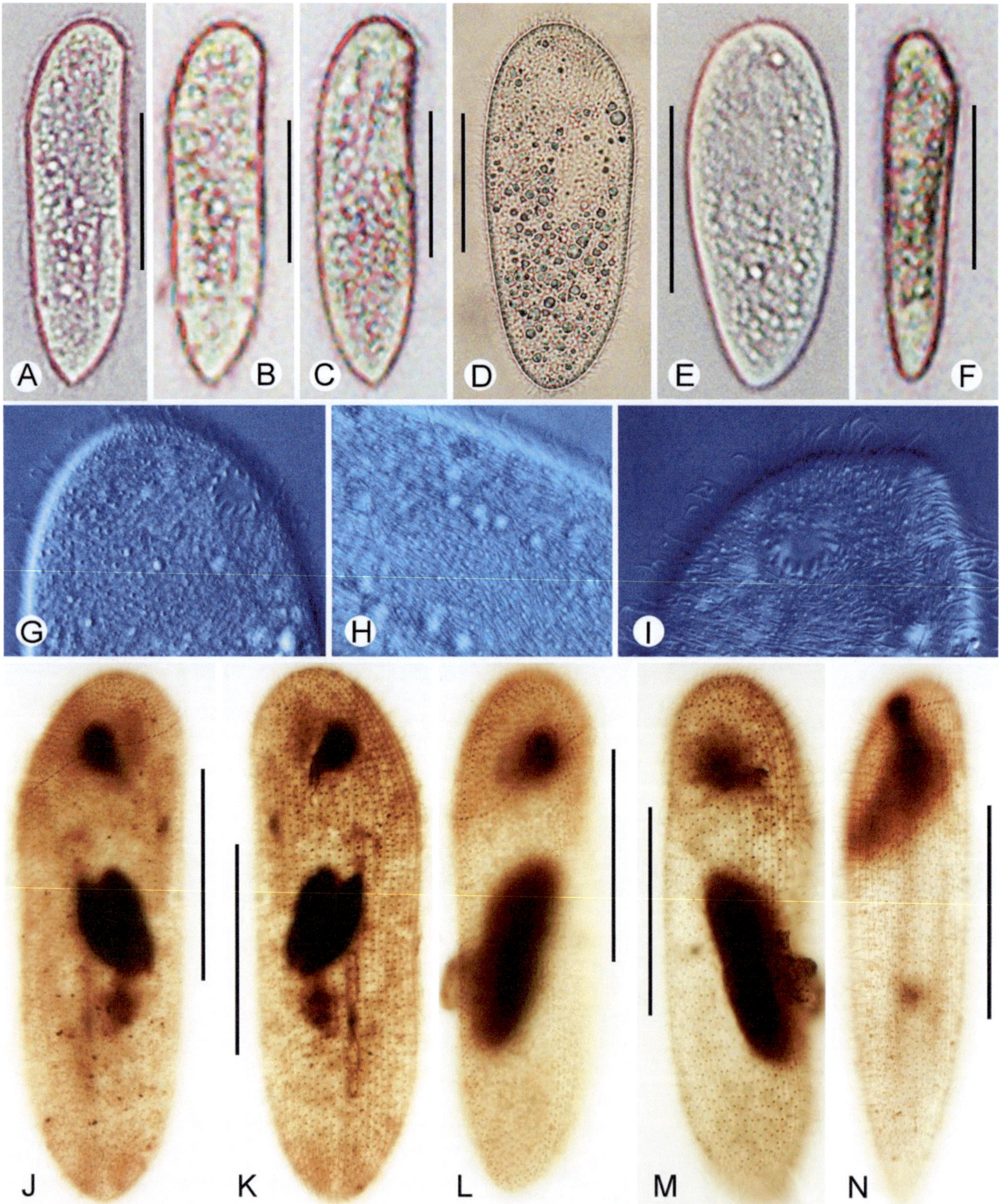


**Fig. 8B** *Zosterodasys jankowskii*
(**a**–**f**) Ventral views of live cells; (**g**, **i**) ventral views of anterior part of cell; (**h**) cortical granules; (**j**–**n**) ciliature on ventral (**j**, **l**, **n**) and dorsal (**k**, **m**) sides. Scale bars: 30 μm

## References

Aliev AR (1990) New species of infusoria (Synhymeniida) of the genus *Zosterodasys* from natural-water bodies of Azerbaijan. Zool Zh 69:13–23

Petz W, Song WB, Wilbert N (1995) Taxonomy and ecology of the ciliate fauna (Protozoa, Ciliophora) in the endopagial and pelagial of the Weddell Sea, Antarctica. Stapfia 40:1–223

Class Nassophorea Small & Lynn, 1981
Order Synhymeniida de Puytorac et al. in Deroux, 1978
Family Orthodonellidae Jankowski, 1968
*Zosterodasys shumerica* Aliev, 1990

**Ecological feature:** brackish water
**Water temperature:** no data available
**pH:** no data available
**Salinity:** 10.0‰

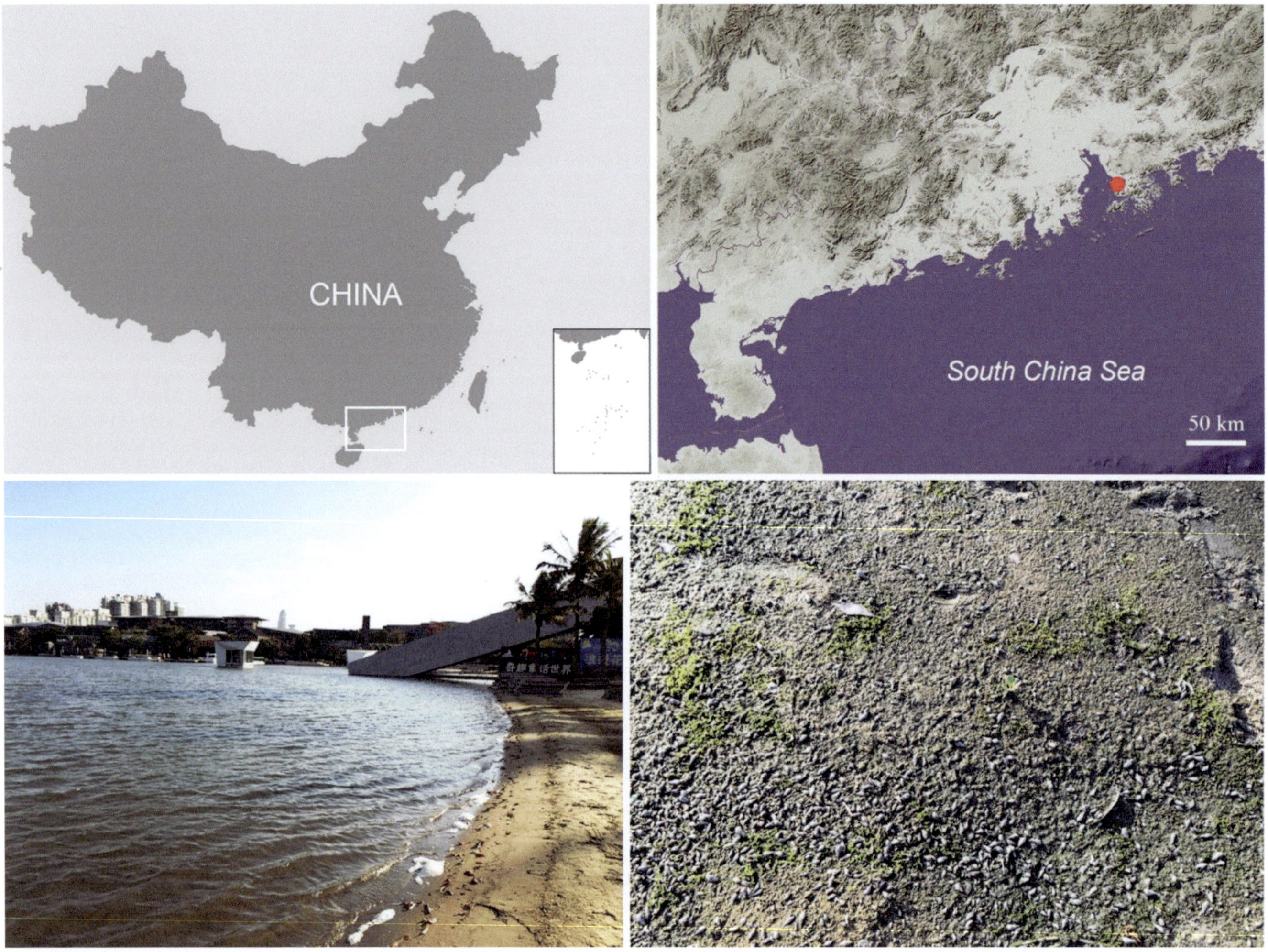


**Fig. 9A** Sample site

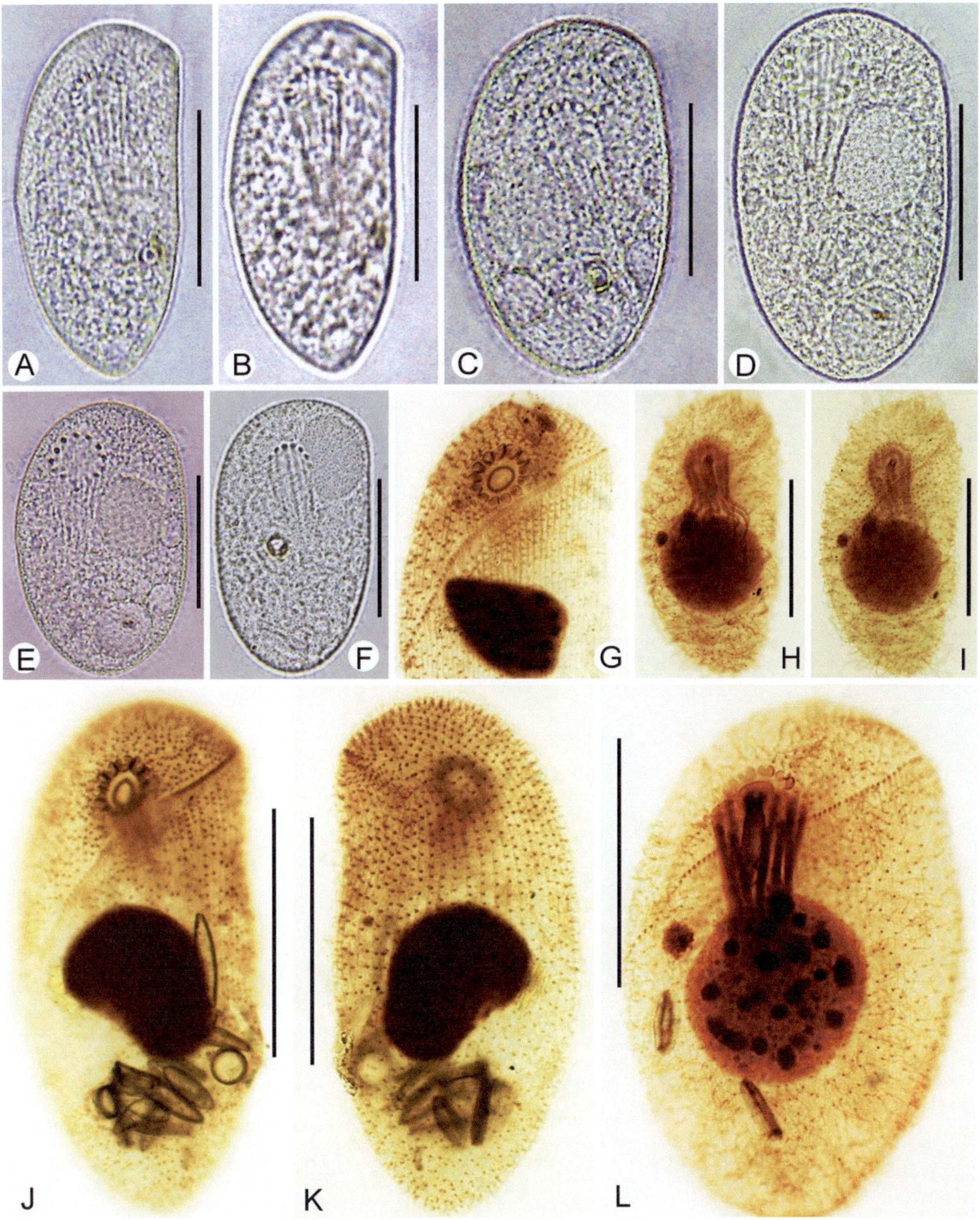


**Fig. 9B** *Zosterodasys shumerica*
(**a–f**) Ventral views of live cells; (**g–l**) ciliature of ventral (**g–j**, **l**) and dorsal (**k**) sides. Scale bars: 40 μm

## Reference

Aliev AR (1990) New species of infusoria (Synhymeniida) of the genus *Zosterodasys* from natural-water bodies of Azerbaijan. Zool Zh 69:13–23

Petz W, Song WB, Wilbert N (1995) Taxonomy and ecology of the ciliate fauna (Protozoa, Ciliophora) in the endopagial and pelagial of the Weddell Sea, Antarctica. Stapfia 40:1–223

Class Nassophorea Small & Lynn, 1981
Order Synhymeniida de Puytorac et al. in Deroux, 1978
Family Scaphidiodontidae Deroux in Corliss, 1979
*Chilodontopsis* sp.

**Ecological feature:** mangrove
**Water temperature:** 24.8 °C
**pH:** 6.7
**Salinity:** 17.0‰

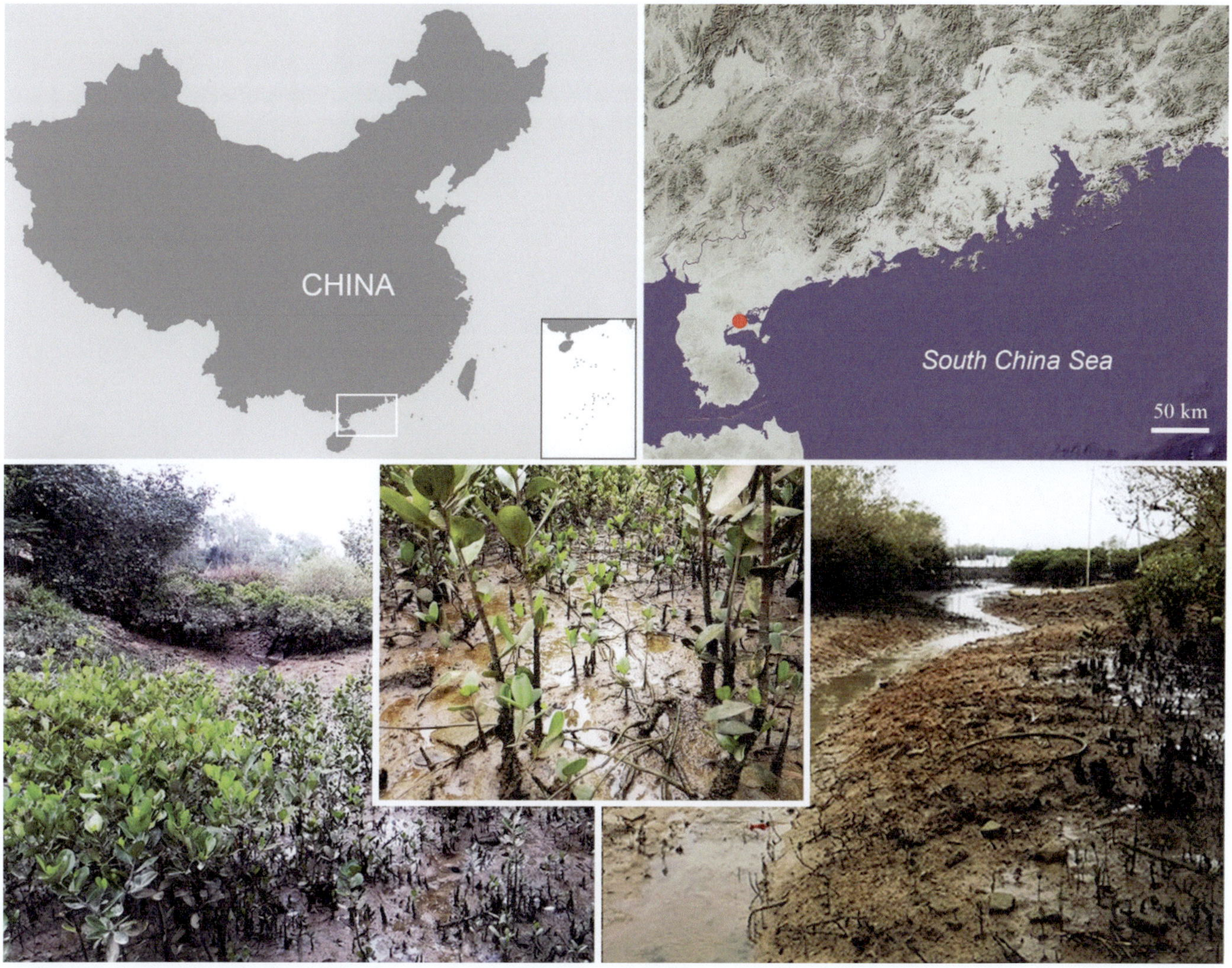


**Fig. 10A** Sample site

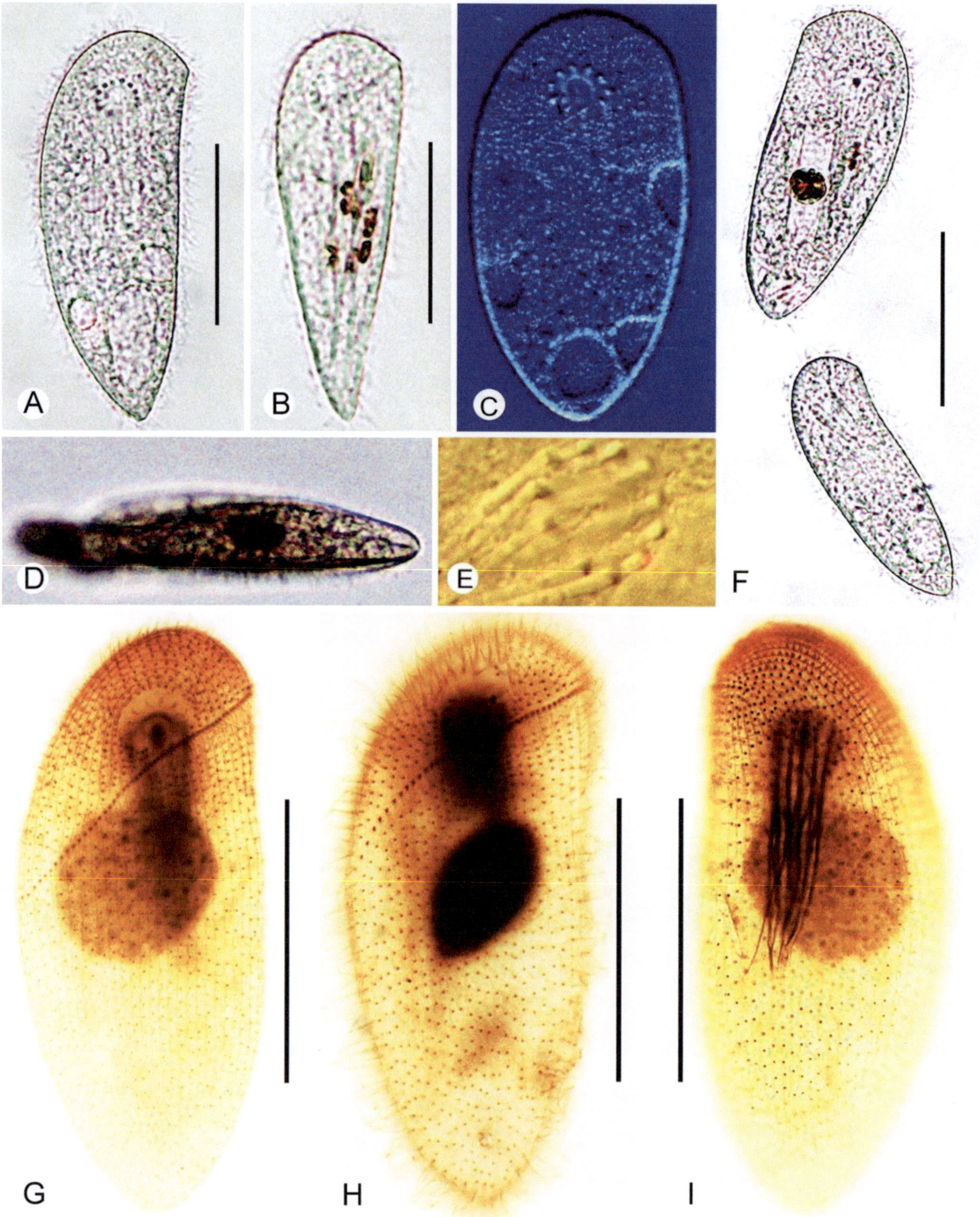


**Fig. 10B** *Chilodontopsis* sp.
(**a**–**c**, **f**) Ventral views of live cells; (**d**) lateral view of live cell; (**e**) nematodesmal rods; (**g**–**i**) ciliature of ventral (**g**, **h**) views and dorsal (**i**) view sides. Scale bars: 30 μm

## References


Petz W, Song WB, Wilbert N (1995) Taxonomy and ecology of the ciliate fauna (Protozoa, Ciliophora) in the endopagial and pelagial of the Weddell Sea, Antarctica. Stapfia 40:1–223

Wang L, Zhao Y, Gong J (2012) Redescription of two synhymeniid ciliates, *Chilodontopsis simplex* Ozaki & Yagiu, 1941 and *Zosterodasys transverses* (Kahl, 1928) Foissner et al., 1994 (Alveolata, Ciliophora, Phyllopharyngea). Zootaxa 3167:45–52

# Class Phyllopharyngea

Xiaozhong Hu, Xiaofeng Lin, and Weibo Song

Class Phyllopharyngea de Puytorac et al., 1974
Order Chlamydodontida Deroux, 1976
Family Chilodonellidae Deroux, 1970
*Chilodonella acuta* Kahl, 1931

**Ecological feature:** brackish water
**Water temperature:** no data available
**pH:** no data available
**Salinity:** 5.0‰

X. Hu (✉)
Fisheries College, Ocean University of China,
Qingdao, Shandong, China
e-mail: xiaozhonghu@ouc.edu.cn

X. Lin
College of Life Science, South China Normal University,
Guangzhou, Guangdong, China
e-mail: xlin@oucscnu.edu.cn

W. Song
Institute of Evolution & Marine Biodiversity,
Ocean University of China, Qingdao, Shandong, China
e-mail: wsong@ouc.edu.cn


X. Hu et al. (eds.), *Ciliate Atlas: Species Found in the South China Sea*,
https://doi.org/10.1007/978-981-13-5901-9_7

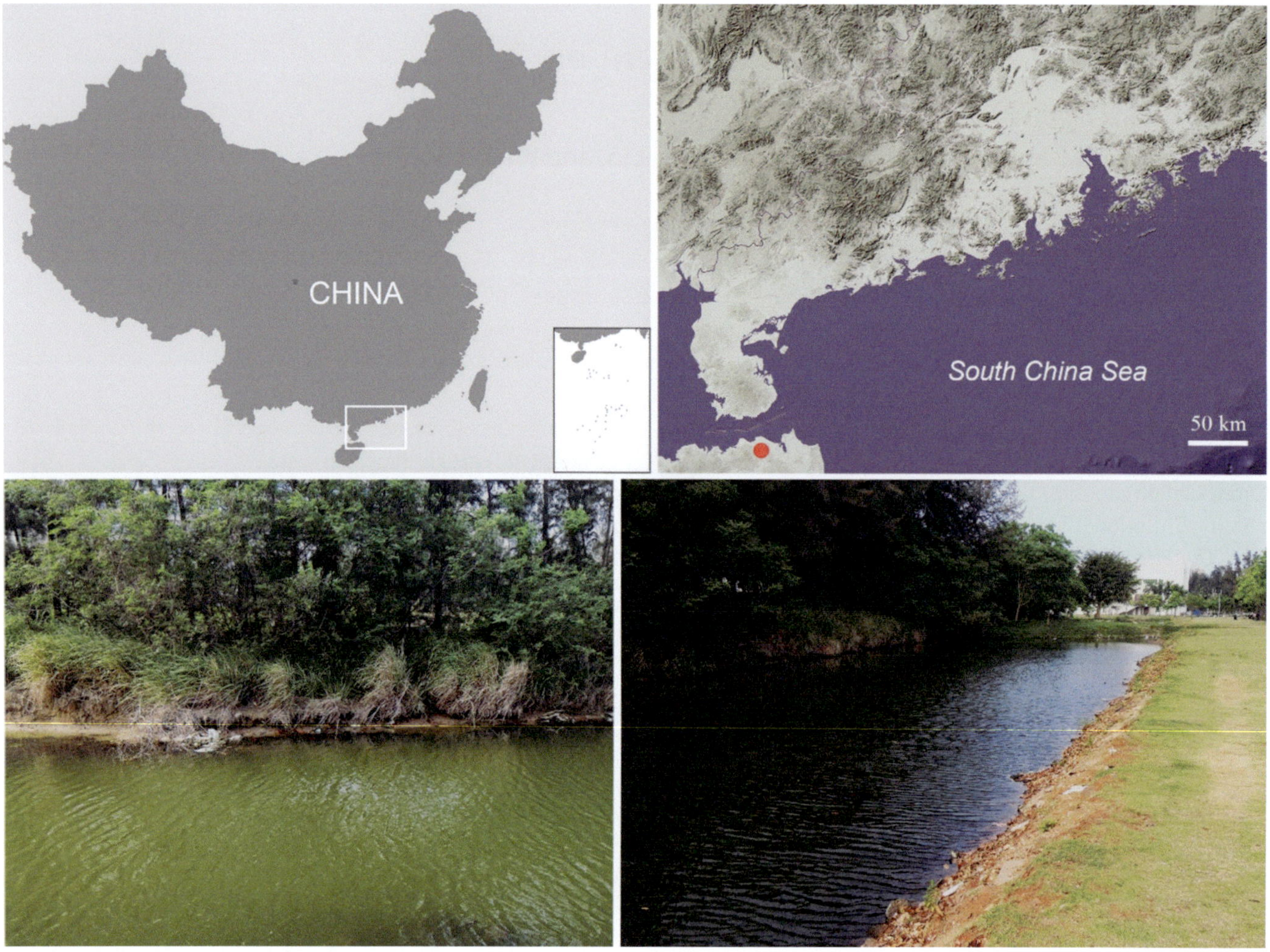


**Fig. 1A** Sample site

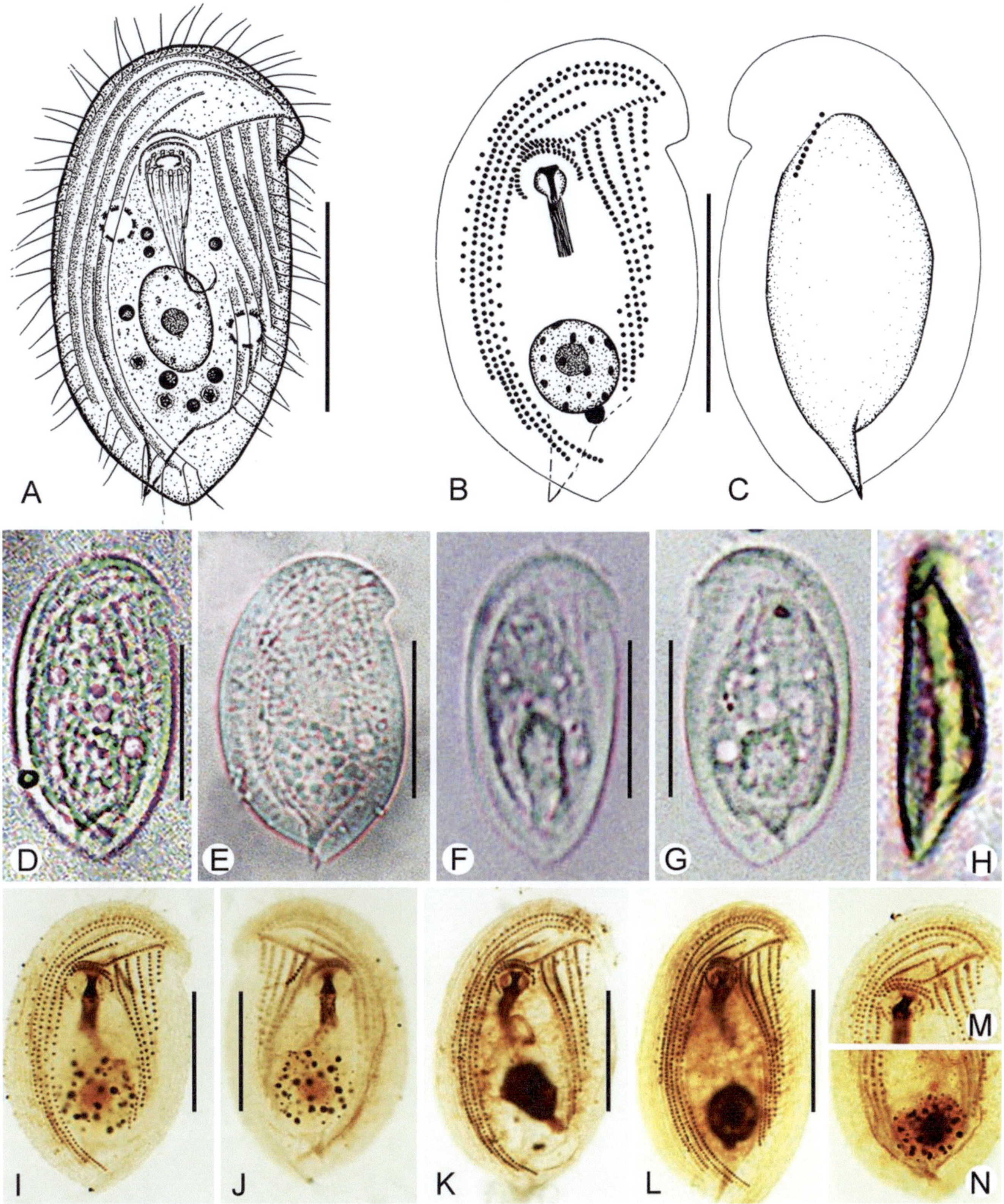


**Fig. 1B** *Chilodonella acuta*
(**a**, **d–f**) Ventral views of live individuals; (**b**, **c**, **i–n**) ciliature of ventral (**b**, **i**, **k–n**) and dorsal (**c**, **j**) sides; (**g**) dorsal view of live cell; (**h**) lateral view of live cell. Scale bars: 20 μm

## Reference

Fan X, Ma R, Al-Farraj SA, Gu F (2014) Morphological and molecular characterization of *Parafurgasonia zhangi* spec. nov. and *Chilodonella acuta* Kahl, 1931 (Protozoa, Ciliophora) from soil habitat of Saudi Arabia. Int J Syst Evol Microbiol 64:2385–2394

Class Phyllopharyngea de Puytorac et al., 1974
Order Chlamydodontida Deroux, 1976
Family Chilodonellidae Deroux, 1970
*Chilodonella uncinata* Ehrenberg, 1938

**Ecological feature:** estuary area
**Water temperature:** 29.0–31.0 °C
**pH:** 7.9–8.6
**Salinity:** 2.5‰–5.0‰

**Fig. 2A** Sample site

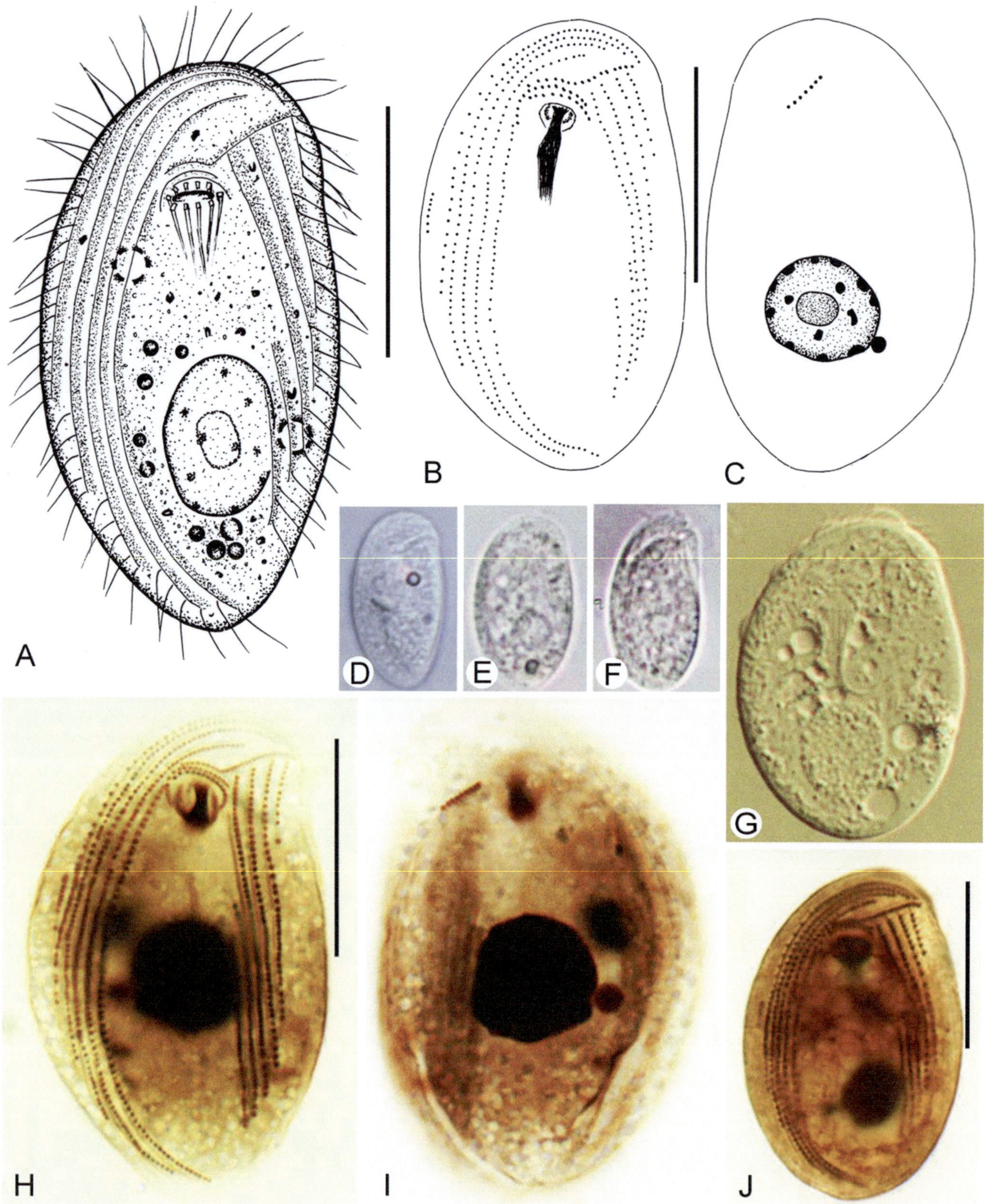


**Fig. 2B** *Chilodonella uncinata*
(**a**, **d–g**) Ventral views of live cells; (**b**, **c**, **h–j**) ciliature of ventral (**b**, **h**, **j**) and dorsal (**c**, **i**) sides. Scale bars: 20 μm

## References

Foissner W (1981) Morphologie und Taxonomie einiger neuer und wenig bekannter kinetofragminophorer Ciliaten (Protozoa: Ciliphora) aus alpinen Böden. Zool Jb Syst 108:264–297

Qu Z, Pan H, Hu X, Li J, Al-Farraj SA, Al-Rasheid KAS, Yi Z (2015) Morphology and molecular phylogeny of three cyrtophorid ciliates (Protozoa, Ciliophora) from China, including two new species, *Chilodonella parauncinata* sp. n. and *Chlamydonella irregularis* sp. n. J Eukaryot Microbiol 62:267–279

Class Phyllopharyngea de Puytorac et al., 1974
Order Chlamydodontida Deroux, 1976
Family Chilodonellidae Deroux, 1970
*Pseudochilodonopsis fluviatilis* Foissner, 1988

**Ecological feature:** mangrove
**Water temperature:** 20.7 °C
**pH:** 7.5
**Salinity:** 13.8‰

**Fig. 3A** Sample site

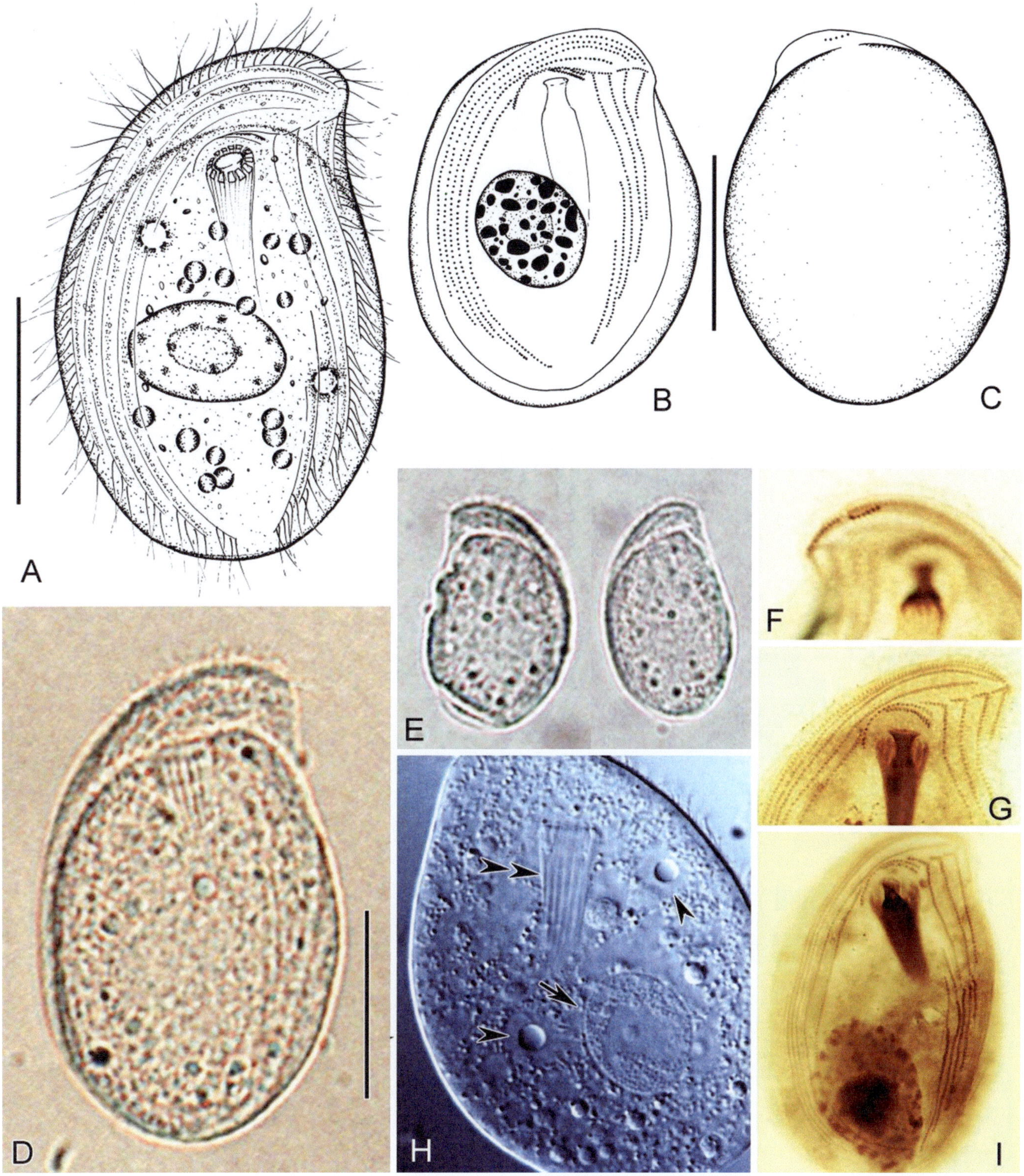


**Fig. 3B** *Pseudochilodonopsis fluviatilis*
(**a**, **d**) Ventral views of live cells; (**b**, **c**) ciliature of ventral (**b**) and dorsal (**c**) sides; (**e**) body shape variants; (**f**, **g**) ciliature of dorsal (**f**) and ventral (**g**) sides; (**h**) details of cytoplasm, arrowheads mark contractile vacuoles, arrow points to macronucleus, double arrowheads indicate nematodesmal rods; (**i**) Ventral ciliature of a dividing specimen. Scale bars: 30 μm

## References

Foissner W (1988) Taxonomie und Ökologie einiger Ciliaten (Protozoa, Ciliophora) des Saprobiensystems. II. Familie Chilodonellidae. Hydrobiologia 162:21–45

Qu Z, Pan H, Qu ZS, Al-Rasheid KAS, Hu X, Gao S (2015) Morphological and phylogenetic studies on three members of the genus *Pseudochilodonopsis* (Ciliophora, Cyrtophoria) isolated from brackish waters in China, including a novel species, *Pseudochilodonopsis quadrivacuolata* sp. nov. Int J Syst Evol Microbiol 65:4323–4334

Class Phyllopharyngea de Puytorac et al., 1974
Order Chlamydodontida Deroux, 1976
Family Chilodonellidae Deroux, 1970
*Pseudochilodonopsis mutabilis* Foissner, 1981

**Ecological feature:** mangrove
**Water temperature:** 24.2 °C
**pH:** 7.0
**Salinity:** 17.0‰

**Fig. 4A** Sample site

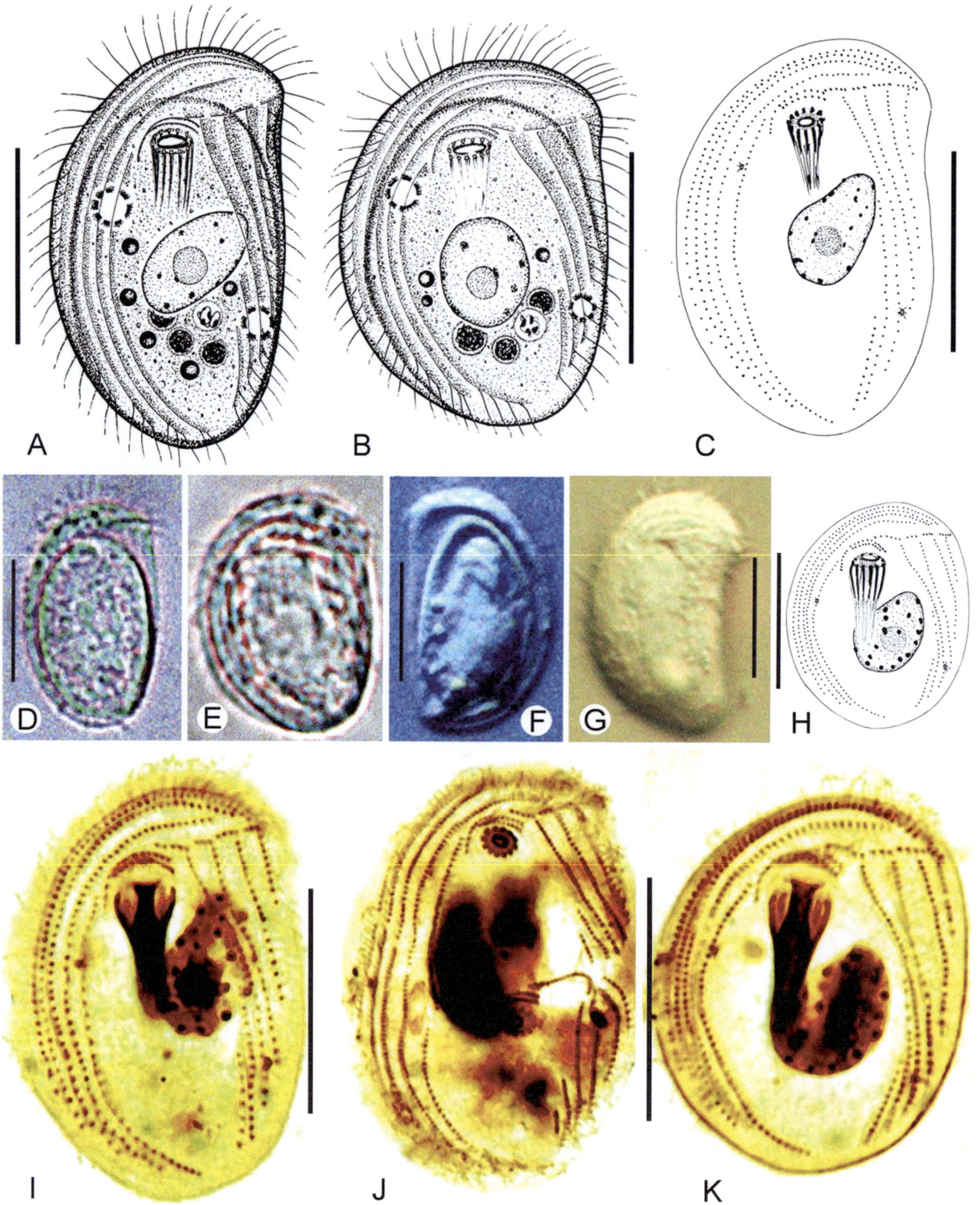


**Fig. 4B** *Pseudochilodonopsis mutabilis*
(**a**, **b**, **d**, **e**, **g**) Ventral views of live cells; (**c**, **h**, **i**, **k**) ventral views of ciliature; (**f**) dorsal view of live cell; (**j**) ventral ciliature of late divider. Scale bars: 20 μm

## Reference

Foissner W (1981) Morphologie und Taxonomie einiger neuer und wenig bekannter kinetofragminophorer Ciliaten (Protozoa: Ciliophora) aus alpine Böden. Zool Jb Syst 108:264–297

Class Phyllopharyngea de Puytorac et al., 1974
Order Chlamydodontida Deroux, 1976
Family Chilodonellidae Deroux, 1970
*Pseudochilodonopsis quadrivacuolata* Qu et al., 2015

**Ecological feature:** estuary area
**Water temperature:** 29.2 °C
**pH:** 7.9
**Salinity:** 2.5‰

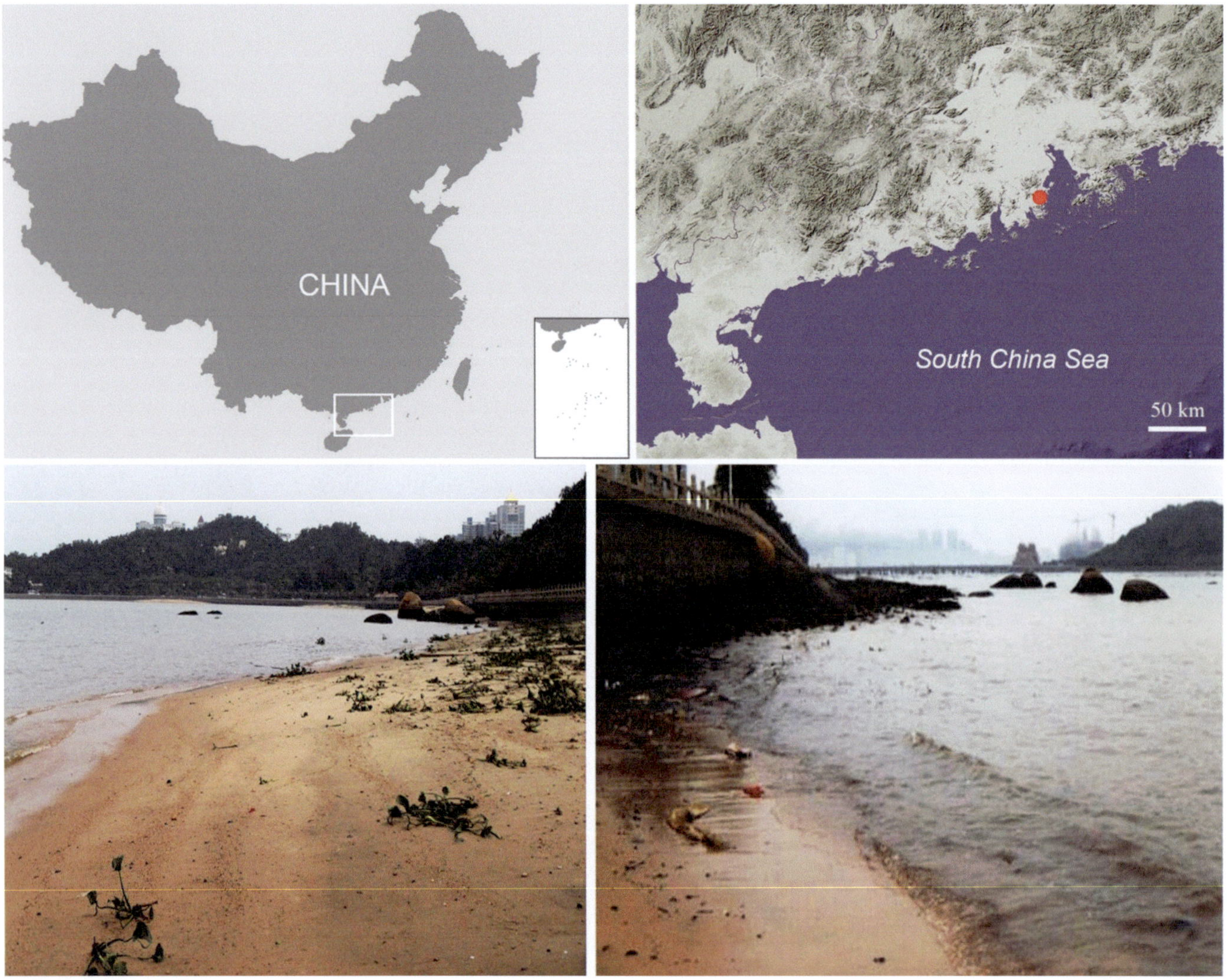


**Fig. 5A** Sample site

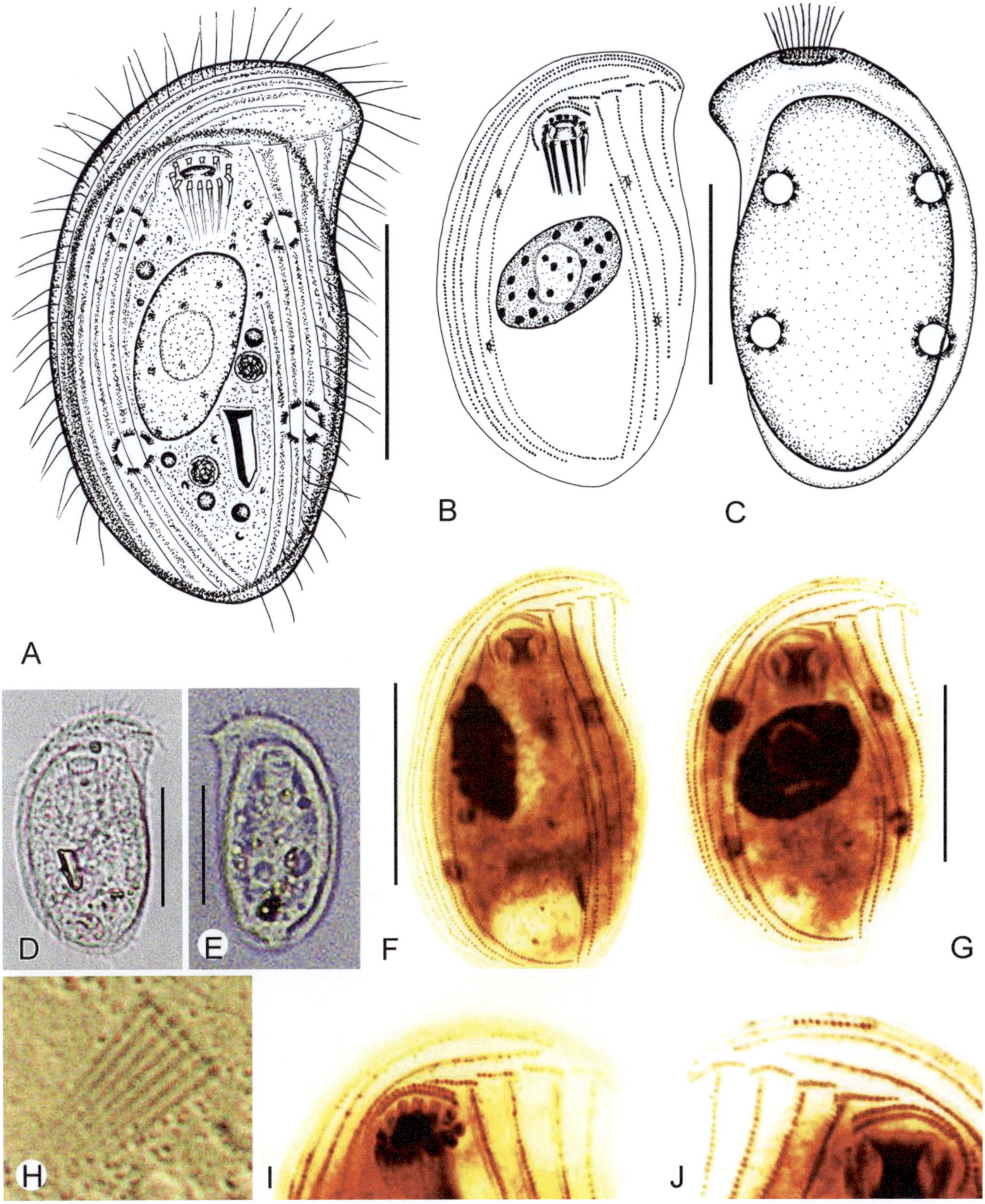


**Fig. 5B** *Pseudochilodonopsis quadrivacuolata*
(**a**, **d**) Ventral views of live cells; (**b**, **f**, **g**) ventral views of ciliature; (**c**, **e**) dorsal views of live cells; (**h**) nematodesmal rods; (**i**) oral ciliature; (**j**) terminal fragment. **a**–**c** from Qu et al. (2015). Scale bars: 20 μm

## Reference

Qu Z, Pan H, Al-Rasheid KAS, Hu X, Gao S (2015) Morphological and phylogenetic studies on three *Pseudochilodonopsis* (Ciliophora, Cyrtophoria) isolated from brackish waters in China, including a new species, *P. quadrivacuolata* spec. nov. Int J Syst Evol Microbiol 65:4323–4334

Class Phyllopharyngea de Puytorac et al., 1974
Order Chlamydodontida Deroux, 1976
Family Chilodonellidae Deroux, 1970
*Pseudochilodonopsis* sp.

**Ecological feature:** sediment (sand)
**Water temperature:** no data available
**pH:** no data available
**Salinity:** 10.0‰

**Fig. 6A** Sample site

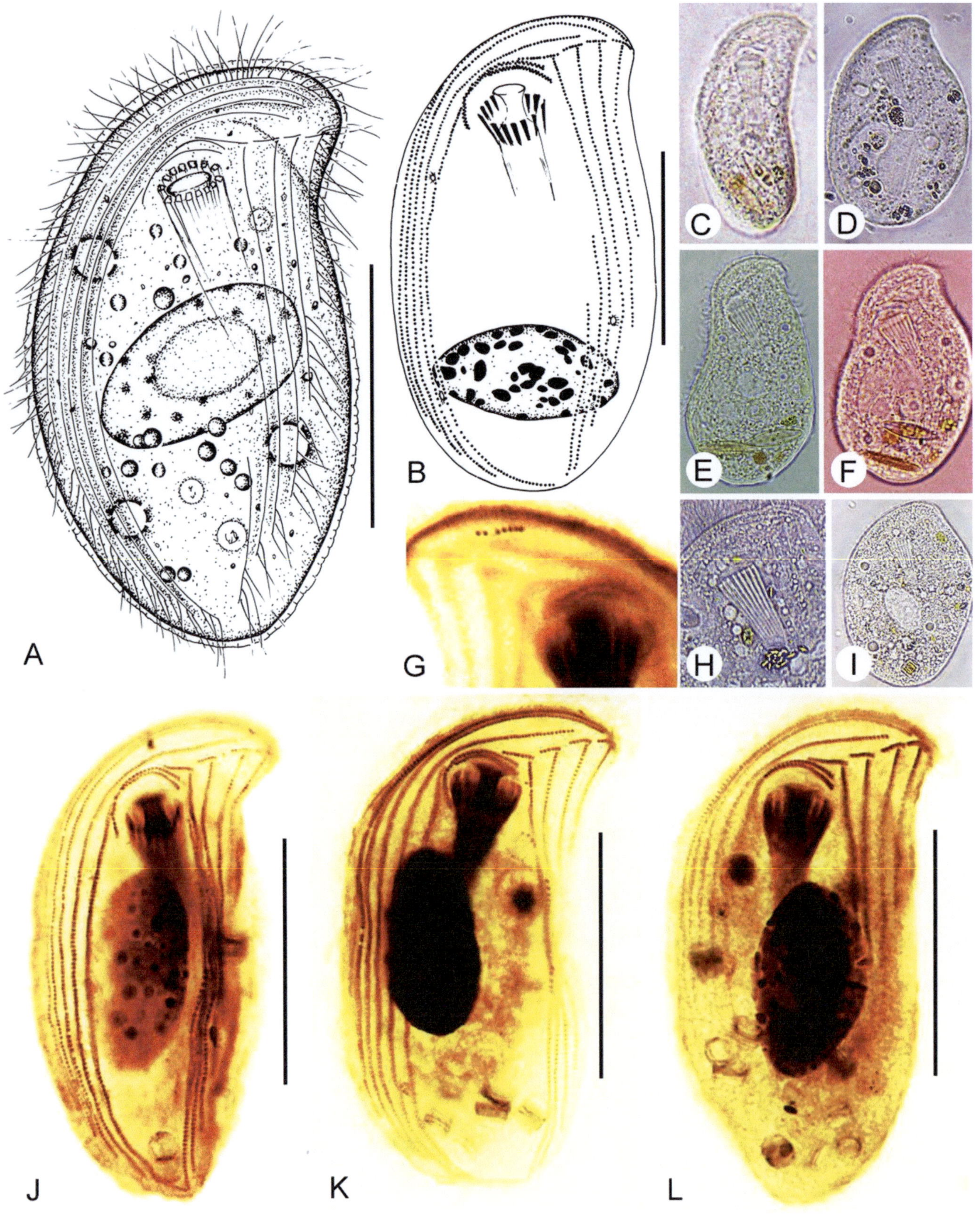


**Fig. 6B** *Pseudochilodonopsis* sp.
(**a**, **c–f**, **i**) Ventral views of live cells; (**b**, **j–l**) ventral views of ciliature; (**g**) dorsal view of anterior part of body, showing terminal fragment; (**h**) nematodesmal rods. Scale bars: 30 μm

## Reference

Pan H (2011) Diversity of marine cyrtophorid and pleurostomatid ciliates in China seas. Ocean University of China, Ph. D. thesis.

Class Phyllopharyngea de Puytorac et al., 1974
Order Chlamydodontida Deroux, 1976
Family Chilodonellidae Deroux, 1970
*Trithigmostoma cucullulus* (Müller, 1786) Jankowski, 1967

**Ecological feature:** estuary
**Water temperature:** 27.0 °C
**pH:** no data available
**Salinity:** 4.3‰

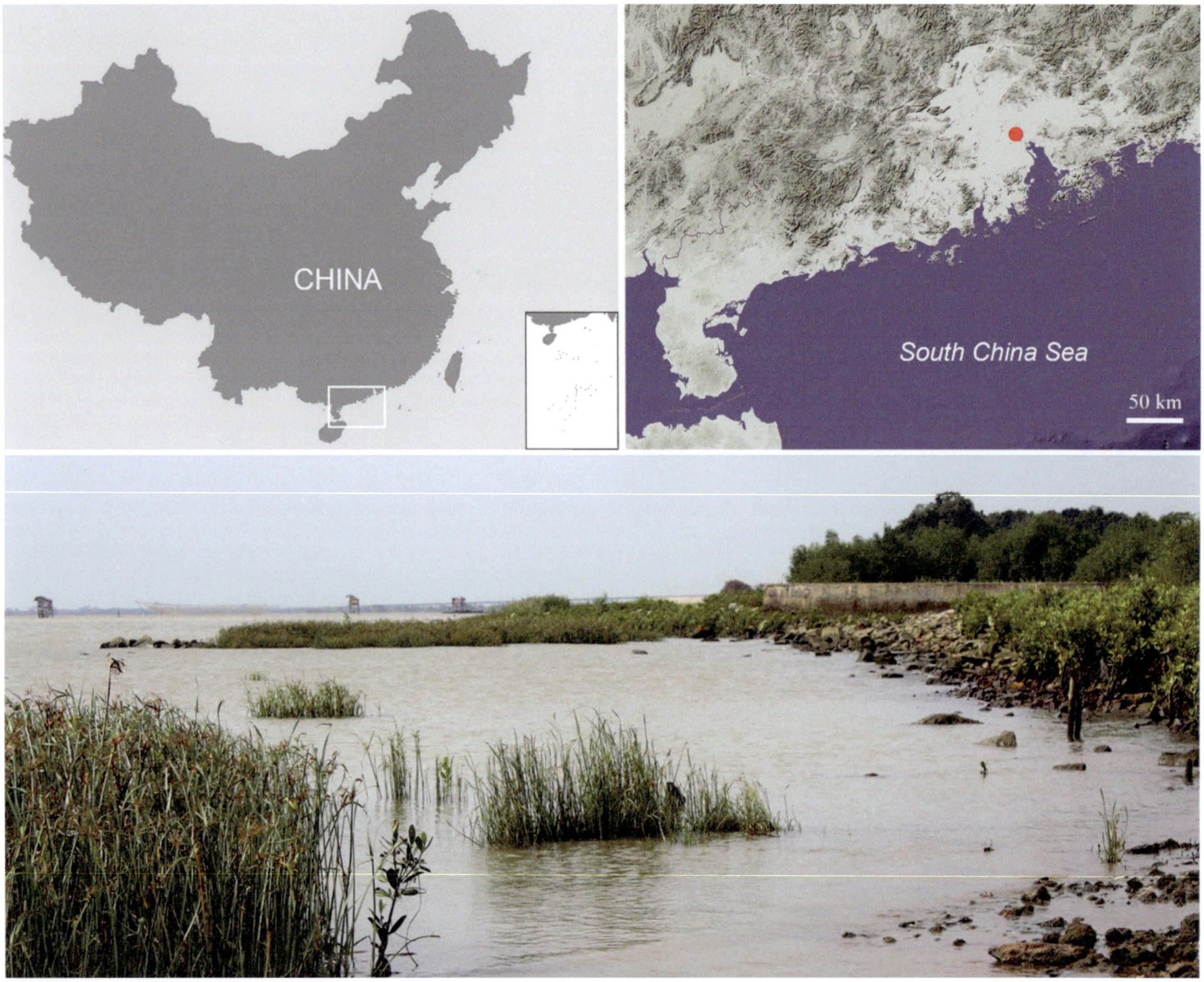


**Fig. 7A** Sample site

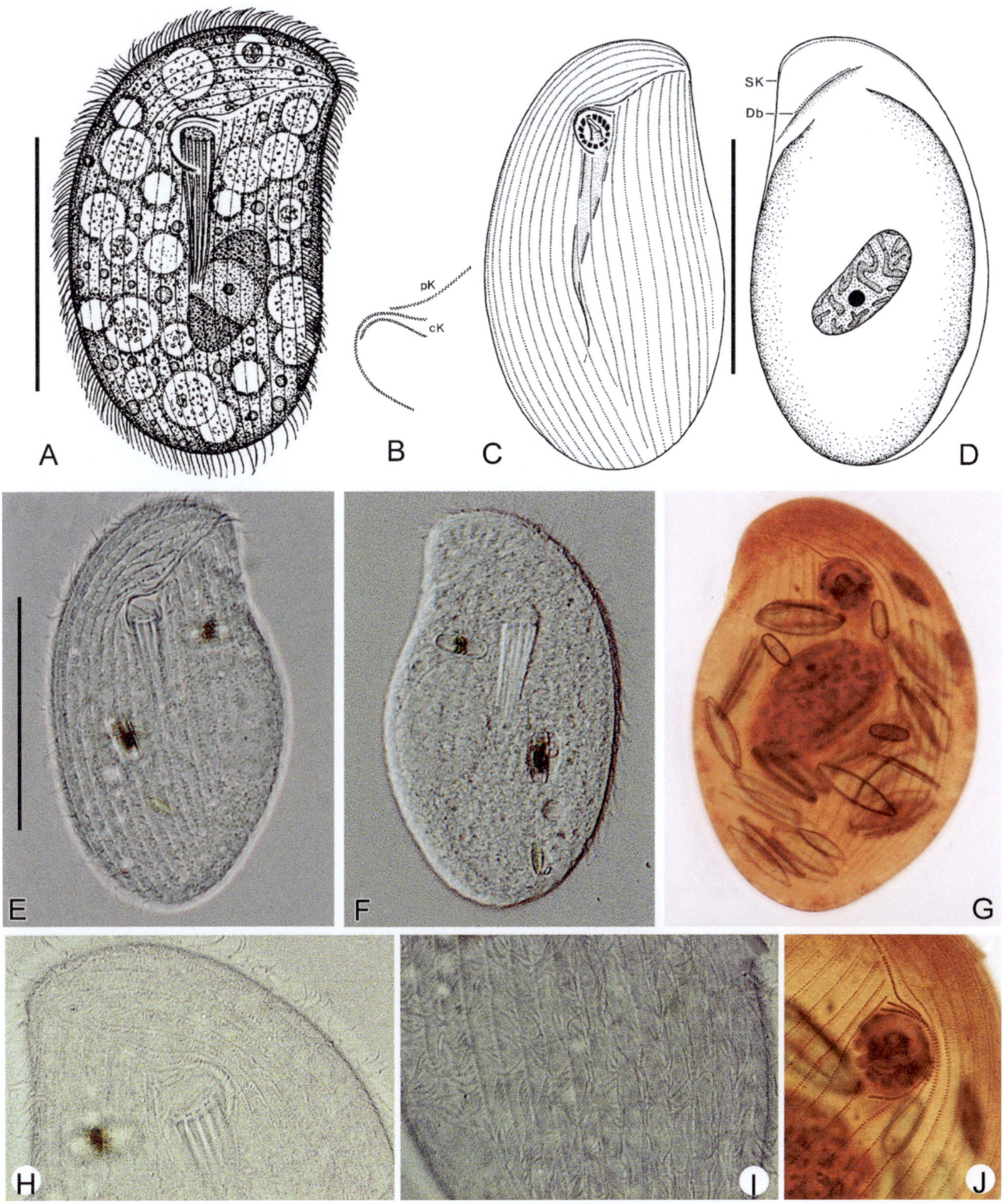


**Fig. 7B** *Trithigmostoma cucullulus*
(**a**, **e**) Ventral views of live cells; (**b**) oral ciliature; (c, **d**) infraciliature of ventral and dorsal sides; (**f**) dorsal view of live cell, showing nematodesmal rods and macronucleus; (**g**, **j**) ventral views of ciliature; (**h**) oral cilia and nematodesmal rods; (**i**) somatic cilia on ventral side. (**a–d**) from Foissner (1988). *cK* circumoral kinety, *Db* dorsal brush, *pK* perioral kinety, *SK* somatic kinety. Scale bars: 70 μm

## Reference

Foissner W (1988) Taxonomie und Ökologie einiger Ciliaten (Protozoa, Ciliophora) des Saprobiensystems. II. Familie Chilodonellidae. Hydrobiologia 162:21–45

Class Phyllopharyngea de Puytorac et al., 1974
Order Chlamydodontida Deroux, 1976
*Incertae sedis* in Family Chilodonellidae
*Odontochlamys alpestris biciliata* Foissner et al., 2002

**Ecological feature:** intertidal zone
**Water temperature:** 25.8 °C
**pH:** 8.9
**Salinity:** 12.2‰

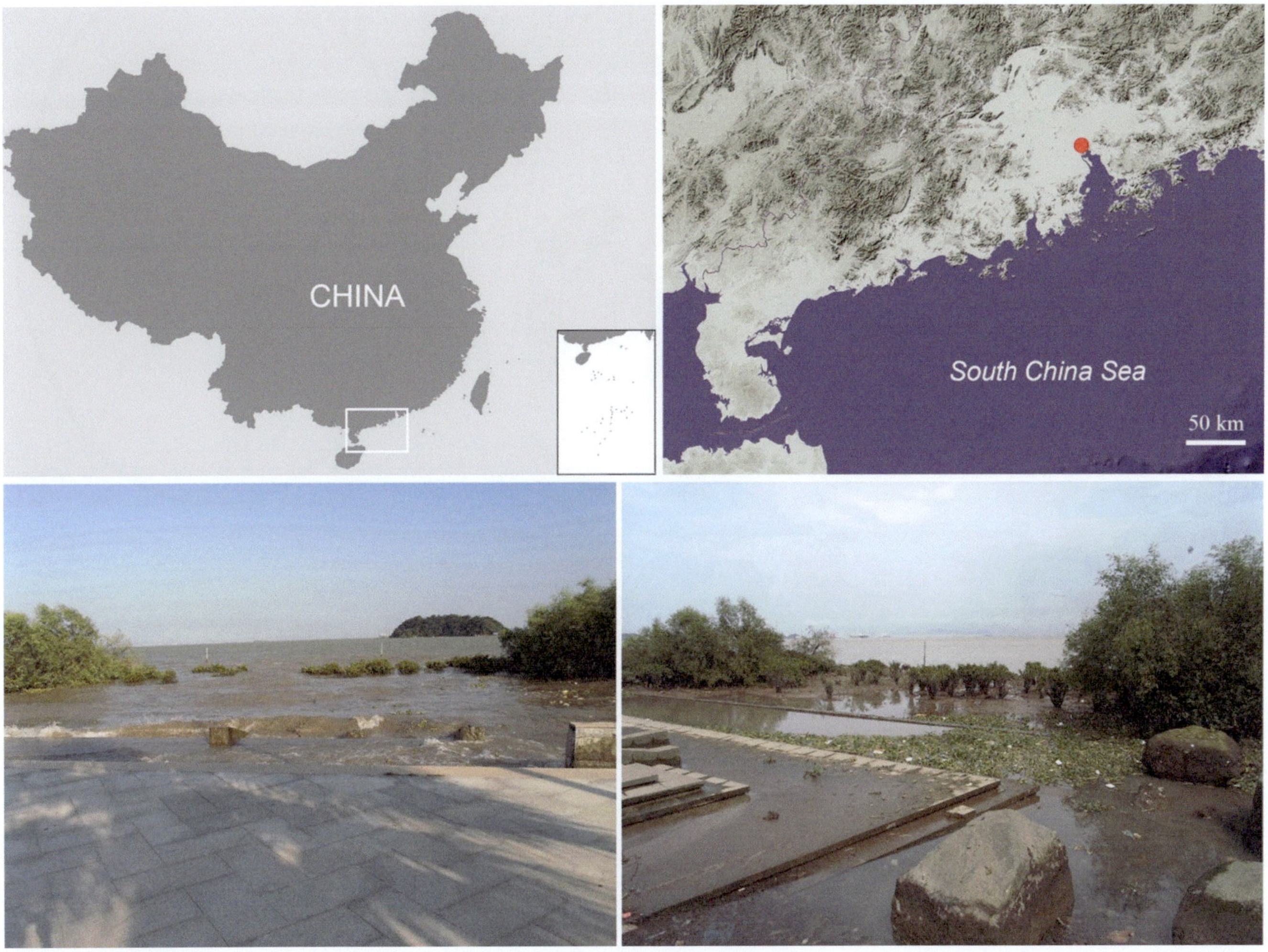


**Fig. 8A** Sample site

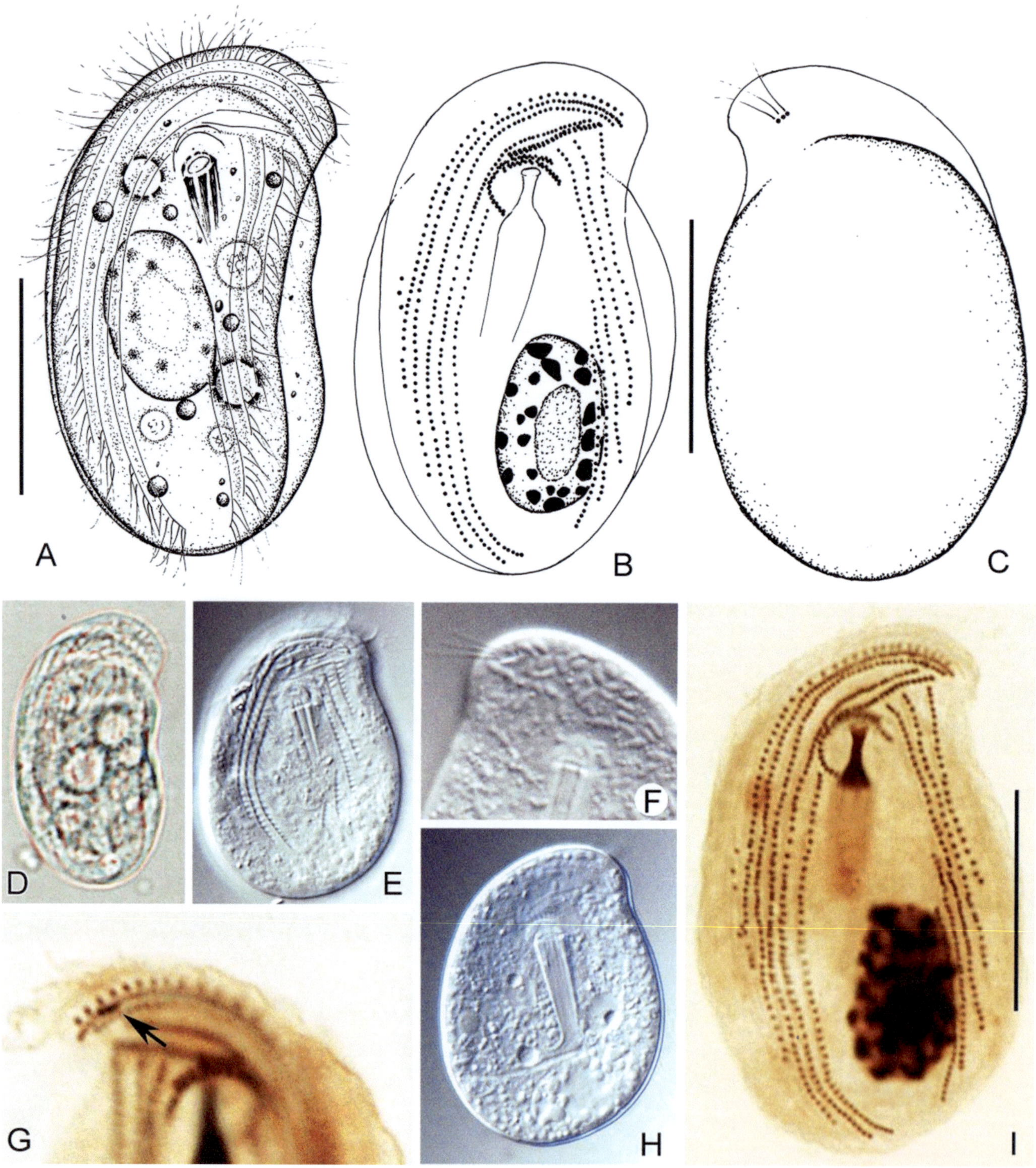


**Fig. 8B** *Odontochlamys alpestris biciliata*
(**a**, **d**, **e**, **h**) Ventral view of live individuals; (**b**, **c**, **i**) ventral (**b**, **i**) and dorsal (**c**) views of ciliature; (**f**, **g**) dorsal view of anterior portion of the body, arrow in **g** refers to terminal fragment. Scale bars: 20 μm

## Reference

Pan H, Jiang J, Fan X, Al-Farraj SA, Gao S (2017) Phylogeny and taxonomy of five poorly known species of cyrtophorian ciliates (Protozoa: Ciliophora: Phyllopharyngea) from China seas. Zool J Linn Soc 180:475–492

Class Phyllopharyngea de Puytorac et al., 1974
Order Chlamydodontida Deroux, 1976
Family Chlamydodontidae Stein, 1859
*Chlamydodon bourlandi* Qu et al., 2018

**Ecological feature:** mangrove
**Water temperature:** 24.5 °C
**pH:** 6.8
**Salinity:** 17.0‰

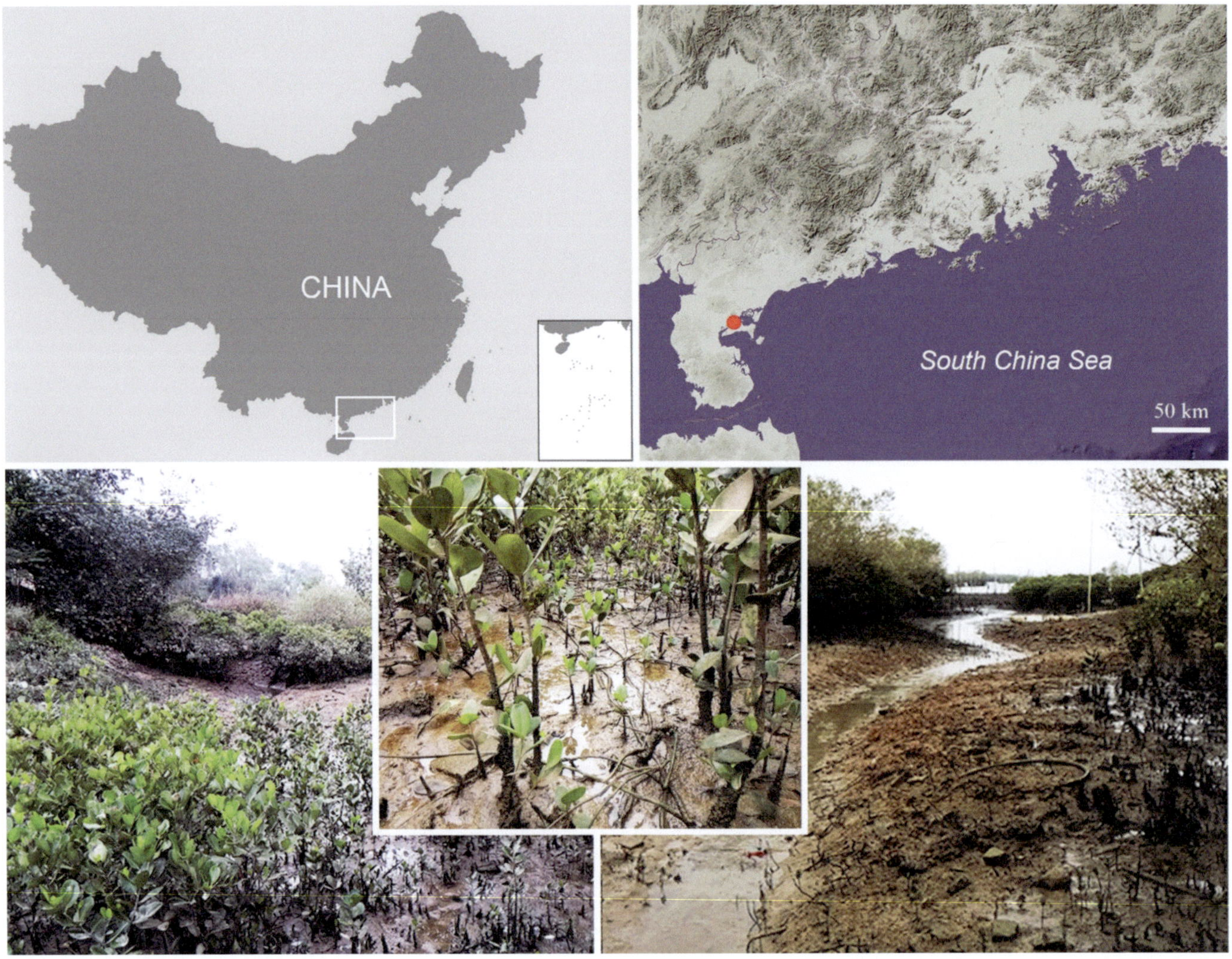


**Fig. 9A** Sample site

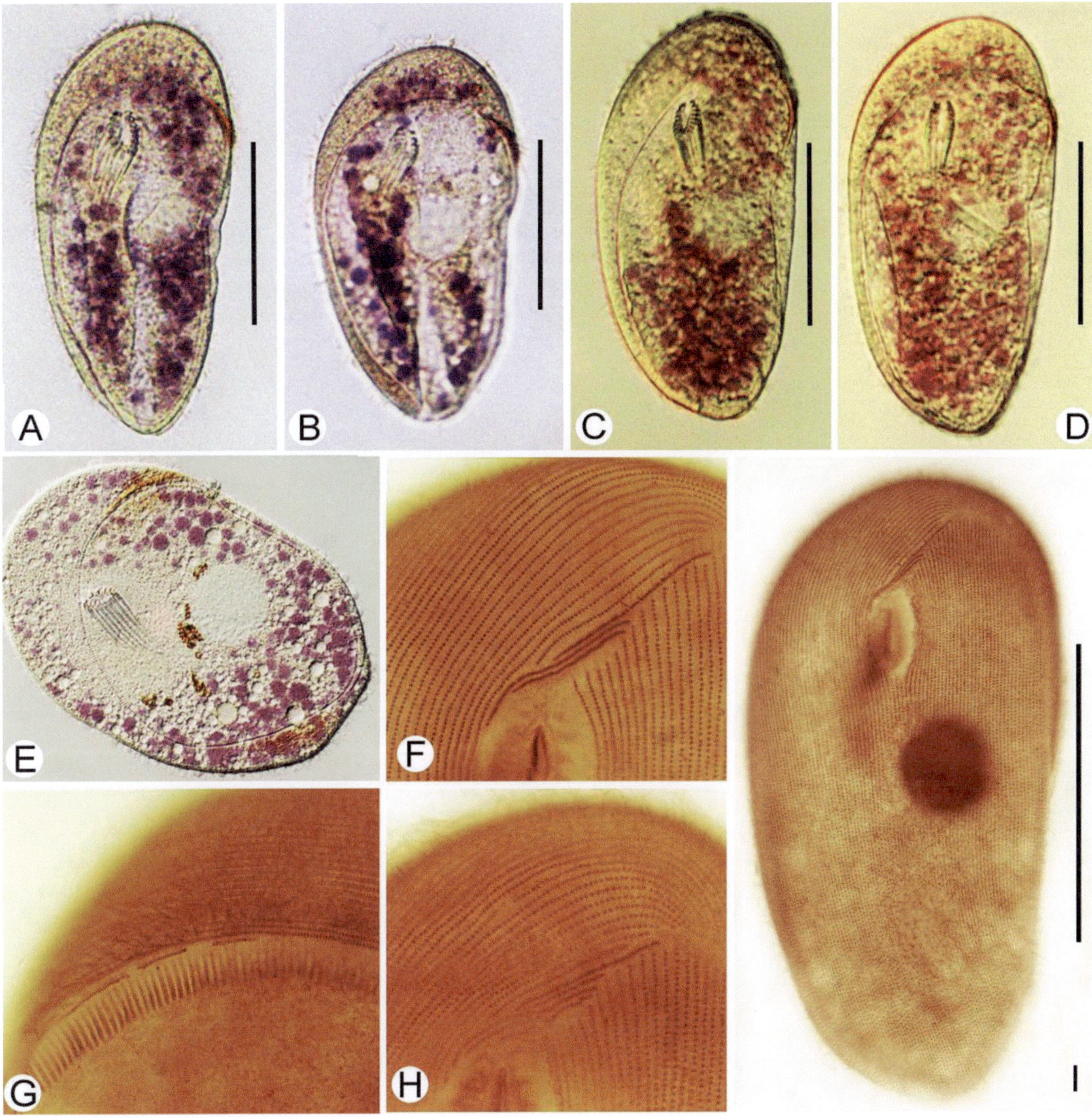


**Fig. 9B** *Chlamydodon bourlandi*
(**a–d**) Ventral views of live individuals; (**e**) ventral view of live cell, showing nematodesmal rods; (**f**, **h**) oral ciliature; (**g**) terminal fragments; (**i**) ventral view of ciliature. Scale bars: 80 μm

## Reference

Qu Z, Li L, Lin X, Stoeck T, Pan H, Al-Rasheid KAS, Song WB (2018) Diversity of the cyrtophorid genus *Chlamydodon* (Protista, Ciliophora): its systematics and geographic distribution, with taxonomic descriptions of three species, Syst Biodivers 16:497–511

Class Phyllopharyngea de Puytorac et al., 1974
Order Chlamydodontida Deroux, 1976
Family Chlamydodontidae Stein, 1859
*Chlamydodon caudatus* Pan et al., 2013

**Ecological feature:** intertidal zone
**Water temperature:** 25.8 °C
**pH:** 8.9
**Salinity:** 12.2‰

**Fig. 10A** Sample site

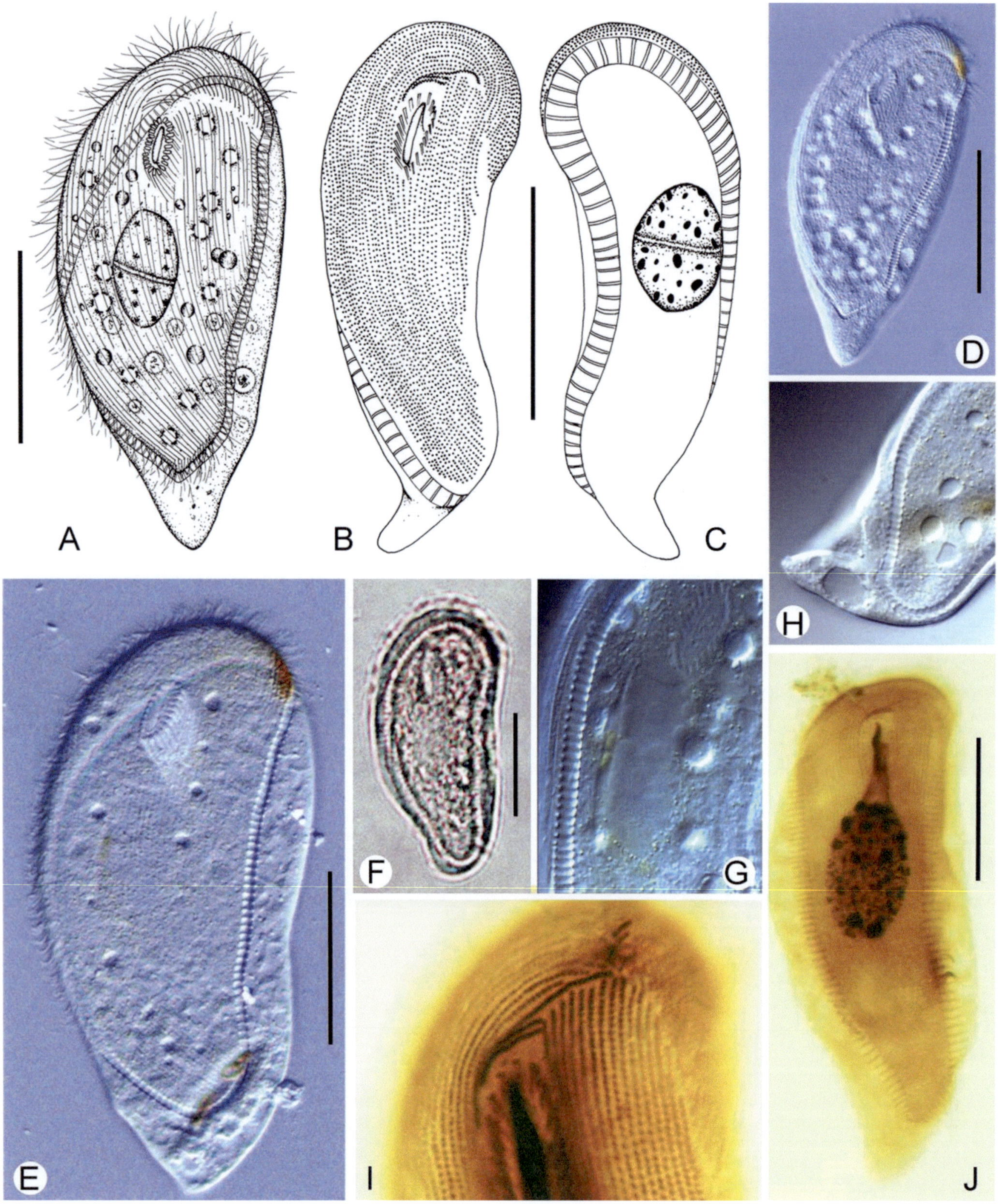


**Fig. 10B** *Chlamydodon caudatus*
(**a**, **d**–**f**) Ventral views of live individuals; (**b**, **c**, **j**) ventral (**b**, **j**) and dorsal (**c**) views of ciliature; (**g**) part of cross-striated band; (**h**) tapered posterior end of body; (**i**) oral ciliature. Scale bars: 60 μm

## Reference

Pan H, Li L, AL-Rasheid KAS, Song WB (2013) Morphological and molecular description of three new species of the cyrtophorid genus *Chlamydodon* (Ciliophora, Cyrtophoria). J Eukaryot Microbiol 60:2–12

Class Phyllopharyngea de Puytorac et al., 1974
Order Chlamydodontida Deroux, 1976
Family Chlamydodontidae Stein, 1859
*Chlamydodon mnemosyne* Ehrenberg, 1835

**Ecological feature:** offshore water
**Water temperature:** no data available
**pH:** 8.5
**Salinity:** 6.3‰

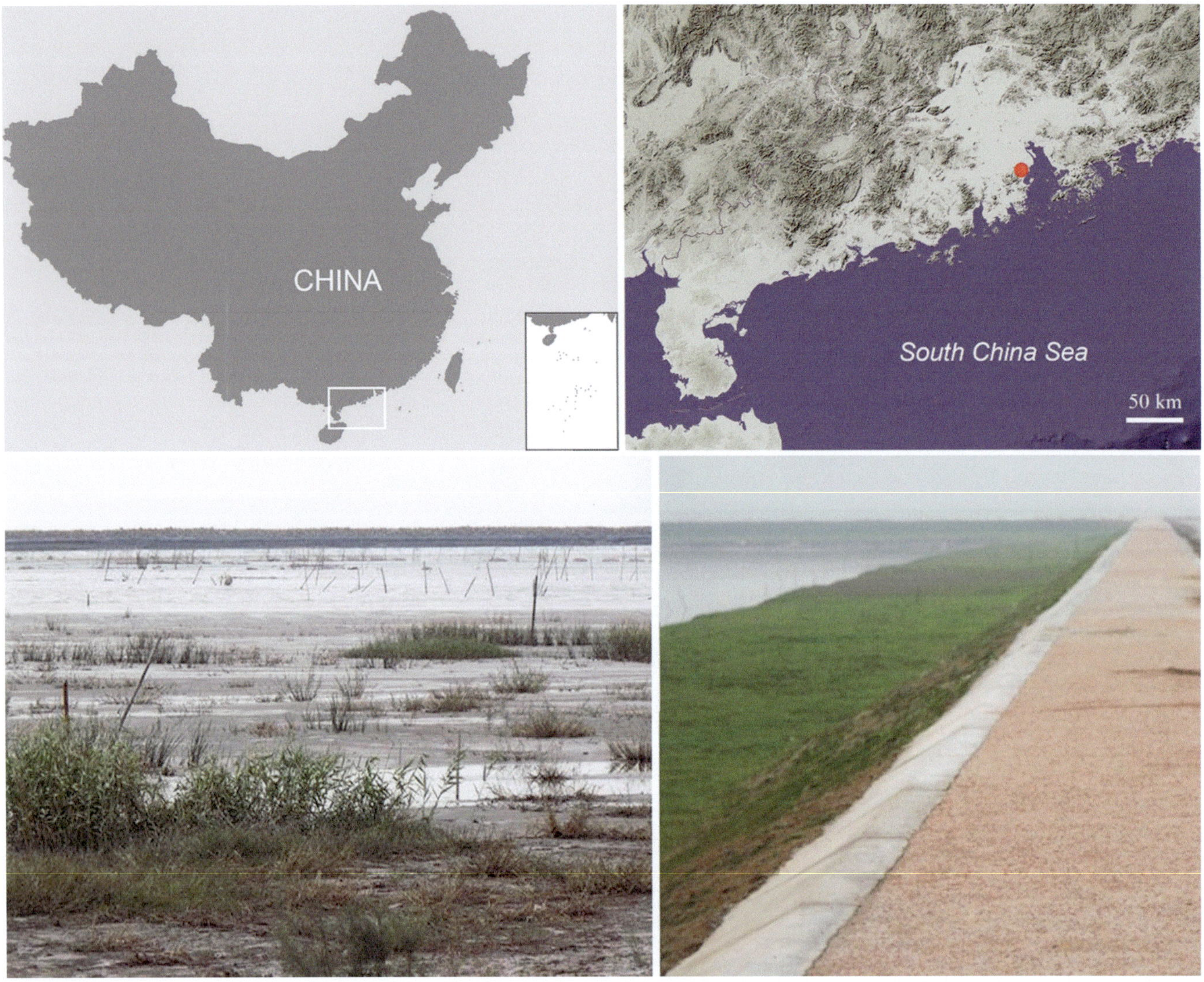


**Fig. 11A** Sample site

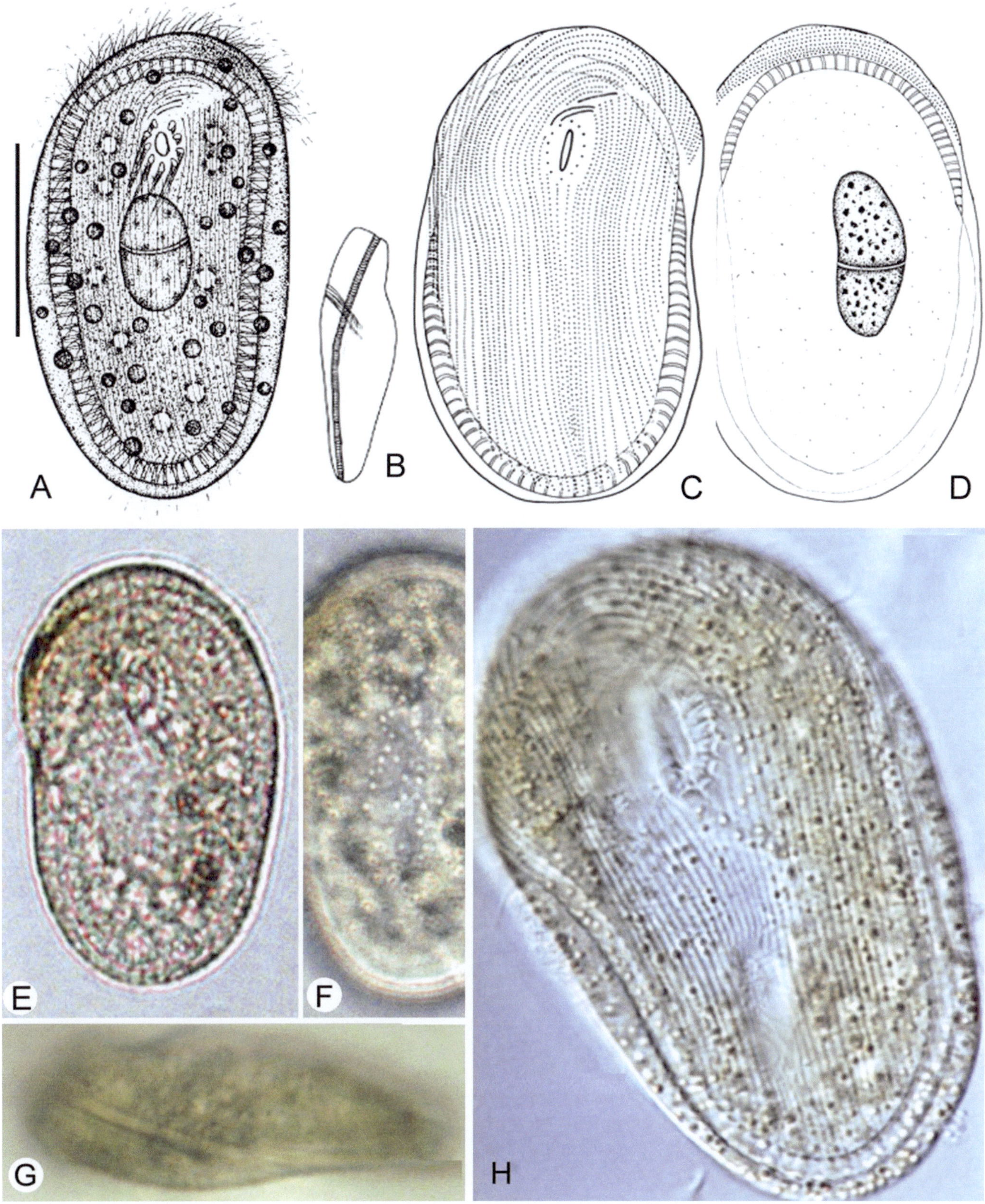


**Fig. 11B** *Chlamydodon mnemosyne*
(**a**) Ventral view of a live cell; (**b**, **g**) lateral view showing the cross-striated band; (**c**, **d**) ventral (**c**) and dorsal (**d**) views of ciliature; (**e**) a live individual showing cytostome; (**f**) dorsal view indicating the brownish cortical granules; (**h**) ventral view, to show the cross-striated band. Scale bars: 30 μm

## Reference

Gong J, Song WB, Warren A (2005) Updating the ciliate genus *Chlamydodon* Ehrenberg, 1835, with redescriptions of three species (Ciliophora, Cyrtophorida). Acta Protozool 44:19–32

Class Phyllopharyngea de Puytorac et al., 1974
Order Chlamydodontida Deroux, 1976
Family Chlamydodontidae Stein, 1859
*Chlamydodon obliquus* Kahl, 1931

**Ecological feature:** offshore water
**Water temperature:** no data available; 21.0 °C
**pH:** 8.0; 7.5
**Salinity:** 29.0‰; 30.0‰

**Fig. 12A** Sample site

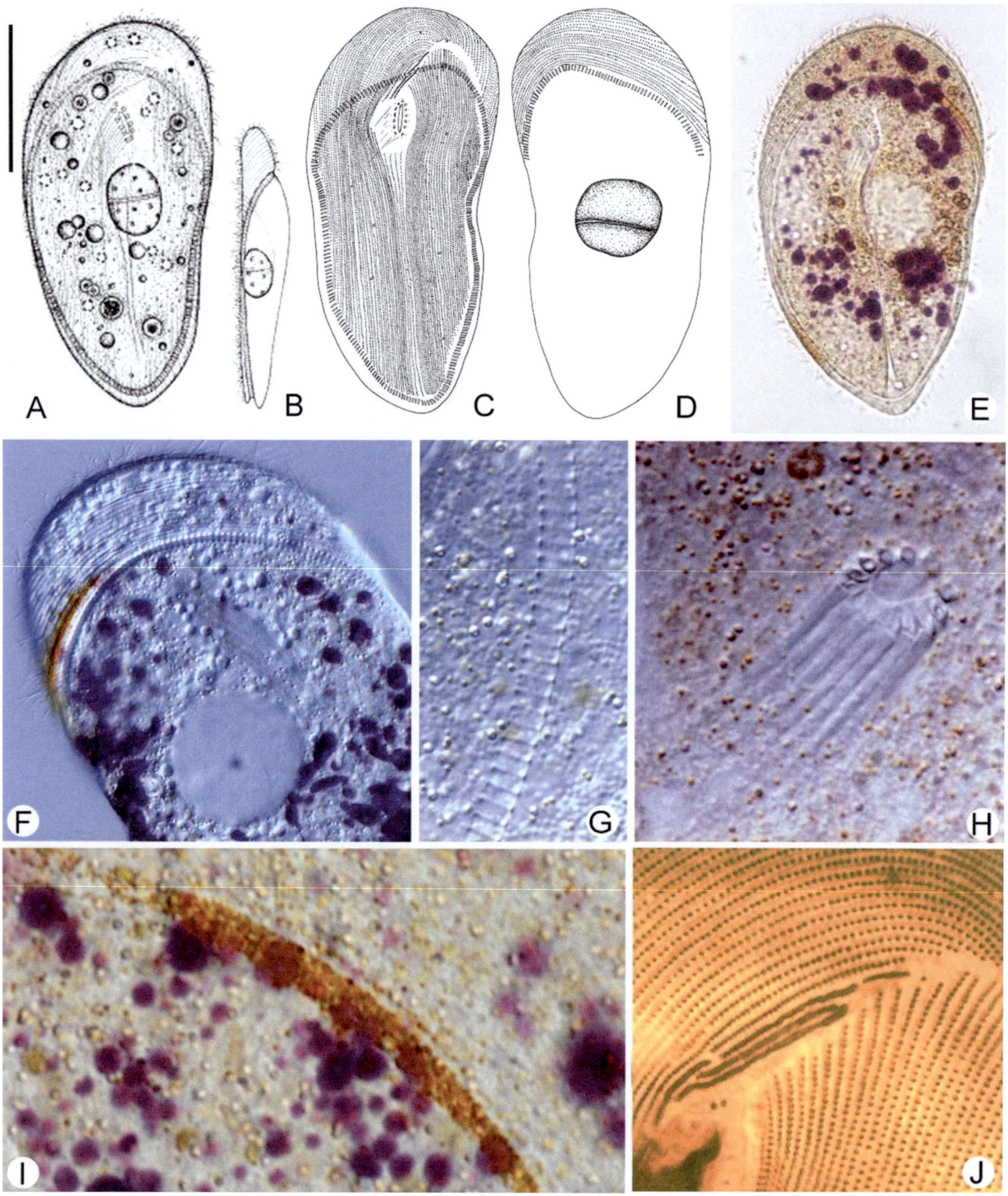


**Fig. 12B** *Chlamydodon obliquus*
(**a**, **e**) Ventral views of a live individual; (**b**) lateral view of a live cell, showing the cross-striated band; (**c**, **d**) ciliature of ventral (**c**) and dorsal (**d**) sides; (**f**, **g**) to show the cross-striated band; (**h**) nematodesmal rods; (**i**) the yellow colored granules in anterior portion of cell; (**j**) buccal apparatus. Scale bar: 50 μm

## Reference

Gong J, Song WB, Warren A (2005) Updating the ciliate genus *Chlamydodon* Ehrenberg, 1835, with redescriptions of three species (Ciliophora, Cyrtophorida). Acta Protozool 44:19–32

Class Phyllopharyngea de Puytorac et al., 1974
Order Chlamydodontida Deroux, 1976
Family Chlamydodontidae Stein, 1859
*Chlamydodon paramnemosyne* Pan et al., 2013

**Ecological feature:** shrimp-farming pond
**Water temperature:** 23.8 °C
**pH:** 7.4
**Salinity:** 24.8‰

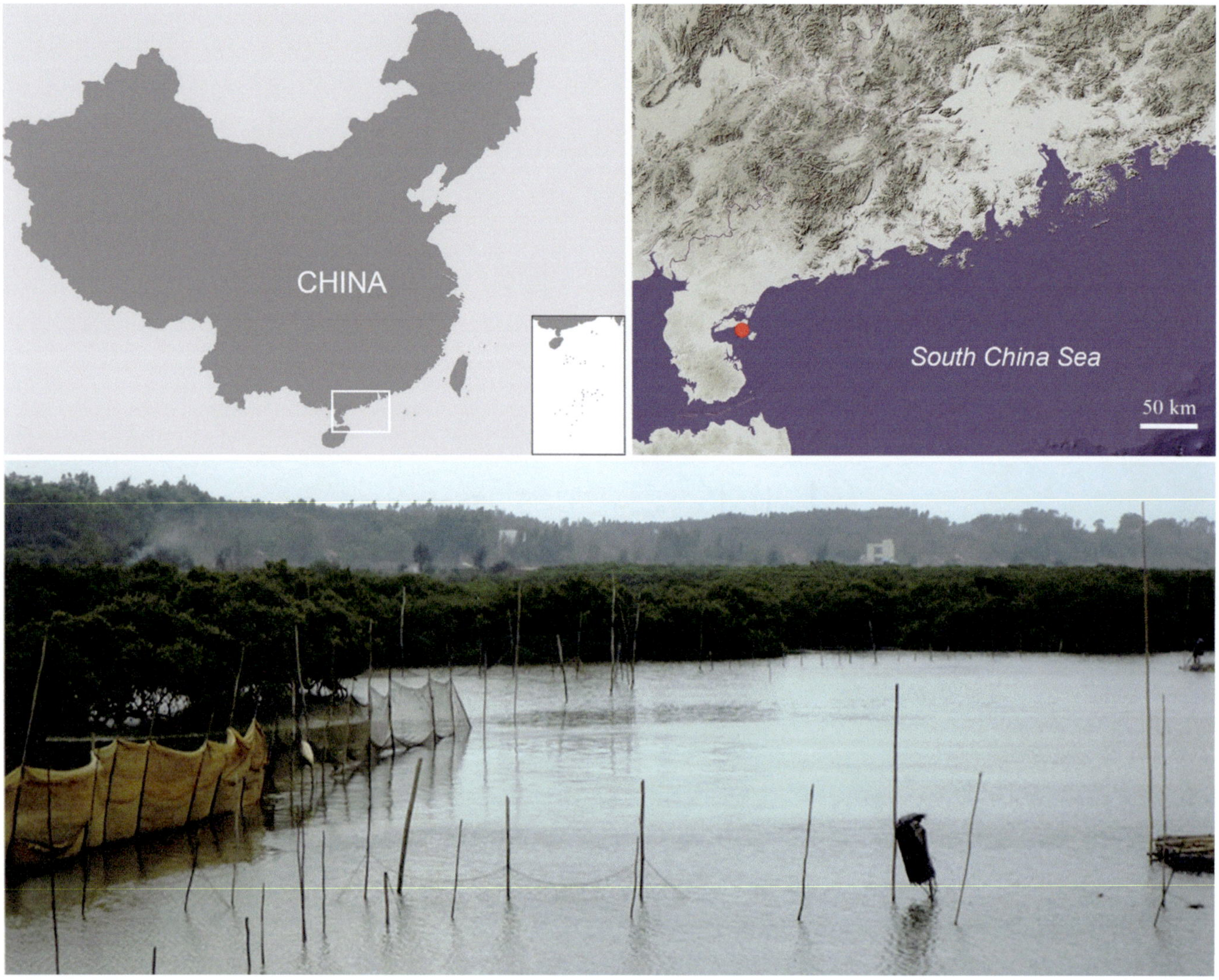


**Fig. 13A** Sample site

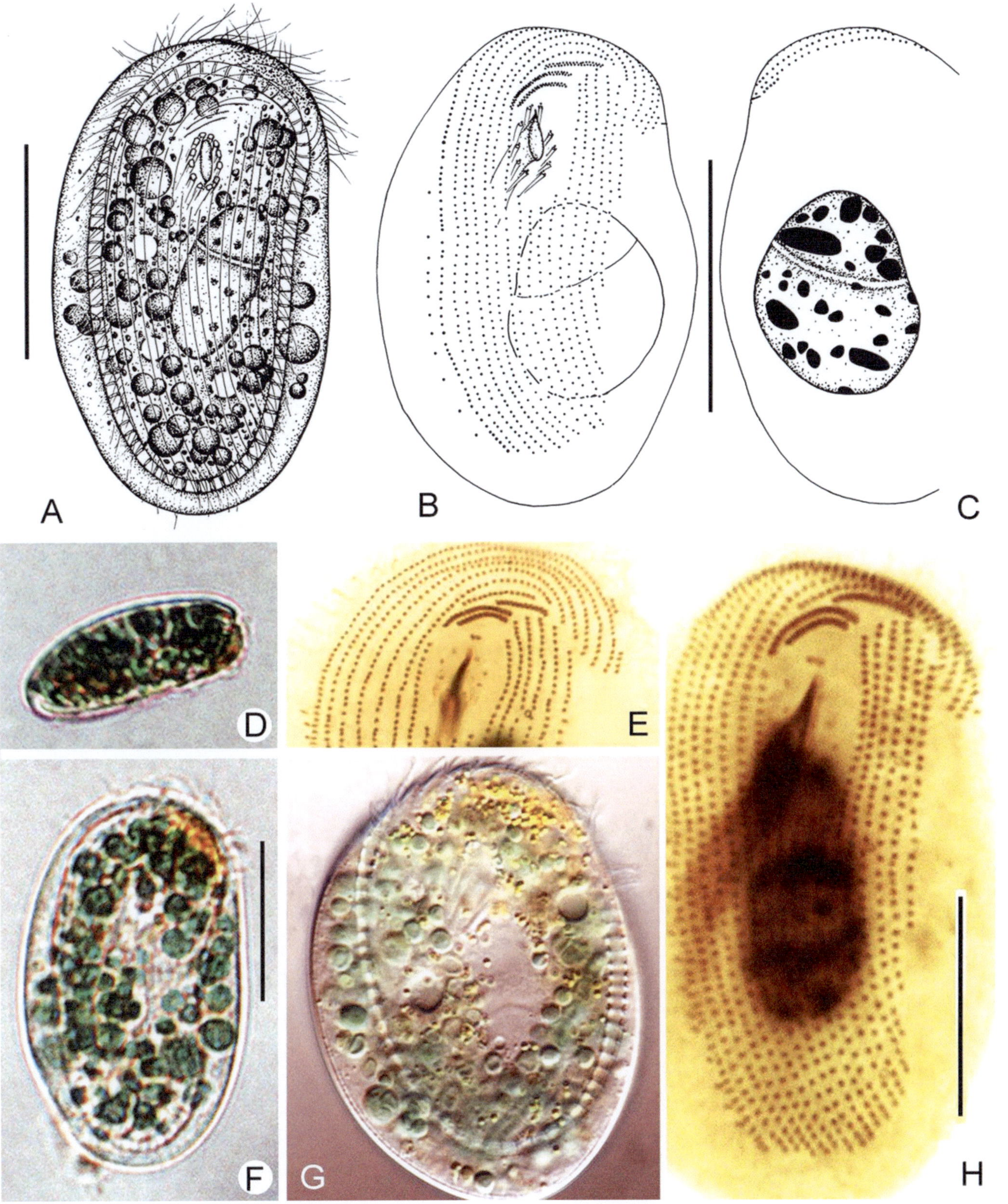


**Fig. 13B** *Chlamydodon paramnemosyne*
(**a**, **f**, **g**) Ventral views of live cells; (**b**, **c**, **h**) ventral (**b**, **h**) and dorsal (**c**) views of ciliature; (**d**) lateral view of live cell; (**e**) oral ciliature. Scale bars: 30 μm

## Reference

Pan H, Li L, AL-Rasheid KAS, Song WB (2013) Morphological and molecular description of three new species of the cyrtophorid genus *Chlamydodon* (Ciliophora, Cyrtophoria). J Eukaryot Microbiol 60:2–12

Class Phyllopharyngea de Puytorac et al., 1974
Order Chlamydodontida Deroux, 1976
Family Chlamydodontidae Stein, 1859
*Chlamydodon similis* Qu et al., 2018

**Ecological feature:** brackish water
**Water temperature:** no data available
**pH:** no data available
**Salinity:** 10.0‰

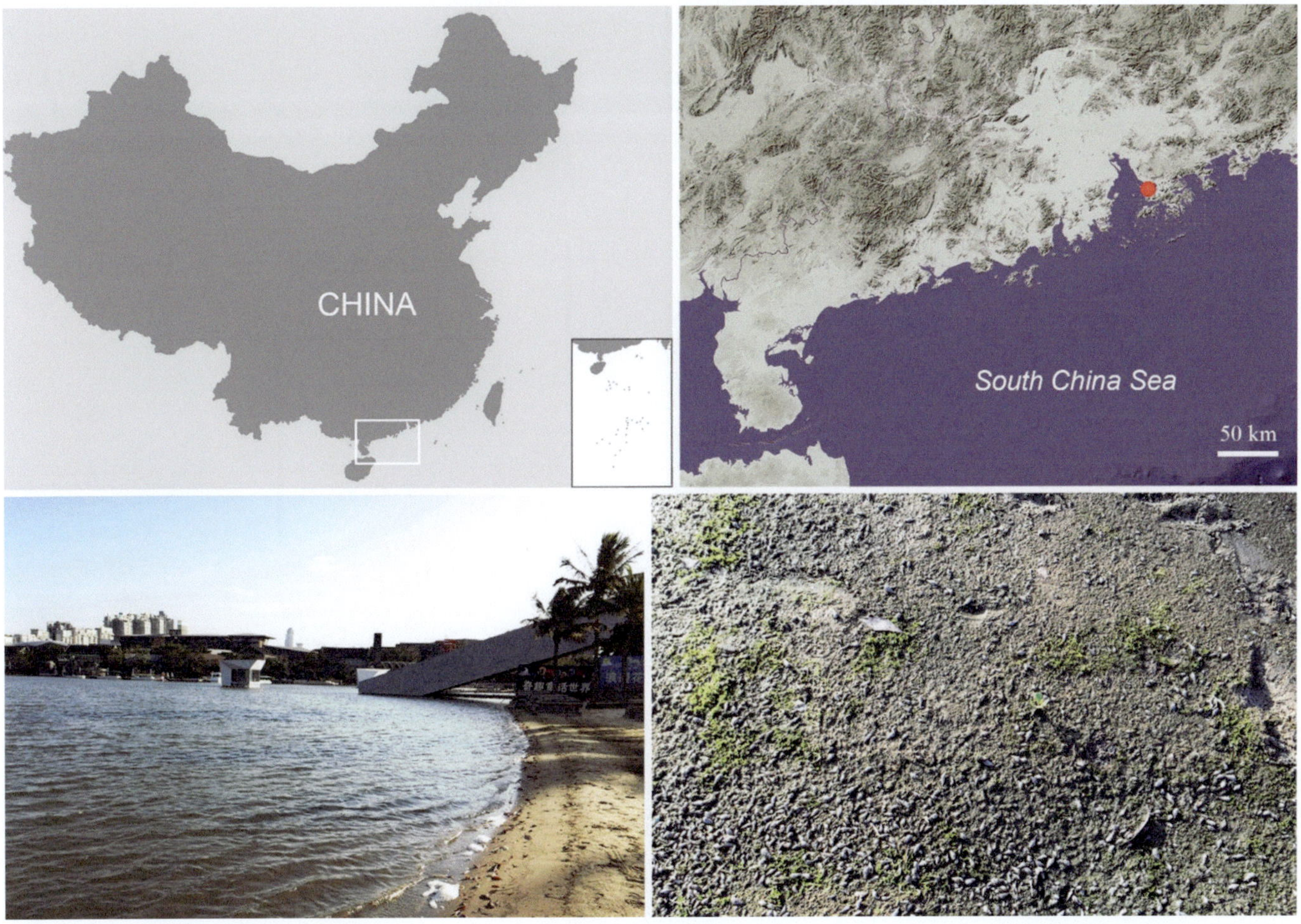


**Fig. 14A** Sample site

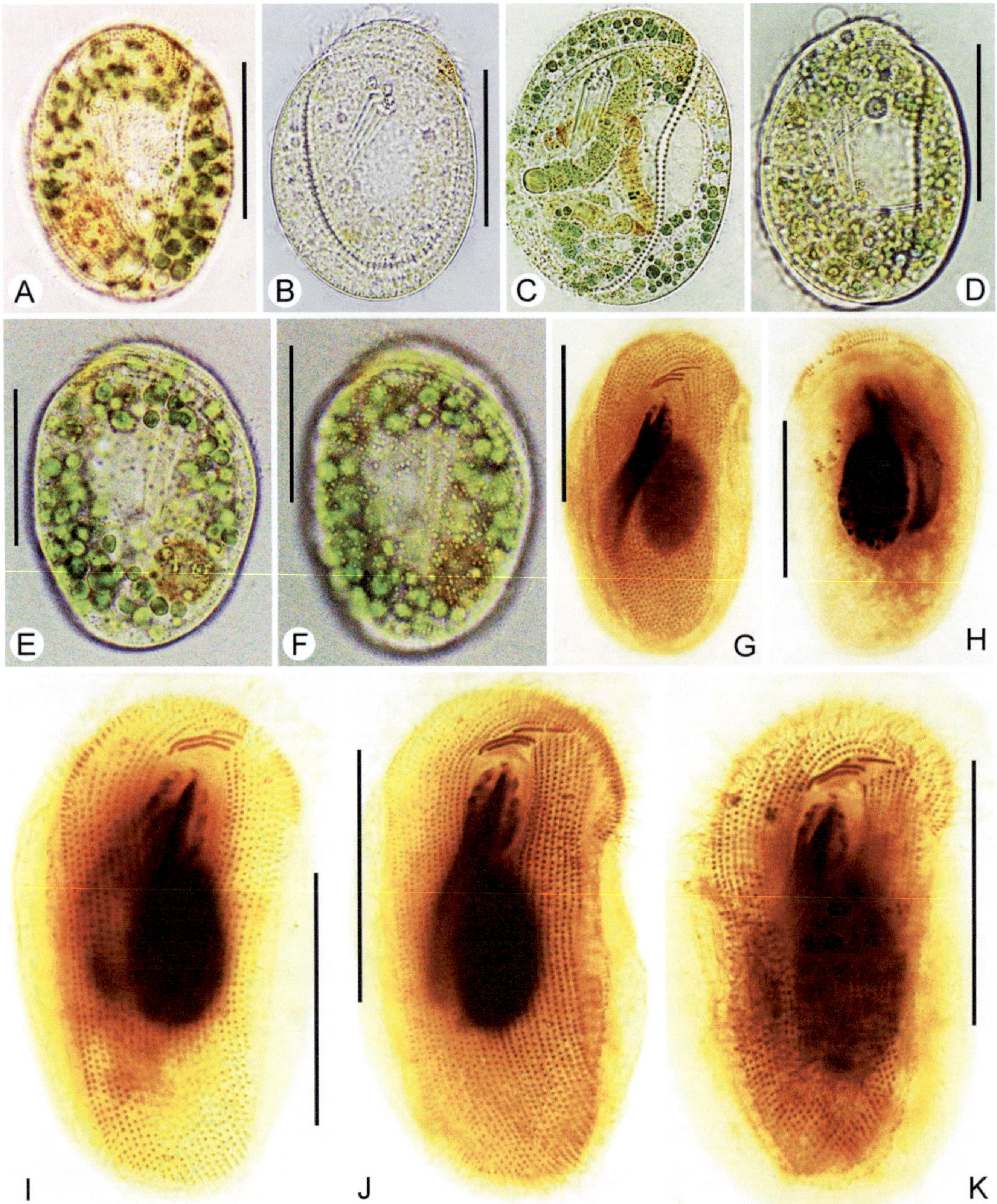


**Fig. 14B** *Chlamydodon similis*
(**a–d**) Ventral views of compressed individuals; (**e**) dorsal view of live cell showing algae in cytoplasm; (**f**) dorsal views of live cell showing the regular-arranged cortical granules; (**g–k**) ciliature on ventral (**g**, **i–k**) and dorsal (**h**) sides. Scale bars: 40 μm

## Reference

Qu Z, Pan H, Lin X, Li L, Aleidan AMA, Al-Farraj SA, Stoeck T, Hu X (2018) A contribution to the morphology and phylogeny of *Chlamydodon*, with three new species from China (Ciliophora, Cyrtophoria). J Eukaryot Microbiol 65:236–249

Class Phyllopharyngea de Puytorac et al., 1974
Order Chlamydodontida Deroux, 1976
Family Chlamydodontidae Stein, 1859
*Chlamydodon wilberti* Qu et al., 2018

**Ecological feature:** mangrove
**Water temperature:** 23.0–34.7 °C
**pH:** 7.2–7.3
**Salinity:** 16.6‰–17.3‰

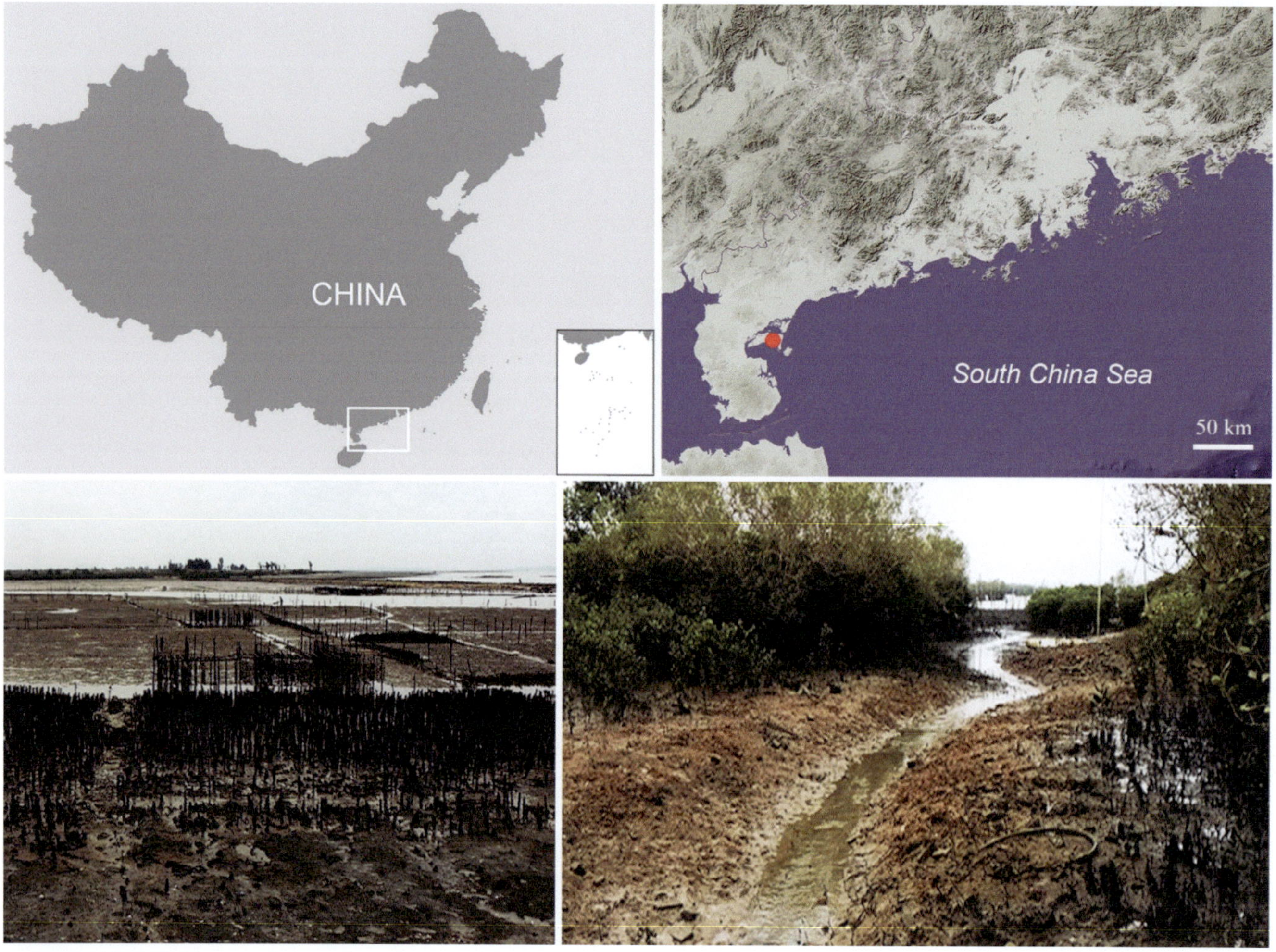


**Fig. 15A** Sample site

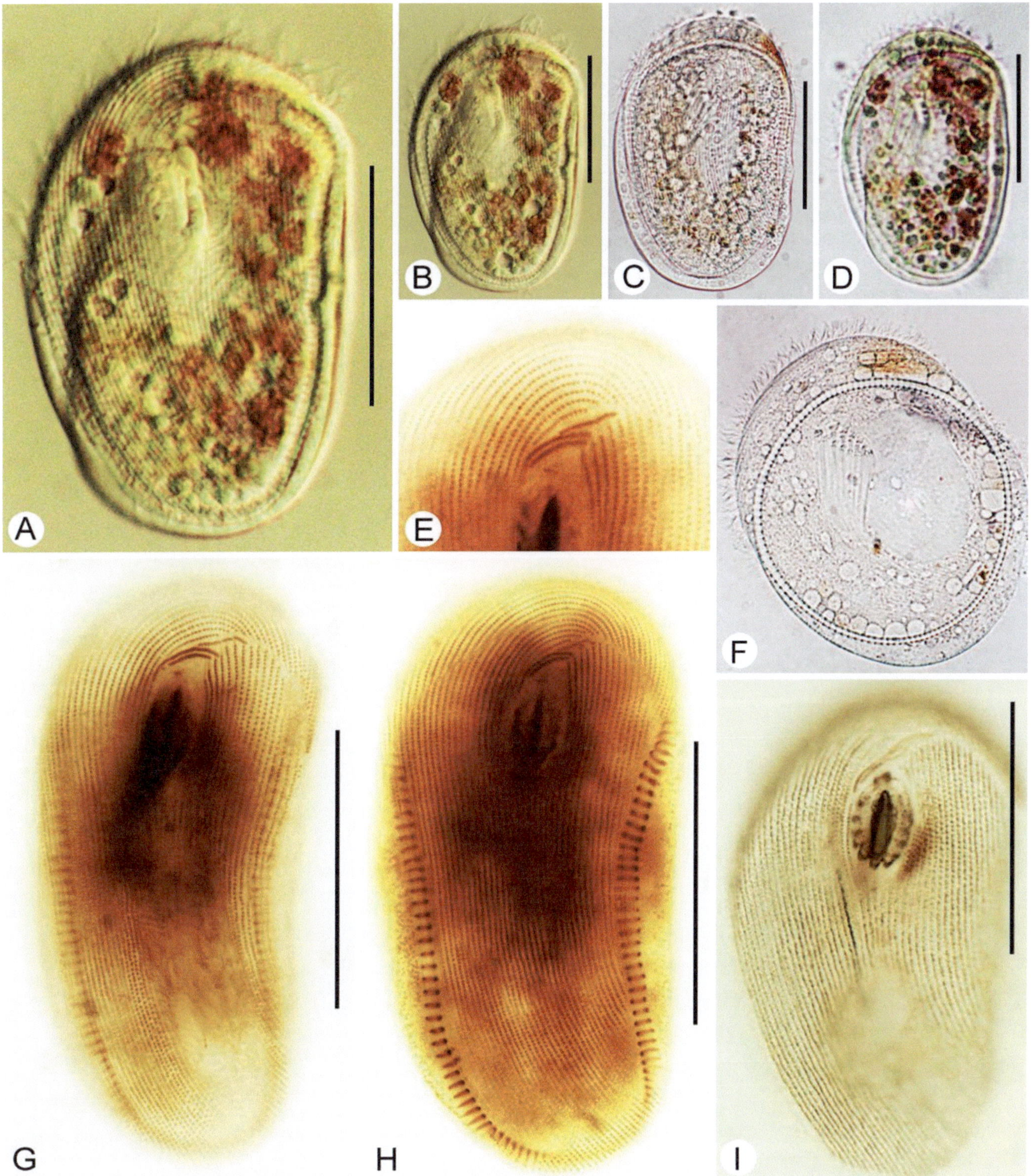


**Fig. 15B** *Chlamydodon wilberti*
(**a–d**) Ventral views of individuals in different body shapes; (e) anterior part of the ventral side, showing oral ciliature; (**f**) a compressed individual showing the complete cross-striated band; (**g–i**) ventral views of ciliature. Scale bars: 80 μm

## Reference

Qu Z, Li L, Lin X, Stoeck T, Pan H, Al-Rasheid KAS, Song WB (2018) Diversity of the cyrtophorid genus *Chlamydodon* (Protista, Ciliophora): its systematics and geographic distribution, with taxonomic descriptions of three species. Syst Biodivers 16:497–511

Class Phyllopharyngea de Puytorac et al., 1974
Order Chlamydodontida Deroux, 1976
Family Chlamydodontidae Stein, 1859
*Paracyrtophoron tropicum* Chen et al., 2012

**Ecological feature:** mangrove
**Water temperature:** 25.0 °C
**pH:** 8.2
**Salinity:** 16.0‰

**Fig. 16A** Sample site

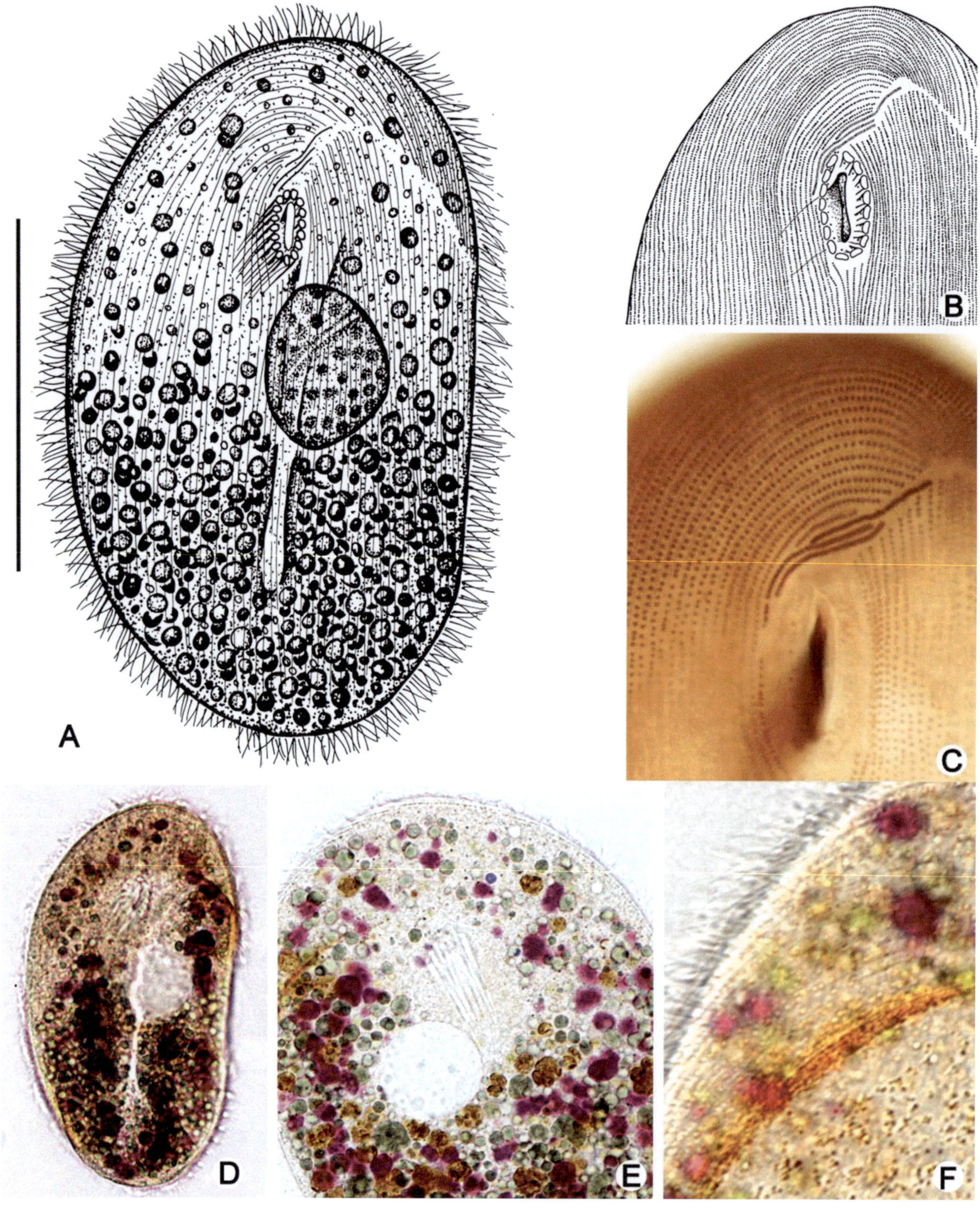


**Fig. 16B** *Paracyrtophoron tropicum*
(**a**, **d**) Ventral views of live individuals; (**b**, **c**) oral ciliature; (**e**) dorsal view of live cell showing macronucleus and nematodesmal rods; (**f**) dorsal view of live cell showing the yellow belt, combined by mass of small cortical granules. Scale bar: 100 μm

## Reference

Chen XR, Hu X, Gong J, Al-Rasheid KAS, Al-Farraj SA, Song WB (2012) Morphology and infraciliature of two new marine cyrtophorid ciliates, *Paracyrtophoron tropicum* nov. gen., nov. spec. and *Aegyria rostellum* nov. spec. (Ciliophora, Cyrtophorida), isolated from tropical waters in southern China. Eur J Protistol 48:63–72

Class Phyllopharyngea de Puytorac et al., 1974
Order Chlamydodontida Deroux, 1976
Family Lynchellidae Jankowski, 1968
*Atopochilodon distichum* Deroux, 1976

**Ecological feature:** intertidal zone
**Water temperature:** 25.0 °C
**pH:** no data available
**Salinity:** 16.0‰

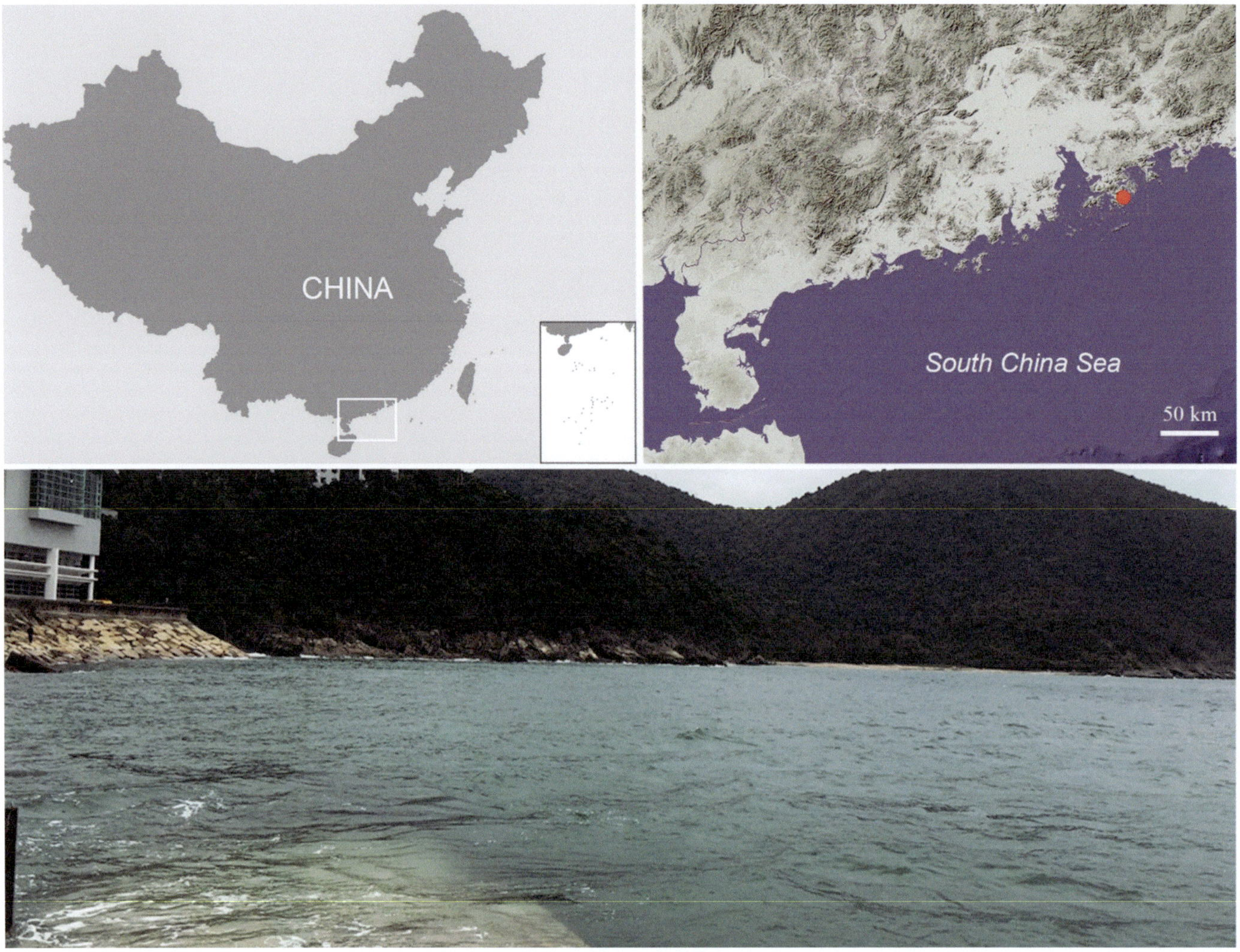


**Fig. 17A** Sample site

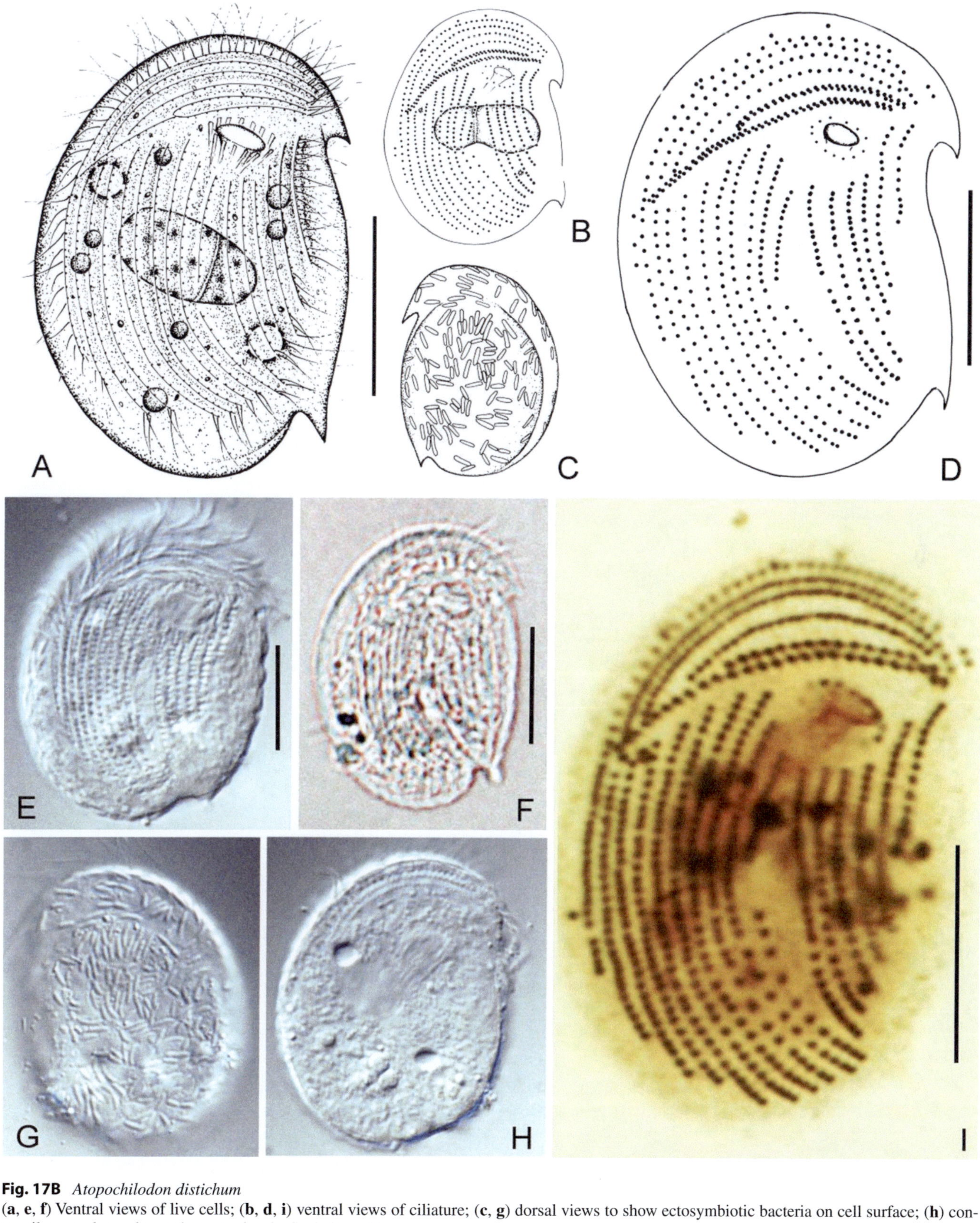


**Fig. 17B** *Atopochilodon distichum*
(**a**, **e**, **f**) Ventral views of live cells; (**b**, **d**, **i**) ventral views of ciliature; (**c**, **g**) dorsal views to show ectosymbiotic bacteria on cell surface; (**h**) contractile vacuoles and cytopharyngeal rods. Scale bars: 20 μm

## Reference

Pan H, Jiang J, Fan X, Al-Farraj SA, Gao S (2017) Phylogeny and taxonomy of five poorly known species of cyrtophorian ciliates (Protozoa, Ciliophora, Phyllopharyngea) from China Seas. Zool J Linn Soc 180:475–492

Class Phyllopharyngea de Puytorac et al., 1974
Order Chlamydodontida Deroux, 1976
Family Lynchellidae Jankowski, 1968
*Chlamydonella derouxi* Song, 2003

**Ecological feature:** mangrove
**Water temperature:** 23.9 °C; no data available
**pH:** 7.1; 7.1
**Salinity:** 15.0‰; 15.0‰

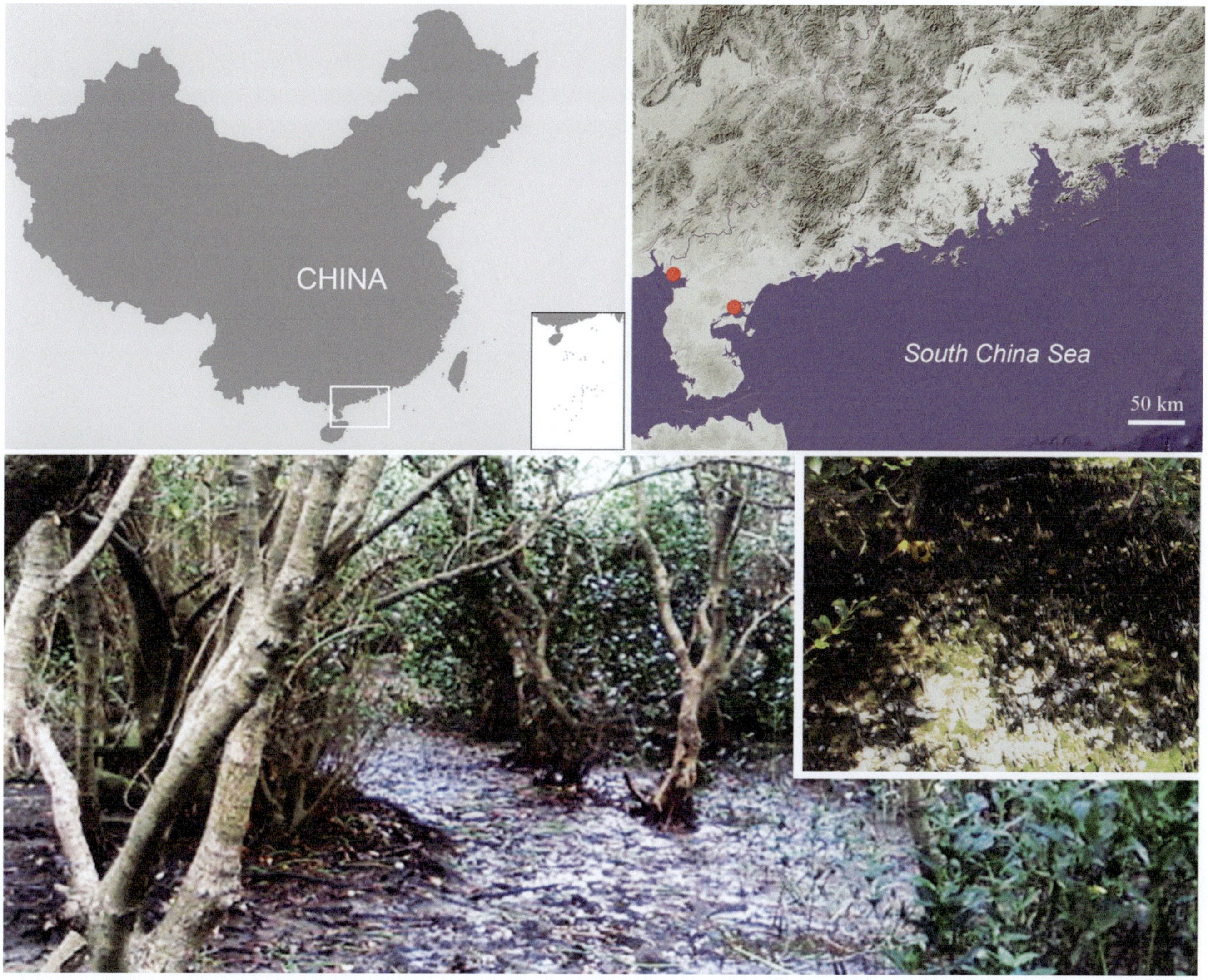


**Fig. 18A** Sample site

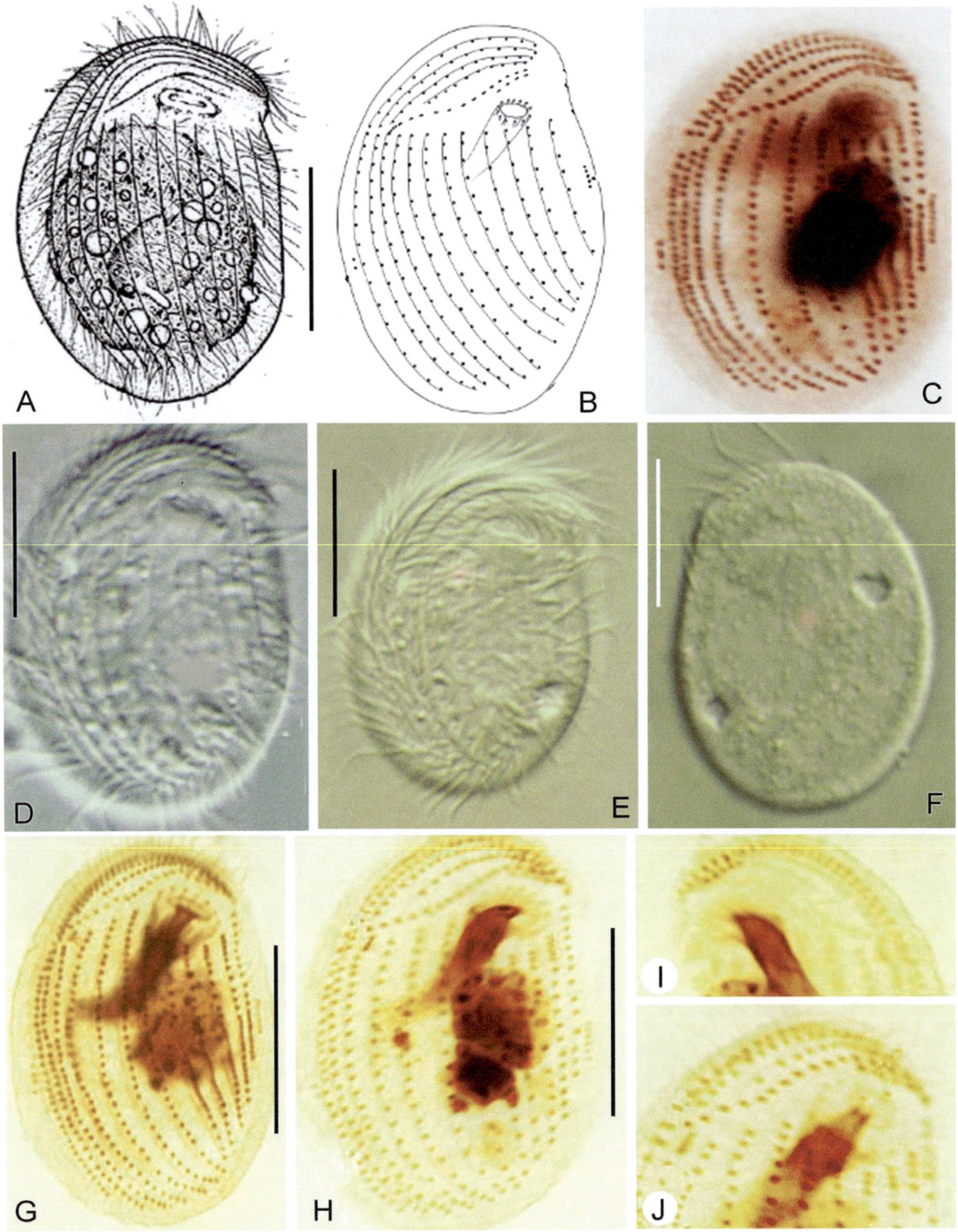


**Fig. 18B** *Chlamydonella derouxi*
(**a**, **d**, **e**) Ventral views of live cells; (**b**, **c**, **g**, **h**) ventral views of ciliature; (**f**, **i**) terminal fragment; (**j**) oral kineties. Scale bars: 15 μm

## Reference

Song WB (2003) Two marine cyrtophorid ciliates from China, *Chlamydonella derouxi* nov. spec. and *Orthotrochilia pilula* (Deroux, 1976) nov. comb., with reestablishment of the genus *Orthotrochilia* nov. gen. (Protozoa, Ciliophora, Cyrtophorida). Hydrobiologia 499:169–177

Class Phyllopharyngea de Puytorac et al., 1974
Order Chlamydodontida Deroux, 1976
Family Lynchellidae Jankowski, 1968
*Chlamydonella pseudochilodon* (Deroux, 1970) Petz et al., 1995

**Ecological feature:** mangrove
**Water temperature:** 23.9 °C
**pH:** 6.6
**Salinity:** 13‰

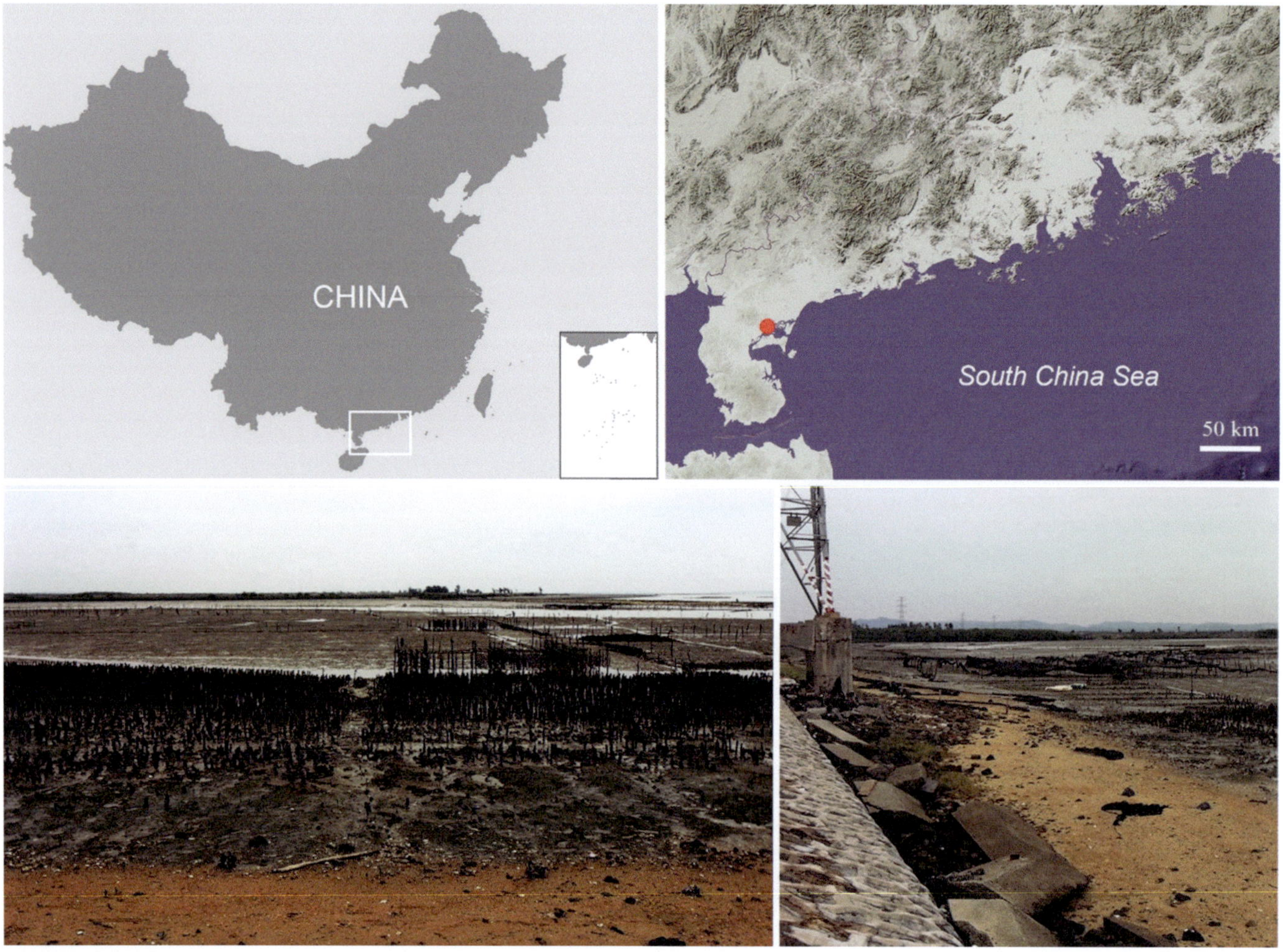


**Fig. 19A** Sample site

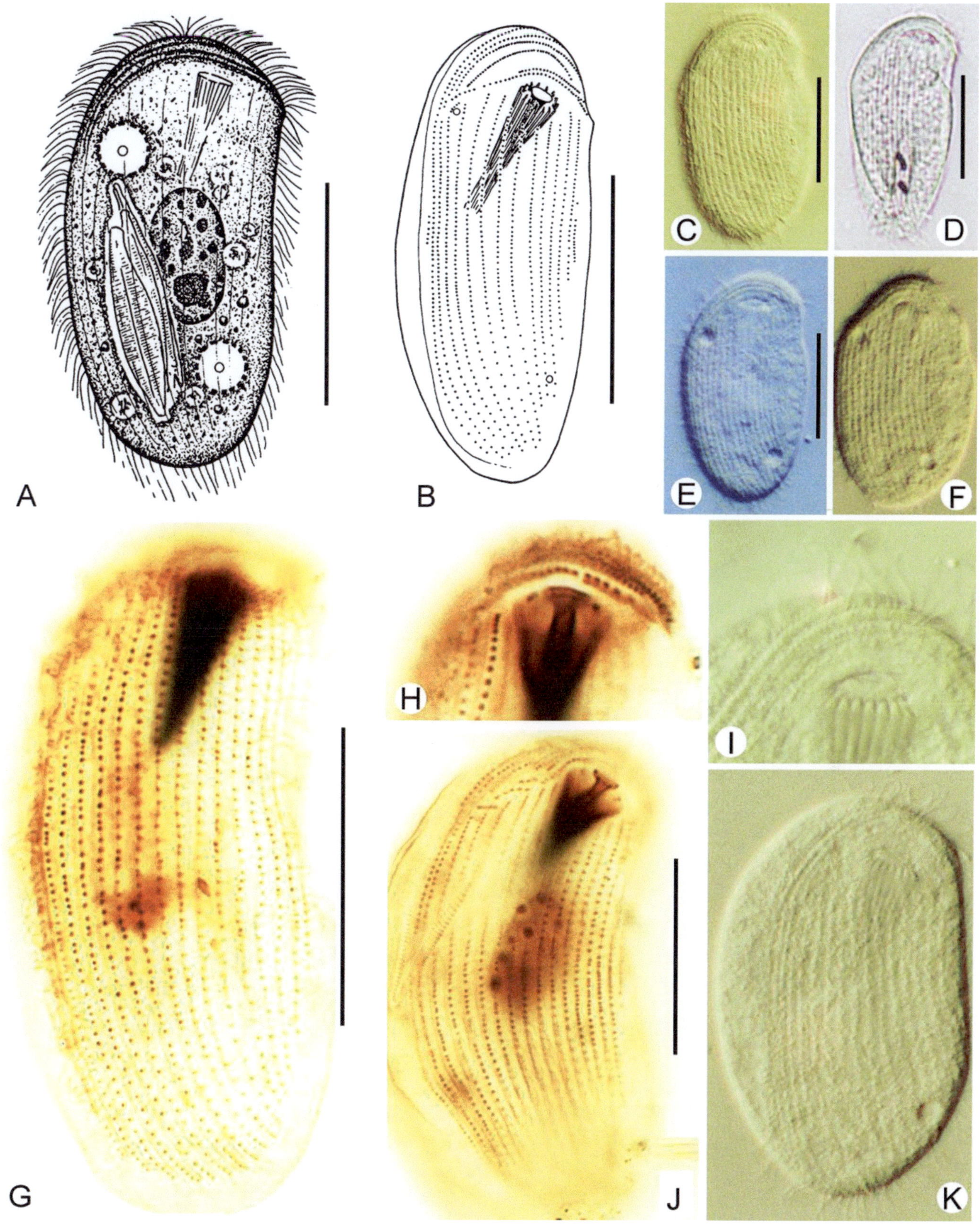


**Fig. 19B** *Chlamydonella pseudochilodon*
(**a**, **c**–**f**, **k**) Ventral views of live cells; (**b**, **g**, **j**) ventral views of ciliature; (**h**) oral ciliature; (**i**) cytopharyngeal rods. **a** and **b** from Petz et al. (1995). Scale bars: 30 μm

## References

Gong J, Song WB (2006) Redescriptions of three cyrtophorid ciliates from marine biofilm, with establishment of a new genus, *Wilbertella* nov. gen. (Ciliophora: Cyrtophorida: Lynchellidae). Acta Protozool 45:153–165

Petz W, Song WB, Wilbert N (1995) Taxonomy and ecology of the ciliate fauna (Protozoa, Ciliophora) in the endopagial and pelagial of the Weddell Sea, Antarctica. Stapfia 40:1–223

Class Phyllopharyngea de Puytorac et al., 1974
Order Chlamydodontida Deroux, 1976
Family Lynchellidae Jankowski, 1968
*Coeloperix sleighi* Gong & Song, 2004

**Ecological feature:** intertidal zone
**Water temperature:** 24.0 °C
**pH:** no data available
**Salinity:** 27.0‰

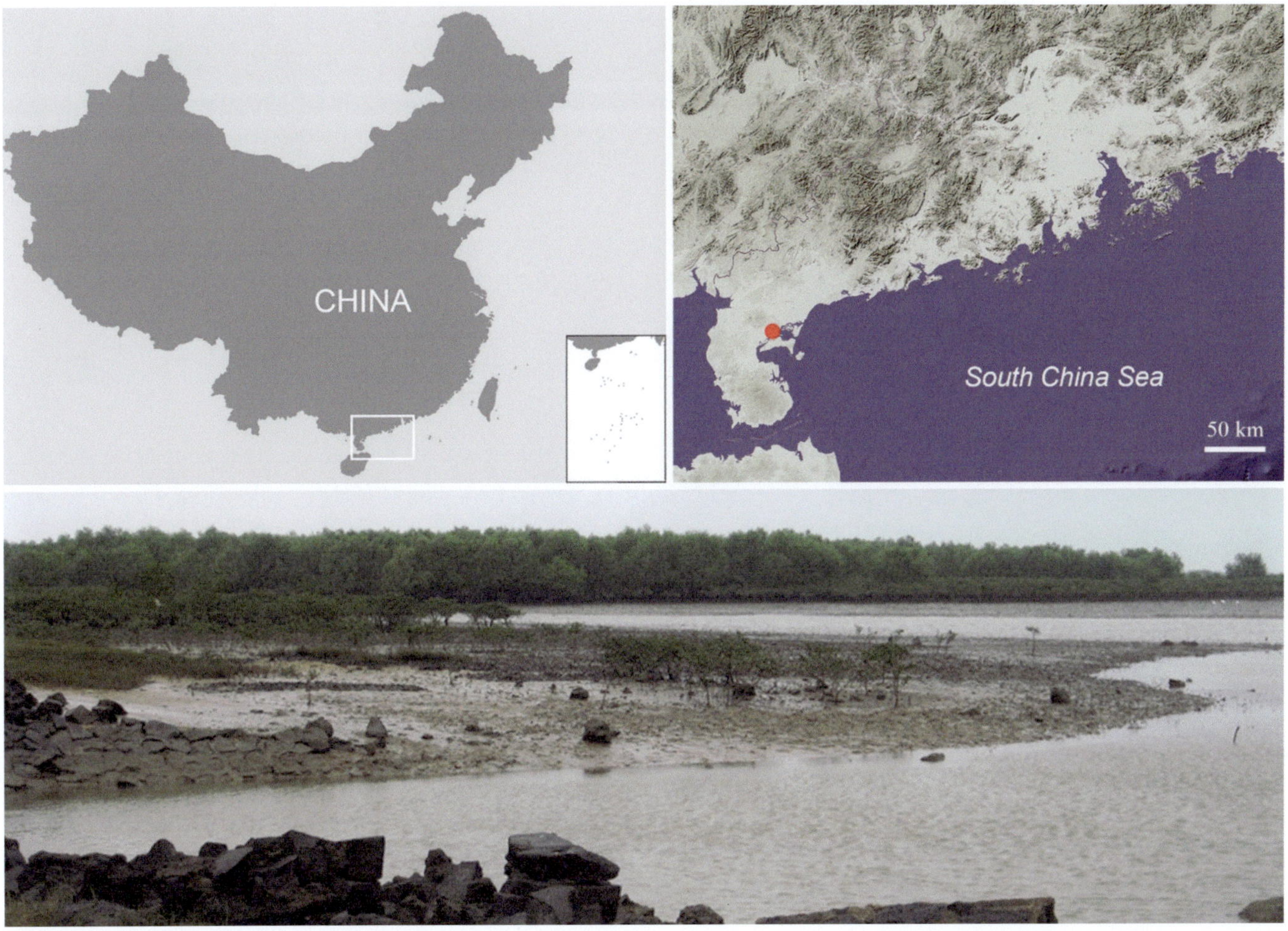


**Fig. 20A** Sample site

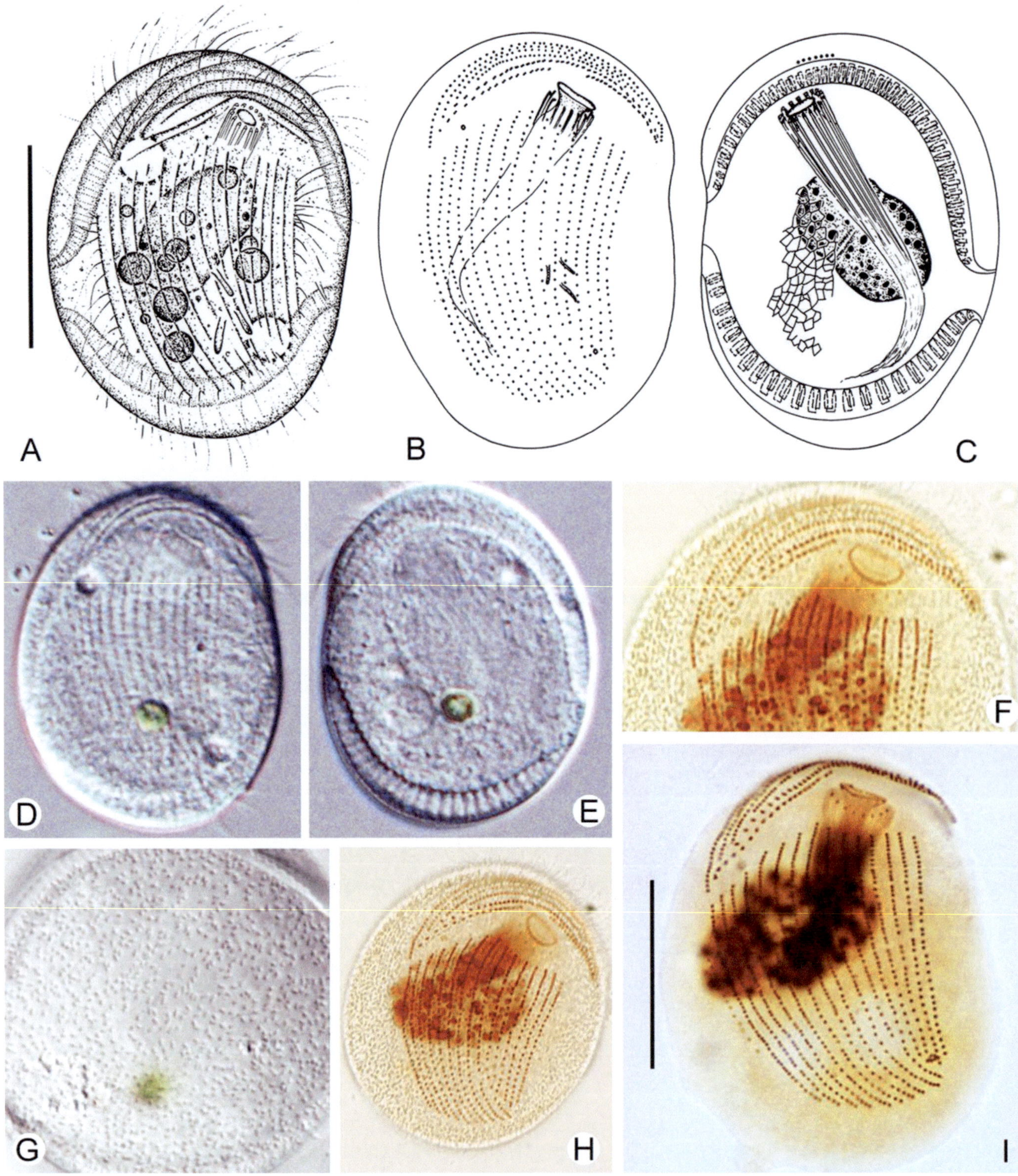


**Fig. 20B** *Coeloperix sleighi*
(**a**, **d**) Ventral views of live individuals; (**b**, **c**, **h**, **i**) ventral (**b**, **h**, **i**) and dorsal (**c**) views of ciliature; (**e**) dorsal view of live cell, showing cross striated band; (**f**) oral ciliature; (**g**) cortical granules. Scale bars: 20 μm

## References

Gong J, Song WB (2004) Reestablishment of the cyrtophorid genus *Coeloperix* Deroux, nov. gen.,with a description of *Coeloperix sleighi* nov. spec. (Protozoa, Ciliophora, Cyrtophorida). Eur J Protistol 40:175–181

Zhao L, Pan H, Ma H, Xu H (2012) Morphological redescription of five marine cyrtophorid ciliates (Protozoa, Ciliophora). Period Ocean Univ Chin 44:52–56 (in Chinese with English abstract)

Class Phyllopharyngea de Puytorac et al., 1974
Order Chlamydodontida Deroux, 1976
Family Lynchellidae Jankowski, 1968
*Lynchella minuta* Qu et al., 2017

**Ecological feature:** estuary area
**Water temperature:** 29.2 °C
**pH:** 7.9
**Salinity:** 2.5‰

**Fig. 21A** Sample site

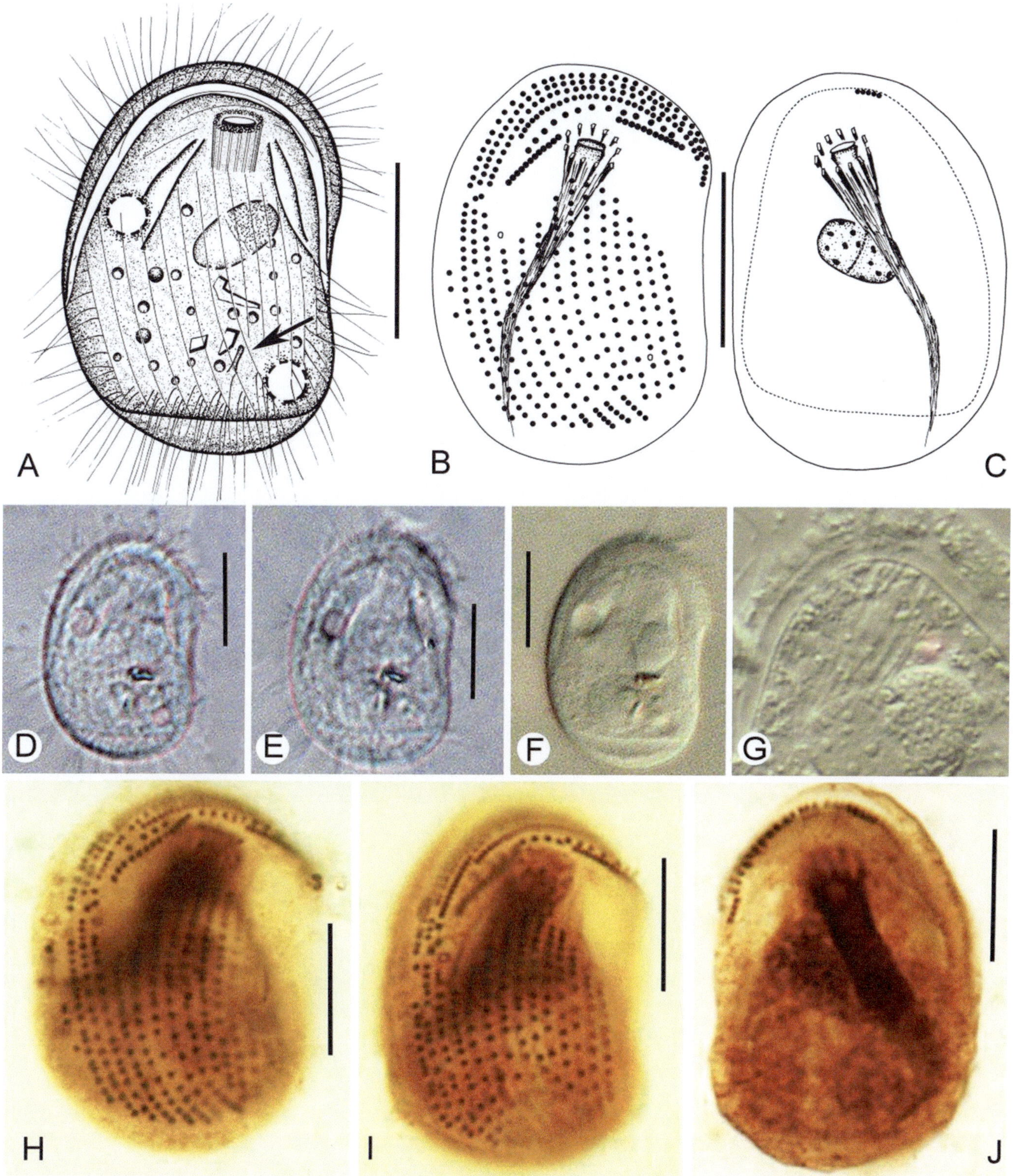


**Fig. 21B** *Lynchella minuta*
(**a**, **d**–**f**) Ventral views of live cells; (**b**, **h**, **i**) ventral views of ciliature; (**c**, **j**) dorsal views, showing terminal fragments; (**g**) showing the perimeter groove. Scale bars: 10 μm

## Reference

Qu Z, Ma H, Al-Farraj SA, Lin X, Hu X (2017) Morphology and molecular phylogeny of *Aegyria foissneri* sp. n. and *Lynchella minuta* sp. n. (Ciliophora, Cyrtophoria) from brackish waters of southern China. Eur J Protistol 57:50–60

Class Phyllopharyngea de Puytorac et al., 1974
Order Dysteriida Deroux, 1976
Family Dysteriidae Claparède and Lachmann, 1858
*Agnathodysteria littoralis* Deroux, 1976

**Ecological feature:** shrimp-farming pond
**Water temperature:** 23.8 °C
**pH:** 7.4
**Salinity:** 24.8‰

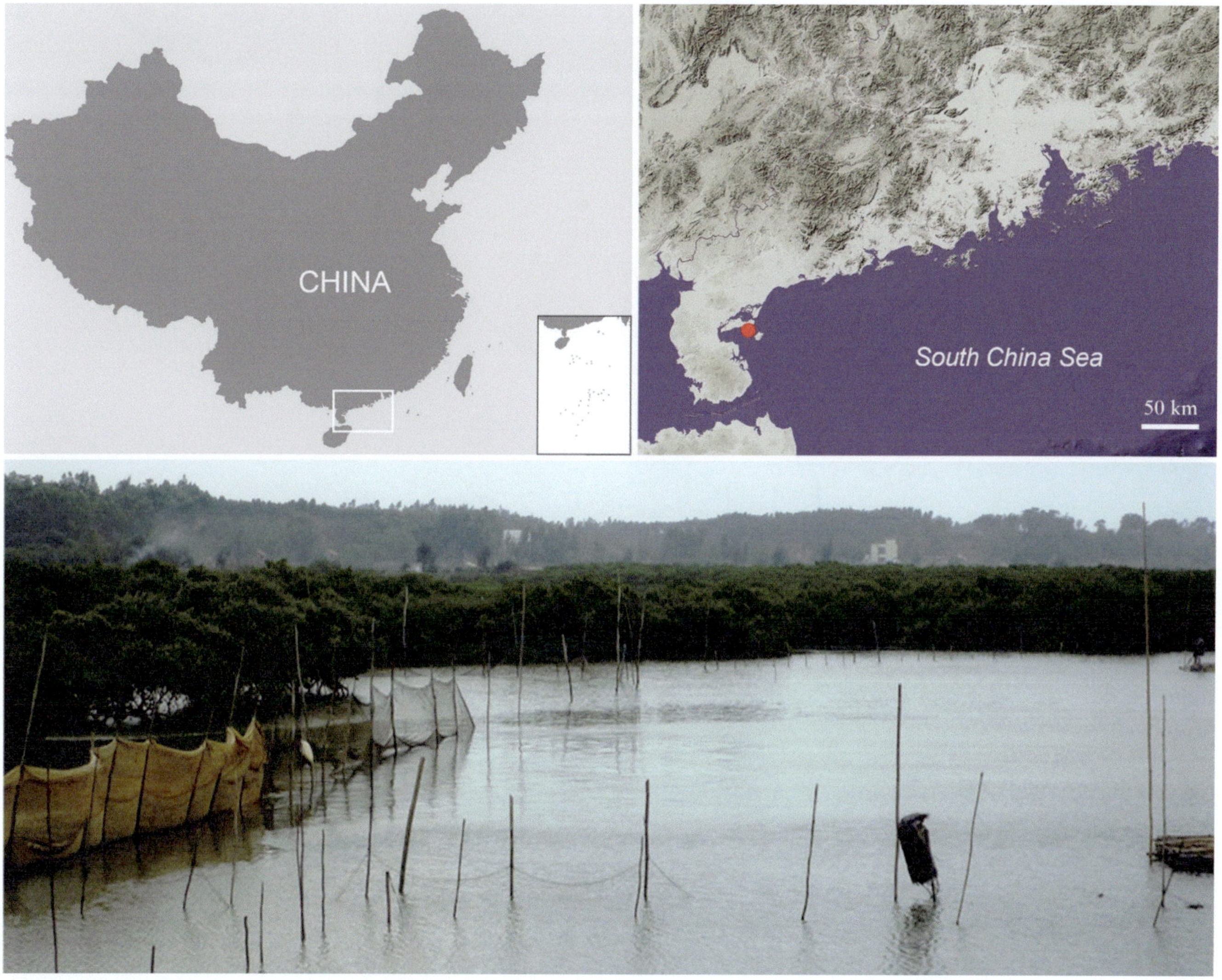


**Fig. 22A** Sample site

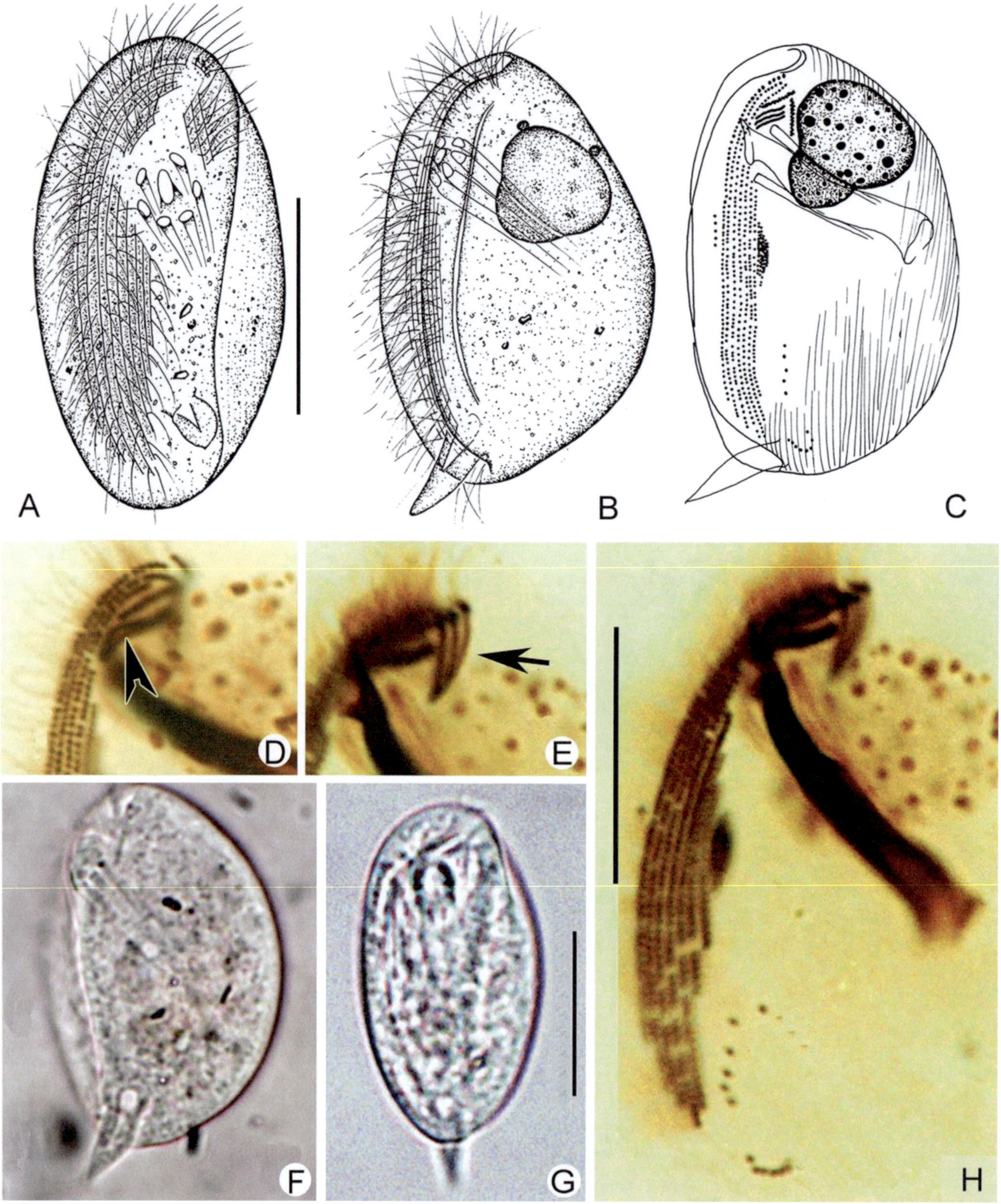


**Fig. 22B** *Agnathodysteria littoralis*
(**a**, **g**) Ventral views of live individuals; (**b**, **f**) left-lateral views of live cells; (**c**, **h**) ciliature; (**d**, **e**) anterior portion of cell, arrowhead refers to circumoral kineties, arrow points to left frontal kineties. Scale bars: 20 μm

## Reference

Zhao L, Pan H, Ma H, Xu H (2014) Morphological redescription of five marine cyrtophorid ciliates (Protozoa, Ciliophora). Period Ocean Univ Chin 44(1):52–56 (in Chinese with English abstract)

Class Phyllopharyngea de Puytorac et al., 1974
Order Dysteriida Deroux, 1976
Family Dysteriidae Claparède & Lachmann, 1858
*Dysteria brasiliensis* Da Cunha et al., 1925

**Ecological feature:** mangrove
**Water temperature:** 24.8 °C
**pH:** 6.7
**Salinity:** 17.0‰

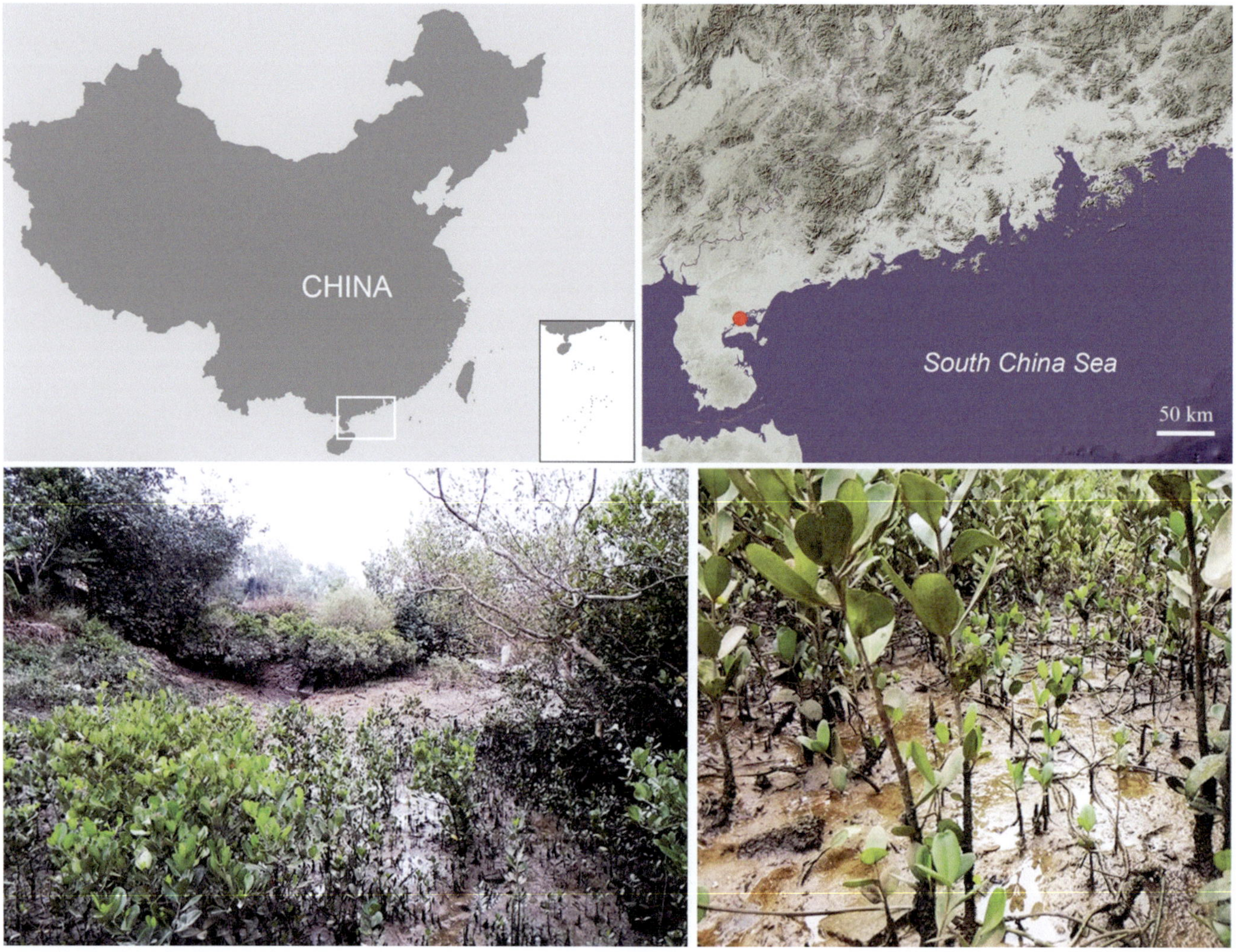


**Fig. 23A** Sample site

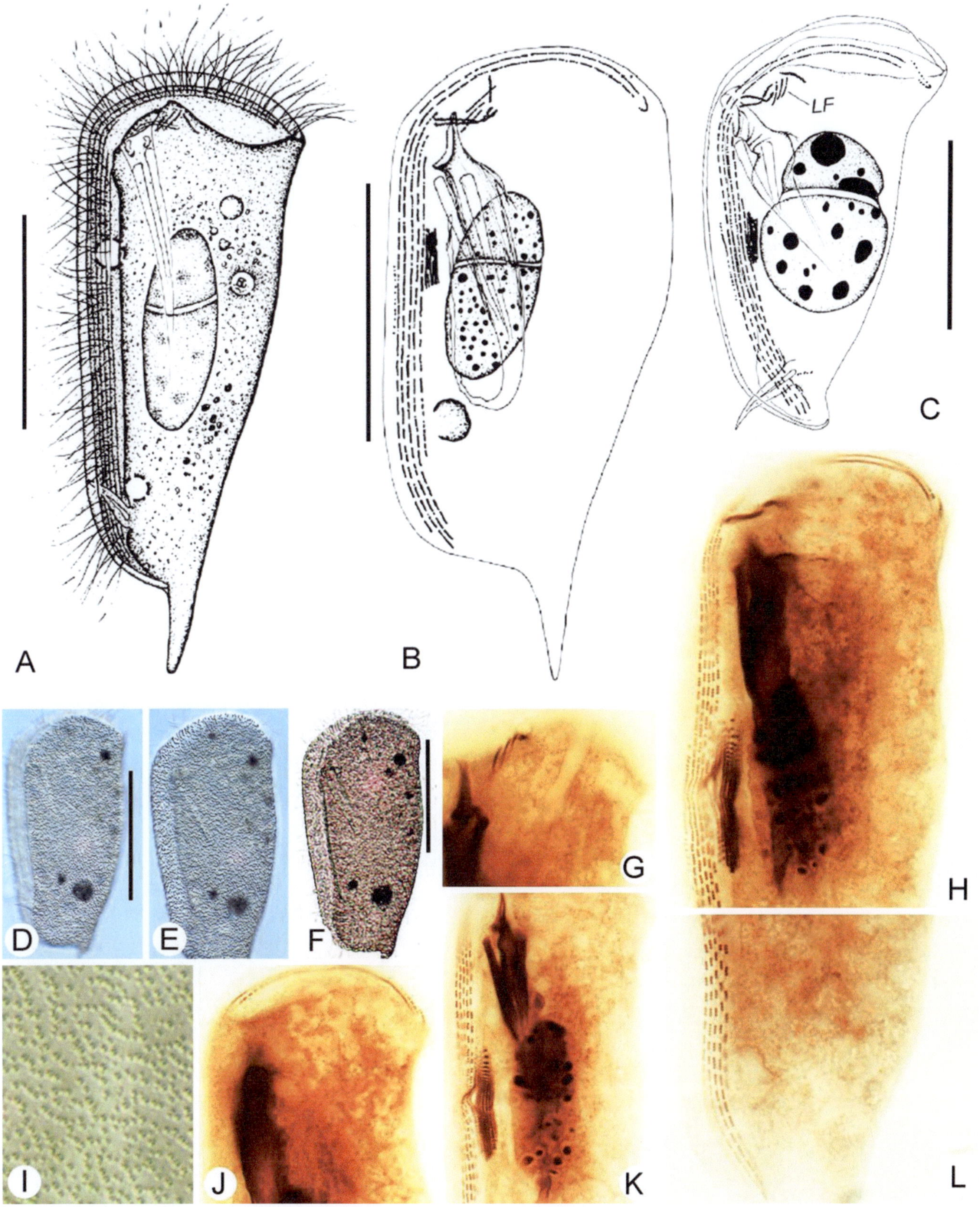


**Fig. 23B** *Dysteria brasiliensis*
(**a**, **d**–**f**) Left-lateral views of live cells; (**b**, **c**, **h**, **j**–**l**) left-lateral views of ciliature; (**g**) left frontal kineties; (**i**) cortical granules. Scale bars: 50 μm

## References

Gong J, Song WB, Warren A, Lin X, Roberts D (2007) Microscopical observations on four marine *Dysteria* species (Ciliophora, Cyrtophorida). Eur J Protistol 43:147–161

Qu Z, Wang C, Gao F, Li J, Al-Rasheid KAS, Hu X (2015) Taxonomic studies on seven species of *Dysteria* (Ciliophora, Cyrtophoria), including a description of *Dysteria paraprocera* sp. n. Eur J Protistol 51:241–258

Class Phyllopharyngea de Puytorac et al., 1974
Order Dysteriida Deroux, 1976
Family Dysteriidae Claparède & Lachmann, 1858
*Dysteria compressa* (Gourret & Roeser, 1886) Kahl, 1931

**Ecological feature:** intertidal zone; mangrove
**Water temperature:** 26.0 °C; 23.0 °C
**pH:** 8.9; 7.3
**Salinity:** 25.9‰; 31‰

**Fig. 24A** Sample site

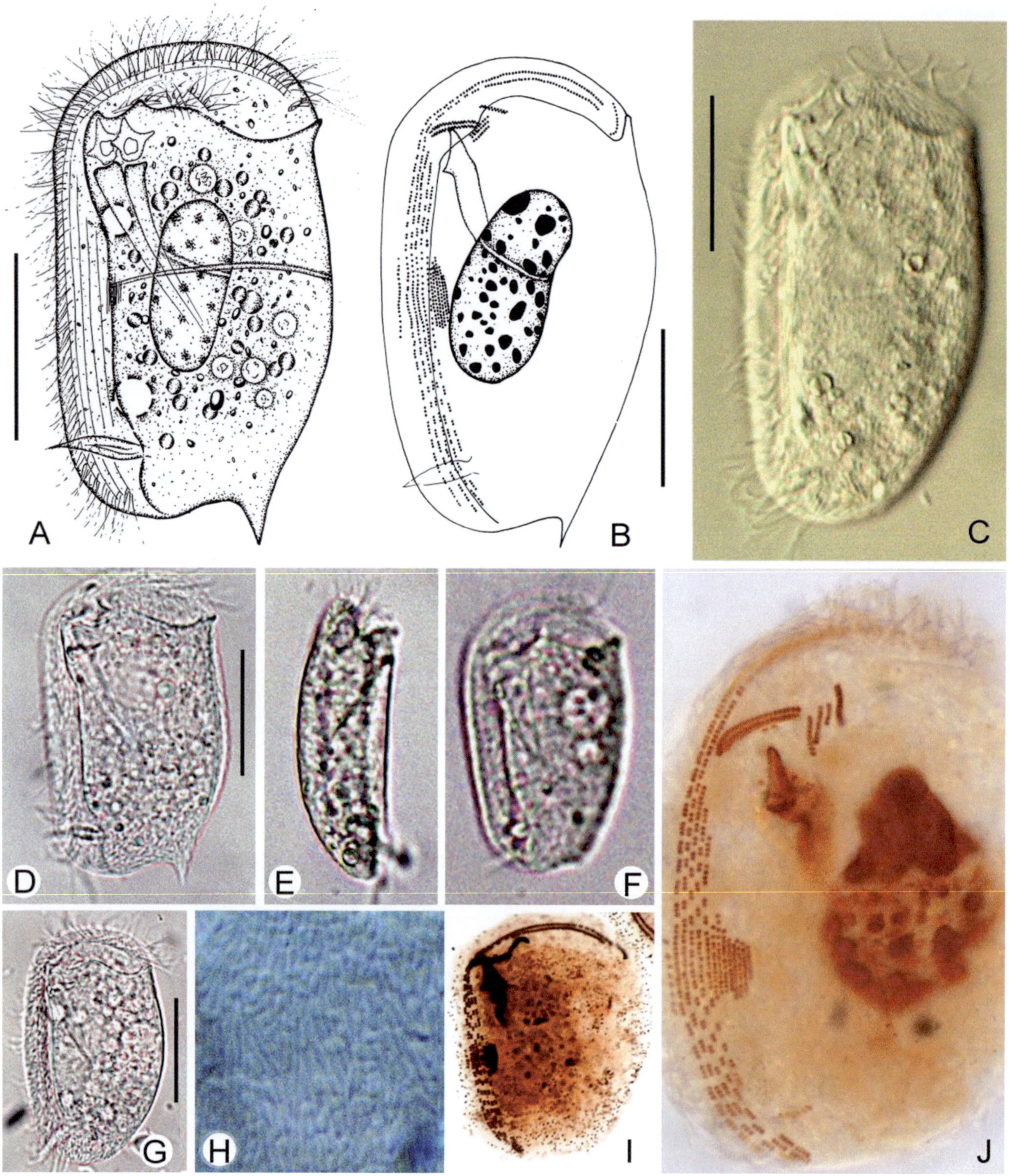


**Fig. 24B** *Dysteria compressa*
(a, **c**, **d**, **f**, **g**) Lateral views of live individuals; (**b**, **i**, **j**) lateral views of ciliature; (**e**) dorsal view of live cell; (**h**) ectosymbiotic bacteria on the cell surface. Scale bars: 30 μm

## Reference

Pan H, Jiang J, Fan X, Al-Farraj SA, Gao S (2017) Phylogeny and taxonomy of five poorly known species of cyrtophorian ciliates (Protozoa, Ciliophora, Phyllopharyngea) from China seas. Zool J Linn Soc 180:475–492

Class Phyllopharyngea de Puytorac et al., 1974
Order Dysteriida Deroux, 1976
Family Dysteriidae Claparède & Lachmann, 1858
*Dysteria crassipes* Claparède & Lachmann, 1859

**Ecological feature:** mangrove (Zhanjiang); offshore water (Daya bay)
**Water temperature:** 24.5 °C; 19.4 °C
**pH:** 6.8; 8.2
**Salinity:** 17.0‰; 32.6‰

**Fig. 25A** Sample site

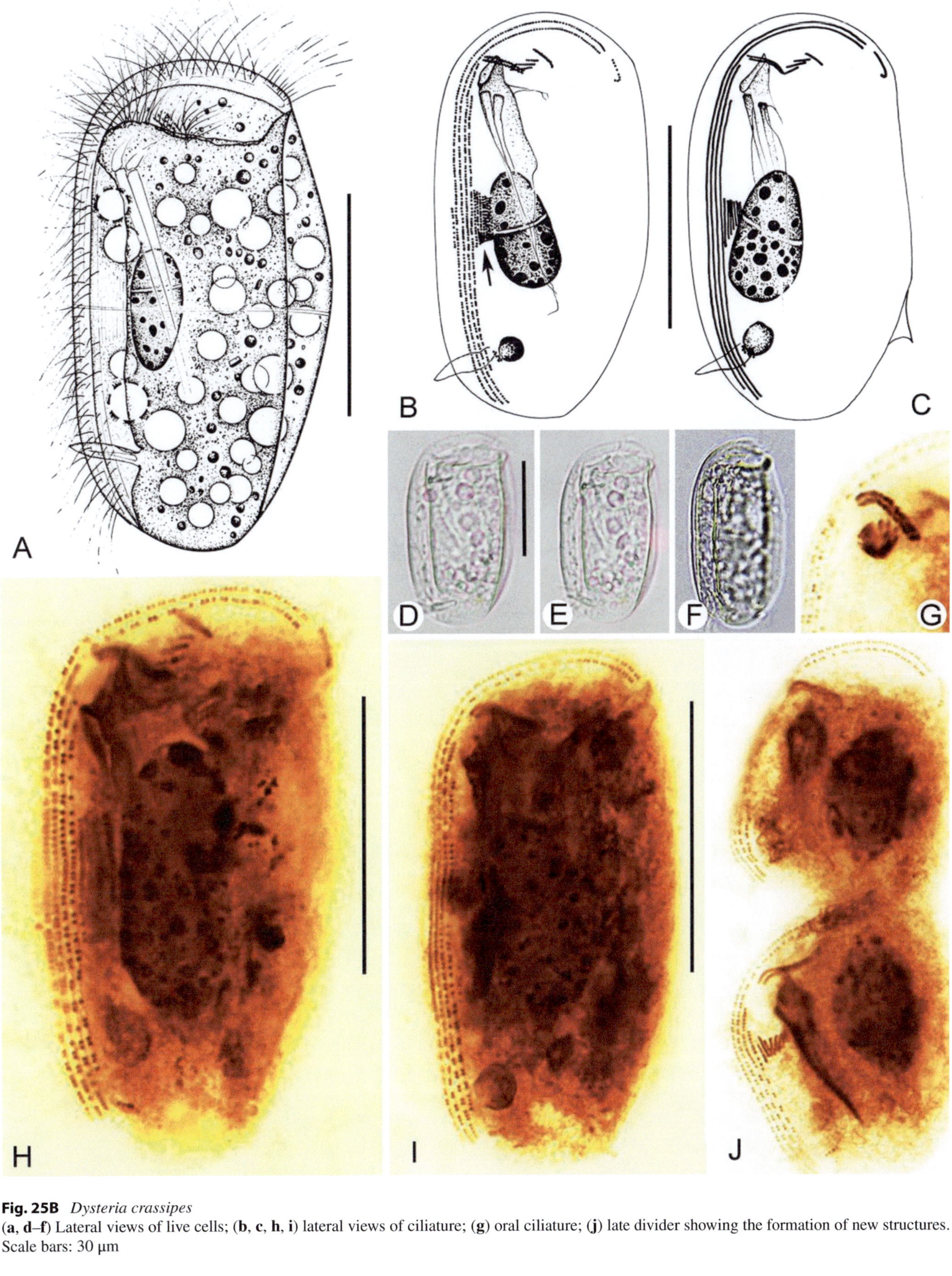


**Fig. 25B** *Dysteria crassipes*
(**a**, **d–f**) Lateral views of live cells; (**b**, **c**, **h**, **i**) lateral views of ciliature; (**g**) oral ciliature; (**j**) late divider showing the formation of new structures. Scale bars: 30 μm

## References

Gong J, Song WB, Warren A, Lin X, Roberts D (2007) Microscopical observations on four marine *Dysteria* species (Ciliophora, Cyrtophorida). Eur J Protistol 43:147–161

Kahl A (1931) Urtiere oder Protozoa I: Wimpertiere oder Ciliata (Infusoria) 2. Holotricha außer den im 1. Teil behandelten Prostomata. Tierwelt Dtl 21:181–398

Class Phyllopharyngea de Puytorac et al., 1974
Order Dysteriida Deroux, 1976
Family Dysteriidae Claparède & Lachmann, 1858
*Dysteria cristata* (Gourret & Roeser, 1886) Kahl, 1931

**Ecological feature:** estuary area
**Water temperature:** 29.2 °C
**pH:** 7.9
**Salinity:** 2.5‰

**Fig. 26A** Sample site

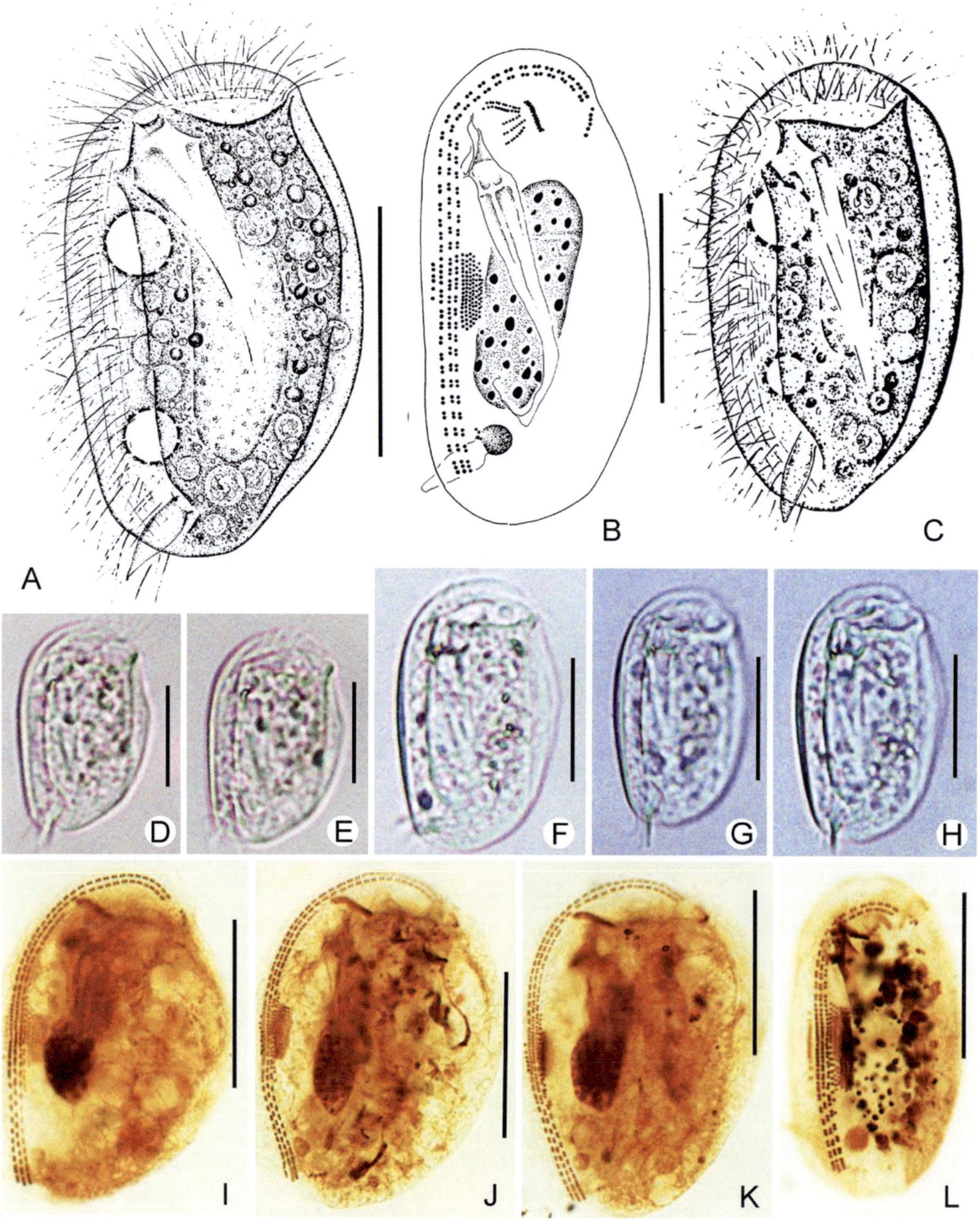


**Fig. 26B.** *Dysteria cristata*
(**a**, **c**, **d**–**h**) Lateral views of live cells; (**b**, **i**–**l**) lateral views of ciliature. Scale bars: 20 μm

## Reference

Gong J, Song WB, Warren A (2002) Redescriptions of two marine cyrtophorid ciliates, *Dysteria cristata* (Gourret and Roeser, 1888) and *Dysteria monostyla* (Ehrenberg, 1838) Kahl, 1931 (Protozoa, Ciliophora, Cyrtophorida), from China. Eur J Protistol 38:213–222

Class Phyllopharyngea de Puytorac et al., 1974
Order Dysteriida Deroux, 1976
Family Dysteriidae Claparède & Lachmann, 1858
*Dysteria derouxi* Gong & Song, 2004

**Ecological feature:** coastal water
**Water temperature:** 21.0 °C; 19.4 °C
**pH:** 7.5; 8.2
**Salinity:** 30.0‰; 32.6‰

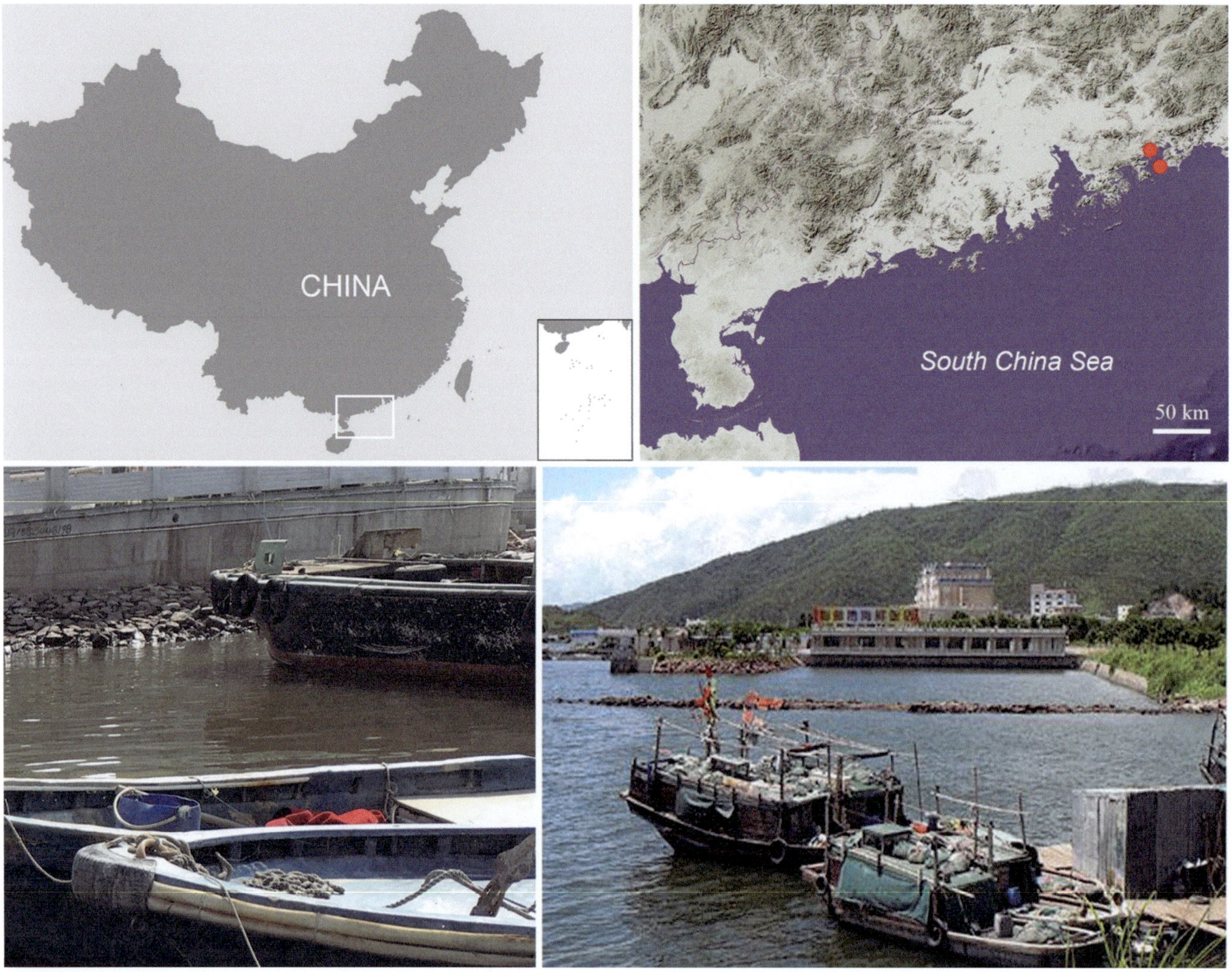


**Fig. 27A** Sample site

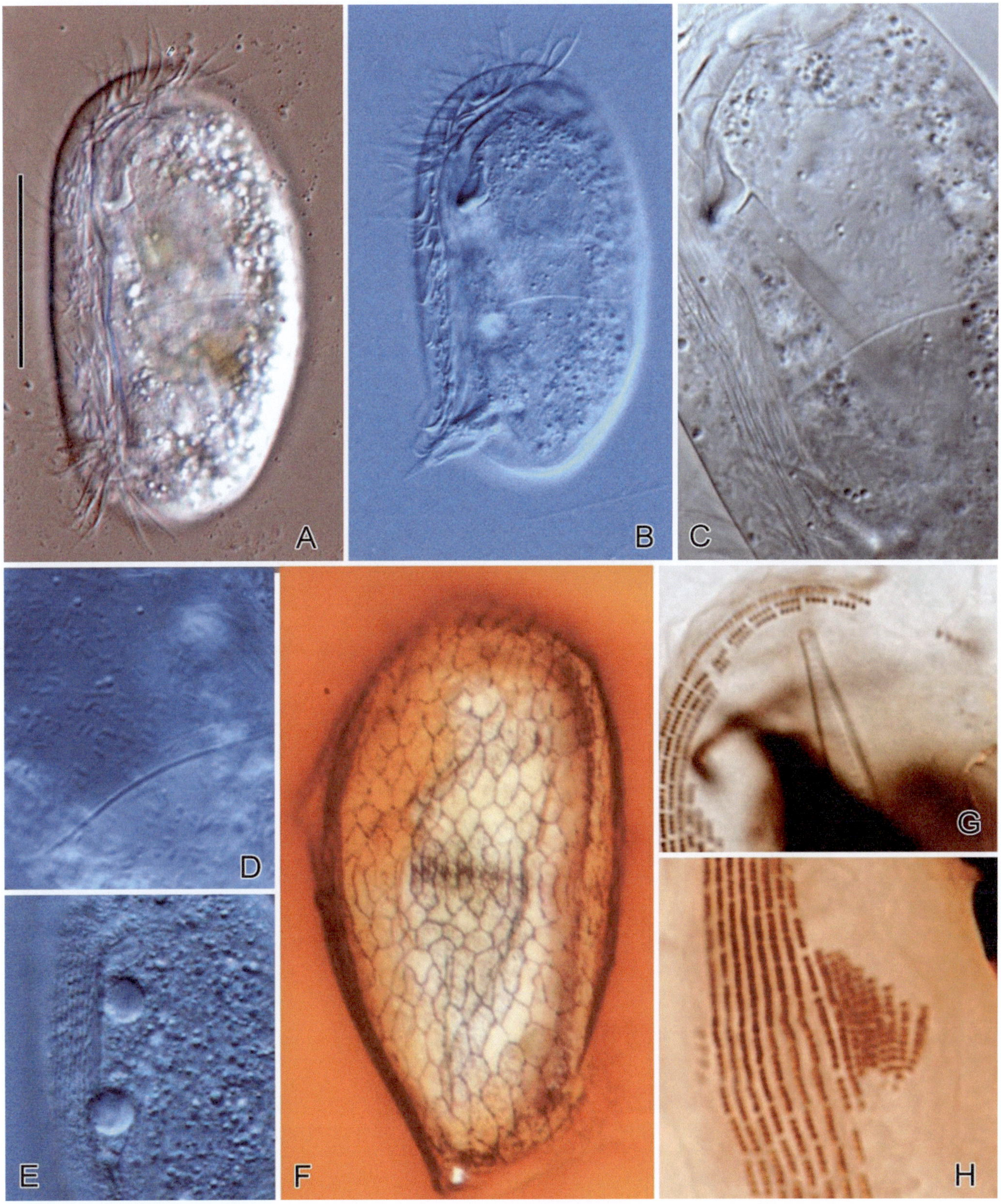


**Fig. 27B** *Dysteria derouxi*
(**a**, **b**) Lateral views of live cells, showing cilia and podite; (**c**) cytopharyngeal rods; (**d**) transverse stripe at equatorial portion; (**e**) two contractile vacuoles; (**f**) right-lateral view of argyrome; (**g**) terminal fragment and right kineties; (**h**) equatorial fragment and left kineties. Scale bar: 100 μm

## Reference

Gong J, Song WB (2004) Description of a new marine cyrtophorid ciliate, *Dysteria derouxi* nov. spec., with an updated key to 12 known *Dysteria* species (Protozoa, Ciliophora). Eur J Protistol 40:13–19

Class Phyllopharyngea de Puytorac et al., 1974
Order Dysteriida Deroux, 1976
Family Dysteriidae Claparède & Lachmann, 1858
*Dysteria monostyla* (Ehrenberg, 1838) Kahl, 1931

**Ecological feature:** mangrove (Zhanjiang); brackish water (Shenzhen)
**Water temperature:** 23 °C; 31.5 °C
**pH:** 7.2; 7.7
**Salinity:** 10‰; 27.0‰

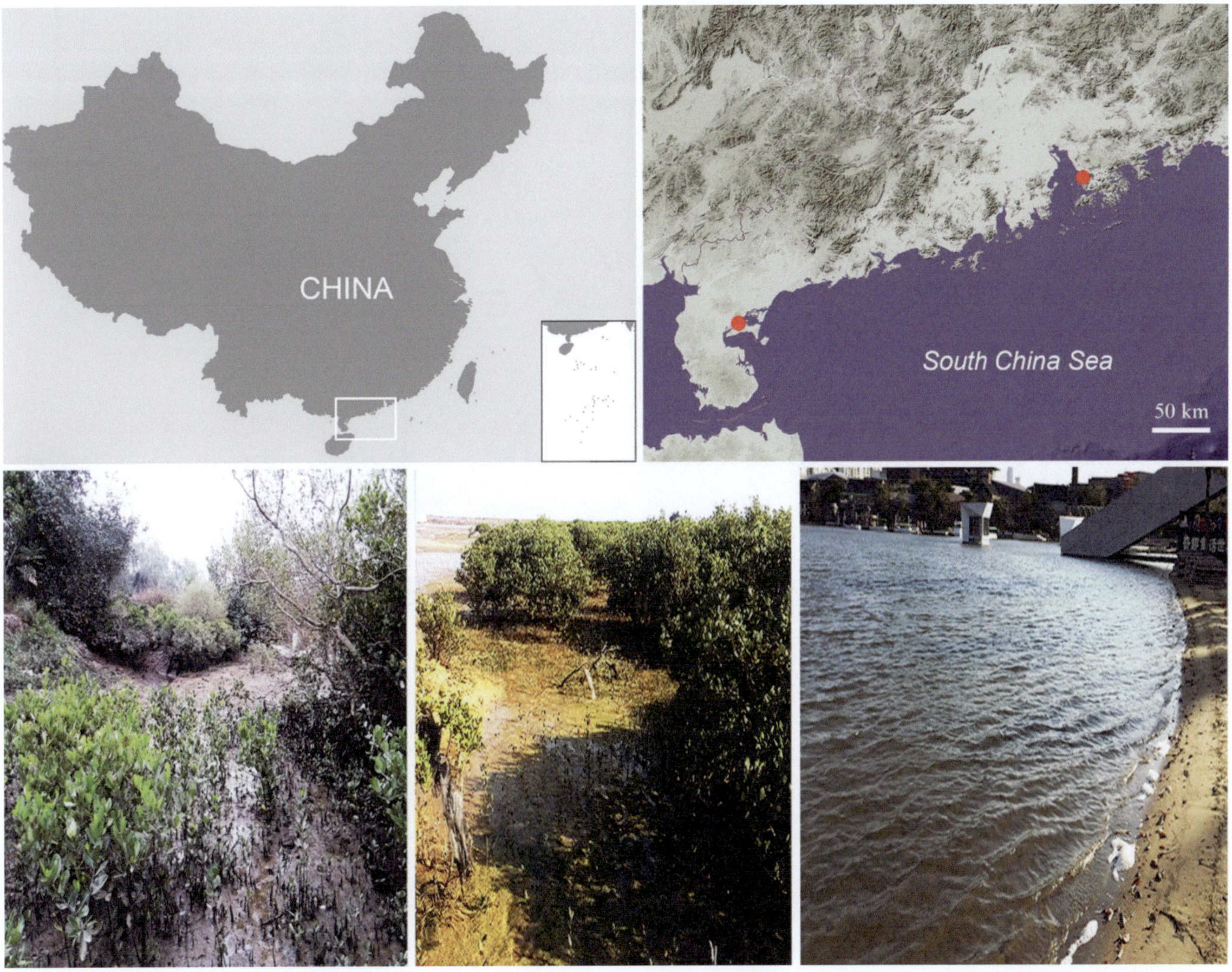


**Fig. 28A** Sample site

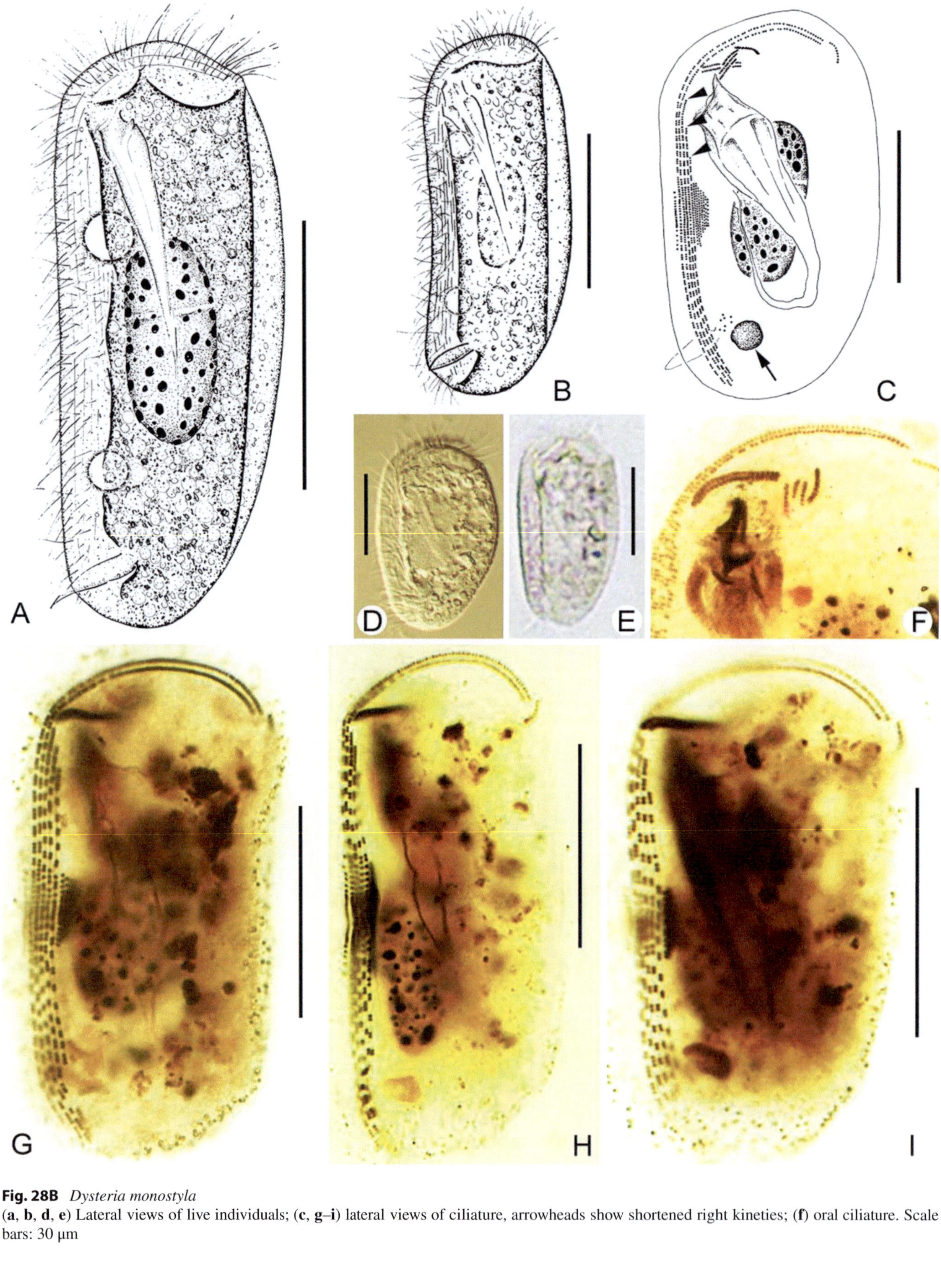


**Fig. 28B** *Dysteria monostyla*
(**a**, **b**, **d**, **e**) Lateral views of live individuals; (**c**, **g–i**) lateral views of ciliature, arrowheads show shortened right kineties; (**f**) oral ciliature. Scale bars: 30 μm

## Reference

Gong J, Song WB, Warren A (2002) Redescriptions of two marine cyrtophorid ciliates, *Dysteria cristata* (Gourret and Roeser, 1888) and *Dysteria monostyla* (Ehrenberg, 1838) Kahl, 1931 (Protozoa, Ciliophora, Cyrtophorida), from China. Eur J Protistol 38:213–222

Class Phyllopharyngea de Puytorac et al., 1974
Order Dysteriida Deroux, 1976
Family Dysteriidae Claparède & Lachmann, 1858
*Dysteria nabia* Park & Min, 2014

**Ecological feature:** mangrove (Zhanjiang); offshore water (Daya Bay)
**Water temperature:** 25.3 °C; 23.1 °C
**pH:** 6.0; 8.2
**Salinity:** 4.0‰; 28.6‰

**Fig. 29A** Sample site

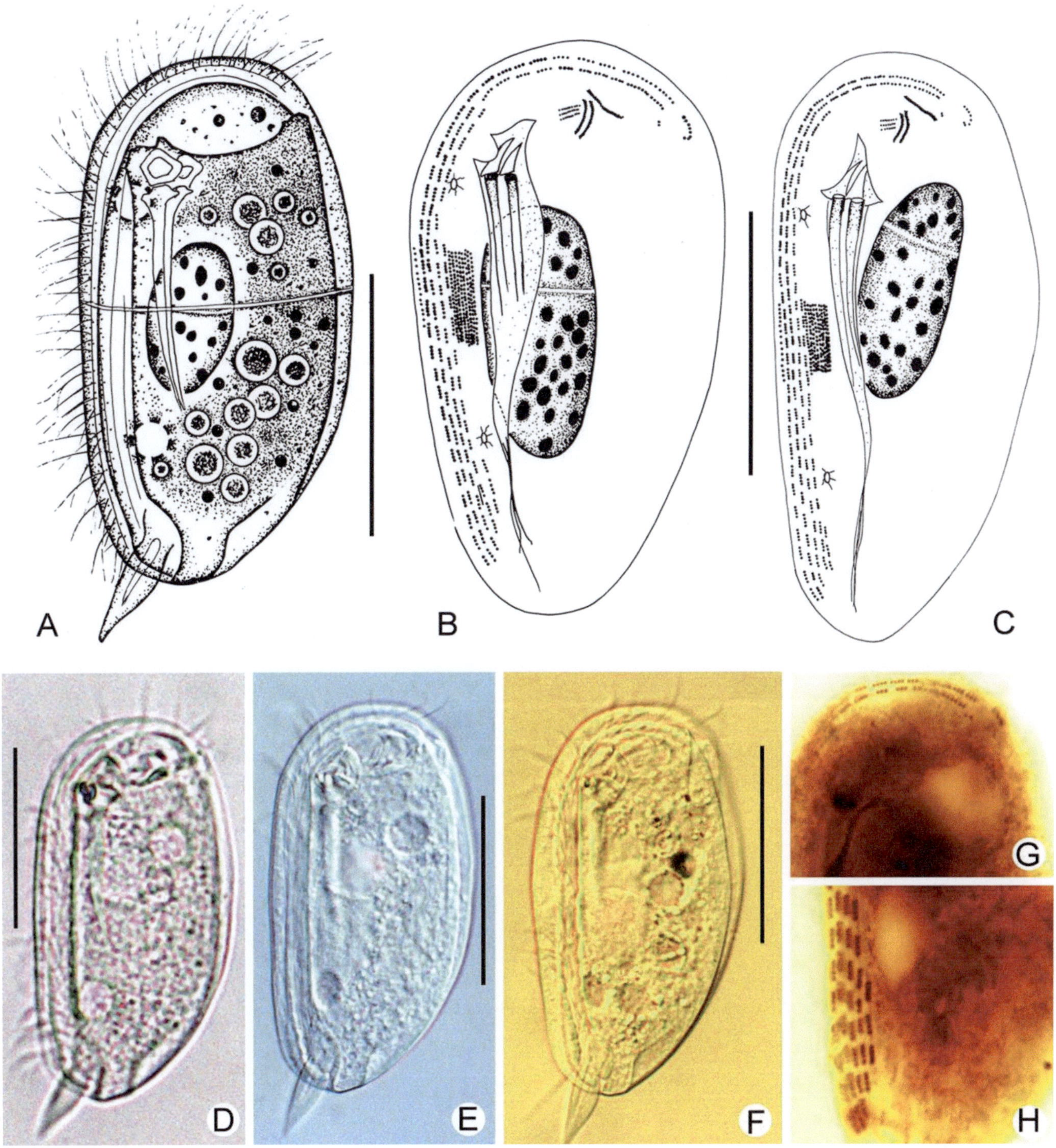


**Fig. 29B** *Dysteria nabia*
(**a**, **d–f**) Left-lateral views of live cells; (**b**, **c**) left-lateral views of ciliature; (**g**) two rightmost kineties extending to anterior end of the body; (**h**) posterior portion of right kineties. Scale bars: 40 μm

## References

Park M, Min G (2014) A new marine cyrtophorid ciliate, *Dysteria nabia* nov. spec. (Ciliophora: Phyllopharyngea: Cyrtophorida: Dysteriidae), from South Korea. Acta Protozool 53:257–268

Qu Z, Wang C, Gao F, Li J, Al-Rasheid KAS, Hu X (2015) Taxonomic studies on seven species of *Dysteria* (Ciliophora, Cyrtophoria), including a description of *Dysteria paraprocera* sp. n. Eur J Protistol 51:241–258

Class Phyllopharyngea de Puytorac et al., 1974
Order Dysteriida Deroux, 1976
Family Dysteriidae Claparède & Lachmann, 1858
*Dysteria paraprocera* Qu et al., 2015

**Ecological feature:** mangrove
**Water temperature:** 17.5 °C
**pH:** 8.0
**Salinity:** 27‰

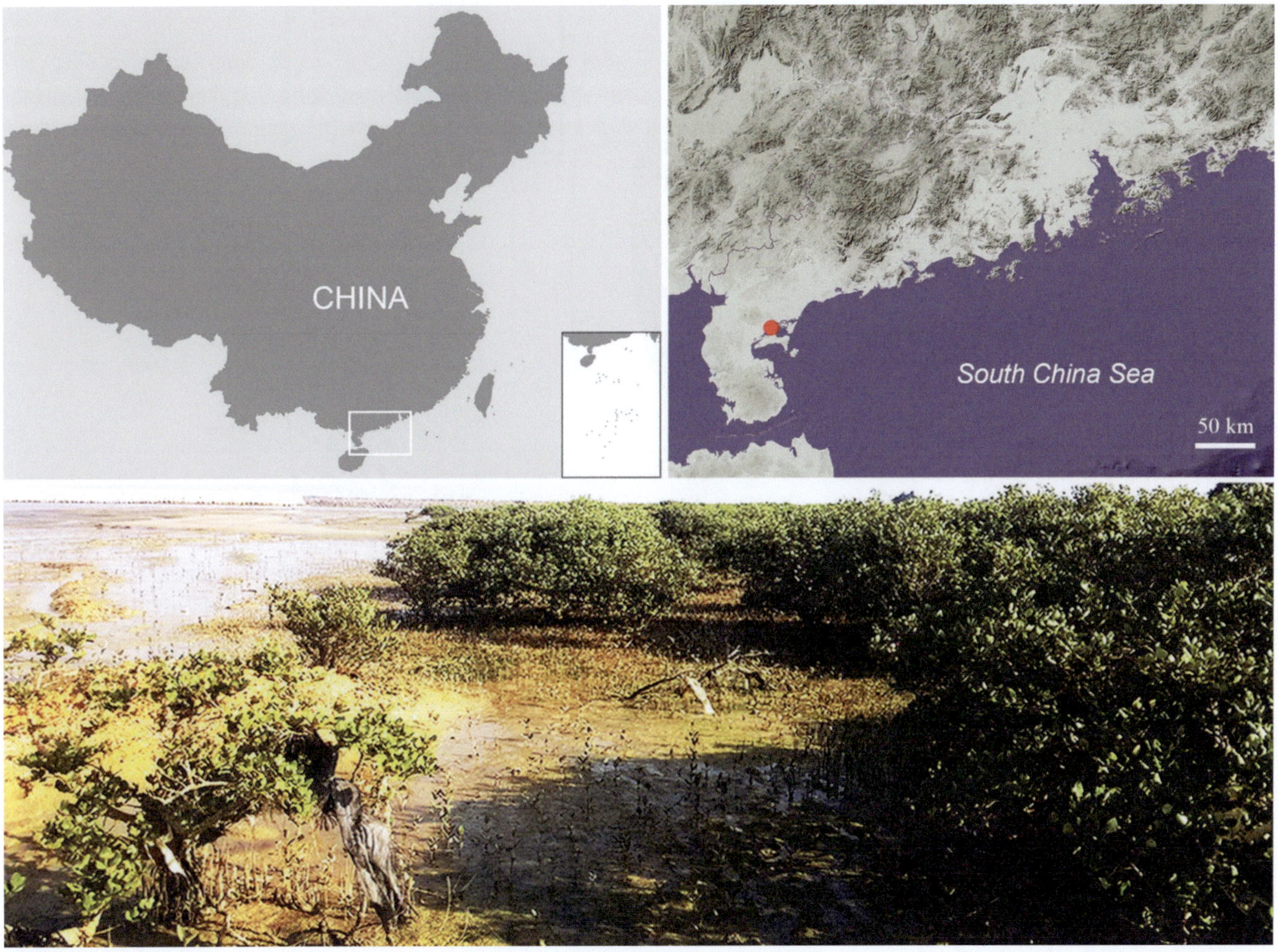


**Fig. 30A** Sample site

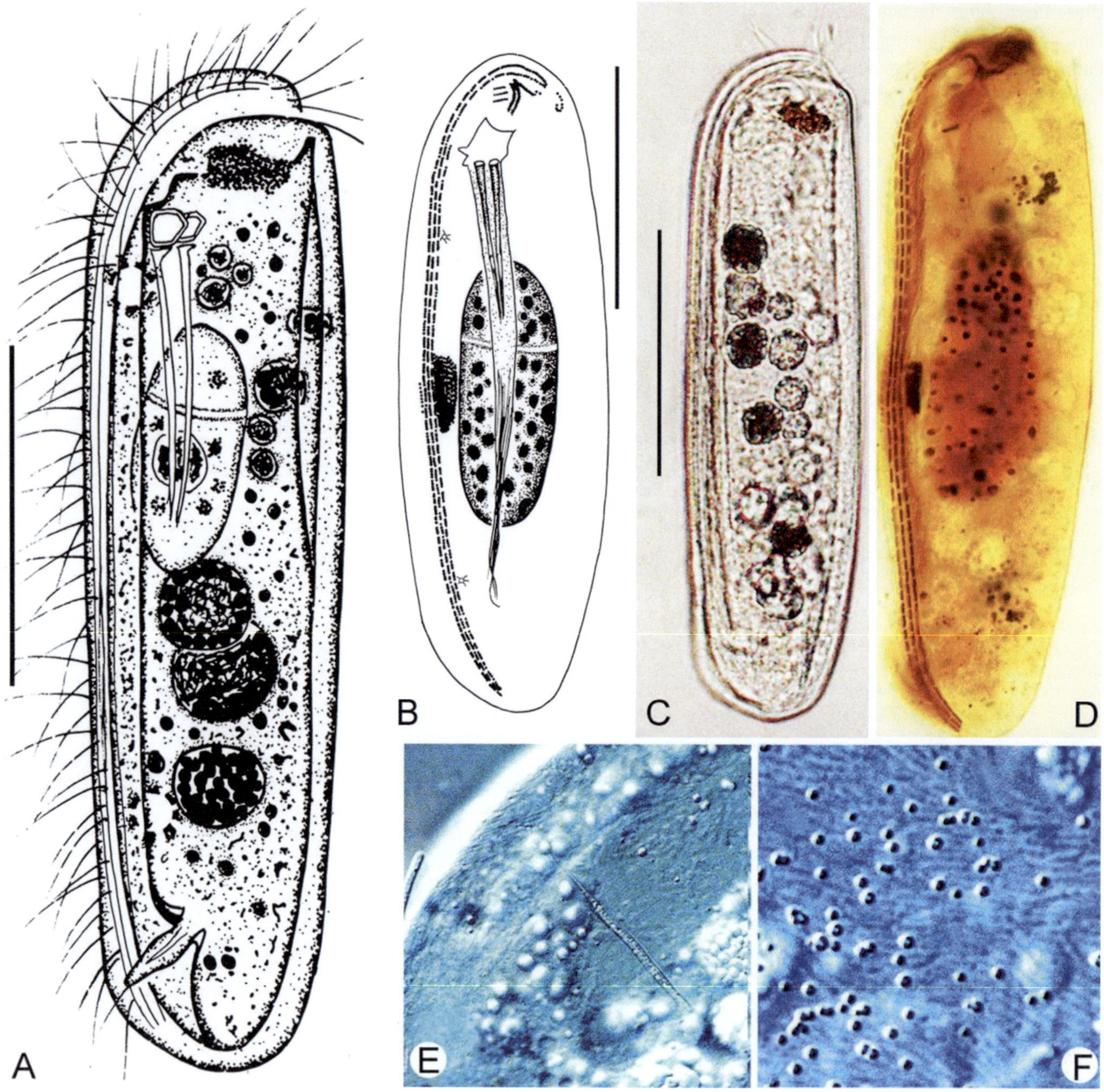


**Fig. 30B** *Dysteria paraprocera*
(**a**, **c**) Left-lateral views of live cells; (**b**, **d**) left-lateral views of stained cells, showing ciliature and macronucleus; (**e**) central part of the body, showing the transverse stripe; (**f**) cortical granules. Scale bars: 40 μm

## Reference

Qu Z, Wang C, Gao F, Li J, Al-Rasheid KAS, Hu X (2015) Taxonomic studies on seven species of *Dysteria* (Ciliophora, Cyrtophoria), including a description of *Dysteria paraprocera* sp. n. Eur J Protistol 51:241–258

Class Phyllopharyngea de Puytorac et al., 1974
Order Dysteriida Deroux, 1976
Family Dysteriidae Claparède & Lachmann, 1858
*Dysteria pectinata* (Nowlin, 1913) Kahl, 1931

**Ecological feature:** offshore water
**Water temperature:** 23.1 °C
**pH:** 8.2
**Salinity:** 28.6‰

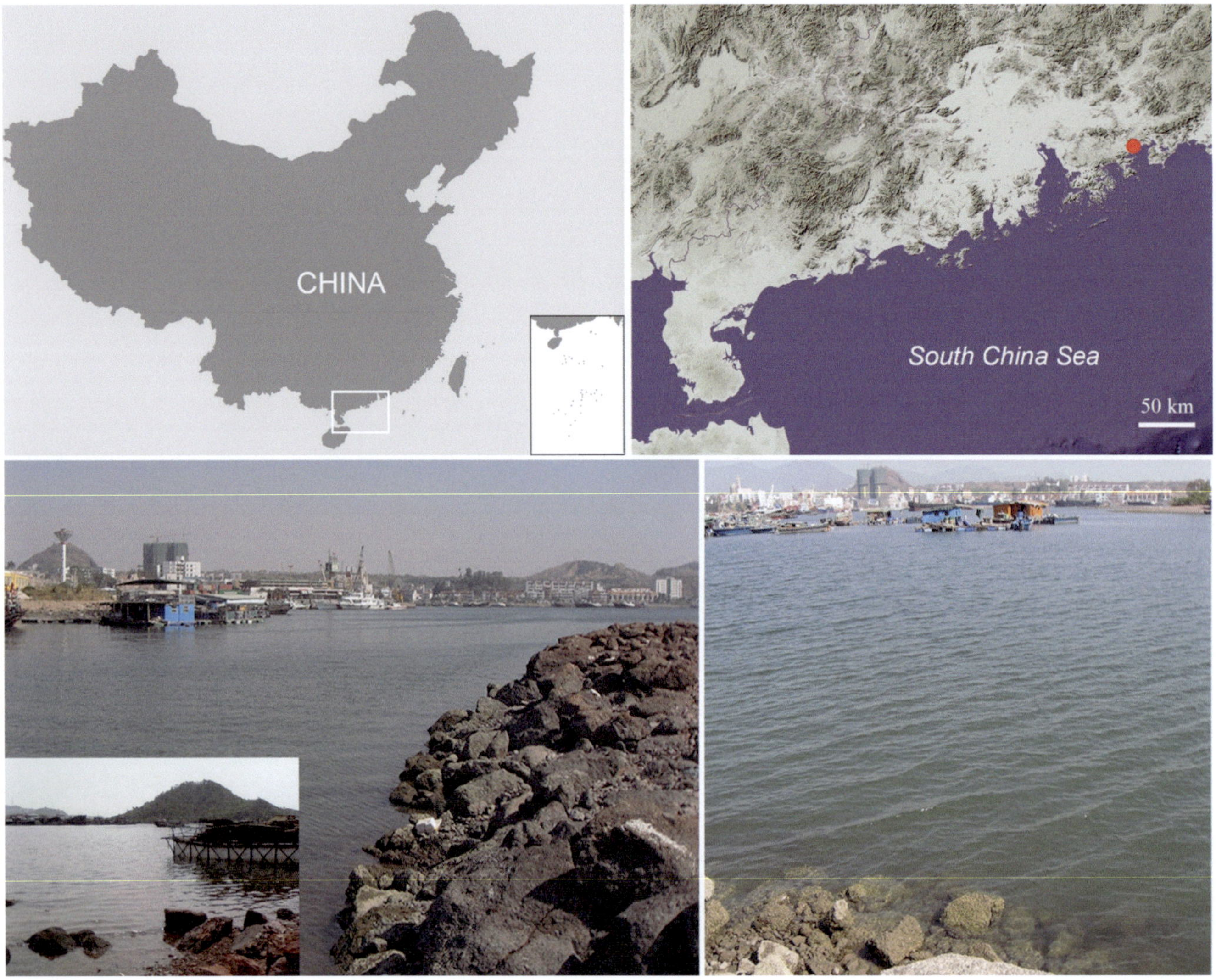


**Fig. 31A** Sample site

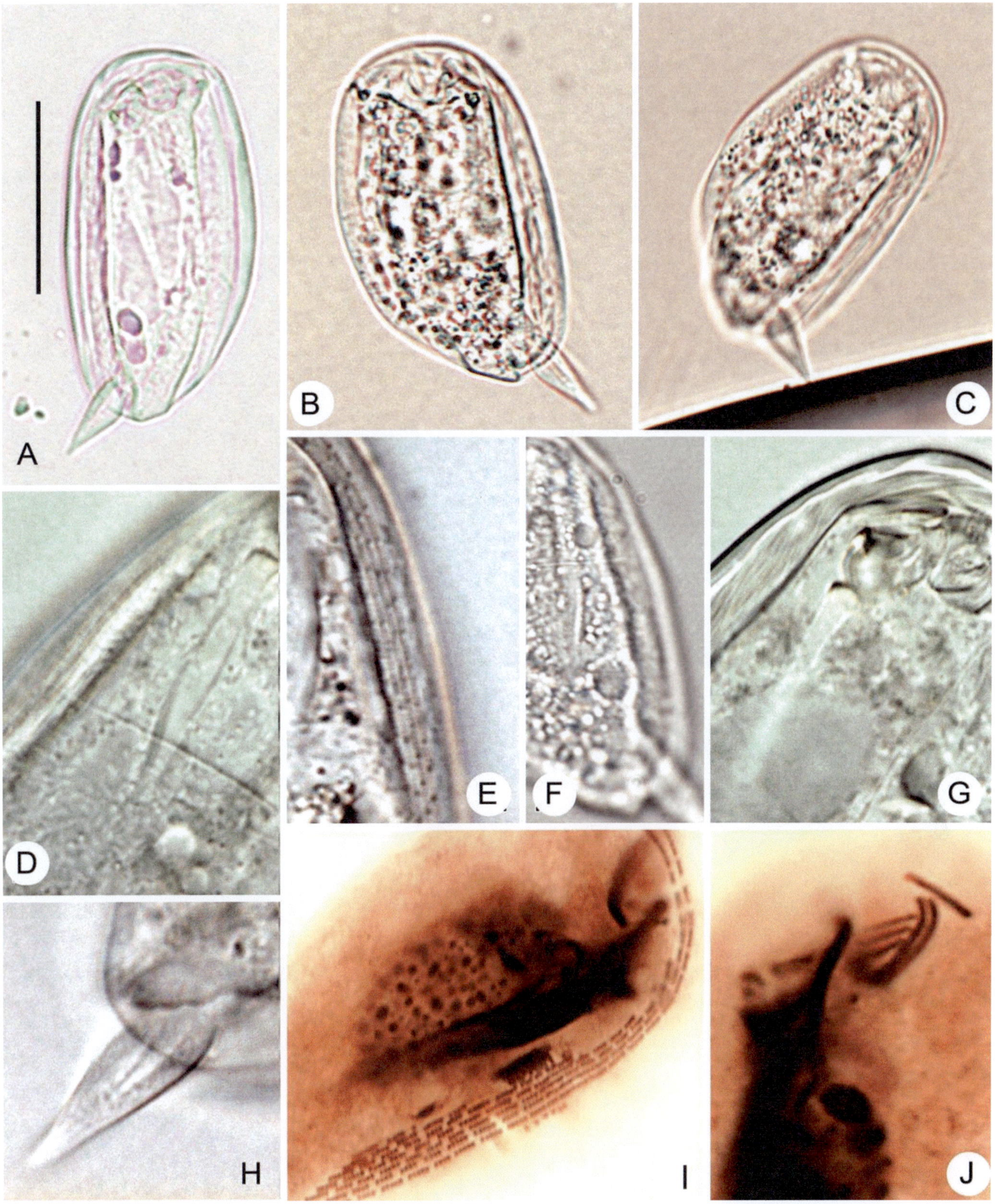


**Fig. 31B** *Dysteria pectinata*
(**a**–**c**) Left-lateral views *of live cells*; (**d**, **g**) cytopharyngeal rods and equatorial transverse stripe; (**e**) ventral groove; (**f**) two contractile vacuoles; (**h**) posterior end of live cell showing podite; (**i**) somatic kineties and cytopharynx; (**j**) oral ciliature. Scale bars: 30 μm

## Reference

Gong J, Song WB, Warren A, Lin X, Roberts D (2007) Microscopical observations on four marine *Dysteria* species (Ciliophora, Cyrtophorida). Eur J Protistol 43:147–161

Class Phyllopharyngea de Puytorac et al., 1974
Order Dysteriida Deroux, 1976
Family Dysteriidae Claparède and Lachmann, 1858
*Dysteria subtropica* Qu et al., 2015

**Ecological feature:** offshore water
**Water temperature:** 19.4 °C
**pH:** 8.2
**Salinity:** 32.6‰

**Fig. 32A** Sample site

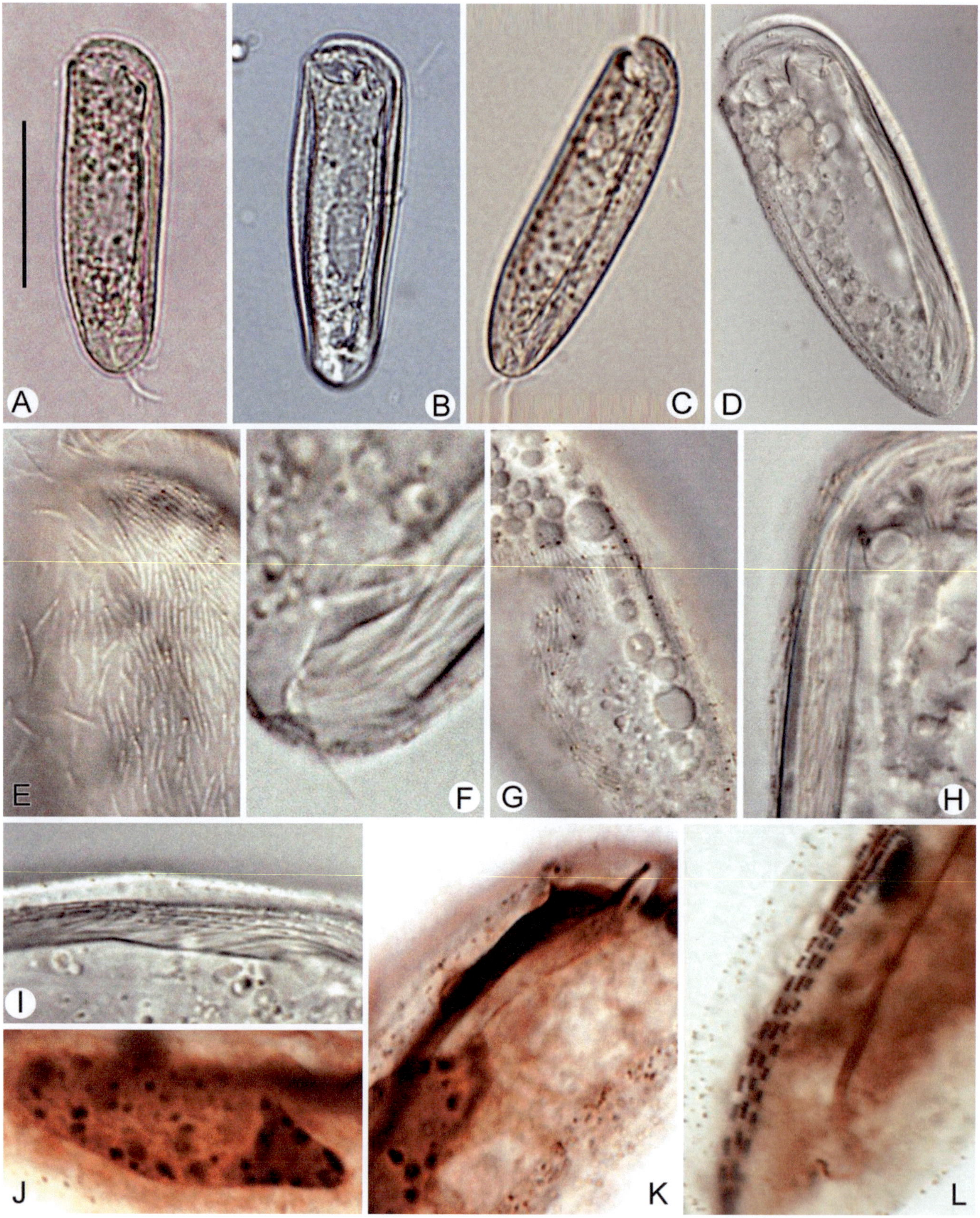


**Fig. 32B** *Dysteria subtropica*
(**a–d**) Lateral views of live cells; (**e**) lateral surface, showing ectosymbiotic bacteria; (**f**) posterior end of cell showing podite; (**g**) contractile vacuoles; (**h**) cytopharyngeal rods and ventral groove; (**i**) ventral groove; (**j**) macronucleus; (**k**) cytopharynx; (**l**) ciliature. Scale bar: 40 μm

## References

Liu W, Gong J, Lin X, Shen Z, Li J, Zhu M, Wang M, Song WB (2008) Redescriptions of two cyrtophorid ciliates, *Dysteria procera* Kahl, 1931 and *Pseudochilodonopsis marina* Song, 1991 (Protozoa, Ciliophora). Acta Hydrobiol Sin 32:84–89 (in Chinese with English abstract)

Qu Z, Wang C, Gao F, Li J, Al-Rasheid KAS, Hu X (2015) Taxonomic studies on seven species of *Dysteria* (Ciliophora, Cyrtophoria), including a description of *Dysteria paraprocera* sp. n. Eur J Protistol 51:241–258

Class Phyllopharyngea de Puytorac et al., 1974
Order Dysteriida Deroux, 1976
Family Dysteriidae Claparède & Lachmann, 1858
*Trochilia alveolata* Pan et al., 2012

**Ecological feature:** intertidal zone
**Water temperature:** 23.0 °C
**pH:** no data available
**Salinity:** 16.0‰

**Fig. 33A** Sample site

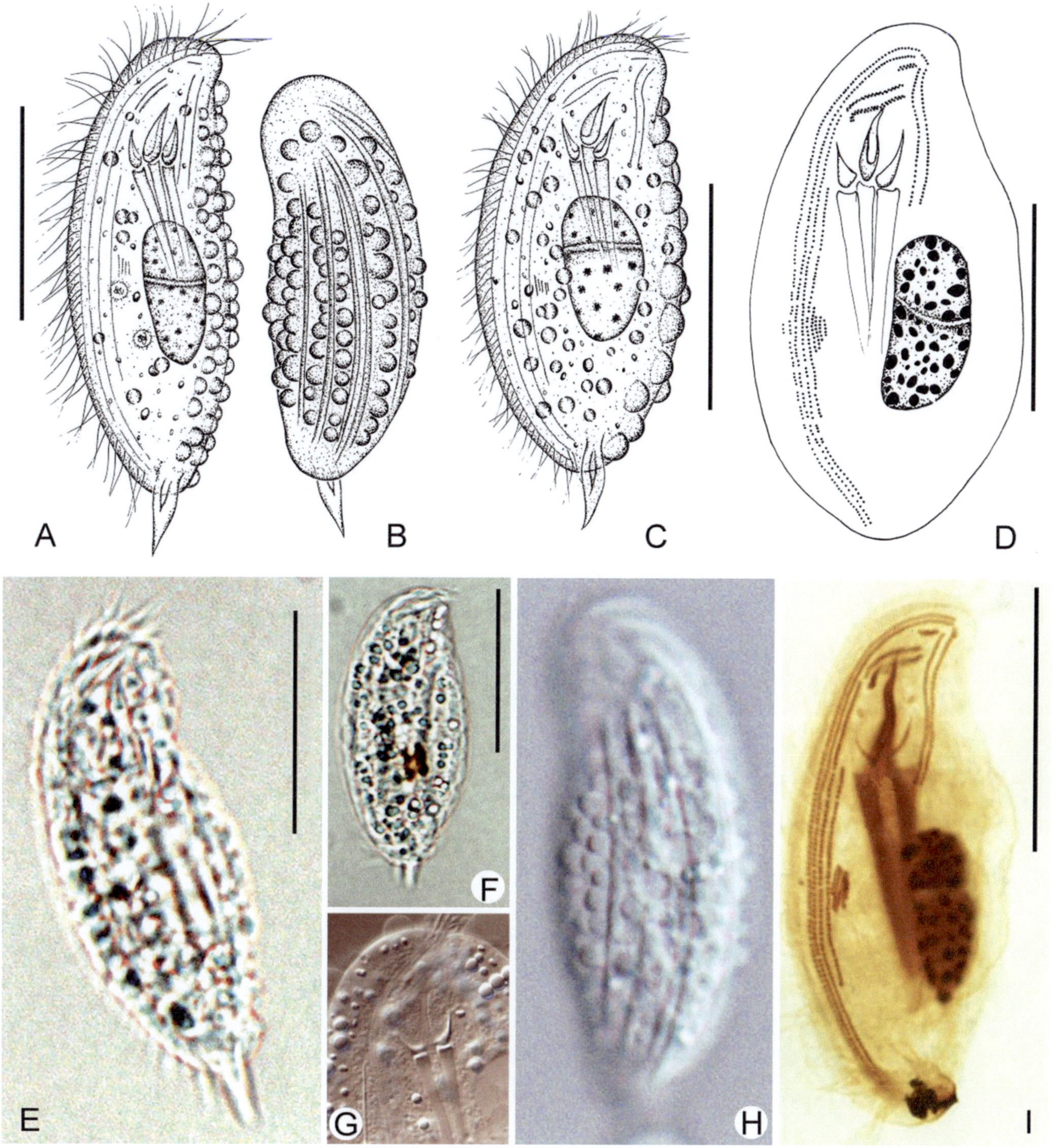


**Fig. 33B** *Trochilia alveolata*
(**a**, **c**, **e**, **f**) Left-lateral views of live cells; (**b**, **h**) right-lateral views of live cells, to show ridges and well-developed alveoli; (**d**, **i**) left-lateral views of ciliature; (**g**) cytopharyngeal rods. Scale bars: 30 μm

## Reference

Pan H, Lin X, Gong J, Al-Rasheid KAS, Song WB (2012) Taxonomy of five species of cyrtophorids (Protozoa: Ciliophora) including consideration of the phylogeny of two new genera. Zool J Linn Soc 164:1–17

Class Phyllopharyngea de Puytorac et al., 1974
Order Dysteriida Deroux, 1976
Family Dysteriidae Claparède & Lachmann, 1858
*Trochilia* sp. 1

**Ecological feature:** mangrove
**Water temperature:** 24.5 °C
**pH:** 6.8
**Salinity:** 17.0‰

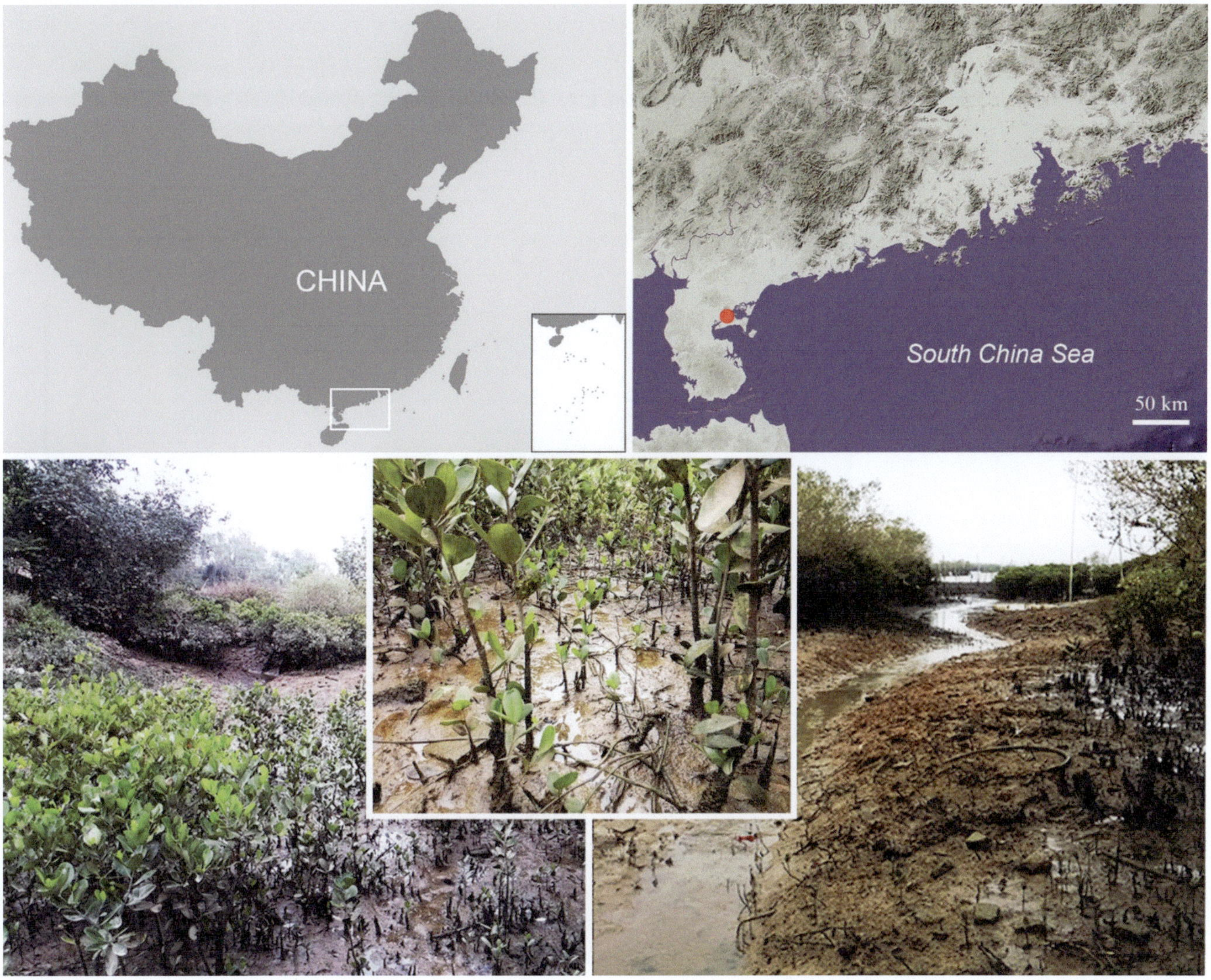


**Fig. 34A** Sample site

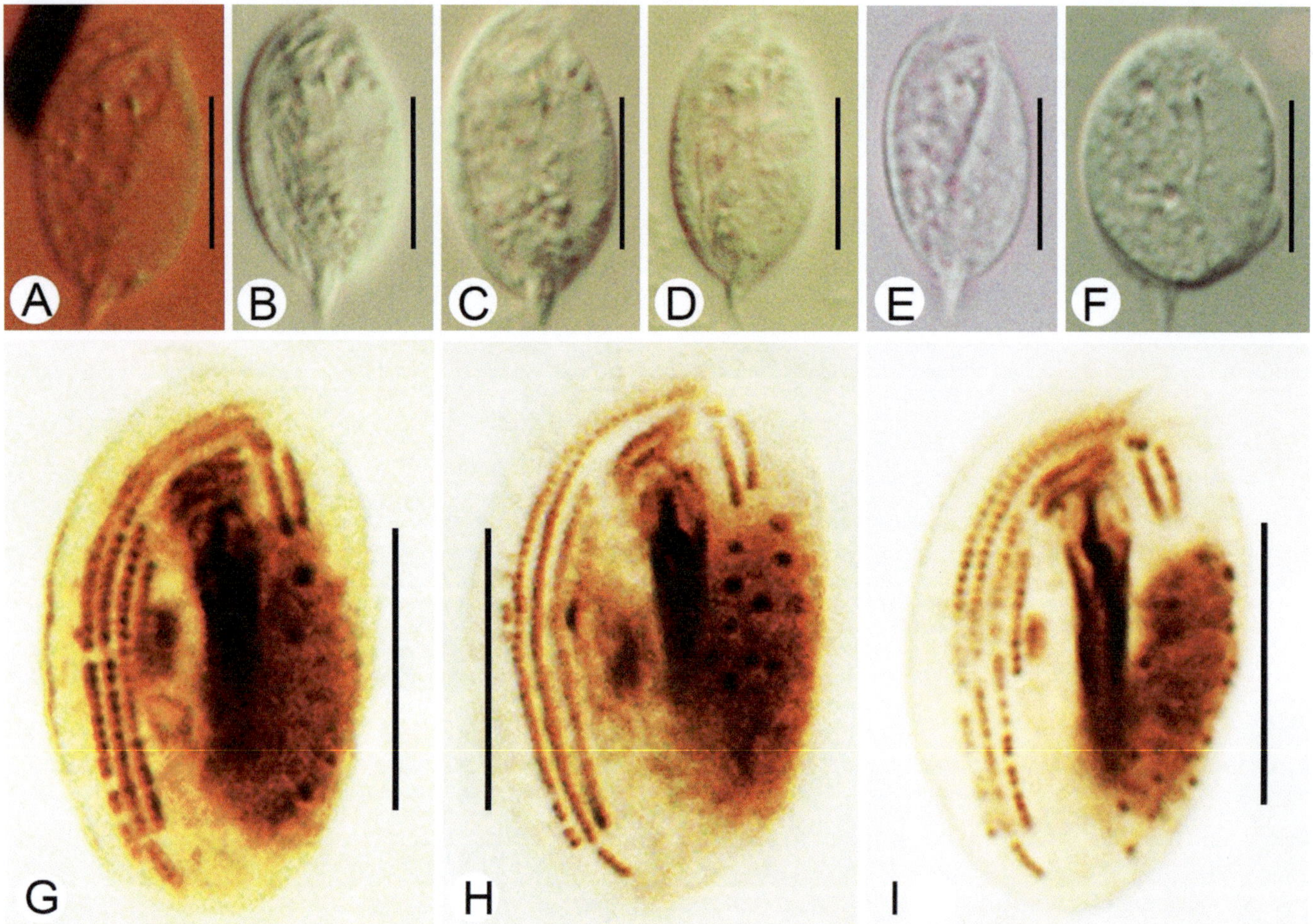


**Fig. 34B** *Trochilia* sp. 1
(**a**–**f**) Ventral views of live cells, showing variations of body shapes; (**g**–**i**) ventral views of ciliature. Scale bars: 10 μm

## Reference

Song WB, Warren A, Hu X (2009) Free-living ciliates in the Bohai and Yellow Seas, China. Science Press, Beijing

Class Phyllopharyngea de Puytorac et al., 1974
Order Dysteriida Deroux, 1976
Family Dysteriidae Claparède & Lachmann, 1858
*Trochilia* sp. 2

**Ecological feature:** brackish water
**Water temperature:** no data available
**pH:** no data available
**Salinity:** 5.0‰

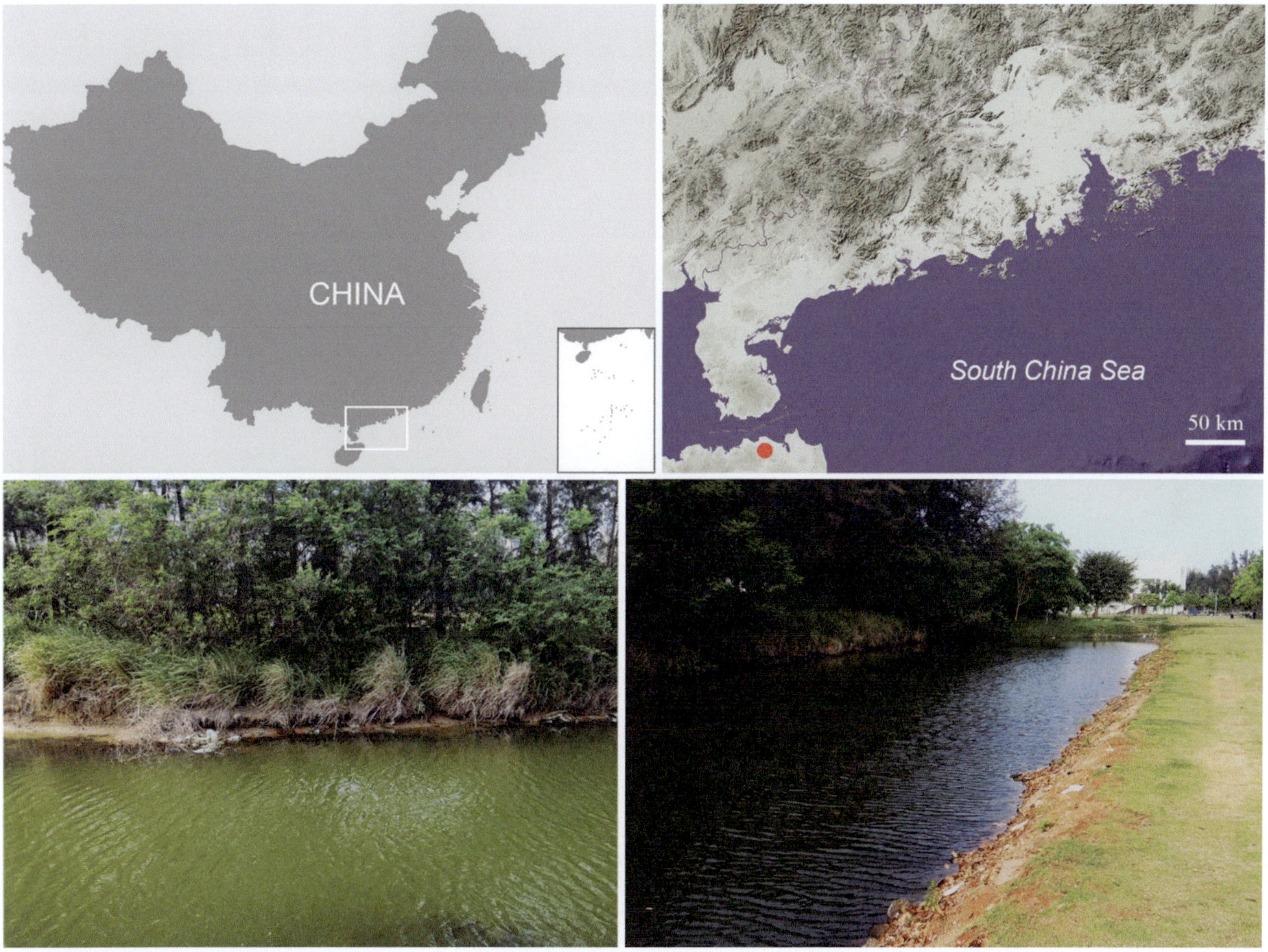


**Fig. 35A** Sample site

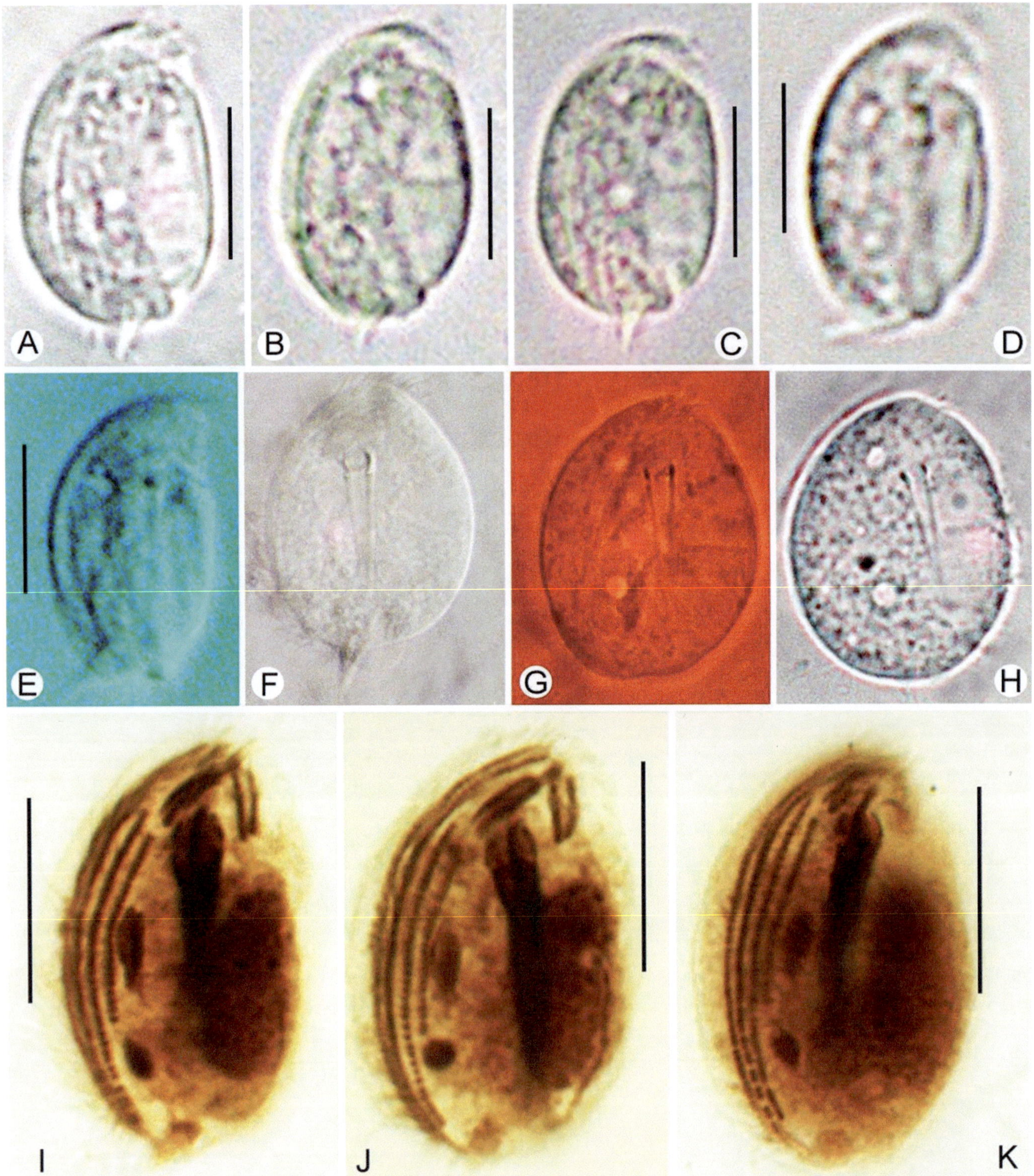


**Fig. 35B** *Trochilia* sp. 2
(**a**–**h**) Left-lateral views of live cells, showing body shapes, cytopharyngeal rods, and contractile vacuoles; (**i**–**k**) left-lateral views of ciliature. Scale bars: 10 μm

## References

Gong J (2005) Taxonomy of cyrtophorid ciliates (Protozoa, Ciliophora) from coastal waters off Qingdao, with systematic revisions on some families and genera. Ocean University of China, Ph. D. thesis

Pan H (2011) Diversity of marine cyrtophorid and pleurostomatid ciliates in China seas. Ocean University of China, Ph. D. thesis

Song W B, Warren A, Hu X (2009) Free-living ciliates in the Bohai and Yellow Seas, China. Science Press, Beijing

Class Phyllopharyngea de Puytorac et al., 1974
Order Dysteriida Deroux, 1976
Family Hartmannulidae Poche, 1913
*Aegyria foissneri* Qu et al., 2017

**Ecological feature:** mangrove
**Water temperature:** 24.8 °C
**pH:** 6.7
**Salinity:** 17.0‰

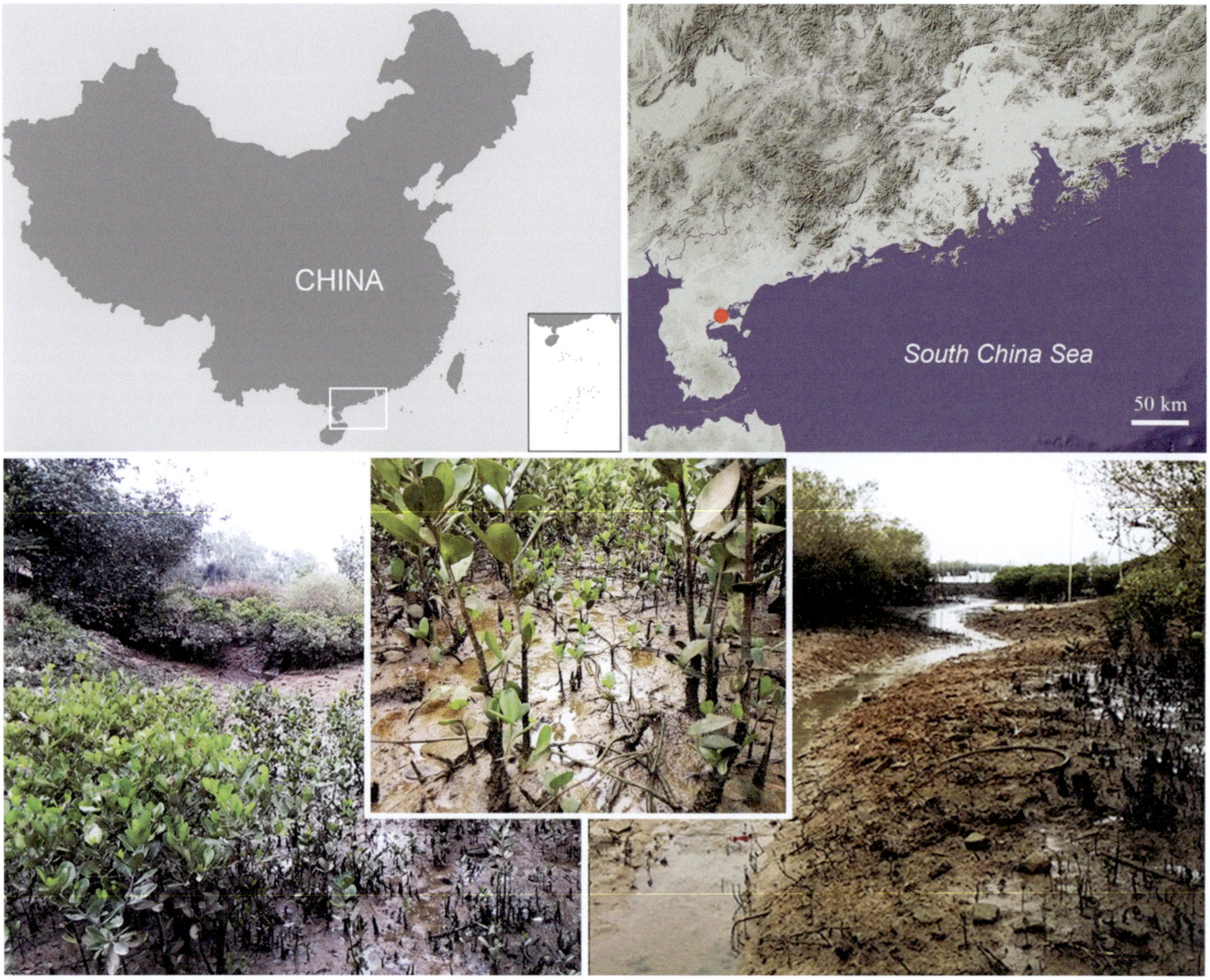


**Fig. 36A** Sample site

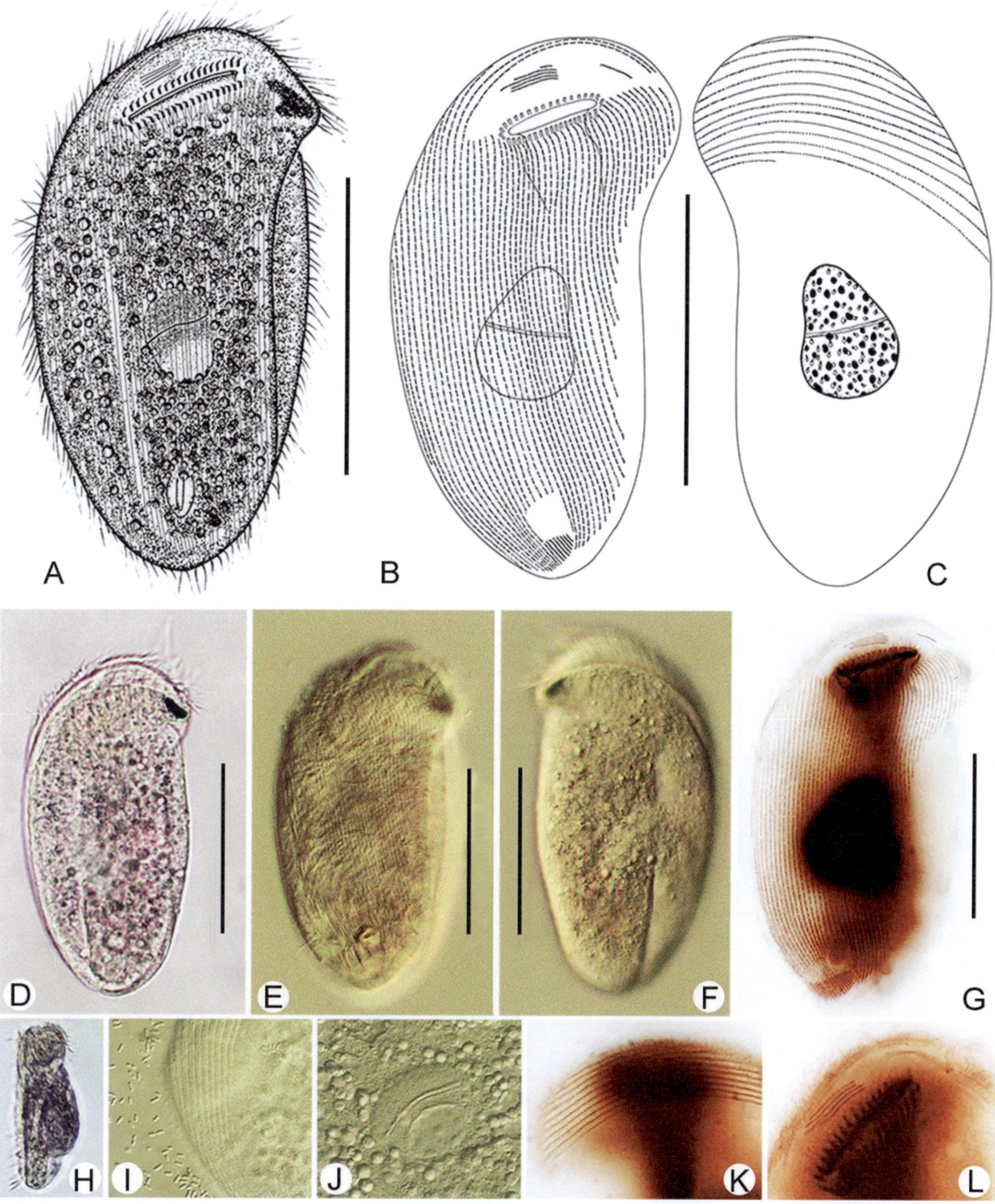


**Fig. 36B** *Aegyria foissneri*
(**a**, **d**, **e**) Ventral views of live individuals; (**b**, **g**, **k**) ventral views of ciliature; (**c**) dorsal view of ciliature; (**f**) dorsal view of live cell; (**h**) lateral view of live cell; (**i**) bar-shaped bacteria; (**j**) macronucleus; (**l**) oral kineties. Scale bars: 50 μm

## Reference

Qu Z, Ma H, Al-Farraj S, Lin X, Hu X (2017) Morphology and molecular phylogeny *Aegyria foissneri* sp. n. and *Lynchella minuta* sp. n. (Ciliophora, Cyrtophoria) from brackish waters of southern China. Eur J Protistol 57:50–60

Class Phyllopharyngea de Puytorac et al., 1974
Order Dysteriida Deroux, 1976
Family Hartmannulidae Poche, 1913
*Aegyria oliva* Claparède & Lachmann, 1859

**Ecological feature:** mangrove (Zhanjiang); offshore water (Huizhou)
**Water temperature:** 17.5 °C; 24.5 °C
**pH:** 8.0; 8.4
**Salinity:** 27.0‰; 28.2‰

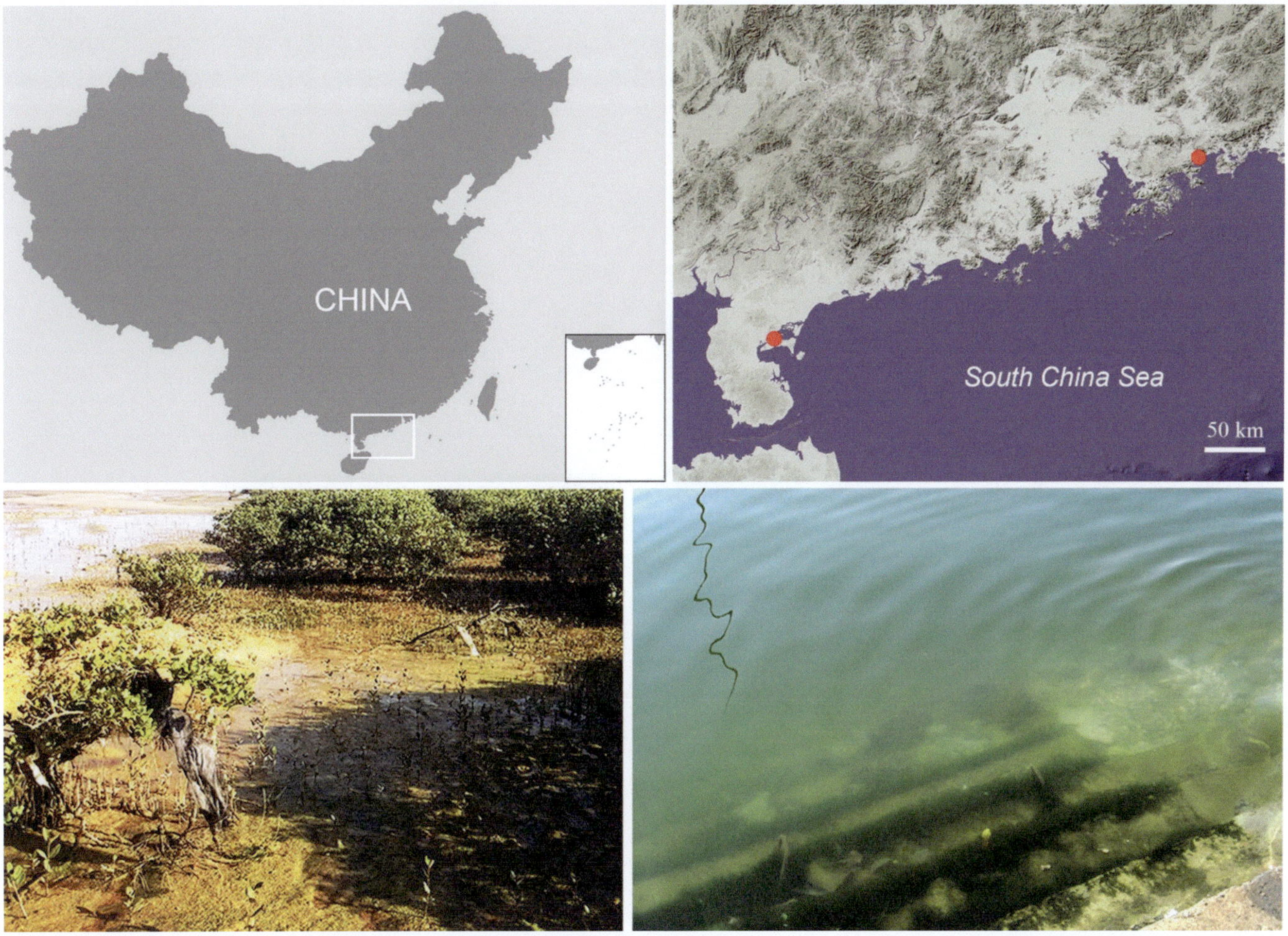


**Fig. 37A** Sample site

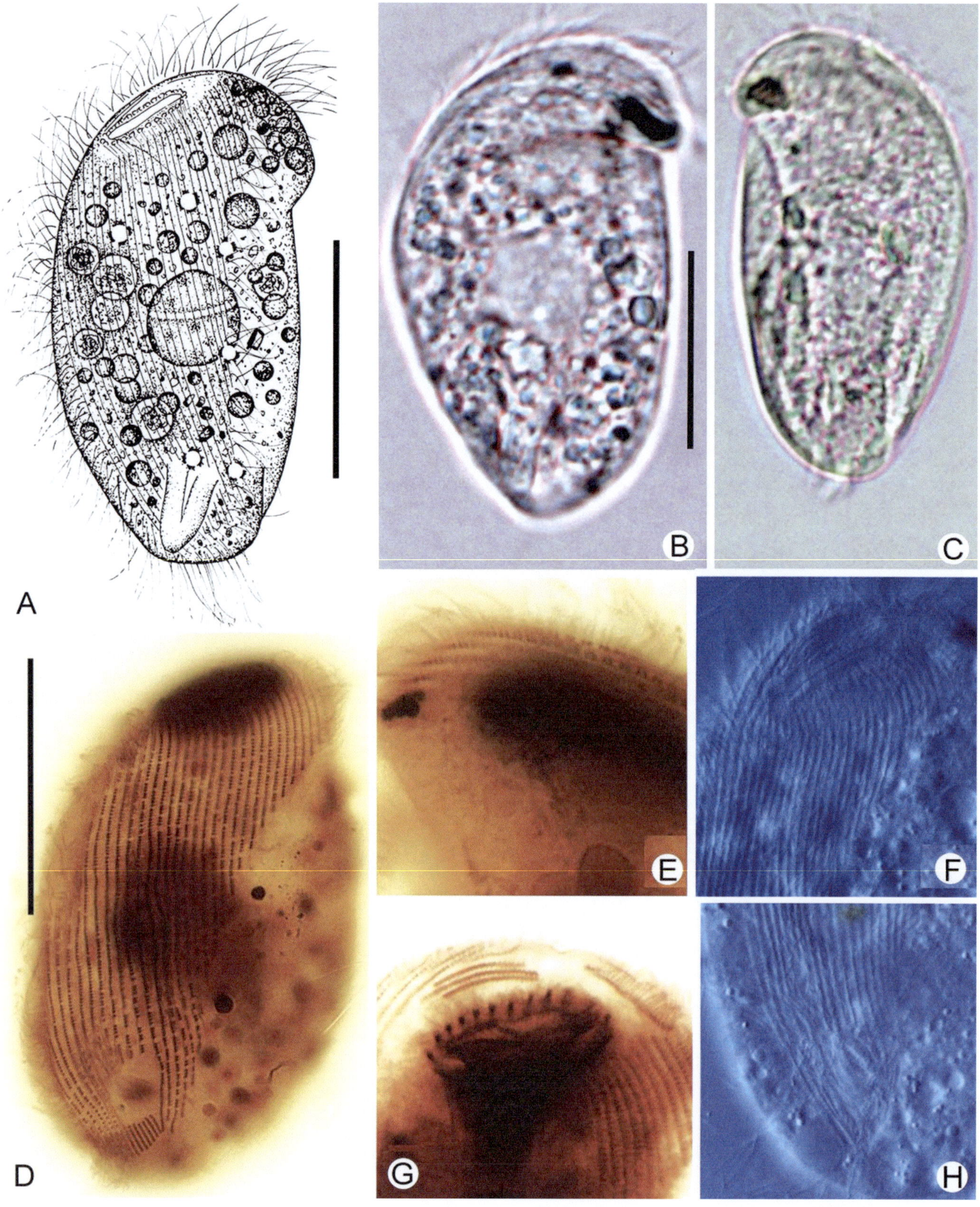


**Fig. 37B** *Aegyria oliva*
(**a**, **b**) Ventral views of live individuals; (**c**) dorsal view of live cell; (**d**) ventral view of ciliature; (**e**) dorsal view of anterior part of ciliature; (**f**) oral split; (**g**) oral kineties; (**h**) podite in posterior end of cell. Scale bars: 40 μm (**a**, **d**); 30 μm (**b**)

## Reference

Deroux G, Dragesco J (1968) Nouvelles données sur quelques Ciliés Holotriches Cyrtophores a Ciliature Ventrale. Protistologica 4:365–404

Song WB, Warren A, Hu X (2009) Free living ciliates in the Bohai and Yellow Seas, China. Science Press, Beijing

Class Phyllopharyngea de Puytorac et al., 1974
Order Dysteriida Deroux, 1976
Family Hartmannulidae Poche, 1913
*Aegyria rostellum* Chen et al., 2012

**Ecological feature:** mangrove
**Water temperature:** 25.0 °C
**pH:** 8.2
**Salinity:** 16.0‰

**Fig. 38A** Sample site

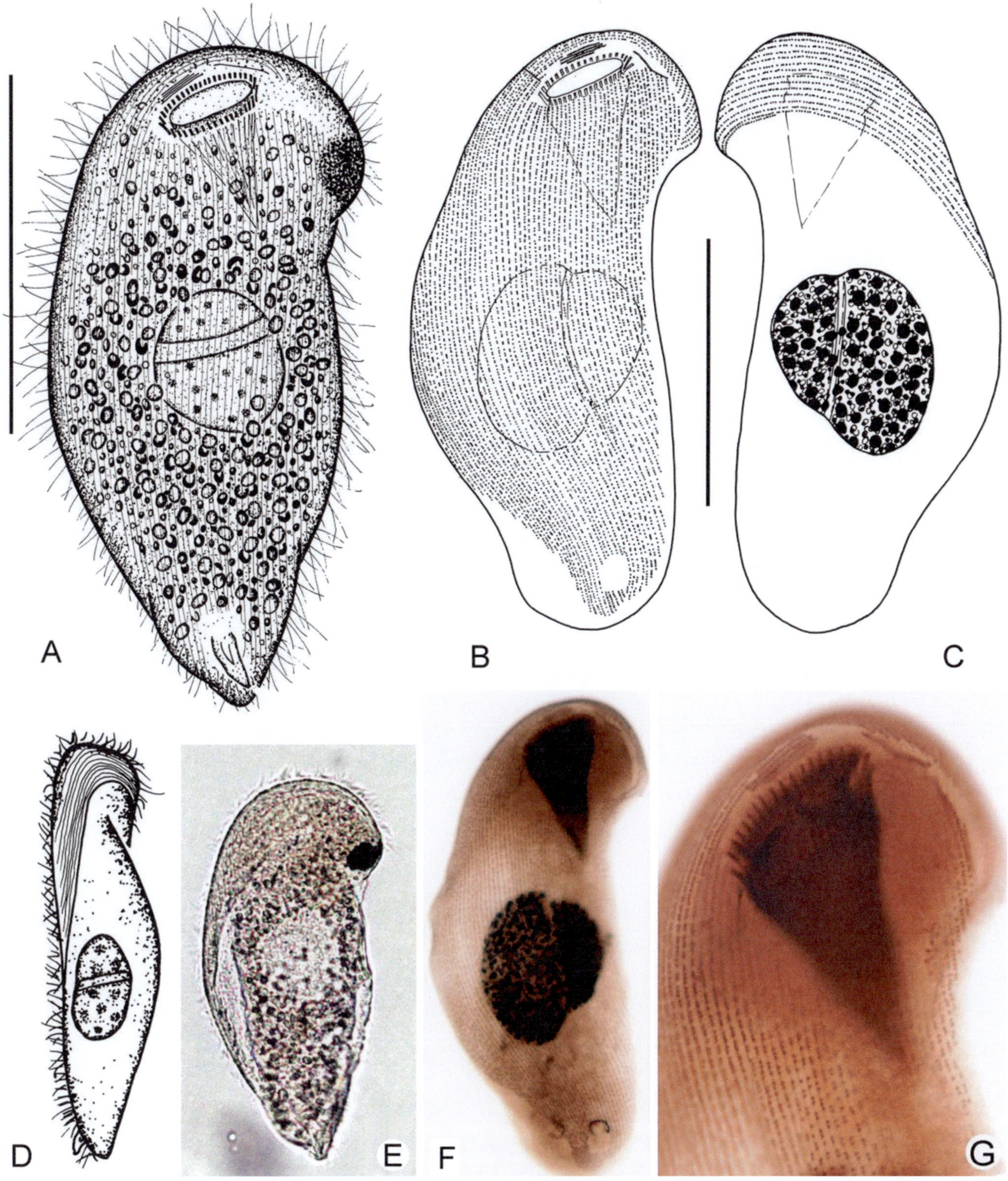


**Fig. 38B** *Aegyria rostellum*
(**a**, **e**) Ventral views of live individuals; (**b**, **c**, **f**) ventral (**b**, **f**) and dorsal (**c**) views of ciliature; (**d**) left-lateral view of live cell; (**g**) oral kineties and nematodesmal rods. Scale bars: 100 μm

## Reference

Chen X, Hu X, Gong J, Al-Rasheid KAS, Al-Farraj SA, Song WB (2012) Morphology and infraciliature of two new marine cyrtophorid ciliates, *Paracyrtophoron tropicum* nov. gen., nov. spec. and *Aegyria rostellum* nov. spec. (Ciliophora, Cyrtophorida), isolated from tropical waters in southern China. Eur J Protistol 48:63–72

Class Phyllopharyngea de Puytorac et al., 1974
Order Dysteriida Deroux, 1976
Family Hartmannulidae Poche, 1913
*Aporthotrochilia pulex* (Deroux, 1976) Pan et al., 2012

**Ecological feature:** intertidal zone
**Water temperature:** 24.0 °C
**pH:** no data available
**Salinity:** 27.0‰

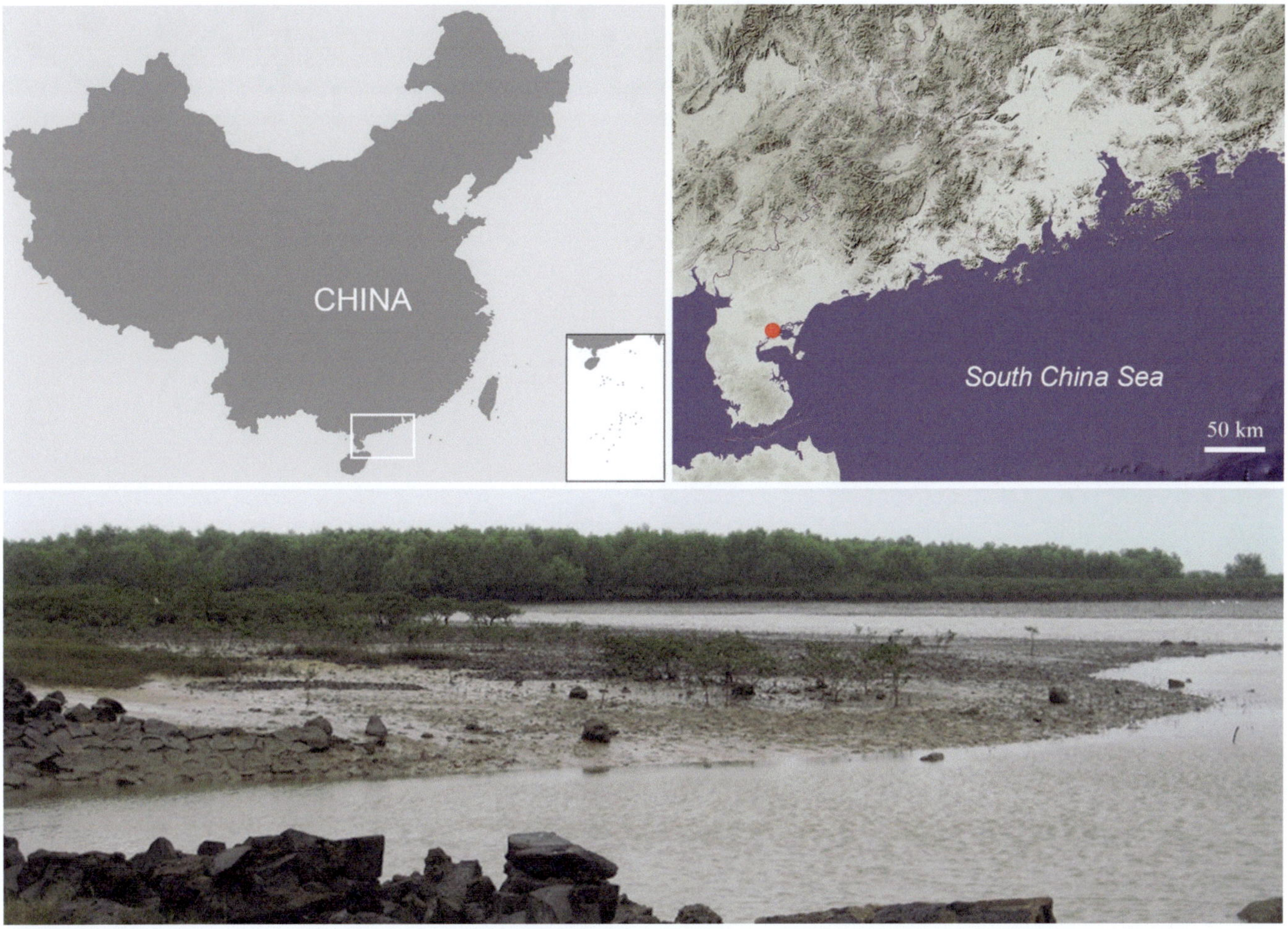


**Fig. 39A** Sample site

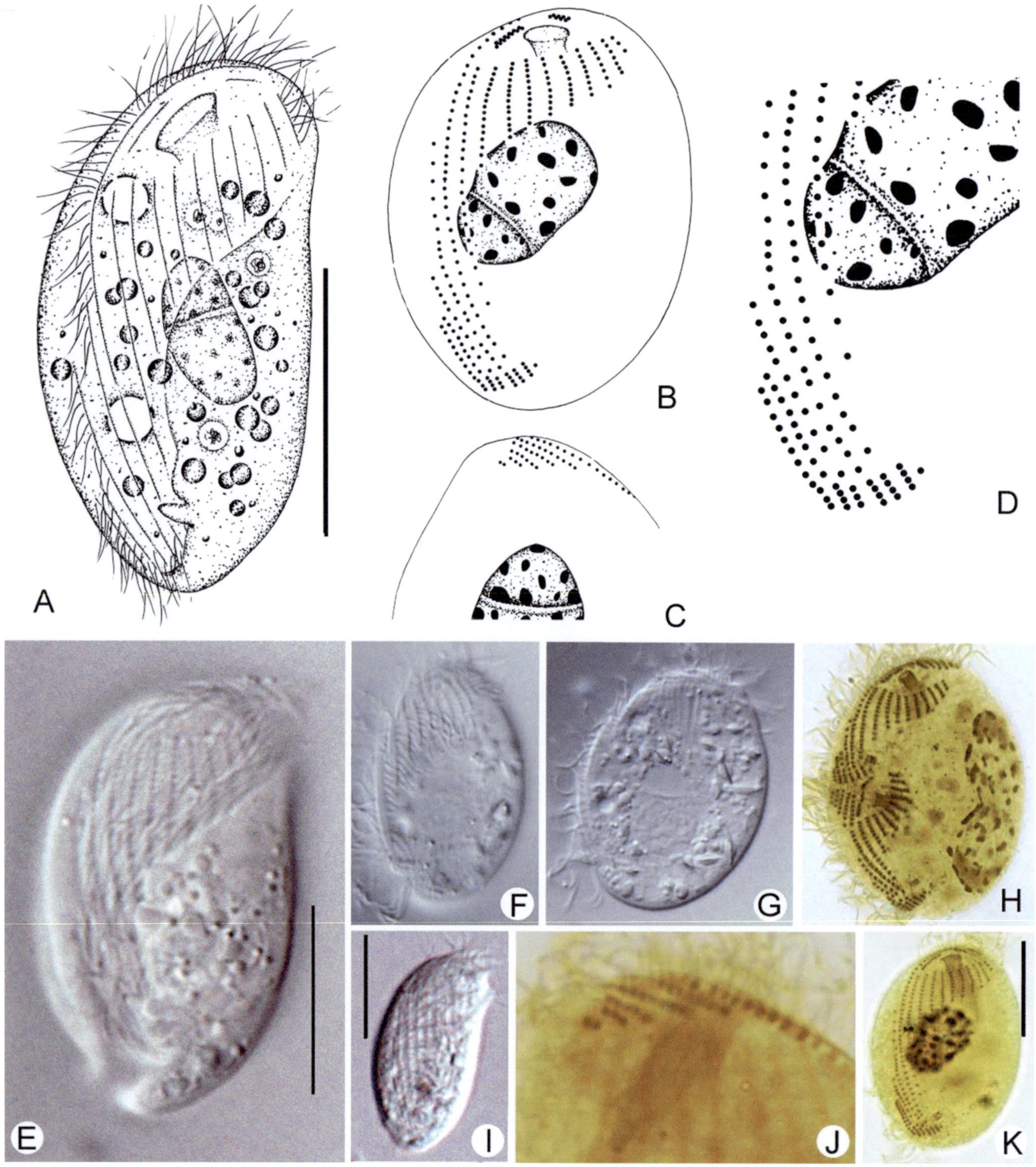


**Fig. 39B** *Aporthotrochilia pulex*
(**a**, **e**, **f**, **i**) Ventral views of live cells; (**b**, **k**) ventral views of ciliature; (**c**, **j**) dorsal views of anterior portion of cell; (**d**) posterior part of somatic kineties; (**g**) cytopharynx, macronucleus, and contractile vacuoles; (**h**) ventral view of ciliature of a divider. Scale bars: 20 μm

## Reference

Pan H, Lin X, Gong J, Al-Rasheid KAS, Song WB (2012) Taxonomy of five species of cyrtophorids (Protozoa: Ciliophora) including consideration of the phylogeny of two new genera. Zool J Linn Soc 164:1–17

Class Phyllopharyngea de Puytorac et al., 1974
Order Dysteriida Deroux, 1976
Family Hartmannulidae Poche, 1913
*Brooklynella sinensis* Gong & Song, 2006

**Ecological feature:** mangrove (Zhanjiang); intertidal zone (Daya Bay)
**Water temperature:** no data available; 30.2 °C
**pH:** no data available; 8.3
**Salinity:** no data available; 31.8‰

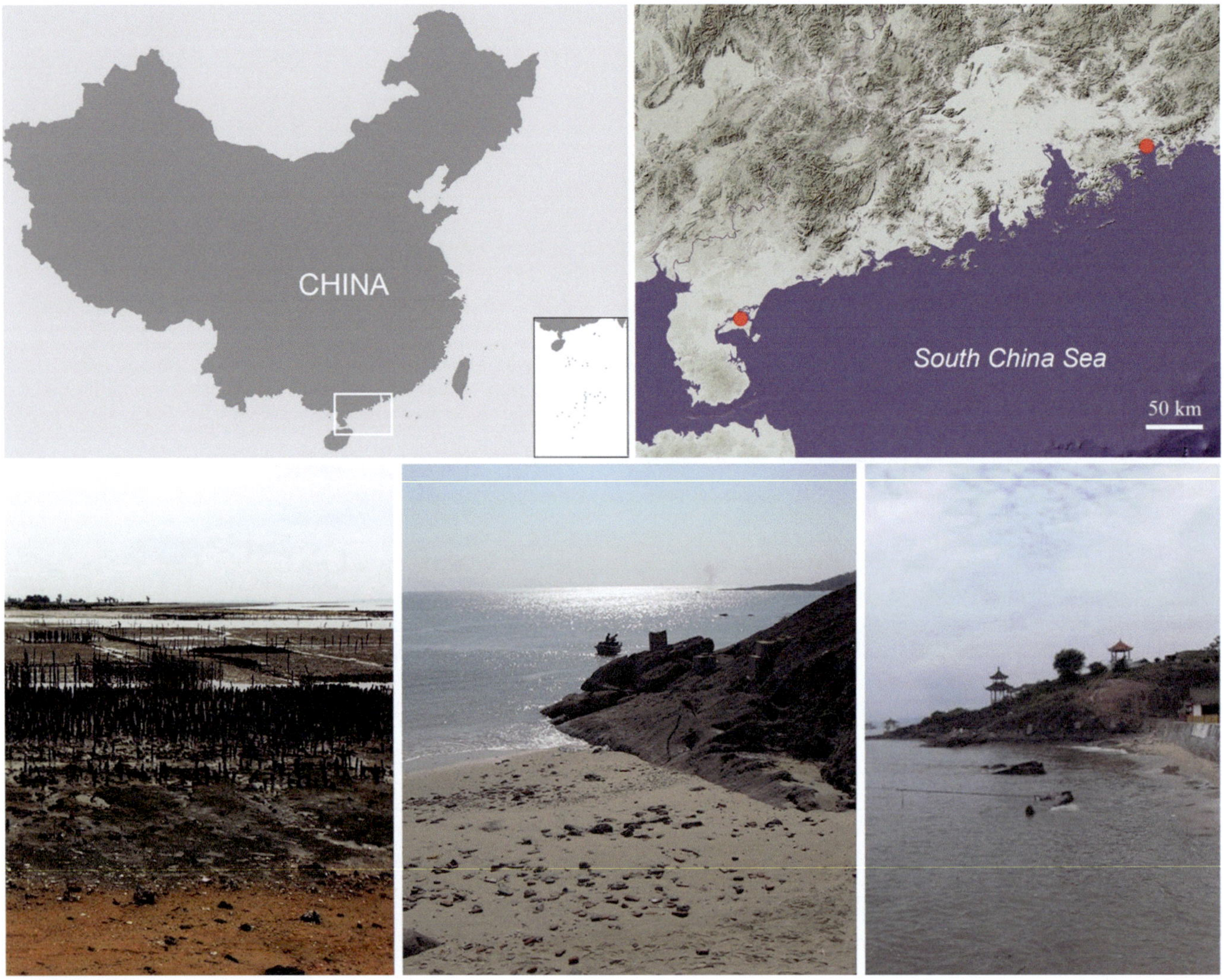


**Fig. 40A** Sample site

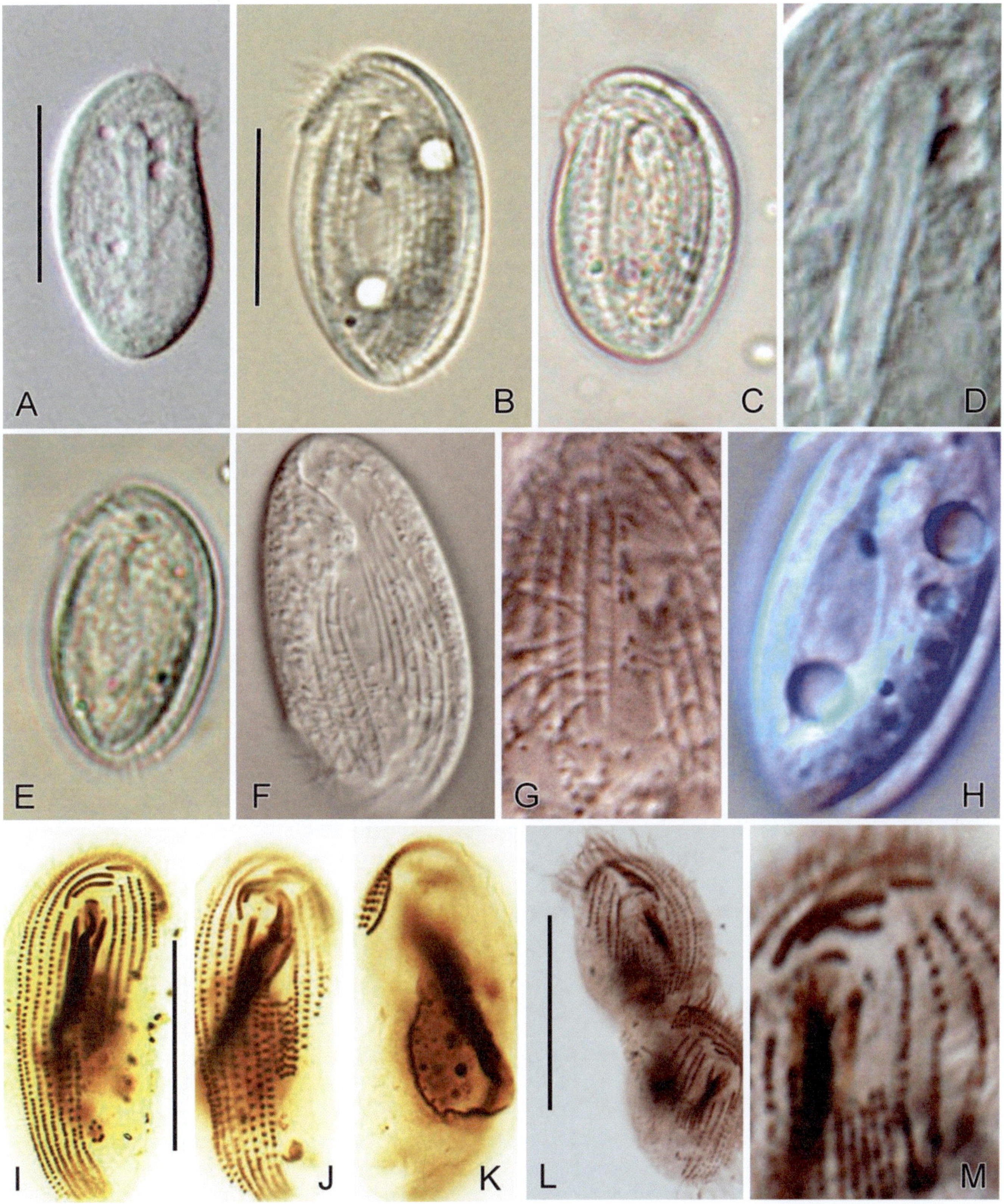


**Fig. 40B** *Brooklynella sinensis*
(**a**, **f**) Ventral views of live individuals; (**b**, **c**, **e**) dorsal views of live cells, showing contractile vacuoles; (**d**) cytopharyngeal rods; (**g**) somatic cilia; (**h**) two contractile vacuoles; (**i**–**k**) ventral (**i**, **j**) and dorsal (**k**) views of ciliature; (**l**) ventral view of ciliature of a late divider; (**m**) oral kineties. Scale bars: 20 μm

## Reference

Gong J, Song WB (2006) Description of a new marine cyrtophorid ciliate, *Brooklynella sinensis* n. sp. from the China Sea with a new definition of the genus *Brooklynella* (Protozoa, Ciliophora, Cyrtophorida). Zootaxa 1113:41–49

Class Phyllopharyngea de Puytorac et al., 1974
Order Dysteriida Deroux, 1976
Family Hartmannulidae Poche, 1913
*Heterohartmannula fangi* Pan et al., 2012

**Ecological feature:** intertidal zone
**Water temperature:** 24.0 °C
**pH:** no data available
**Salinity:** 27.0‰

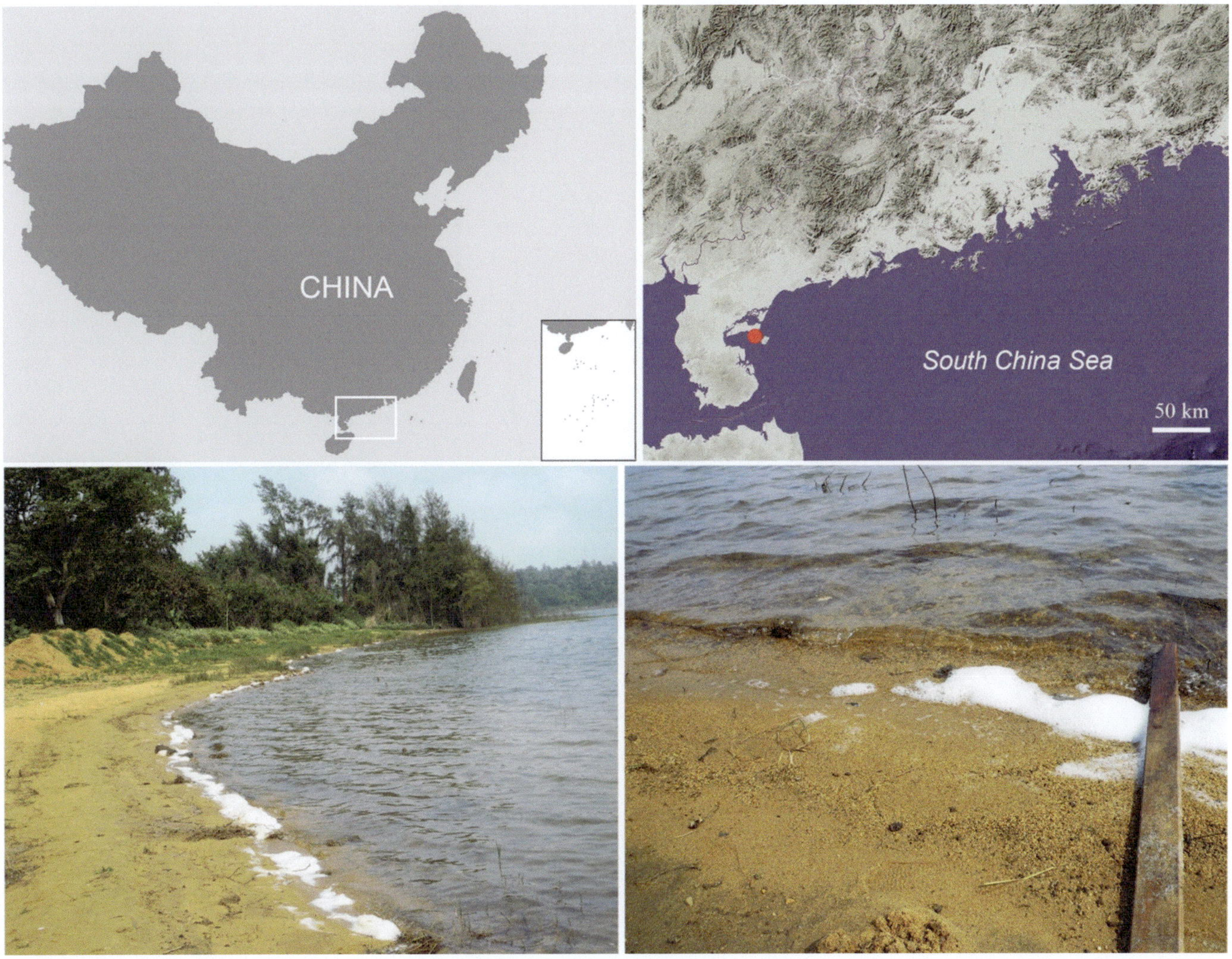


**Fig. 41A** Sample site

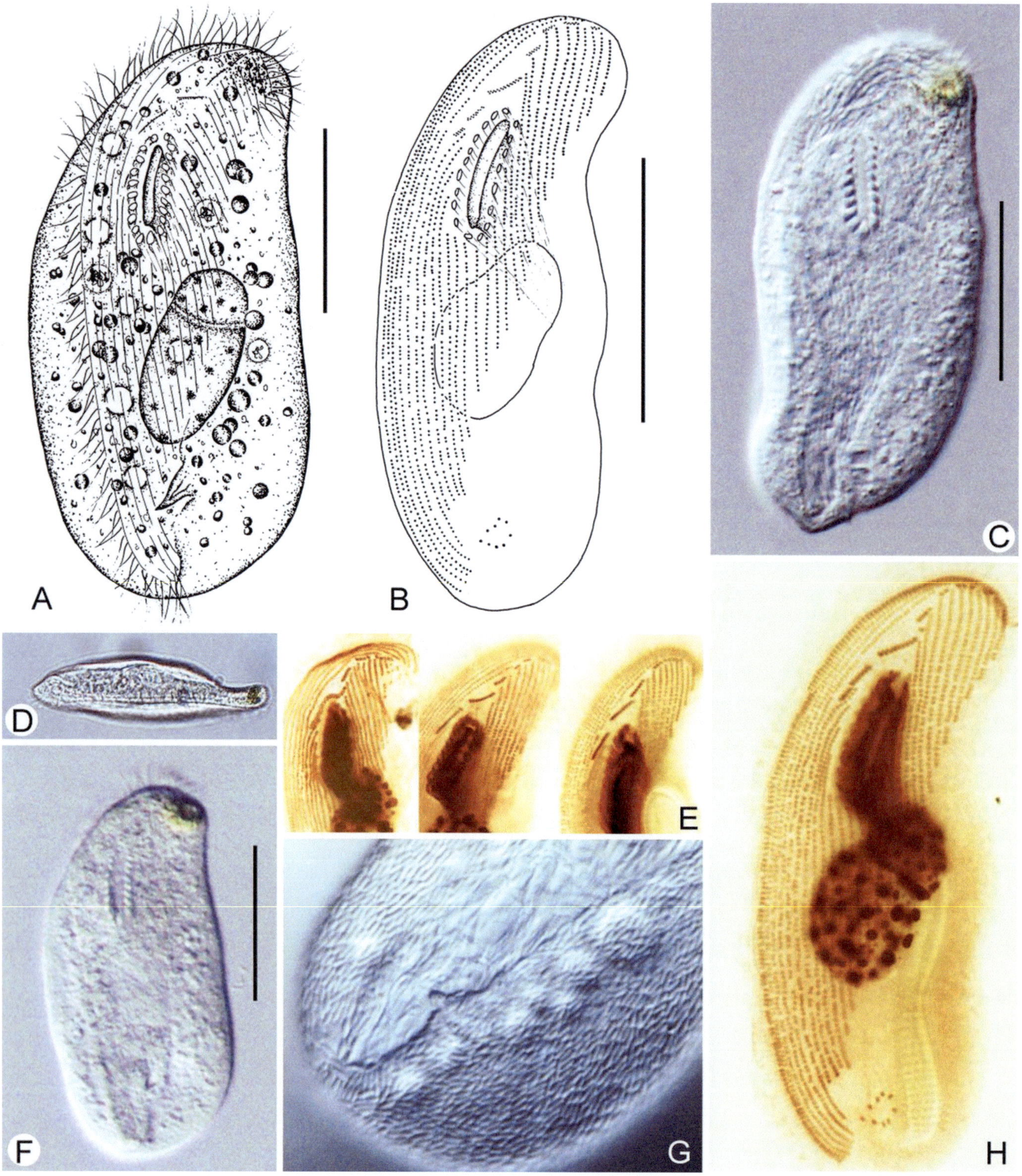


**Fig. 41B** *Heterohartmannula fangi*
(**a**, **c**, **f**) Ventral views of live cells; (**b**, **h**) ventral ciliature; (**d**) lateral view of live cell; (**e**) anterior portion of different individuals, to show oral ciliature; (**g**) posterior ventral view of live cell, showing ectosymbiotic bacteria. Scale bars: 40 μm

## Reference

Pan H, Lin X, Gong J, Al-Rasheid KAS, Song WB (2012) Taxonomy of five species of cyrtophorids (Protozoa: Ciliophora) including consideration of the phylogeny of two new genera. Zool J Linn Soc 164:1–17

Class Phyllopharyngea de Puytorac et al., 1974
Order Hypocomatida Deroux, 1976
Family Hypocomidae Bütschli, 1889
*Hypocoma* sp.

**Ecological feature:** mangrove
**Water temperature:** no data available
**pH:** no data available
**Salinity:** 5.0‰

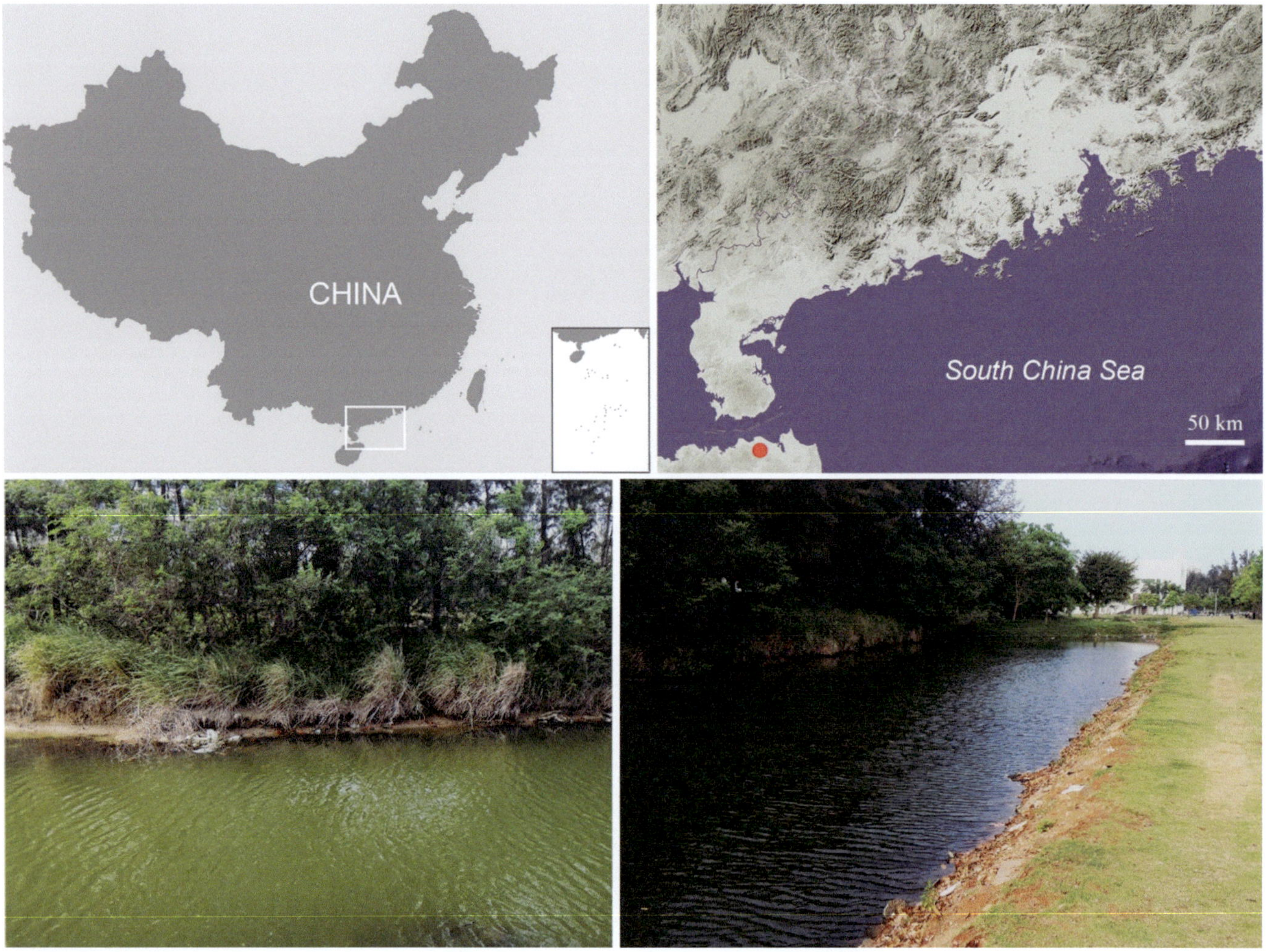


**Fig. 42A** Sample site

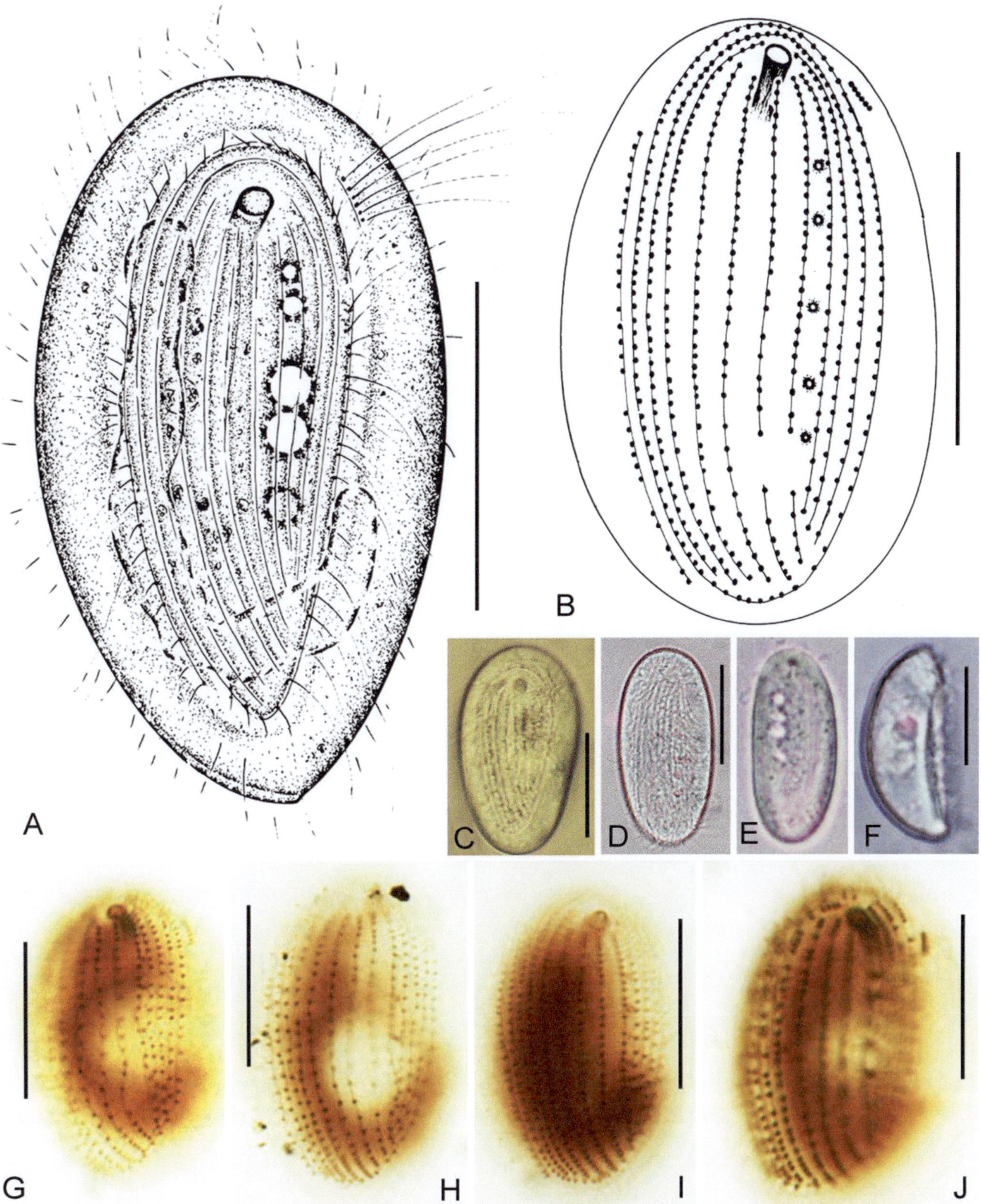


**Fig. 42B** *Hypocoma* sp.
(**a**, **c**, **d**) Ventral views of live cells; (**b**, **g–j**) ventral views of ciliature; (**e**) dorsal view of live cell; (**f**) lateral view of live cell. Scale bars: 15 μm

## References

Chen X (2009) Biodiversity of the marine ciliates: Gymnostomata, Hypostomata, Suctoria, Spirotrichea. Ocean University of China, Ph. D. thesis

Pan H (2011) Diversity of marine cyrtophorid and pleurostomatid ciliates in China seas. Ocean University of China, Ph. D. thesis

Class Phyllopharyngea de Puytorac et al., 1974
Order Endogenida Collin, 1912
Family Acinetidae Stein, 1859
*Acineta tuberosa* Ehrenberg, 1833

**Ecological feature:** offshore water
**Water temperature:** 26.0 °C
**pH:** 8.0
**Salinity:** 31.0‰

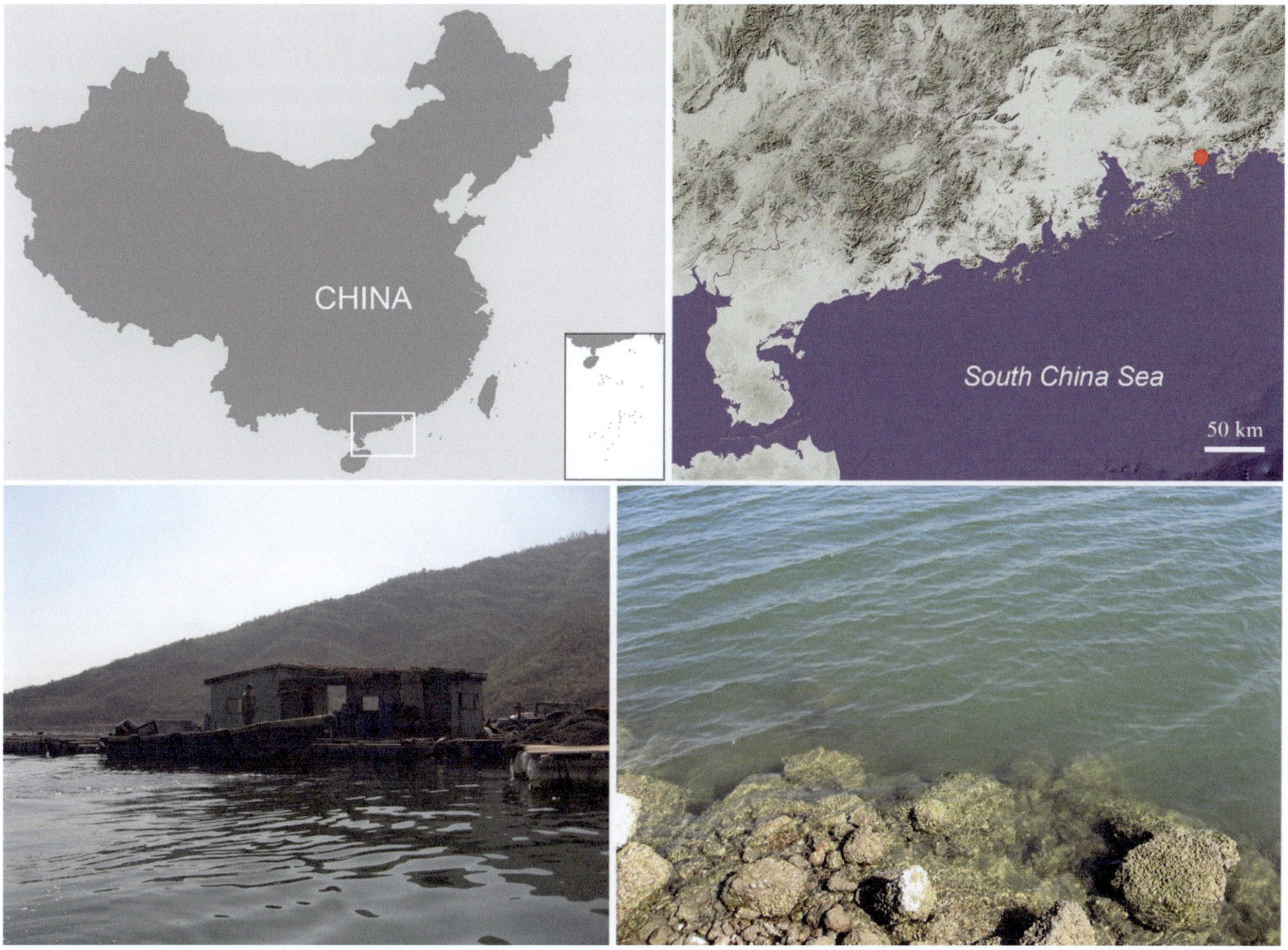


**Fig. 43A** Sample site

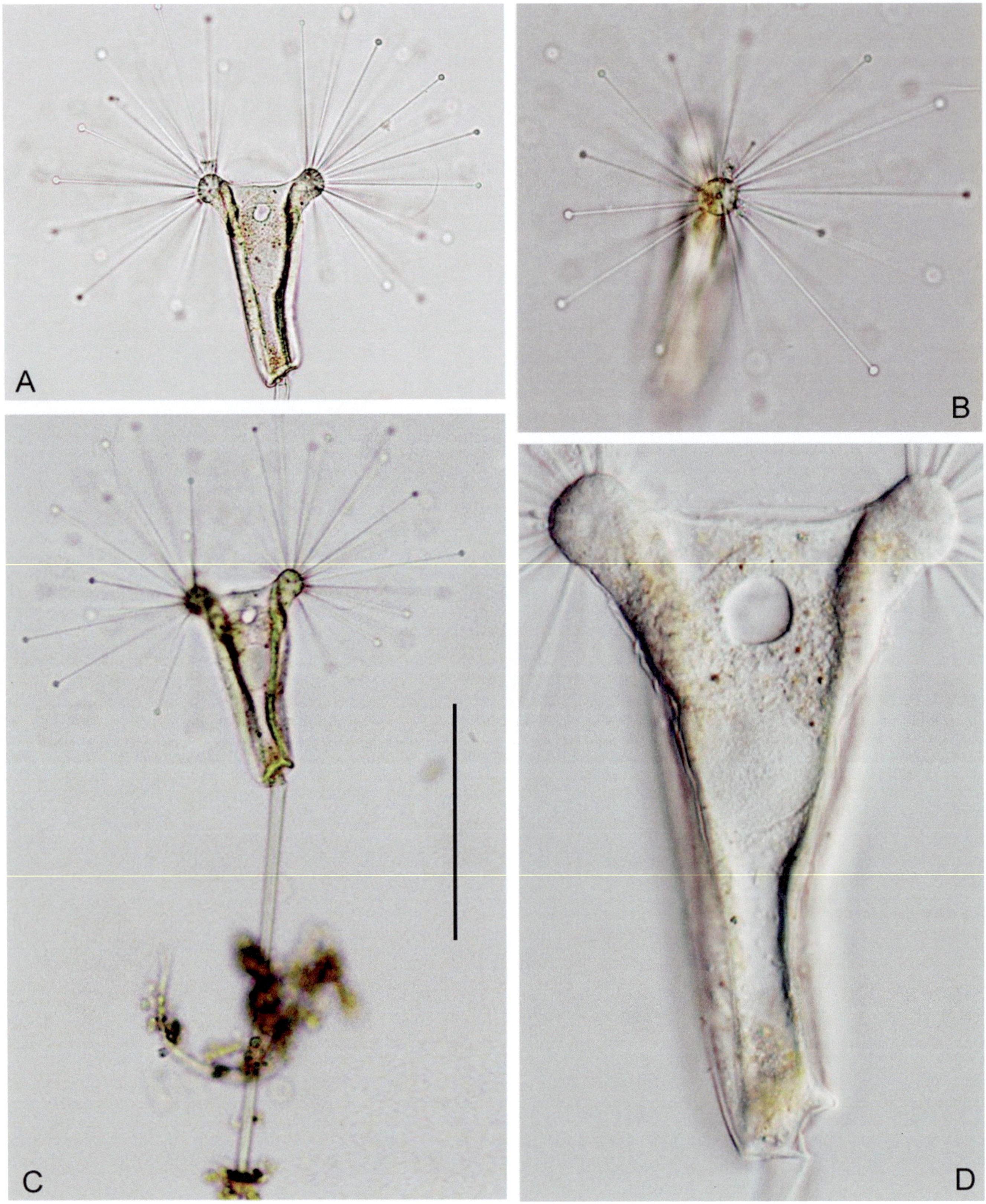


**Fig. 43B** *Acineta tuberosa*
(**a**, **c**) Lateral views of live individuals; (**b**) showing tentacles; (**d**) contractile vacuole, macronucleus, and lorica. Scale bar: 20 μm

## Reference

Chen X, Song WB, Hu X (2005) Morphology of suctorid ciliates from coastal water off Qingdao, with description of a new genus and a new species (Protozoa, Ciliophora). Acta Zootaxonom Sin 30:493–500

# Class Prostomatea

Xiaozhong Hu, Xiaofeng Lin, and Weibo Song

Class Prostomatea Schewiakoff, 1896
Order Prostomatida Schewiakoff, 1896
Family Colepidae Ehrenberg, 1838
*Coleps hirtus hirtus* (Müller, 1786) Nitzsch, 1827

**Ecological feature:** mangrove
**Water temperature:** 21.0 °C
**pH:** 7.0
**Salinity:** 27.0‰

X. Hu (✉)
Fisheries College, Ocean University of China,
Qingdao, Shandong, China
e-mail: xiaozhonghu@ouc.edu.cn

X. Lin
College of Life Science, South China Normal University,
Guangzhou, Guangdong, China
e-mail: xlin@oucscnu.edu.cn

W. Song
Institute of Evolution & Marine Biodiversity,
Ocean University of China, Qingdao, Shandong, China
e-mail: wsong@ouc.edu.cn


X. Hu et al. (eds.), *Ciliate Atlas: Species Found in the South China Sea*,
https://doi.org/10.1007/978-981-13-5901-9_8

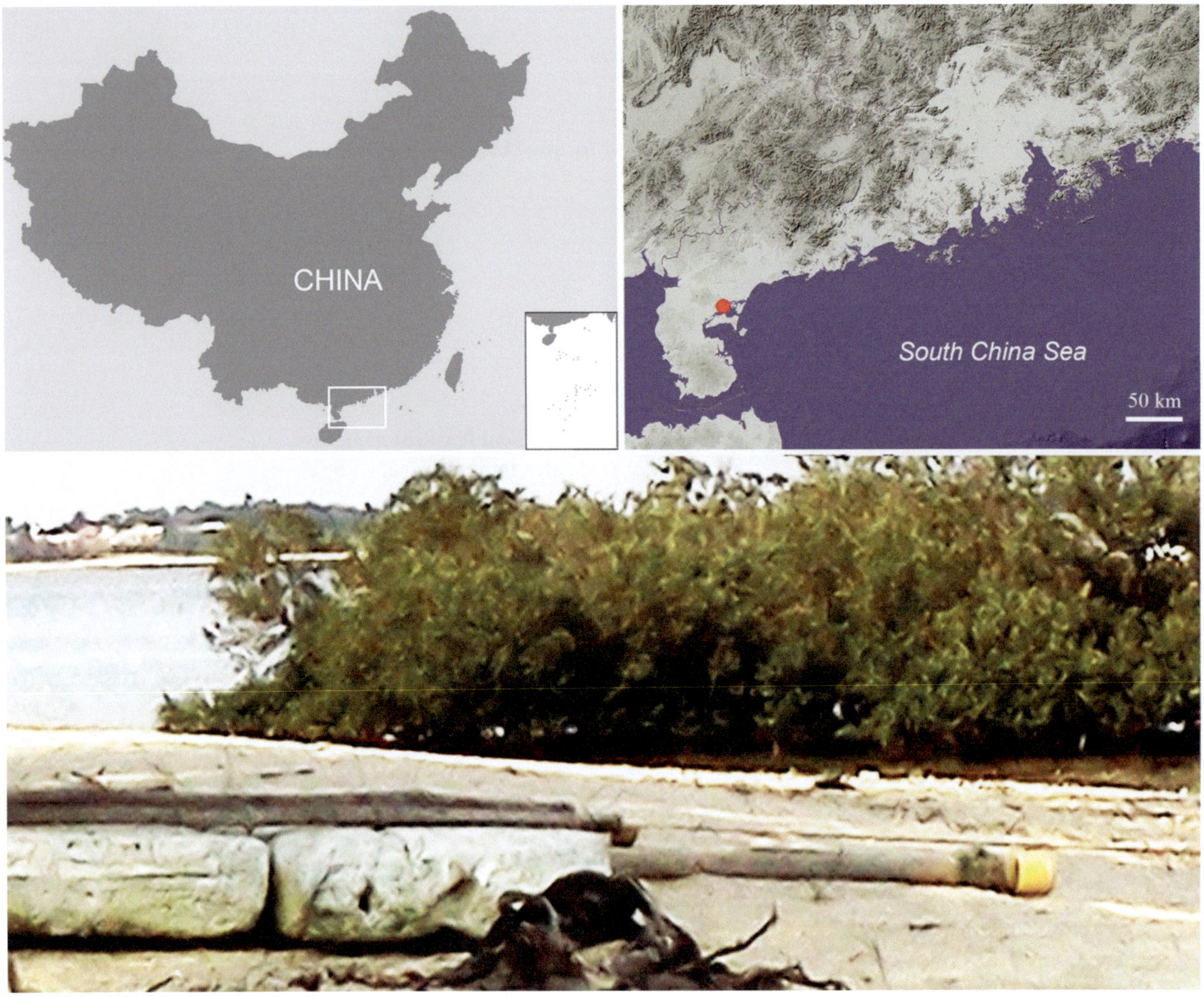


**Fig. 1A** Sample site

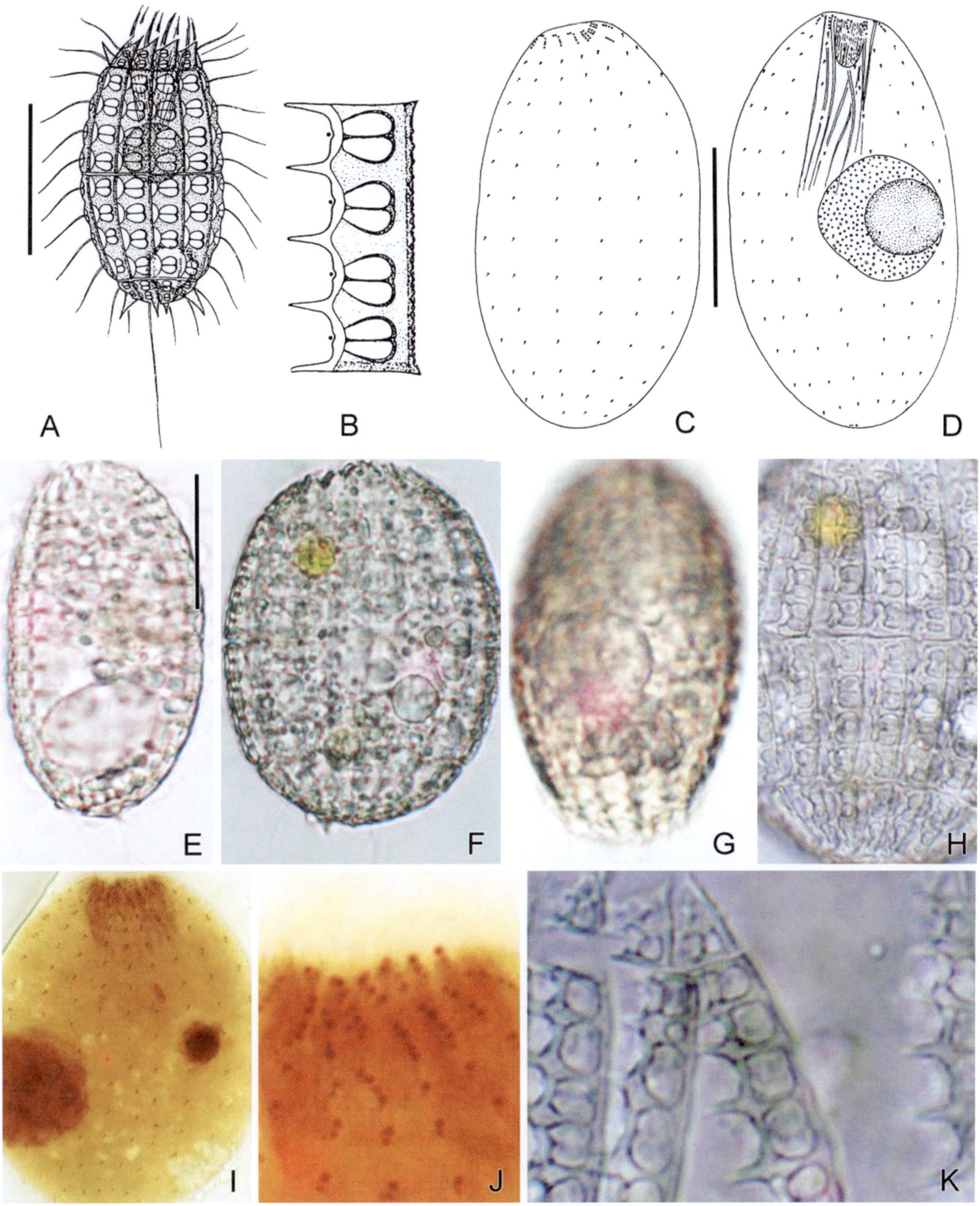


**Fig. 1B** *Coleps hirtus hirtus*
(**a**) Lateral view of live cell; (**b**) details of anterior main plate; (**c**, **d**) ciliature and macronucleus; (**e–g**) different sizes of body in vivo; (**h**) plates on cell surface; (**i**) ciliature; (**j**) buccal apparatus; (**k**) details of anterior main plate. Scale bars: 30 μm

## Reference

Foissner W (1994) Progress in taxonomy of planktonic freshwater ciliates. Mar Microbial Food Webs 8:9–35

Class Prostomatea Schewiakoff, 1896
Order Prostomatida Schewiakoff, 1896
Family Colepidae Ehrenberg, 1838
*Pinacocoleps* sp.

**Ecological feature:** mangrove
**Water temperature:** 26.1 °C
**pH:** 7.5
**Salinity:** 10.0‰

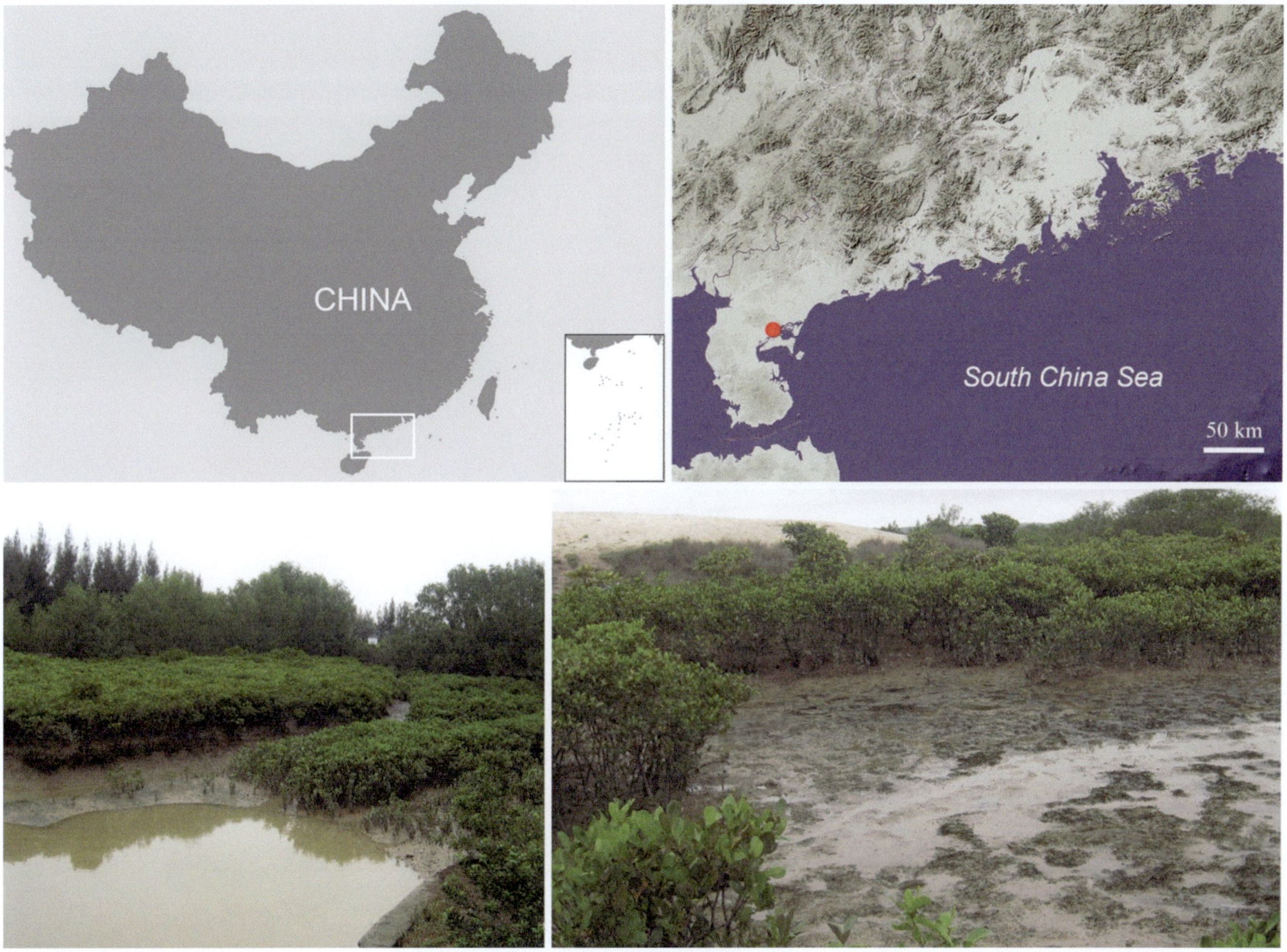


**Fig. 2A** Sample site

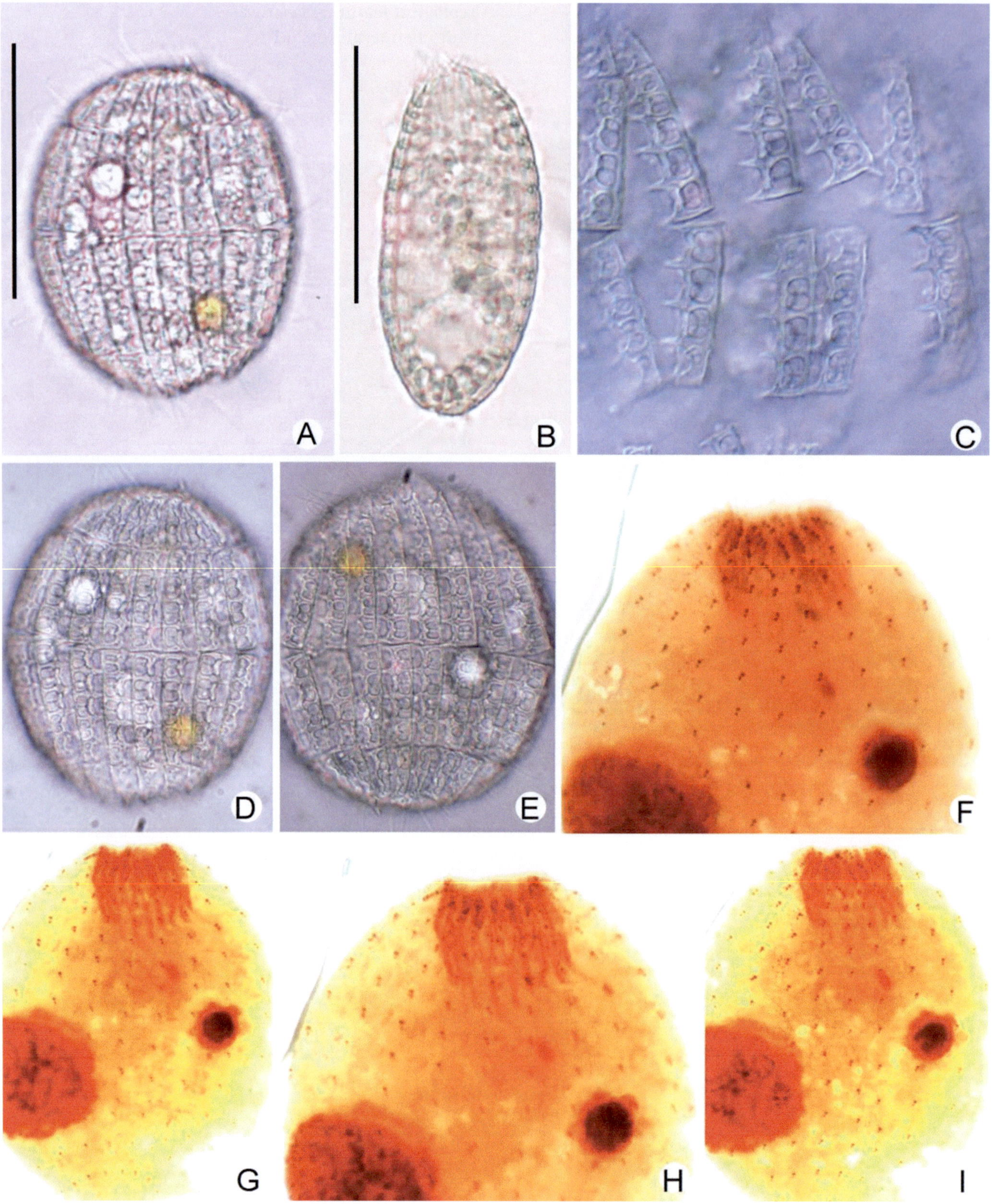


**Fig. 2B** *Pinacocoleps* sp.
(**a**, **d**, **e**) Lateral views of live individuals; (**b**) shape variant; (**c**) armor plates; (**f–i**) ciliary pattern. Scale bars: 50 μm

## References

Chen XR, Wang Y, Long H, Al-Rasheid KAS, Warren A, Song WB (2010). Morphological studies on two marine colepid ciliates from Qingdao, China, *Nolandia orientalis* spec. nov. and *Pinacocoleps similis* (Kahl, 1933) comb. nov. (Ciliophora, Colepidae). Eur J Protistol 46:254–262

Foissner W, Kusuoka Y, Shimano S (2008) Morphology and gene sequence of *Levicoleps biwae* n. gen., n. sp. (Ciliophora, Prostomatida), a proposed endemic from the ancient Lake Biwa, Japan. J Eukaryot Microbiol 55:185–200

Class Prostomatea Schewiakoff, 1896
Order Prostomatida Schewiakoff, 1896
Family Colepidae Ehrenberg, 1838
*Tiarina fusa* (Claparède & Lachmann, 1857) Bergh, 1880

**Ecological feature:** intertidal zone
**Water temperature:** 29.7 °C
**pH:** 8.4
**Salinity:** 28.7‰

**Fig. 3A** Sample site

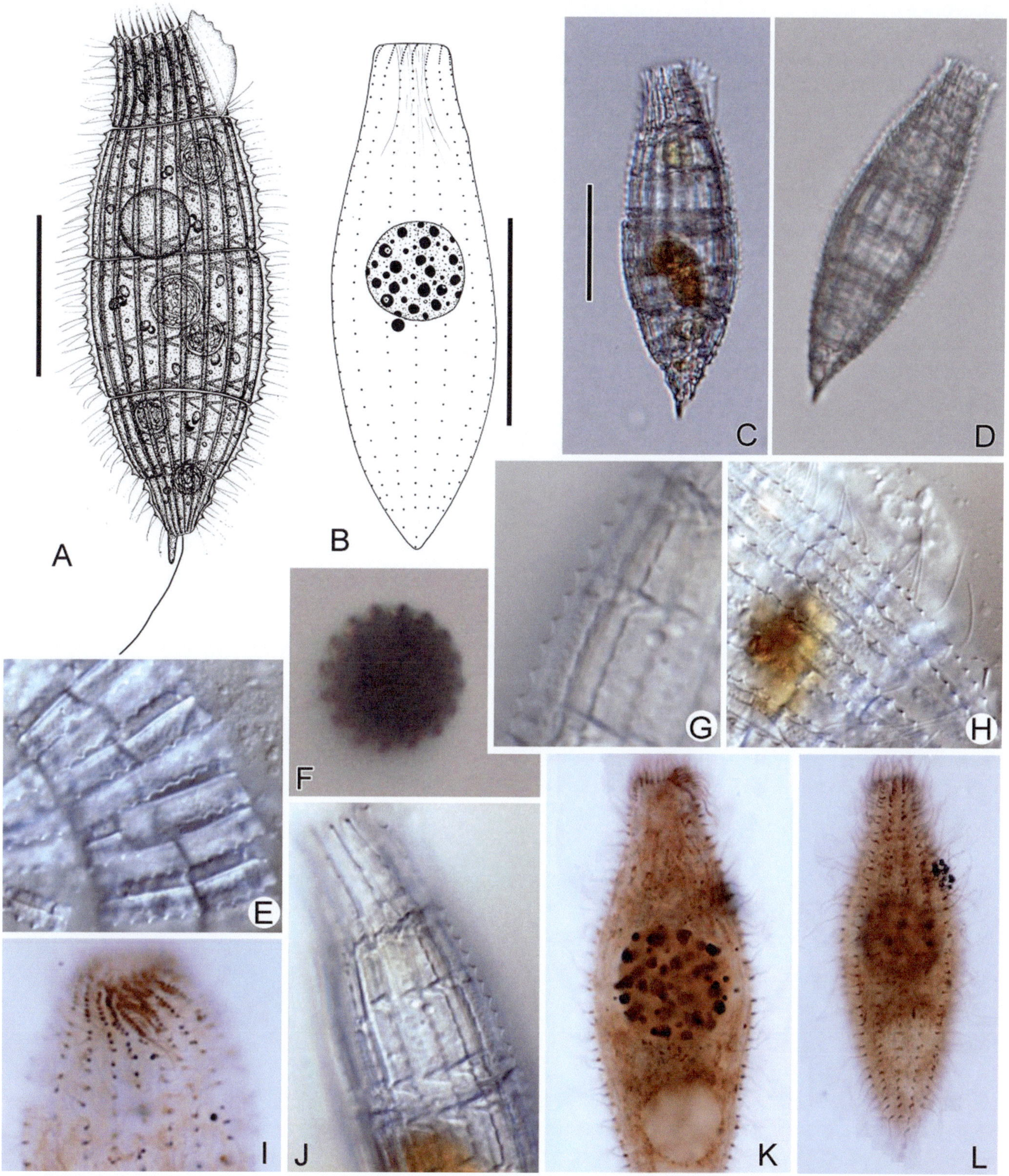


**Fig. 3B** *Tiarina fusa*
(**a**, **c**, **d**) Lateral views of live cells, showing body shape; (**b**, **l**) ciliature; (**e**, **g**, **h**) detail of plates, showing the ridge and serrated left margin; (**f**) transverse view, showing the circular cross-section; (**i**) detail of anterior end, showing ciliary pattern and adoral organelle; (**j**) anterior portion of the body, showing armor plates; (**k**) macronucleus and contractile vacuole. Scale bars: 30 μm

## Reference

Chen XR, Gao S, Liu W, Song WB, Al-Rasheid KAS, Warren A (2012) Taxonomic descriptions of three marine colepid ciliates, *Nolandia sinica* spec. nov., *Apocoleps caoi* spec. nov. and *Tiarina fusa* (Clapare`de & Lachmann, 1858) Bergh, 1881 (Ciliophora, Prorodontida). Int J Syst Evol Microbiol 62:735–744

Class Prostomatea Schewiakoff, 1896
Order Prostomatida Schewiakoff, 1896
Family Placidae Small & Lynn, 1985
*Spathidiopsis socialis* Fabre-Domergue, 1889

**Ecological feature:** mangrove
**Water temperature:** 24.0 °C
**pH:** 6.7
**Salinity:** 12.0‰

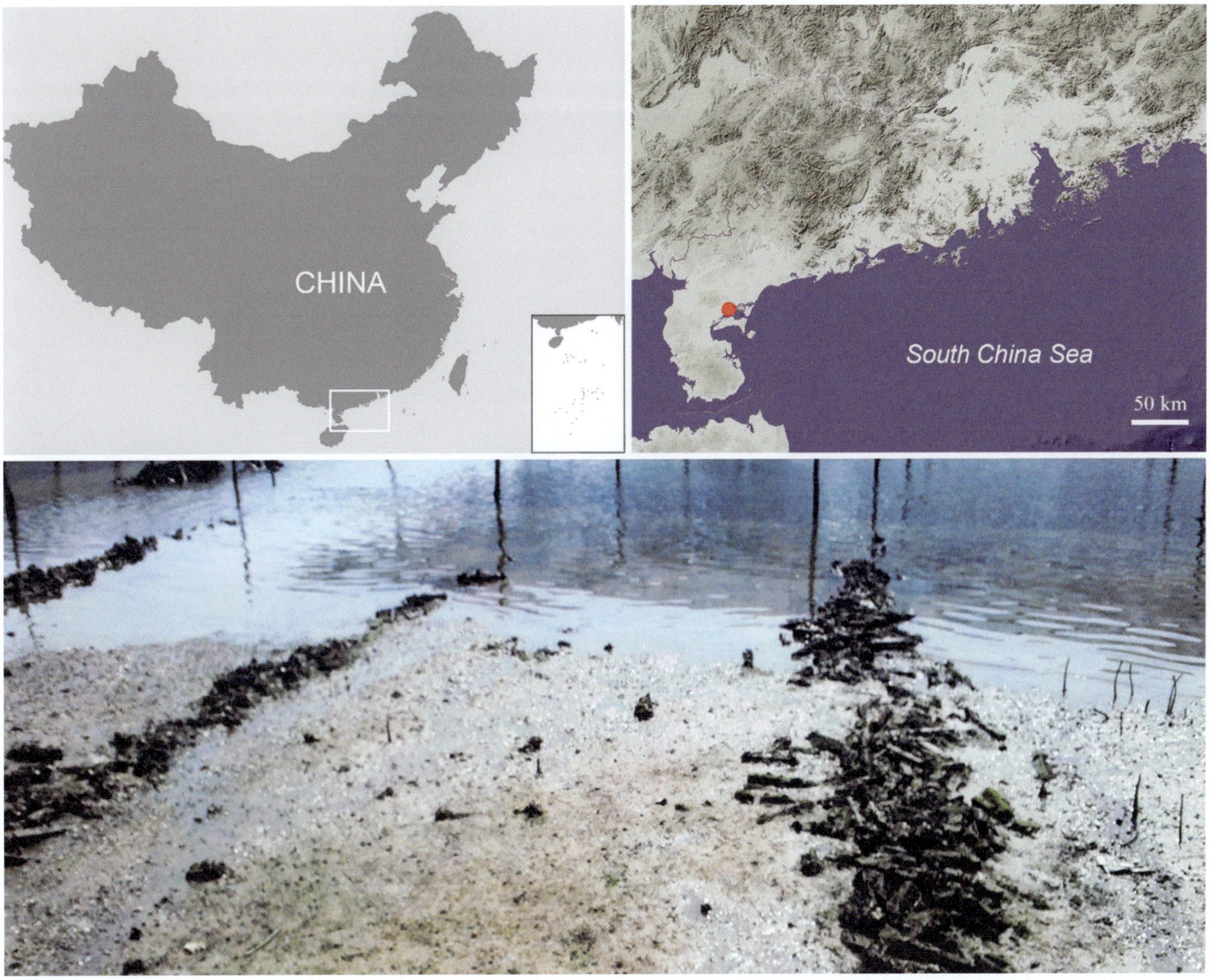


**Fig. 4A** Sample site

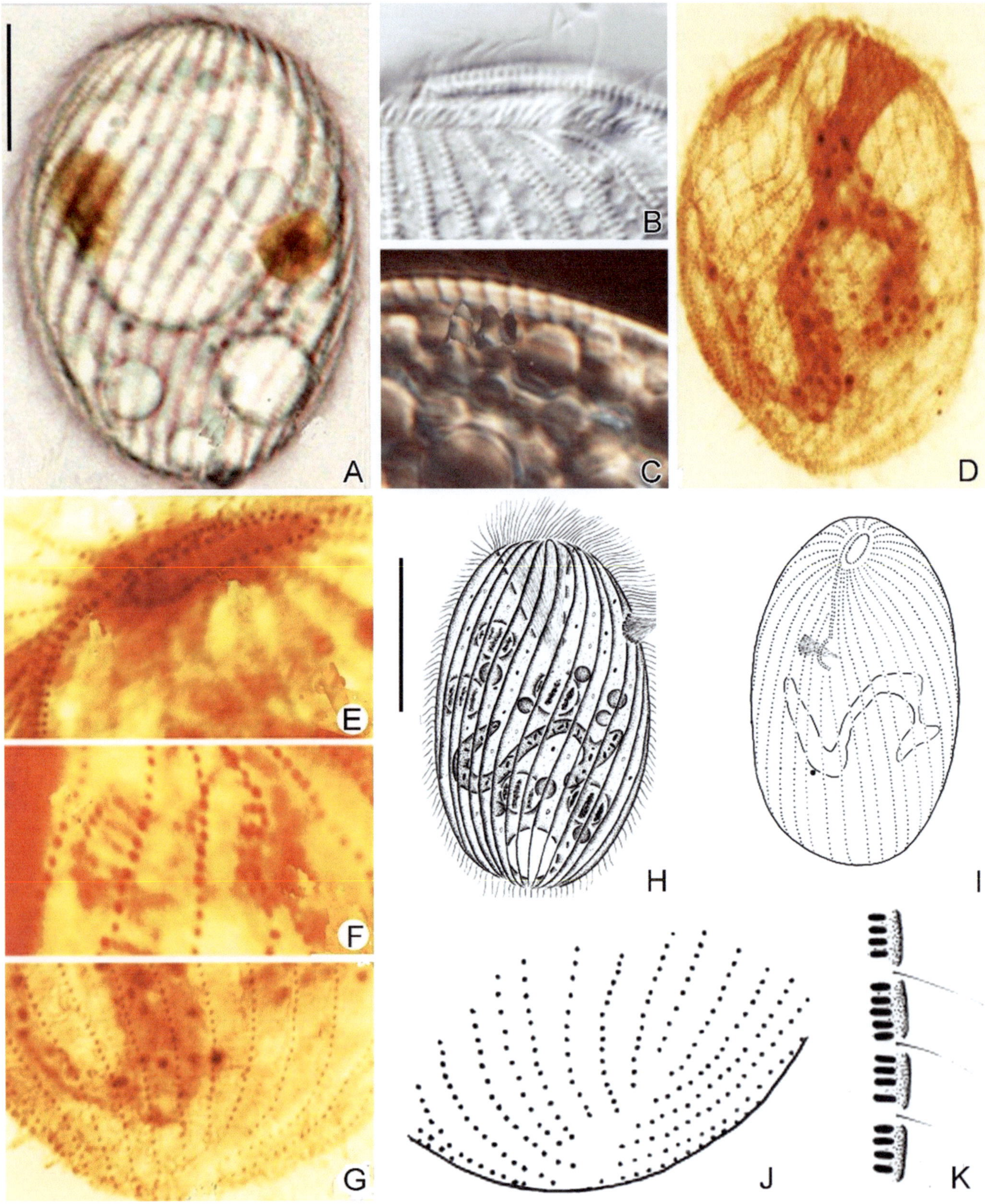


**Fig. 4B** *Spathidiopsis socialis*
(**a**, **h**) Lateral views of live cells; (**b**) buccal region of live cell, showing oral cilia; (**c**, **k**) portion of pellicle; (**d**, **i**) ciliature and nuclear apparatus; (**e**) oral ciliature; (**f**) fragment kinety; (**g**, **j**) subcaudal ciliature, showing suture. Scale bars: 20 μm

## References

Lipscomb DL, Riordan GP (2012) The ultrastructure of *Placus striatus* and a revision of the family Placidae (Ciliophora). J Eukaryot Microbiol 59:407–422

Xu D, Song WB, Hu X (2005) Notes on two marine ciliates from the Yellow Sea, China: *Placus salinus* and *Strombidium apolatum* (Protozoa, Ciliophora). J Ocean Univ China 4:137–144

# Class Oligohymenophorea

Xiaozhong Hu, Xiaofeng Lin, and Weibo Song

Class Oligohymenophorea de Puytorac et al., 1974
Order Peniculida Fauré-Fremiet in Corliss, 1956
Family Frontoniidae Kahl, 1926
*Frontonia canadensis* Roque & Puytorac, 1972

**Ecological feature:** culturing water
**Water temperature:** 16.0 °C
**pH:** 7.4
**Salinity:** 16.0‰

X. Hu (✉)
Fisheries College, Ocean University of China,
Qingdao, Shandong, China
e-mail: xiaozhonghu@ouc.edu.cn

X. Lin
College of Life Science, South China Normal University,
Guangzhou, Guangdong, China
e-mail: xlin@oucscnu.edu.cn

W. Song
Institute of Evolution & Marine Biodiversity,
Ocean University of China, Qingdao, Shandong, China
e-mail: wsong@ouc.edu.cn


X. Hu et al. (eds.), *Ciliate Atlas: Species Found in the South China Sea*,
https://doi.org/10.1007/978-981-13-5901-9_9

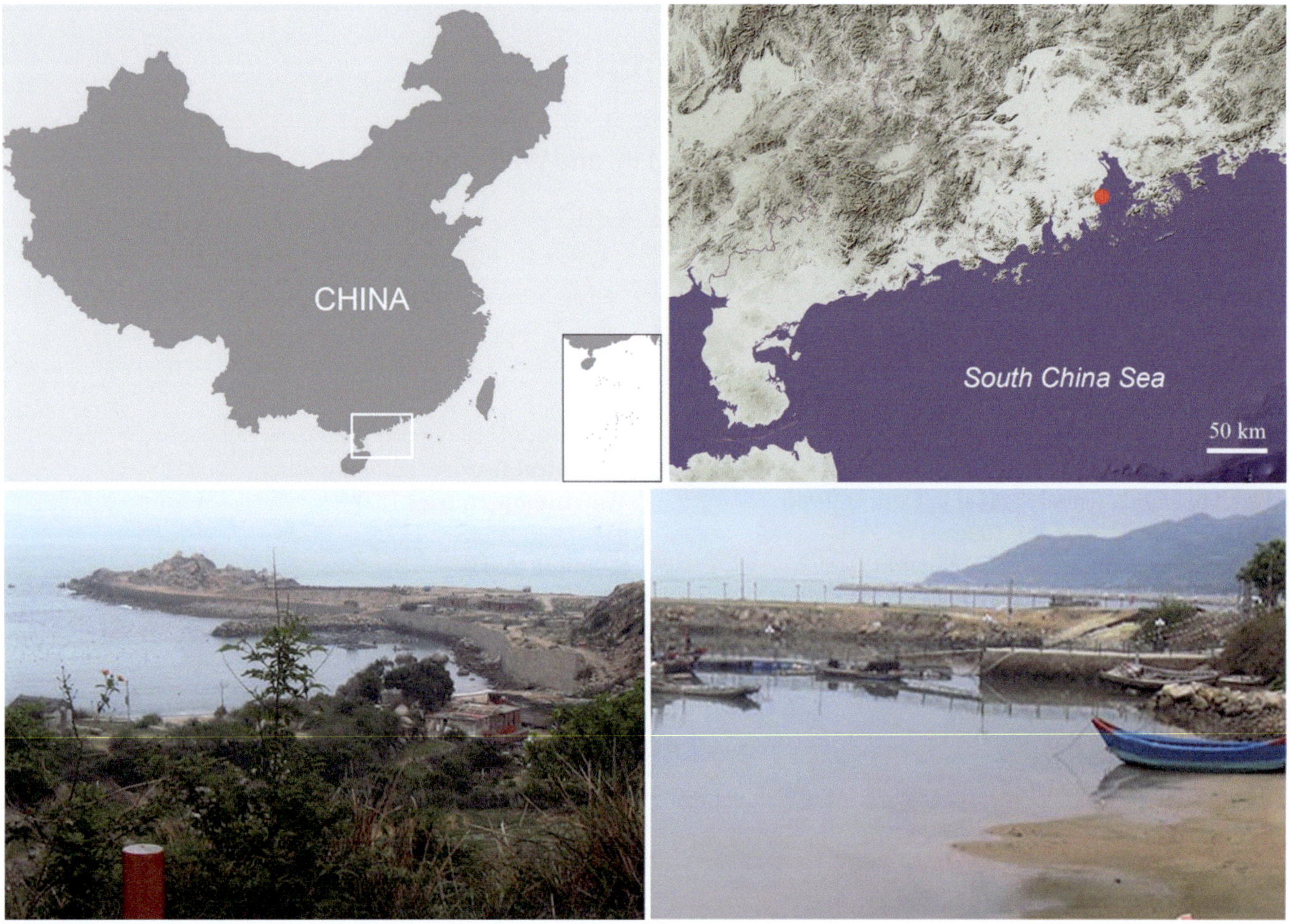


**Fig. 1A** Sample site

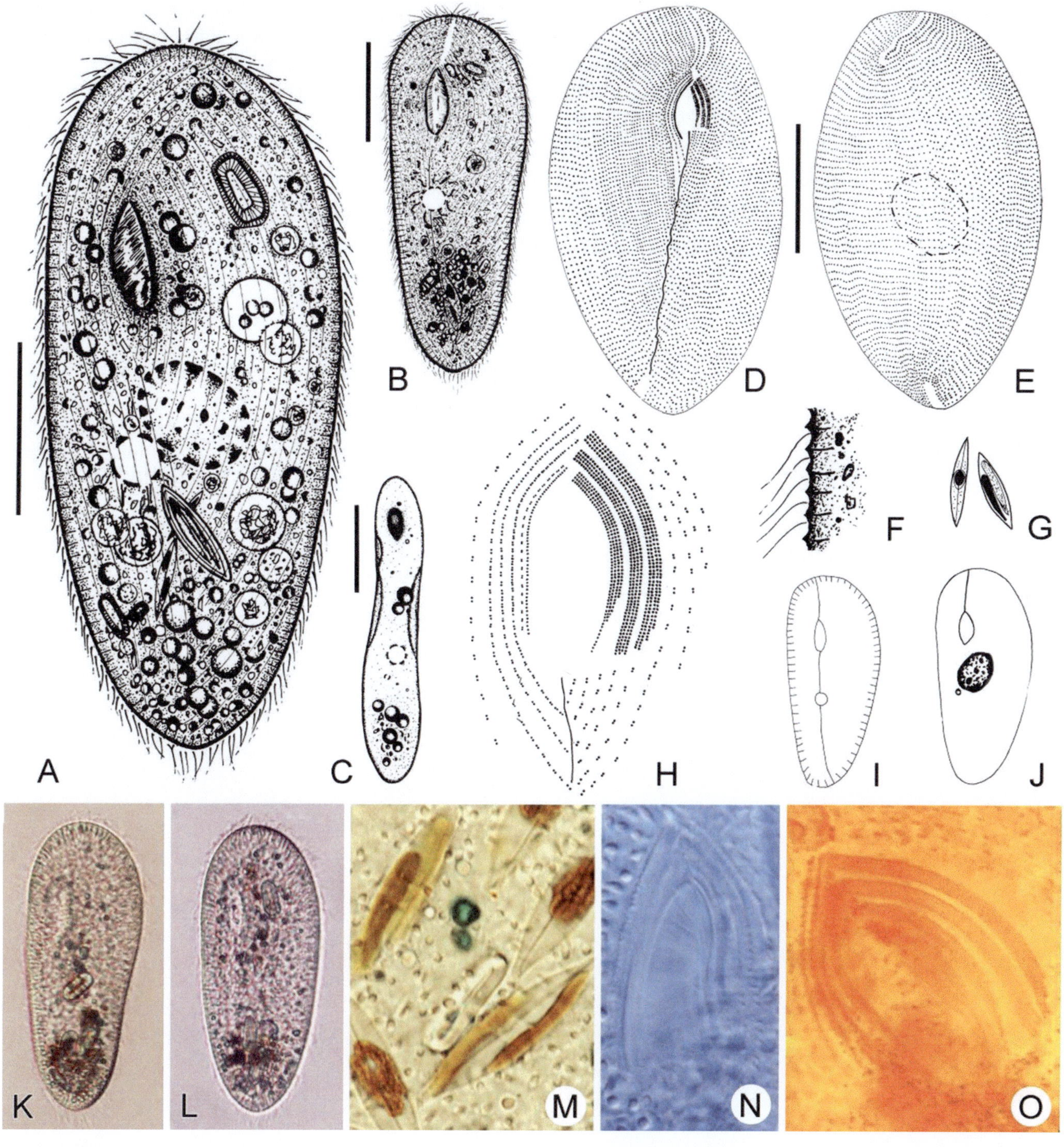


**Fig. 1B** *Frontonia canadensis*
(**a**, **b**, **k**, **l**) Ventral views of live cells, showing body shapes; (**c**) lateral view of live cell; (**d**, **e**) ciliature of ventral (**d**) and dorsal (**e**) sides; (**f**) extrusomes beneath pellicle; (**g**, **m**). Ingested algae; (**h**, **n**, **o**) structures of oral apparatus; (**i**, **j**) ventral views showing arrangement of extrusomes, sutures, contractile vacuole, and nuclear apparatus. Scale bars: 40 μm (**a–c**); 50 μm (**d**, **e**)

## References

Fan X, Lin X, Liu W, Xu Y, Al-Farraj SA, Al-Rasheid KAS, Warren A (2013) Morphology and phylogeny of three new marine *Frontonia* species (Ciliophora; Peniculia) from coastal waters of China. Eur J Protistol 49:312–323

Pan X, Gao F, Liu W, Fan X, Warren A, Song WB (2013) Morphology and SSU rRNA gene sequences of three *Frontonia* species, including a description of *F. subtropica* spec. nov. (Ciliophora, Peniculida). Eur J Protistol 49:67–77

Class Oligohymenophorea de Puytorac et al., 1974
Order Peniculida Fauré-Fremiet in Corliss, 1956
Family Frontoniidae Kahl, 1926
*Frontonia elegans* Fan et al., 2013

**Ecological feature:** shrimp-culturing pond
**Water temperature:** 20.0 °C
**pH:** 8.2
**Salinity:** 9.0‰

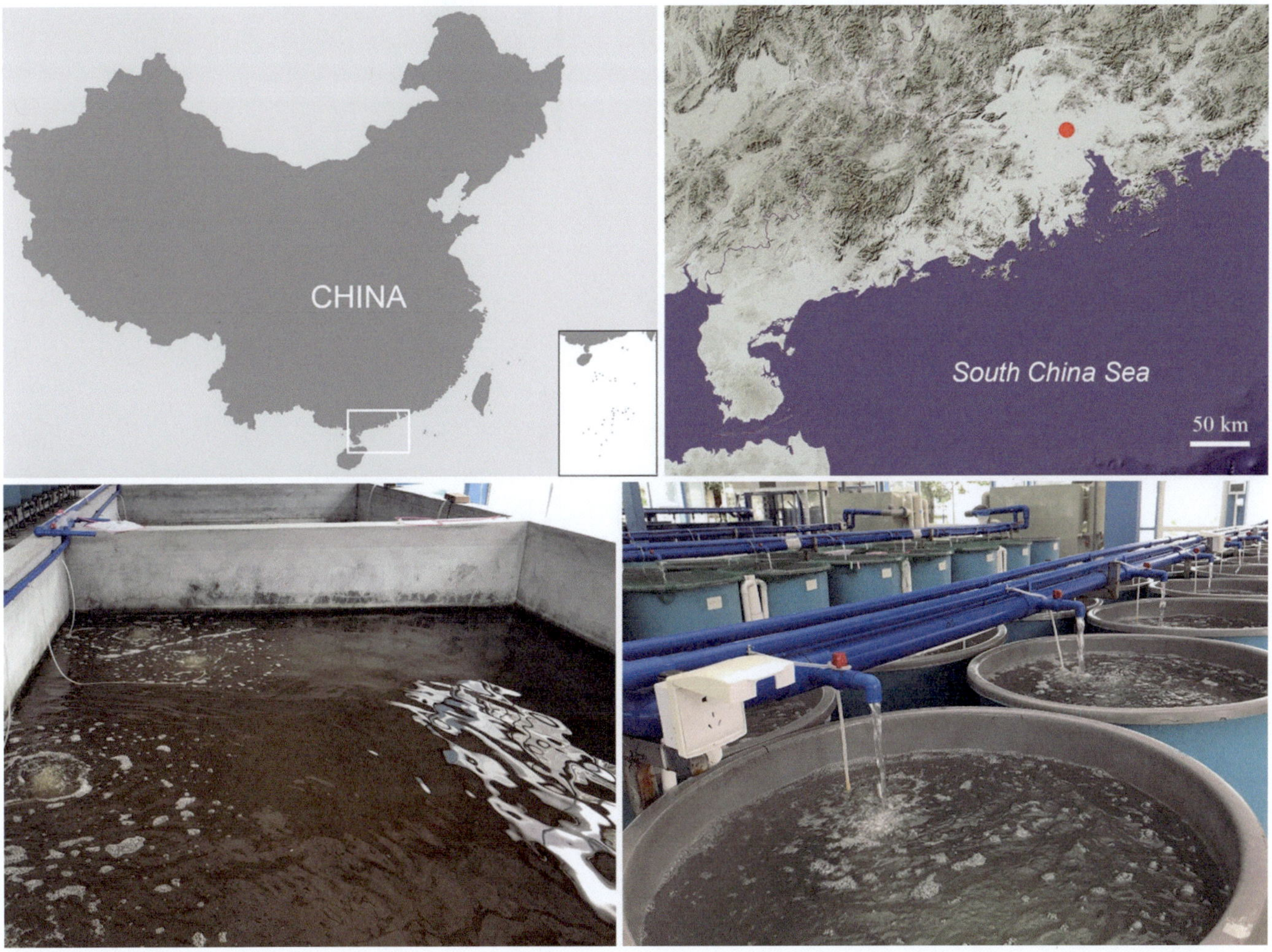


**Fig. 2A** Sample site

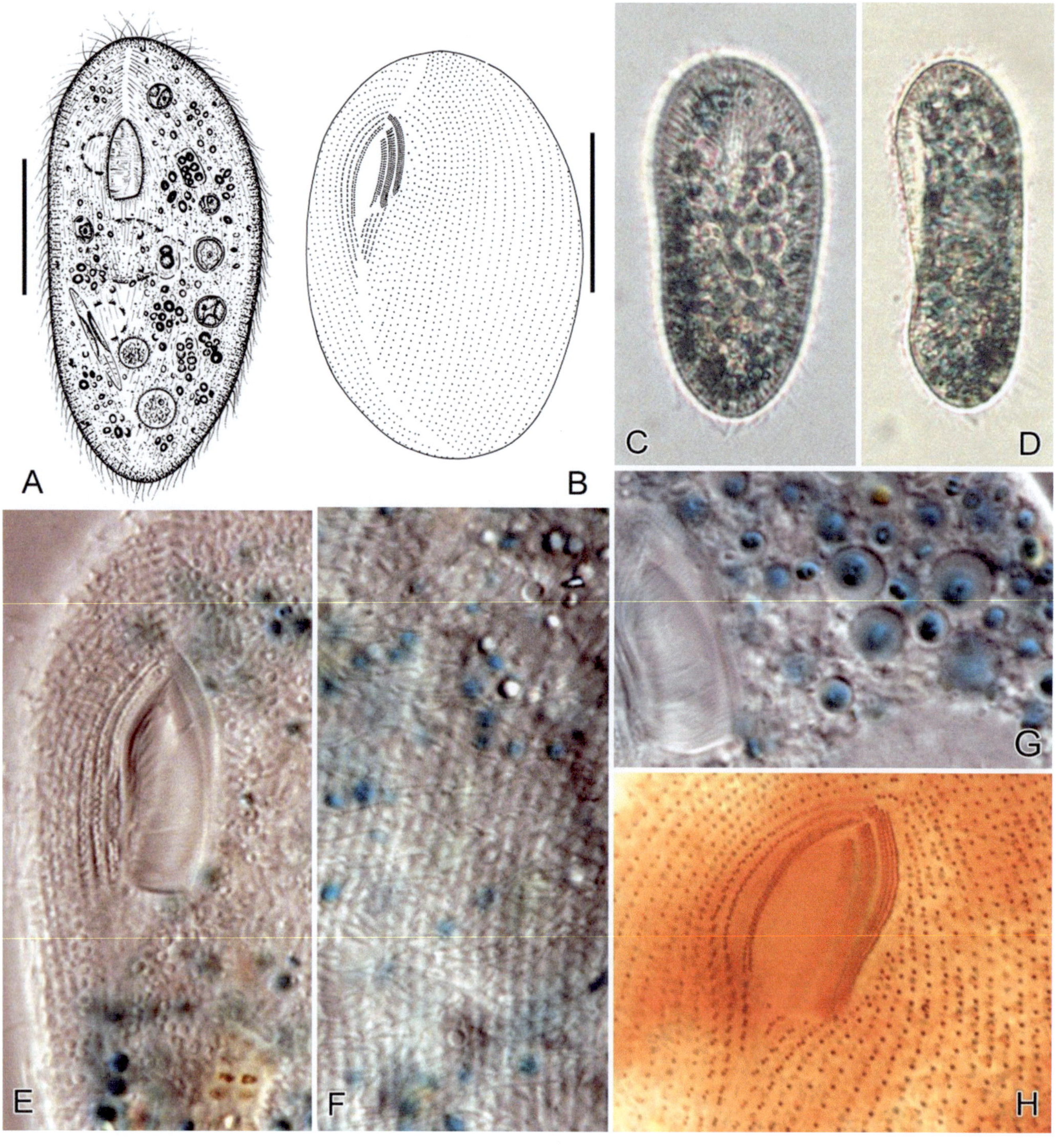


**Fig. 2B** *Frontonia elegans*
(**a**, **c**) Ventral views of live individuals; (**b**) ventral view of ciliature; (**d**) lateral view of live cell; (**e**) details of buccal field; (**f**) excretory pores of two contractile vacuoles; (**g**) food vacuoles containing blue-colored algae; (**h**) oral apparatus. Scale bar: 40 μm

## Reference

Fan X, Lin X, Liu W, Xu Y, Al-Farraj SA, Warren A (2013) Morphology and phylogeny of three new marine *Frontonia* species (Ciliophora; Peniculida) from coastal waters of China. Eur J Protistol 49:312–323

Class Oligohymenophorea de Puytorac et al., 1974
Order Peniculida Fauré-Fremiet in Corliss, 1956
Family Frontoniidae Kahl, 1926
*Frontonia guangdongensis* Pan et al., 2013

**Ecological feature:** culturing water
**Water temperature:** 24.0 °C
**pH:** 6.5
**Salinity:** 28.3‰

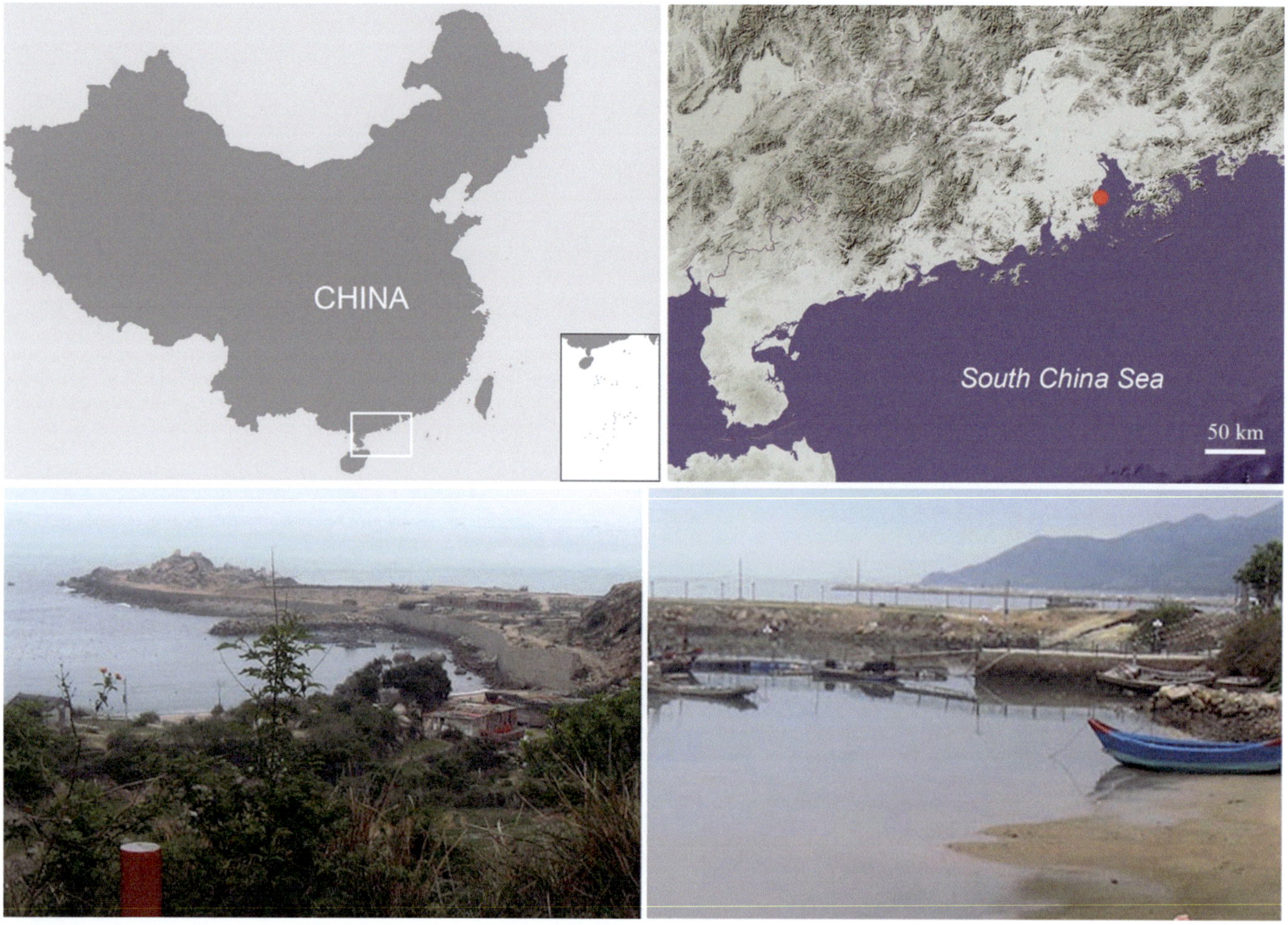


**Fig. 3A** Sample site

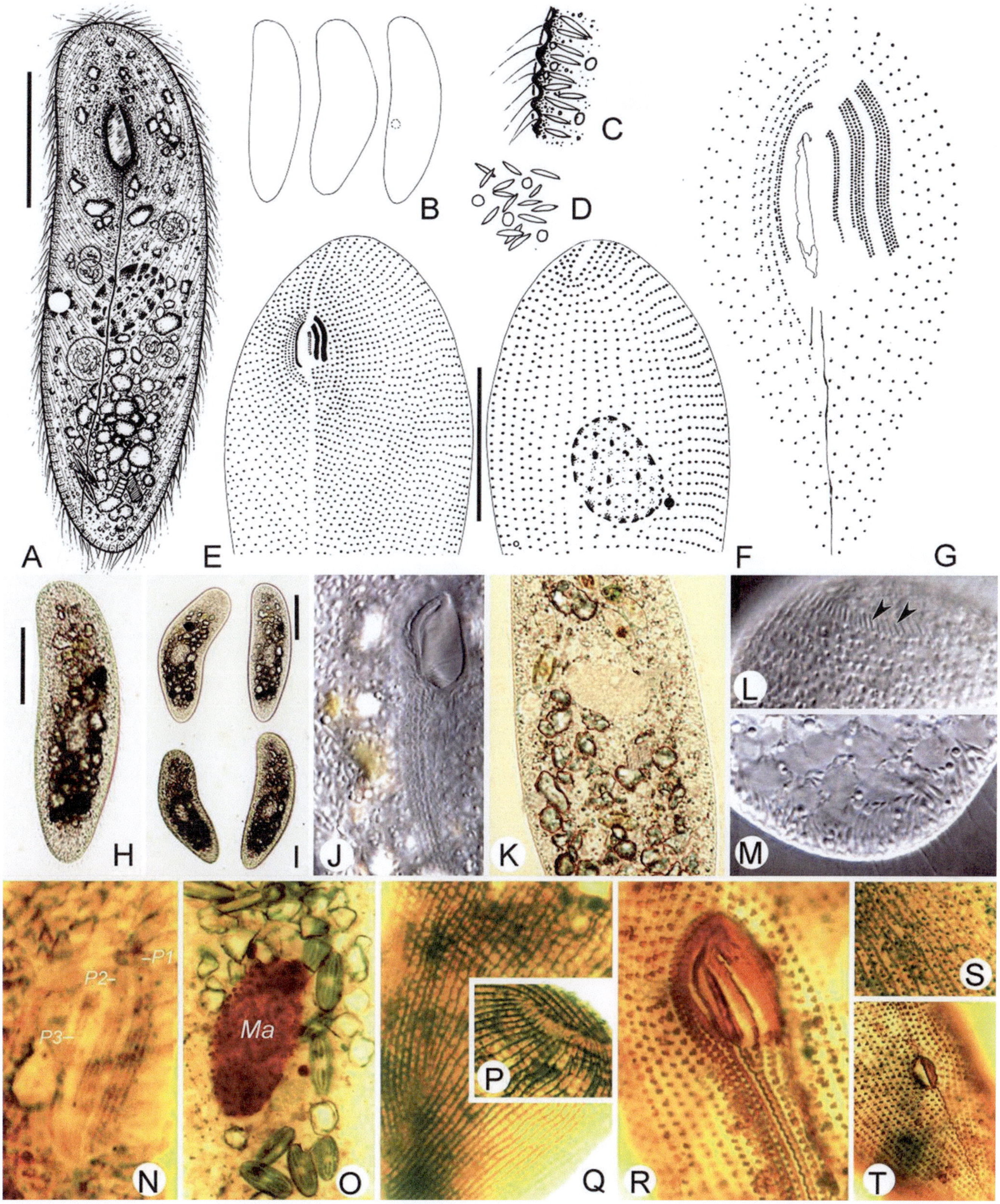


**Fig. 3B** *Frontonia guangdongensis*
(**a**, **b**, **h**, **i**) Views of live cells, showing body shapes; (**c**, **d**) extrusomes; (**e**, **f**) ciliature; (**g**, **n**, **r**) oral apparatus; (**j**) cytostome; (**k**, **o**) nucleus; (**l**, **p**, **q**, **t**) sutures (arrowheads in **l**); (**m**) posterior part of cell; (**s**) somatic kineties. Scale bars: 50 μm. P1-3, peniculi 1, 2, 3; Ma, macronucleus

## Reference

Pan X, Liu W, Yi ZZ, Fan XP, Al-Rasheid KAS, Lin X (2013) Studies on three diverse *Frontonia* species (Ciliophora, Peniculida), with brief notes on 14 marine or brackish congeners. Acta Protozool 52:35–49

Class Oligohymenophorea de Puytorac et al., 1974
Order Peniculida Fauré-Fremiet in Corliss, 1956
Family Frontoniidae Kahl, 1926
*Frontonia magna* Fan et al., 2011

**Ecological feature:** mangrove
**Water temperature:** 25.0 °C; 24.0 °C
**pH:** no data available; 8.1
**Salinity:** 16.0‰; 9.0‰

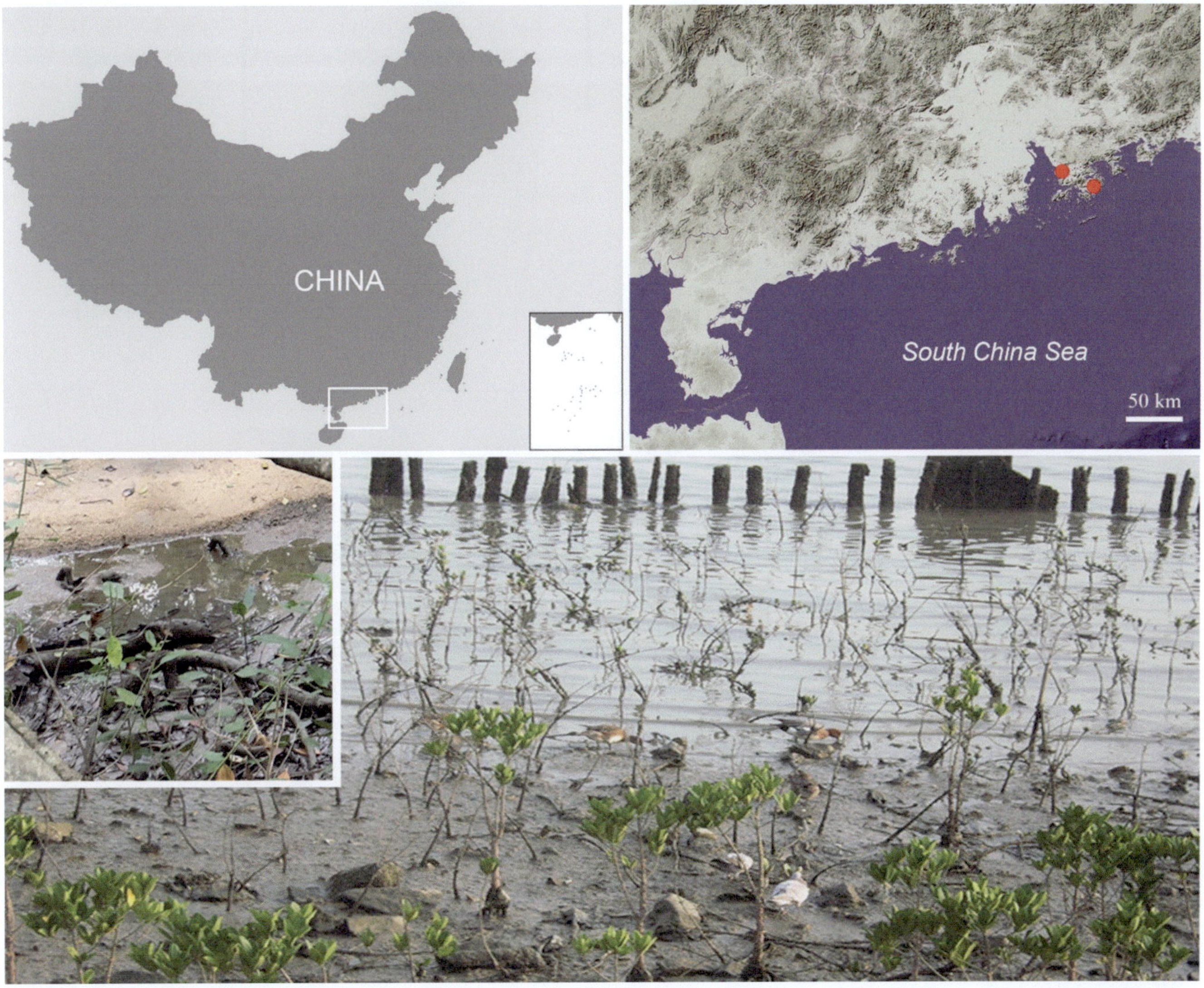


**Fig. 4A** Sample site

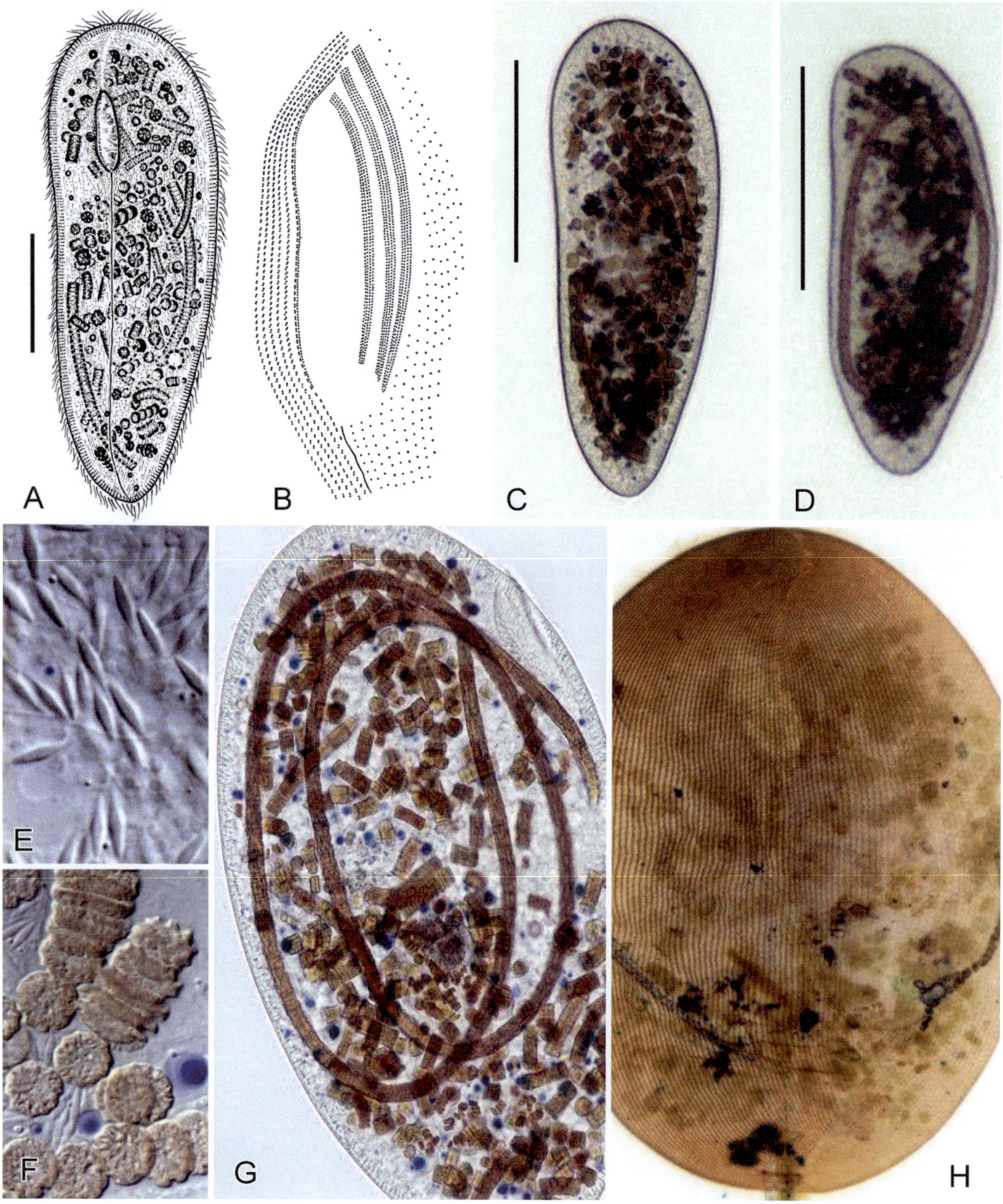


**Fig. 4B** *Frontonia magna*
(**a**, **c**, **d**)Ventral views of live individuals, showing body shapes; (**b**) oral apparatus; (**e**) resting trichocysts; (**f**, **g**) Showing the algae food (**f**) and well-fed individual (**g**); (**h**) dorsal view, showing the tightly arranged somatic kineties. Scale bars: 100 μm (**a**); 200 μm (**c**, **d**)

## Reference

Fan X, Chen X, Song WB, Al-Rasheid KAS, Warren A (2011). Two novel marine *Frontonia* species, *F. mengi* spec. nov. and *F. magna* spec. nov. (Protozoa; Ciliophora), with notes on their phylogeny based on SSU rRNA gene sequence data. Int J Syst Evol Microbiol 61:1476–1486

Class Oligohymenophorea de Puytorac et al., 1974
Order Peniculida Fauré-Fremiet in Corliss, 1956
Family Frontoniidae Kahl, 1926
*Frontonia ocularis* Bullington, 1939

**Ecological feature:** mangrove
**Water temperature:** 17.0 °C
**pH:** 8.1
**Salinity:** 26.0‰

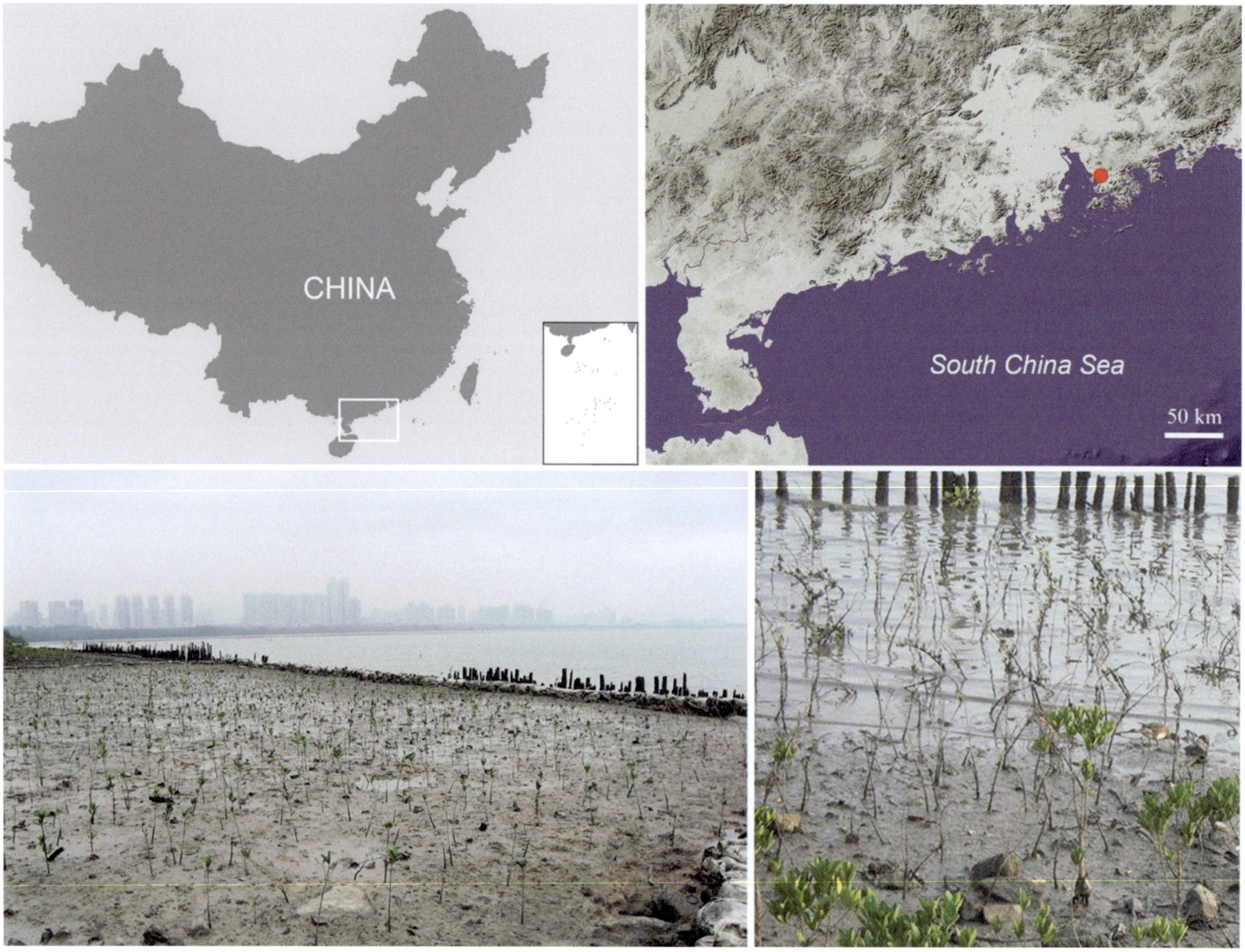


**Fig. 5A** Sample site

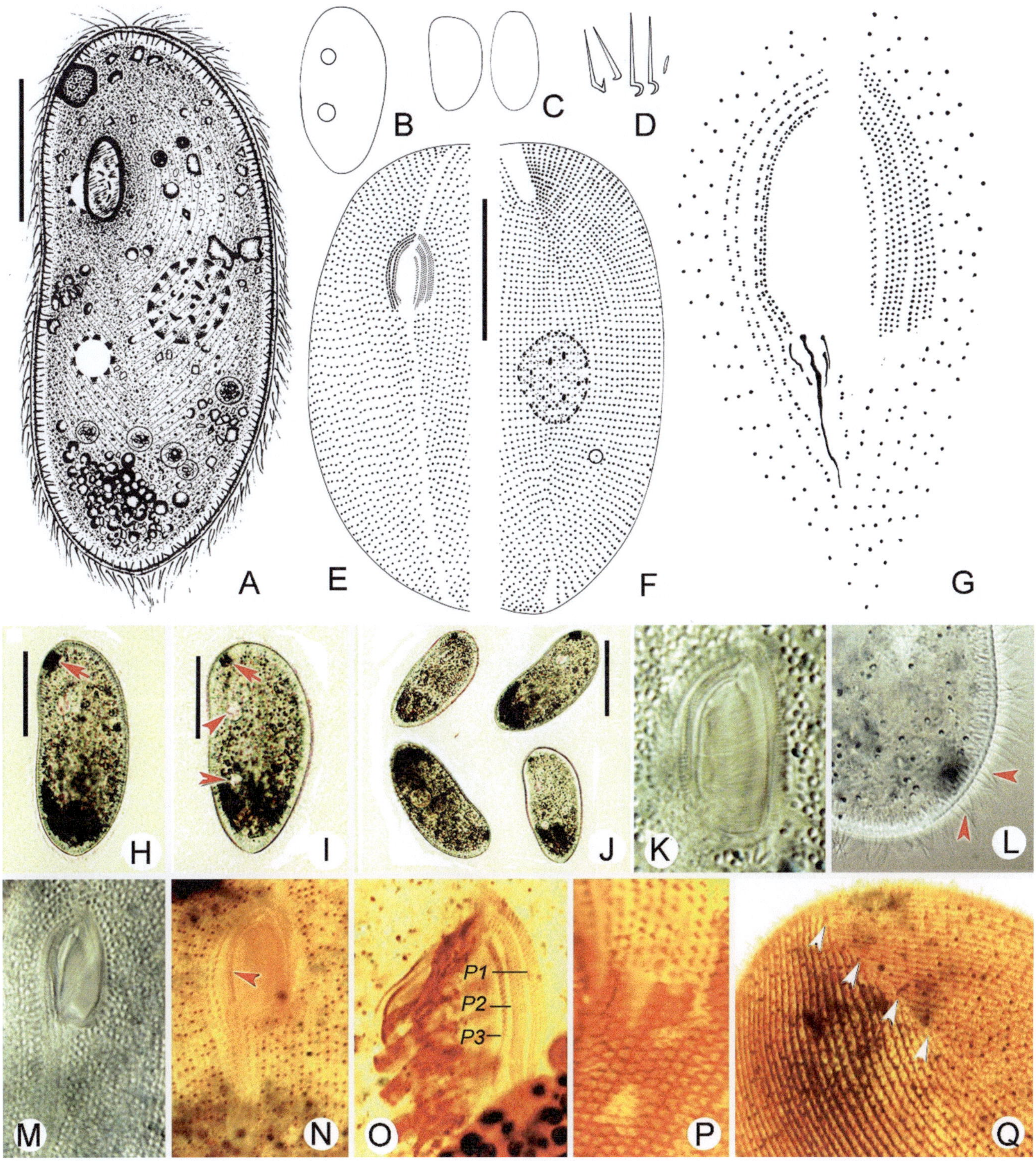


**Fig. 5B** *Frontonia ocularis*
(**a**, **c**, **h**–**j**) Ventral views of live cells, arrows show the black brown pigment spot, arrowheads mark contractile vacuole; (**b**) position of contractile vacuoles; (**d**) ejected extrusomes; (**e**, **f**) ciliature of ventral and dorsal sides; (**g**, **n**, **o**) oral apparatus, arrowhead shows paroral membrane; (**k**, **m**) buccal area; (**l**) posterior part of the cell, arrowheads indicate caudal cilia; (**p**) part of argyrome; (**q**) anterior suture (arrowheads). P1-3, peniculi 1, 2, 3. Scale bars: 50 µm (**a**, **e**, **f**); 60 µm (**h**, **i**); 80 µm (**j**)

## Reference

Pan X, Liu W, Yi Z, Fan X, Al-Rasheid KAS, Lin X (2013) Studies on three diverse *Frontonia* species (Ciliophora, Peniculida), with brief notes on 14 marine or brackish congeners. Acta Protozool 52:35–49

Class Oligohymenophorea de Puytorac et al., 1974
Order Peniculida Fauré-Fremiet in Corliss, 1956
Family Frontoniidae Kahl, 1926
*Frontonia pusilla* Fan et al., 2013

**Ecological feature:** mangrove
**Water temperature:** 19.0 °C
**pH:** no data available
**Salinity:** 25.0‰

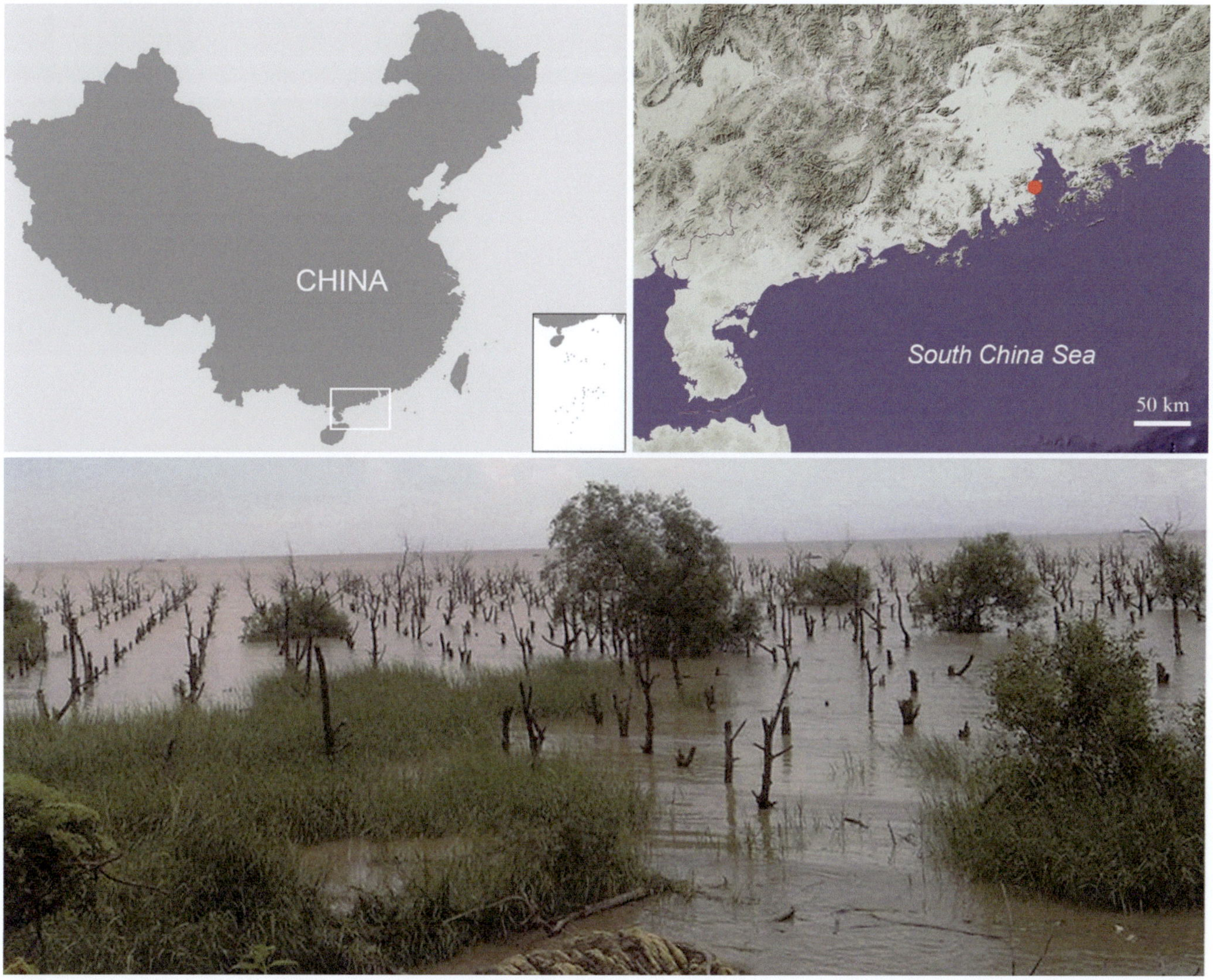


**Fig. 6A** Sample site

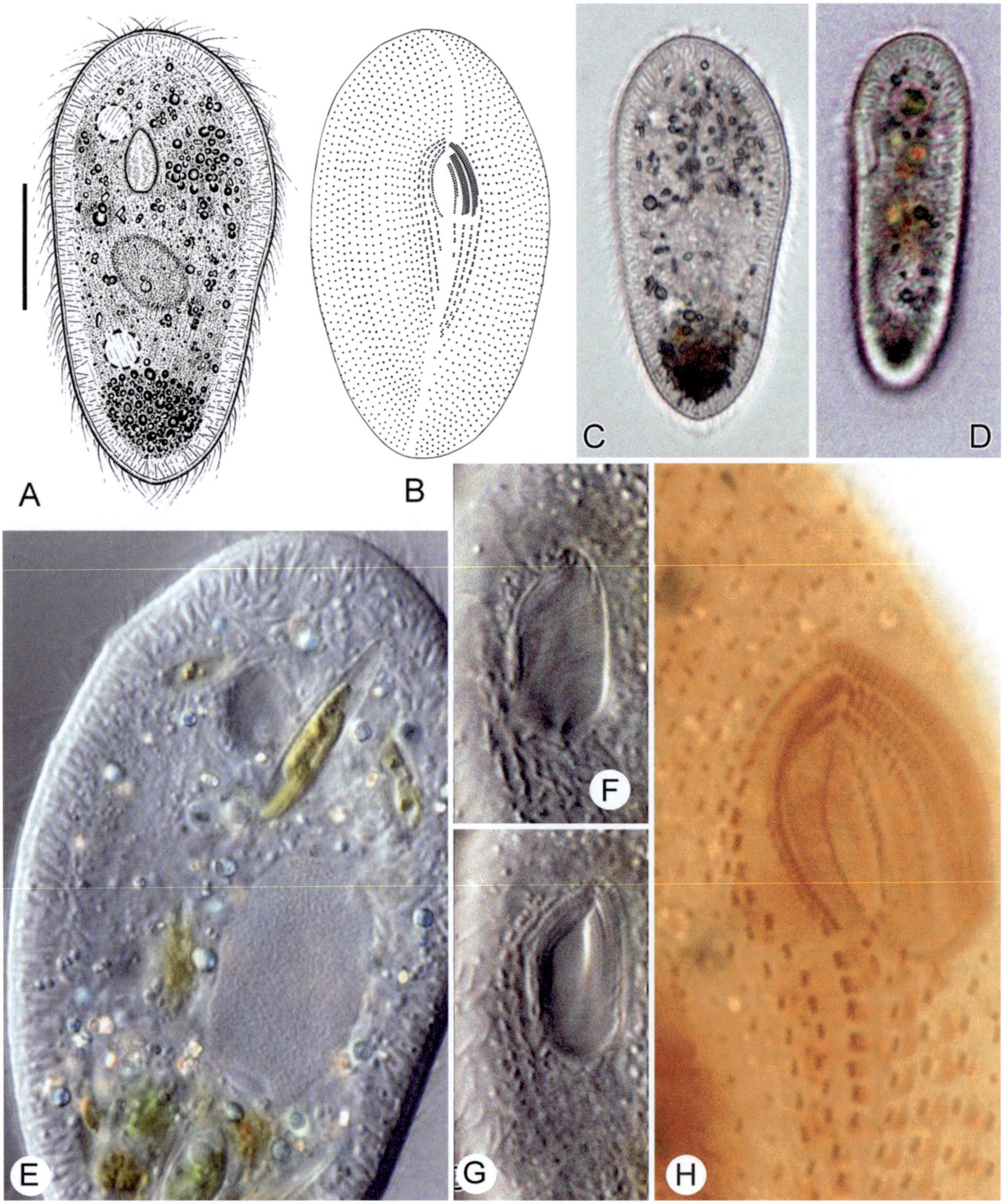


**Fig. 6B** *Frontonia pusilla*
(**a**, **c**) Ventral views of live individuals; (**b**) ventral ciliature; (**c**, **d**) lateral view of live cell, to show buccal field; (**e**) to show extrusomes and macronucleus; (**f**, **g**) buccal area; (**h**) oral apparatus. Scale bar: 30 μm

## Reference

Fan X, Lin X, Liu W, Xu Y, Al-Farraj SA,Warren A (2013) Morphology and phylogeny of three new marine *Frontonia* species (Ciliophora; Peniculida) from coastal waters of China. Eur J Protistol 49:312–323

Class Oligohymenophorea de Puytorac et al., 1974
Order Peniculida Fauré-Fremiet in Corliss, 1956
Family Frontoniidae Kahl, 1926
*Frontonia schaefferi* Bullington, 1939

**Ecological feature:** mangrove
**Water temperature:** 25.0 °C
**pH:** 7.8
**Salinity:** 21.0‰

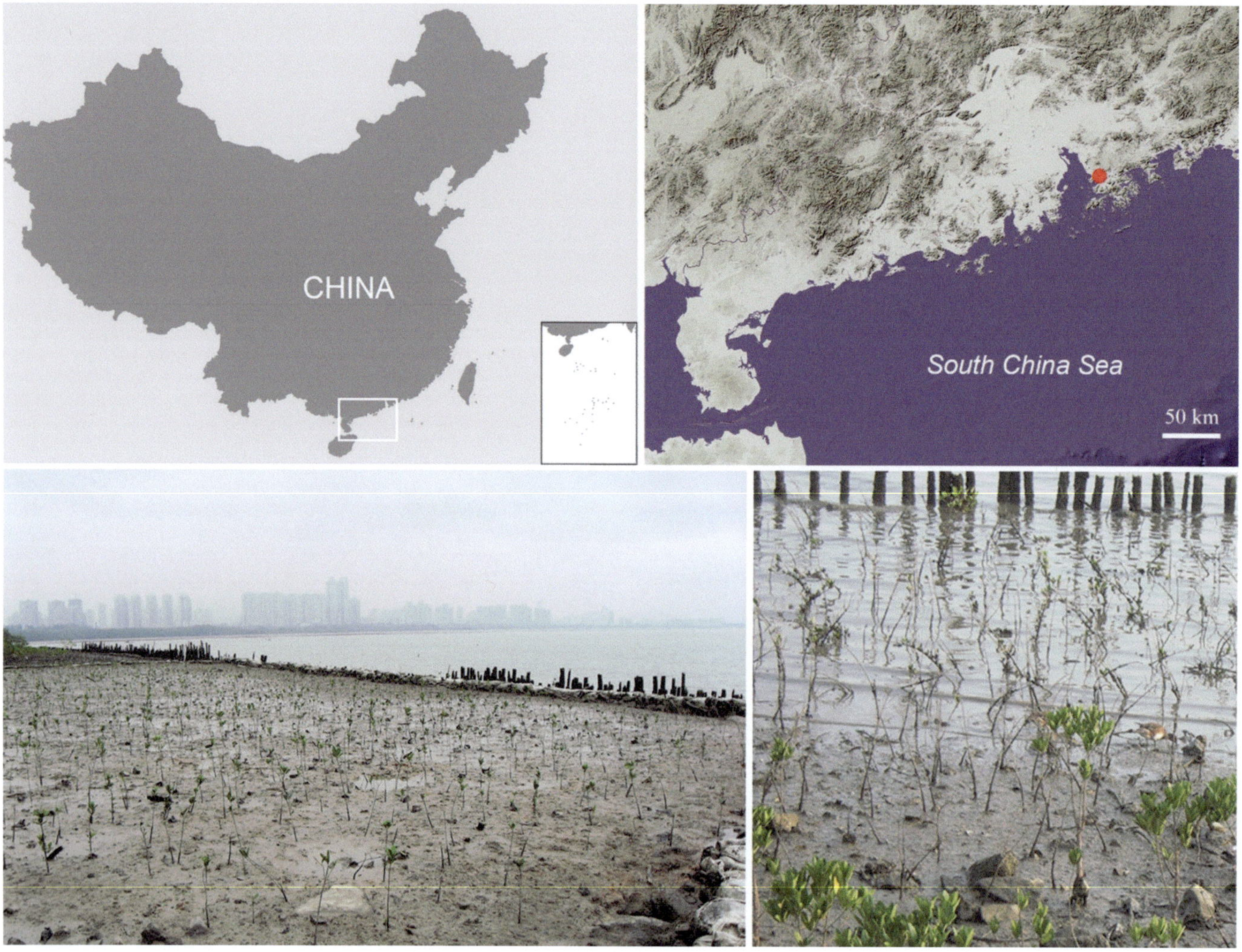


**Fig. 7A** Sample site

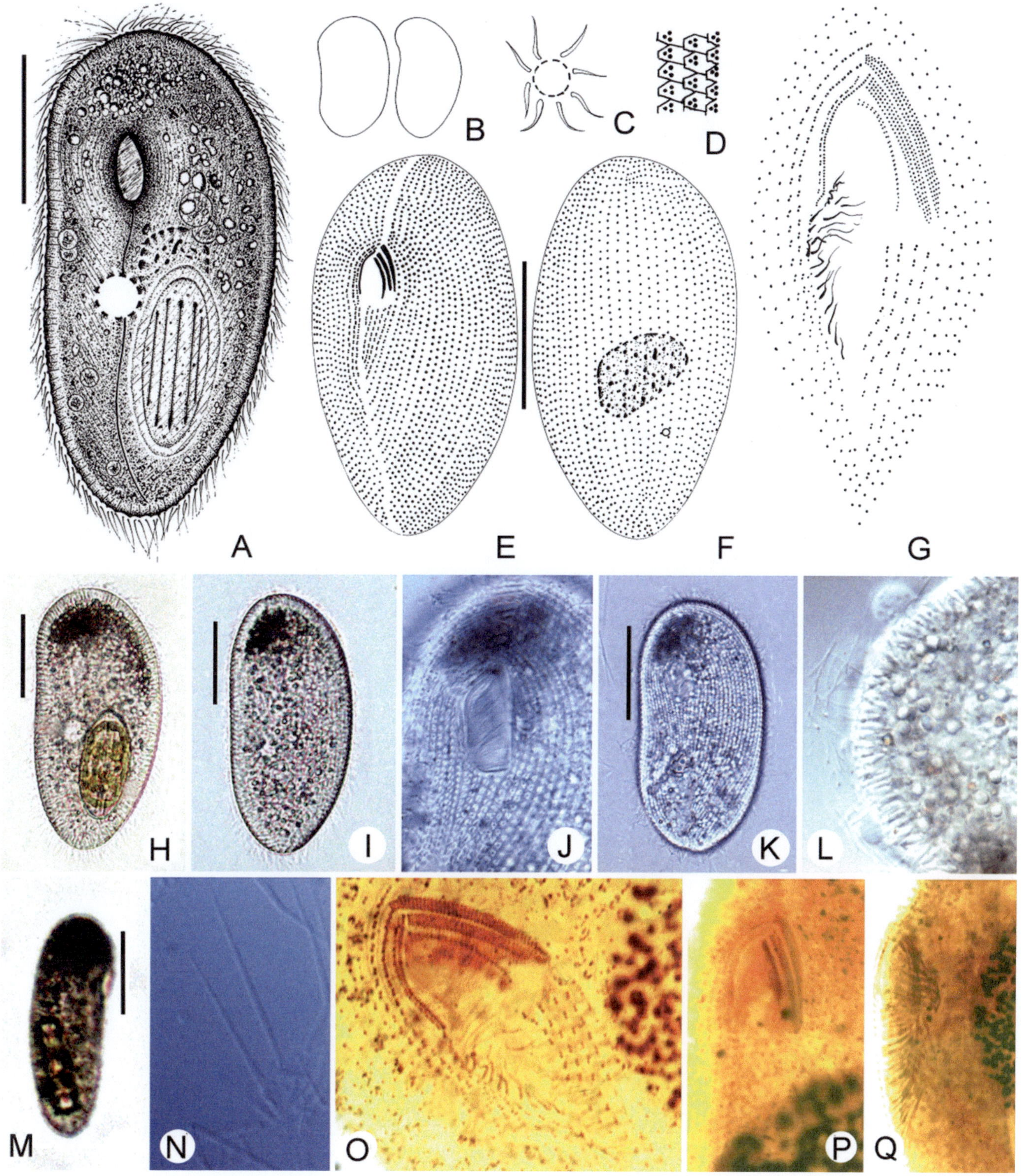


**Fig. 7B** *Frontonia schaefferi*
(**a**, **b**, **h**, **i**, **k**) Ventral views of live individuals, showing different body shapes; (**c**) contractile vacuole with about eight long collecting canals; (**d**) part of argyrome; (**e**, **f**) ciliature of ventral (**e**) and dorsal (**f**) sides; (**g**, **o**, **p**) **o**ral apparatus; (**j**) buccal area; (**l**) extrusomes; (**m**) lateral view of live cell; (**n**) ejected extrusomes; (**q**) cytopharyngeal fibers. Scale bars: 40 μm (**a**); 70 μm (**e, f, h, i, k, m**)

## Reference

Pan XM, Liu W, Yi Z, Fan X, Al-Rasheid KAS, Lin X (2013) Studies on three diverse *Frontonia* species (Ciliophora, Peniculida), with brief notes on 14 marine or brackish congeners. Acta Protozool 52:35–49

Class Oligohymenophorea de Puytorac et al., 1974
Order Peniculida Fauré-Fremiet in Corliss, 1956
Family Frontoniidae Kahl, 1926
*Frontonia subtropica* Pan et al., 2013

**Ecological feature:** culturing area
**Water temperature:** 16.0 °C
**pH:** 7.4
**Salinity:** 16.0‰

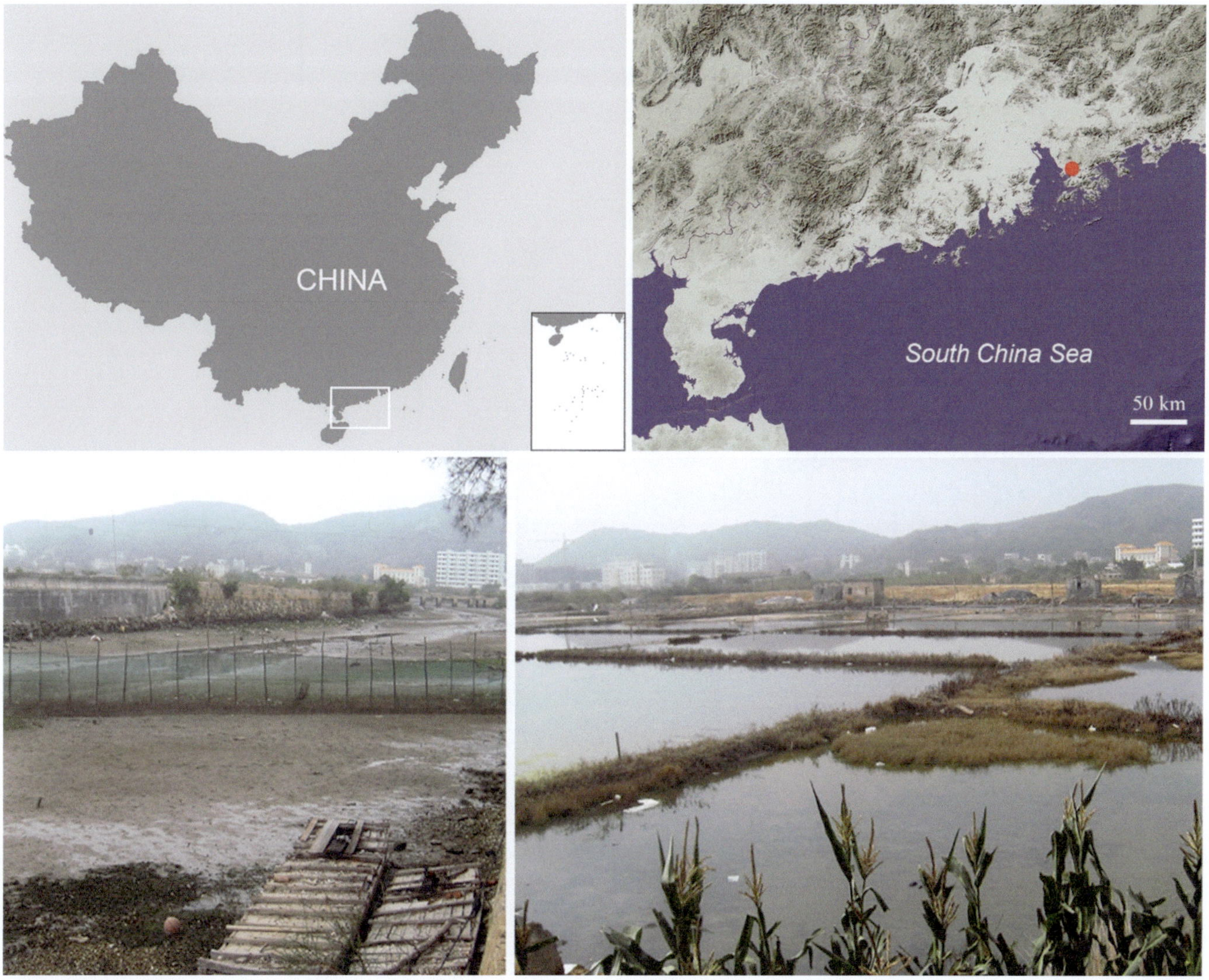


**Fig. 8A** Sample site

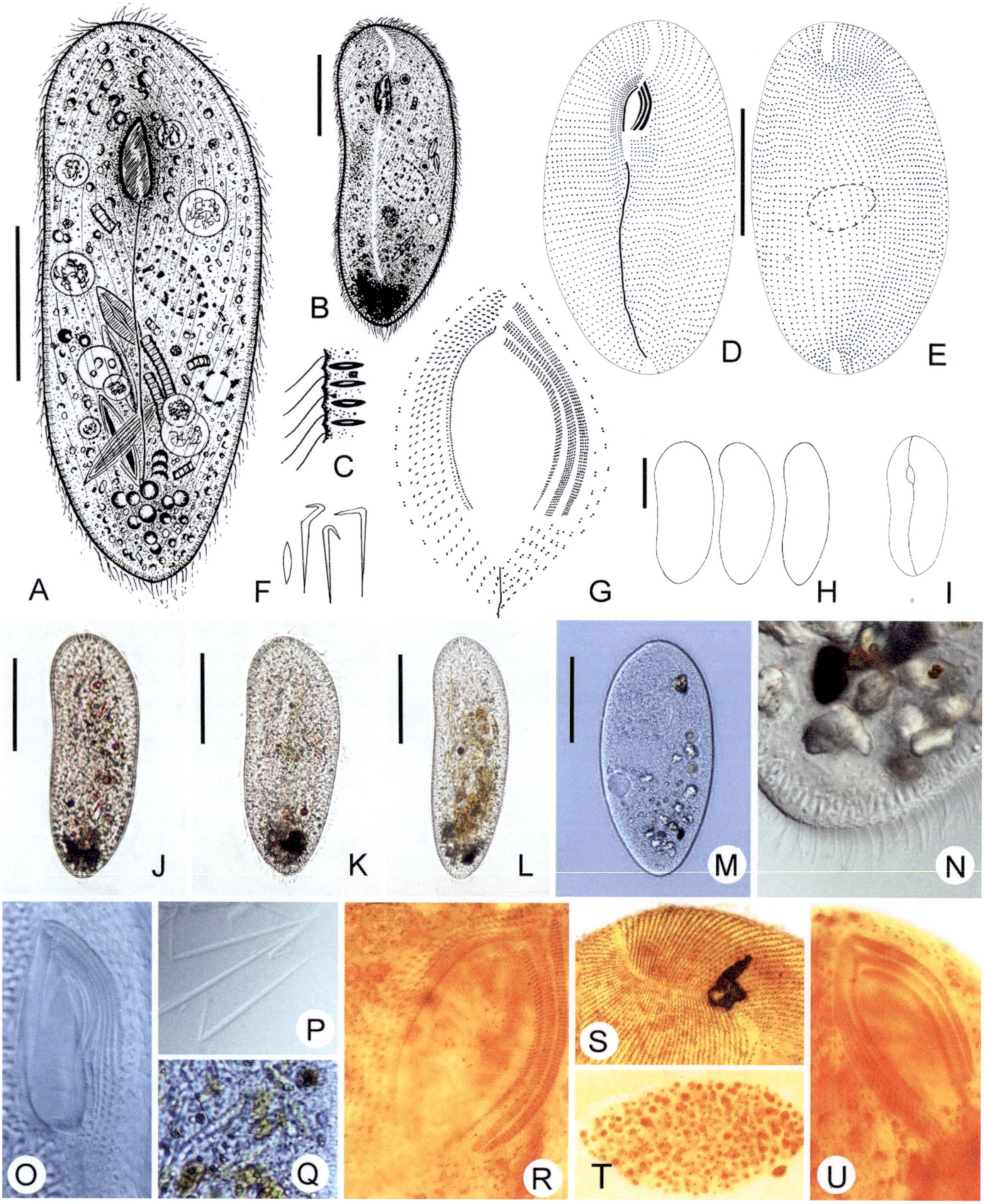


**Fig. 8B** *Frontonia subtropica*
(**a**, **b**, **h**, **j–l**) Ventral views of live cells; (**c**, **n**) extrusomes; (**d**, **e**) ciliature of ventral (**d**) and dorsal (**e**) sides; (**f**, **p**) ejected extrusomes; (**g**, **o**, **r**, **u**) oral ciliature; (**i**) cytostome and suture; (**m**) contractile vacuole; (**q**) cytoplasm with food vacuoles; (**s**) anterior suture; (**t**) nuclei. Scale bar: 100 µm

## Reference

Pan X, Gao F, Liu W, Fan X, Warren A, Song WB (2013) Morphology and SSU rRNA gene sequences of three *Frontonia* species, including a description of *F. subtropica* spec.nov. (Ciliophora, Peniculida). Eur J Protistol 49:67–77

Class Oligohymenophorea de Puytorac et al., 1974
Order Peniculida Fauré-Fremiet in Corliss, 1956
Family Maritujidae Jankowski in Small & Lynn, 1985
*Marituja* cf. *caudata*

**Ecological feature:** culturing area
**Water temperature:** 23.0 °C
**pH:** 7.9
**Salinity:** 19.0‰

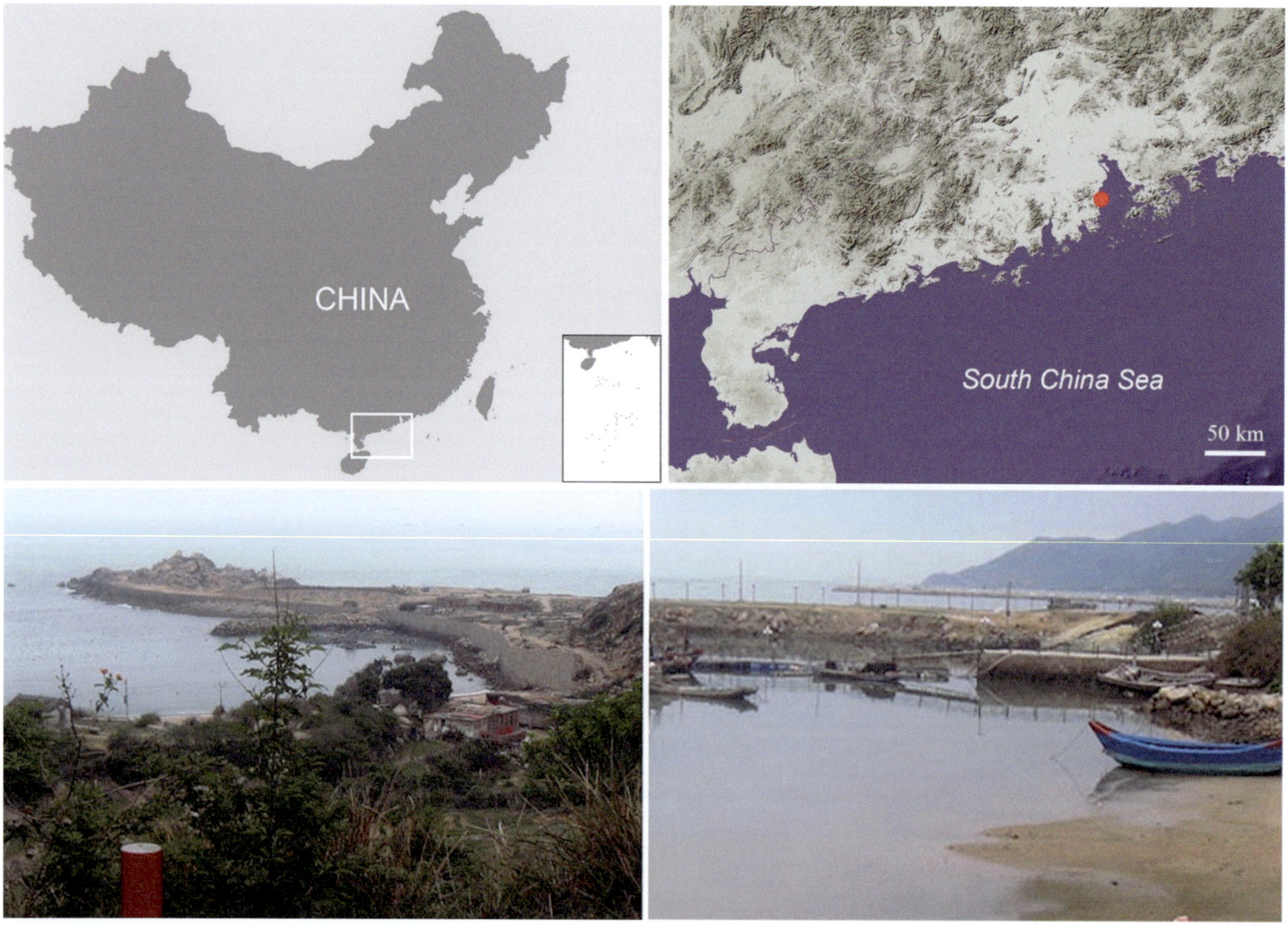


**Fig. 9A** Sample site

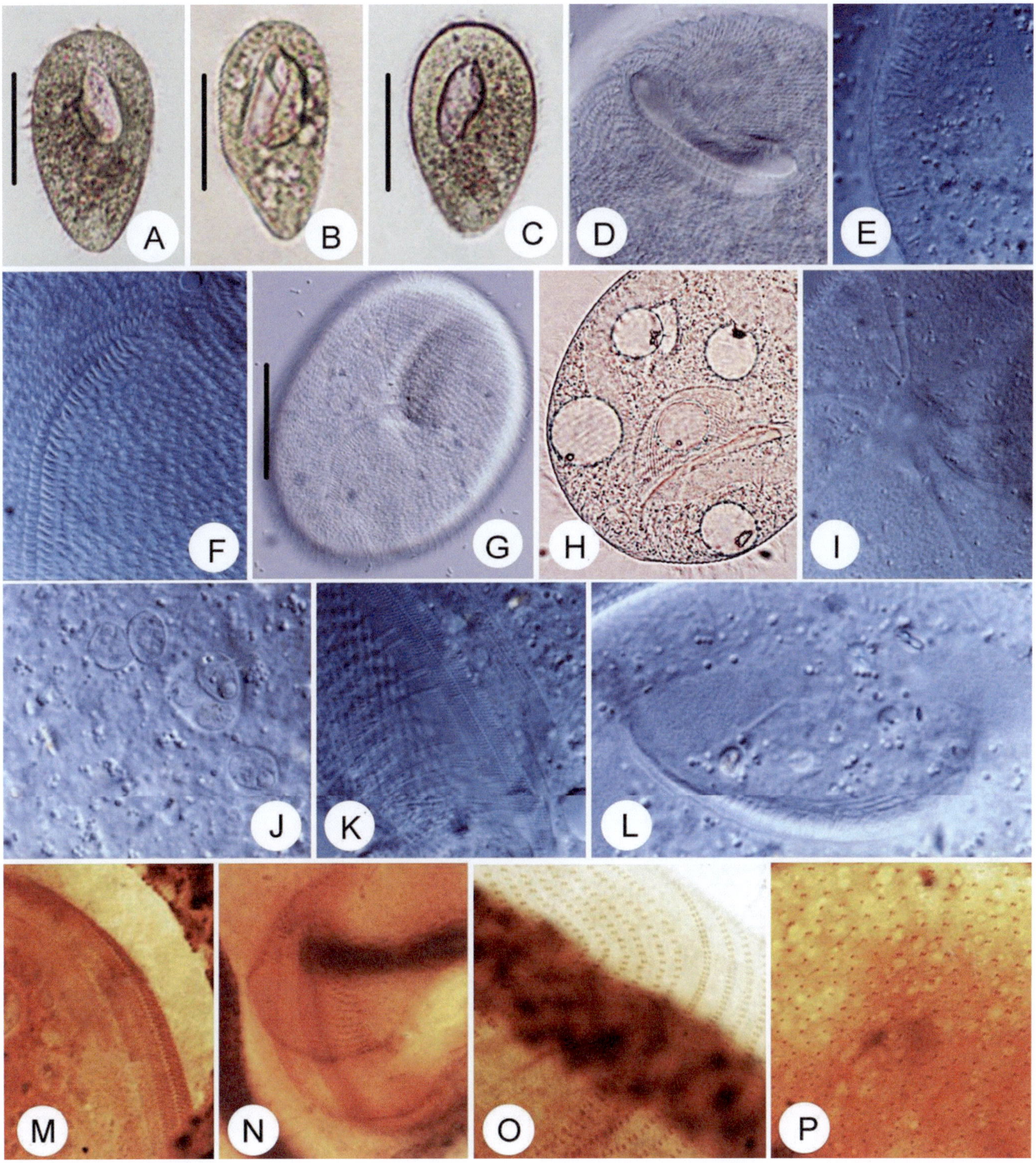


**Fig. 9B** *Marituja* cf. *caudata*
(**a–c**) Ventral views of live individuals, showing different body shapes; (**d**) buccal field; (**e**) part of pellicle, showing extrusomes; (**f, o**) anterior suture; (**g**) dorsal view of live cell; (**h, j**) food vacuoles; (**i**) contractile vacuole; (**k–n**) buccal structure; (**p**) part of somatic kineties. Scale bar: 50μm

## Reference

Xu Y, Gao F, Fan X (2018) Reconsideration of the systematics of Peniculida (Protista, Ciliophora) based on SSU rRNA gene sequences and new morphological features of Marituja and Disematostoma. Hydrobiologia 806:313–331

Class Oligohymenophorea de Puytorac et al., 1974
Order Peniculida Fauré-Fremiet in Corliss, 1956
Family Maritujidae Jankowski in Small & Lynn, 1985
*Marituja pelagica* Gajewskaja, 1928

**Ecological feature:** mangrove
**Water temperature:** 18.0 °C
**pH:** 6.0
**Salinity:** 4.4‰

**Fig. 10A** Sample site

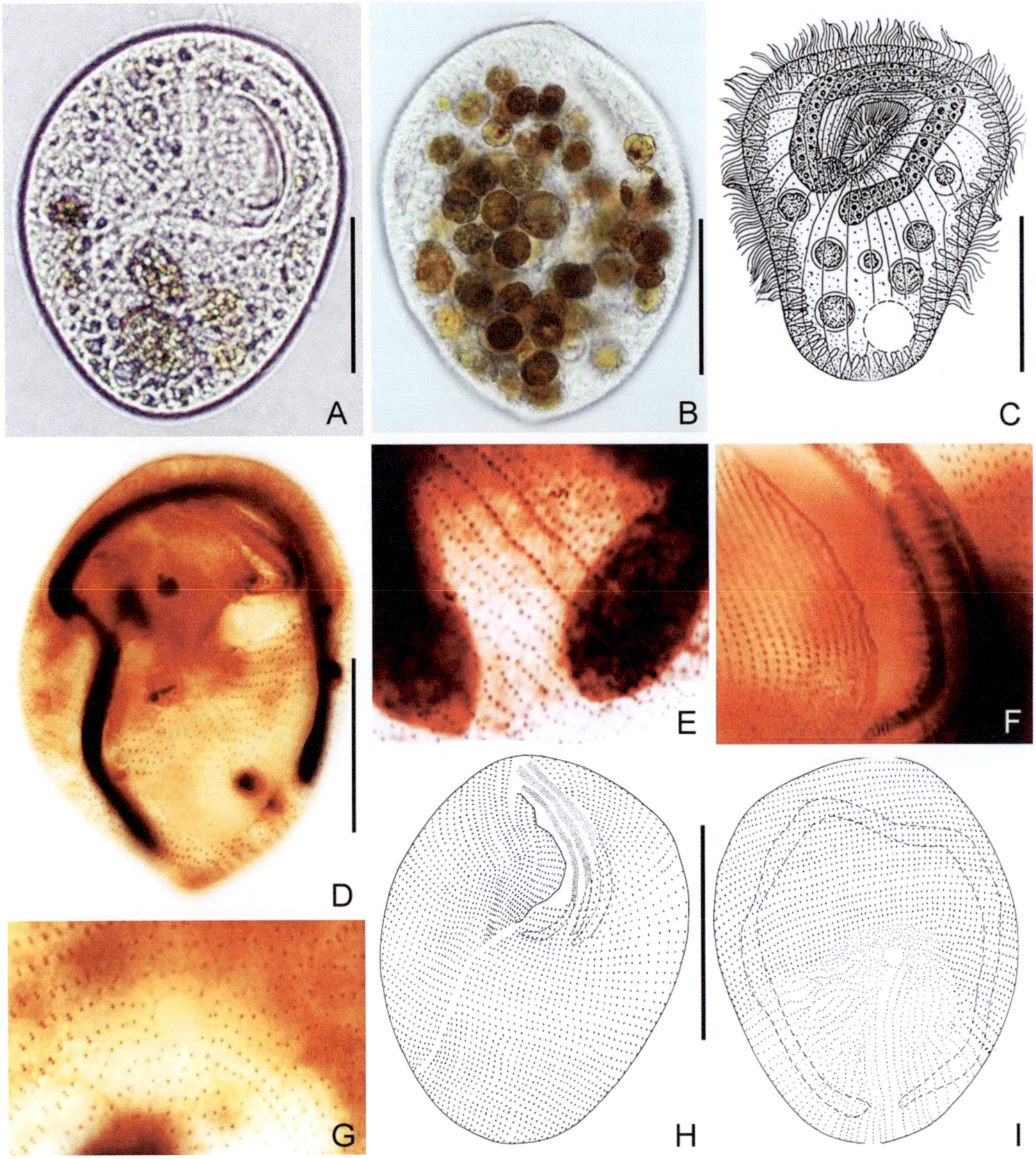


**Fig. 10B** *Marituja pelagica*
(**a–c**) Lateral views of live individuals, (**c**) from Foissner et al. (1999, p434); (**d**, **h**, **i**) ciliature of ventral (**d**, **h**) and dorsal (**i**) sides; (**e**) scutica; (**f**) oral structure; (**g**) contractile vacuole pore. Scale bar: 50 μm

## References

Foissner W, Berger H, Schaumburg J (1999) Identification and ecology of limnetic plankton ciliates. Bavarian State Office of Water Management, Munich, pp 793

Gajewskaja N (1933) Zur Okologie, Morphologie und Systematik der Infusorien des Baikalsees. Zoologica 83:1–298

Class Oligohymenophorea de Puytorac et al., 1974
Order Philasterida Small, 1967
Family Cohnilembidae Kahl, 1933
*Cohnilembus verminus* (Müller, 1786)

**Ecological feature:** mangrove
**Water temperature:** 19.0 °C; 23.0 °C; 20.0 °C
**pH:** 6.7; 6.9; 7.2
**Salinity:** 10.0‰; 10.0‰; 16.0‰

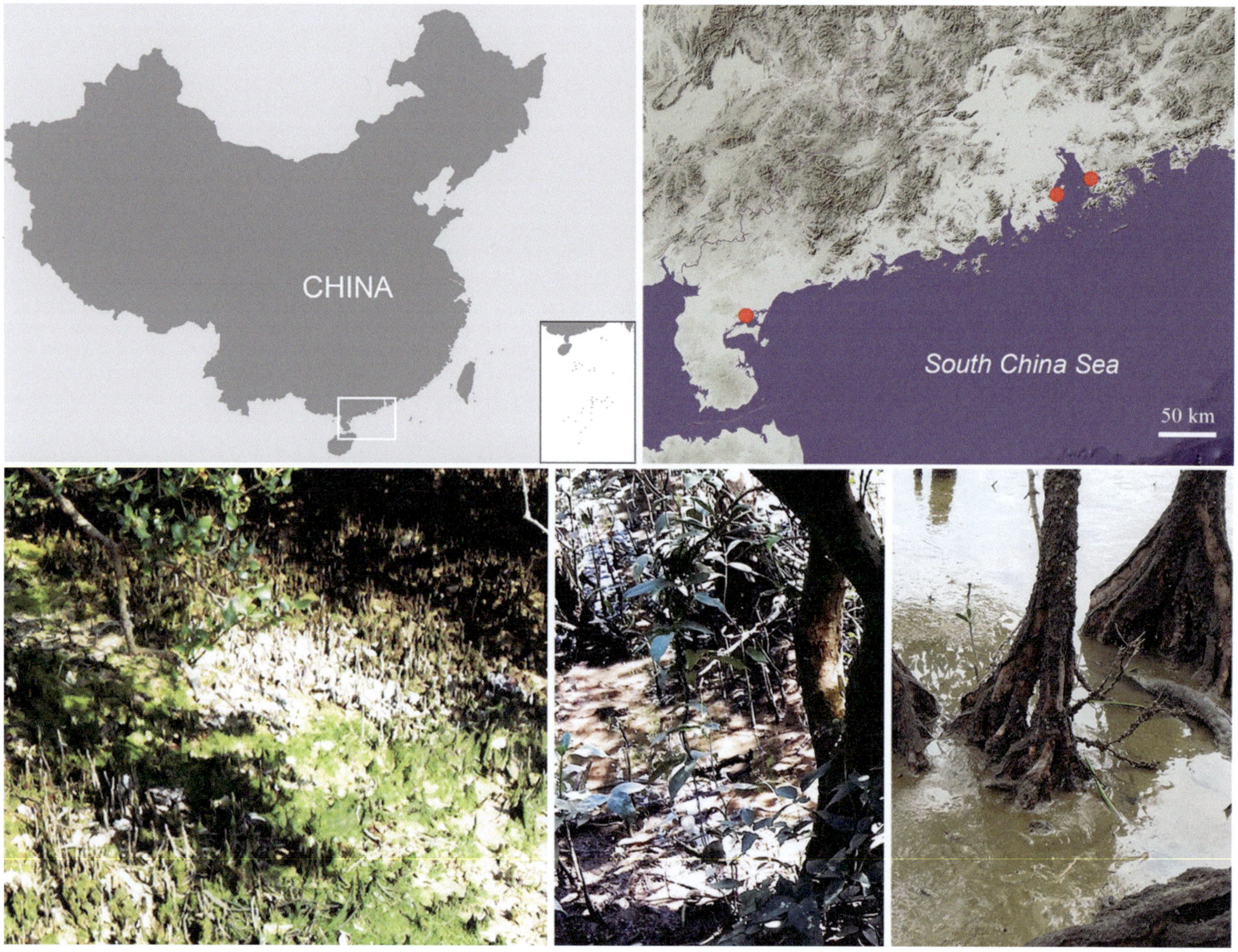


**Fig. 11A** Sample site

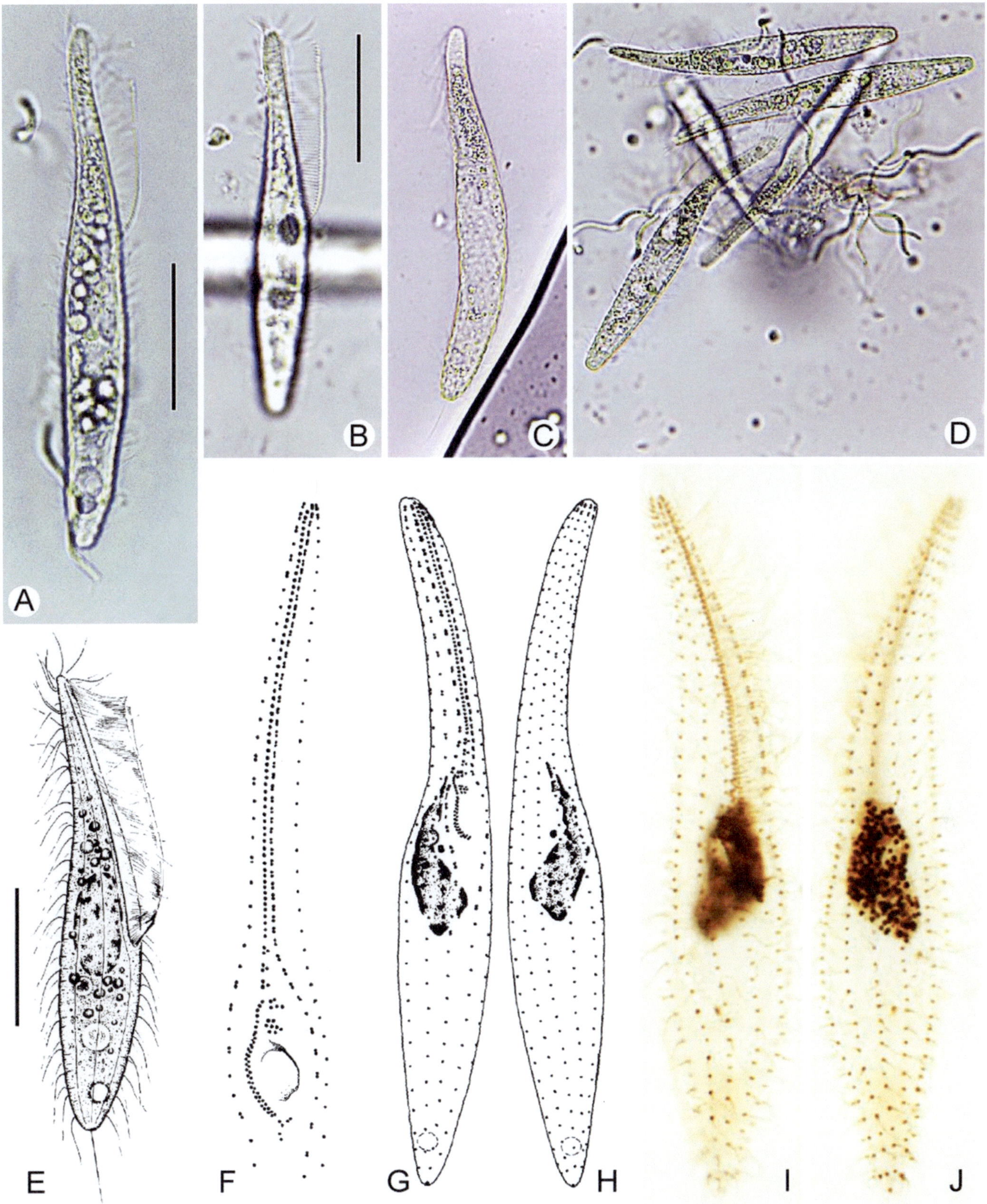


**Fig. 11B** *Cohnilembus verminus*
(**a–e**) Lateral views of live individuals, showing different body shapes and sizes; (**f**) oral apparatus; (**g–j**) ventral (**g**, **i**) and dorsal (**h**, **j**) views of ciliature, showing ciliature and macronucleus. Scale bar: 15 μm

## Reference

Song WB (2000) Morphological and taxonomical studies on some marine scuticociliates from China Sea, with description of two new species, *Philasterides armatalis* sp. n. and *Cyclidium varibonneti* sp. n. (Protozoa: Ciliophora: Scuticociliatida). Acta Protozool 39:295–322

Class Oligohymenophorea de Puytorac et al., 1974
Order Philasterida Small, 1967
Family Loxocephalidae Jankowski, 1964
*Paratetrahymena wassi* Thompson, 1963

**Ecological feature:** offshore water
**Water temperature:** 19.4 °C
**pH:** 8.2
**Salinity:** 32.6‰

**Fig. 12A** Sample site

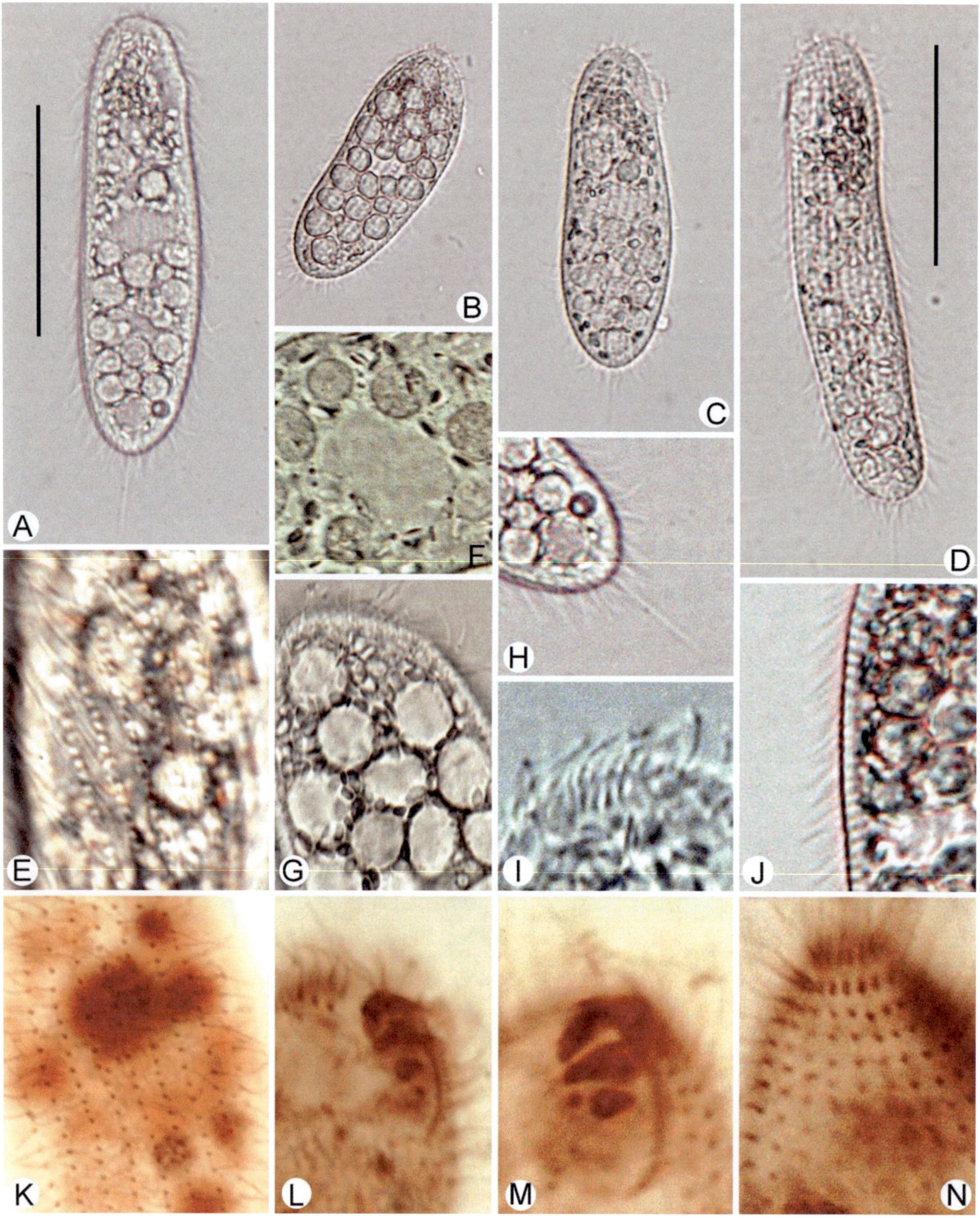


**Fig. 12B** *Paratetrahymena wassi*
(**a–d**) Different body shapes; (**e**, **i**, **j**) extrusomes; (**f**) macronucleus; (**g**) food granules in anterior portion of the body; (**h**) caudal cilium and contractile vacuole; (**k**) somatic kineties; (**l**, **m**) oral apparatus; (**n**) two basal body pairs at the anterior end of each somatic kinety. Scale bar: 50 μm

## Reference

Li J, Lin X, Yi Z, Clamp J, Liu W, Al-Rasheid KAS (2010) Molecules or morphogenesis: how to determine the phylogenetic assignment of *Paratetrahymena* (Protista, Ciliophora, Oligohymenophorea). Zool Scr 39:499–510

Class Oligohymenophorea de Puytorac et al., 1974
Order Philasterida Small, 1967
Family Orchitophryidae Cépède, 1910
*Mesanophrys carcini* (Grolière & Leglise, 1977)

**Ecological feature:** intertidal zone
**Water temperature:** 20.0 °C
**pH:** no data available
**Salinity:** 31.0‰

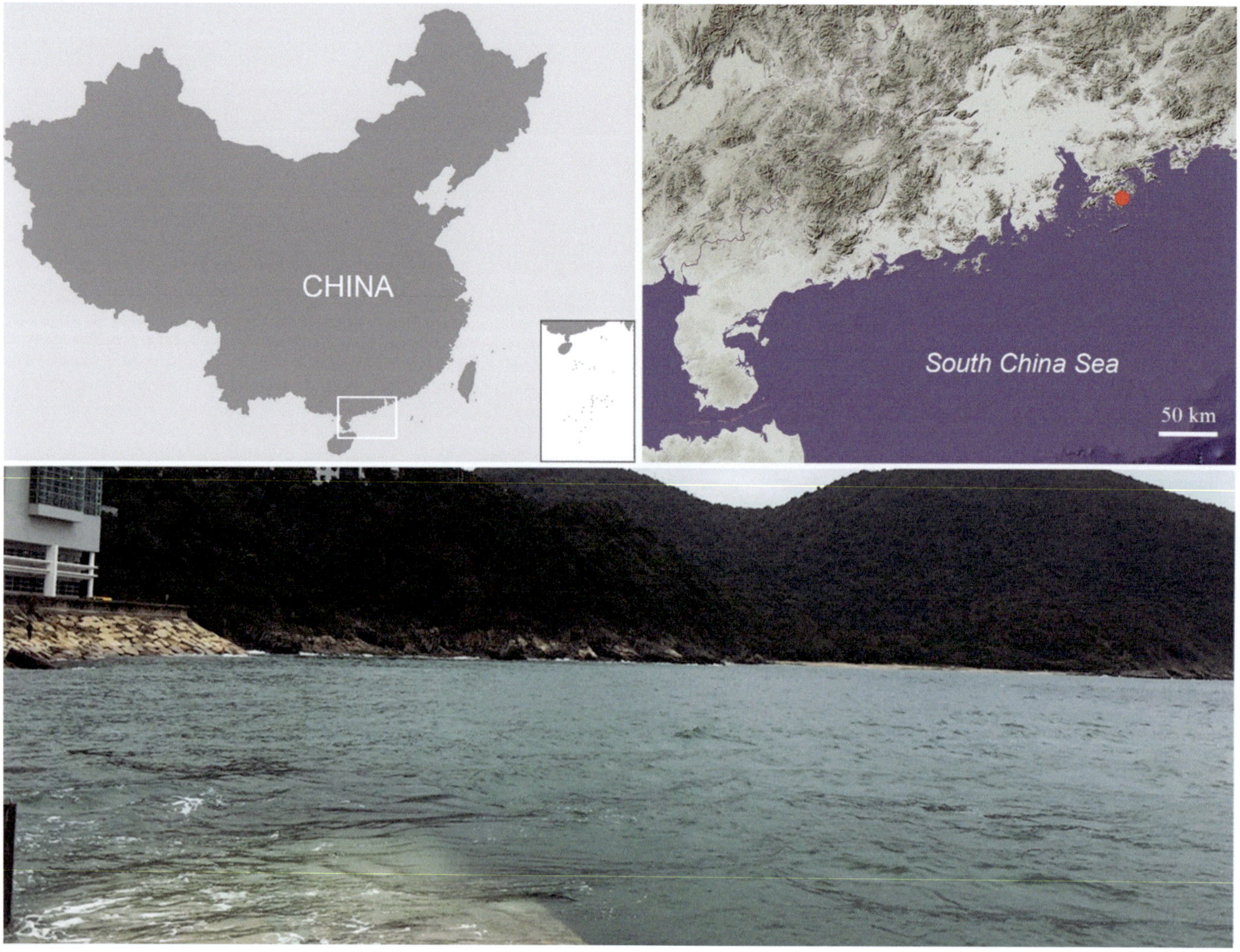


**Fig. 13A** Sample site

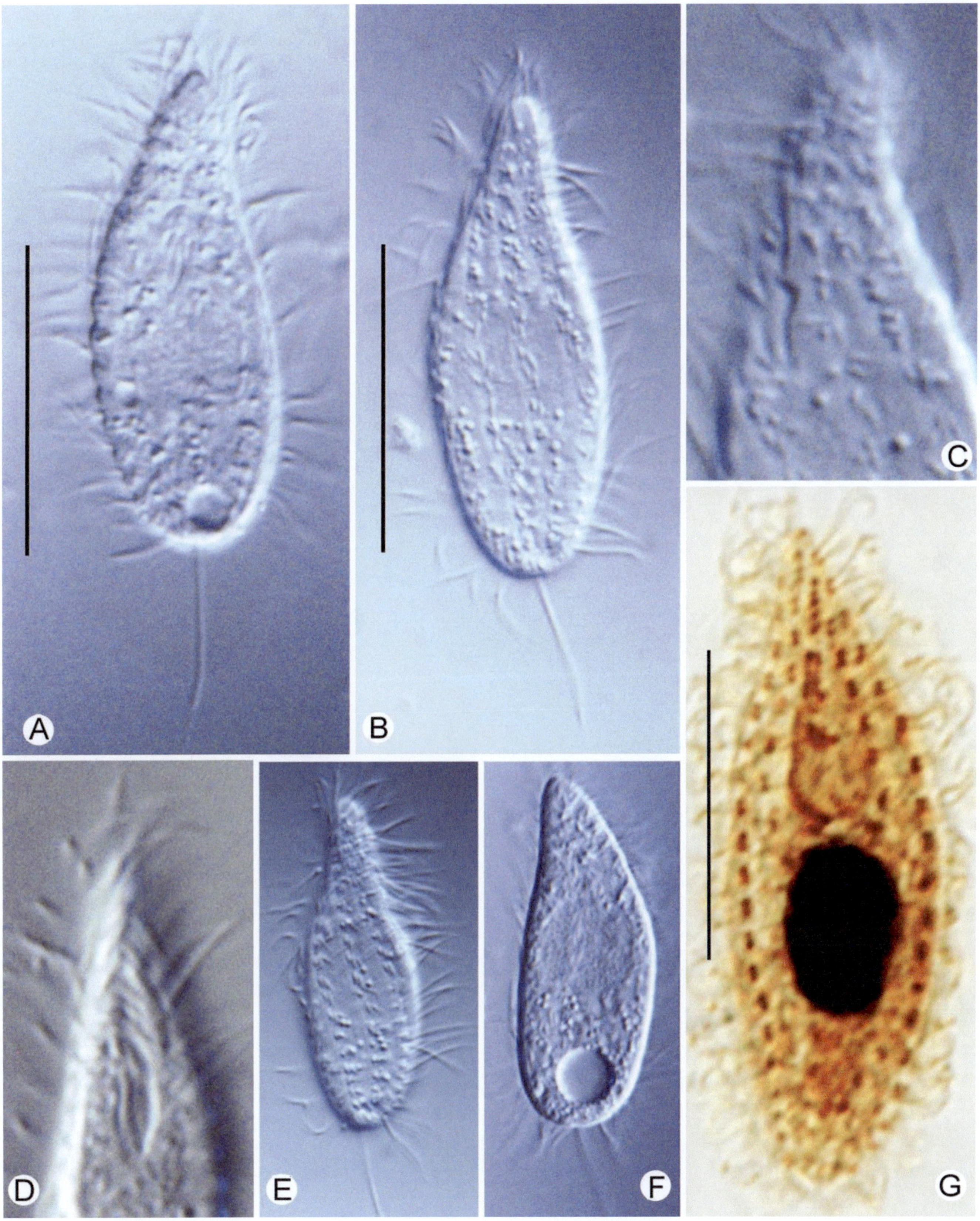


**Fig. 13B** *Mesanophrys carcini*
(**a**, **b**) Lateral views of live cells, showing body shapes; (**c**, **d**) buccal area; (**e**) distribution pattern of cytoplasm granules; (**f**) contractile vacuole; (**g**) ventral ciliature. Scale bar: 15 μm

## References

Qin L, Wang Y, Zhang L, Dai J (2007) Histopathology of turbot associated with *Mesanophrys carcini* parasite. Acta Hydrobiol Sin 31:618–628

Zhao Y, Fan X, Xu Y, Hu X, Ma H (2011) Morphological studies on eight marine scuticociliates (Protozoa, Ciliophora) from China. Acta Hydrobiol Sin 35:929–939 (In Chinese with English summary)

Class Oligohymenophorea de Puytorac et al., 1974
Order Philasterida Small, 1967
Family Orchitophryidae Small & Lynn, 1985
*Metanophrys* cf. *sinensis*

**Ecological feature:** intertidal zone
**Water temperature:** 16.0 °C
**pH:** 7.0
**Salinity:** 22.0‰

**Fig. 14A** Sample site

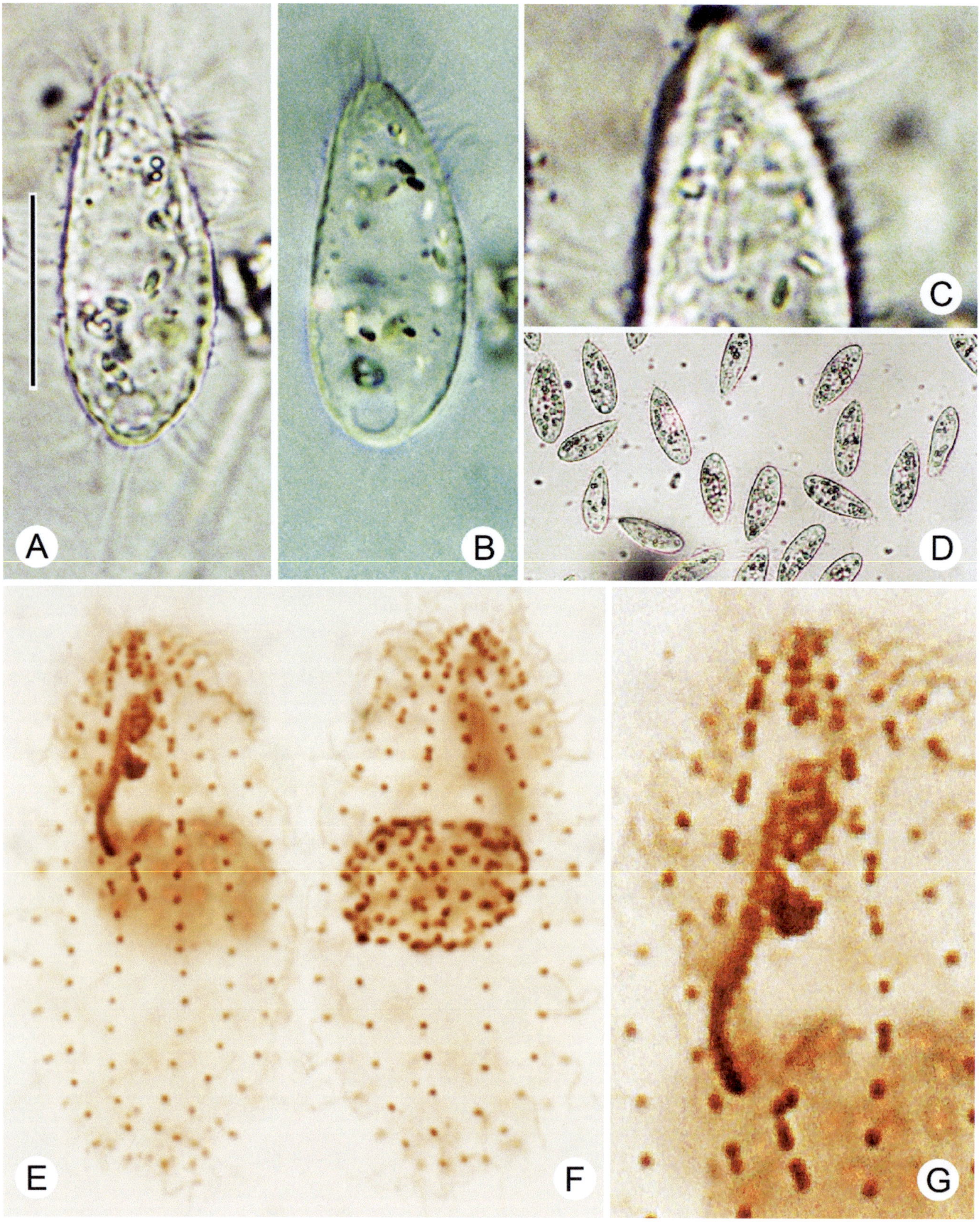


**Fig. 14B** *Metanophrys* cf. *sinensis*
(**a–c**) Views of live cells, showing notched pellicle; (**d**) different body shapes and sizes; (**e**, **f**) ventral (**e**) and dorsal (**f**) views of ciliature; (**g**) oral apparatus. Scale bar: 15 μm

## References

Pan X, Yi Z, Li J, Ma H, Al-Farraj SA, Al-Rasheid KAS (2015) Biodiversity of marine scuticociliates (Protozoa, Ciliophora) from China:description of seven morphotypes including a new species, *Philaster sinensis* spec. nov. Eur J Protistol 51:142–157

Song WB, Wilbert N (2000) Redefinition and redescription of some marine scuticociliates from China, with report of a new species, *Metanophrys sinensis* nov. spec. (Ciliophora, Scuticociliatida). Zool Anz 239:45–74

Class Oligohymenophorea de Puytorac et al., 1974
Order Philasterida Small, 1967
Family Orchitophryidae Cépède, 1910
*Metanophrys similis* Song et al., 2002

**Ecological feature:** mangrove
**Water temperature:** 29.0 °C
**pH:** 7.6
**Salinity:** 20.0‰

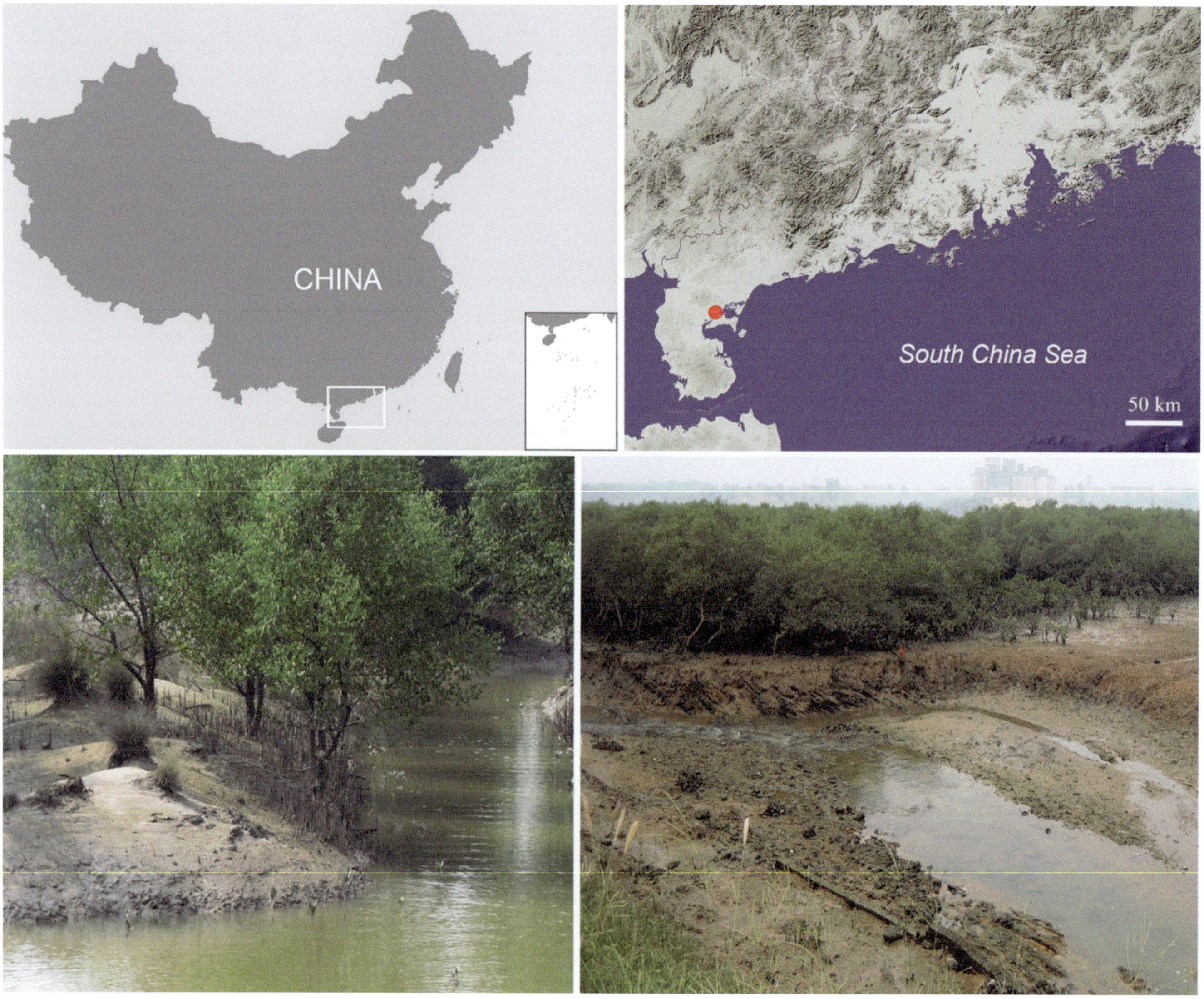


**Fig. 15A** Sample site

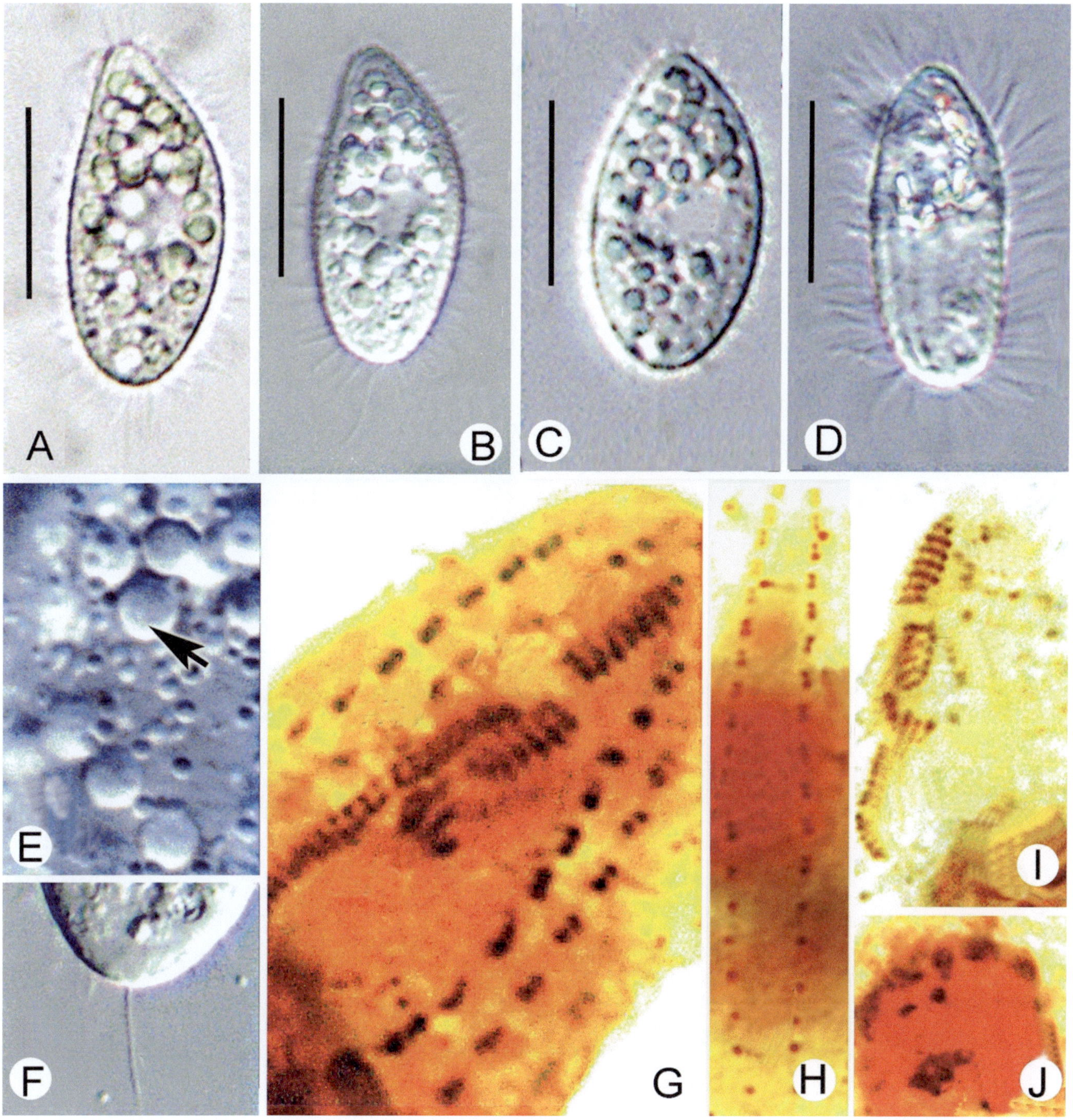


**Fig. 15B** *Metanophrys similis*
(**a–d**) Lateral views of live individuals, showing cilia, macronucleus, and inclusions; (**e**) food vacuole (arrow); (**f**) caudal cilium; (**g**, **i**) **o**ral apparatus; (**h**) dorsal view of ciliature, showing dikinetids and monokinetids; (**j**) macronucleus. Scale bar: 30 μm

## Reference

Song WB, Shang H, Chen Z, Ma HG (2002) Comparison of some closely-related *Metanophrys*-taxa with description of a new species *Metanophrys similis* nov. spec. (Ciliophora, Scuticociliatida). Eur J Protistol 38:45–53

Class Oligohymenophorea de Puytorac et al., 1974
Order Philasterida Small, 1967
Family Orchitophryidae Cépède, 1910
*Metanophrys sinensis* Song & Wilbert, 2000

**Ecological feature:** mangrove
**Water temperature:** 26.0 °C
**pH:** 7.3
**Salinity:** 21.0‰

**Fig. 16A** Sample site

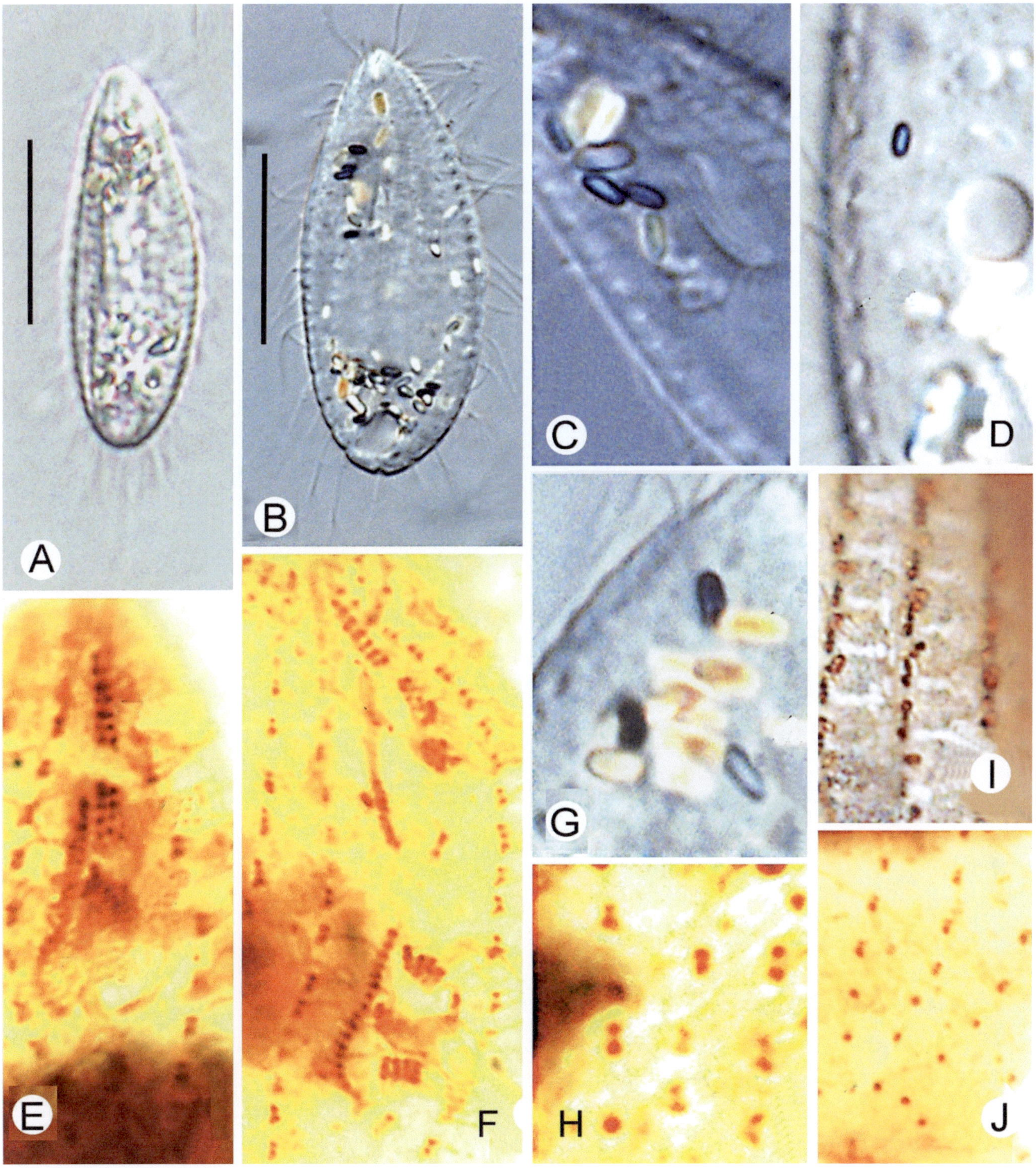


**Fig. 16B** *Metanophrys sinensis*
(**a**, **b**) Ventral views of live individuals; (**c**) buccal field; (**d**) notched pellicle; (**e**) oral apparatus; (**f**) late stage of morphogenesis, to show new oral apparatus; (**g**) the bar-shaped crystals; (**h**) to show basal bodies of scutica; (**i**) part of argyrome; (**j**) posterior portion of somatic kineties. Scale bar: 15 μm

## Reference

Song WB, Wilbert N (2000) Redefinition and redescription of some marine scuticociliates from China, with report of a new species, *Metanophrys sinensis* nov. spec. (Ciliophora, Scuticociliatida). Zool Anz 239:45–74

Class Oligohymenophorea de Puytorac et al., 1974
Order Philasterida Small, 1967
Family Orchitophryidae Cépède, 1910
*Paramesanophrys typicus* Pan et al., 2016

**Ecological feature:** coastal water
**Water temperature:** 24.0 °C
**pH:** 8.2
**Salinity:** 32.0‰

**Fig. 17A** Sample site

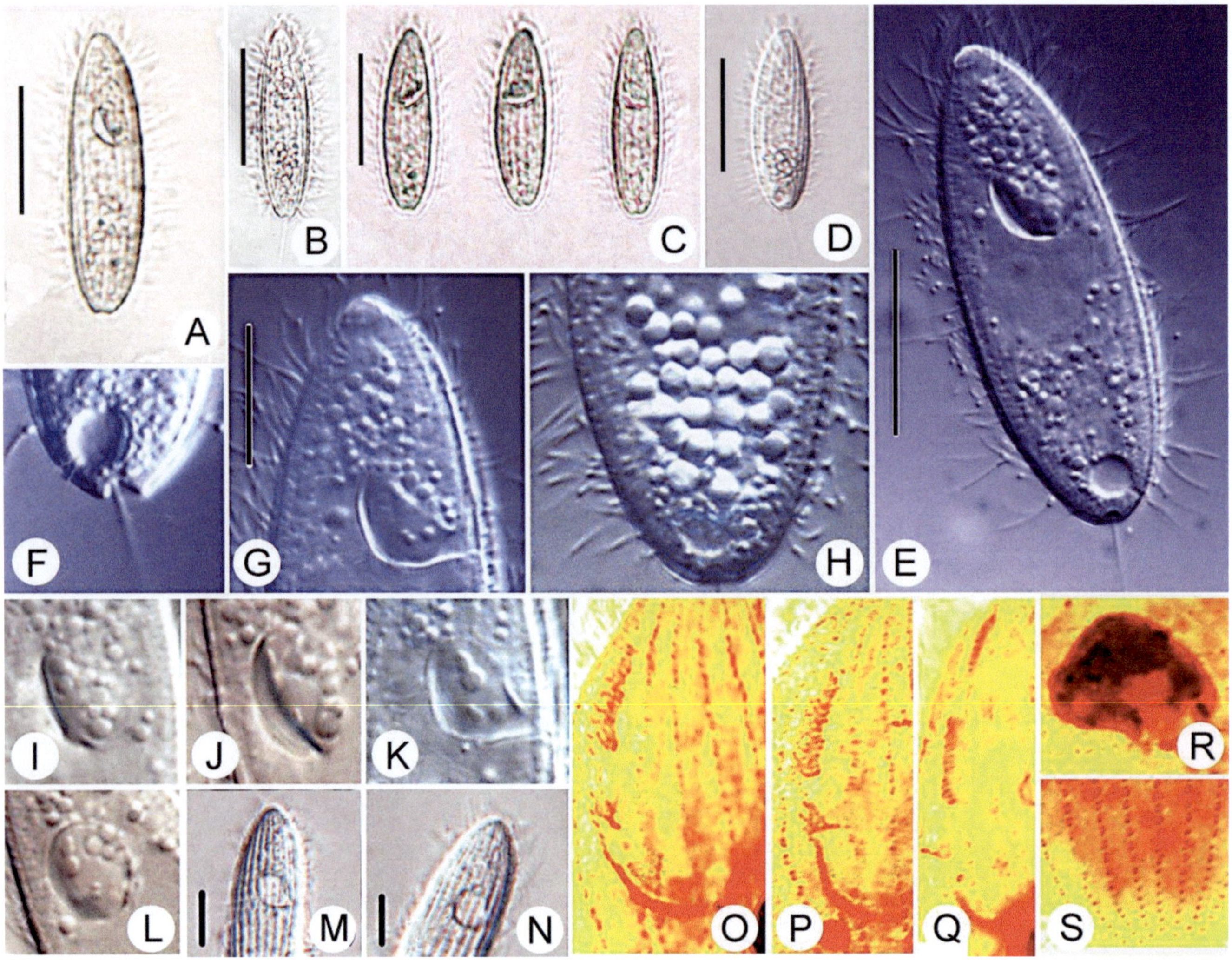


**Fig. 17B** *Paramesanophrys typicus*
(**a–d**) Ventral views of live individuals; (**e**) ventral view of live cell, showing cytostome, macronucleus, contractile vacuole, and caudal cilium; (**f**) posterior region of live cell, showing contractile vacuole and caudal cilium; (**g**) buccal region; (**h**) food vacuoles in posterior portion of cell; (**i–n**) ventral views of buccal area; (**o–q**) buccal apparatus; (**r**) macronucleus; (**s**) part of somatic kineties. Scale bars: 10 μm (**e, m, n**); 40 μm (**a, g**); 80 μm (**b–d**)

## Reference

Pan XM, Fan XP, Al-Farraj SA, Gao S, Chen Y (2016) Taxonomy and morphology of four "ophrys-related" scuticociliates (Ciliophora, Scuticociliatia), with description of a new genus, *Paramesanophrys* gen. nov. Eur J Taxon 191:1–18

Class Oligohymenophorea de Puytorac et al., 1974
Order Philasterida Small, 1967
Family Parauronematidae Small & Lynn, 1985
*Parauronema longum* Song, 1995

**Ecological feature:** mangrove area
**Water temperature:** 25.0 °C
**pH:** 7.8
**Salinity:** 21.0‰

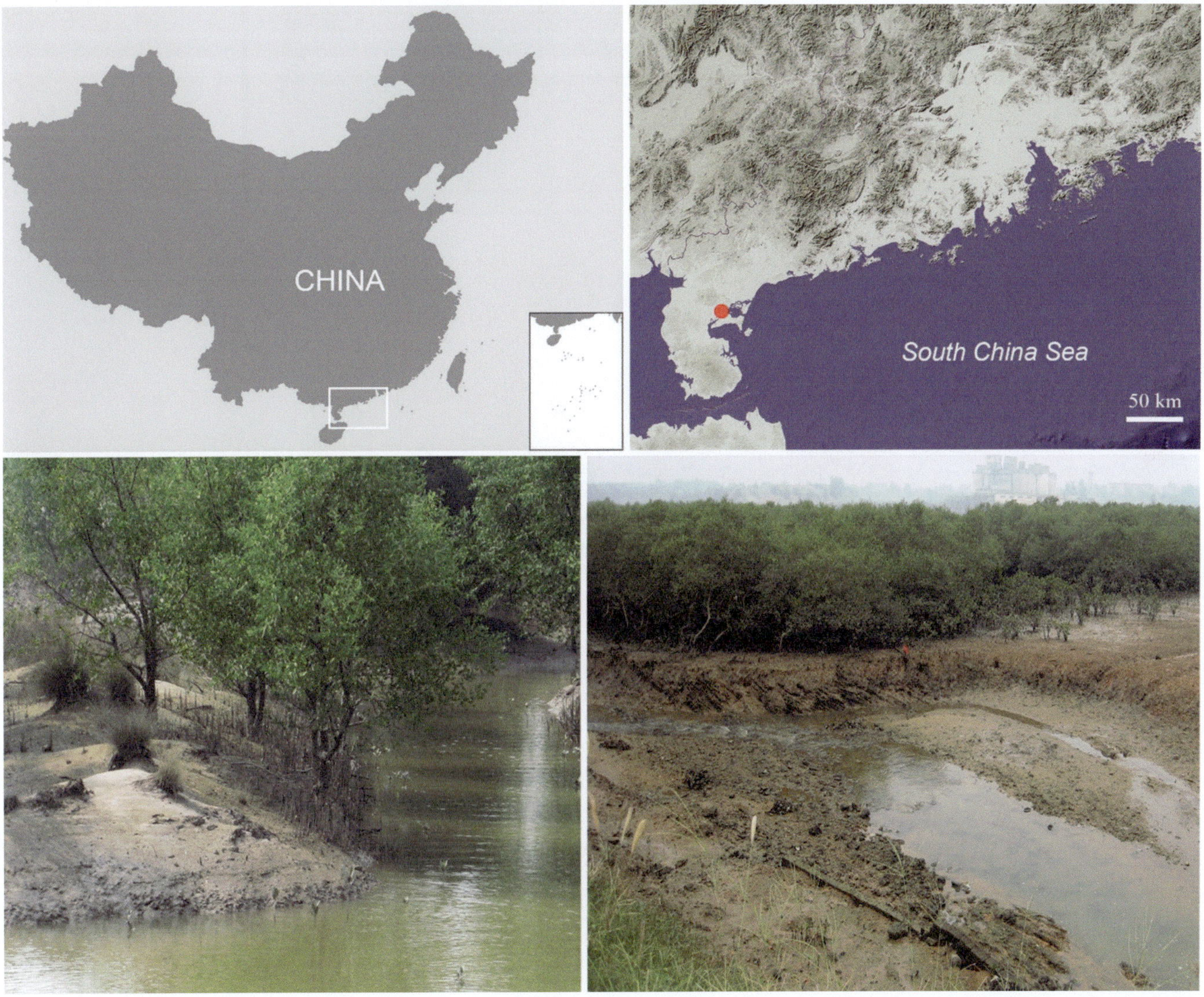


**Fig. 18A** Sample site

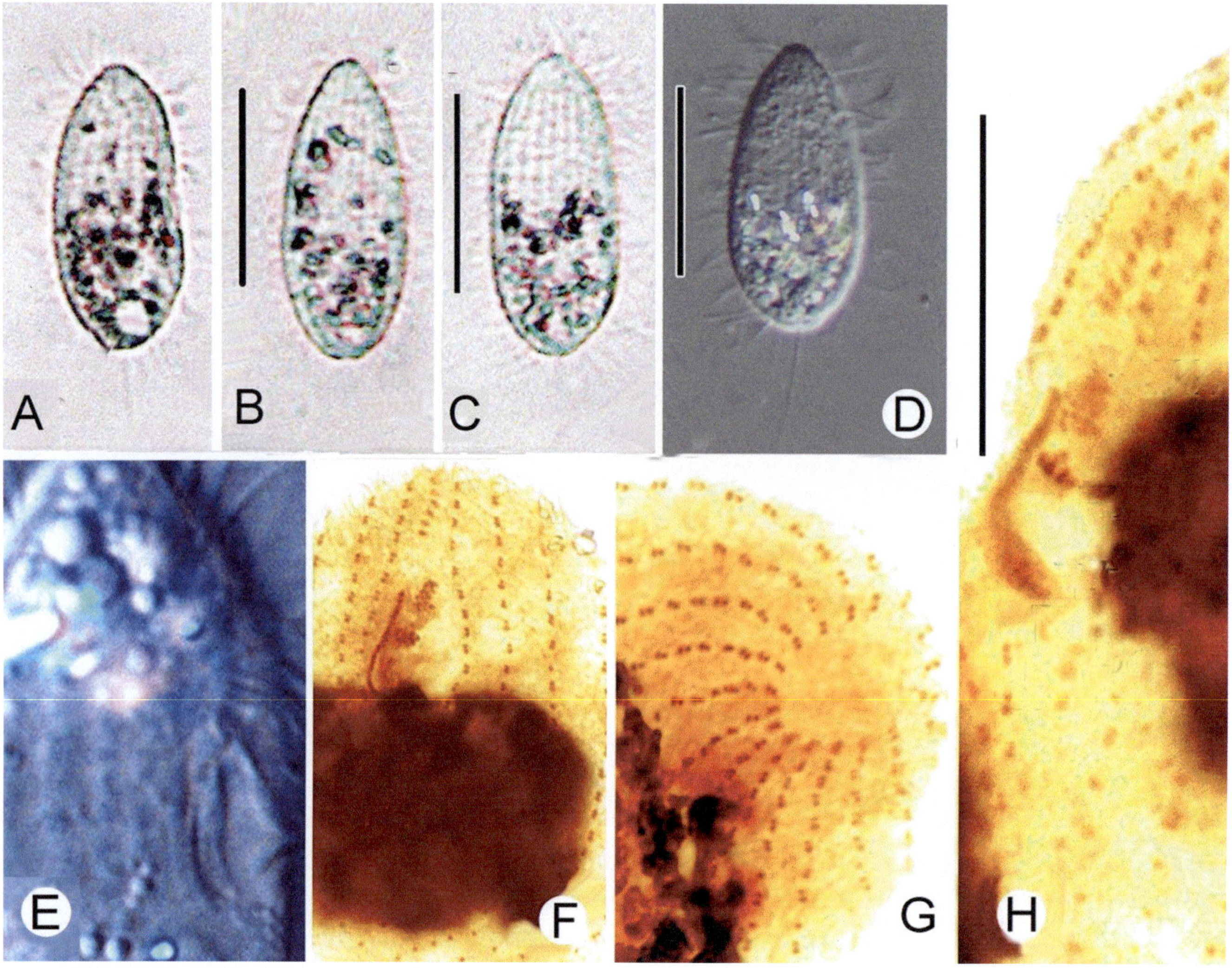


**Fig. 18B** *Parauronema longum*
(**a–d**) Lateral views of live individuals; (**e**) buccal field; (**f**, **h**) oral structure; (**g**) anterior part of the body, to show a large apical plate. Scale bars: 10 μm (**h**); 40 μm (**b–d**)

## Reference

Pan X, Yi Z, Huang J, Li J, Ma H, Al-Farraj SA, Al-Rasheid KAS (2015) Biodiversity of marine scuticociliates (Protozoa, Ciliophora) from China: description of seven morphotypes including a new species, *Philaster sinensis* spec. nov. Eur J Protistol 51:142–157

Class Oligohymenophorea de Puytorac et al., 1974
Order Philasterida Small, 1967
Family Philasteridae Kahl, 1931
*Philaster hiatti* Thompson, 1969

**Ecological feature:** mollusc-farming pond
**Water temperature:** 19.4 °C
**pH:** 8.2
**Salinity:** 32.6‰

**Fig. 19A** Sample site

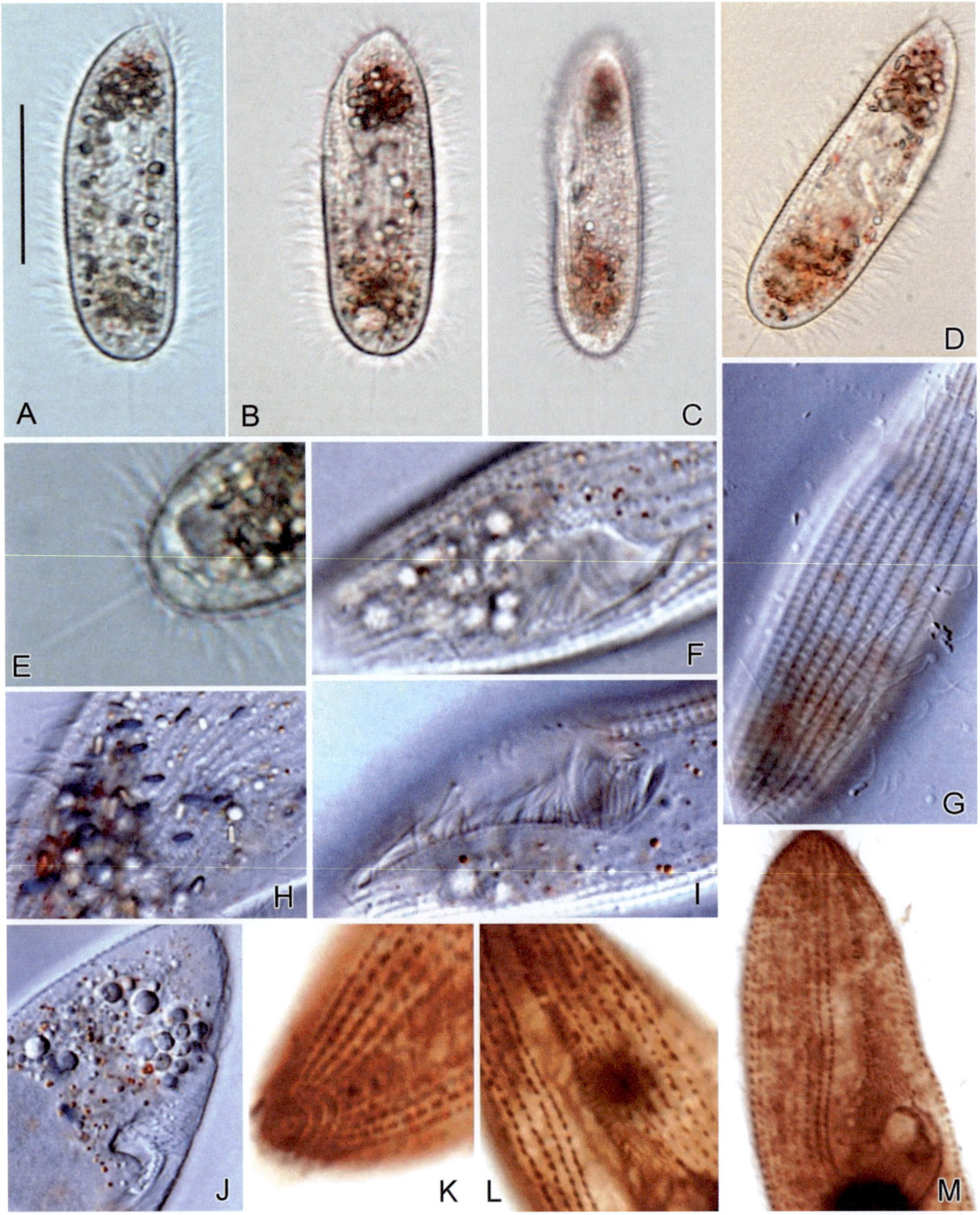


**Fig. 19B** *Philaster hiatti*
(**a**–**d**) Lateral views of live cells; (**e**) posterior end of cell showing caudal cilium and contractile vacuoles; (**f**, **i**) buccal field and membranelles; (**g**) somatic cilia; (**h**, **j**) cytoplasm showing granules; (**k**) ciliature of anterior portion of the body; (**l**, **m**) oral apparatus. Scale bar: 30 μm

## References

Small E, Lynn D (1985) Phylum Ciliophora. In: Lee J, Hunter S, Bovee E (eds.) An Illustrated Guide to the Protozoa. Society of Protozoologists, Lawrence, pp 393–575

Thompson JC (1969) *Philaster hiatti* n. sp., a holotrichous ciliate from Hawaii. J Protozool 16:81–83

Class Oligohymenophorea de Puytorac et al., 1974
Order Philasterida Small, 1967
Family Philasteridae Kahl, 1931
*Philasterides* cf. *armatalis*

**Ecological feature:** intertidal zone
**Water temperature:** 16.0 °C
**pH:** 7.0
**Salinity:** 22.0‰

**Fig. 20A** Sample site

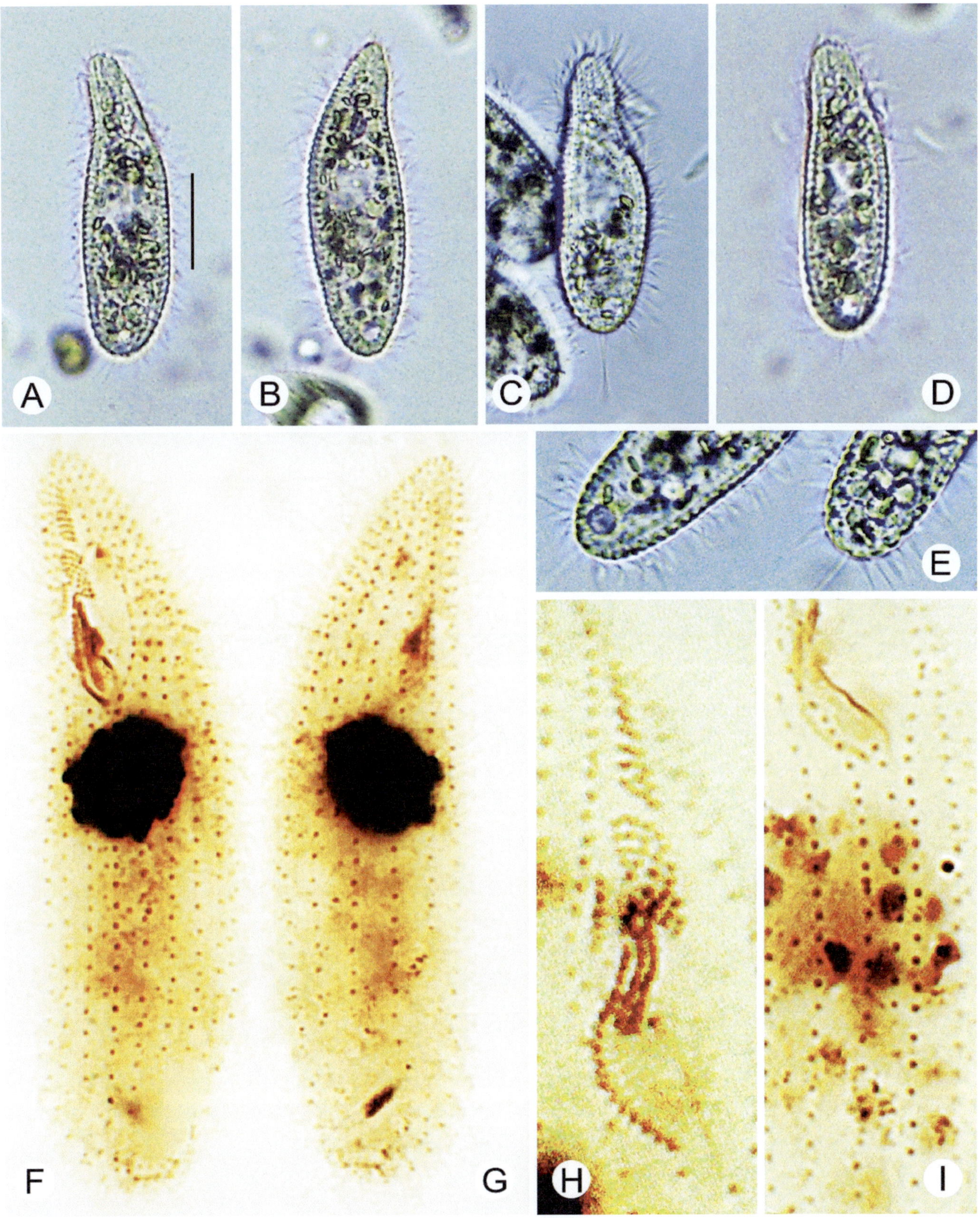


**Fig. 20B** *Philasterides* cf. *armatalis*
(**a–d**) Views of live cells, showing different body shapes and sizes; (**e**) posterior portion of two cells, showing notched pellicle; (**f**, **g**) ventral (**f**) and dorsal (**g**) views of ciliature; (**h**) oral apparatus; (**i**) Scutica. Scale bar: 20 μm

## References

Song WB (2000) Morphological and taxonomical studies on some marine scuticociliates from China Sea, with description of two new species, *Philasterides armatalis* sp. n. and *Cyclidium varibonneti* sp. n. (Protozoa: Ciliophora: Scuticociliatida). Acta Protozool 39:295–322

Song WB, Zhao Y, Xu K, Hu X, Gong J (2003) Pathogenic protozoa in mariculture. Science Press, Beijing

Class Oligohymenophorea de Puytorac et al., 1974
Order Philasterida Small, 1967
Family Pseudocohnilembidae Evans & Thompson, 1964
*Pseudocohnilembus hargisi* Evans & Thompson, 1964

**Ecological feature:** mangrove; intertidal zone (Daya Bay)
**Water temperature:** 20.0 °C; 30.2 °C
**pH:** 7.2; 8.3
**Salinity:** 16.0‰; 31.8‰

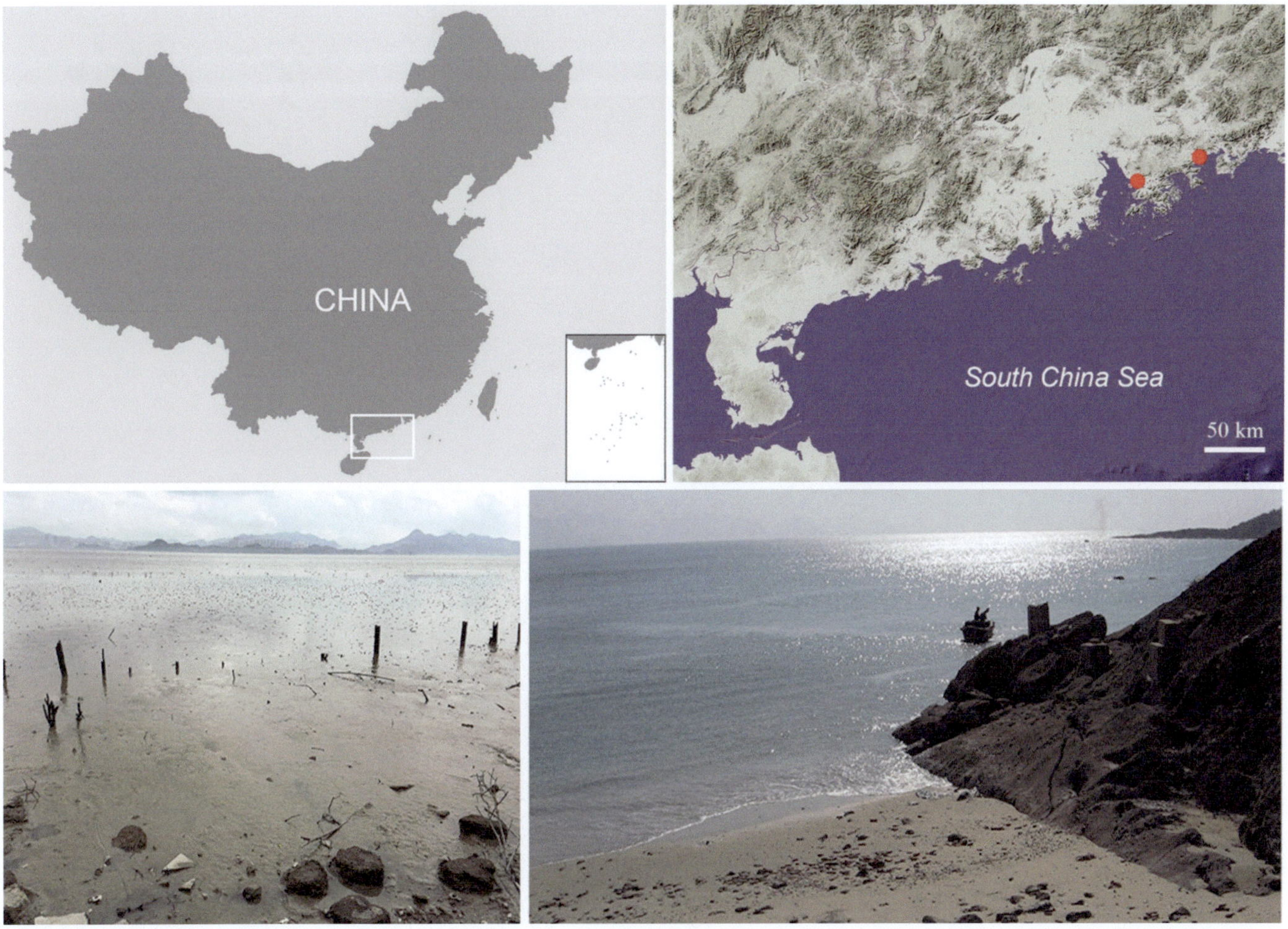


**Fig. 21A** Sample site

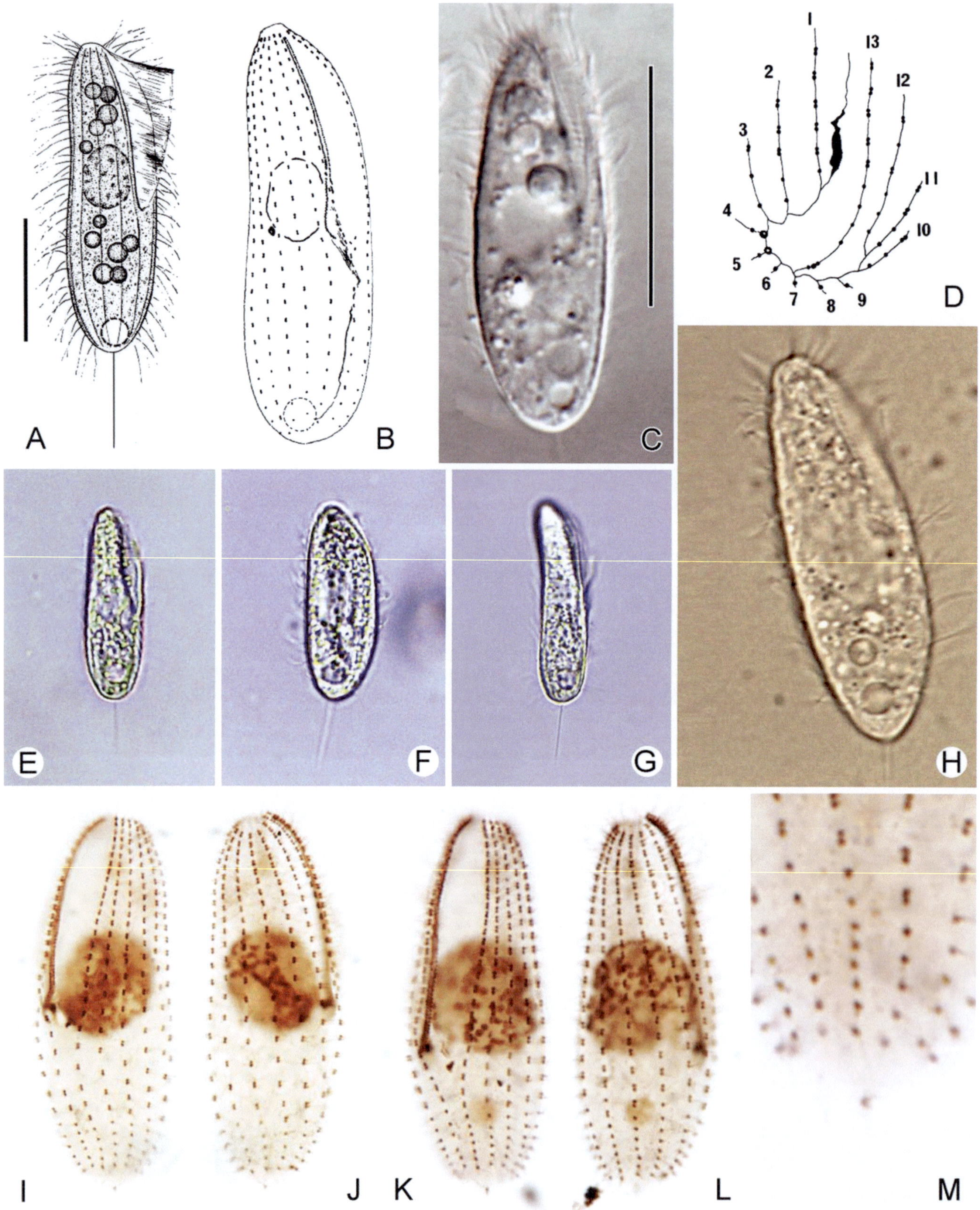


**Fig. 21B** *Pseudocohnilembus hargisi*
(**a**, **c**, **e–h**) Views of live cells; (**b**, **i–l**) lateral views of ciliature; (**d**) caudal view of posterior portion, showing caudal complex and contractile vacuole pores; (**m**) posterior portion of a cell, indicating basal body of caudal cilium. 1–13, somatic kineties 1–13. Scale bars: 15 μm (**a**); 30 μm (**c**)

## References

Pan X, Yi Z, Li J, Ma H, Al-Farraj SA, Al-Rasheid KAS (2015) Biodiversity of marine scuticociliates (Protozoa, Ciliophora) from China: description of seven morphotypes including a new species, *Philaster sinensis* spec. nov. Eur J Protistol 51:142–157

Song WB, Wilbert N (2002) Reinvestigations of three "well-known" marine scuticociliates: *Uronemella filificum* (Kahl, 1931) nov. gen., nov. comb., *Pseudocohnilembus hargisi* Evans & Thompson, 1964 and *Cyclidium citrullus* Cohn 1865, with description of the new genus *Uronemella* (Protozoa, Ciliophora, Scuticociliatida). Zool Anz 241:317–331

Class Oligohymenophorea de Puytorac et al., 1974
Order Philasterida Small, 1967
Family Pseudocohnilembidae Evans & Thompson, 1964
*Pseudocohnilembus persalinus* Evans & Thompson, 1964

**Ecological feature:** mangrove
**Temperature:** no data available
**pH:** 7.5
**Salinity:** 18.0‰

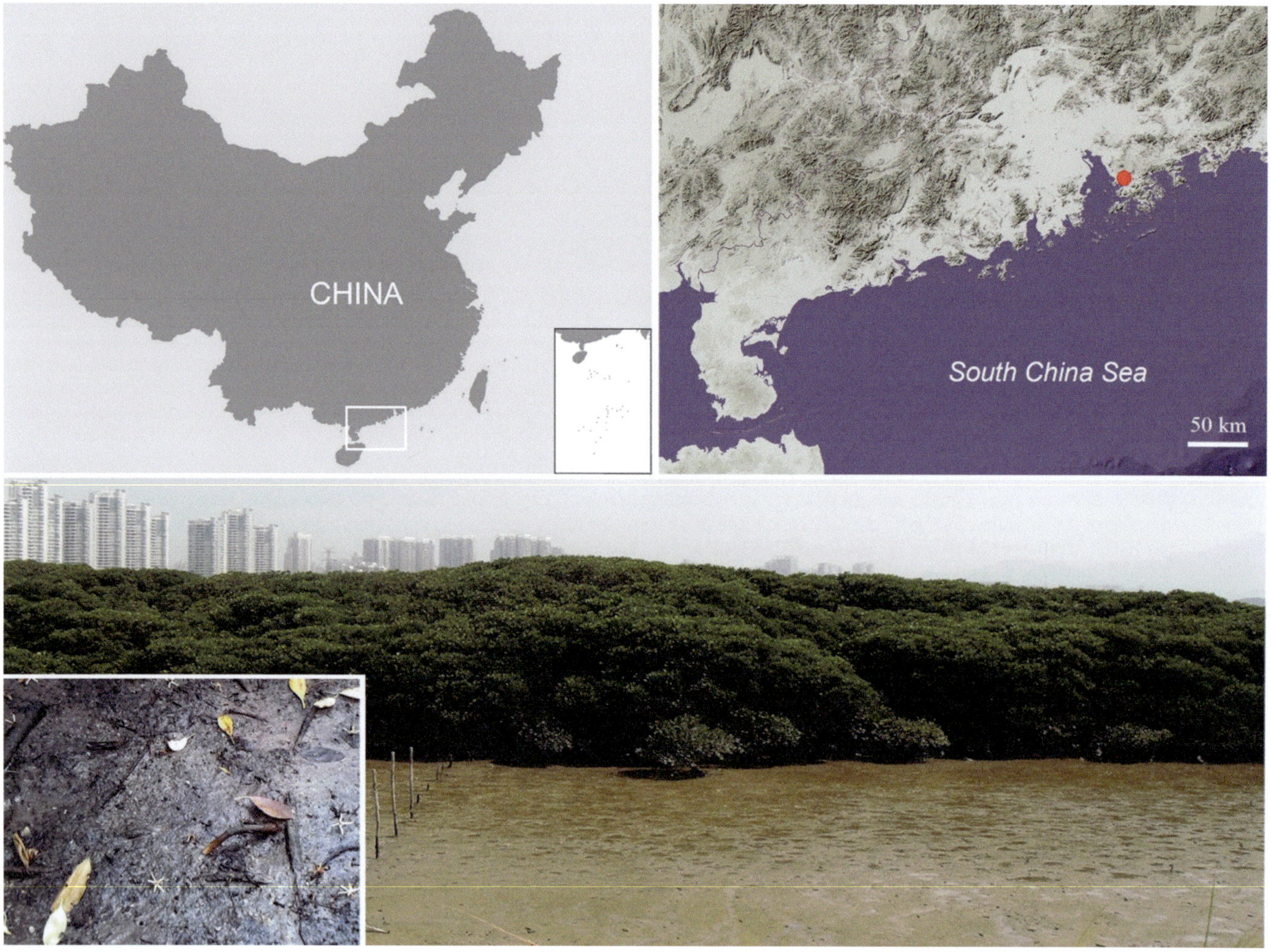


**Fig. 22A** Sample site

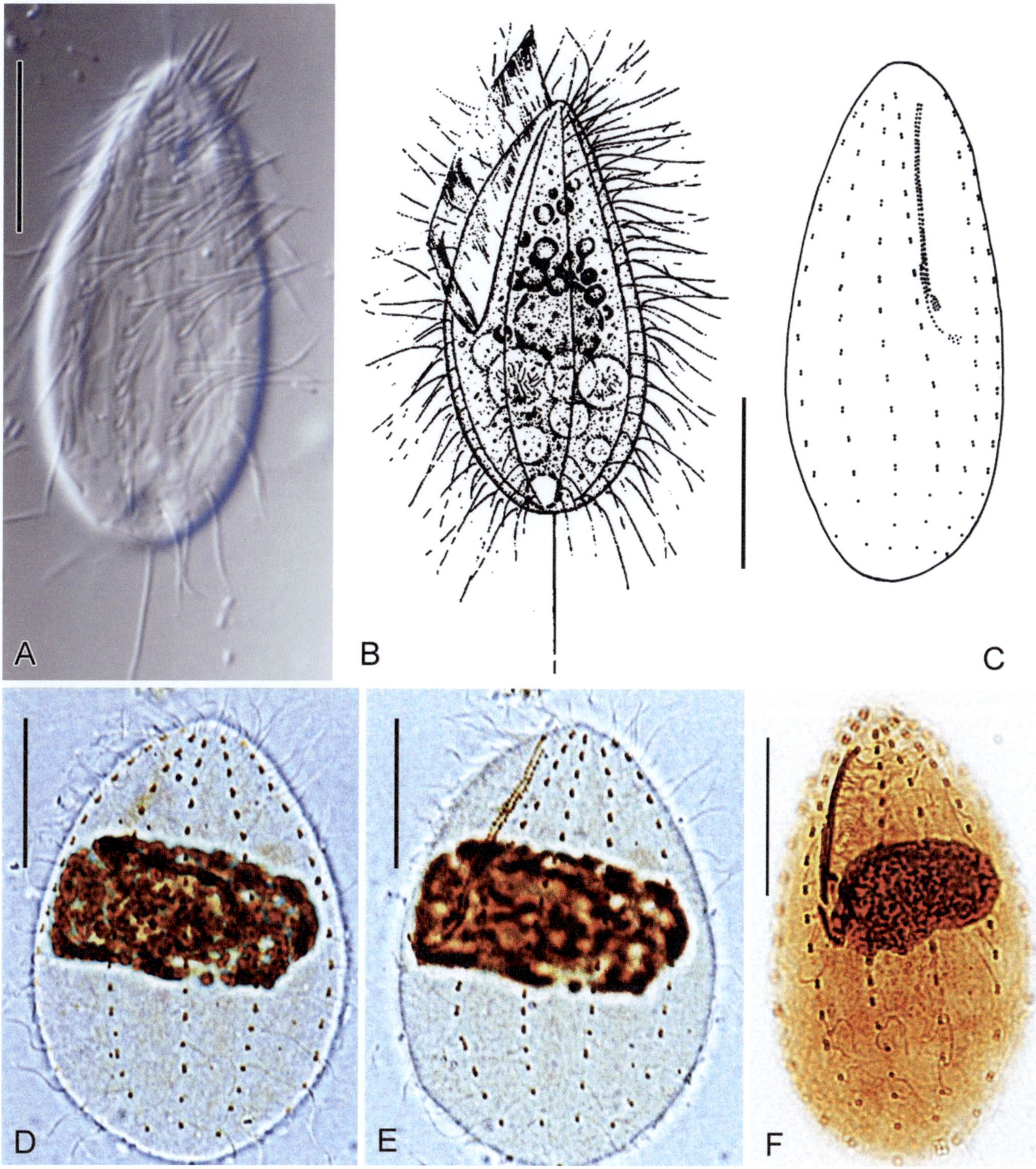


**Fig. 22B** *Pseudocohnilembus persalinus*
(**a**, **b**) Ventral views of live cells; (**c–f**) ventral (**c**, **e**, **f**) and dorsal (**d**) views of ciliature. Scale bar: 15 μm

## References

Pomp R, Wilbert N (1988) Taxonomic and ecological studies of ciliates from Australian saline soils: colpodids and hymenostomate ciliates. Mar Freshwater Res 39:479–495

Song WB (2000) Morphological and taxonomical studies on some marine scuticociliates from China Sea, with description of two new species, *Philasterides armatalis* sp. n. and *Cyclidium varibonneti* sp. n. (Protozoa: Ciliophora: Scuticociliatida). Acta Protozool 39:295–322

Class Oligohymenophorea de Puytorac et al., 1974
Order Philasterida Small, 1967
Family Uronematidae Thompson, 1964
*Uronema apomarinum* Liu et al., 2018

**Ecological feature:** mangrove; brackish water pond
**Water temperature:** 15.0 °C; 15.0 °C
**pH:** 7.8; 7.1
**Salinity:** 21.0‰; 10.0‰

**Fig. 23A** Sample site

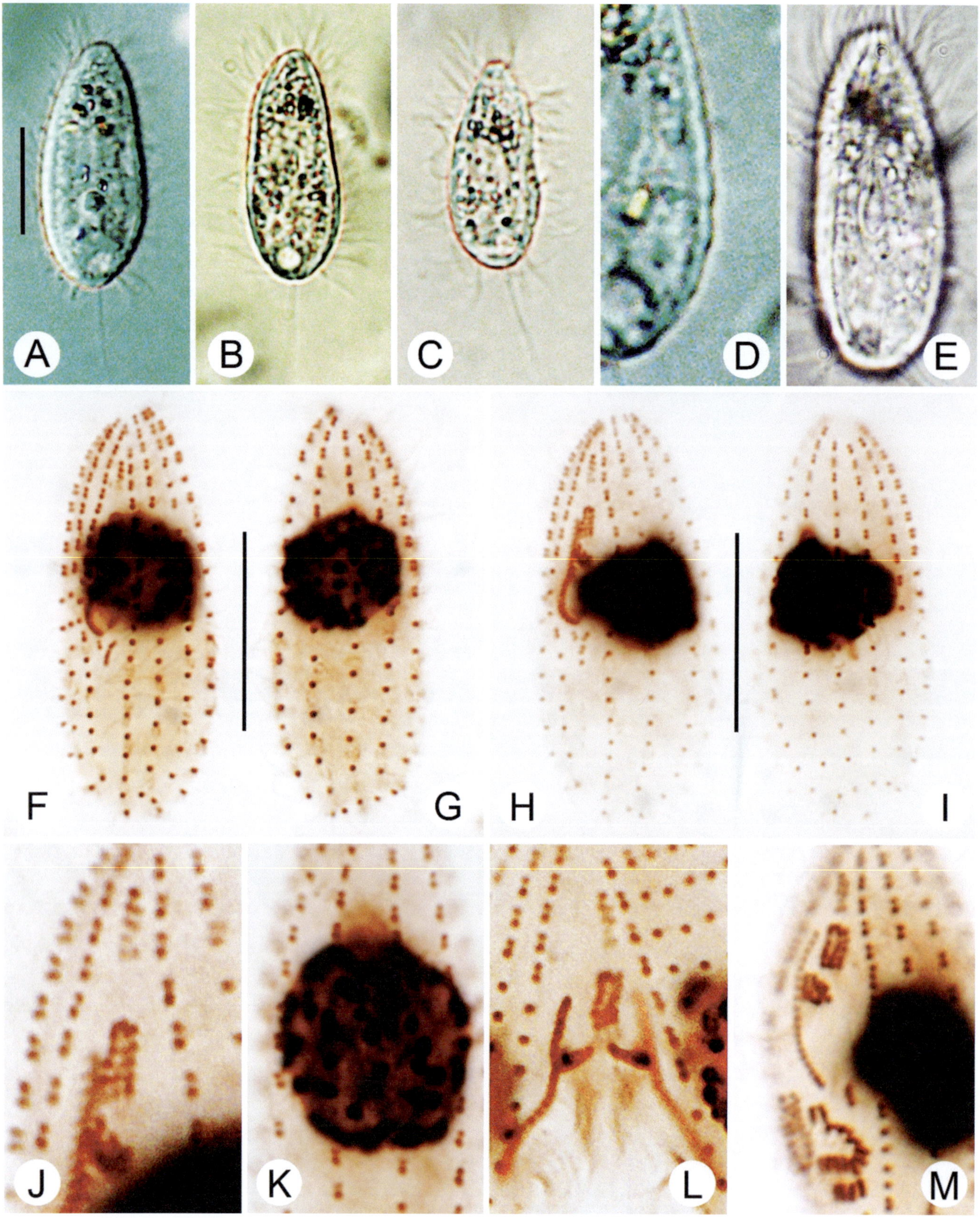


**Fig. 23B** *Uronema apomarinum*
(**a–c**, **e**) Views of live cells; (**d**) slightly notched pellicle; (**f–i**) ventral (**f**, **h**) and dorsal (**g**, **i**) views of ciliature; (**j**) oral apparatus; (**k**) nuclear apparatus; (**l**) oral structure when two cells conjugated; (**m**) oral apparatus during division. Scale bar: 15 μm

## Reference

Liu M, Li LF, Zhang TT, Fan X, Yi Z, Lin X (2018) Taxonomy of seven scuticociliates from coastal waters of southern China including a new species: *Uronema apomarinum* n. sp. (Protozoa, Ciliophora). Eur J Protistol (accepted)

Class Oligohymenophorea de Puytorac et al., 1974
Order Philasterida Small, 1967
Family Uronematidae Thompson, 1964
*Uronema elegans* Maupas, 1883

**Ecological feature:** mangrove
**Water temperature:** 24.0 °C
**pH:** 6.5
**Salinity:** 10.0‰

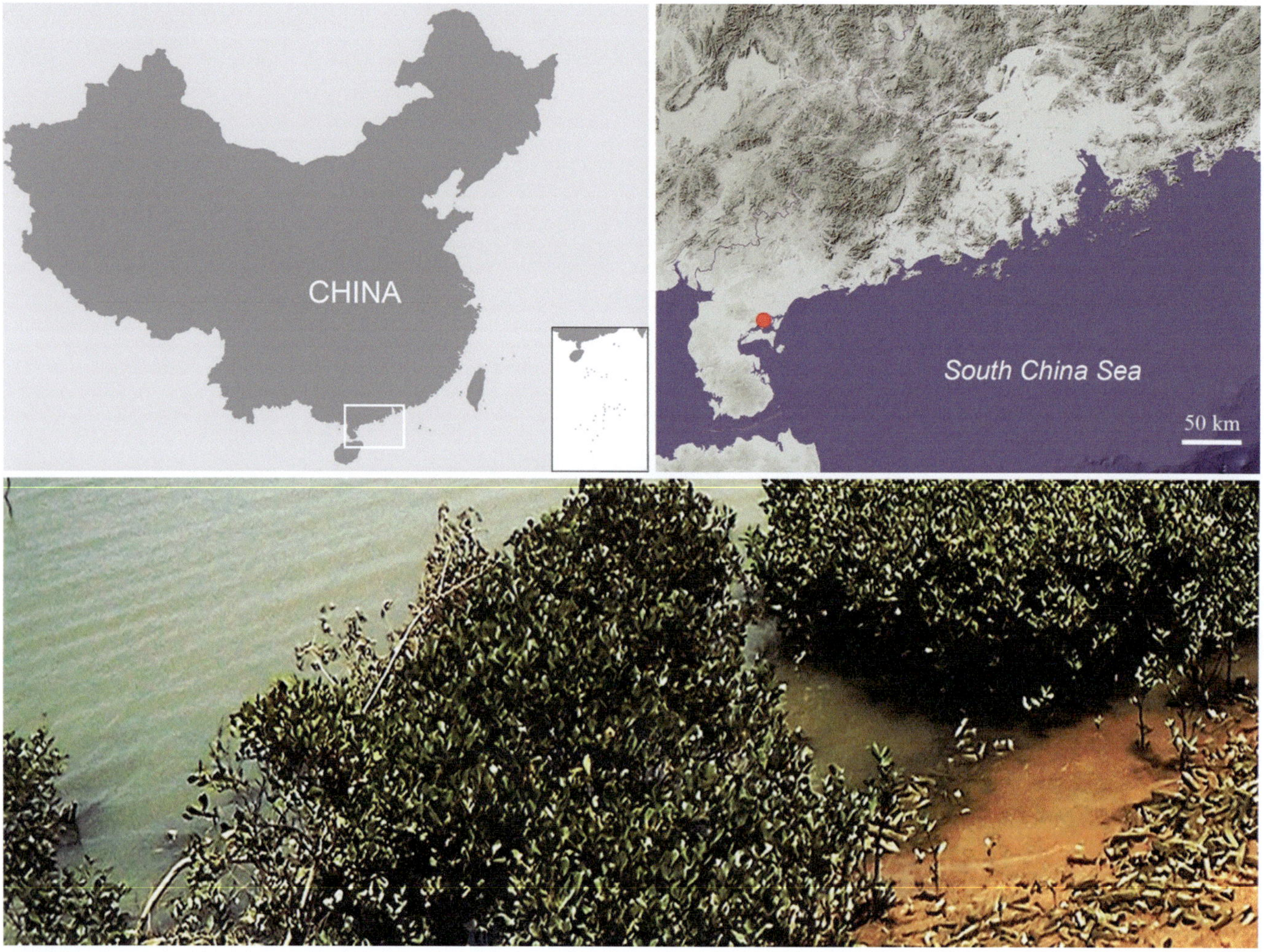


**Fig. 24A** Sample site

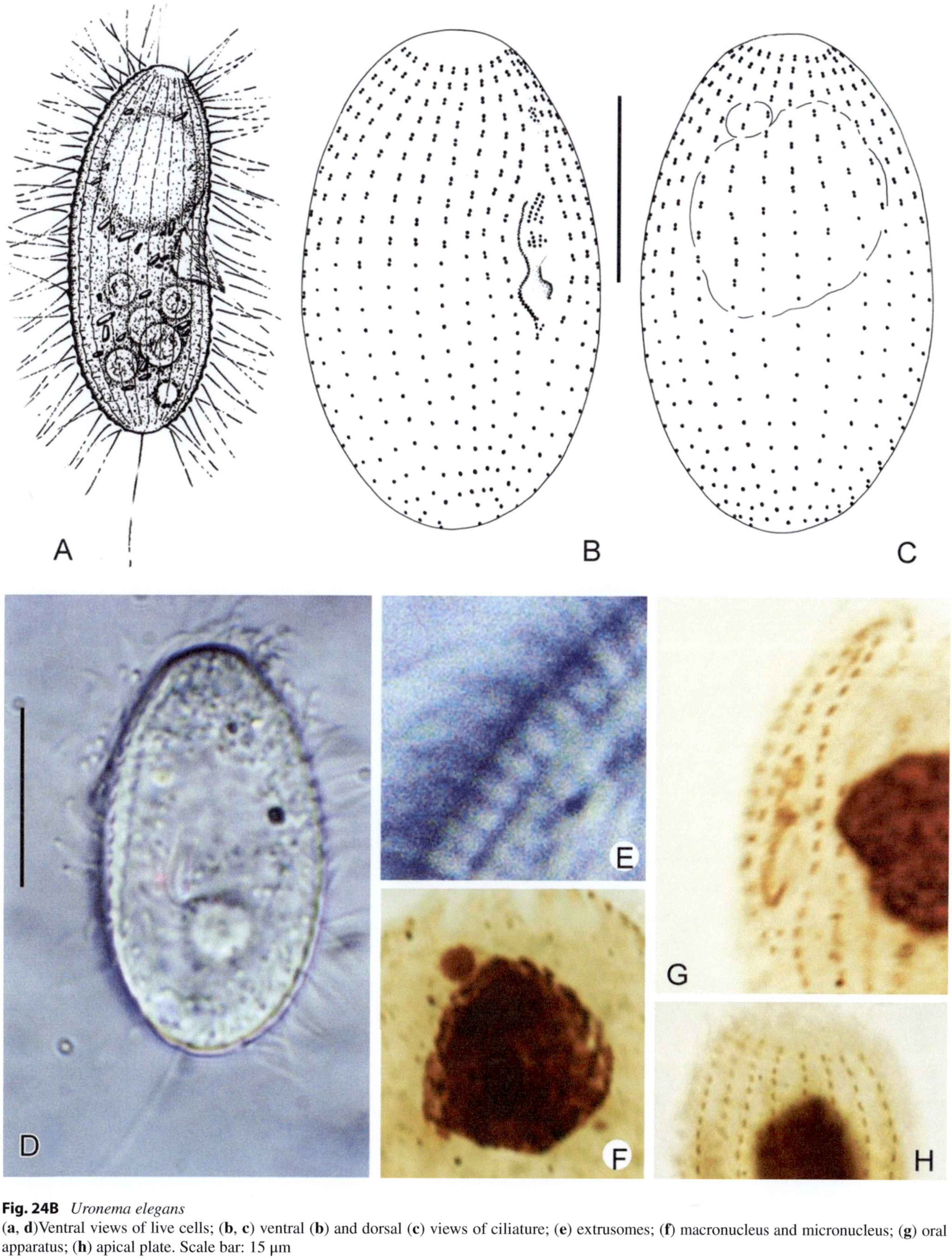


**Fig. 24B** *Uronema elegans*
(**a**, **d**)Ventral views of live cells; (**b**, **c**) ventral (**b**) and dorsal (**c**) views of ciliature; (**e**) extrusomes; (**f**) macronucleus and micronucleus; (**g**) oral apparatus; (**h**) apical plate. Scale bar: 15 μm

## Reference

Song WB, Ma H, Wang M, Zhu M (2002) Comparative studies on two closely related species *Uronemella filificum* (Kahl, 1931) and *Uronema elegans* Maupas, 1883 with redescription of *Paranophrys marina* Thompson et Berger, 1965 (Ciliophora: Scuticociliatida) from China seas. Acta Protozool 41:263–278

Class Oligohymenophorea de Puytorac et al., 1974
Order Philasterida Small, 1967
Family Uronematidae Thompson, 1964
*Uronema heteromarinum* Pan et al., 2010

**Ecological feature:** Mangrove; shore wharf
**Water temperature:** 18.0 °C; 23.0 °C
**pH:** 6.0; no data available
**Salinity:** 4.4‰; 32.0‰

**Fig. 25A** Sample site

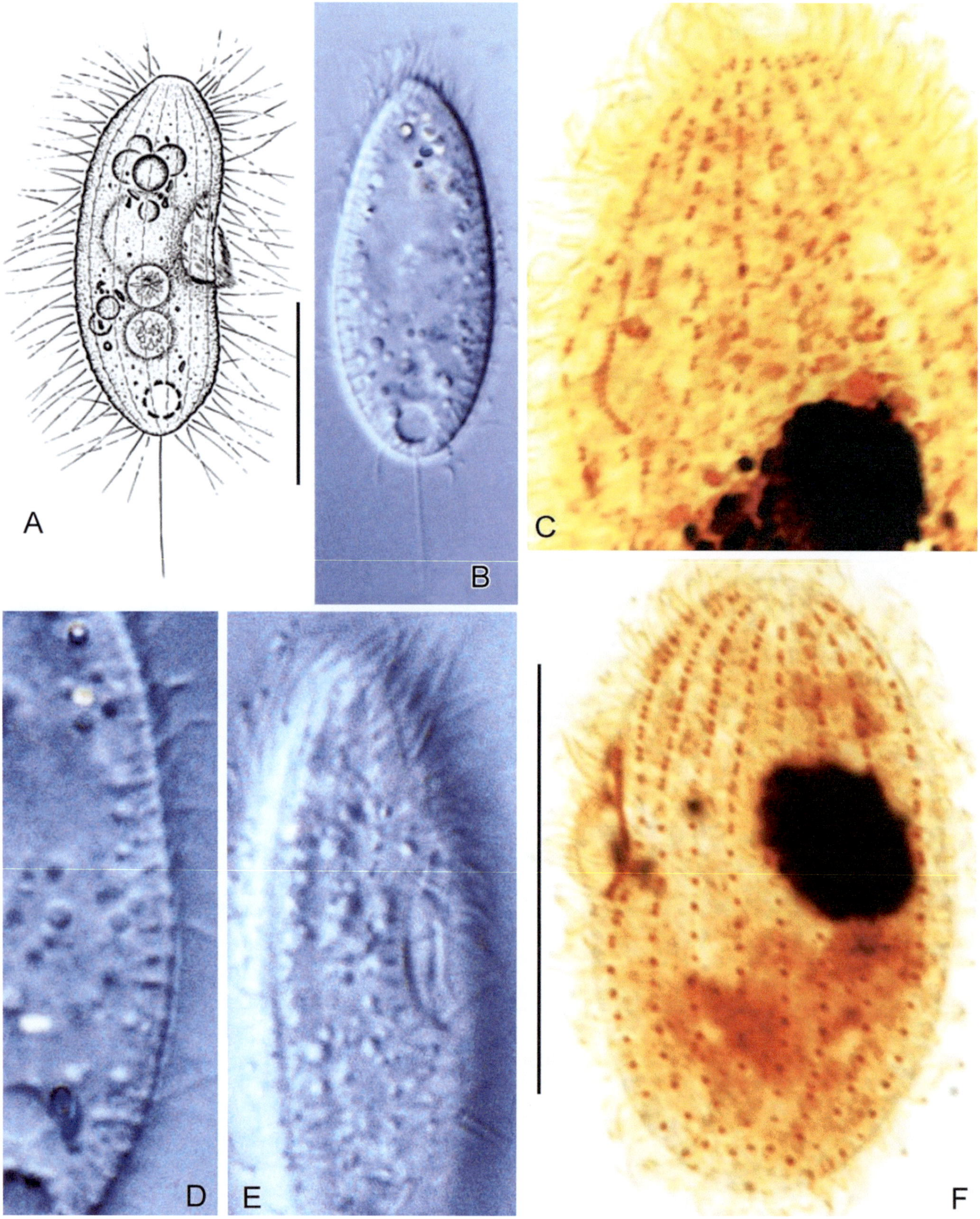


**Fig. 25B** *Uronema heteromarinum*
(**a**, **b**) Lateral views of live individuals; (**c**) oral apparatus; (**d**) extrusomes; (**e**) buccal area; (**f**) ciliature and macronucleus. Scale bar: 25 μm

## References

Pan H, Huang J, Hu X, Fan X, Al-Rasheid KAS, Song WB (2010) Morphology and SSU rRNA gene sequences of three marine ciliates from Yellow Sea, China, including one new species, *Uronema heteromarinum* nov. spec. (Ciliophora, Scuticociliatida). Acta Protozool 49:45–59

Zhao Y, Fan X, Xu Y, Hu X, Ma H (2011) Morphological studies on eight marine scuticociliates (Protozoa, Ciliophora) from China. Acta Hydrobiol Sin 35:929–939 (in Chinese)

Class Oligohymenophorea de Puytorac et al., 1974
Order Philasterida Small, 1967
Family Uronematidae Thompson, 1964
*Uronemita filificum* (Kahl, 1931) Jankowski, 1980

**Ecological feature:** mangrove (Shenzhen); coastal water (Daya Bay)
**Water temperature:** 18.0 °C; 14.0 °C
**pH:** 7.6; 7.2
**Salinity:** 28.0‰; 28.0‰

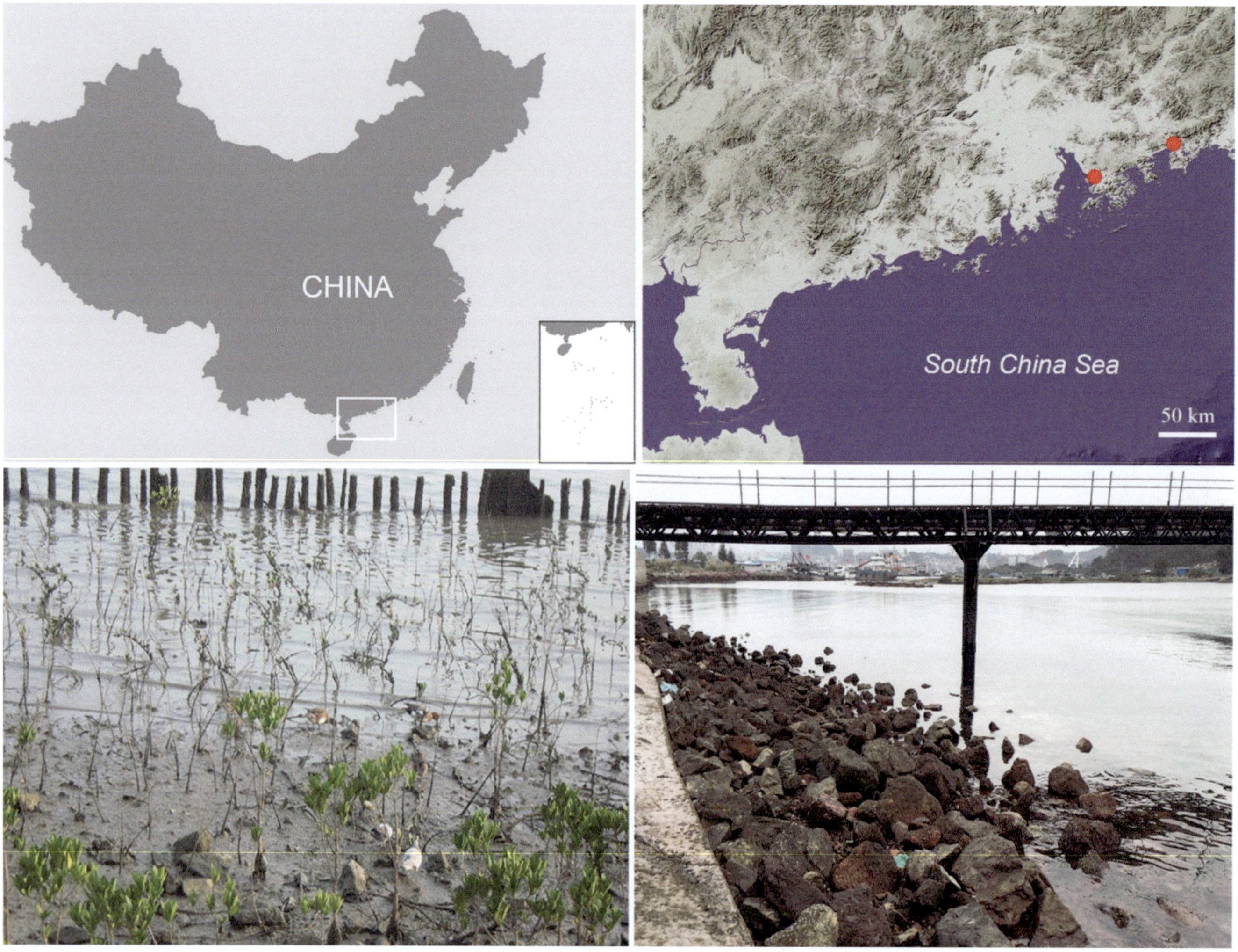


**Fig. 26A** Sample site

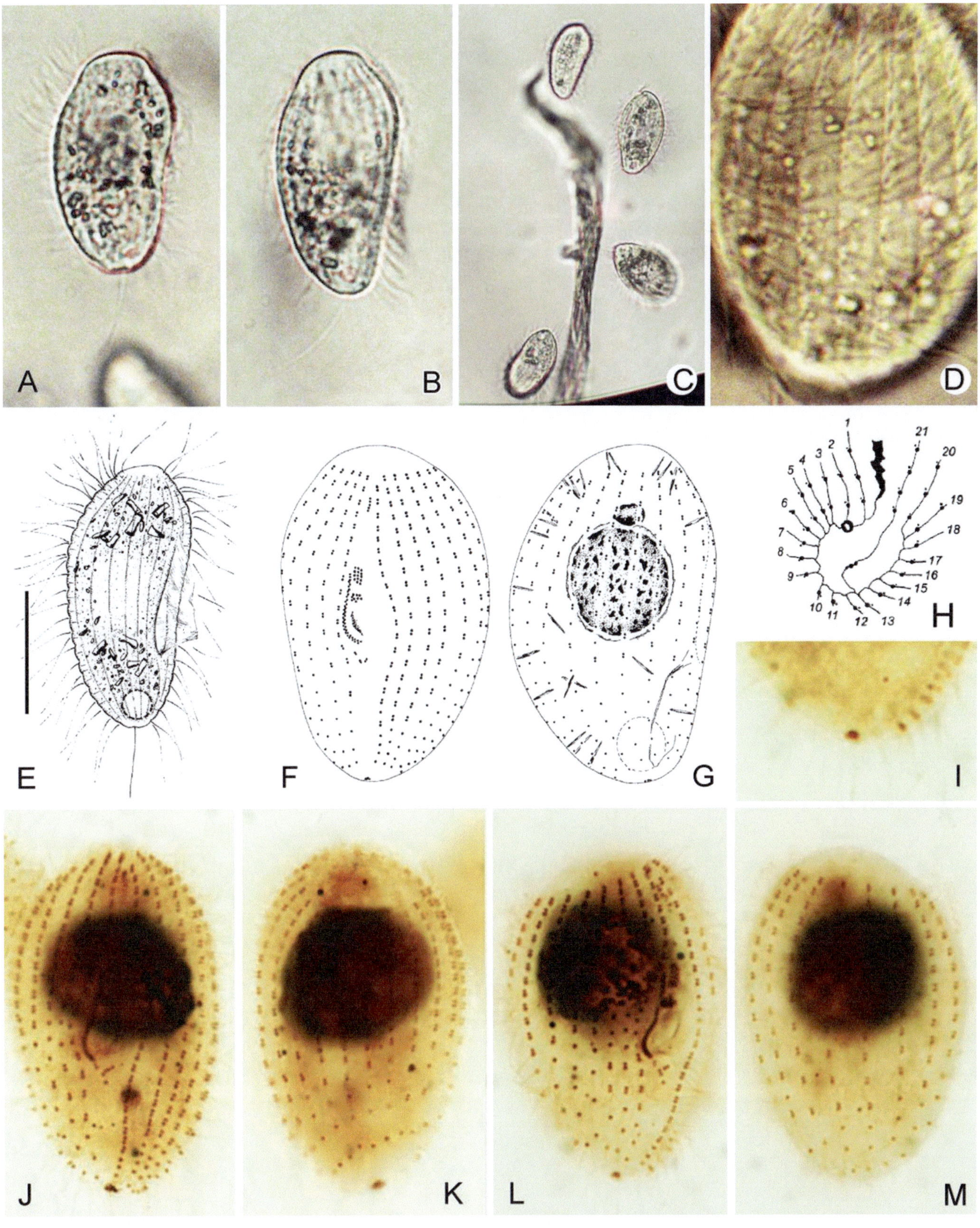


**Fig. 26B** *Uronemita filificum*
(**a–c**, **e**) Lateral views of live individuals; (**d**) somatic cilia; (**f**, **g**, **j–m**) ventral (**f**, **j**, **l**) and dorsal (**g**, **k**, **m**) views of ciliature; (**h**) posterior region of cell, showing silver lines; (**i**) base of caudal cilia. 1–21, somatic kineties 1–21. Scale bar: 20 μm

## Reference

Liu M, Gao F, Al-Farraj SA, Hu X (2016) Morphology and small subunit rRNA gene sequence of *Uronemita parabinucleata* n. sp. (Ciliophora, Uronematidae), with an improved generic diagnosis. Eur J Protistol 51:1–10

Class Oligohymenophorea de Puytorac et al., 1974
Order Pleuronematida Fauré-Fremiet in Corliss, 1956
Family Cyclidiidae Ehrenberg, 1838
*Cyclidium varibonneti* Song, 2000

**Ecological feature:** shrimp-farming pond (Zhanjiang); mangrove (Shenzhen)
**Water temperature:** 18.0 °C; 16.0 °C
**pH:** no data available; 7.2
**Salinity:** 25.0‰; 8.0‰

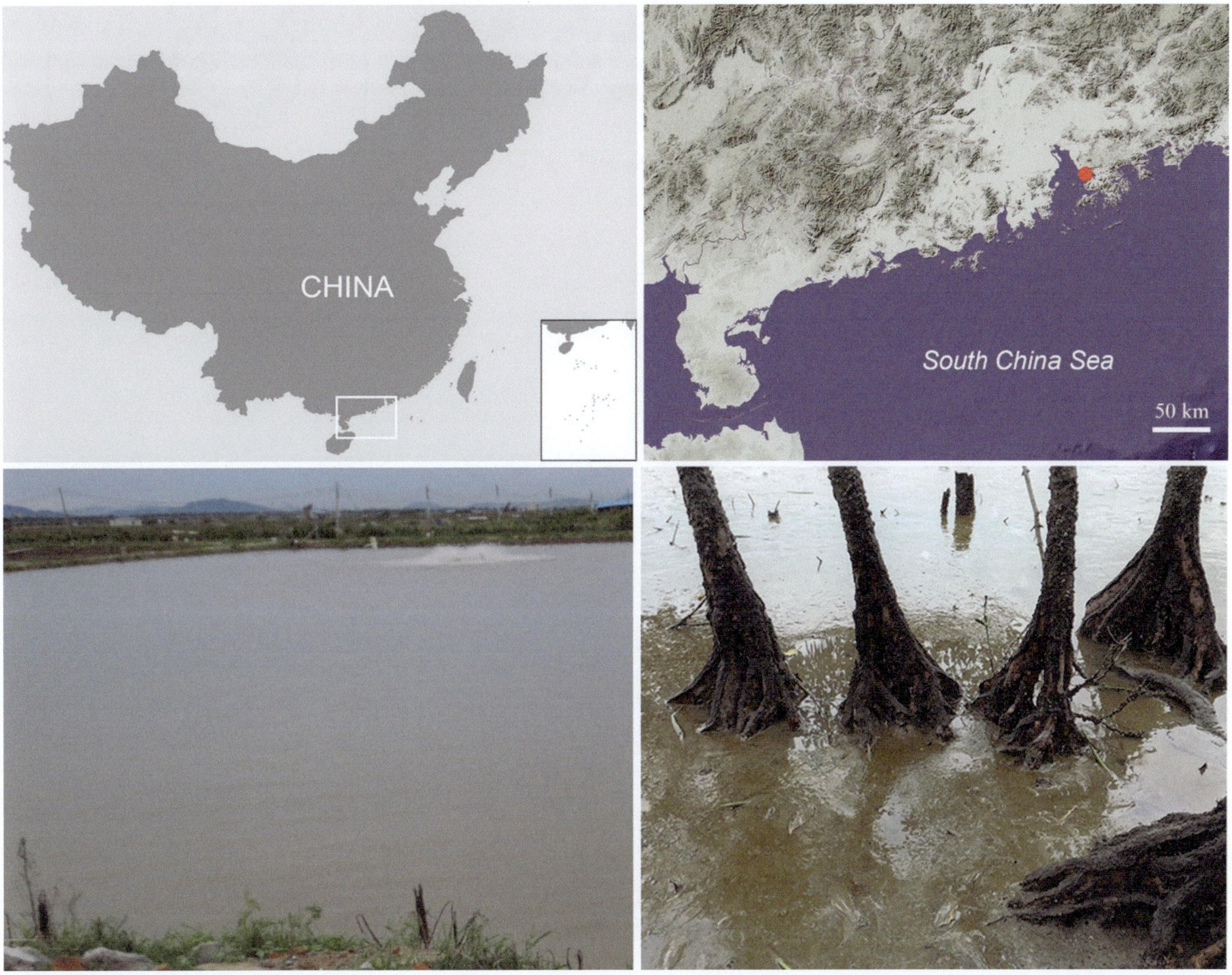


**Fig. 27A** Sample site

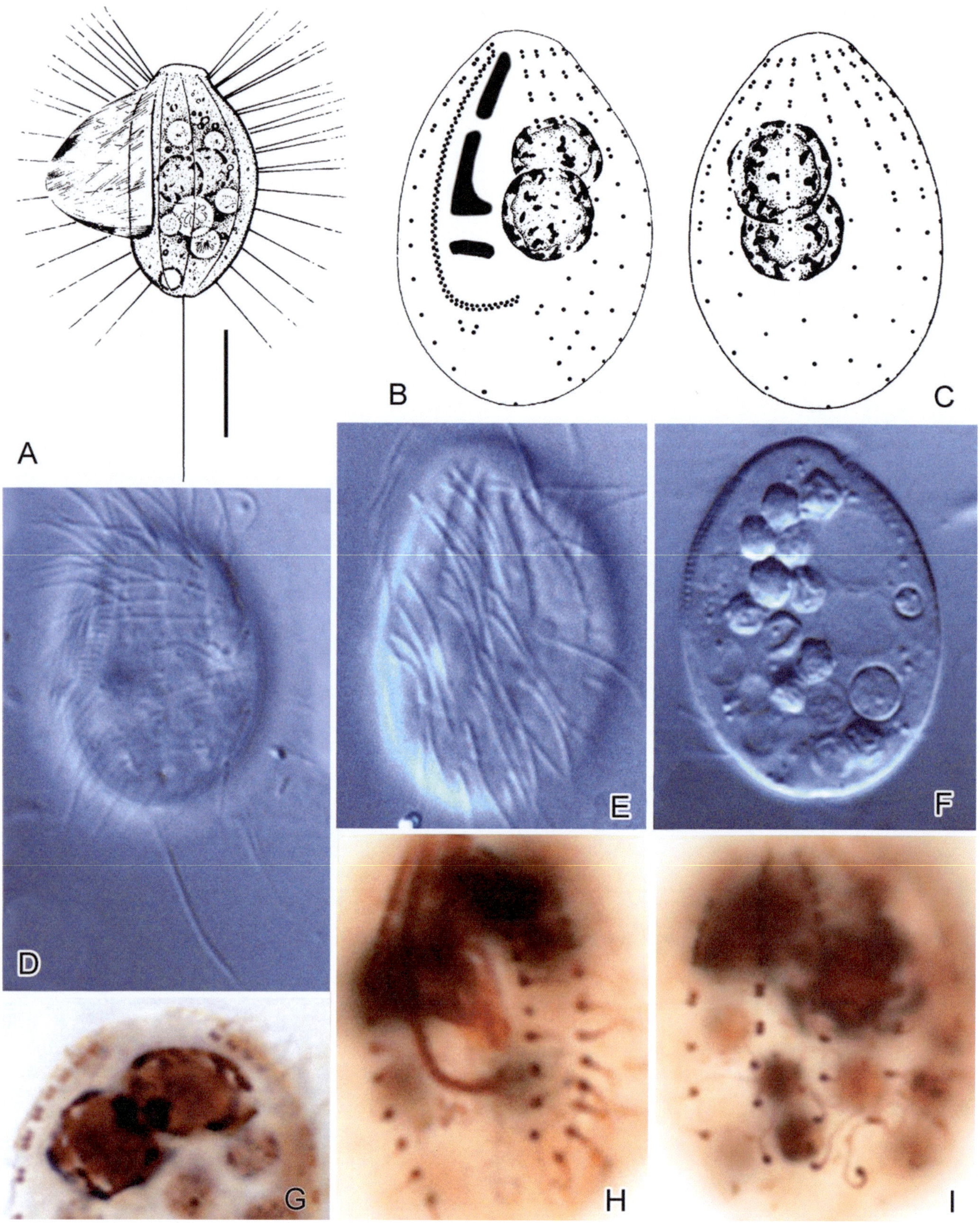


**Fig. 27B** *Cyclidium varibonneti*
(**a**, **d**) Lateral views of live cells; (**b**, **c**, **h**, **i**) ventral (**b**, **h**) and dorsal (**c**, **i**) views of ciliature; (**e**) somatic cilia; (**f**, **g**) two macronuclei. Scale bar: 10 μm

## Reference

Song WB (2000) Morphology and taxonomical studies on some marine scuticociliates from China sea, with description of two new species, *Philasterides armatalis* sp. n and *Cyclidium varibonneti* sp. n. (Protozoa: Ciliophora: Scuticociliatida). Acta Protozool 39:295–322

Class Oligohymenophorea de Puytorac et al., 1974
Order Pleuronematida Fauré-Fremiet in Corliss, 1956
Family Cyclidiidae Rhrenberg, 1838
*Falcicyclidium fangi* Fan et al., 2011

**Ecological feature:** intertidal zone
**Water temperature:** 24.4 °C
**pH:** 8.5
**Salinity:** 29.5‰

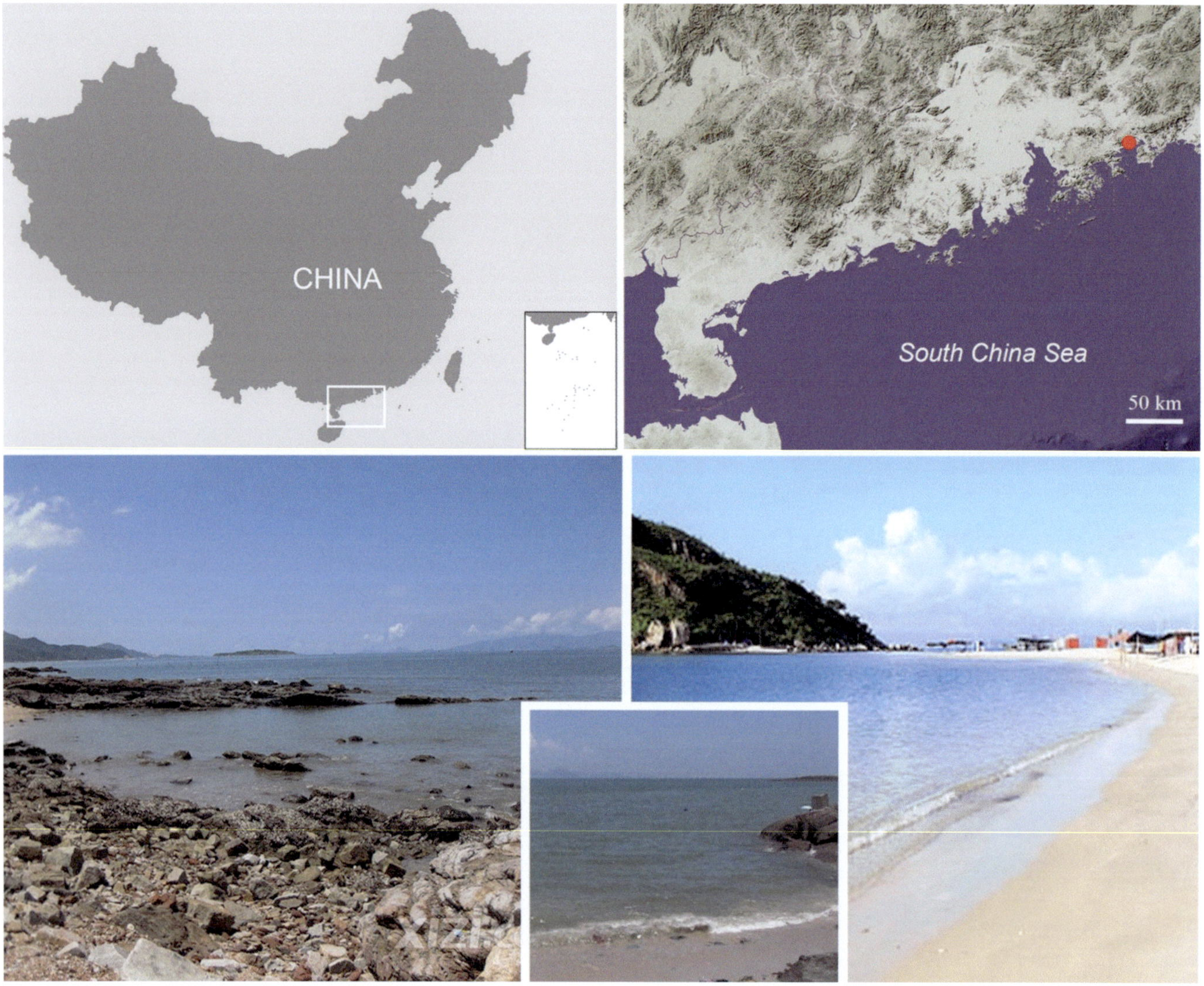


**Fig. 28A** Sample site

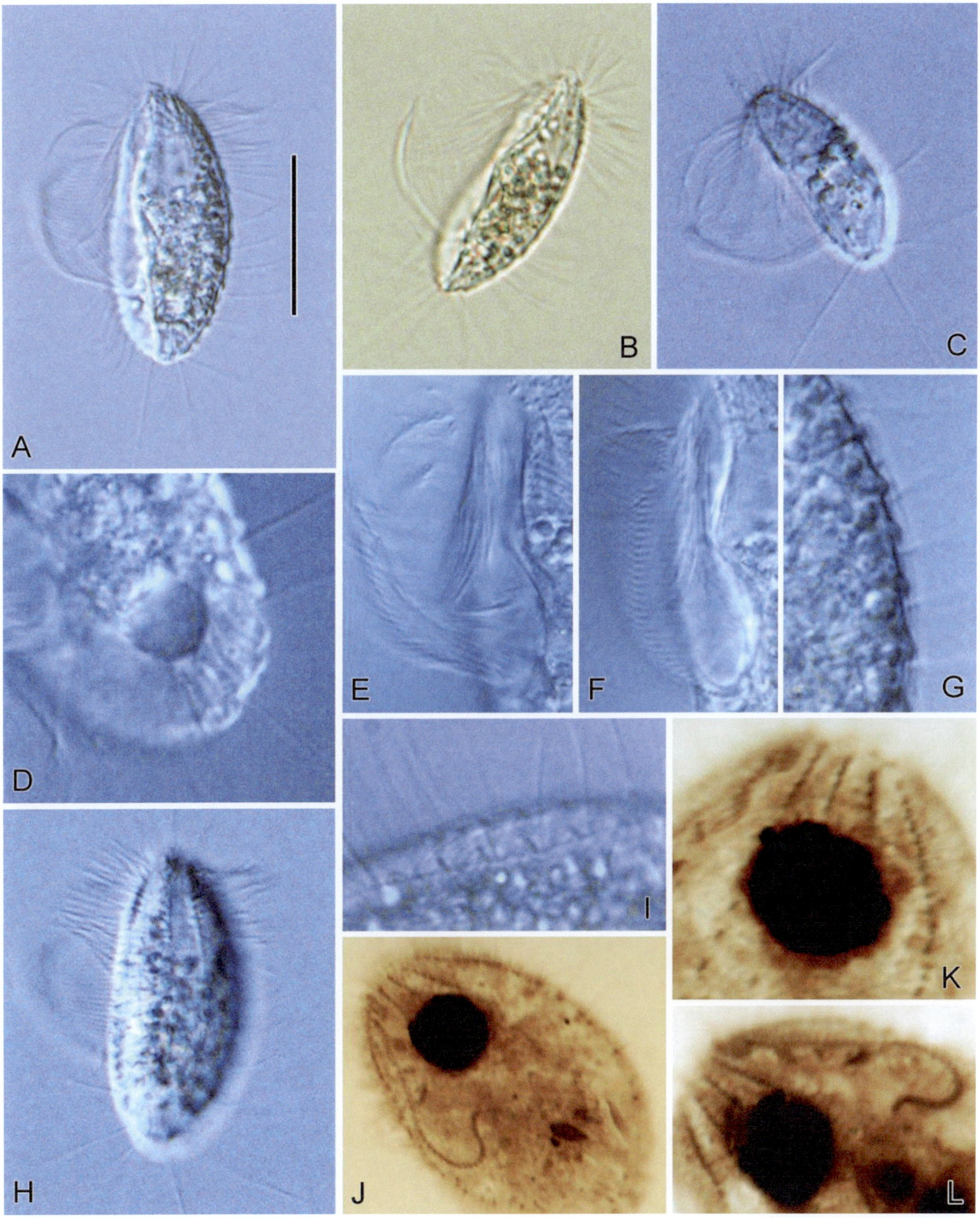


**Fig. 28B** *Falcicyclidium fangi*
(**a–c**, **h**) Lateral views of live cells; (**d**) posterior end of cell showing caudal cilium and contractile vacuole; (**e**, **f**) buccal membranelles; (**g**, **i**) notched pellicle; (**j–l**) ciliature. Scale bar: 30 μm

## Reference

Fan X, Hu X, Al-Farraj SA, Clamp J, Song WB (2011) Morphological description of three marine ciliates (Ciliophora, Scuticociliatia), with establishment of a new genus and two new species. Eur J Protistol 47:186–196

Class Oligohymenophorea de Puytorac et al., 1974
Order Pleuronematida Fauré-Fremiet in Corliss, 1956
Family Cyclidiidae Ehrenberg, 1838
*Protocyclidium citrullus* (Cohn, 1866) Foissner et al., 2002

**Ecological feature:** mangrove
**Water temperature:** 18.0 °C
**pH:** 7.6
**Salinity:** 10.0‰

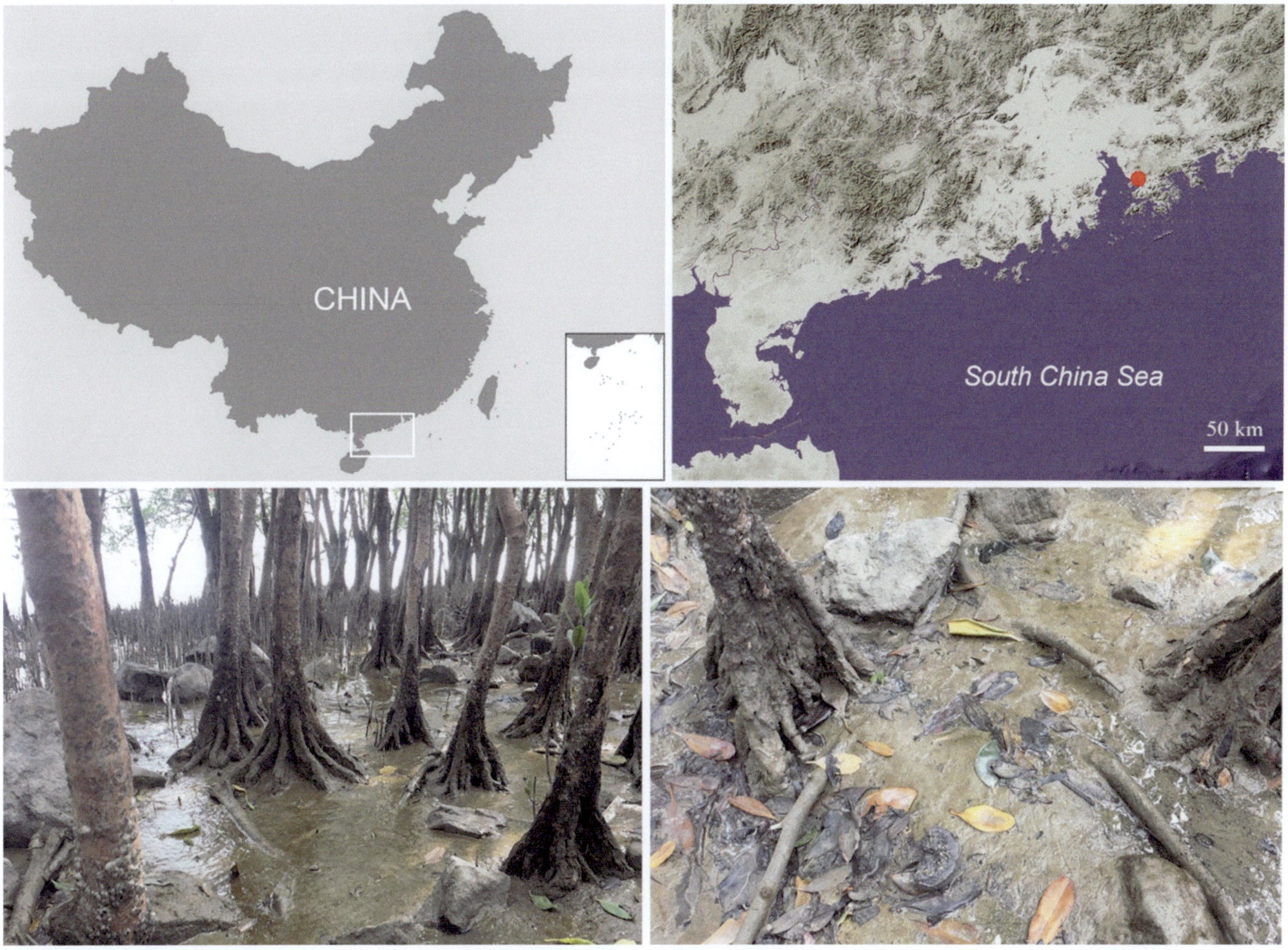


**Fig. 29A** Sample site

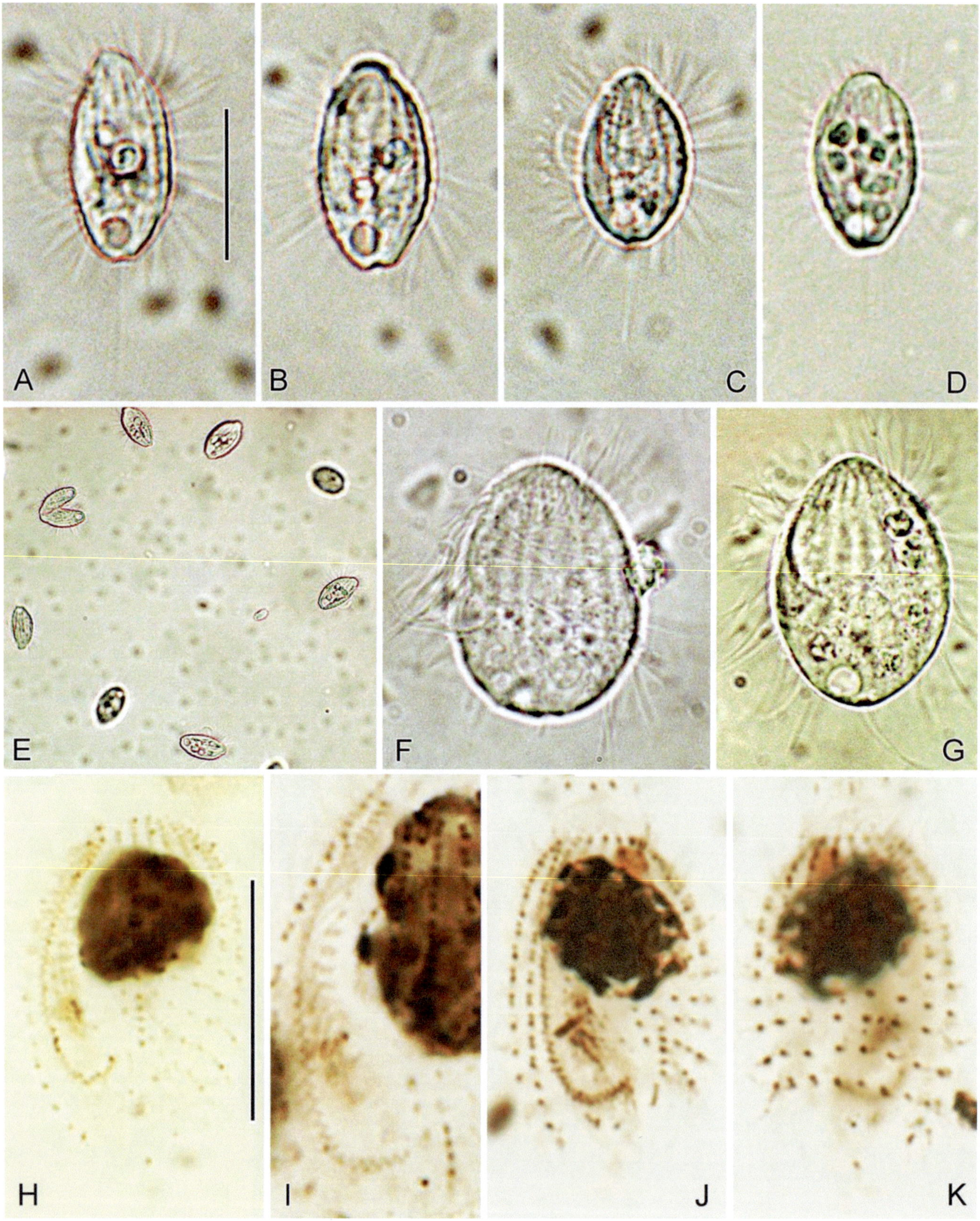


**Fig. 29B** *Protocyclidium citrullus*
(**a–d**) Lateral view of live cells; (**e**) different body sizes and shapes; (**f**, **g**) slightly depressed cell, showing oral cilia and somatic cilia; (**h–j**) oral apparatus; (**k**) dorsal ciliature and nuclear apparatus. Scale bar: 20 μm

## References

Fan X, Lin X, Al-Rasheid KAS, Al-Farraj SA, Warren A, Song WB (2011) The diversity of scuticociliates (Protozoa, Ciliophora): a report on eight marine forms found in coastal waters of China, with a description of one new species. Acta Protozool 50:205–218

Foissner W, Agatha S, Berger H (2002) Soil ciliates (Protozoa, Ciliophora) from Namibia (Southwest southwest Africa), with emphasis on two contrasting environments, the Etosha region and the Namib desert. Denisia 5:1–1459

Class Oligohymenophorea de Puytorac et al., 1974
Order Pleuronematida Fauré-Fremiet in Corliss, 1956
Family Pleuronematidae Kent, 1881
*Pleuronema coronatum* Kent, 1881

**Ecological feature:** mangrove
**Water temperature:** 19.0 °C; 27.0 °C; 21.0 °C
**pH:** 6.7; 7; 7.7
**Salinity:** 17.0‰; 20.0‰; 26.0‰

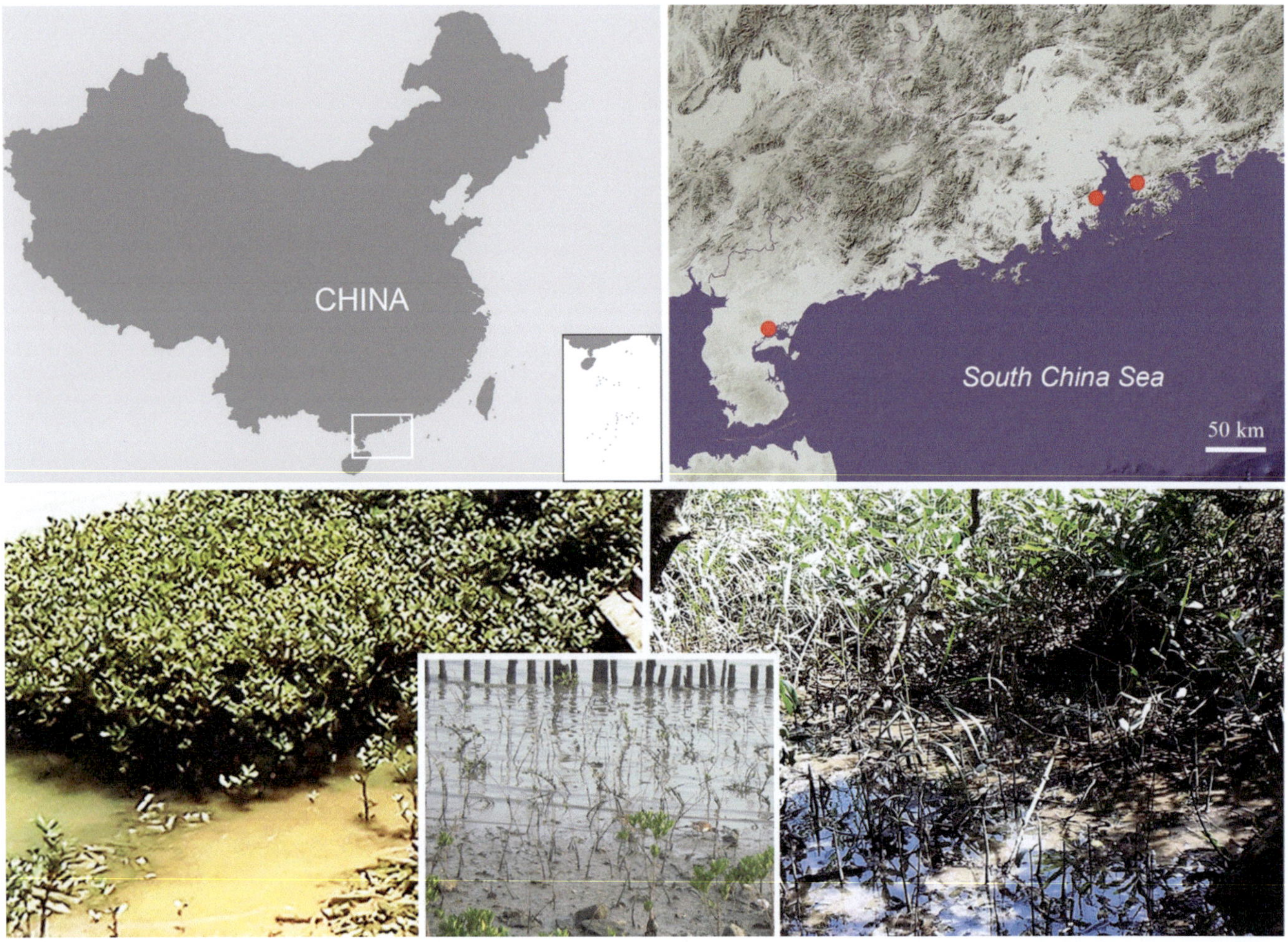


**Fig. 30A** Sample site

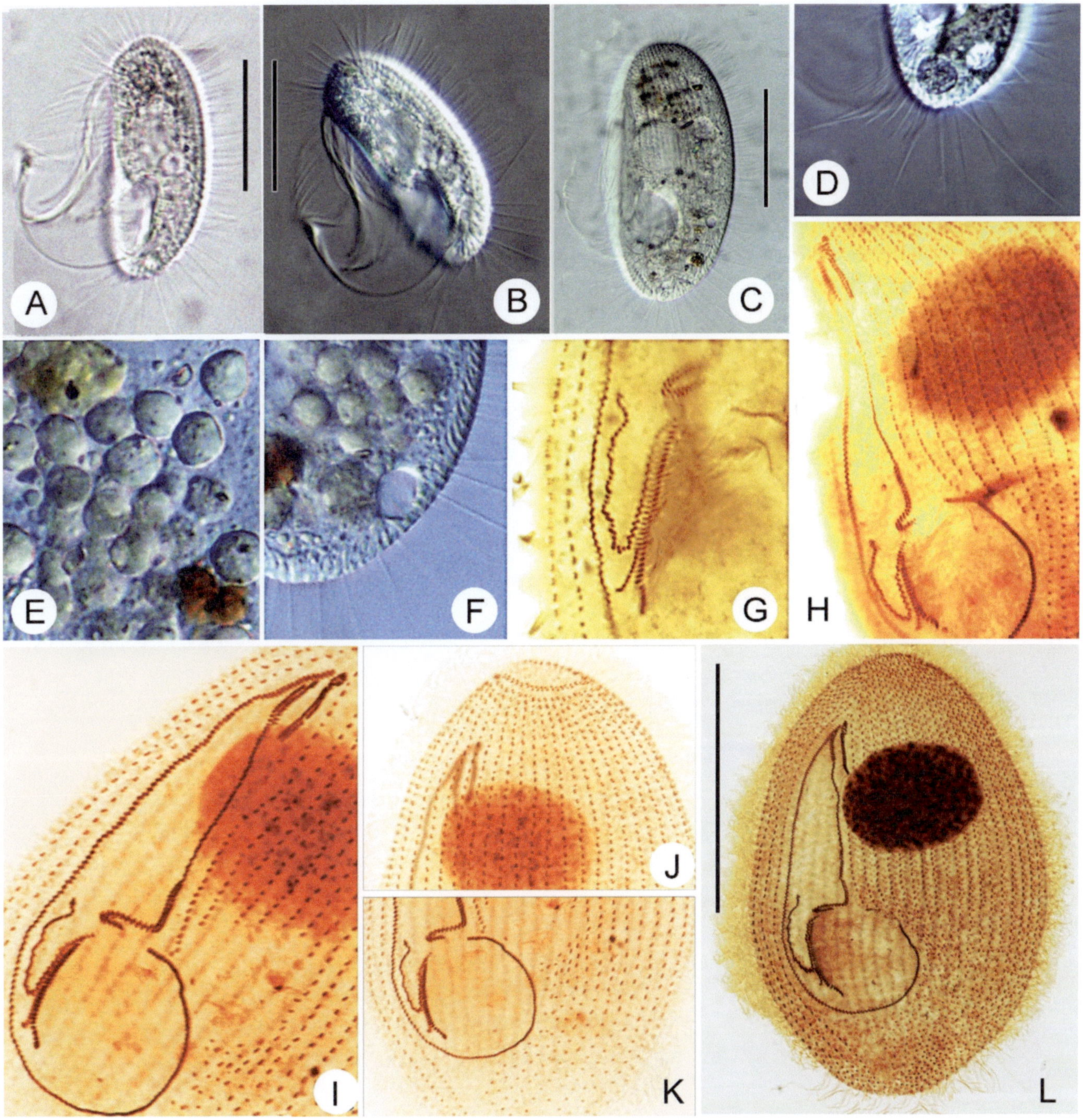


**Fig. 30B** *Pleuronema coronatum*
(**a–d**) Ventral view of live individuals; (**d**) caudal cilia; (**e**) cytoplasmic granules; (**f**) the arrangement of extrusomes and contractile vacuole; (**g**, **k**) detail of posterior part of oral apparatus, (**h**, **i**) oral apparatus. (**j**) anterior view of ciliature, showing the somatic kineties; (**l**) ventral ciliature and macronucleus. Scale bars: 30 μm (**a–c**), 40 μm (**l**)

## References

Song WB (2000) Morphology and taxonomical studies on some marine scuticociliates from China sea, with description of two new species, *Philasterides armatalis* sp. n and *Cyclidium varibonneti* sp. n. (Protozoa: Ciliophora: Scuticociliatida). Acta Protozool 39:295–322

Wang Y, Hu X, Long H, Al-Rasheid KAS, Al-Farraj SA, Song WB (2008) Morphological studies indicate that *Pleuronema grolierei* nov. spec. and *P. coronatum* Kent, 1881 represent different sections of the genus *Pleuronema* (Ciliophora: Scuticociliatida). Eur J Protistol 44:131–140

Class Oligohymenophorea de Puytorac et al., 1974
Order Pleuronematida Fauré-Fremiet in Corliss, 1956
Family Pleuronematidae Kent, 1881
*Pleuronema orientale* Pan et al., 2015

**Ecological feature:** intertidal zone
**Water temperature:** 22.0 °C
**pH:** 7.2
**Salinity:** 7.0‰

**Fig. 31A** Sample site

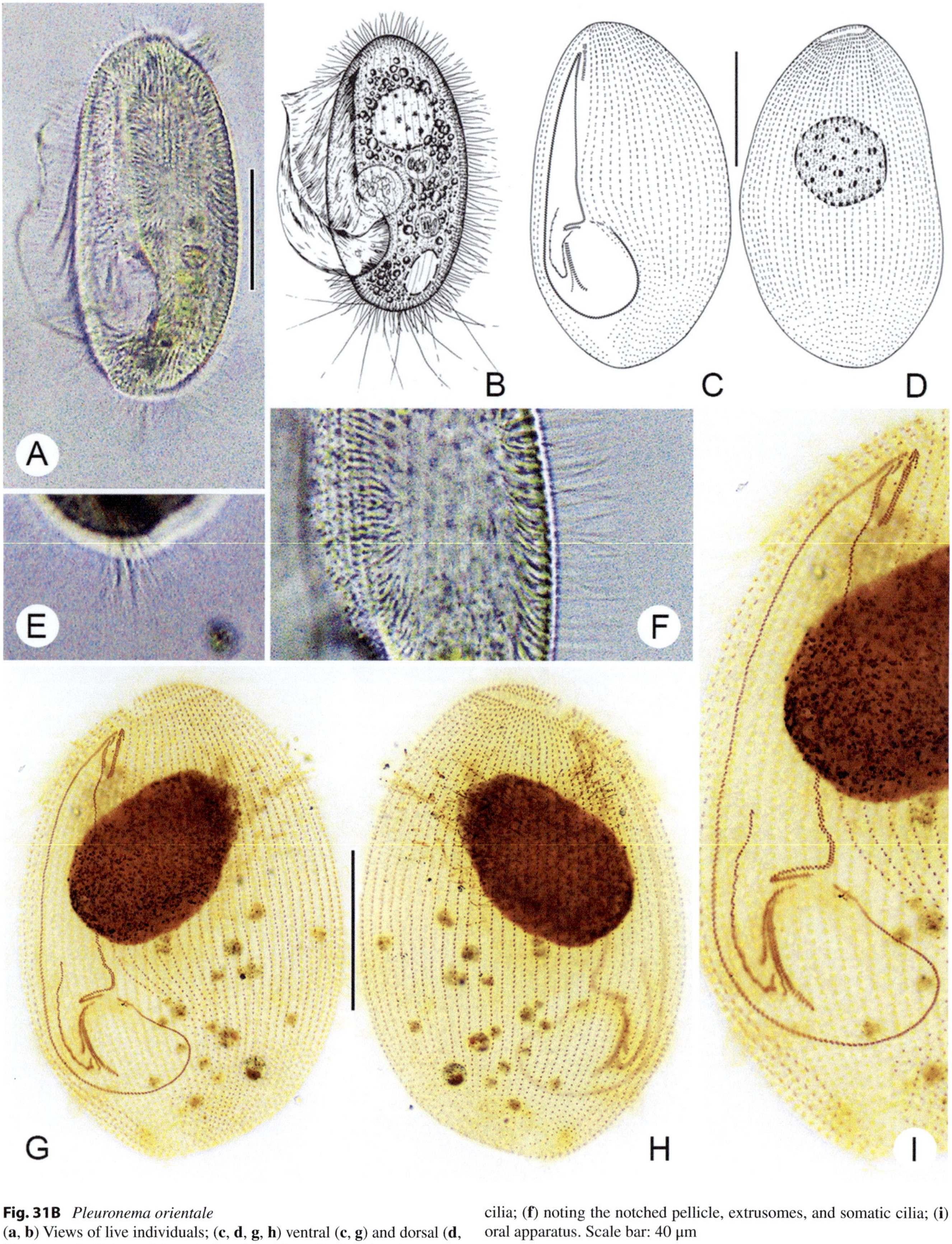


**Fig. 31B** *Pleuronema orientale*
(**a**, **b**) Views of live individuals; (**c**, **d**, **g**, **h**) ventral (**c**, **g**) and dorsal (**d**, **h**) views of ciliature; (**e**) posterior view of a cell, showing the caudal cilia; (**f**) noting the notched pellicle, extrusomes, and somatic cilia; (**i**) oral apparatus. Scale bar: 40 μm

## Reference

Pan H, Hu J, Warren A, Wang L, Jiang J, Hao R (2015) Morphology and molecular phylogeny of *Pleuronema orientale* spec. nov. and *Pleuronema paucisaetosum* spec. nov. (Ciliophora, Scuticociliata) from Hangzhou Bay, China. Int J Syst Evol Microbiol 65:4800–4808

Class Oligohymenophorea de Puytorac et al., 1974
Order Pleuronematida Fauré-Fremiet in Corliss, 1956
Family Pleuronematidae Kent, 1881
*Pleuronema puytoraci* Grolière & Detcheva, 1974

**Ecological feature:** mangrove
**Water temperature:** 18.0 °C
**pH:** 7.6
**Salinity:** 16.0‰

**Fig. 32A** Sample site

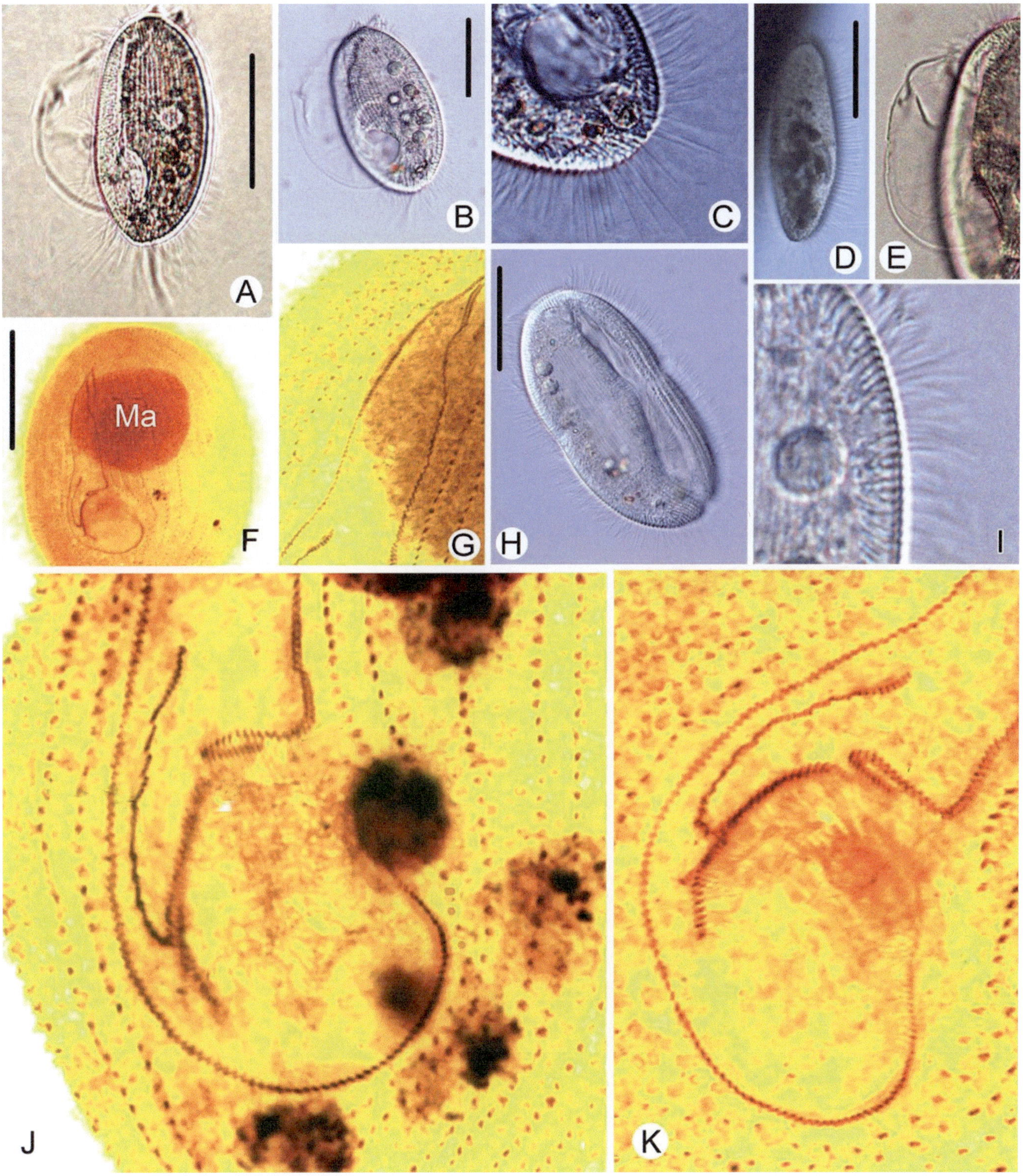


**Fig. 32B** *Pleuronema puytoraci*
(**a**, **b**, **d**) Lateral views of live individuals; (**c**) posterior portion of cell, showing contractile vacuole; (**e**) paroral membrane; (**f**, **g**, **j**, **k**) oral apparatus; (**h**) a dividing cell; (**i**) distribution of extrusomes. *Ma* macronucleus. Scale bar: 50 µm

## Reference

Pan X, Shao C, Ma H, Fan X, Ma H, Al-Rasheid KAS, Al-Farraj SA, Hu X (2011) Redescriptions of two marine scuticociliates from China, with notes on stomatogenesis in *Parauronema longum* (Ciliophora, Scuticociliatida). Acta Protozool 50:301–310

Class Oligohymenophorea de Puytorac et al., 1974
Order Pleuronematida Fauré-Fremiet in Corliss, 1956
Family Pleuronematidae Kent, 1881
*Pleuronema setigerum* Calkins, 1903

**Ecological feature:** mangrove
**Water temperature:** 21.0 °C
**pH:** 8.2
**Salinity:** 19.0‰

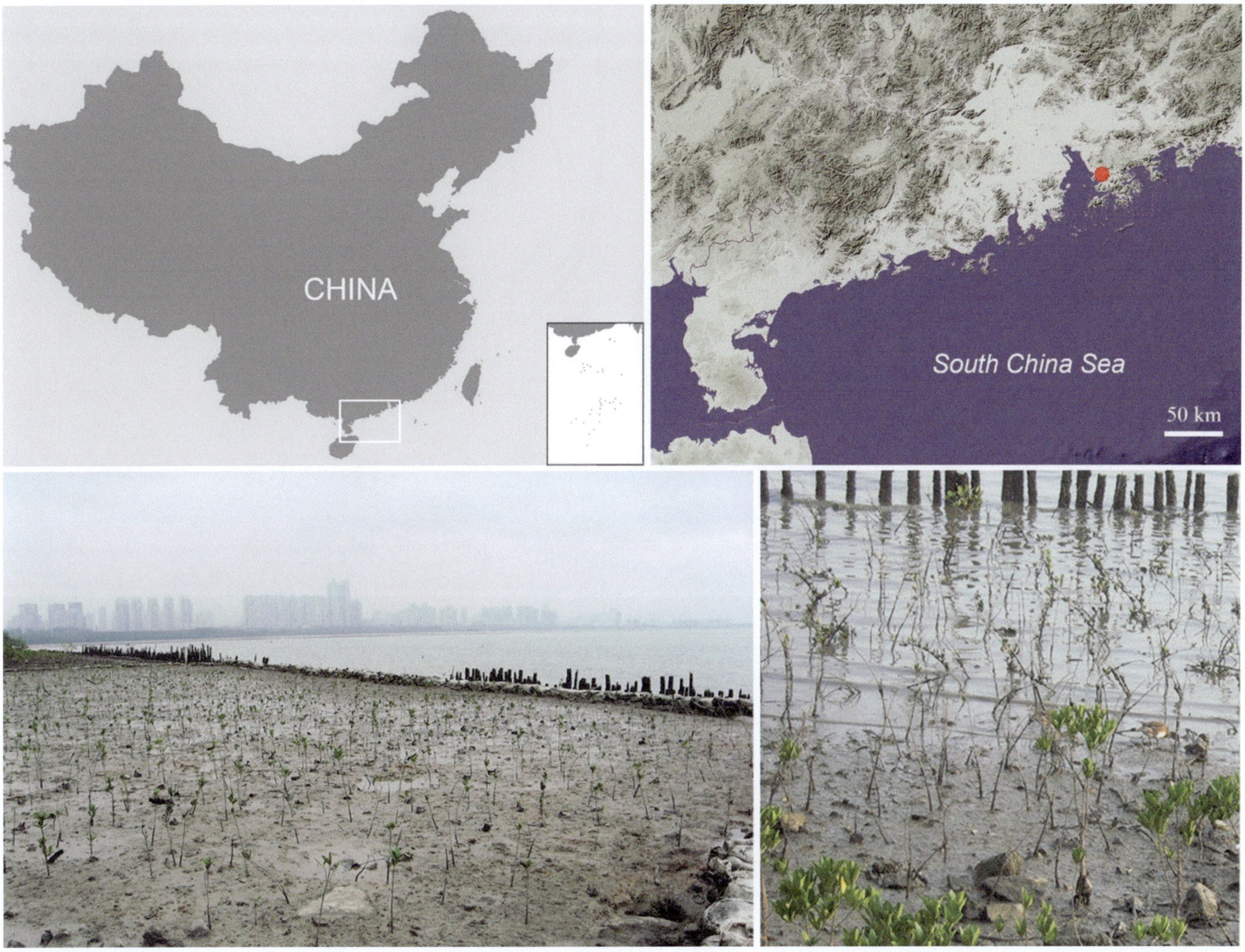


**Fig. 33A** Sample site

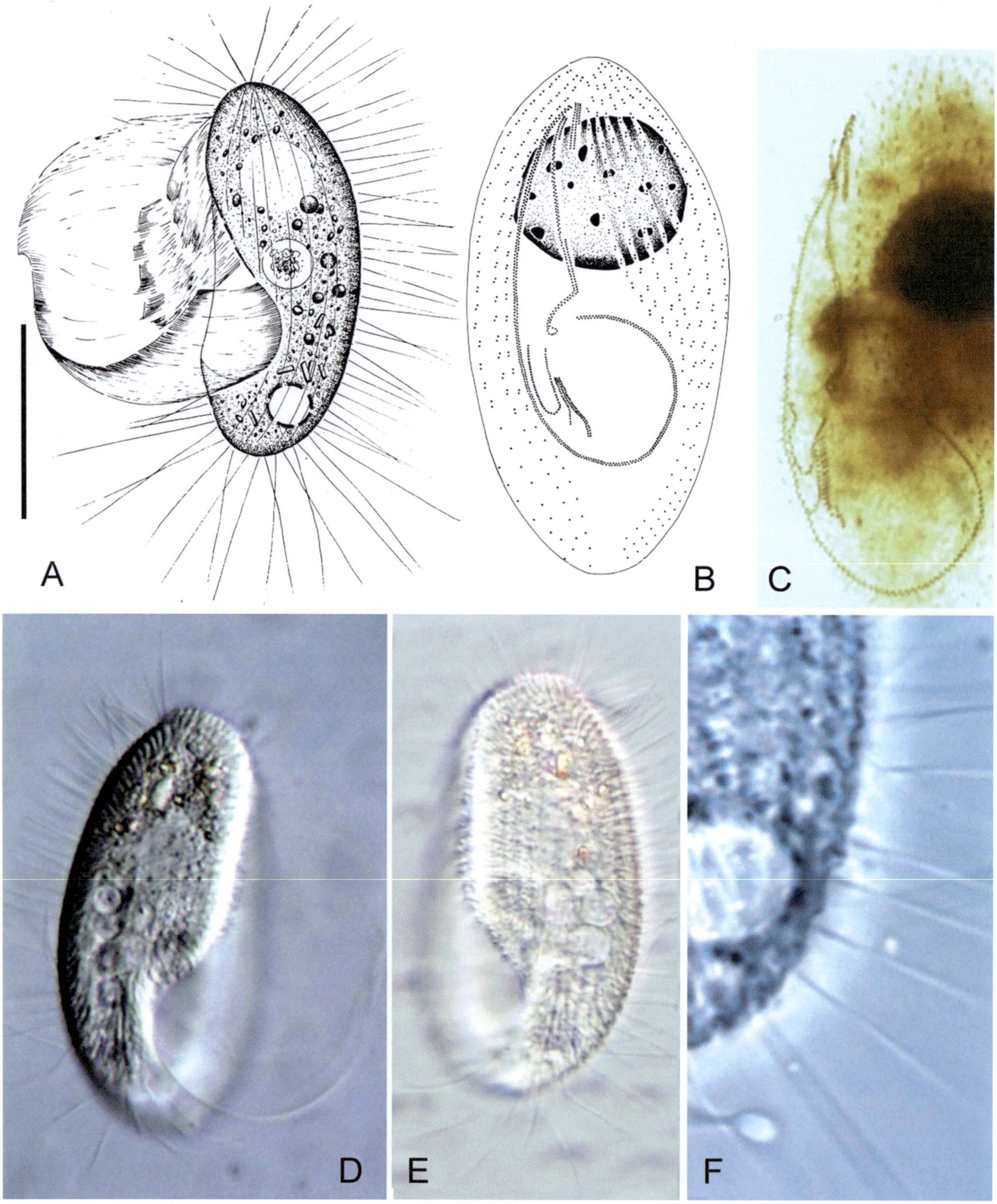


**Fig. 33B** *Pleuronema setigerum*
(**a**, **d**, **e**) Lateral views of live individuals; (**b**) ventral view of ciliature; (**c**) oral apparatus; (**f**) notched pellicle. Scale bar: 30 μm

## Reference

Pan H, Huang J, Hu X, Fan X, Al-Rasheid KAS, Song WB (2010) Morphology and SSU rRNA gene sequences of three marine ciliates from Yellow Sea, China, including one new species, *Uronema heteromarinum* nov. spec. (Ciliophora, Scuticociliatida). Acta Protozool 49:45–59

Class Oligohymenophorea de Puytorac et al., 1974
Order Pleuronematida Fauré-Fremiet in Corliss, 1956
Family Pleuronematidae Kent, 1881
*Pleuronema wiackowskii* Wang et al., 2008

**Ecological feature:** intertidal zone
**Water temperature:** 23 °C
**Salinity:** 32.0‰

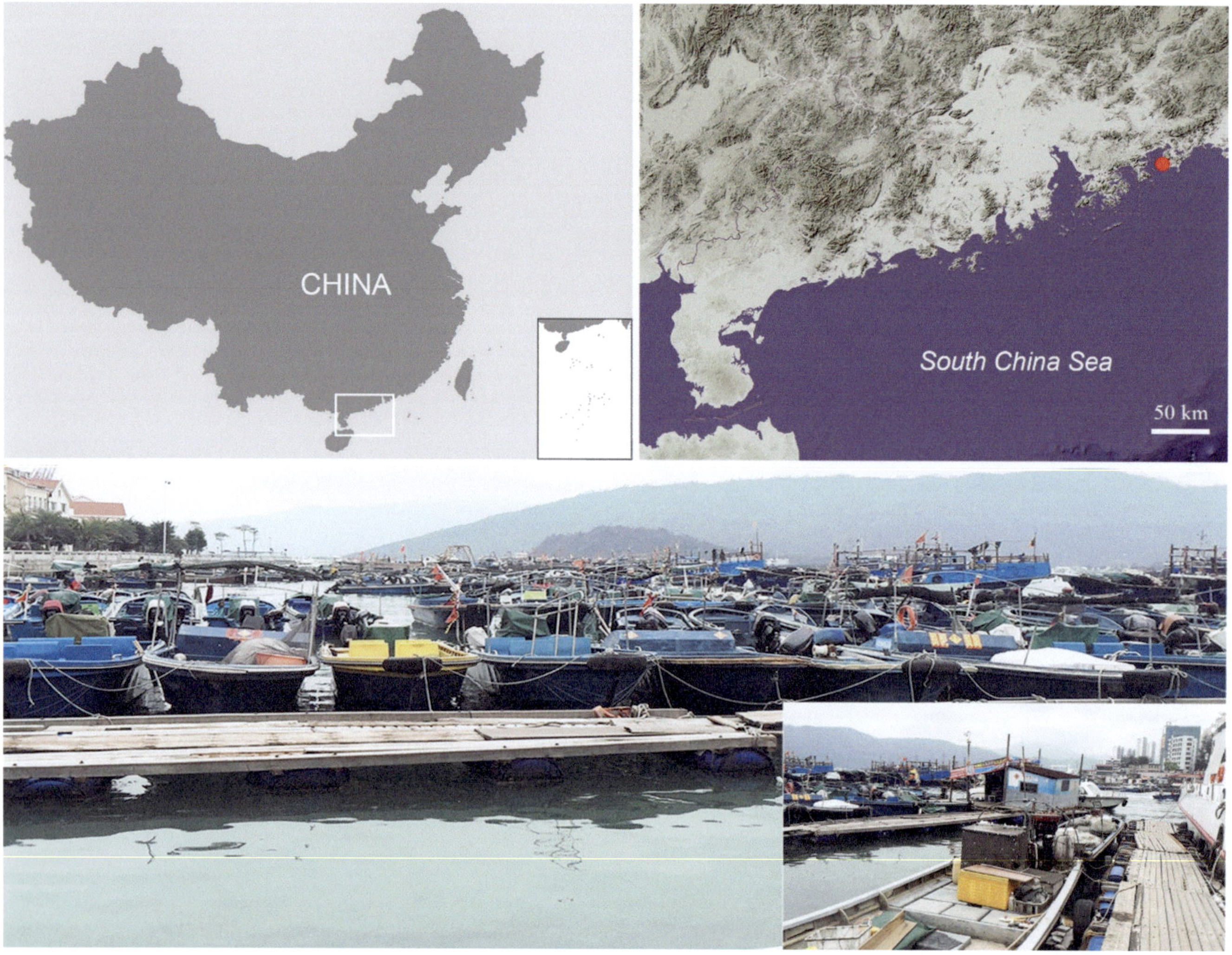


**Fig. 34A** Sample site

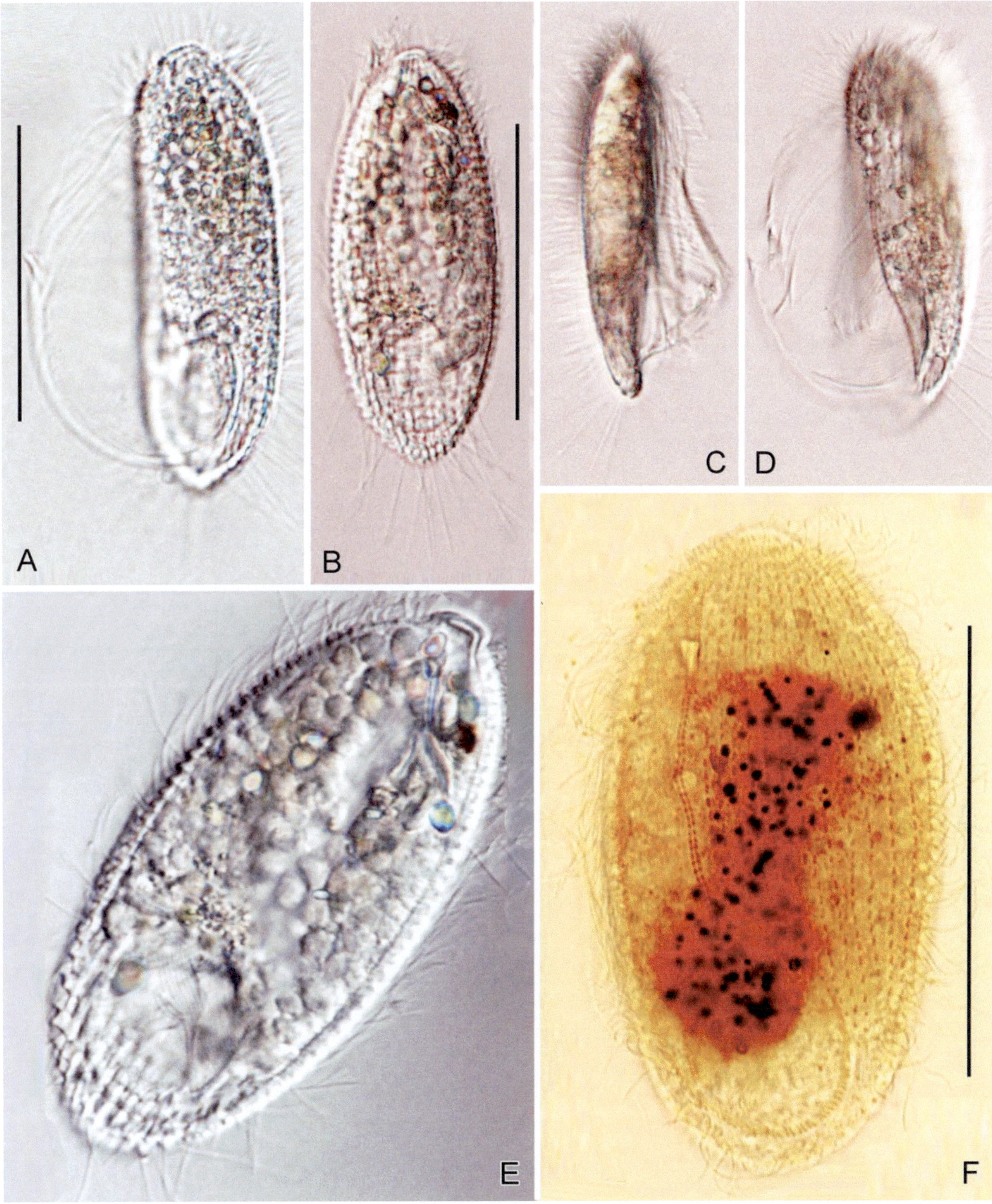


**Fig. 34B** *Pleuronema wiackowskii*
(**a**) Ventral view of live individual; (**b**) dorsal view of live cell, showing notched pellicle and caudal cilia; (**c**, **d**) lateral views of live cells to show paroral membrane; (**e**) notched pellicle and apical plate; (**f**) ventral view of ciliature. Scale bar: 35 μm

## Reference

Wang Y, Song WB, Hu X, Warren A, Al-Rasheid KAS (2008) Descriptions of two new marine *Pleuronema, P. czapikae* sp. n. and *P. wiackowskii* sp. n. (Ciliophora: Scuticociliatida) from the Yellow Sea, north China. Acta Protozool 47:35–45

Class Oligohymenophorea de Puytorac et al., 1974
Order Sessilida Kahl, 1933
Family Epistylididae Kahl, 1933
*Pseudepistylis songi* Shi et al., 2007

**Ecological feature:** estuary area
**Water temperature:** 25.0 °C
**pH:** 7.0
**Salinity:** 6.0‰

**Fig. 35A** Sample site

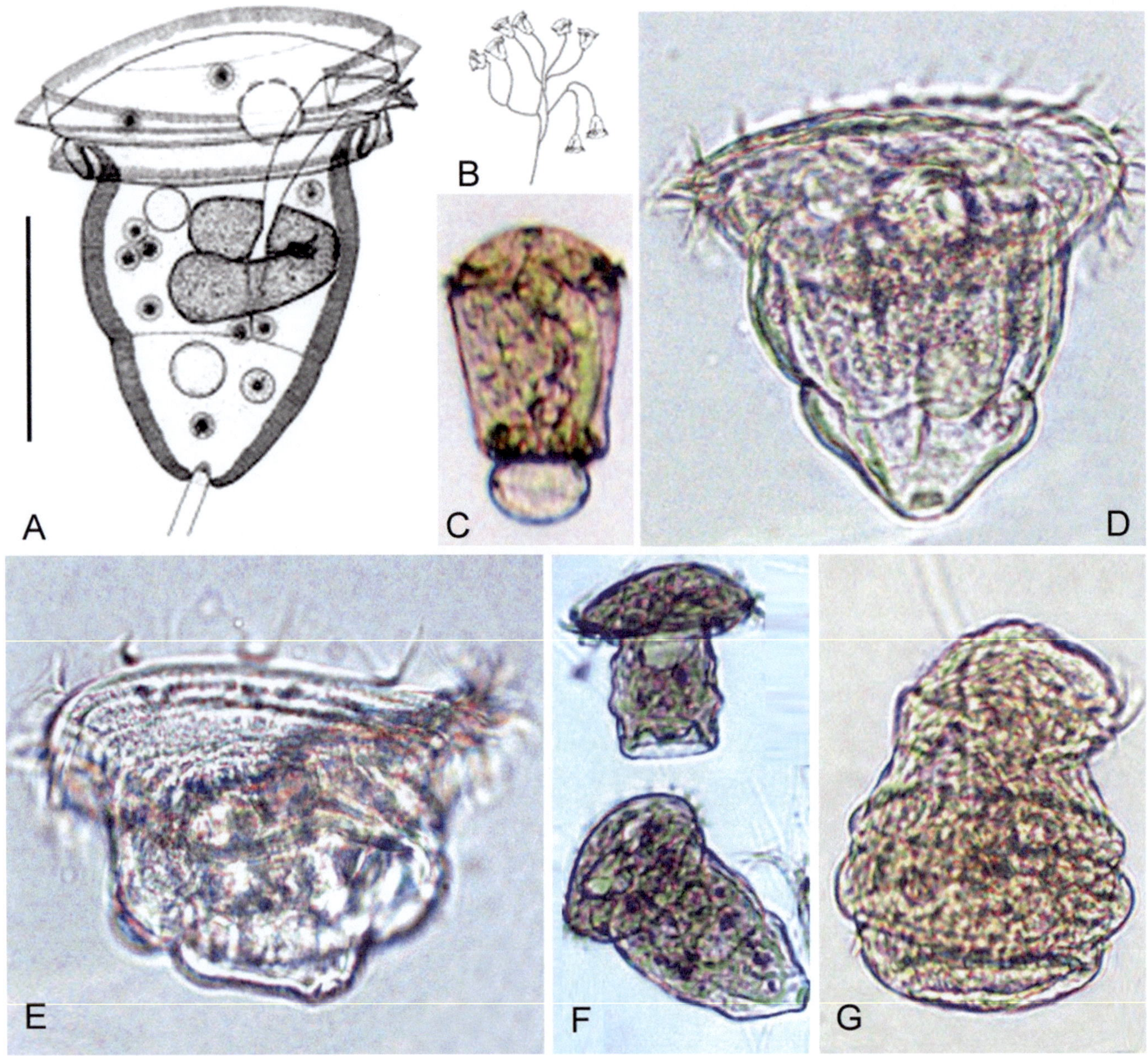


**Fig. 35B** *Pseudepistylis songi*
(**a**) Lateral view of representative zooid (from Peng et al., 2007); (**b**) colony outline; (**c**) telotroch; (**d–g**). To show different body shapes and sizes. Scale bar: 100 μm

## Reference

Peng X, Shi X, Warren A (2007) Establishment of a new peritrich ciliates genus, *Pseudepistylis* n. gen. (Ciliophora: Peritrichia: Epistylididae), with a description of a new freshwater species, *Pseudepistylis songi* n. sp. from Wenzhou, China. Zootaxa 1524:35–45

Class Oligohymenophorea de Puytorac et al., 1974
Order Sessilida Kahl, 1933
Family Vorticellidae Ehrenberg, 1838
*Vorticella lima* Kahl, 1933

**Ecological feature:** mangrove
**Water temperature:** 24.0 °C
**pH:** 7.8
**Salinity:** 23‰

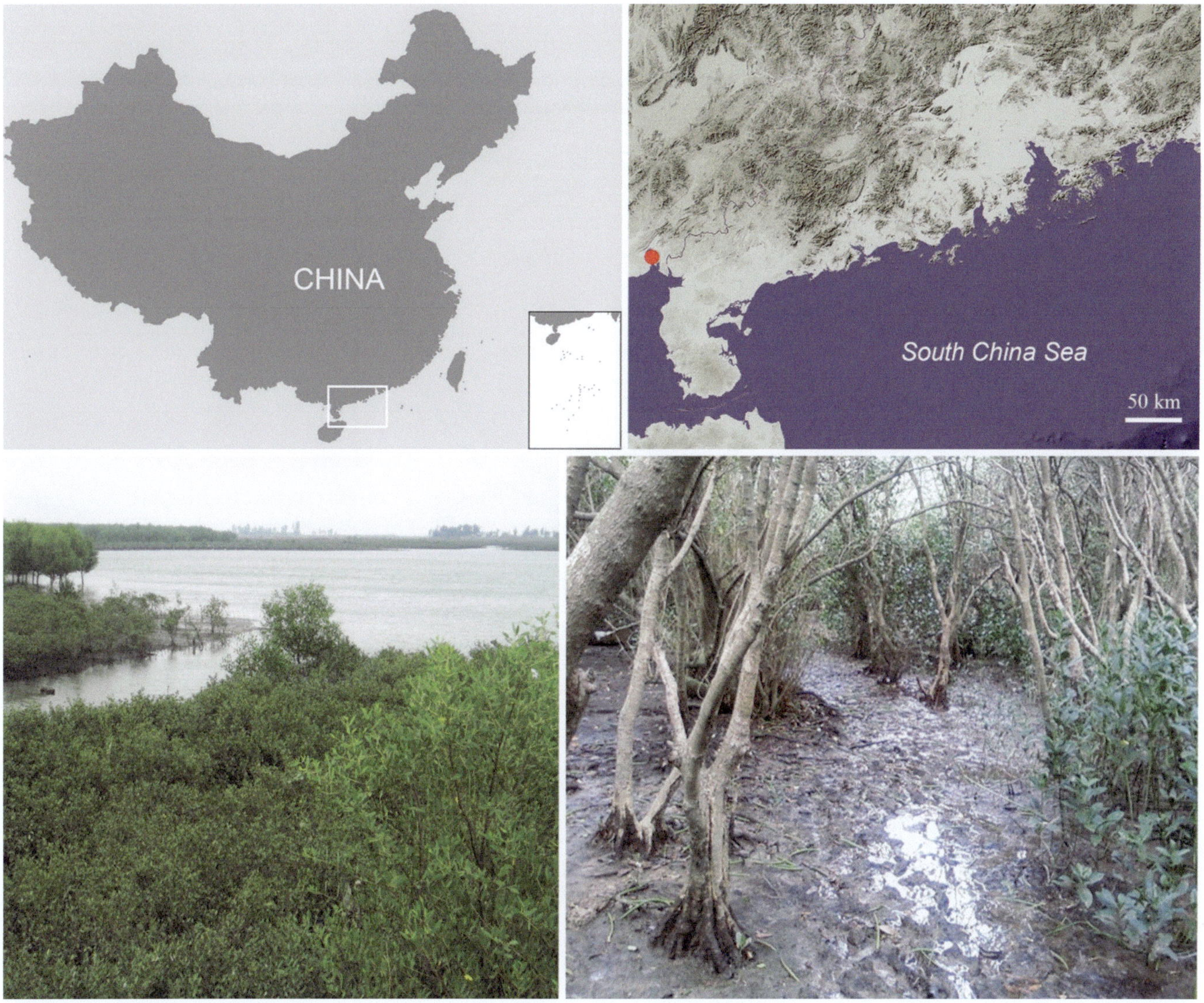


**Fig. 36A** Sample site

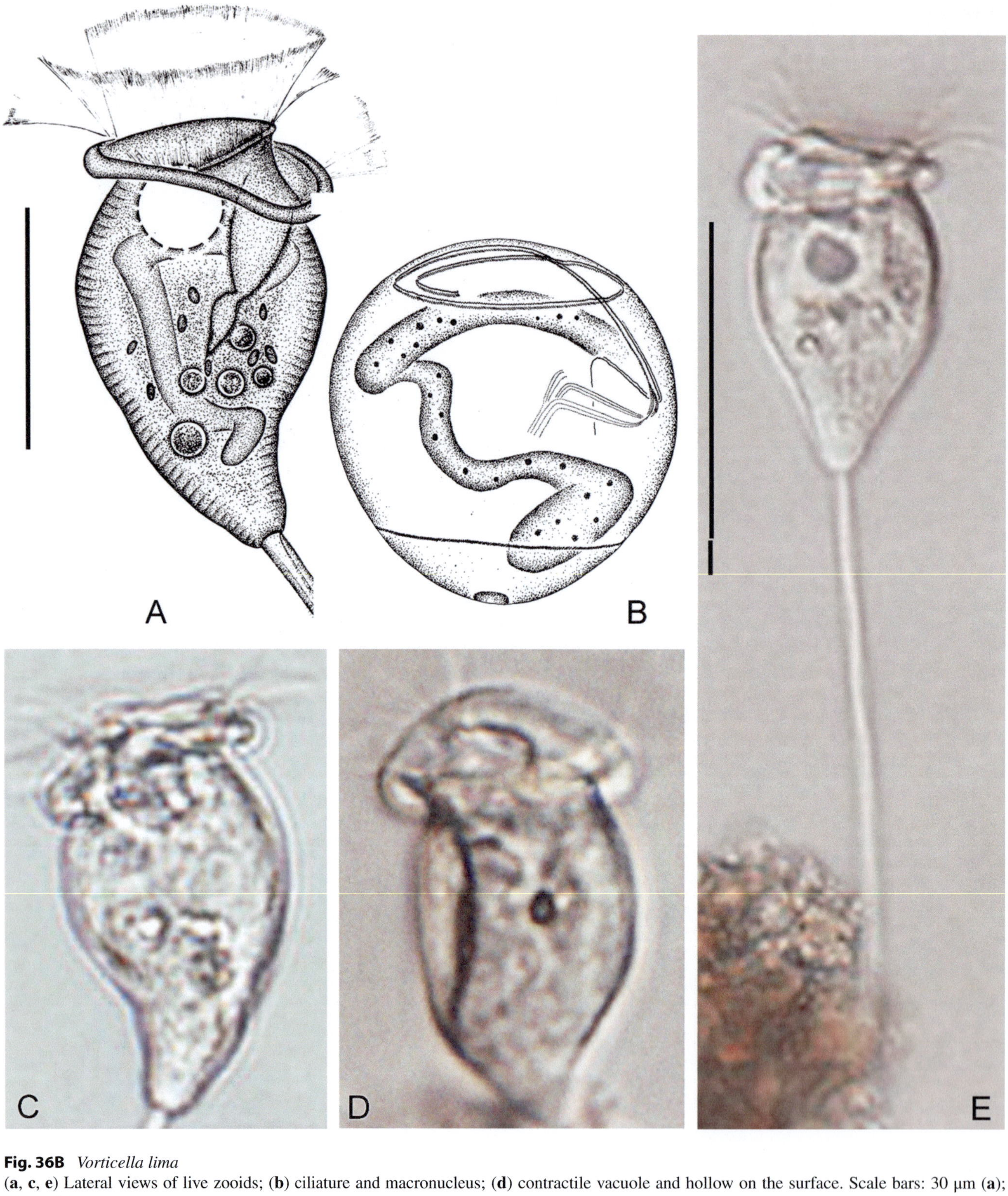


**Fig. 36B** *Vorticella lima*
(**a**, **c**, **e**) Lateral views of live zooids; (**b**) ciliature and macronucleus; (**d**) contractile vacuole and hollow on the surface. Scale bars: 30 μm (**a**); 50 μm (**b**)

## Reference

Liang Z, Shen Z, Zhang Y, Ji D, Warren A, Lin X (2018) Morphology and phylogeny of four new *Vorticella* species (Ciliophora: Peritrichia) from coastal waters of southern China. J Eukaryot Microbiol. (in press)

Class Oligohymenophorea de Puytorac et al., 1974
Order Sessilida Kahl, 1933
Family Vorticellidae Ehrenberg, 1838
*Vorticella minuta* Liang et al., 2018

**Ecological feature:** estuary area
**Water temperature:** 23.0 °C
**pH:** 7.3
**Salinity:** 5‰

**Fig. 37A** Sample site

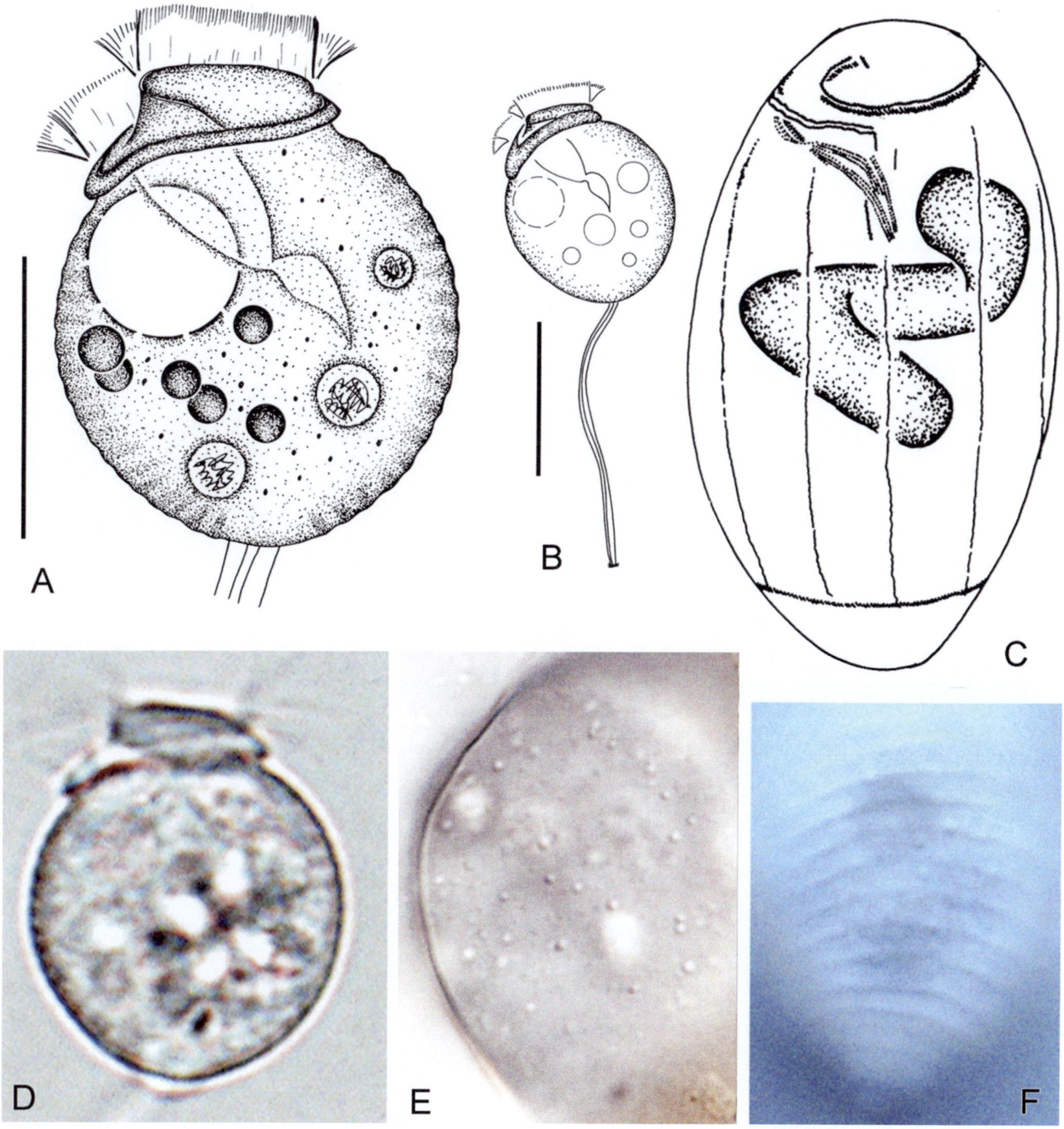


**Fig. 37B** *Vorticella minuta*
(**a**, **b**, **d**) General views of live zooids; (**c**) ciliature, macronucleus, and myoneme system; (**e**) granules on cell surface; (**f**) pellicular striations. Scale bars: 15 μm (**a**); 30 μm (**b**)

## Reference

Liang Z, Shen Z, Zhang Y, Ji D, Warren A, Lin X (2018) Morphology and phylogeny of four New *Vorticella* species (Ciliophora: Peritrichia) from coastal waters of southern China. J Eukaryot Microbiol (in press)

Class Oligohymenophorea de Puytorac et al., 1974
Order Sessilida Kahl, 1933
Family Vorticellidae Ehrenberg, 1838
*Vorticella parachiangi* Liang et al., 2018

**Ecological feature:** mangrove
**Water temperature:** 24.0 °C
**pH:** 7.8
**Salinity:** 12‰

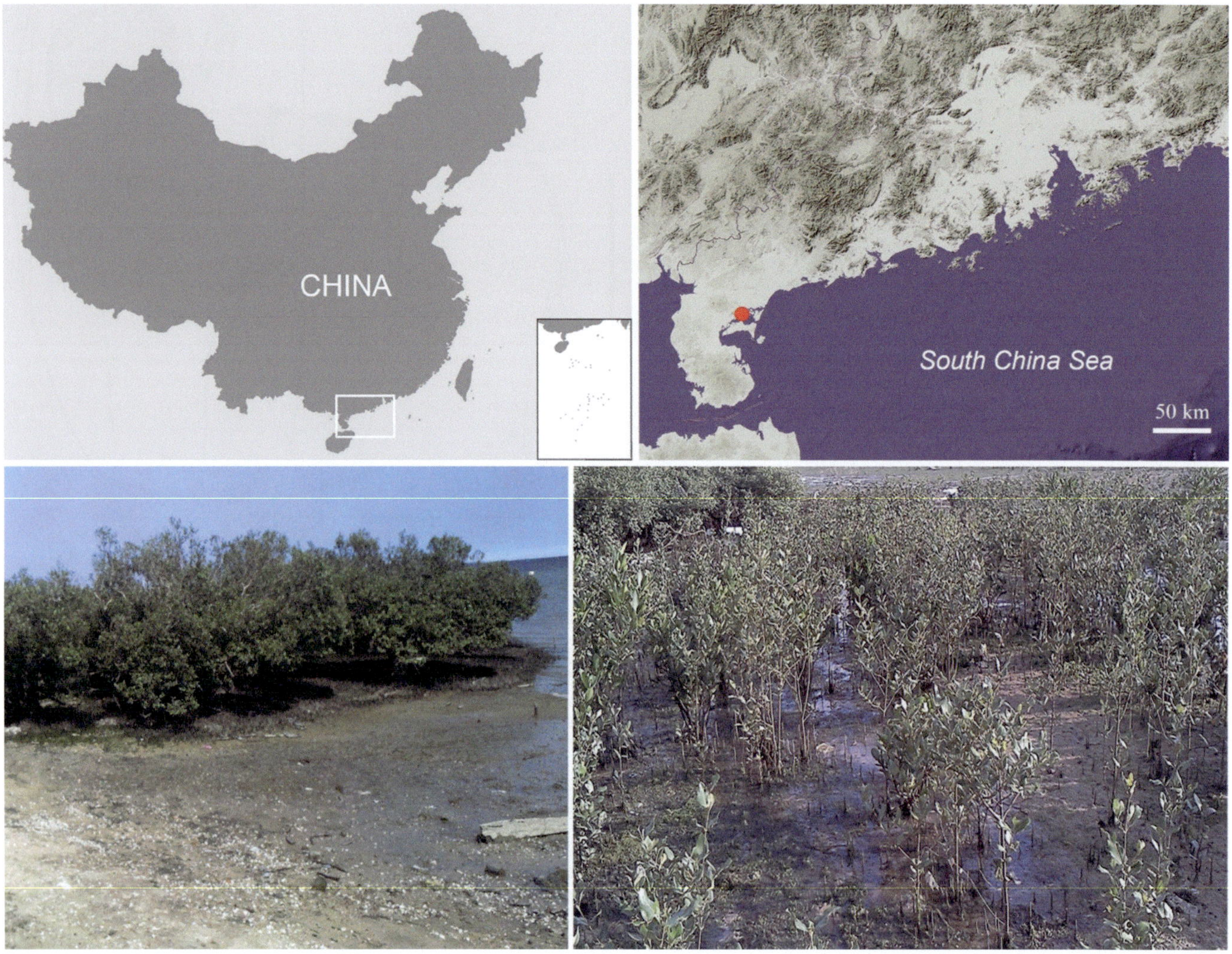


**Fig. 38A** Sample site

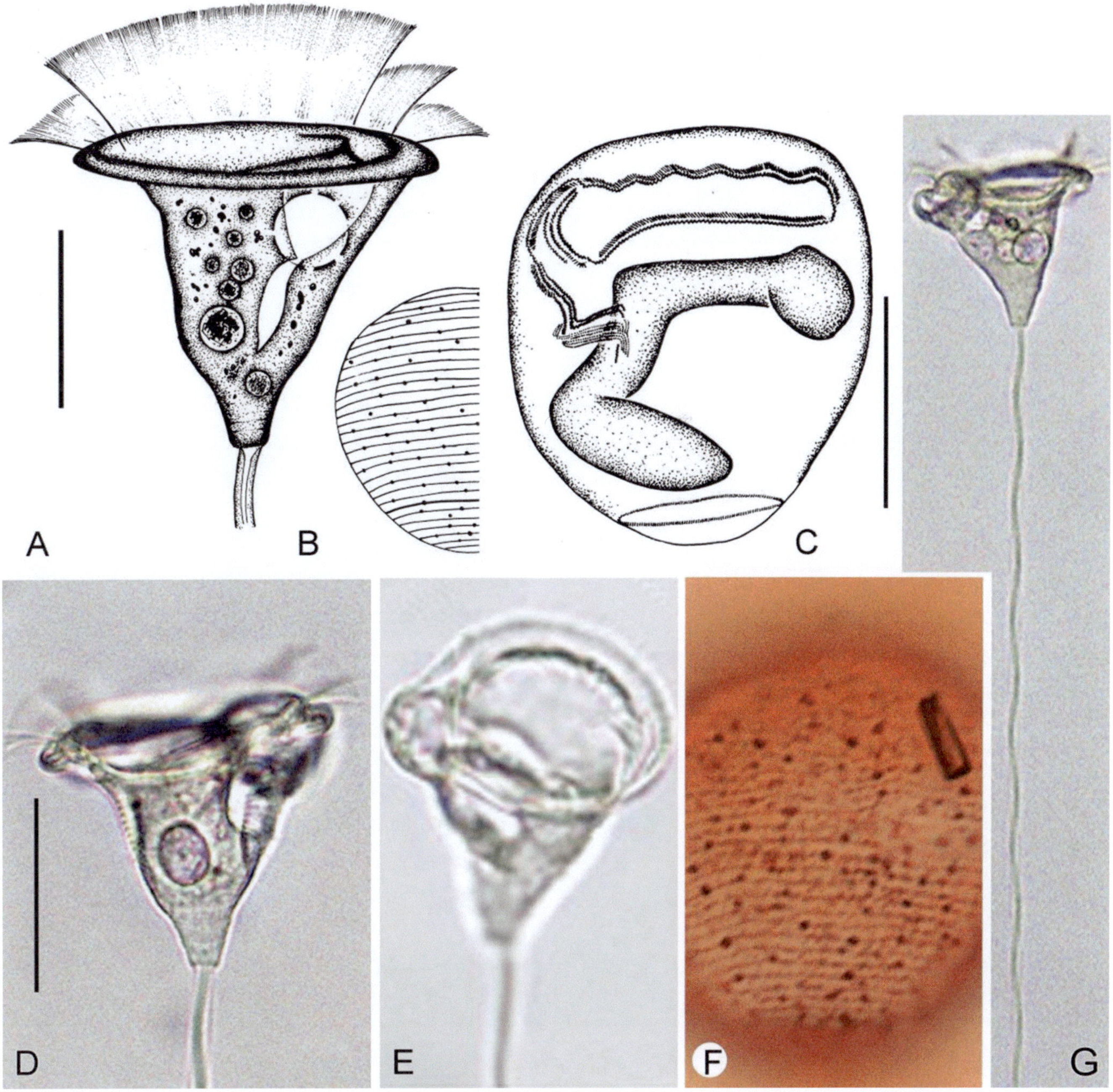


**Fig. 38B** *Vorticella parachiangi*
(**a**, **d**) General views of live zooids; (**b**) part of silverline system; (**c**) ciliature and macronucleus; (**e**) edge of peristomial lip; (**f**) silverline system; (**g**) a live zooid at low magnification. Scale bar: 40 μm (**a**, **c**, **d**)

## Reference

Liang Z, Shen Z, Zhang Y, Ji D, Warren A, Lin X (2018) Morphology and phylogeny of four New *Vorticella* species (Ciliophora: Peritrichia) from coastal waters of southern China. J Eukaryot Microbiol. (in press)

Class Oligohymenophorea de Puytorac et al., 1974
Order Sessilida Kahl, 1933
Family Vorticellidae Ehrenberg, 1838
*Vorticella scapiformis* Liang et al., 2018

**Ecological feature:** fish-farming pond
**Water temperature:** 25.0 °C
**pH:** 7.2
**Salinity:** 22‰

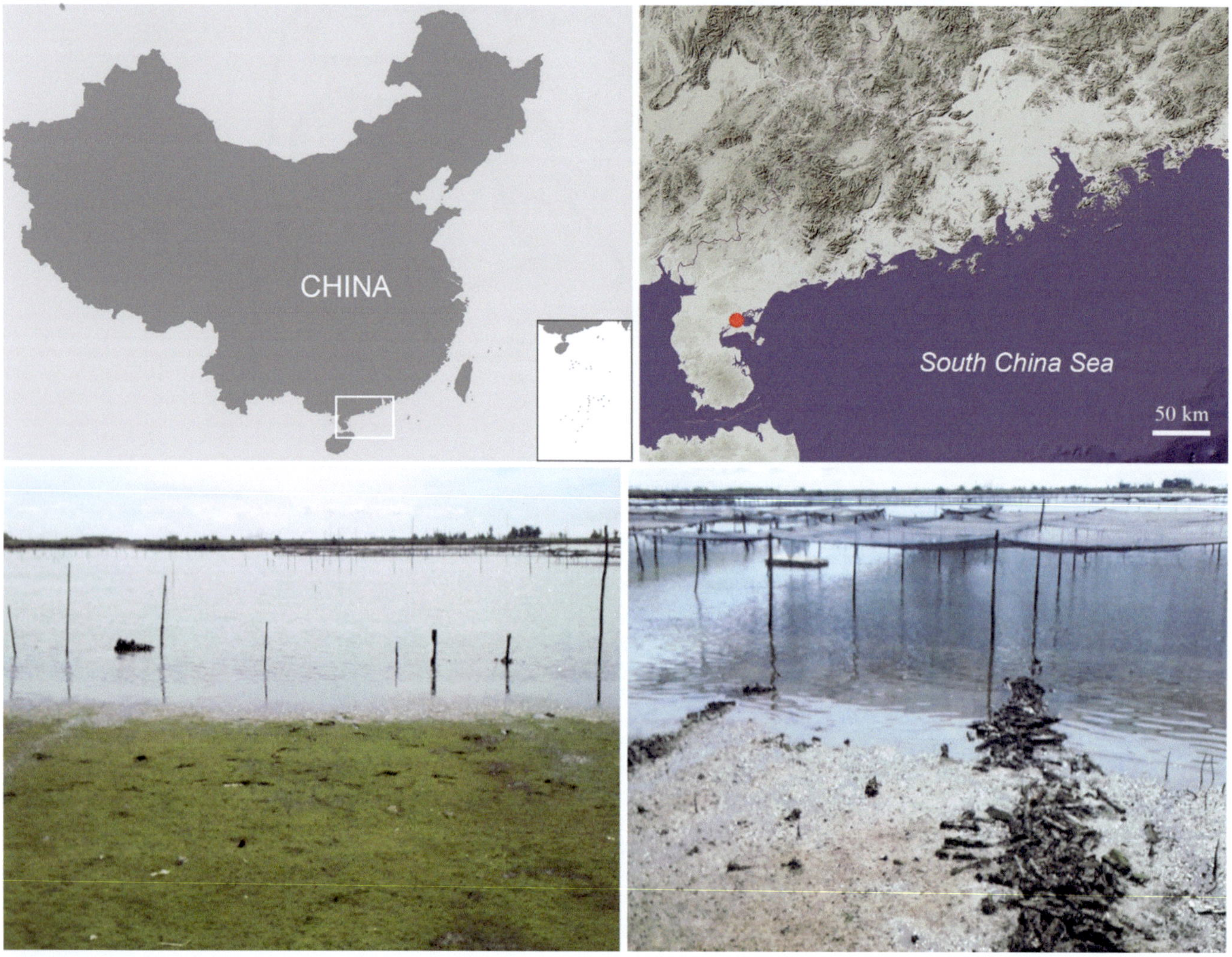


**Fig. 39A** Sample site

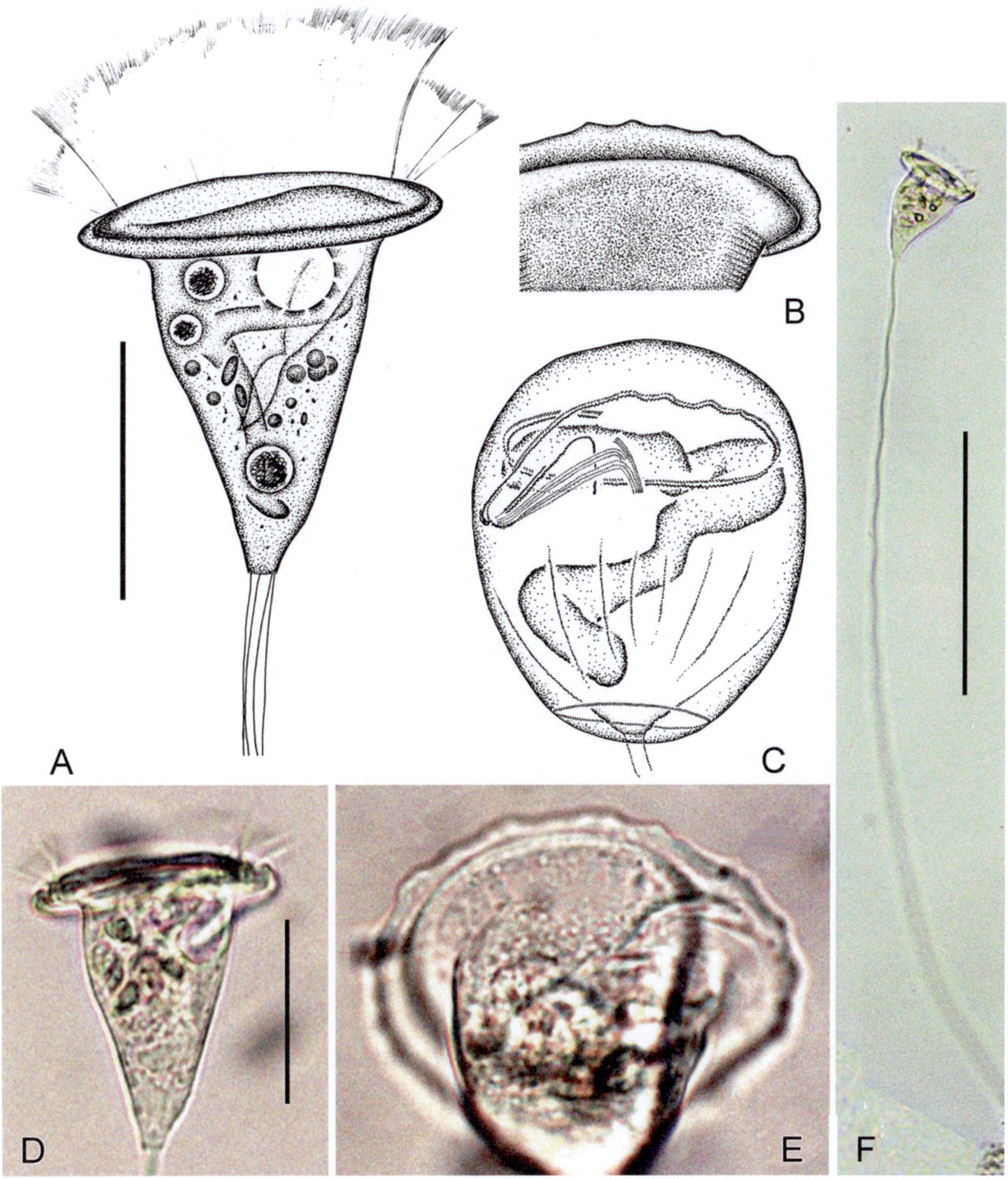


**Fig. 39B** *Vorticella scapiformis*
(**a**, **d**) General views of representative zooids; (**b**, **e**) edge of peristomial lip; (**c**) ciliature, macronucleus, and myoneme system; (**f**) a live zooid at low magnification. Scale bars: 50 μm (**a**, **d**); 180 μm (**f**)

## Reference

Liang Z, Shen Z, Zhang Y, Ji D, Warren A, Lin X (2018) Morphology and phylogeny of four New *Vorticella* species (Ciliophora: Peritrichia) from coastal waters of southern China. J Eukaryot Microbiol. (in press)

Class Oligohymenophorea de Puytorac et al., 1974
Order Sessilida Kahl, 1933
Family Vorticellidae Ehrenberg, 1838
*Vorticella utriculus* (Stokes, 1885)

**Ecological feature:** fish-farming pond
**Water temperature:** 25.0 °C
**pH:** 7.2
**Salinity:** 22.0‰

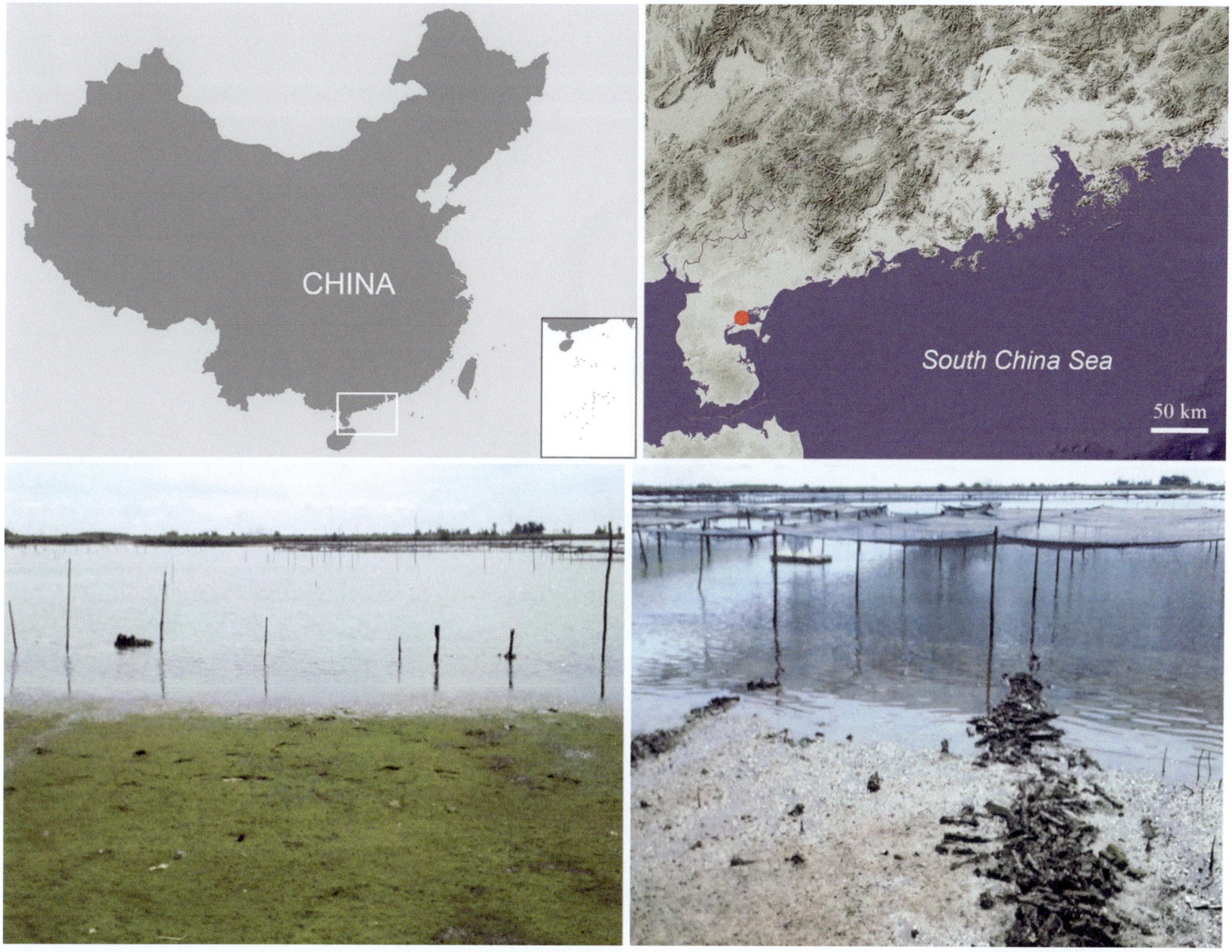


**Fig. 40A** Sample site

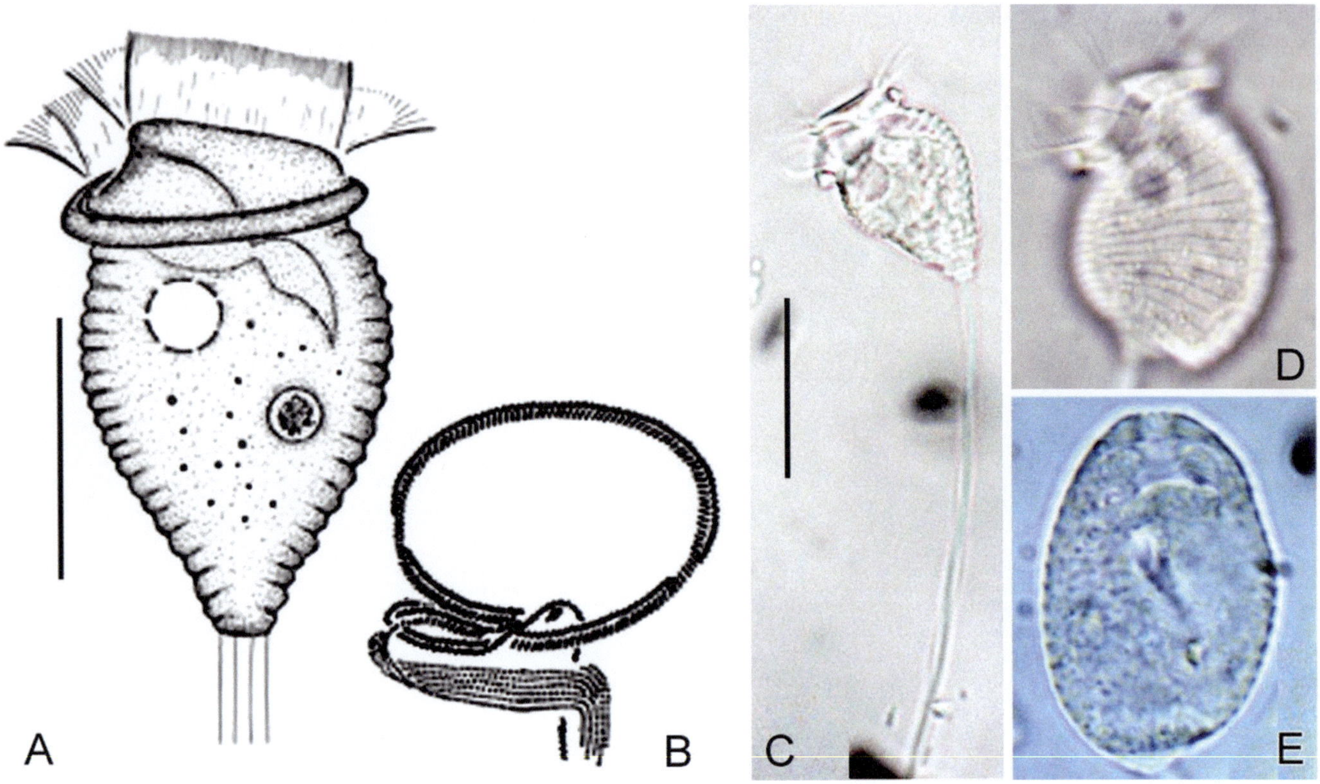


**Fig. 40B** *Vorticella utriculus*
(**a**) General view of live zooid, showing elevated peristomial disc and ventrally located contractile vacuole; (**b**) oral ciliature; (**c**) a live zooid at low magnification; (d) zooids at high magnification, showing pellicular striations; (**e**) shrinked body. Scale bars: 15 μm (**a**); 30 μm (**c**)

## Reference

Stokes AC (1885) On some new infusoria from American freshwaters. Ann Nat Hist Series 5(15):437–449

Class Oligohymenophorea de Puytorac et al., 1974
Order Sessilida Kahl, 1933
Family Zoothamniidae Sommer, 1951
*Zoothamnium bucciniiformum* Shen et al., 2017

**Ecological feature:** offshore water
**Water temperature:** 26.0 °C
**pH:** 7.0
**Salinity:** 30.0‰

**Fig. 41A** Sample site

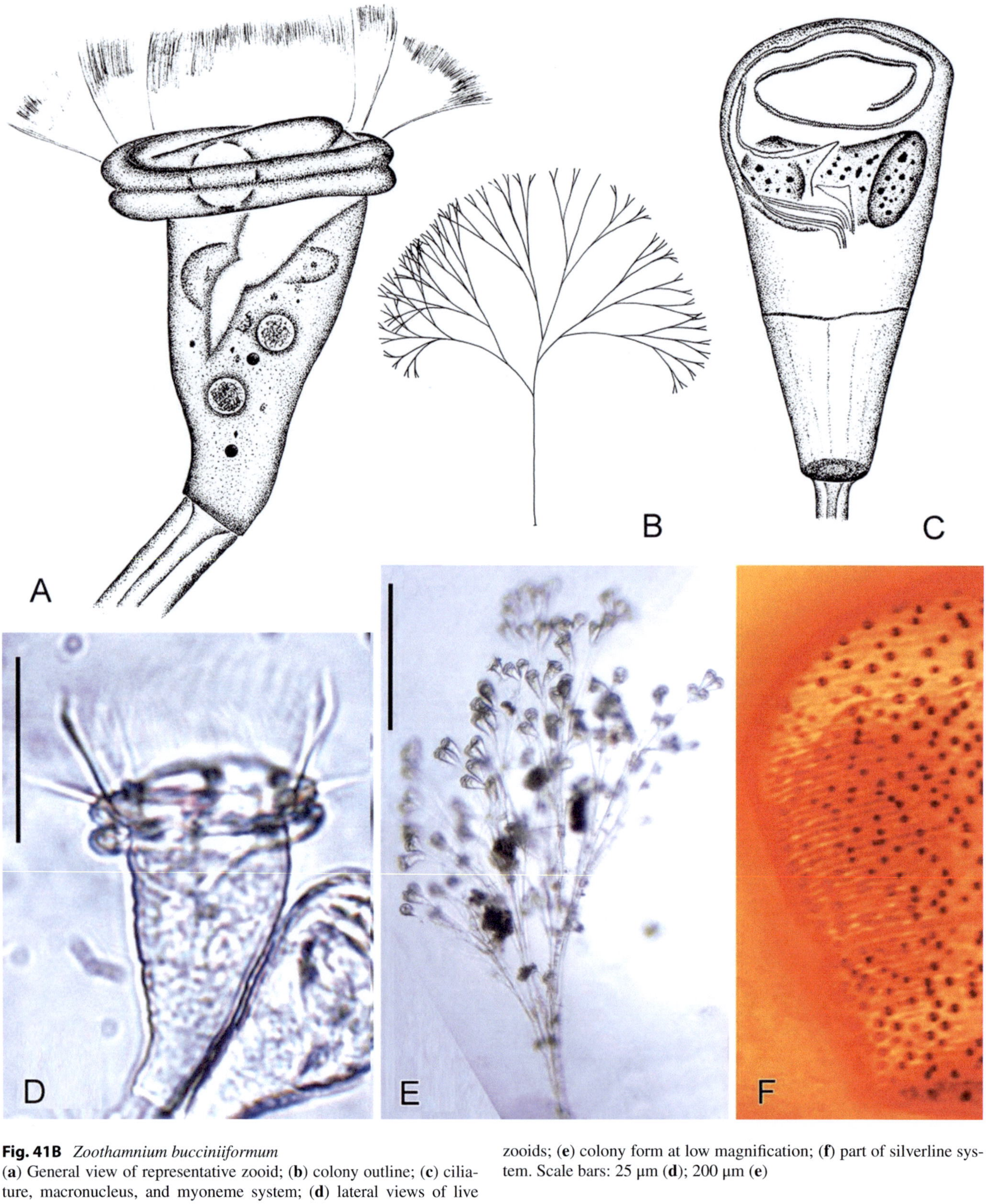


**Fig. 41B** *Zoothamnium bucciniiformum*
(**a**) General view of representative zooid; (**b**) colony outline; (**c**) ciliature, macronucleus, and myoneme system; (**d**) lateral views of live zooids; (**e**) colony form at low magnification; (**f**) part of silverline system. Scale bars: 25 μm (**d**); 200 μm (**e**)

## Reference

Shen Z, Ji D, Yi Z, Al-Rasheid KAS, Lin X (2017) Morphology and phylogenetic placement of three new *Zoothamnium* species (Ciliophora: Peritrichia) from coastal waters of southern China. J Eukaryot Microbiol 64:266–277

Class Oligohymenophorea de Puytorac et al., 1974
Order Sessilida Kahl, 1933
Family Zoothamniidae Sommer, 1951
*Zoothamnium candelabrum* Wailes, 1932

**Ecological feature:** mangrove
**Water temperature:** 22.0 °C
**pH:** 7.5
**Salinity:** 10.0‰

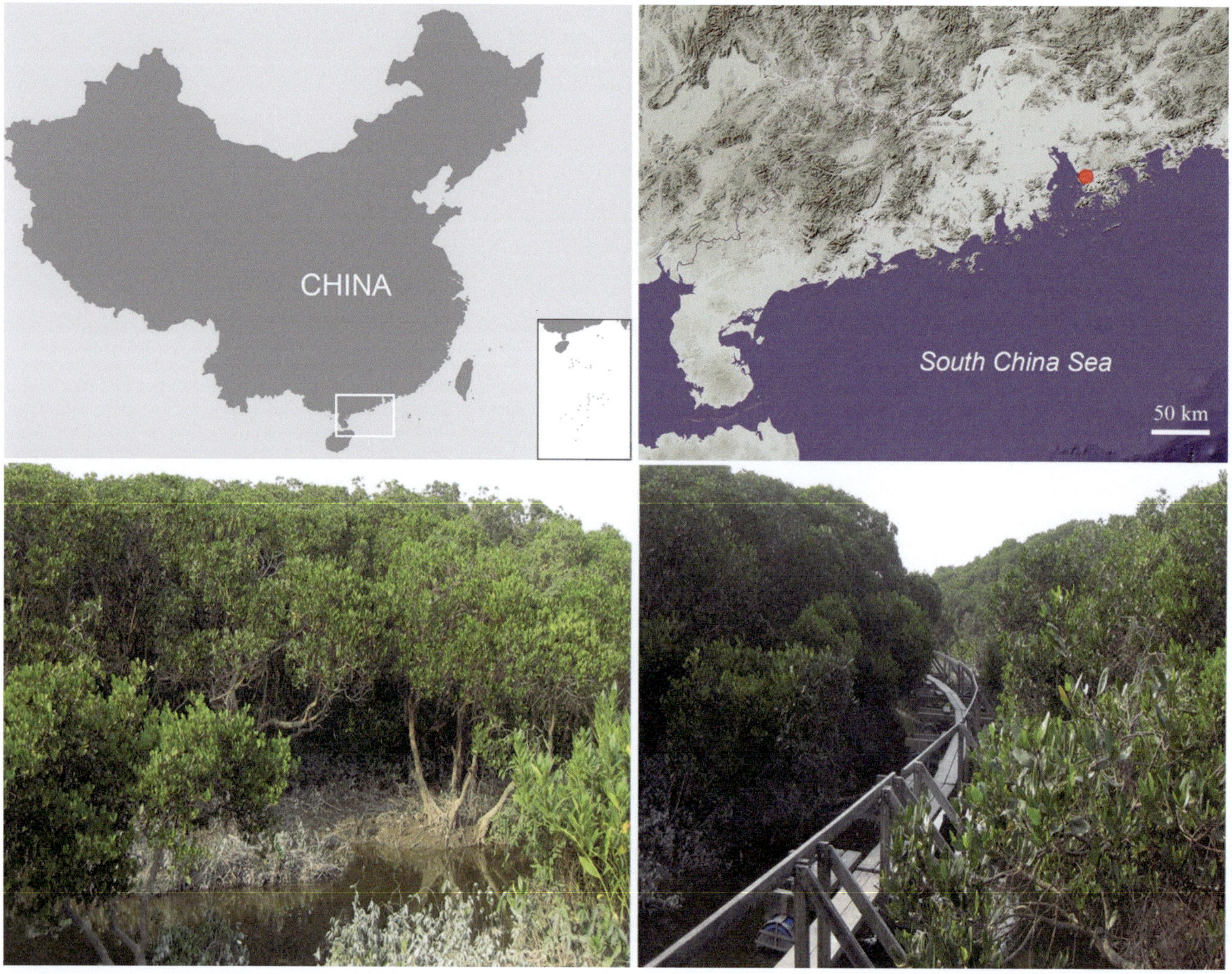


**Fig. 42A** Sample site

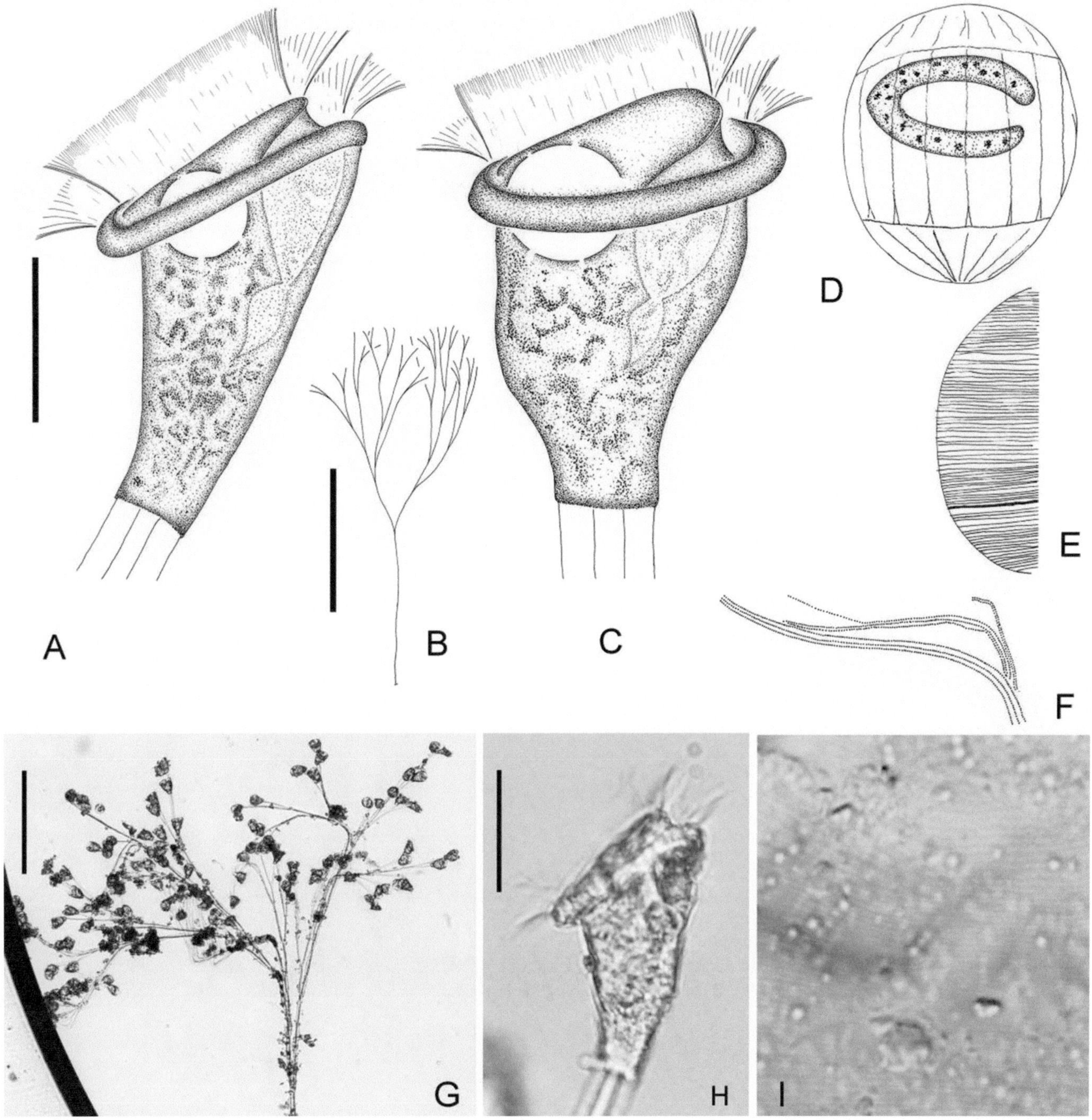


**Fig. 42B** *Zoothamnium candelabrum*
(**a**, **c**, **h**) General views of live zooids; (**b**) colony outline; (**d**) macronucleus and myoneme system; (**e**) part of silverline system; (**f**) peniculi 1–3; (**g**) colony form at low magnification; (**i**) pellicular striations. Scale bars: 15 μm (**a**); 600 μm (**b**); 300 μm (**g**); 20 μm (**h**)

## Reference

Wailes GH (1932) Description of new species of Protozoa from British Columbia. Contr Can Biol Fish 7:213–219

Class Oligohymenophorea de Puytorac et al., 1974
Order Sessilida Kahl, 1933
Family Zoothamniidae Sommer, 1951
*Zoothamnium florens* Shen et al., 2017

**Ecological feature:** mangrove
**Water temperature:** 25.0 °C
**pH:** 7.6
**Salinity:** 13.0‰

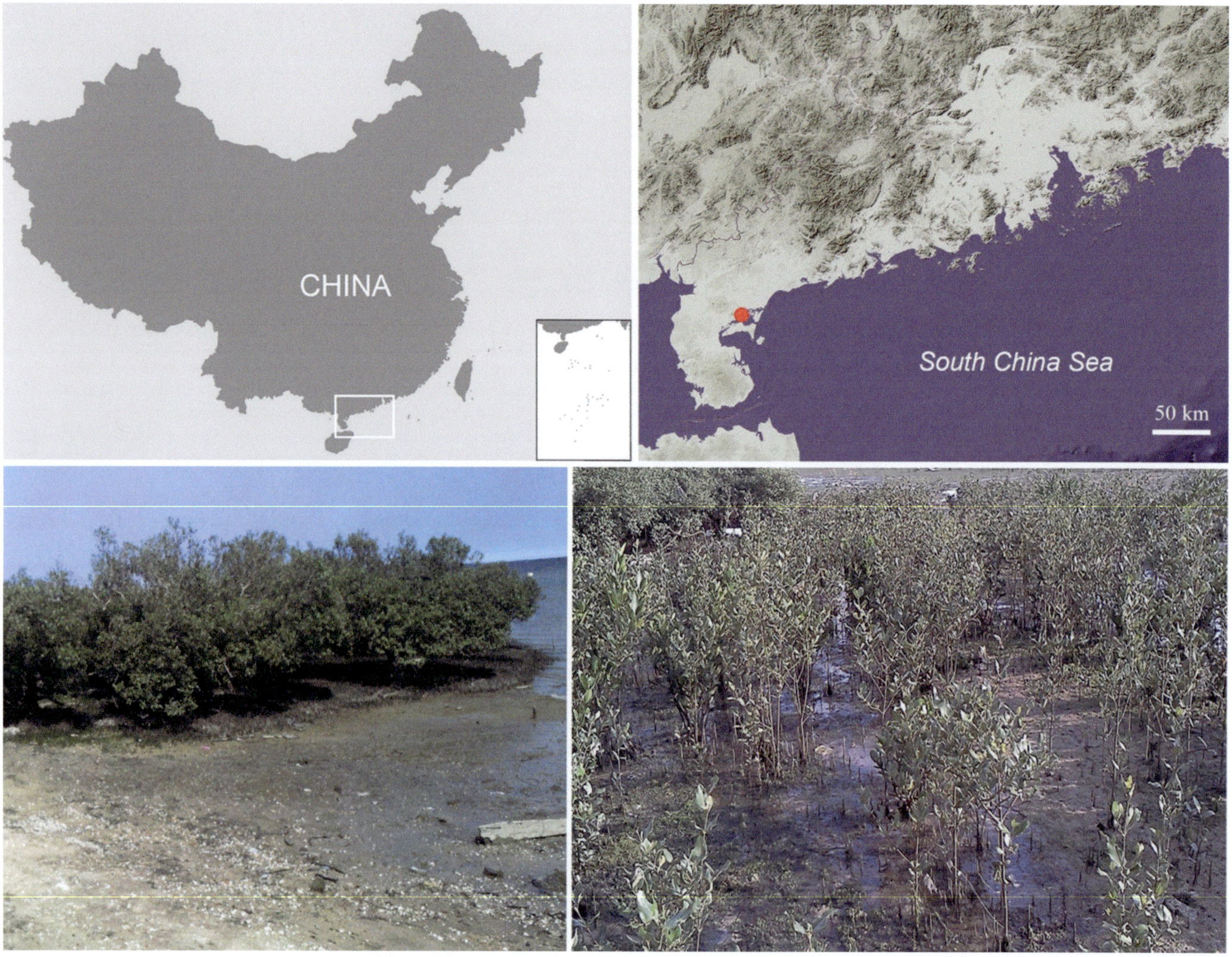


**Fig. 43A** Sample site

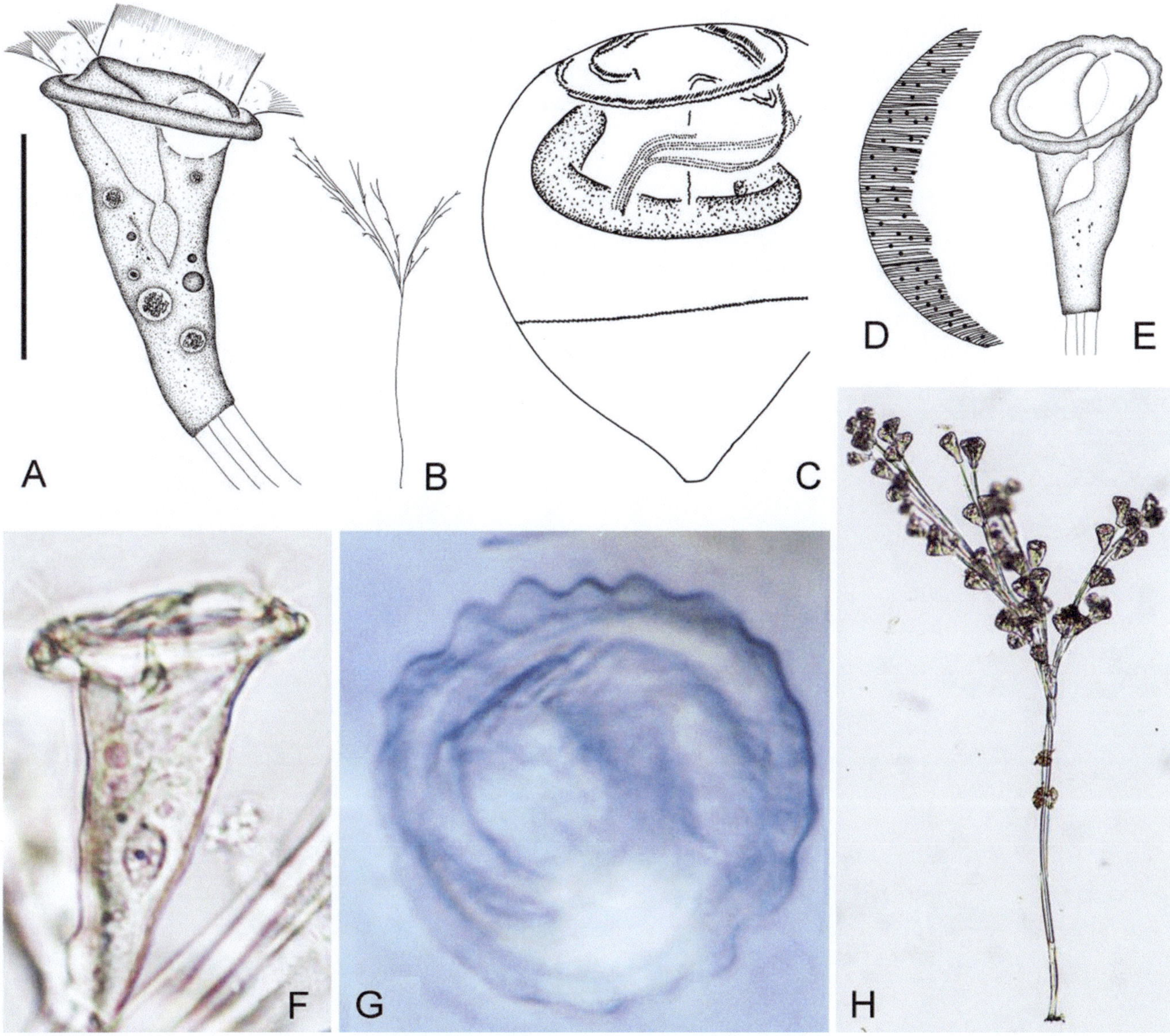


**Fig. 43B** *Zoothamnium florens*
(**a**, **f**) General views of live zooids; (**b**) colony outline; (**c**) ciliature and macronucleus; (**d**) part of silverline system; (**e**) apical view of live zooid, showing peristomial lip; (**g**) oral view, indicating the edge of peristomial lip; (**h**) colony form at low magnification. Scale bar: 50 μm

## Reference

Shen Z, Ji D, Yi Z, Al-Rasheid KAS, Lin X (2017) Morphology and phylogenetic placement of three new *Zoothamnium* species (Ciliophora: Peritrichia) from coastal waters of southern China. J Eukaryot Microbiol 64:266–277

Class Oligohymenophorea de Puytorac et al., 1974
Order Sessilida Kahl, 1933
Family Zoothamniidae Sommer, 1951
*Zoothamnium hentscheli* Kahl, 1933

**Ecological feature:** mangrove
**Water temperature:** 25.0 °C
**pH:** 8.0
**Salinity:** 10.0‰

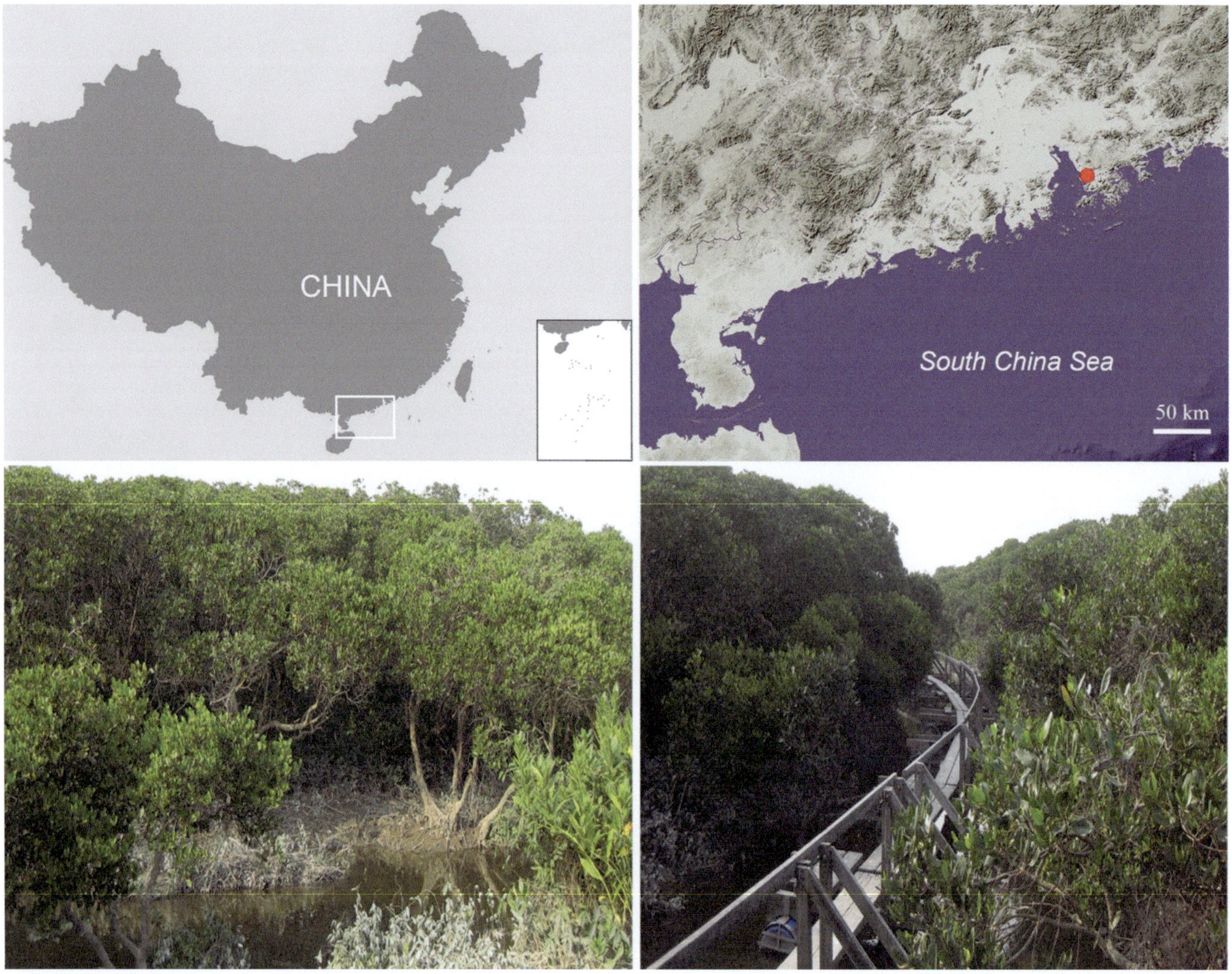


**Fig. 44A** Sample site

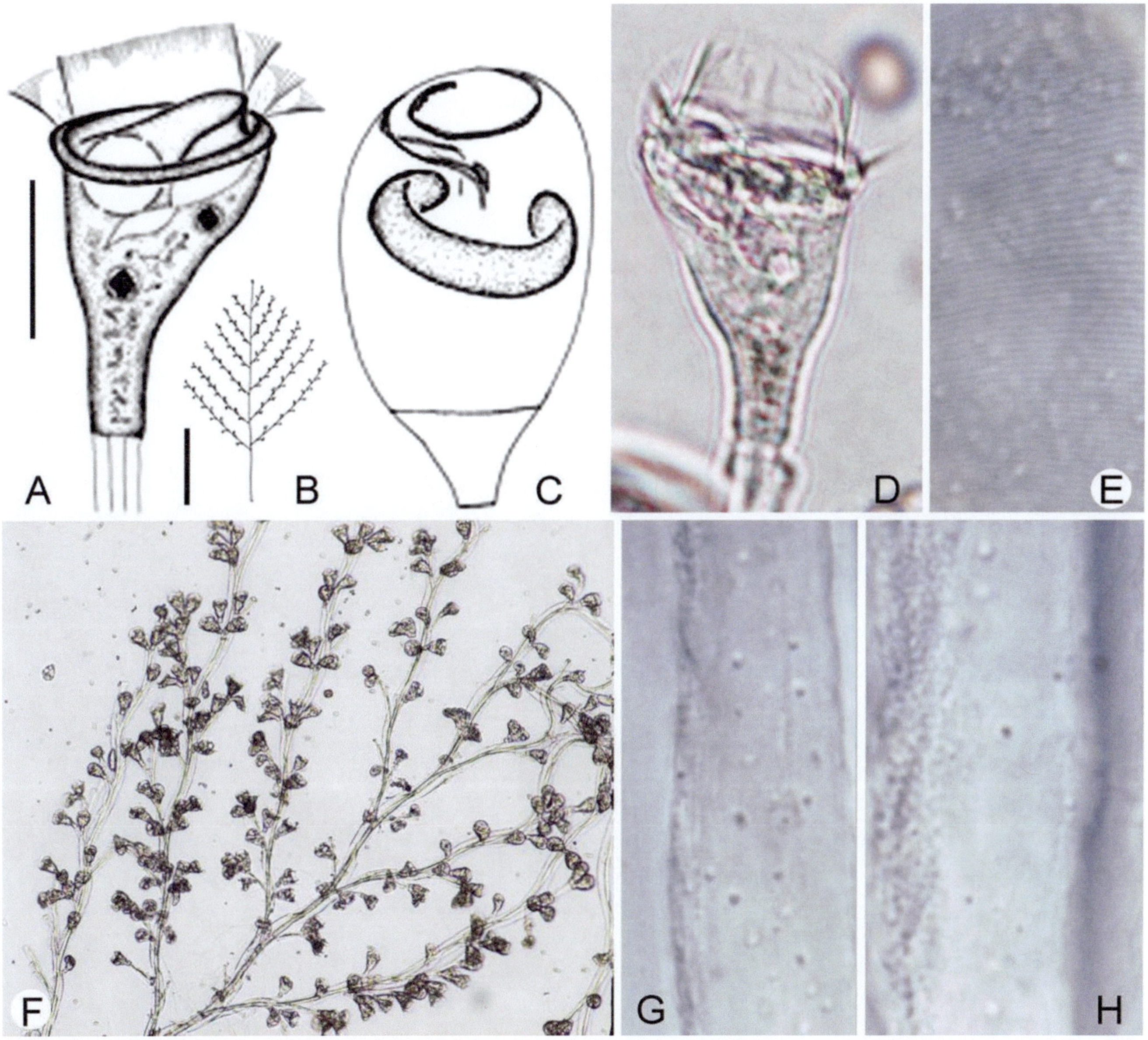


**Fig. 44B** *Zoothamnium hentscheli*
(**a**, **d**) General views of typical zooids; (**b**) colony form; (**c**) ciliature and macronucleus; (**e**) zooid at high magnification, to show pellicular striations; (**f**) colony form at low magnification; (**g**) smaller the coplasmic granules in the spasmoneme; (**h**) bigger the coplasmic granules in the spasmoneme. Scale bars: 25 μm (**a**); 1000 μm (**b**)

## Reference

Kahl A (1935) Urtiere oder Protozoa I: Wimpertiere oder Ciliata (Infusoria) 4. Peritricha und Chonotricha. Tierwelt Dtl 30:651–886

Class Oligohymenophorea de Puytorac et al., 1974
Order Sessilida Kahl, 1933
Family Zoothamniidae Sommer, 1951
*Zoothamnium paraentzii* Song, 1991

**Ecological feature:** mangrove
**Water temperature:** 25.0 °C
**pH:** 7.0
**Salinity:** 25.0‰

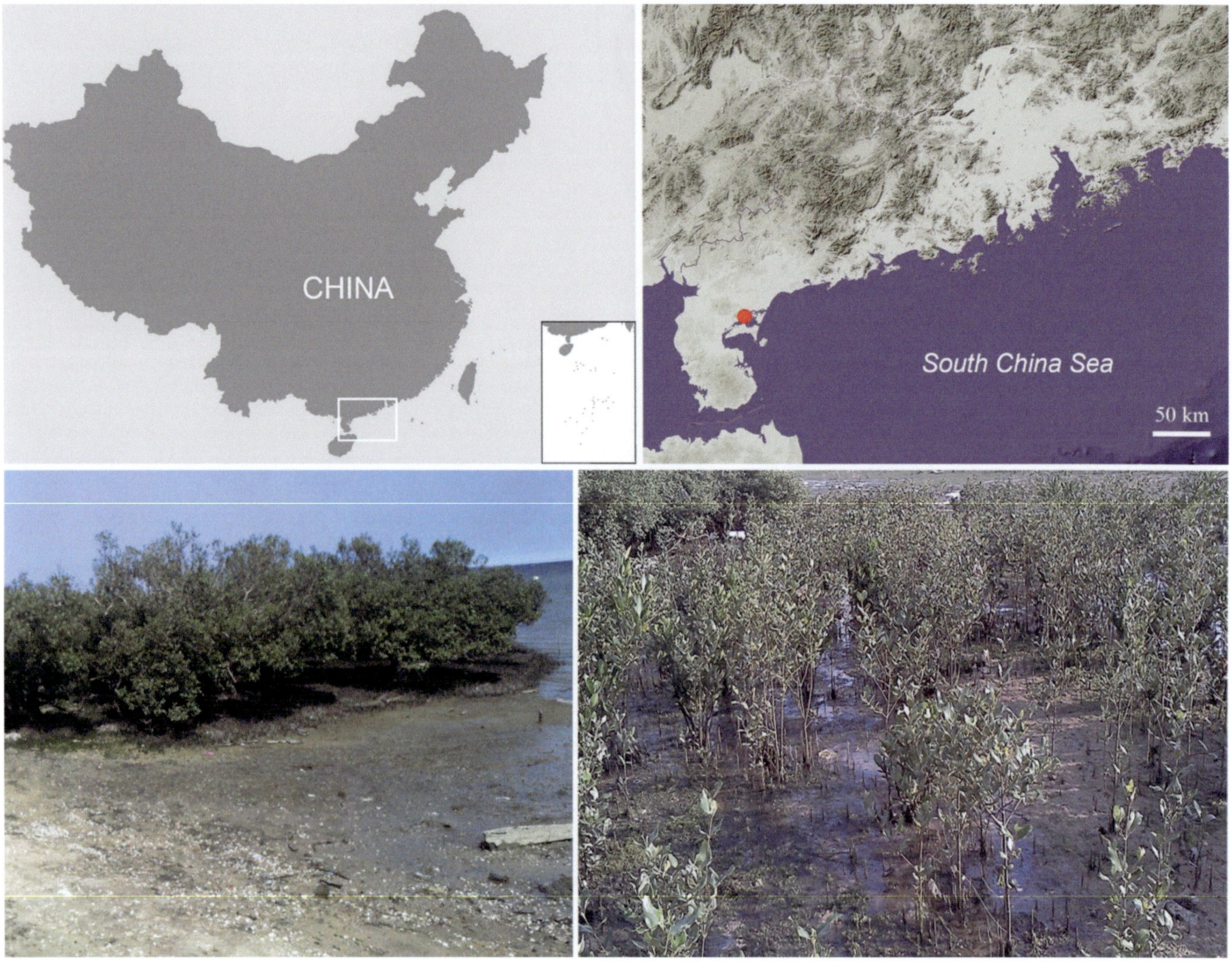


**Fig. 45A** Sample site

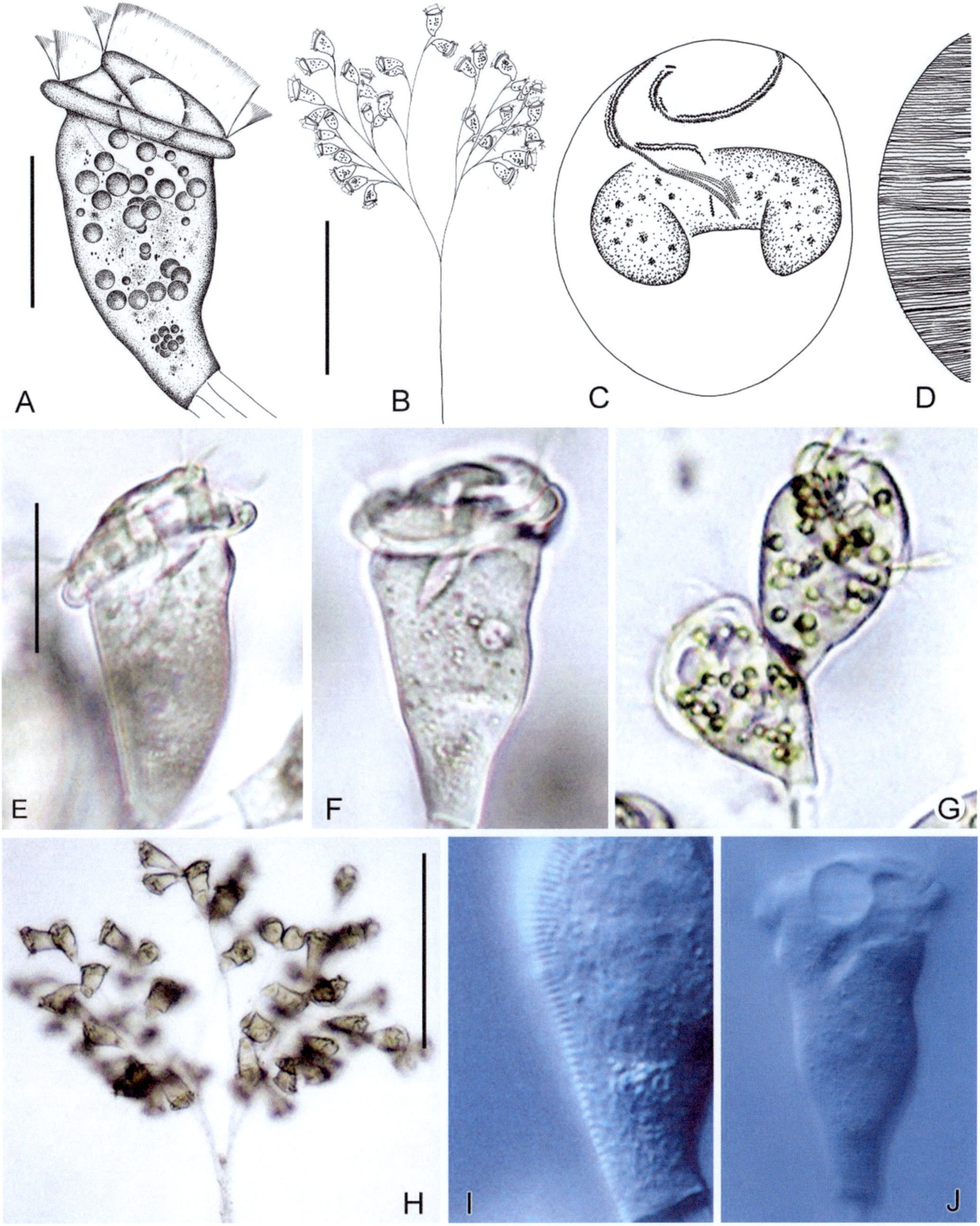


**Fig. 45B** *Zoothamnium paraentzii*
(**a**, **e–g**, **j**) Lateral views of representative zooids; (**b**, **h**) colony form; (**c**) ciliature and macronucleus; (**d**) part of silverline system; (**i**) zooids at high magnification, to show pellicular striations. Scale bars: 40 μm (**a**, **e**); 300 μm (**h**); 450 μm (**b**)

## References

Song WB (1991) A new commensal ciliate, *Zoothamnium paraentzii* (Ciliophora, Peritrichida). Zool Res 12:355–359 (in Chinese with English summary)

Sun P, Ji D (2005) Notes on a new marine peritrichous ciliate (Ciliophora: Peritrichida), *Zoothamnium xuianum* n. sp., with redescription of *Z. paraentzii* Song, 1991 from northern China. Zootaxa 1075:41–53

Class Oligohymenophorea de Puytorac et al., 1974
Order Sessilida Kahl, 1933
Family Zoothamniidae Sommer, 1951
*Zoothamnium plumula* Kahl, 1933

**Ecological feature:** offshore water
**Water temperature:** 26.0 °C
**pH:** 6.8
**Salinity:** 30.0‰

**Fig. 46A** Sample site

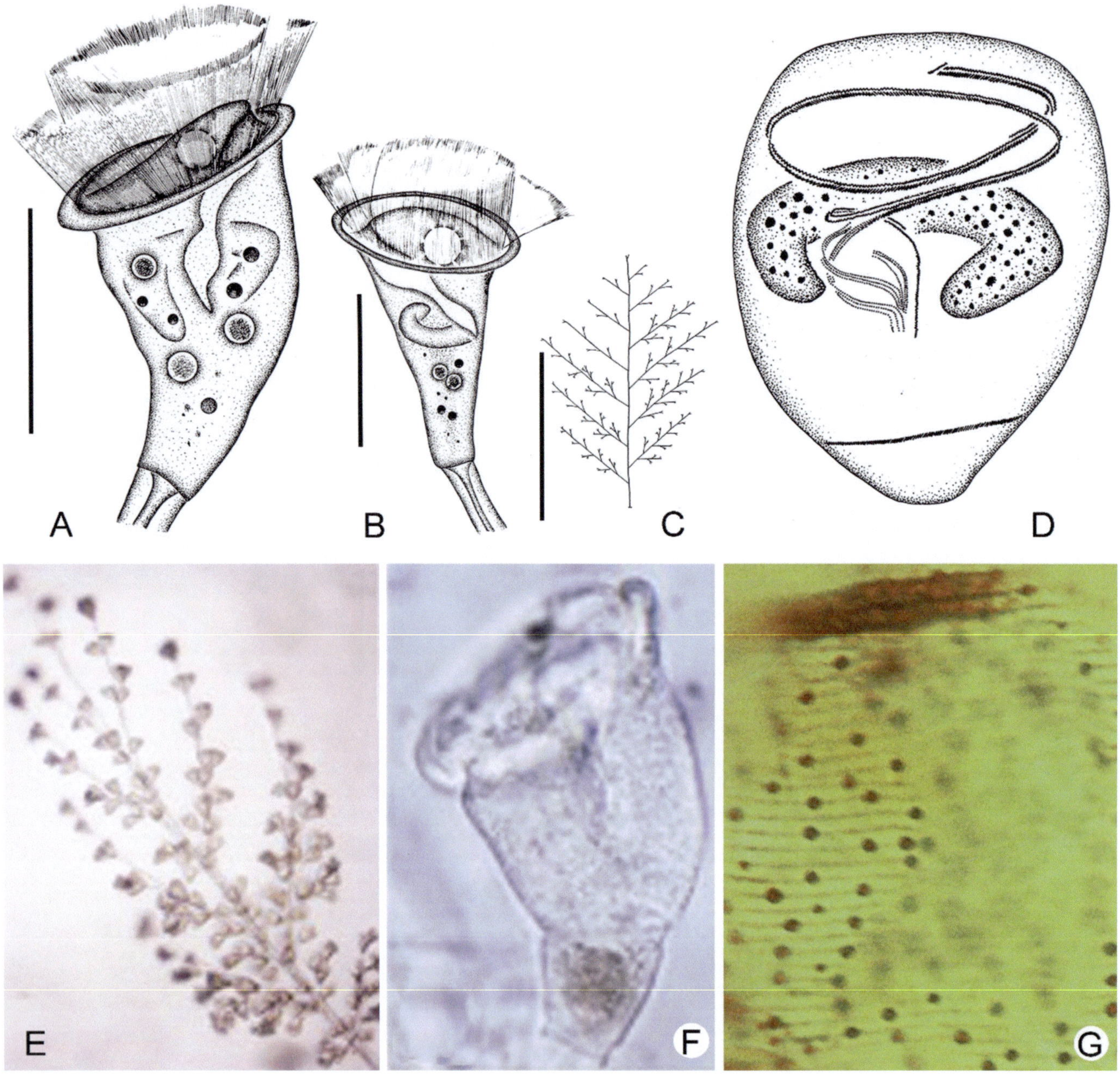


**Fig. 46B** *Zoothamnium plumula*
(**a**, **b**) General views of live zooids, to show bigger (**a**) and smaller (**b**) ones; (**c**) colony outline; (**d**) ciliature and macronucleus; (**e**) colony form at low magnification; (**f**) lateral view of bigger zooid; (**g**) part of silverline system. Scale bars: 50 μm (**a**, **b**); 1500 μm (**c**)

## Reference

Ji D, Shin M, Choi J, Clamp J, Al-Rasheid KAS, Song WB (2011) Redescriptions of five species of marine peritrichs, *Zoothamnium plumula*, *Zoothamnium nii*, *Zoothamnium wang*, *Pseudovorticella bidulphiae*, and *Pseudovorticella marina* (Protista, Ciliophora). Zootaxa 2930:47–59

Class Oligohymenophorea de Puytorac et al., 1974
Order Sessilida Kahl, 1933
Family Zoothamniidae Sommer, 1951
*Zoothamnium zhanjiangense* Shen et al., 2017

**Ecological feature:** mangrove
**Water temperature:** 27.0 °C
**pH:** 6.6
**Salinity:** 10.0‰

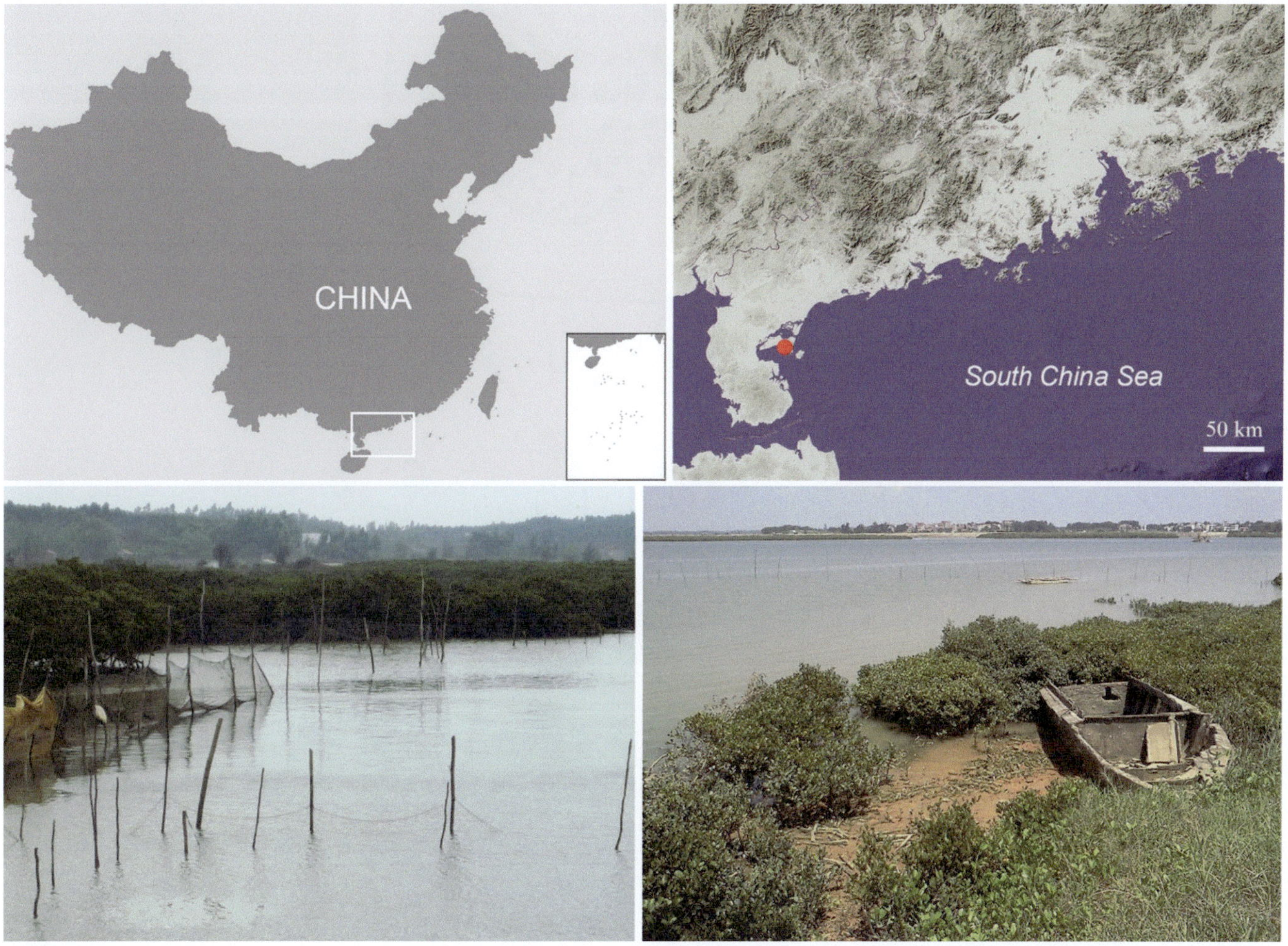


**Fig. 47A** Sample site

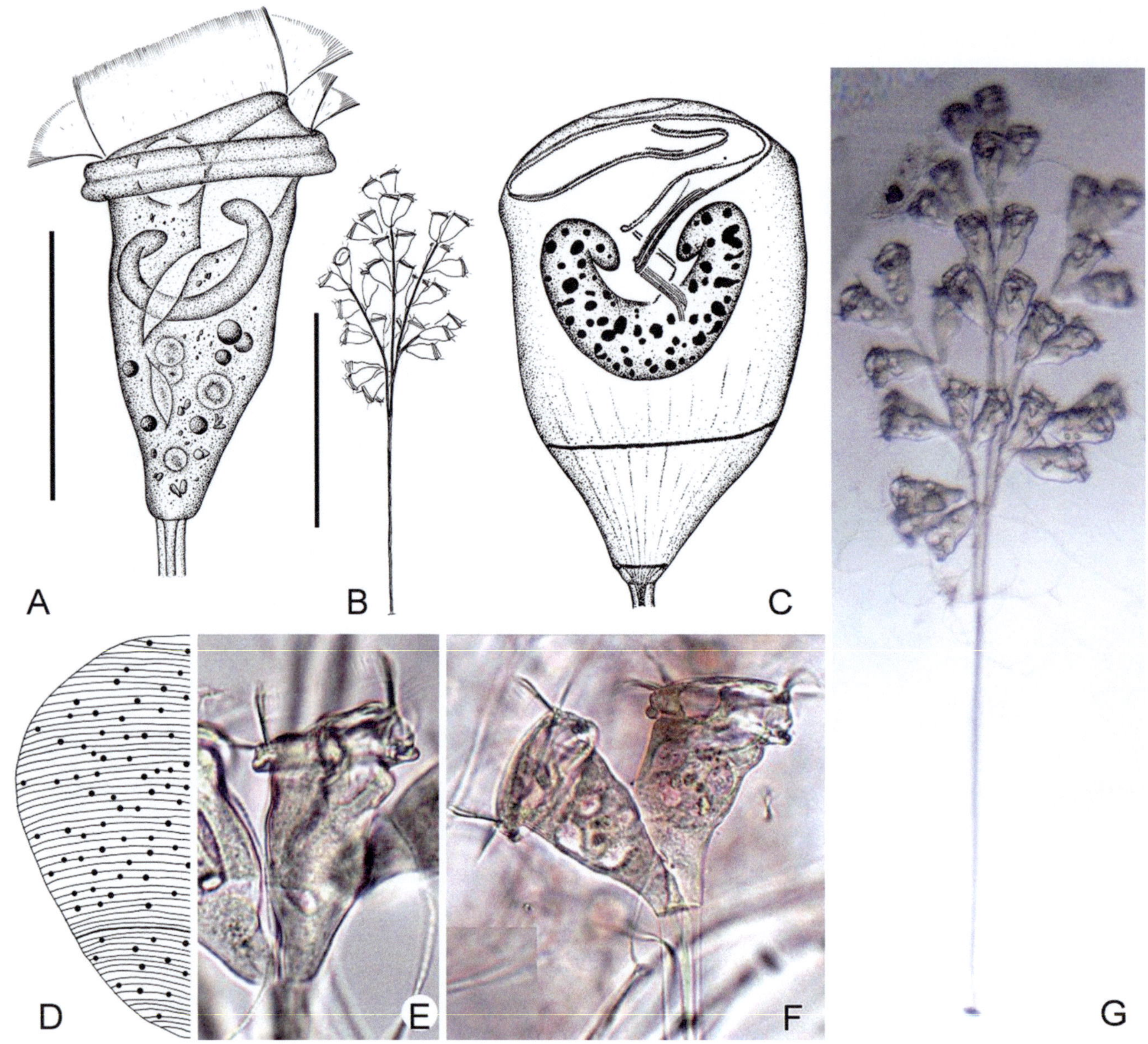


**Fig. 47B** *Zoothamnium zhanjiangense*
(**a**) General view of live zooid; (**b**) colony outline; (**c**) oral ciliature and macronucleus; (**d**) part of silverline system; (**e, f**) representative zooids in vivo; (**g**) colony form at low magnification. Scale bars: 50 μm (**a**); 400 μm (**b**)

## Reference

Shen Z, Ji D, Yi Z, Al-Rasheid KAS, Lin X (2017) Morphology and phylogenetic placement of three new *Zoothamnium* species (Ciliophora: Peritrichia) from coastal waters of southern China. J Eukaryot Microbiol 64:266–277

# Class Protocruziea

Xiaozhong Hu, Xiaofeng Lin, and Weibo Song

Class Protocruziea Gao et al., 2016
Order Protocruziida Jankowski, 1980
Protocruziidae Jankowski, 1980
*Protocruzia* cf. *contrax*

**Ecological feature:** offshore water
**Water temperature:** 24.5 °C
**pH:** 8.4
**Salinity:** 28.2‰

X. Hu (✉)
Fisheries College, Ocean University of China,
Qingdao, Shandong, China
e-mail: xiaozhonghu@ouc.edu.cn

X. Lin
College of Life Science, South China Normal University,
Guangzhou, Guangdong, China
e-mail: xlin@oucscnu.edu.cn

W. Song
Institute of Evolution & Marine Biodiversity,
Ocean University of China, Qingdao, Shandong, China
e-mail: wsong@ouc.edu.cn


X. Hu et al. (eds.), *Ciliate Atlas: Species Found in the South China Sea*,
https://doi.org/10.1007/978-981-13-5901-9_10

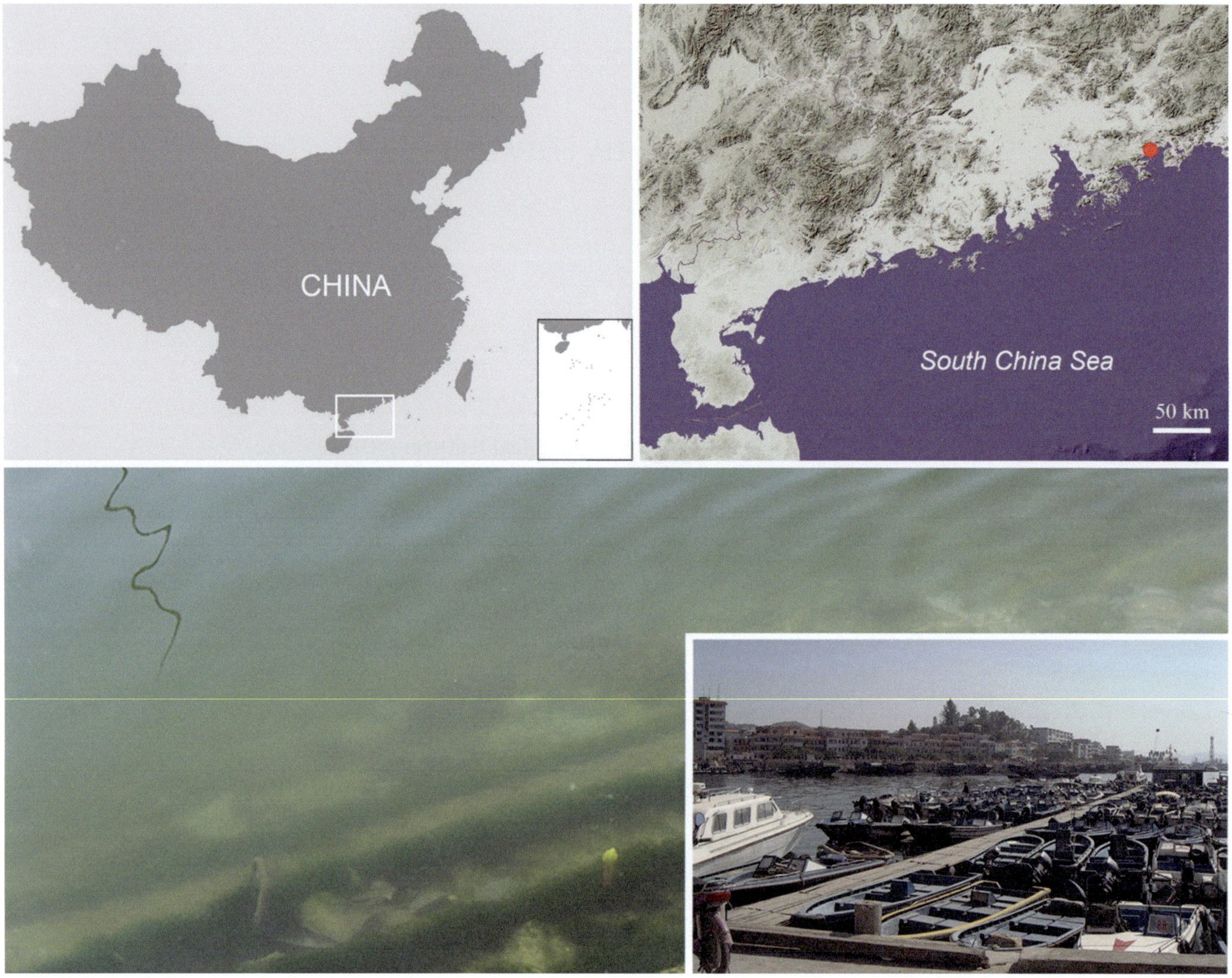


**Fig. 1A** Sample site

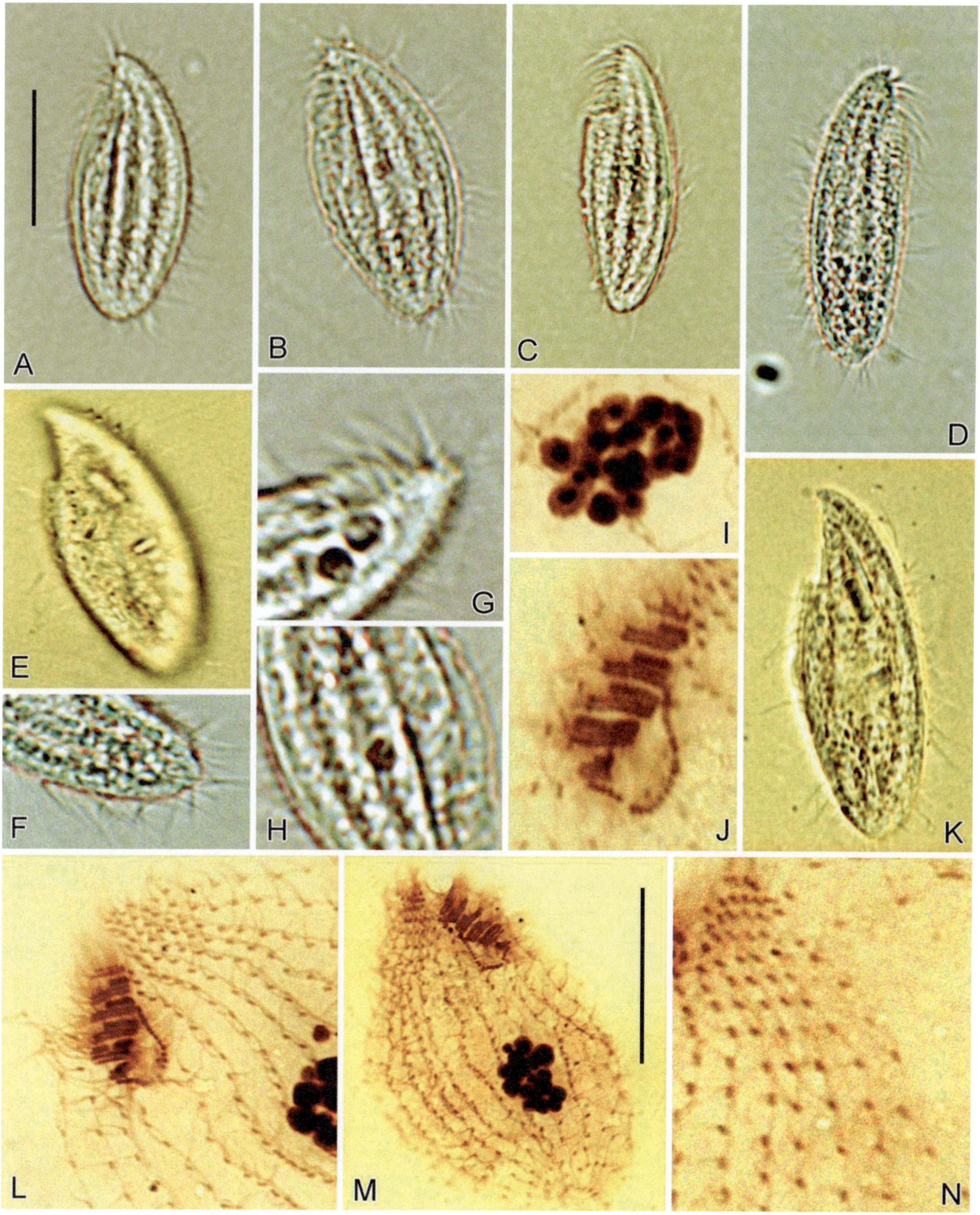


**Fig. 1B** *Protocruzia* cf. *contrax*
(**a–e**, **k**) Lateral views of live cells; (**f**) posterior portion of cell showing somatic cilia; (**g**) anterior portion of cell showing oral cilia; (**h**) ridges on cell surface; (**i**) **m**acronuclear segments; (**j**, **l**) oral apparatus; (**m**) ciliature and nuclei; (**n**) somatic kineties with dikinetids. Scale bar: 20 μm

## Reference

Song WB, Wilbert N (1997) Morphological investigation on some free living ciliates (Protozoa, Ciliophora) from China Sea with description of a new hypotrichous genus, *Hemigastrostyla* nov. gen. Arch Protistenk 148:413–444

Class Protocruziea Gao et al., 2016
Order Protocruziida Jankowski, 1980
Protocruziidae Jankowski, 1980
*Protocruzia granulosa* Kahl, 1933

**Ecological feature:** coastal water
**Water temperature:** 22.0 °C
**pH:** no data available
**Salinity:** 32.0‰

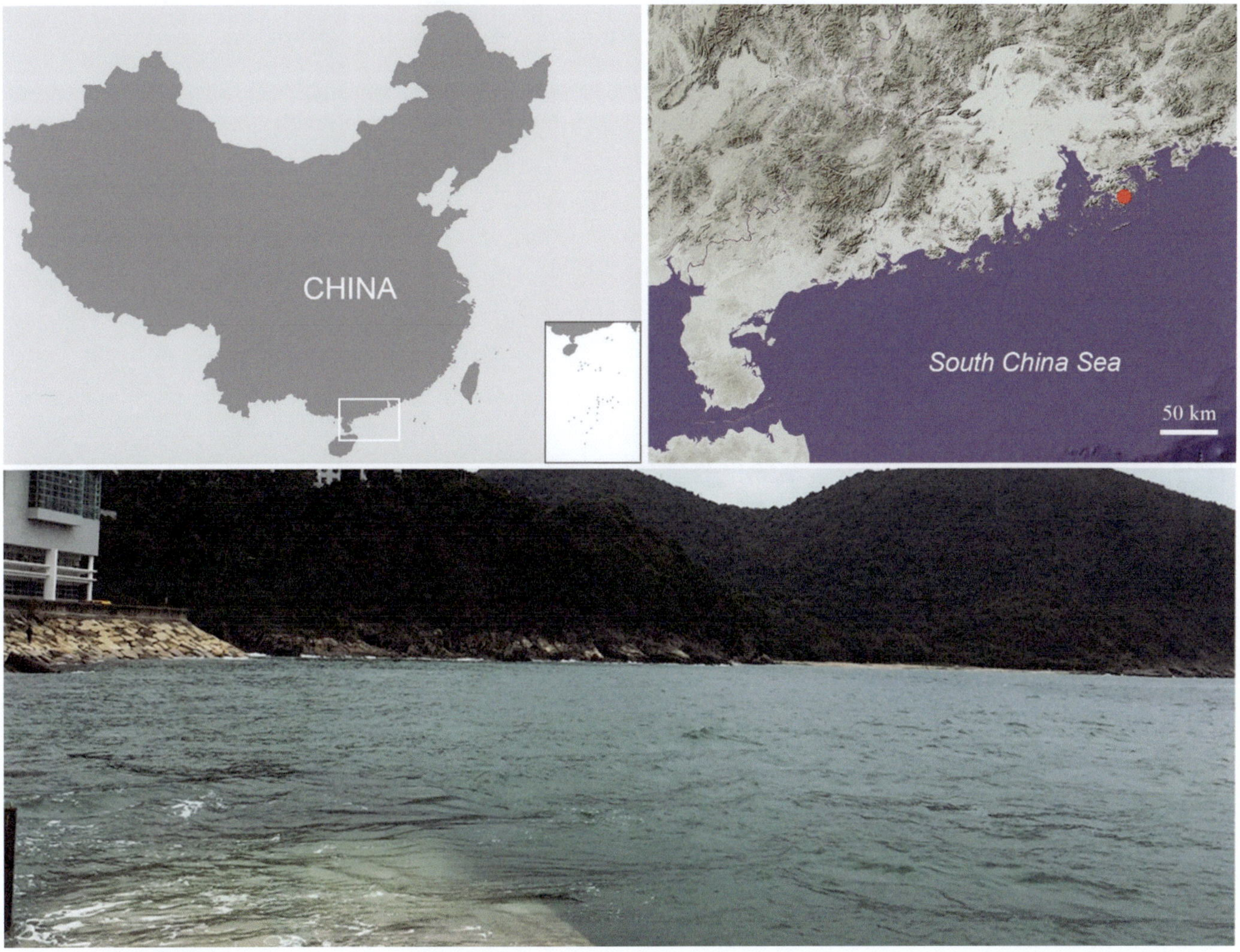


**Fig. 2A** Sample site

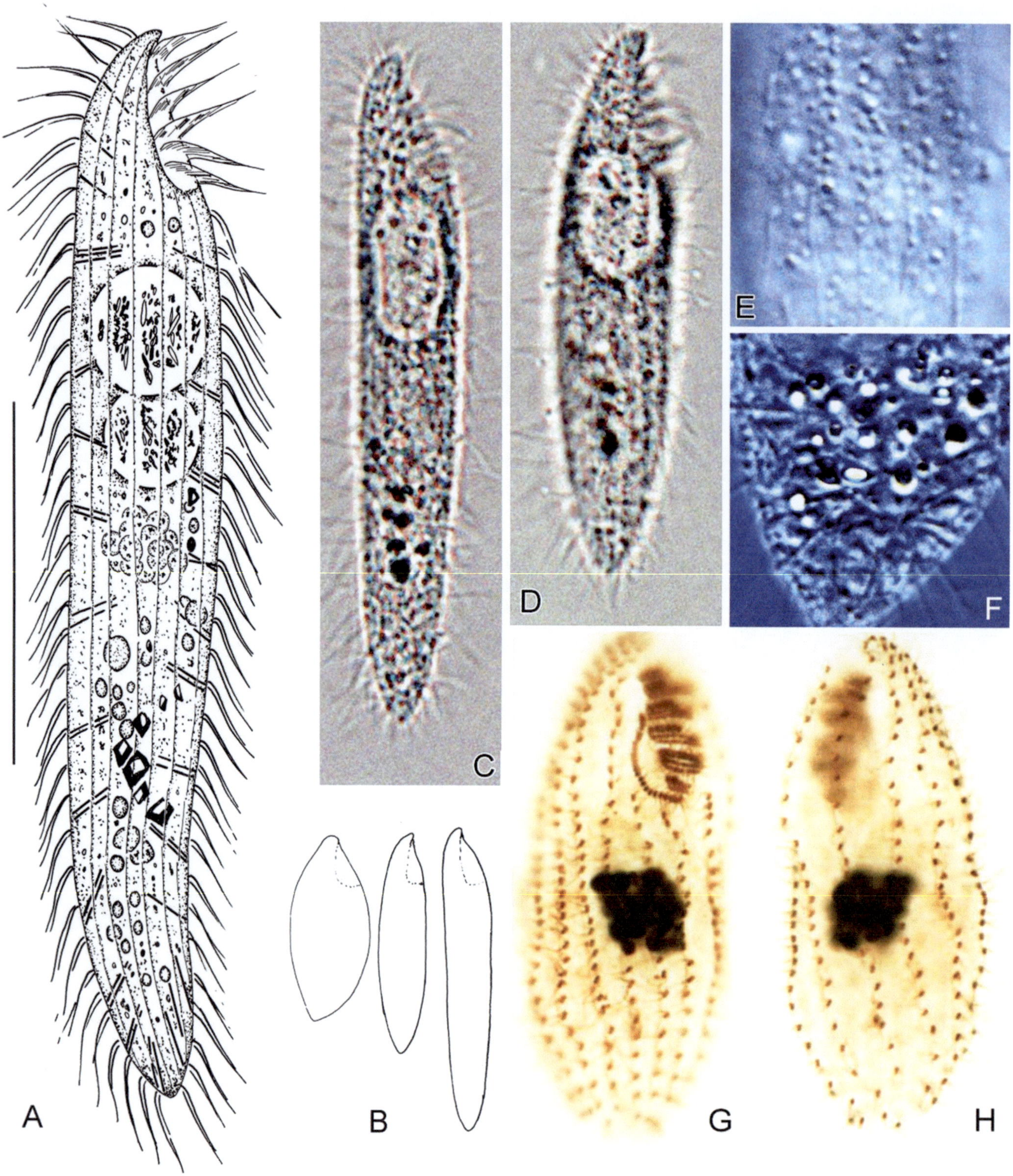


**Fig. 2B** *Protocruzia granulosa*
(**a**) Right view of live cell; (**b**) variation in shape and size of the same individual; (**c**, **d**) right views of the same cell; (**e**) cortical granules; (**f**) extrusomes; (**g**, **h**) right and left views of the same stained specimen, showing ciliature and nuclei. Scale bar: 30 μm

## Reference

Jiang J, Huang J, Al-Farraj SA, Lin X, Hu X (2017) Morphology and molecular phylogeny of two poorly known species of *Protocruzia* (Ciliophora: Protocruziida). J Eukaryot Microbiol 64:144–152

Class Protocruziea Gao et al., 2016
Order Protocruziida Jankowski, 1980
Protocruziidae Jankowski, 1980
*Protocruzia tuzeti* Villeneuve-Brachon, 1940

**Ecological feature:** coastal water
**Water temperature:** 21.0 °C
**pH:** no data available
**Salinity:** 29.0‰

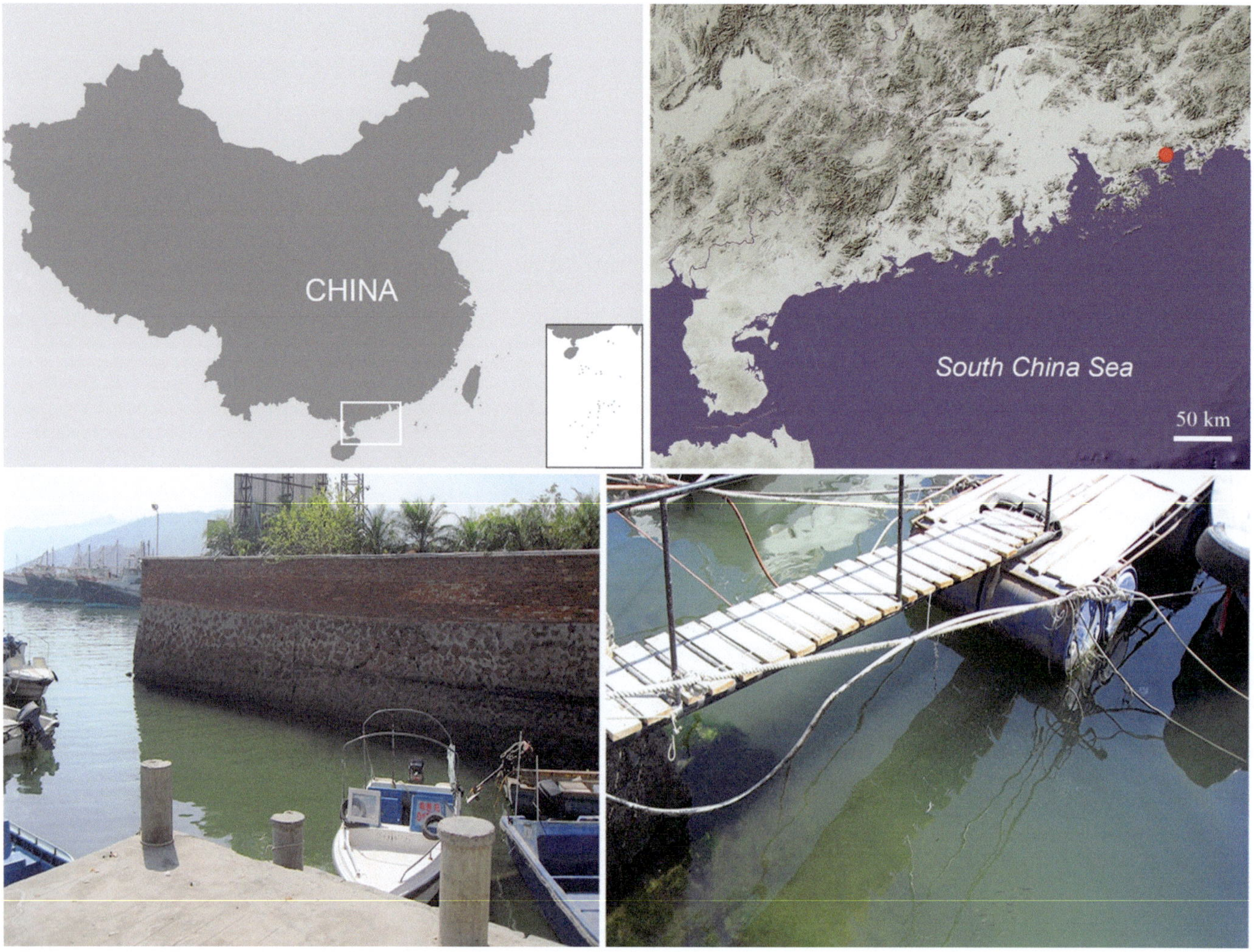


**Fig. 3A** Sample site

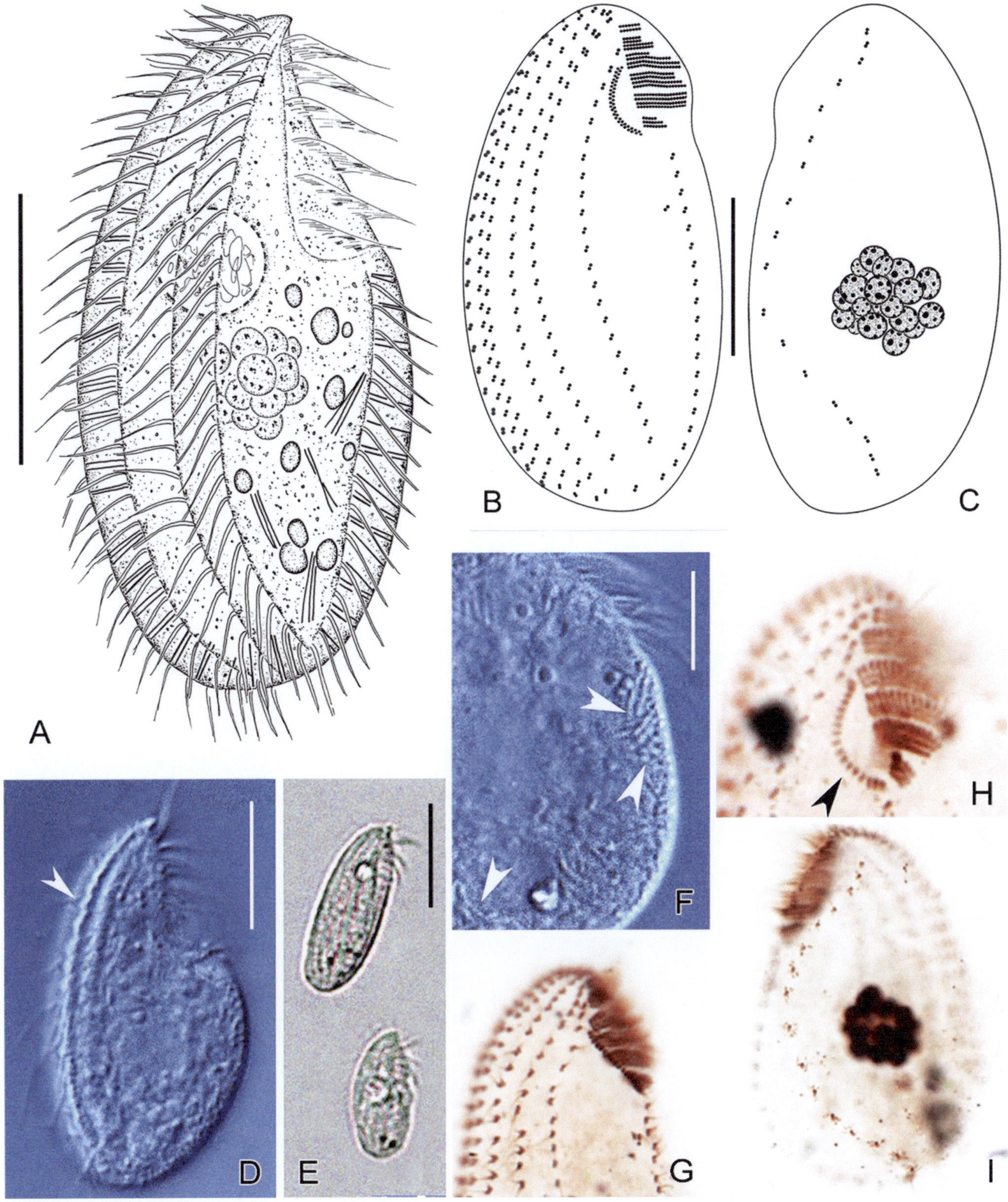


**Fig. 3B** *Protocruzia tuzeti*
(**a**, **d**, **e**) Right-lateral views of live cells, arrowhead shows ciliary furrow; (**b**, **c**, **i**) right (**b**) and left (**c**, **i**) lateral views of ciliature; (**f**) extrusomes (arrowheads); (**g**, **h**) oral apparatus, arrowhead marks paroral membrane. Scale bars: 15 μm (**a–c**, **d**); 30 μm (**e**); 5 μm (**f**)

## Reference

Jiang J, Huang J, Al-Farraj SA, Lin X, Hu X (2017) Morphology and molecular phylogeny of two poorly known species of *Protocruzia* (Ciliophora: Protocruziida). J Eukaryot Microbiol 64:144–152

# Class Spirotrichea

Xiaozhong Hu, Xiaofeng Lin, and Weibo Song

Class Spirotrichea Bütschli, 1889
Order Choreotrichida Small & Lynn, 1985
Family Strobilidiidae Kahl in Doflein & Reichenow, 1929
*Pelagostrobilidium minutum* Liu et al., 2012

**Ecological feature:** intertidal zone
**Water temperature:** 21.7 °C
**pH:** 8.2
**Salinity:** 33.3‰

X. Hu (✉)
Fisheries College, Ocean University of China,
Qingdao, Shandong, China
e-mail: xiaozhonghu@ouc.edu.cn

X. Lin
College of Life Science, South China Normal University,
Guangzhou, Guangdong, China
e-mail: xlin@oucscnu.edu.cn

W. Song
Institute of Evolution & Marine Biodiversity,
Ocean University of China, Qingdao, Shandong, China
e-mail: wsong@ouc.edu.cn


X. Hu et al. (eds.), *Ciliate Atlas: Species Found in the South China Sea*,
https://doi.org/10.1007/978-981-13-5901-9_11

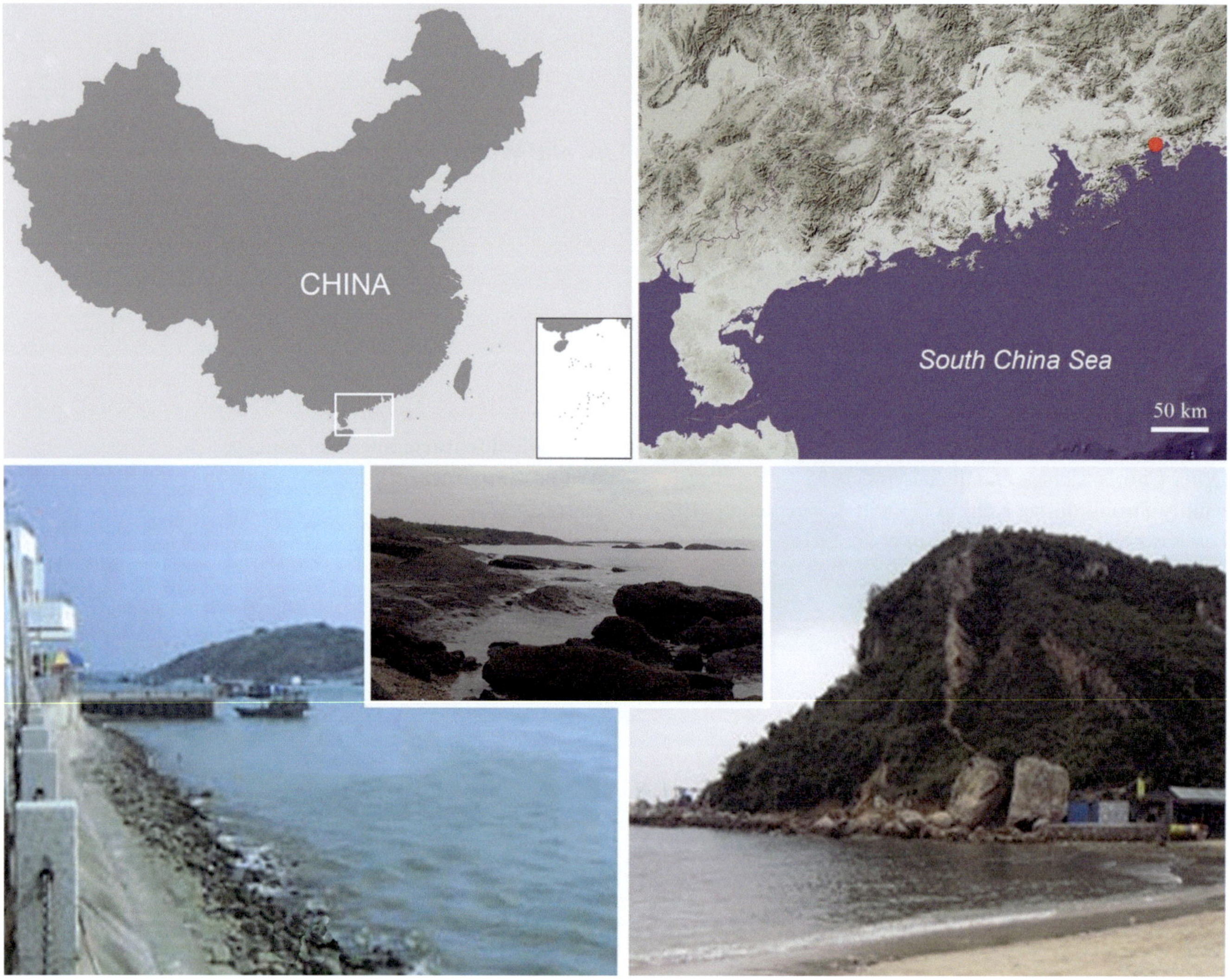


**Fig. 1A** Sample site

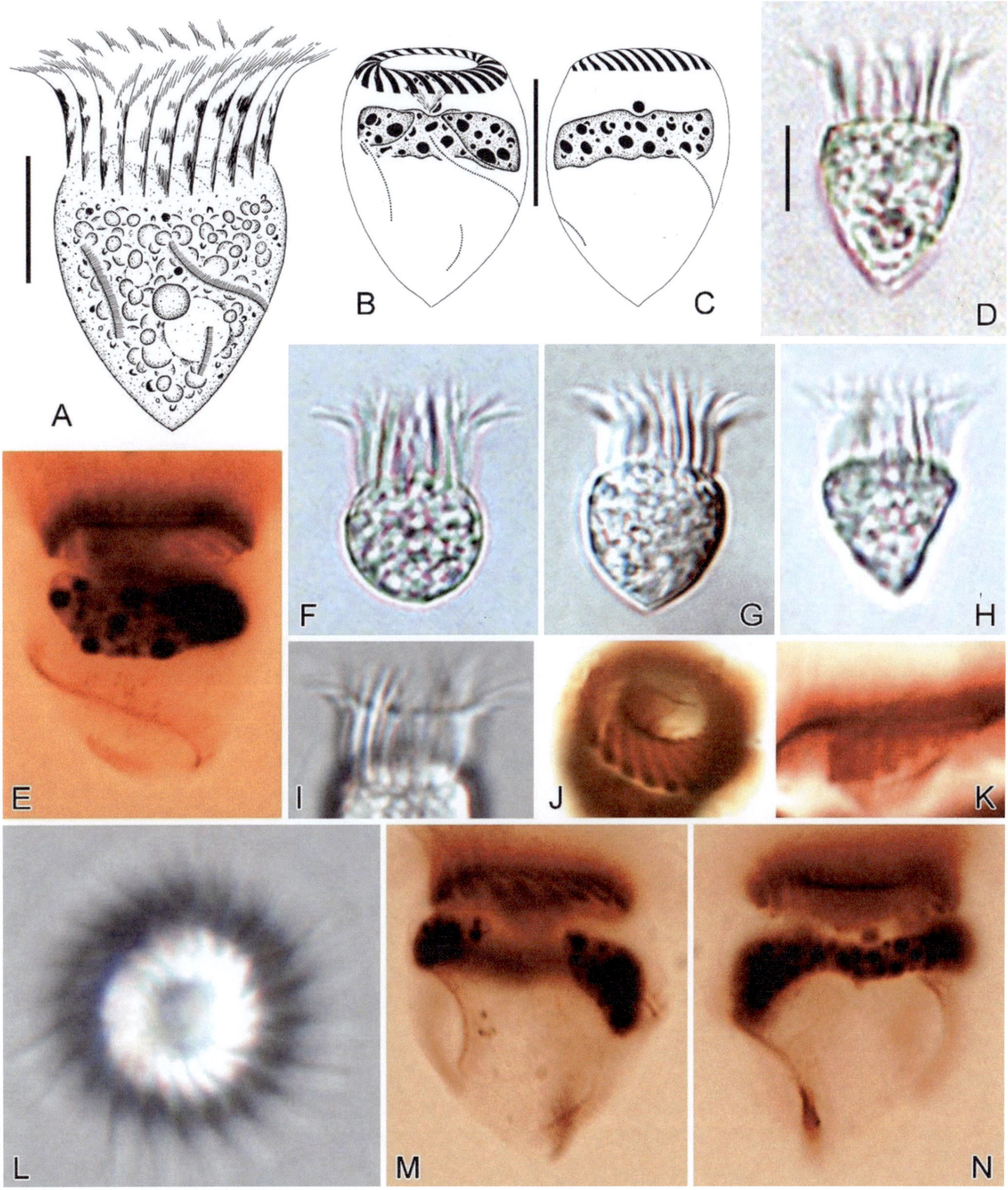


**Fig. 1B** *Pelagostrobilidium minutum*
(**a**, **d**) Lateral views of live cells. (**b**, **c**) Ciliature of the same specimen. (**e**, **m**, **n**) Ciliature and nuclear apparatus. (**f**–**g**) Different body shapes. (**i**) External membranelles in vivo. (**j**) Apical-lateral view to show oral membranelles. (**k**) Internal membranelles. (**l**) Apical view to show oral membranelles. Scale bars: 10 μm

## Reference

Liu W, Yi Z, Lin X, Warren A, Song WB (2012) Phylogeny of three choreotrich genera (Protozoa, Ciliophora, Spirotrichea), with morphological, morphogenetic and molecular investigations on three strobilidiid species. Zool Scr 41:417–434

Class Spirotrichea Bütschli, 1889
Order Choreotrichida Small & Lynn, 1985
Family Strobilidiidae Kahl in Doflein & Reichenow, 1929
*Pelagostrobilidium paraepacrum* Liu et al., 2012

**Ecological feature:** offshore water
**Water temperature:** 20.1 °C
**pH:** 8.1
**Salinity:** 27.8‰

**Fig. 2A** Sample site

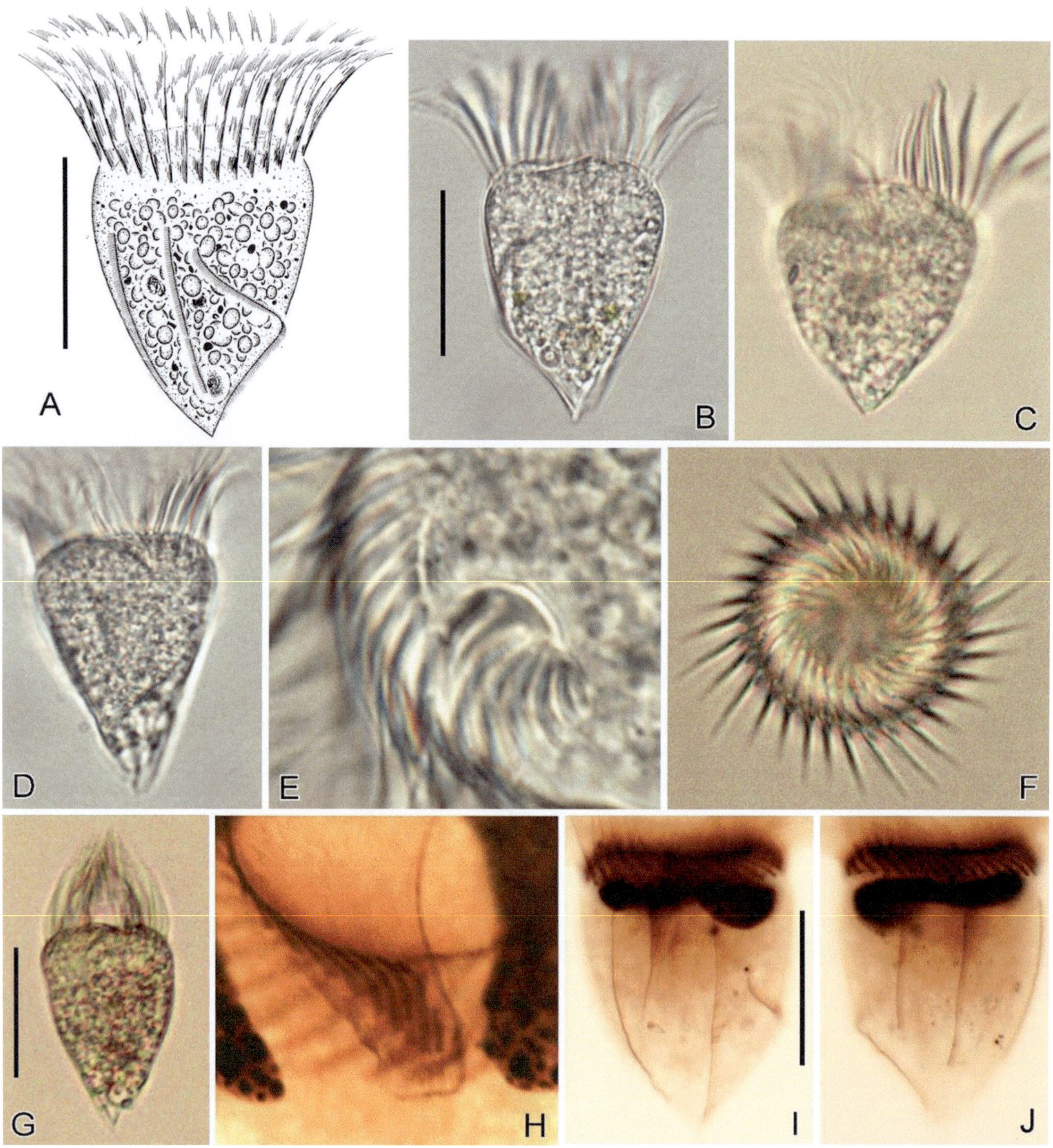


**Fig. 2B** *Pelagostrobilidium paraepacrum*
(**a–d**) Lateral views of live cells showing different body shapes. (**e**) Detail of oral field showing internal membranelles. (**f**) Apical view, to show external membranelles. (**g**) Resting specimen in vivo. (**h**) Internal membranelles and endoral membrane. (**i**, **j**) Lateral views of ciliature. Scale bars: 40 μm

## Reference

Liu W, Yi Z, Lin X, Warren A, Song WB (2012) Phylogeny of three choreotrich genera (Protozoa, Ciliophora, Spirotrichea), with morphological, morphogenetic and molecular investigations on three strobilidiid species. Zool Scr 41:417–434

Class Spirotrichea Bütschli, 1889
Order Choreotrichida Small & Lynn, 1985
Family Strobilidiidae Kahl in Doflein & Reichenow, 1929
*Pelagostrobilidium simile* Song & Bradbury, 1998

**Ecological feature:** offshore water
**Water temperature:** 29.7 °C
**pH:** 8.4
**Salinity:** 28.7‰

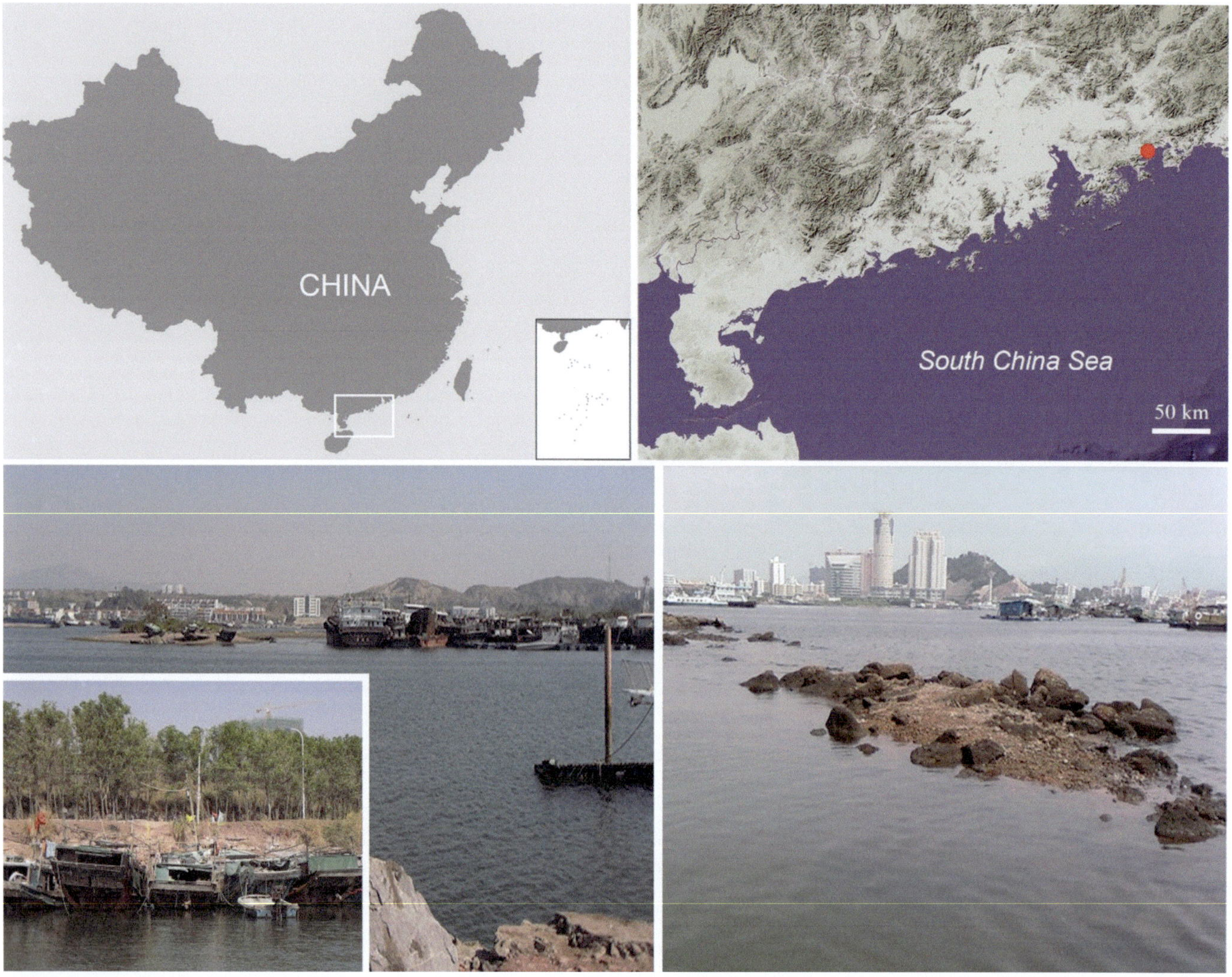


**Fig. 3A** Sample site

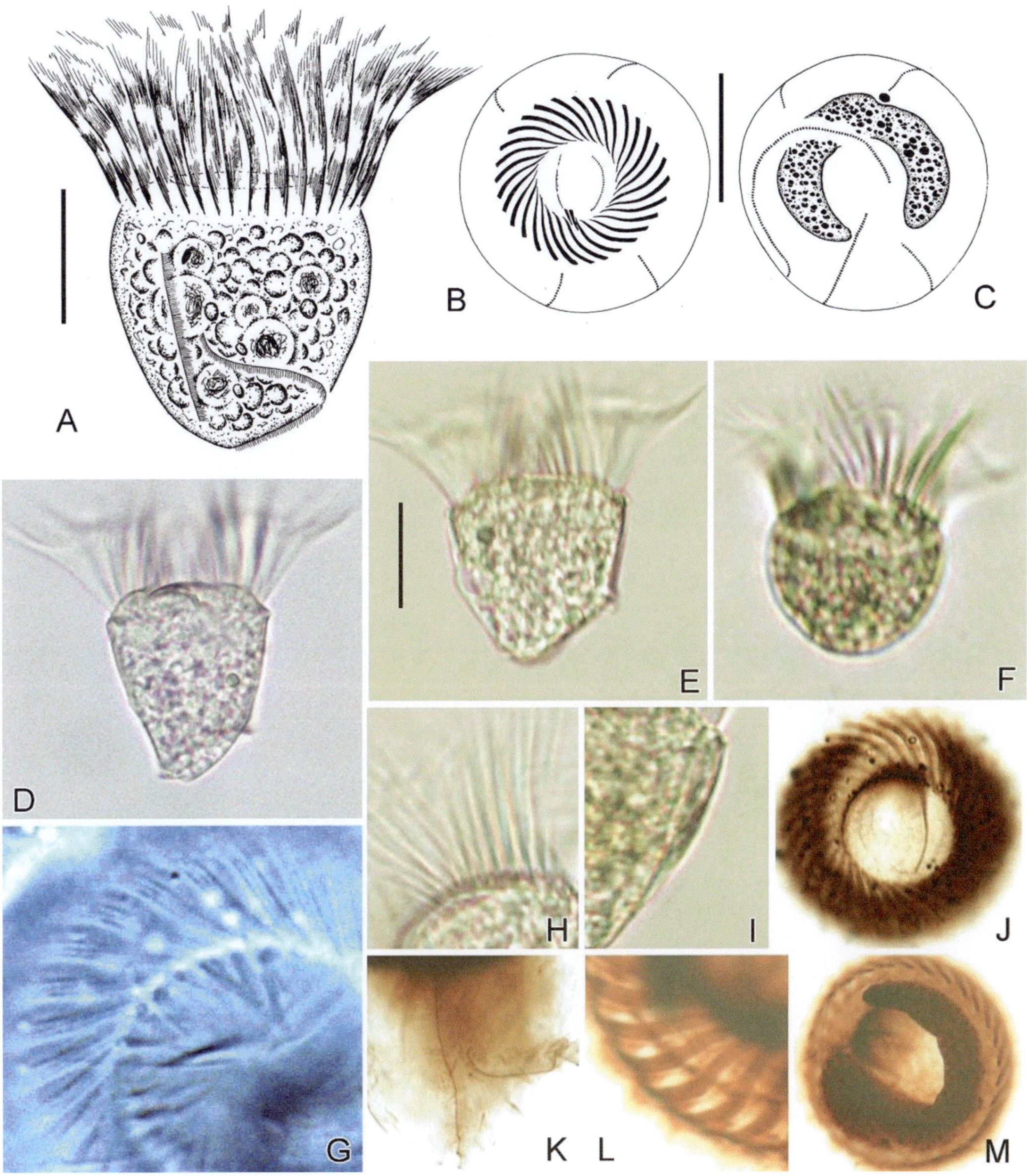


**Fig. 3B** *Pelagostrobilidium simile*
(**a**, **d**–**f**) Lateral views of live cells showing different body shapes. (**b**, **c**) Apical (**b**) and aboral (**c**) views of ciliature. (**g**, **h**, **l**) External membranelles in vivo (**g**, **h**) and after protargol staining (**l**). (**i**) Detail of cell surface showing the kinetal lip. (**j**) Apical view to show oral membranelles. (**k**) Lateral view to show somatic kineties. (**m**) Macronucleus. Scale bars: 20 μm

## Reference

Song WB, Bradbury PC (1998) Studies on some new and rare reported marine planktonic ciliates (Ciliophora: Oligotricha) from coastal waters in North China. J Mar Biol Ass UK 78:767–794

Class Spirotrichea Bütschli, 1889
Order Choreotrichida Small & Lynn, 1985
Family Strobilidiidae Kahl in Doflein & Reichenow, 1929
*Rimostrombidium veniliae* Montagnes & Taylor, 1994

**Ecological feature:** offshore water
**Water temperature:** 24.5 °C
**pH:** 8.4
**Salinity:** 28.2‰

**Fig. 4A** Sample site

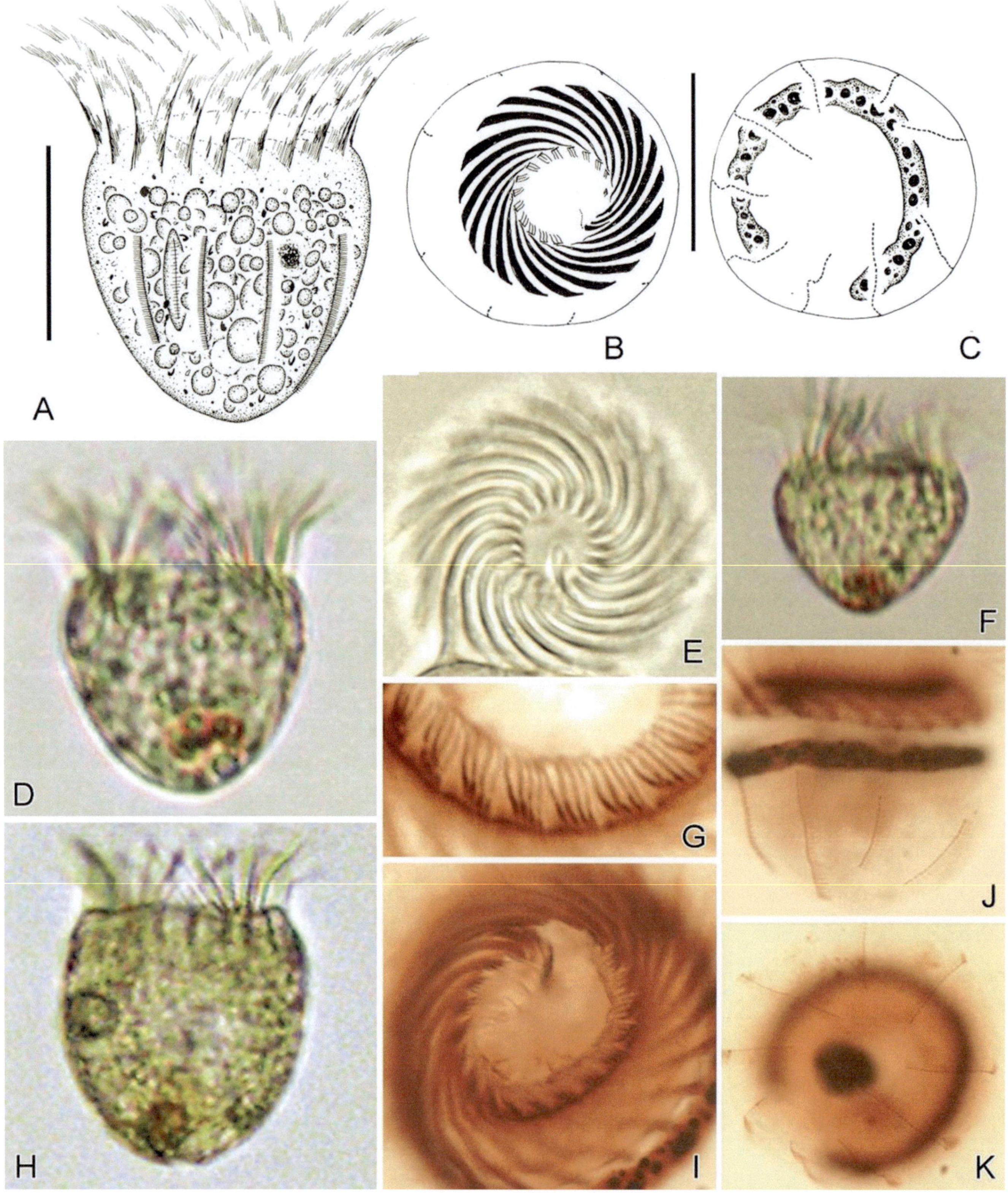


**Fig. 4B** *Rimostrombidium veniliae*
(**a**, **d**, **f**, **h**) Lateral views of live cells showing different body shapes. (**b**, **c**) Apical (**b**) and aboral (**c**) views of ciliature. (**e**, **i**) Apical views in vivo (**e**) and after protargol staining (**i**) to show oral membranelles. (**g**) Detail of oral field to show fiber systems associated with external membranelles. (**j**, **k**) Lateral (**j**) and aboral (**k**) views of somatic kineties. Scale bars: 30 μm

## Reference

Montagnes DJS, Taylor FJR (1994) The salient features of five marine ciliates in the class Spirotrichea (Oligotrichia), with notes on their culturing and behaviour. J Eukaryot Microbiol 41:569–586

Class Spirotrichea Bütschli, 1889
Order Choreotrichida Small & Lynn, 1985
Family Strombidinopsidae Small & Lynn, 1985
*Parastrombidinopsis minima* Tsai et al., 2008

**Ecological feature:** estuary area
**Water temperature:** 29.9 °C
**pH:** 8.8
**Salinity:** 18.2‰

**Fig. 5A** Sample site

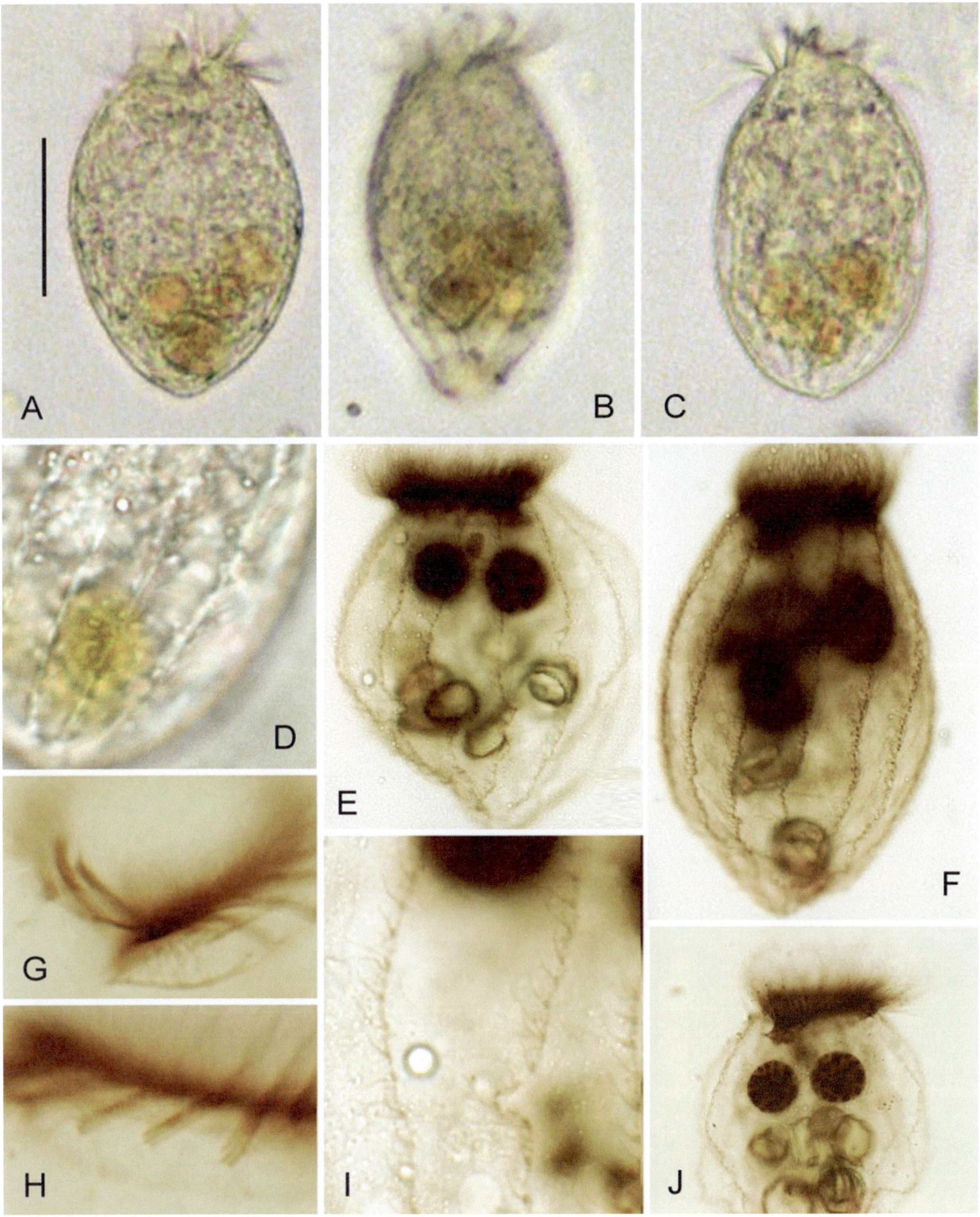


**Fig. 5B** *Parastrombidinopsis minima*
(**a–c**) Lateral views of live individuals. (**d**) Somatic cilia. (**e**, **f**) Lateral views of ciliature. (**g**) Internal membranelles. (**h**) External membranelles. (**i**) Somatic kineties. (**j**) Macronucleus. Scale bar: 30 μm

## Reference

Tsai S, Xu D, Chung C, Chiang K (2008) *Parastrombidinopsis minima* n. sp. (Ciliophora: Oligotrichia) from the coastal waters of northeastern Taiwan: morphology and small subunit ribosomal DNA sequence. J Eukaryot Microbiol 55:567–573

Class Spirotrichea Bütschli, 1889
Order Choreotrichida Small & Lynn, 1985
Family Strombidinopsidae Small & Lynn, 1985
*Parastrombidinopsis* sp.

**Ecological feature:** offshore water
**Water temperature:** 29.2 °C
**pH:** 8.0
**Salinity:** 2.5‰

**Fig. 6A** Sample site

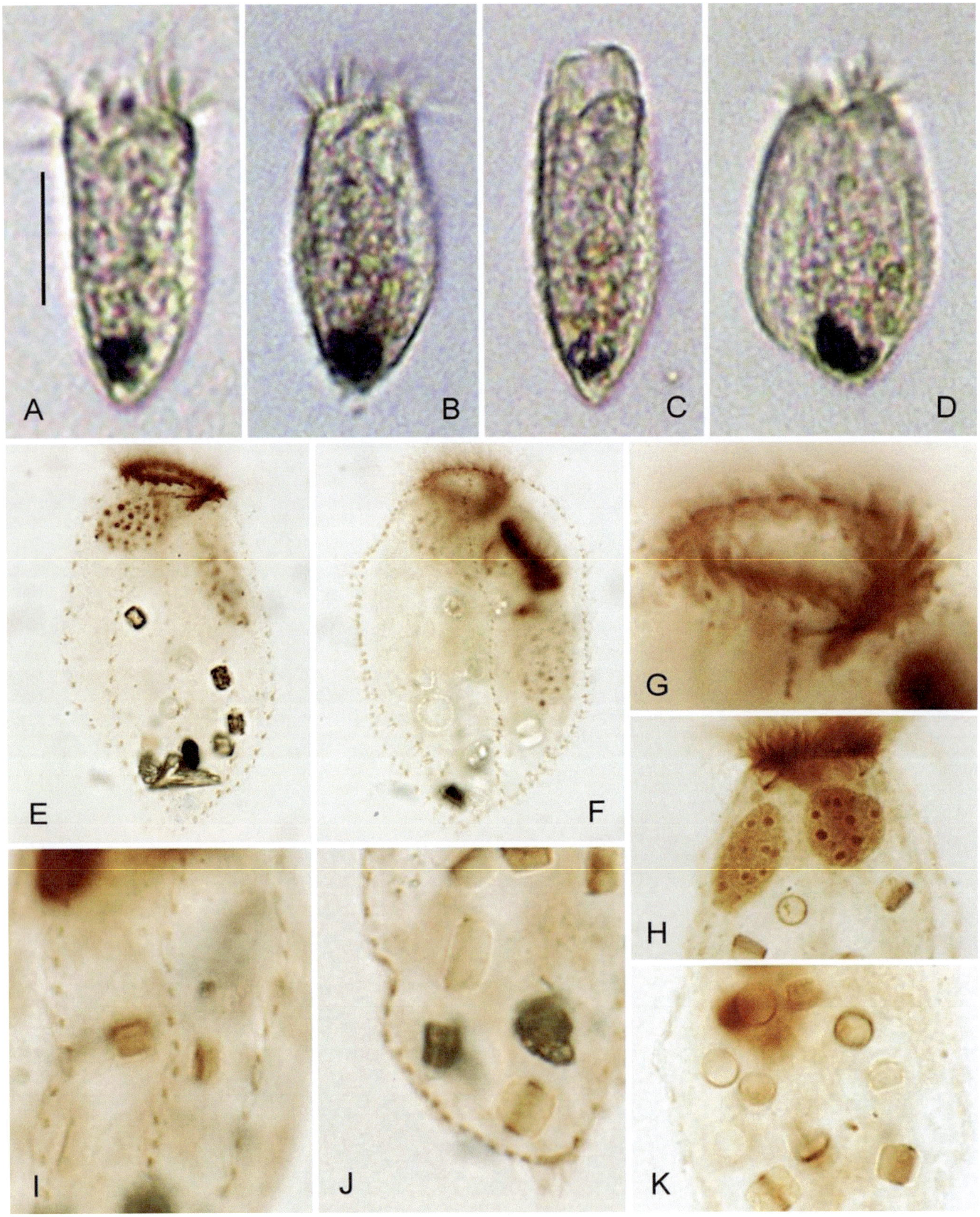


**Fig. 6B** *Parastrombidinopsis* sp.
(**a–d**) Lateral views of live individuals. (**e**, **f**) Lateral views of ciliature. (**g**) Oral apparatus. (**h**) Macronuclei. (**i**, **j**) Detail of somatic kineties. (**k**) Ingested diatom food. Scale bar: 30 μm

## References

Kim J, Jeong H, Strüder-Kypke M, Lynn D, Kim S, Kim J, Lee S (2005) *Parastrombidinopsis shimi* n. gen., n. sp. (Ciliophora: Choreotrichia) from the coastal waters of Korea: morphology and small subunit ribosomal DNA sequence. J Eukaryot Microbiol 52:514–522

Song WB, Warren A, Hu X (2009) Free-living Ciliates in the Bohai and Yellow Seas, China. Science Press, Beijing

Class Spirotrichea Bütschli, 1889
Order Choreotrichida Small & Lynn, 1985
Family Strombidinopsidae Small & Lynn, 1985
*Strombidinopsis batos* Lynn et al., 1991

**Ecological feature:** mangrove
**Water temperature:** 27.0 °C
**pH:** 8.2
**Salinity:** 17.0‰

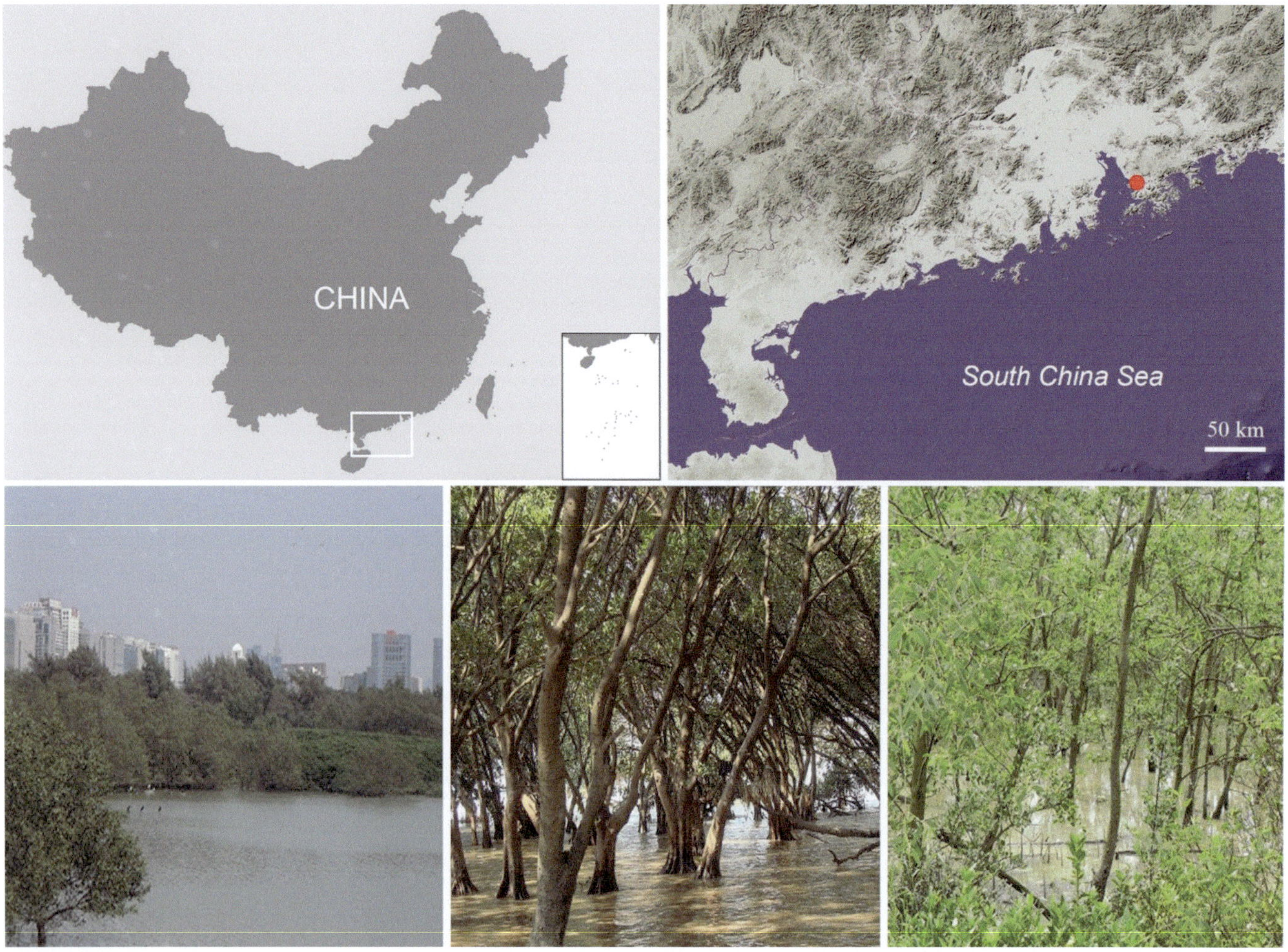


**Fig. 7A** Sample site

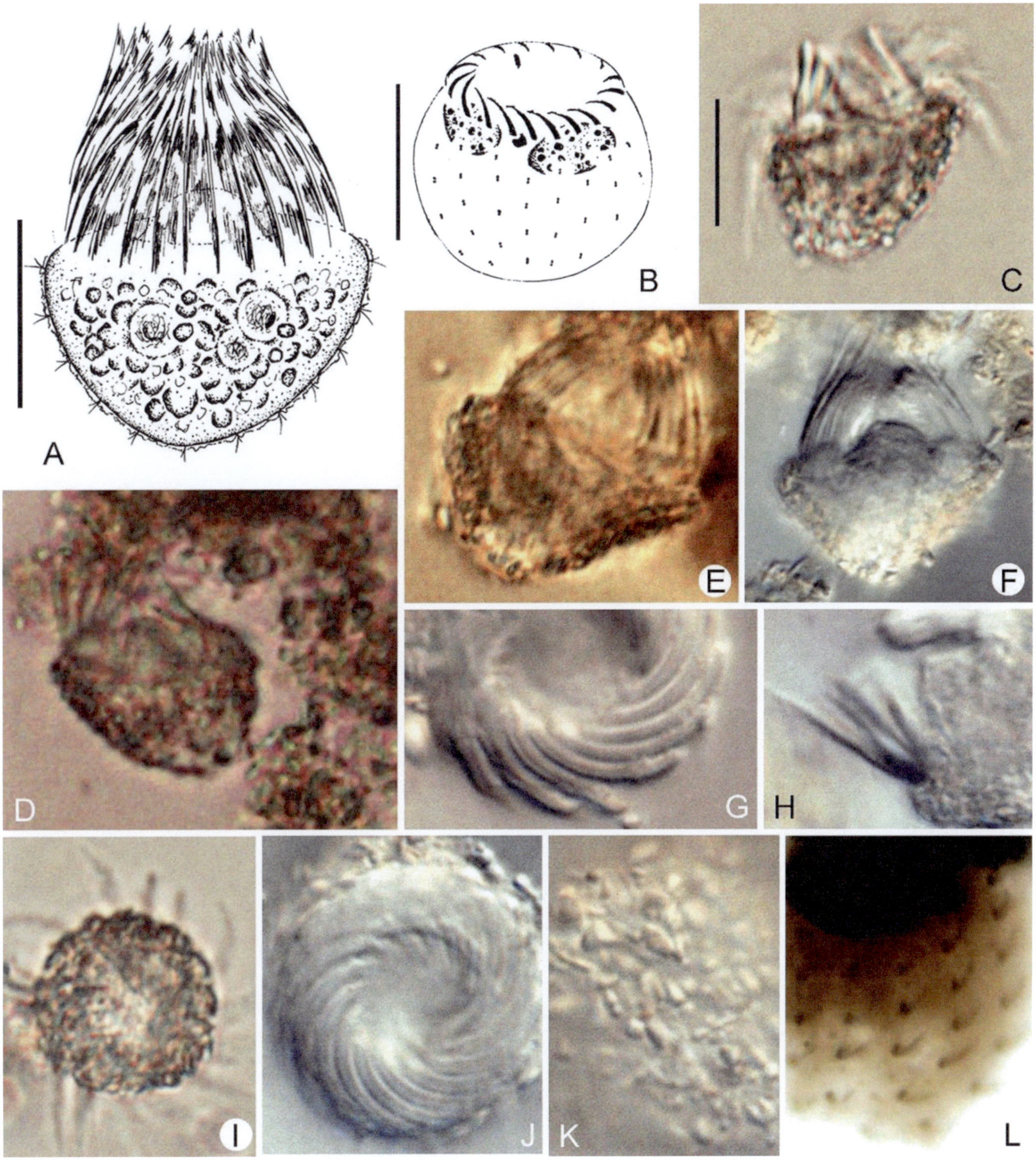


**Fig. 7B** *Strombidinopsis batos*
(**a**) Lateral view of resting specimen. (**b**) Lateral view of ciliature. (**c**) Lateral view of swimming specimen. (**d–f**, **h**) Lateral views of specimens crawling on the sediment. (**g**, **j**) Apical views in vivo to show external membranelles. (**i**, **k**) Cell convex, to show the mineral envelope. (**l**) Somatic kineties. Scale bars: 20 μm

## Reference

Lynn D, Montagnes D, Dale T, Gilron G, Strom S (1991) A reassessment of the genus *Strombidinopsis* (Ciliophora: Choreotrichida) with description of four new planktonic species and remarks on its taxonomy and phylogeny. J Mar Biol Ass UK 71:597–612

Class Spirotrichea Bütschli, 1889
Order Choreotrichida Small & Lynn, 1985
Family Strombidinopsidae Small & Lynn, 1985
*Strombidinopsis cheshiri* Montagnes & Taylor, 1994

**Ecological feature:** intertidal zone
**Water temperature:** 30.2 °C
**pH:** 8.3
**Salinity:** 31.8‰

**Fig. 8A** Sample site

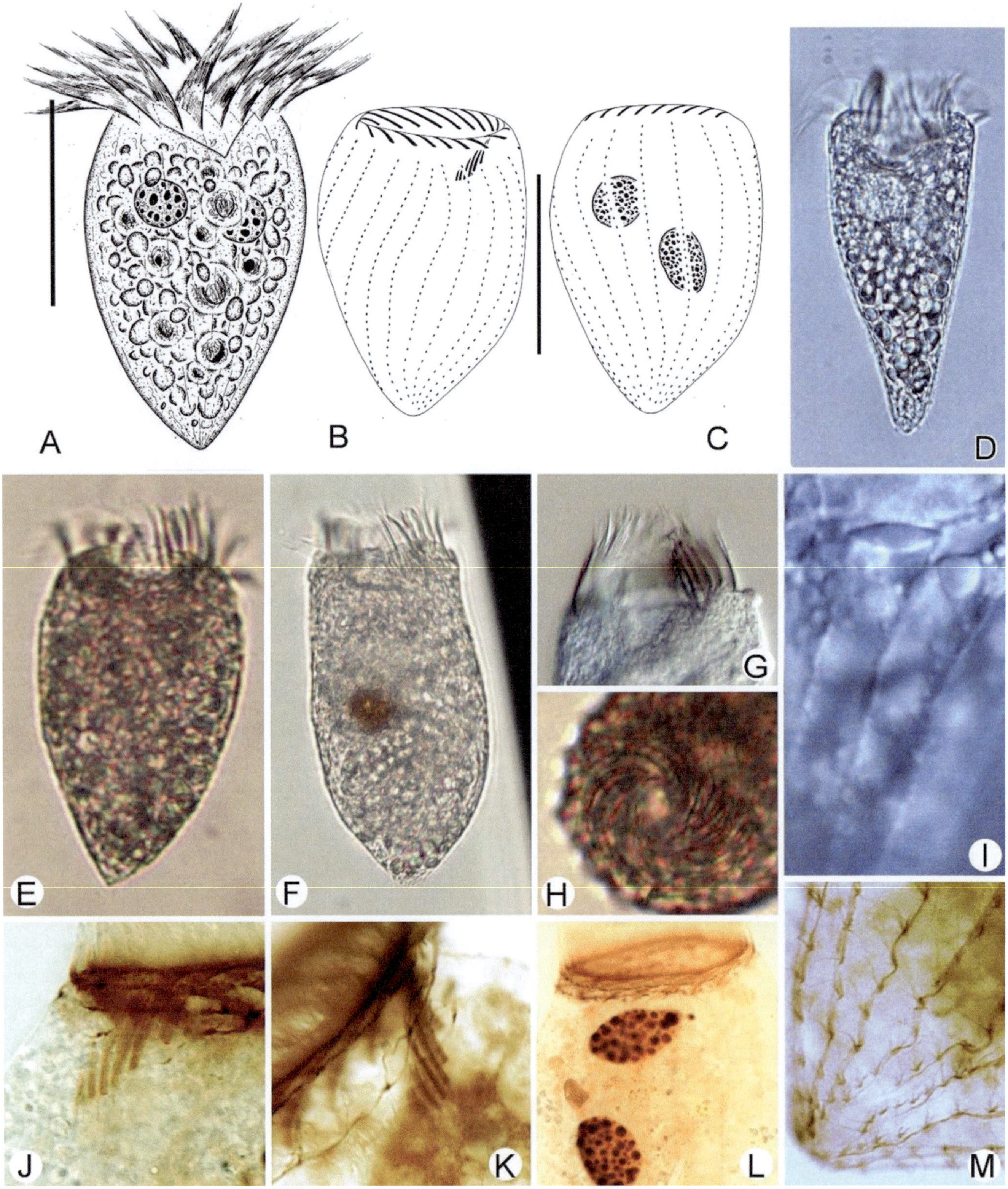


**Fig. 8B** *Strombidinopsis cheshiri*
(**a**, **d–f**) Lateral views of live cells. (**b**, **c**) Lateral views of ciliature. (**g**) Anterior portion of live cell, to show oral cilia when cell rests. (**h**) Apical view of live cell to show external membranelles. (**i**) Somatic cilia. (**j**, **k**) Internal membranelles. (**l**) Macronuclei. (**m**) Posterior portion of somatic kineties. Scale bars: 80 μm

## References

Montagnes D, Taylor F (1994) The salient features of five marine ciliates in the class Spirotrichea (Oligotrichia), with notes on their culturing and behaviour. J Eukaryot Microbiol 41:569–586

Song WB, Warren A, Hu X (2009) Free-living Ciliates in the Bohai and Yellow Seas, China. Science Press, Beijing

Class Spirotrichea Bütschli, 1889
Order Choreotrichida Small & Lynn, 1985
Family Strombidinopsidae Small & Lynn, 1985
*Strombidinopsis minima* (Gruber, 1884) Song & Bradbury, 1998

**Ecological feature:** mangrove
**Water temperature:** 20.8 °C
**pH:** 7.7
**Salinity:** 21.3‰

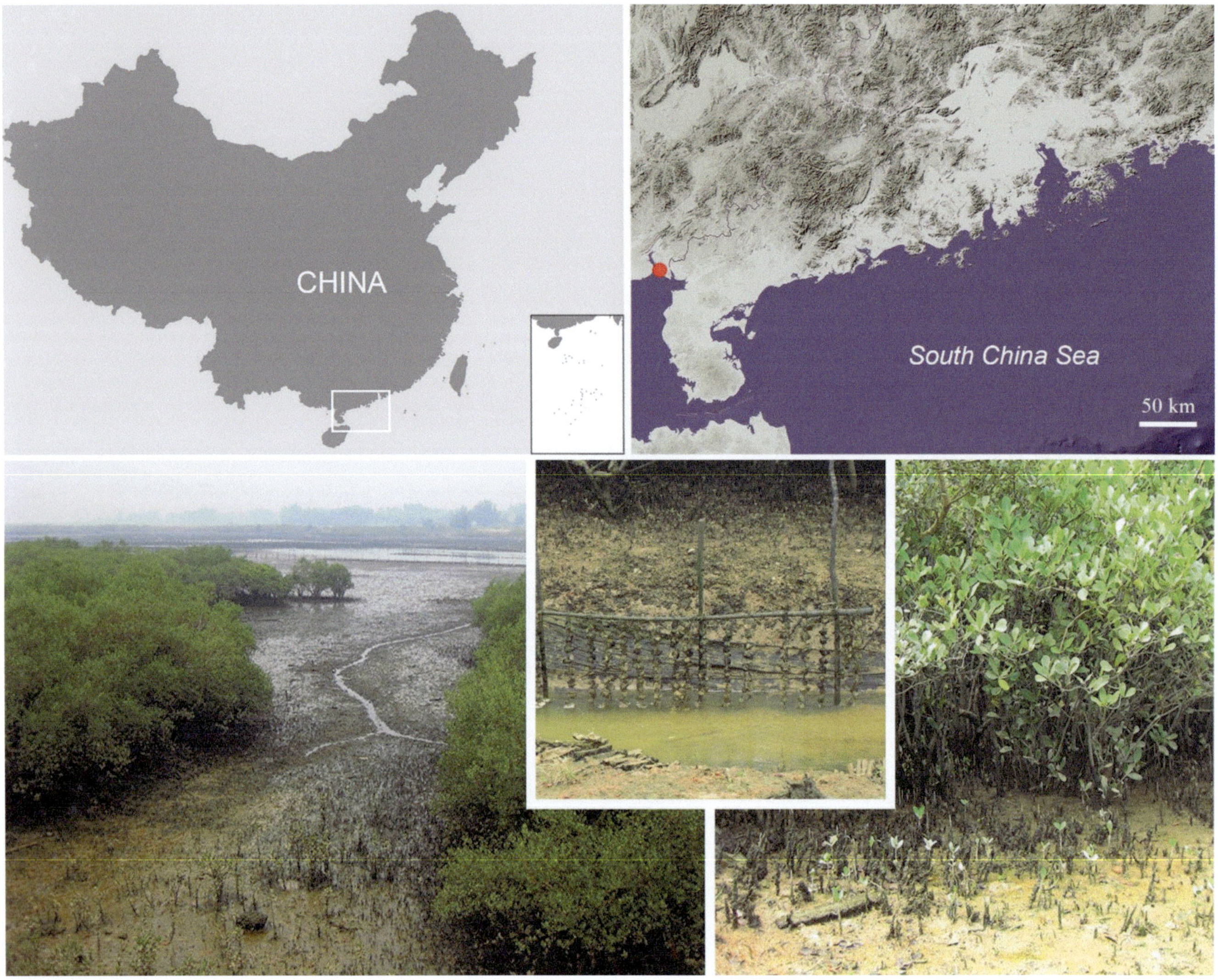


**Fig. 9A** Sample site

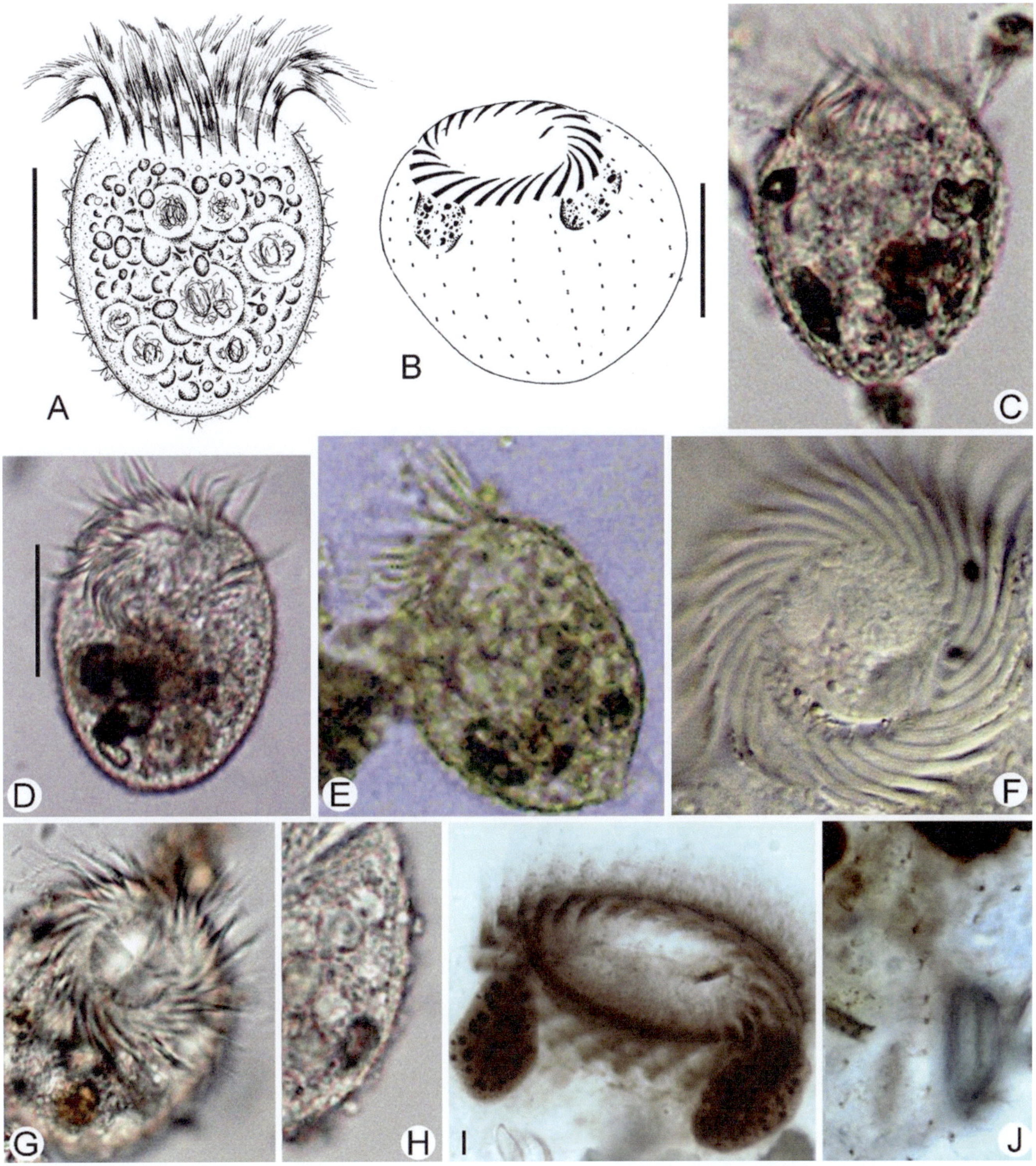


**Fig. 9B** *Strombidinopsis minima*
(**a**, **c**–**e**) Lateral views of live cells. (**b**) Lateral-apical view of ciliature. (**f**) Apical view of live cell, to show external membranelles. (**g**) Lateral-apical view of live cell, showing external membranelles. (**h**) Cell convex, showing mineral envelope. (**i**) Apical-ventral view, to show adoral membranelles and macronucleus. (**j**) Part of somatic kineties. Scale bars: 20 μm

## References

Agatha S (2003) Redescription of *Strombidinopsis minima* (Gruber, 1884) Lynn et al., 1991 (Protozoa, Ciliophora), with notes on its ontogenesis and distribution. Eur J Protistol 39:233–244

Song WB, Bradbury PC (1998) Studies on some new and rare reported marine planktonic ciliates (Ciliophora: Oligotricha) from coastal waters in North China. J Mar Biol Ass UK 78:767–794

Class Spirotrichea Bütschli, 1889
Order Choreotrichida Small & Lynn, 1985
Family Strombidinopsidae Small & Lynn, 1985
*Strombidinopsis sinicum* Liu et al., 2016

**Ecological feature:** mangrove
**Water temperature:** 26.0 °C
**pH:** 8.9
**Salinity:** 25.9‰

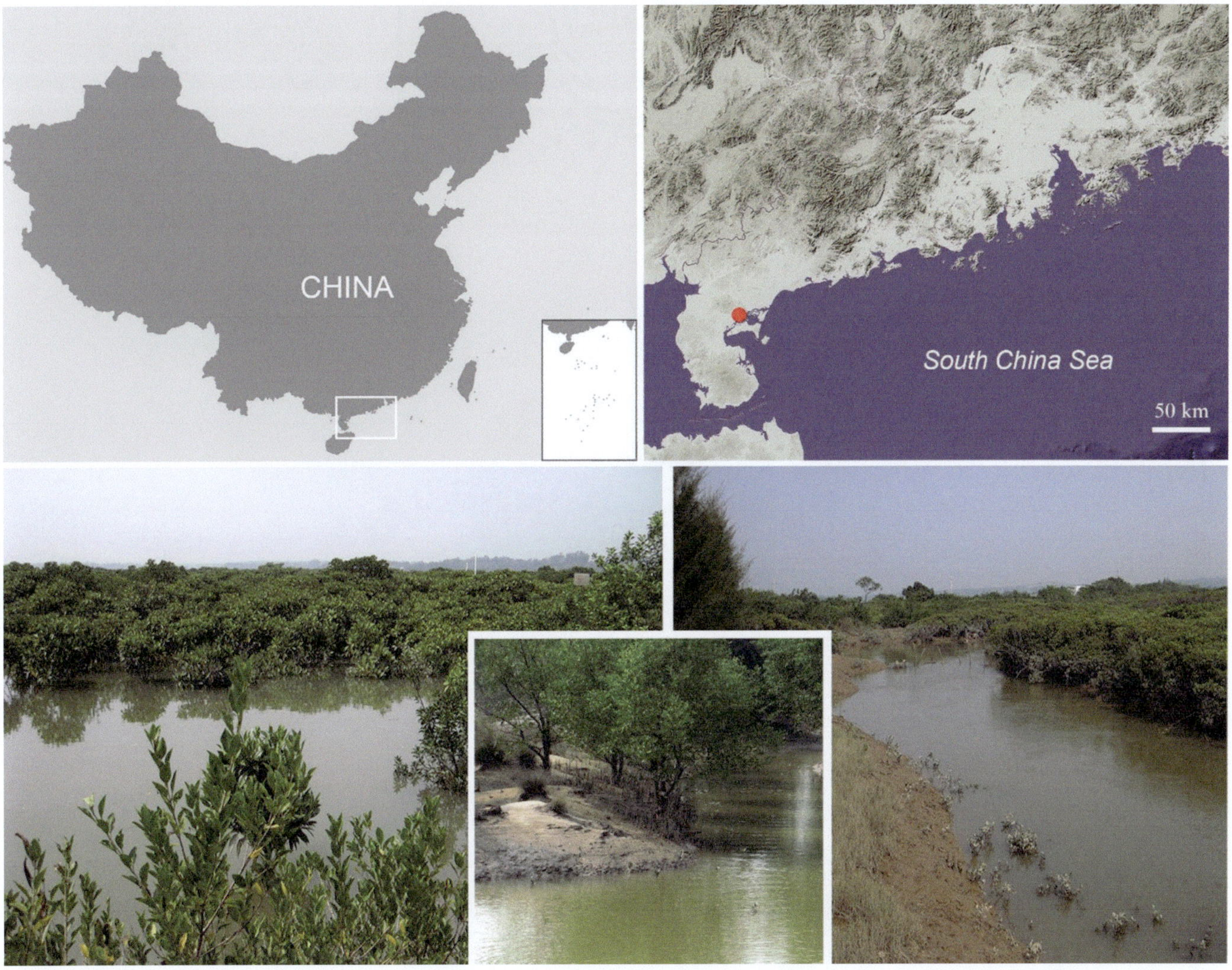


**Fig. 10A** Sample site

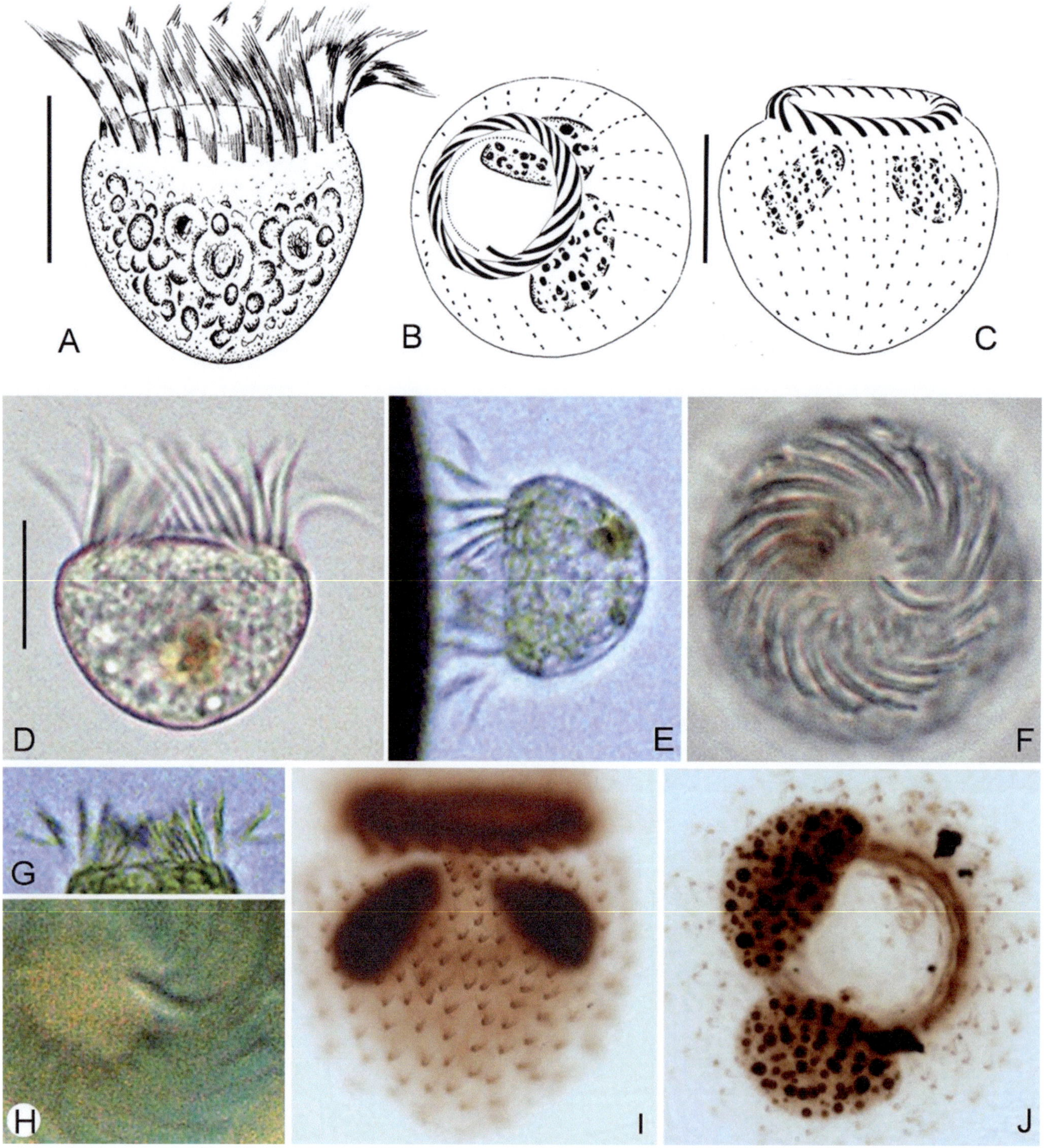


**Fig. 10B** *Strombidinopsis sinicum*
(**a**, **d**) Lateral views of live cells. (**b**, **c**) Apical (**b**) and lateral (**c**) views of ciliature. (**e**) A live cell crawling on sediment. (**f**, **h**) Apical view of live cells showing external membranelles. (**g**) Anterior portion of live cell showing external membranelles. (**i**) Lateral view of ciliature. (**j**) Apical view of stained specimen showing ciliature and macronuclei. Scale bars: 30 μm

## Reference

Liu W, Xu D, Ma H, Al-Farraj SA, Warren A, Yi Z (2016) Taxonomy and molecular systematics of three oligotrich (s. l.) ciliates including descriptions of two new species, *Strombidium guangdongense* sp. nov. and *Strombidinopsis sinicum* sp. nov. (Protozoa, Ciliophora). Syst Biodiver 14:452–465

Class Spirotrichea Bütschli, 1889
Order Choreotrichida Small & Lynn, 1985
Family Strombidinopsidae Small & Lynn, 1985
*Strombidinopsis* sp.

**Ecological feature:** mangrove
**Water temperature:** 34.2 °C
**pH:** 9.0
**Salinity:** 20.0‰

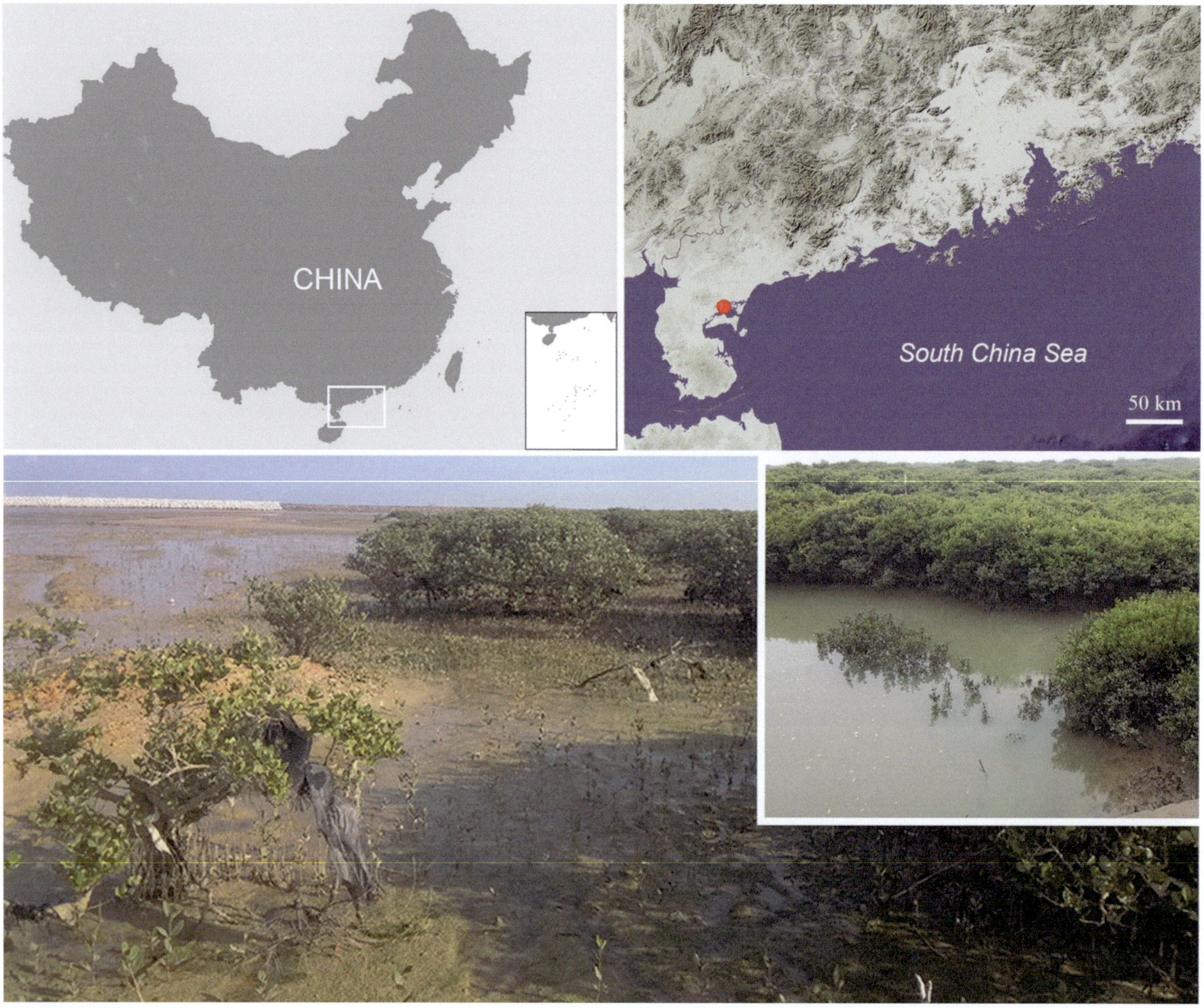


**Fig. 11A** Sample site

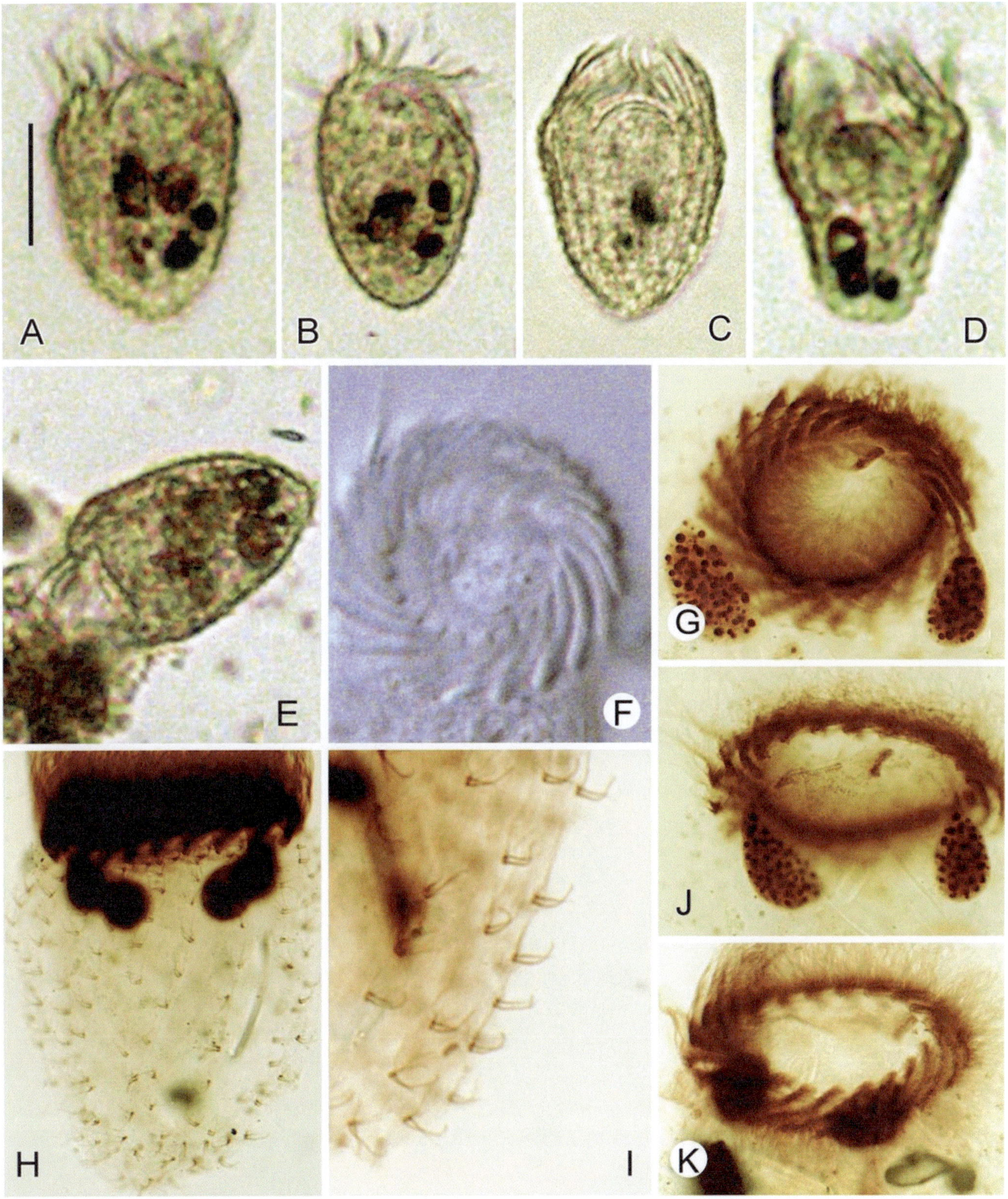


**Fig. 11B** *Strombidinopsis* sp.
(**a–e**) Lateral views of live individuals. (**f**) Apical view of live cell to show external membranelles. (**g**, **j**, **k**) Apical views of buccal field of stained specimens, showing external membranelles. (**h**) Lateral view of ciliature. (**i**) Detail of somatic kineties. Scale bars: 20 μm

## Reference

Song WB, Warren A, Hu X (2009) Free-living Ciliates in the Bohai and Yellow Seas, China. Science Press, Beijing

Class Spirotrichea Bütschli, 1889
Order Halteriida Petz & Foissner, 1992
Family Halteriidae Claparède & Lachmann, 1858
*Pelagohalteria cirrifera* (Kahl, 1932) Foissner et al., 1988

**Ecological feature:** mangrove
**Water temperature:** 18.0 °C
**pH:** 8.1
**Salinity:** 10.4‰

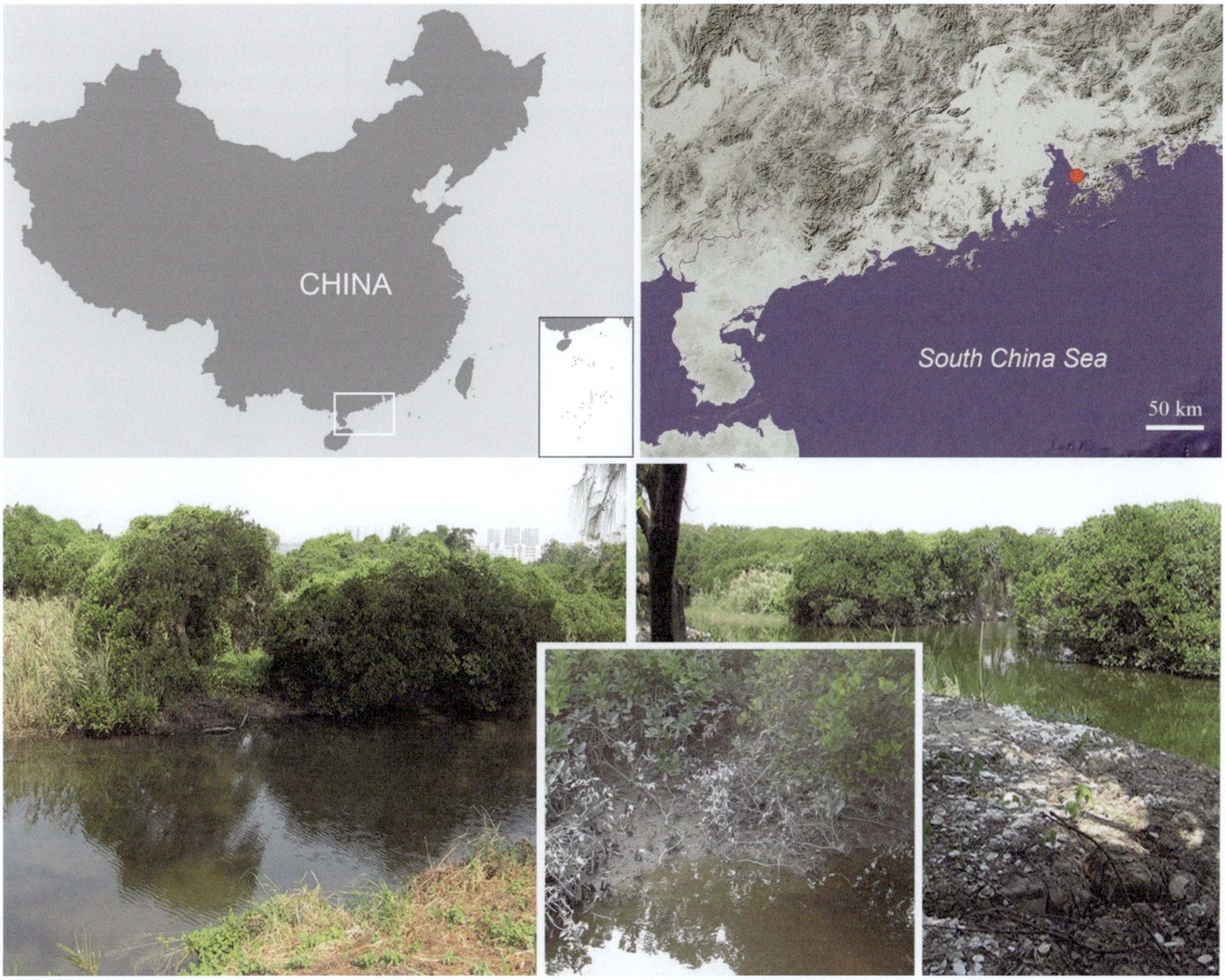


**Fig. 12A** Sample site

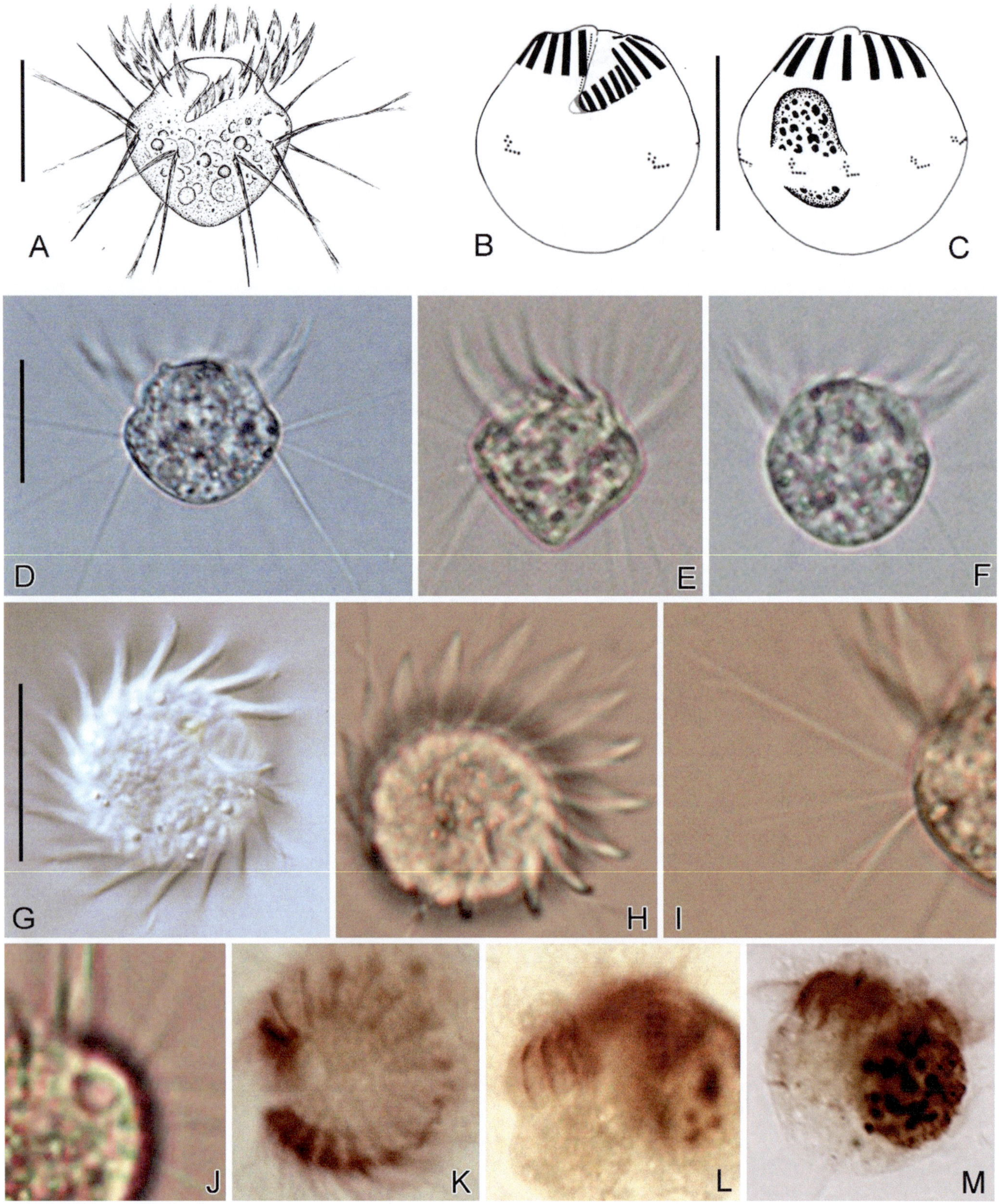


**Fig. 12B** *Pelagohalteria cirrifera*
(**a**, **d**) Ventral views of live cells. (**b**, **c**) Ventral (**b**) and dorsal (**c**) views of ciliature. (**e**, **f**) Different body shapes. (**g**, **h**) Apical views to show adoral membranelles. (**i**) Lateral view to show the bristle complexes. (**j**) Contractile vacuole. (**k**, **l**) Apical (**k**) and ventral (**l**) views to show adoral zone of membranelles. (**m**) Macronucleus and bristle complexes. Scale bars: 20 μm

## Reference

Agatha S, Riedel-Lorjé JC (1997) Morphology, infraciliature, and ecology of halteriids and strombidiids (Ciliophora, Oligotrichea) from coastal brackish water basins. Arch Protistenk 148:445–459

Class Spirotrichea Bütschli, 1889
Order Lynnellida Liu et al., 2015
Family Lynnellidae Liu et al., 2011
*Lynnella semiglobulosa* Liu et al., 2011

**Ecological feature:** offshore water
**Water temperature:** 20.6 °C
**pH:** 7.5
**Salinity:** 28.6‰

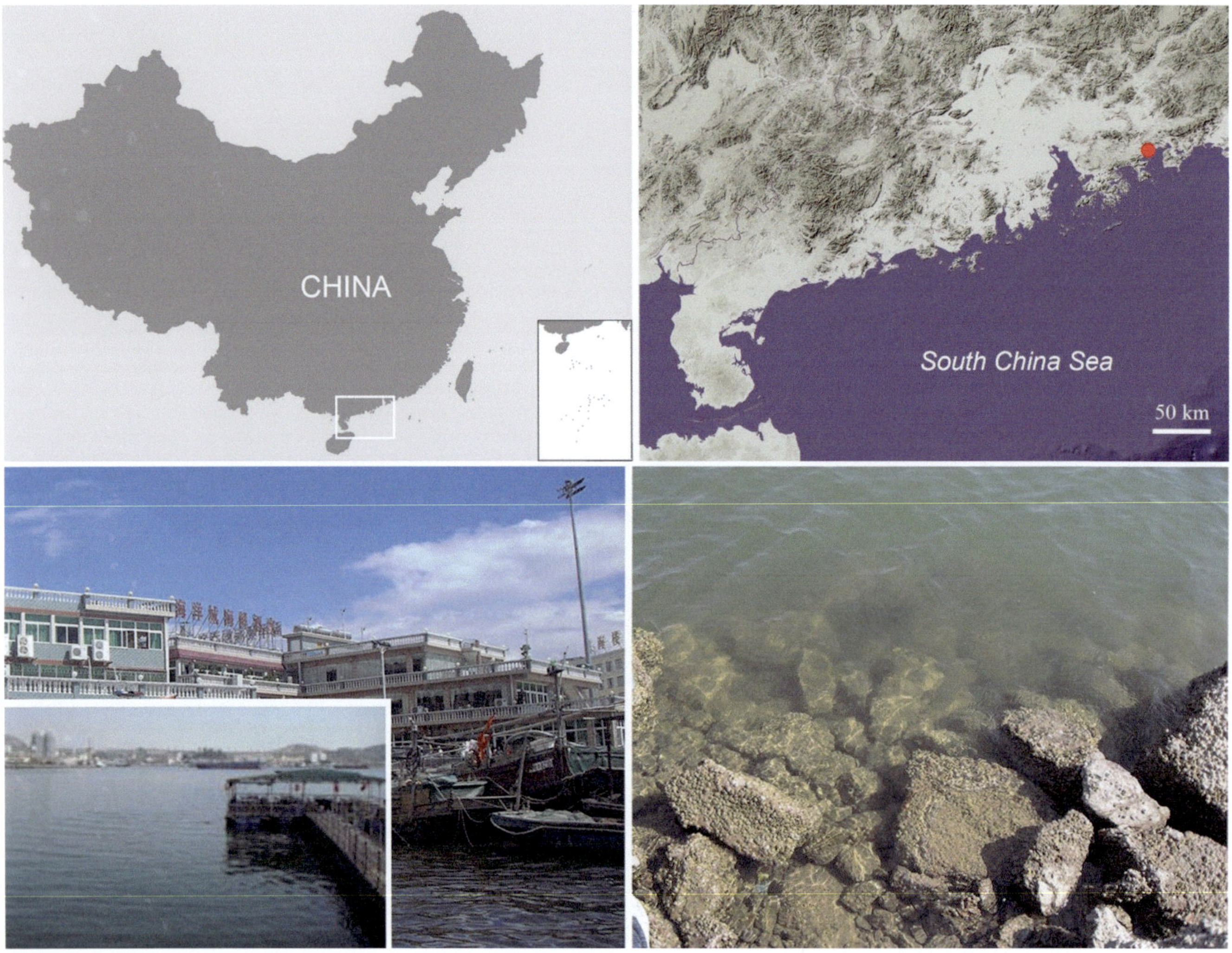


**Fig. 13A** Sample site

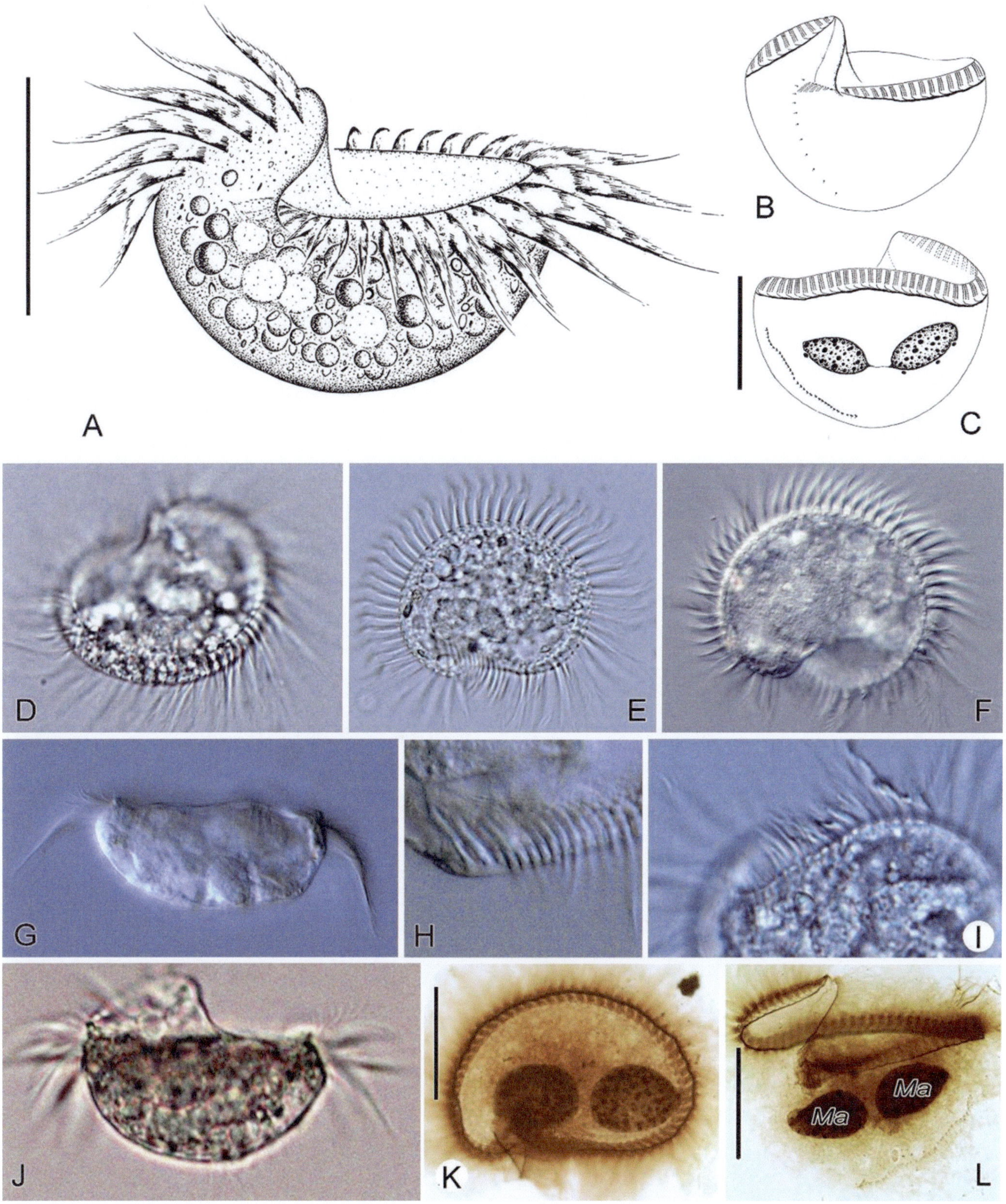


**Fig. 13B** *Lynnella semiglobulosa*
(**a**, **j**) Ventral views of live cells. (**b**, **c**, **l**) Ventral (**b**, **l**) and dorsal (**c**) views of ciliature. (**d**) Apical-dorsal view showing apical protrusion. (**e**, **f**) Adoral membranelles in vivo. (**g**) Lateral view of live cell. (**h**, **i**) Details of proximal end of the adoral zone of membranelles in apical (**h**) and aboral (**i**) views. (**k**) Apical views of adoral membranelles and macronuclei. Scale bars: 30 μm

## Reference

Liu W, Yi Z, Lin X, Al-Rasheid KAS (2011) Morphologic and molecular data suggest that *Lynnella semiglobulosa* n. g., n. sp. represents a new family within the subclass Choreotrichia (Ciliophora, Spirotrichea). J Eukaryot Microbiol 58:43–49

Class Spirotrichea Bütschli, 1889
Order Strombidiida Petz & Foissner, 1992
Family Strombidiidae Fauré-Fremiet, 1970
*Antestrombidium agathae* Liu et al., 2015

**Ecological feature:** mangrove
**Water temperature:** 19.7 °C
**pH:** 7.8
**Salinity:** 23.9‰

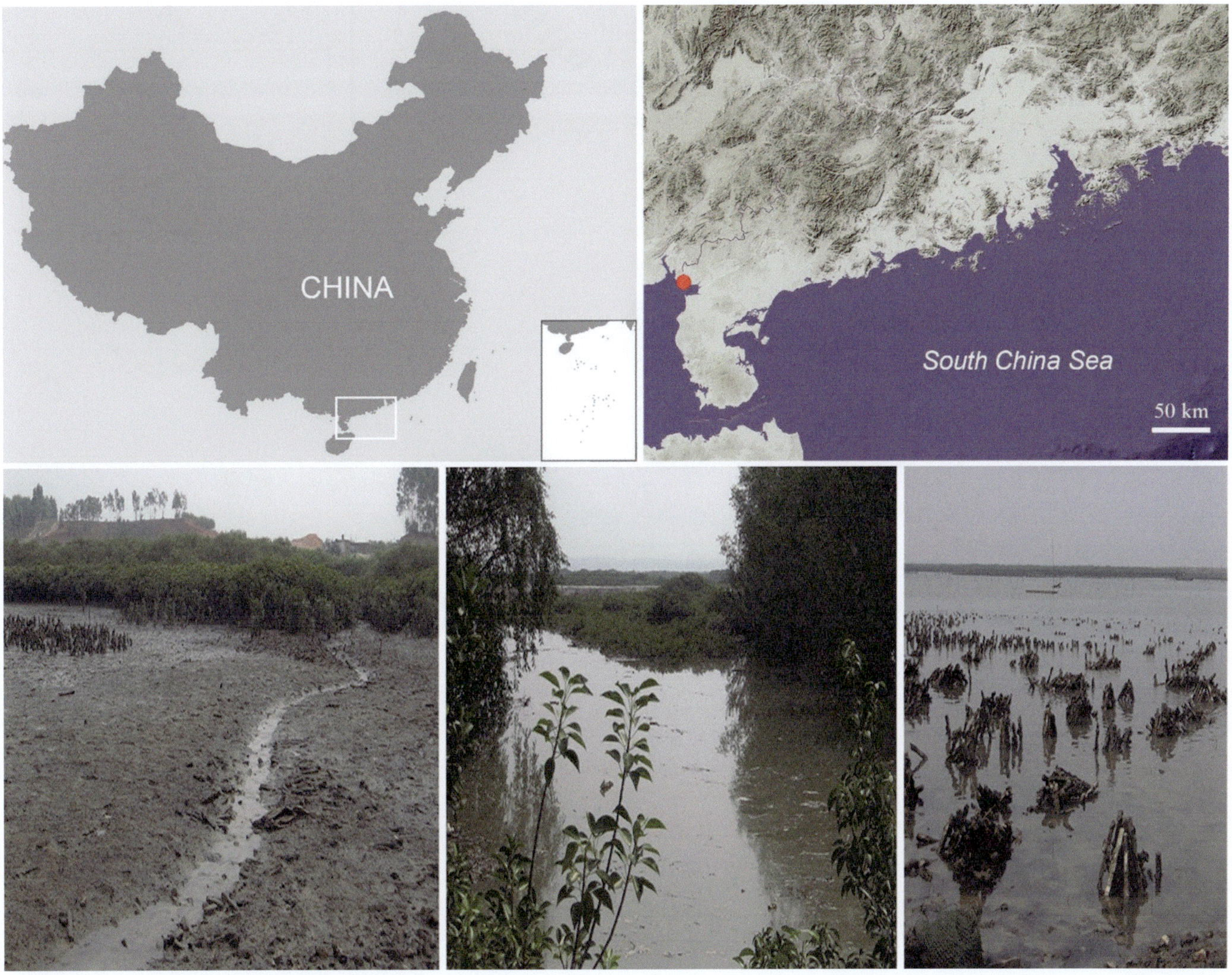


**Fig. 14A** Sample site

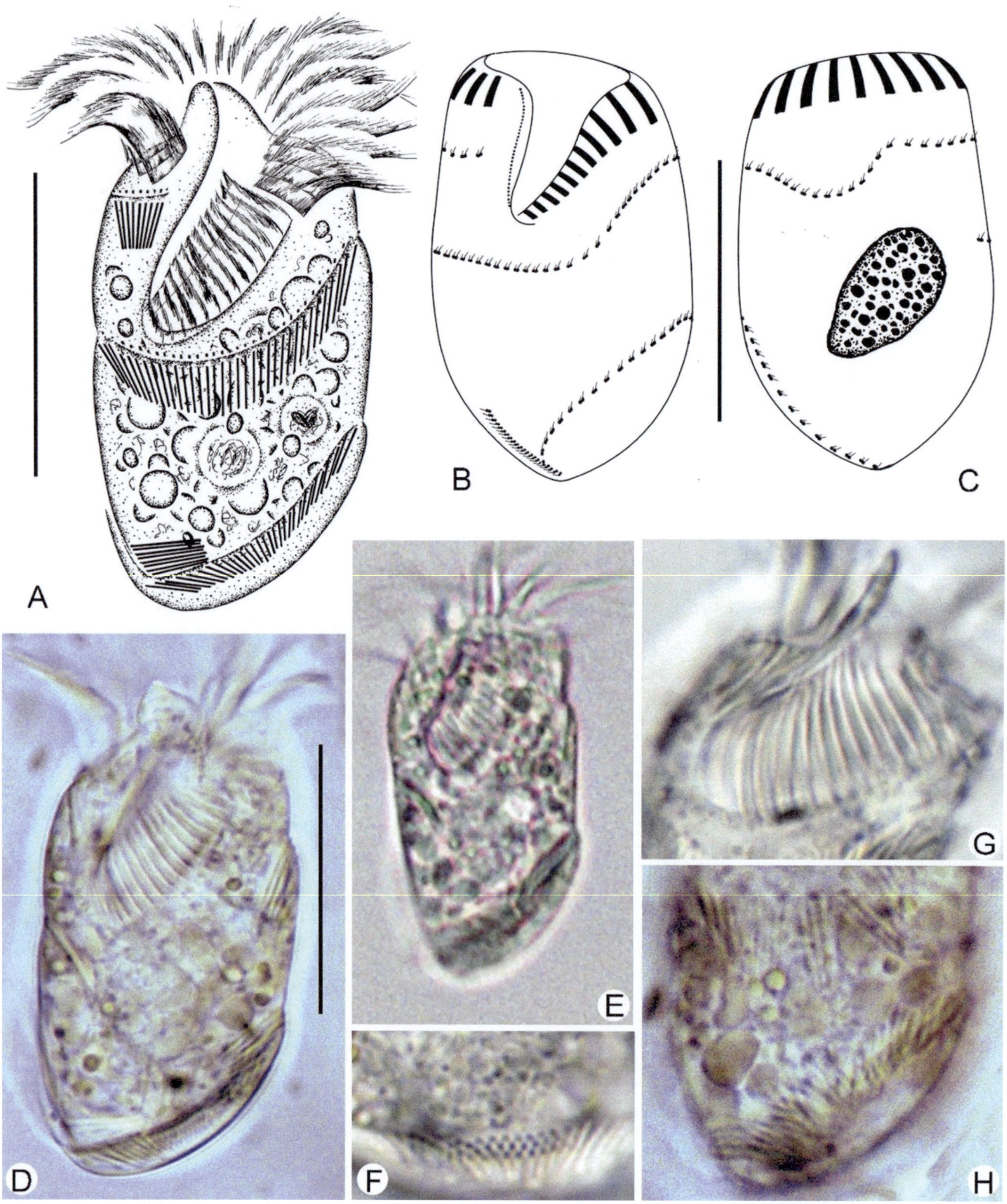


**Fig. 14B** *Antestrombidium agathae*
(**a**, **d**, **e**) Ventral view of live individuals. (**b**, **c**) Ventral (**b**) and dorsal (**c**) views of ciliature. (**f**, **h**) The arrangement of extrusomes. (**g**) Ventral view of anterior portion of cell to show buccal membranelles. Scale bars: 30 μm

## Reference

Liu W, Yi Z, Xu D, Clamp J, Li J, Lin X, Song WB (2015) Two new genera of planktonic ciliates and insights into the evolution of the family Strombidiidae (Protista, Ciliophora, Oligotrichia). PLoS ONE 10(6):e0131726

Class Spirotrichea Bütschli, 1889
Order Strombidiida Petz & Foissner, 1992
Family Strombidiidae Fauré-Fremiet, 1970
*Apostrombidium* sp.

**Ecological feature:** mangrove
**Water temperature:** 24.5 °C
**pH:** 6.8
**Salinity:** 17.0‰

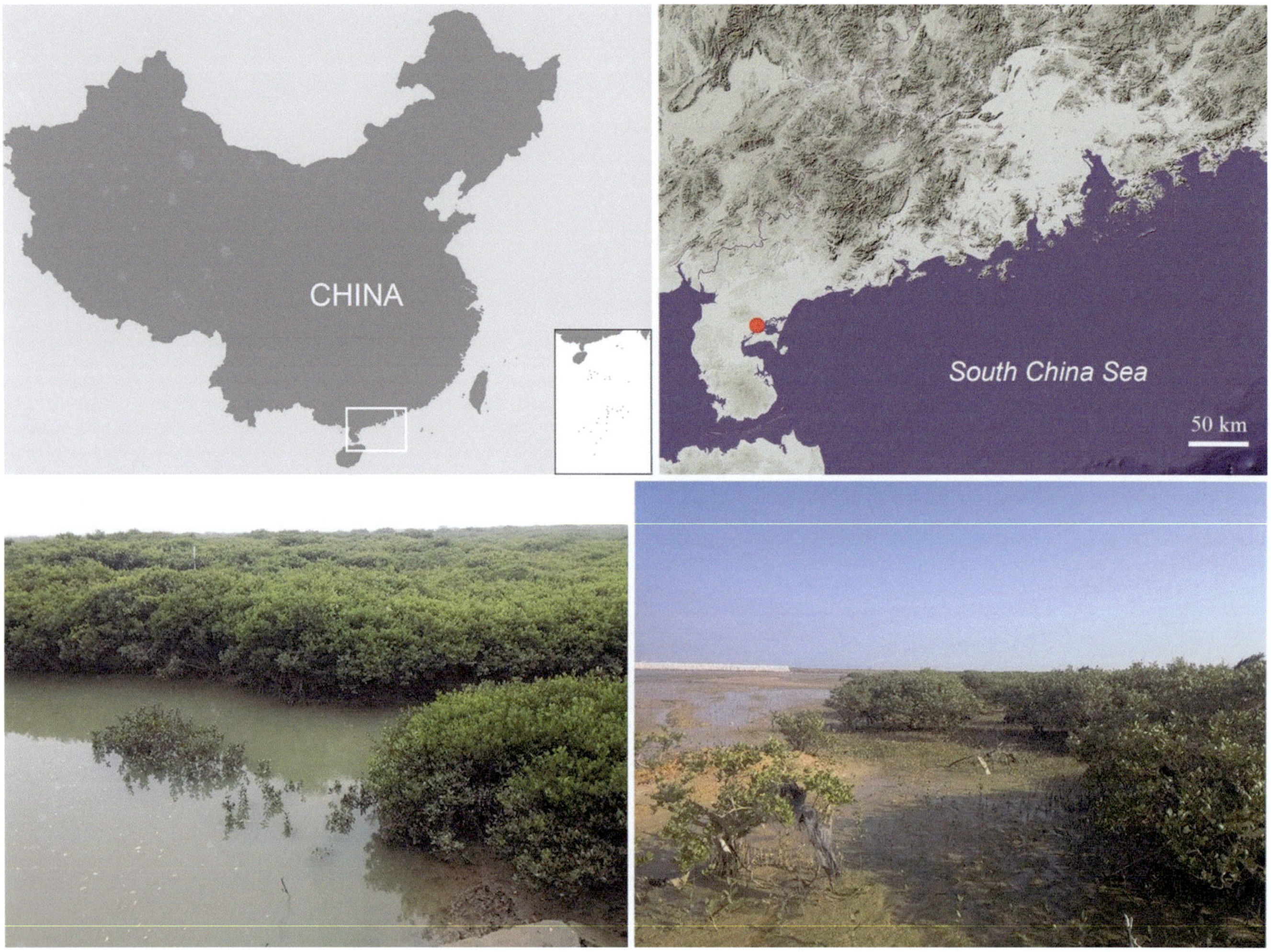


Fig. 15A Sample site

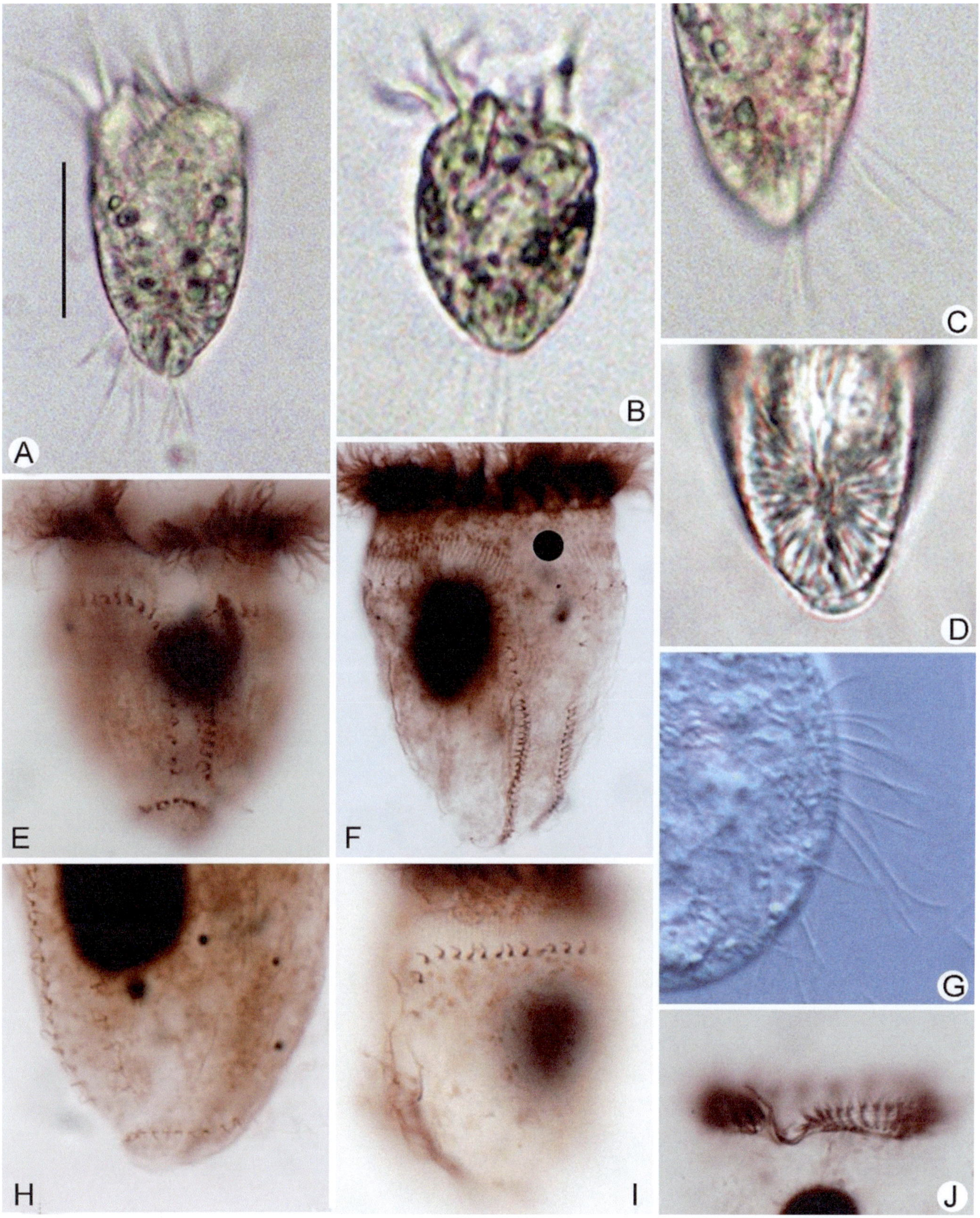


**Fig. 15B** *Apostrombidium* sp
(**a**, **b**) Ventral views of live individuals. (**c**, **g**) Cilia on dorsal side. (**d**) Extrusomes. (**e**, **f**) Lateral views of ciliature. (**h**, **i**) Detail of kineties. (**j**) Buccal field. Scale bar: 30 μm

## References

Song W, Li J, Liu W, Jiang J, Al-Rasheid KAS, Hu X (2013) Taxonomy, morphology and molecular systematics of three oligotrich ciliates, including a description of *Apostrombidium parakielum* spec. nov. (Ciliophora, Oligotrichia). Int J Syst Evol Microbiol 63:1179–1191

Xu D, Warren A, Song WB (2009) Oligotrichs. In: Song WB, Warren A, Hu X (eds.) Free-living Ciliates in the Bohai and Yellow Seas, China. Science Press, Beijing

Class Spirotrichea Bütschli, 1889
Order Strombidiida Petz & Foissner, 1992
Family Strombidiidae Fauré-Fremiet, 1970
*Cyrtostrombidium paralongisomum* Tsai et al., 2015

**Ecological feature:** mangrove
**Water temperature:** 19.7 °C
**pH:** 7.8
**Salinity:** 23.9‰

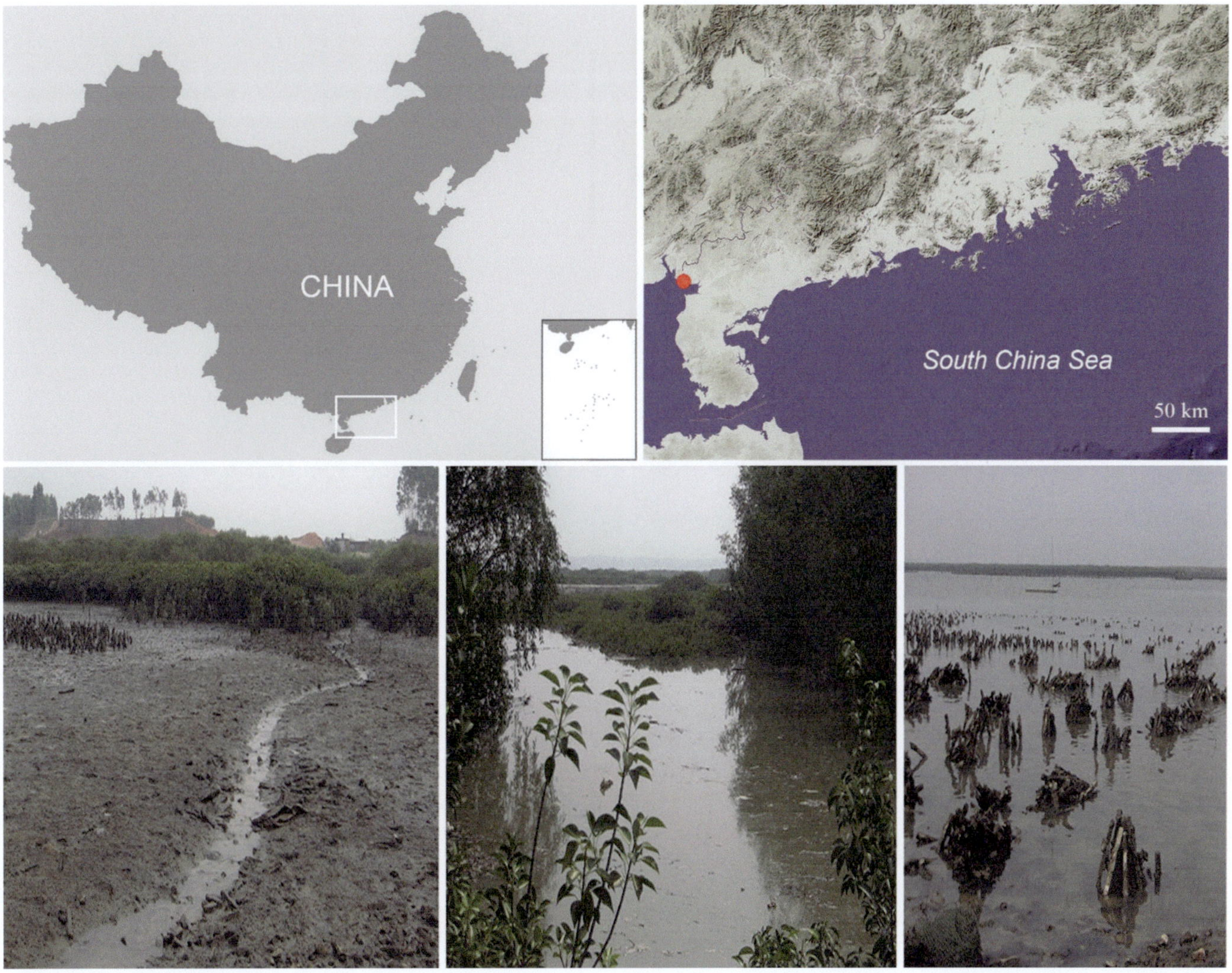


**Fig. 16A** Sample site

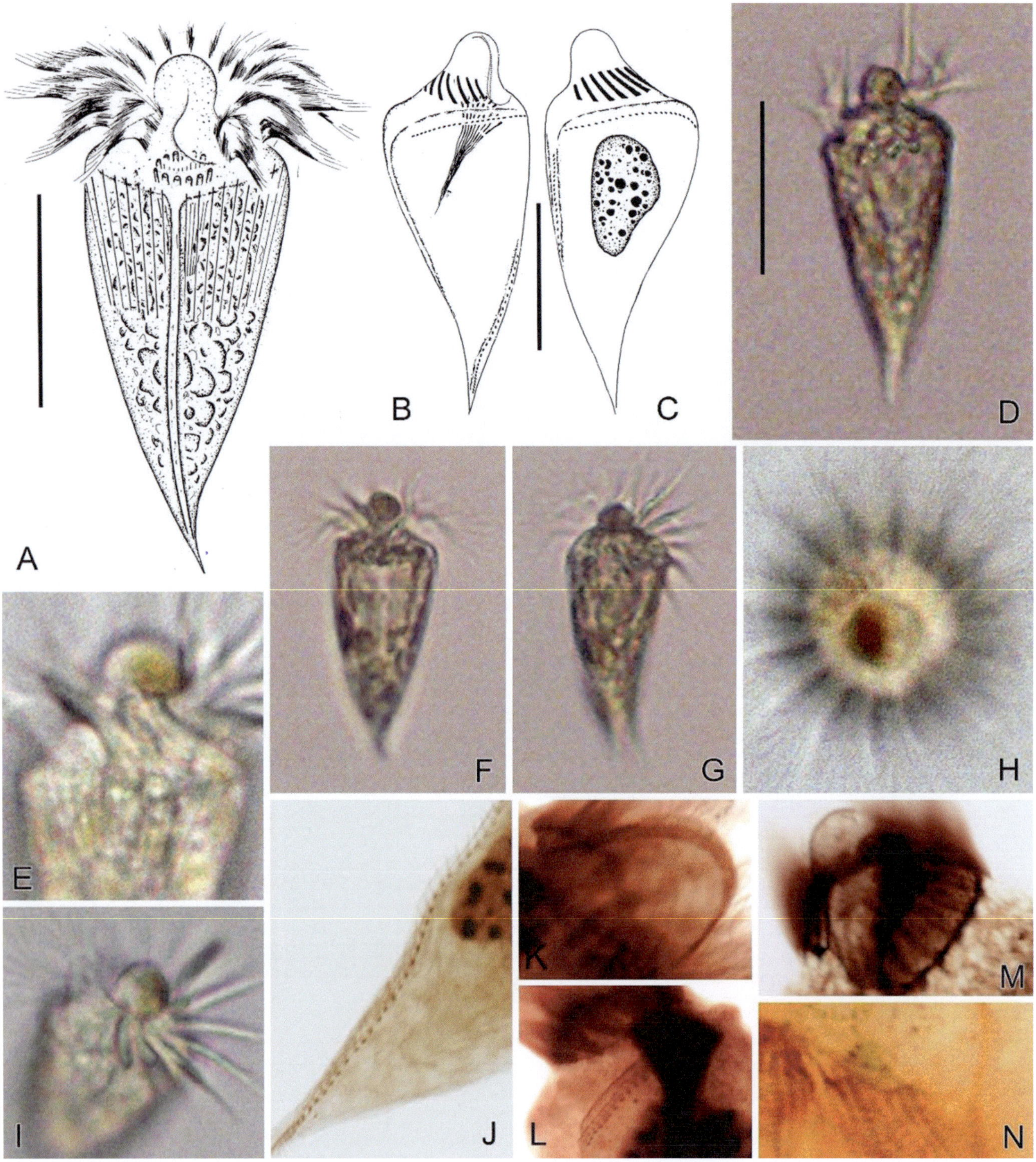


**Fig. 16B** *Cyrtostrombidium paralongisomum*
(**a**, **d**, **f**) Ventral views of live individuals. (**b**, **c**) Right (**b**) and left (**c**) views of ciliature. (**e**) Anterior portion of cell to show apical protrusion and extrusomes. (**g**) Lateral view of live cell. (**h**, **i**) Apical (**h**) and apical-lateral views of anterior portion of cell to show oral membranelles; ventral kinety. (**k**) Endoral membrane. (**m**) Ventral view of anterior portion to show oral membranelles. (**l**) Right view of middle portion showing girdle kinety and cytopharyngeal basket. (**n**) Anterior membranelles. Scale bars: 40 μm

## Reference

Tsai S, Chen W, Chiang K (2015) Phylogenetic position of the genus *Cyrtostrombidium*, with a description of *Cyrtostrombidium paralongisomum* nov. spec. and a redescription of *Cyrtostrombidium longisomum* Lynn & Gilron, 1993 (Protozoa, Ciliophora) based on live observation, protargol staining, and 18S rDNA sequences. J Eukaryot Microbiol 62:239–248

Class Spirotrichea Bütschli, 1889
Order Strombidiida Petz & Foissner, 1992
Family Strombidiidae Fauré-Fremiet, 1970
*Laboea strobila* Lohmann, 1908

**Ecological feature:** offshore water
**Water temperature:** 22.6 °C
**pH:** 7.6
**Salinity:** 22.6‰

**Fig. 17A** Sample site

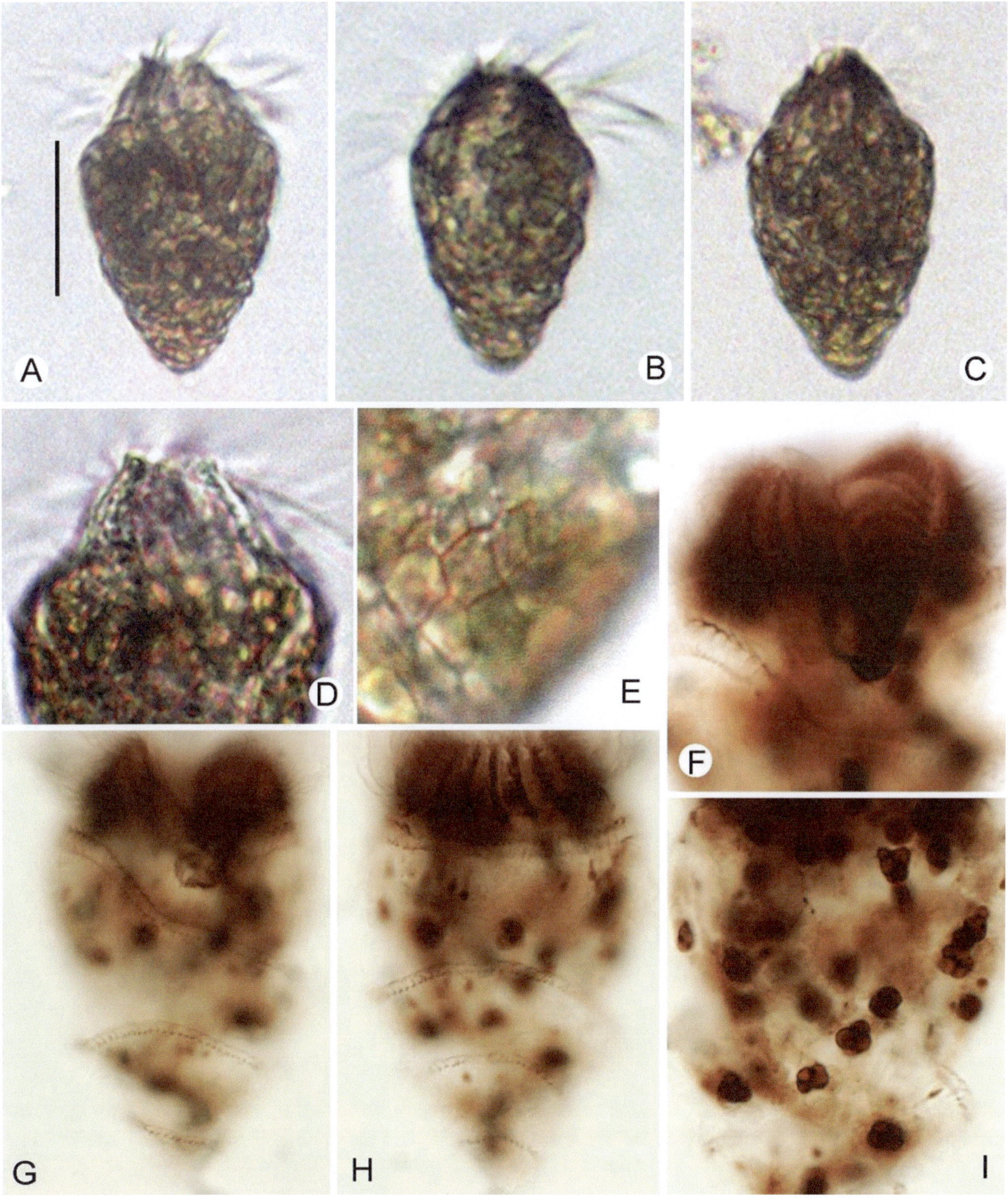


**Fig. 17B** *Laboea strobila*
(**a–c**) Ventral views of live individuals. (**d**) Buccal field. (**e**) Hemitheca. (**f**) Buccal structure. (**g**, **h**) Ventral (**g**) and dorsal (**h**) views of ciliature. (**i**) Macronuclei. Scale bars: 30 μm

## Reference

Agatha S, Strüder-Kypke M C, Beran A (2004) Morphologic and genetic variability in the marine planktonic ciliate *Laboea strobila* Lohmann, 1908 (Ciliophora, Oligotrichea), with notes on its ontogenesis. J Eukaryot Microbiol 51:267–281

Class Spirotrichea Bütschli, 1889
Order Strombidiida Petz & Foissner, 1992
Family Strombidiidae Fauré-Fremiet, 1970
*Novistrombidium apsheronicum* (Alekperov & Asadullayeva, 1997) Agatha, 2003

**Ecological feature:** mangrove; intertidal zone
**Water temperature:** 31.2 °C; 30.2 °C
**pH:** 8.2; 8.3
**Salinity:** 31.8‰; 31.8‰

**Fig. 18A** Sample site

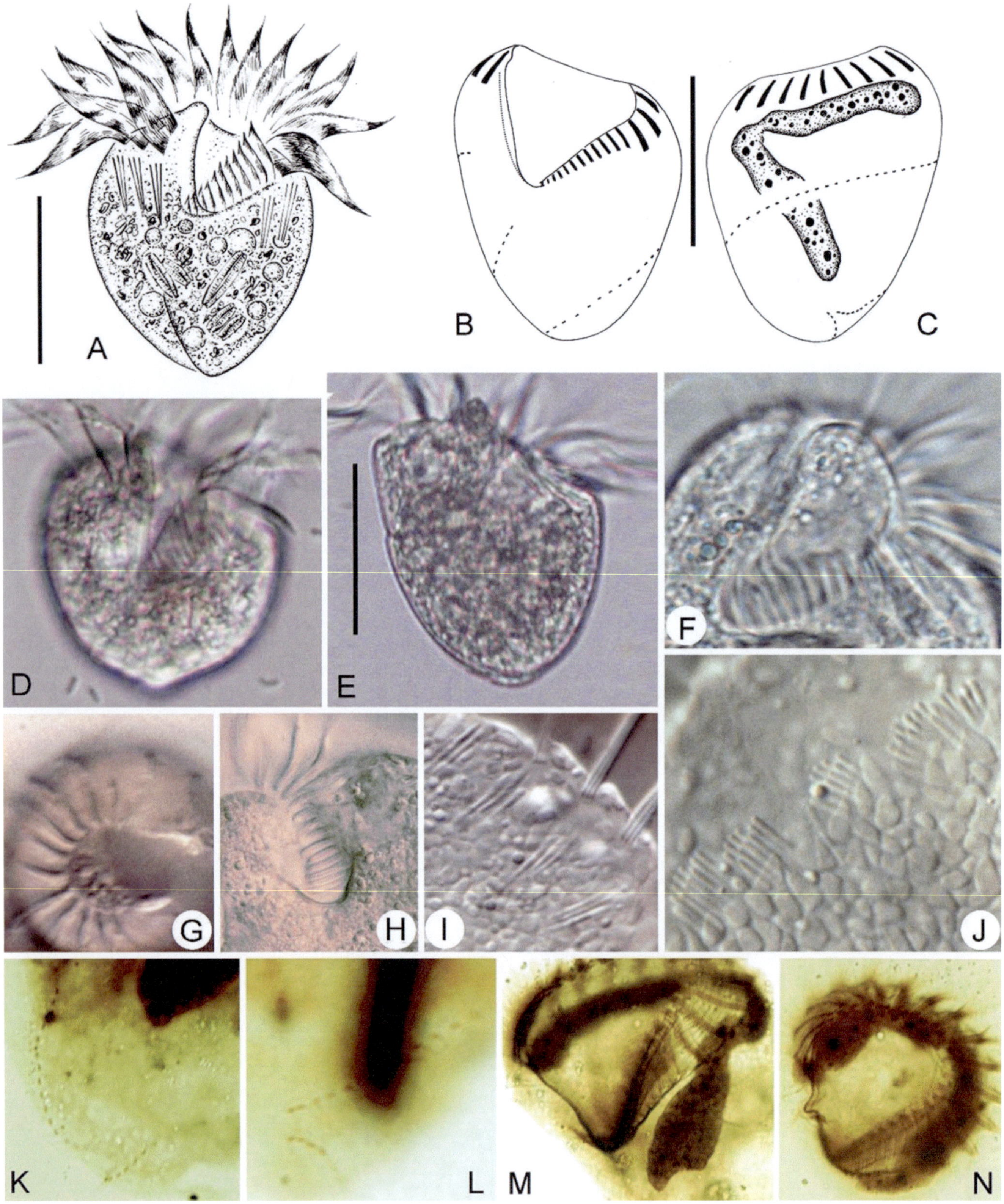

**Fig. 18B** *Novistrombidium apsheronicum*
(**a**, **d**, **e**) Ventral views of live cells. (**b**, **c**) Ventral (**b**) and dorsal (**c**) views of ciliature. (**f**) Buccal field, showing ventral membranelles. (**g**, **h**) Anterior (**g**) and ventral (**h**) membranelles. (**i**, **j**) Grouped extrusomes. (**k**, **l**) Girdle kinety and ventral kinety. (**m**) Macronucleus. (**n**) Adoral zone of membranelles. Scale bars: 30 μm

## References

Agatha S (2003) Morphology and ontogenesis of *Novistrombidium apsheronicum* nov. comb. and *Strombidium arenicola* (Protozoa, Ciliophora): a comparative light microscopical and SEM study. Eur J Protistol 39:245–266

Song W, Li J, Liu W, Jiang J, Al-Rasheid KAS, Hu X (2013) Taxonomy, morphology and molecular systematics of three oligotrich ciliates, including a description of *Apostrombidium parakielum* spec. nov. (Protozoa, Ciliophora, Oligotrichia). Int J Syst Evol Microbiol 63:1179–1191

Class Spirotrichea Bütschli, 1889
Order Strombidiida Petz & Foissner, 1992
Family Strombidiidae Fauré-Fremiet, 1970
*Novistrombidium orientale* Liu et al., 2009

**Ecological feature:** offshore water
**Water temperature:** 22.7 °C
**pH:** 7.9
**Salinity:** 26.0‰

**Fig. 19A** Sample site

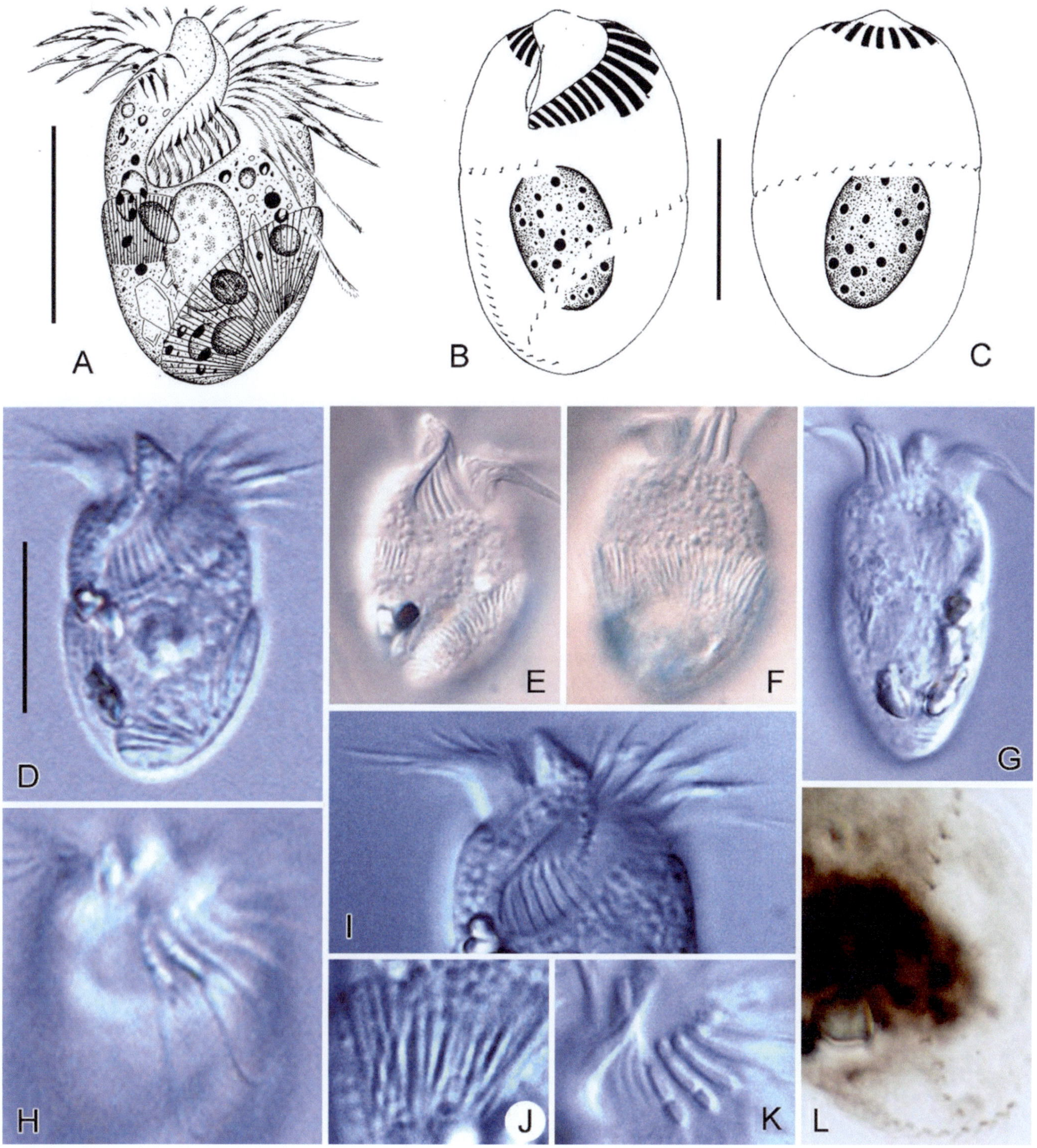


**Fig. 19B** *Novistrombidium orientale*
(**a**, **d**) Ventral views of live cells. (**b**, **c**) Ventral (**b**) and dorsal (**c**) views of ciliature. (**e**, **f**) Ventral (**e**) and dorsal (**f**) views in vivo to show the distribution of extrusomes. (**g**) Lateral view of live cell. (**h**, **i**) Anterior portion of cell to show the oral membranelles. (**j**) Extrusomes. (**k**) Ventral membranelles. (**l**) Somatic kineties. Scale bars: 20 μm

## Reference

Liu W, Xu D, Lin X, Li J, Gong J, Al-Rasheid KAS, Song WB (2009) *Novistrombidium sinicum* n. sp. and *N. orientale* n. sp. (Protozoa: Ciliophora): Two new oligotrich ciliates from a mangrove wetland, south China. J Eukaryot Microbiol 56:459–465

Class Spirotrichea Bütschli, 1889
Order Strombidiida Petz & Foissner, 1992
Family Strombidiidae Fauré-Fremiet, 1970
*Novistrombidium sinicum* Liu et al., 2009

**Ecological feature:** mangrove
**Water temperature:** 17.4 °C
**pH:** 7.5
**Salinity:** 19.8‰

**Fig. 20A** Sample site

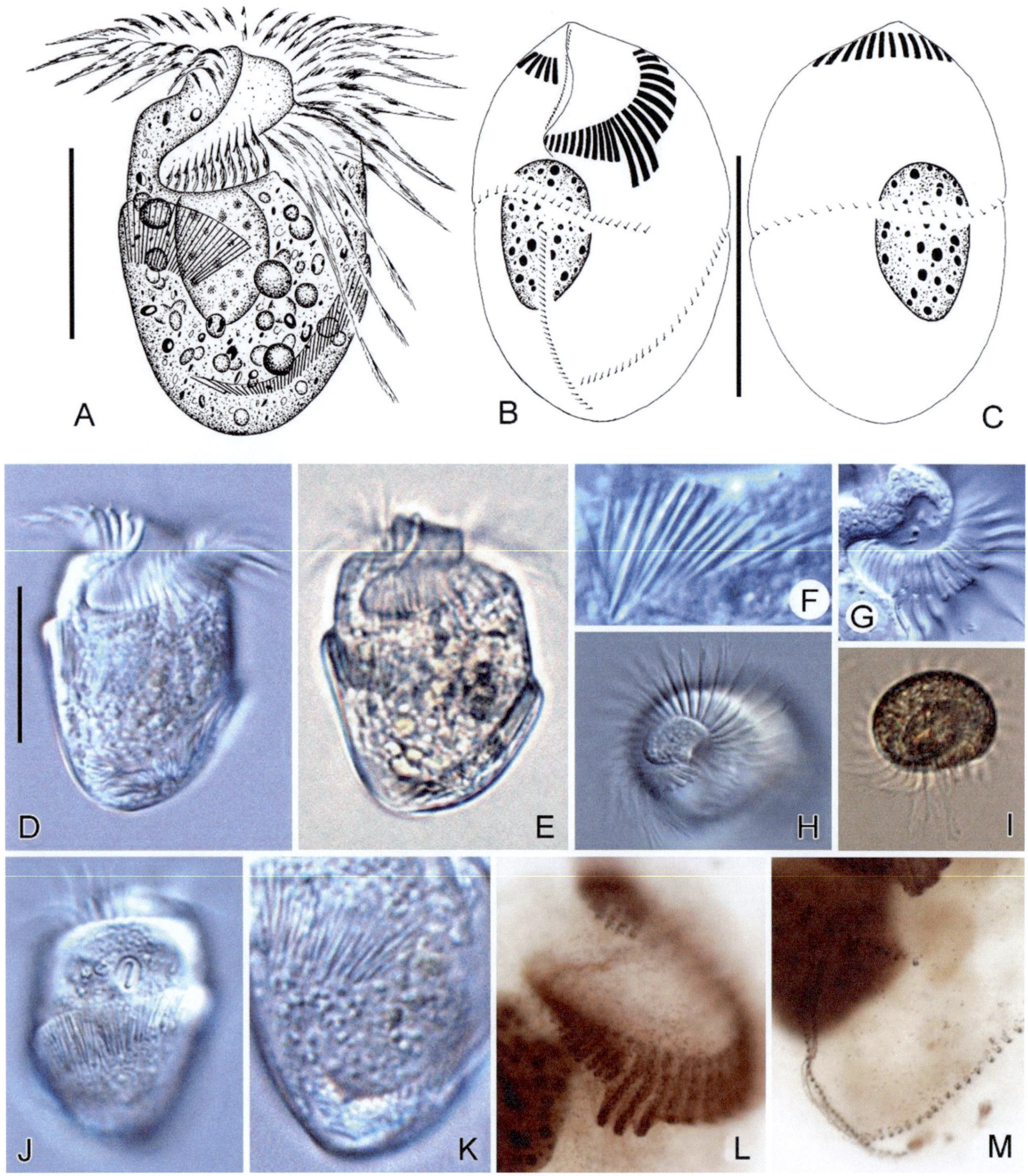


**Fig. 20B** *Novistrombidium sinicum*
(**a**, **d**, **e**) Ventral views of live cells. (**b**, **c**) Ventral (**b**) and dorsal (**c**) views of ciliature. (**f**) Extrusomes. (**g–i**) Apical views of live cells to show oral membranelles. (**j**, **k**) Lateral views to show the arrangement of extrusomes. (**l**) Adoral zone of membranelles. (**m**) Ventral view of posterior portion of cell to show the somatic kineties. Scale bars: 30 μm

## Reference

Liu W, Xu D, Lin X, Li J, Gong J, Al-Rasheid KAS, Song WB (2009) *Novistrombidium sinicum* n. sp. and *N. orientale* n. sp. (Protozoa: Ciliophora): two new oligotrich ciliates from a mangrove wetland, south China. J Eukaryot Microbiol 56:459–465

Class Spirotrichea Bütschli, 1889
Order Strombidiida Petz & Foissner, 1992
Family Strombidiidae Fauré-Fremiet, 1970
*Novistrombidium testaceum* (Anigstein, 1914) Song & Bradbury, 1998

**Ecological feature:** mangrove
**Water temperature:** no data available; 27.0 °C
**pH:** no data available; 8.2
**Salinity:** no data available; 17.0‰

**Fig. 21A** Sample site

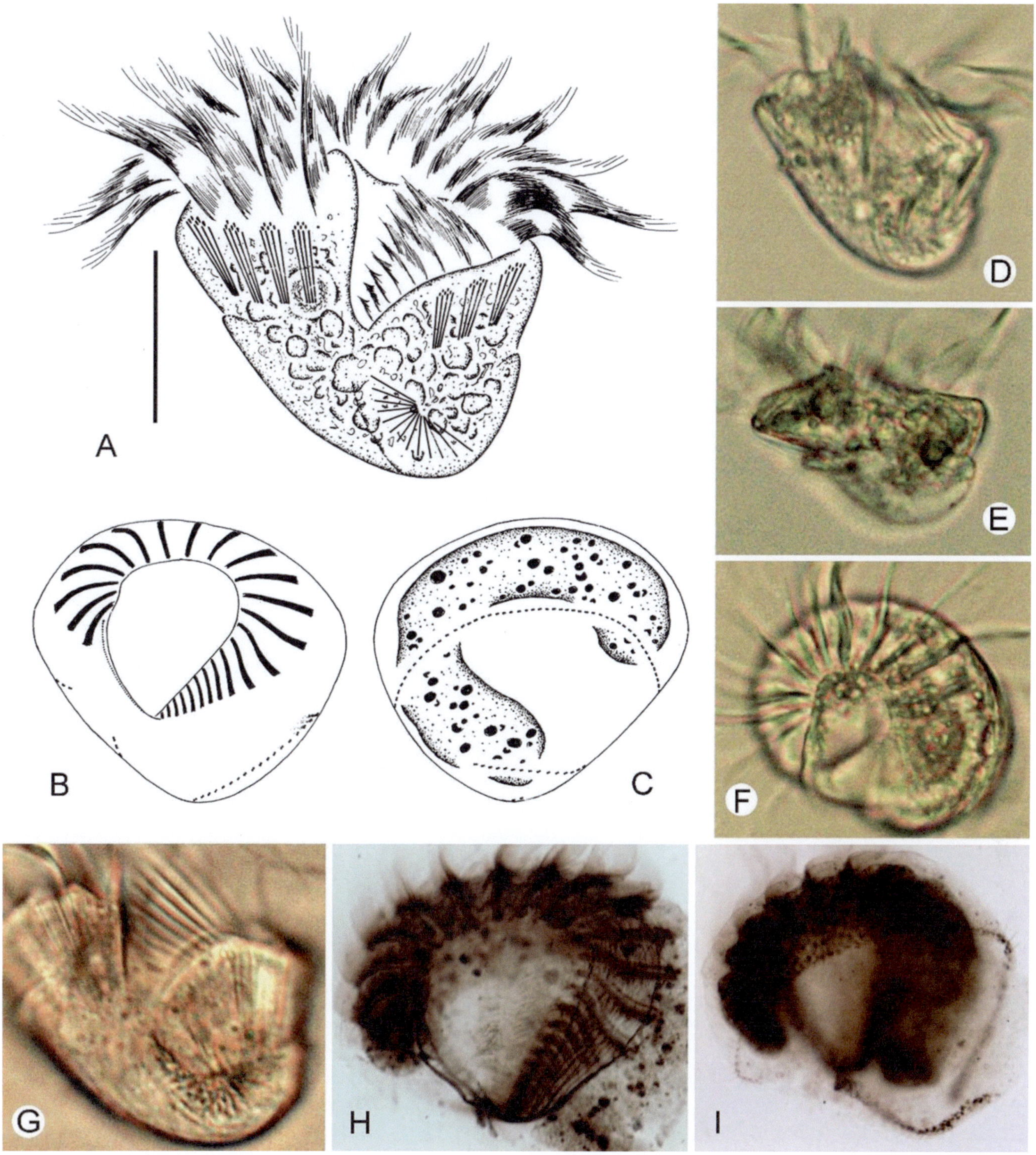


**Fig. 21B** *Novistrombidium testaceum*
(**a**, **d**, **e**) Ventral views of live individuals. (**b**, **c**) Ventral (**b**) and dorsal (**c**) views of ciliature. (**f**) Apical view of live cell, showing adoral membranelles. (**g**) Extrusomes. (**h**) Adoral zone of membranelles. (**i**) Somatic kineties. Scale bars: 20 μm

## References

Song WB, Bradbury P (1998) Studies on some new and rare reported marine planktonic ciliates (Ciliophora: Oligotrichia) from coastal waters in North China. J Mar Biol Ass UK 78:767–794

Song W, Li J, Liu W, Jiang J, Al-Rasheid KAS, Hu X (2013) Taxonomy, morphology and molecular systematics of three oligotrich ciliates, including a description of *Apostrombidium parakielum* spec. nov. (Ciliophora, Oligotrichia). Int J Syst Microbiol 63:1179–1191

Class Spirotrichea Bütschli, 1889
Order Strombidiida Petz & Foissner, 1992
Family Strombidiidae Fauré-Fremiet, 1970
*Omegastrombidium elegans* (Florentin, 1901) Agatha, 2004

**Ecological feature:** offshore water
**Water temperature:** 18.1 °C
**pH:** 8.1
**Salinity:** 32.7‰

**Fig. 22A** Sample site.

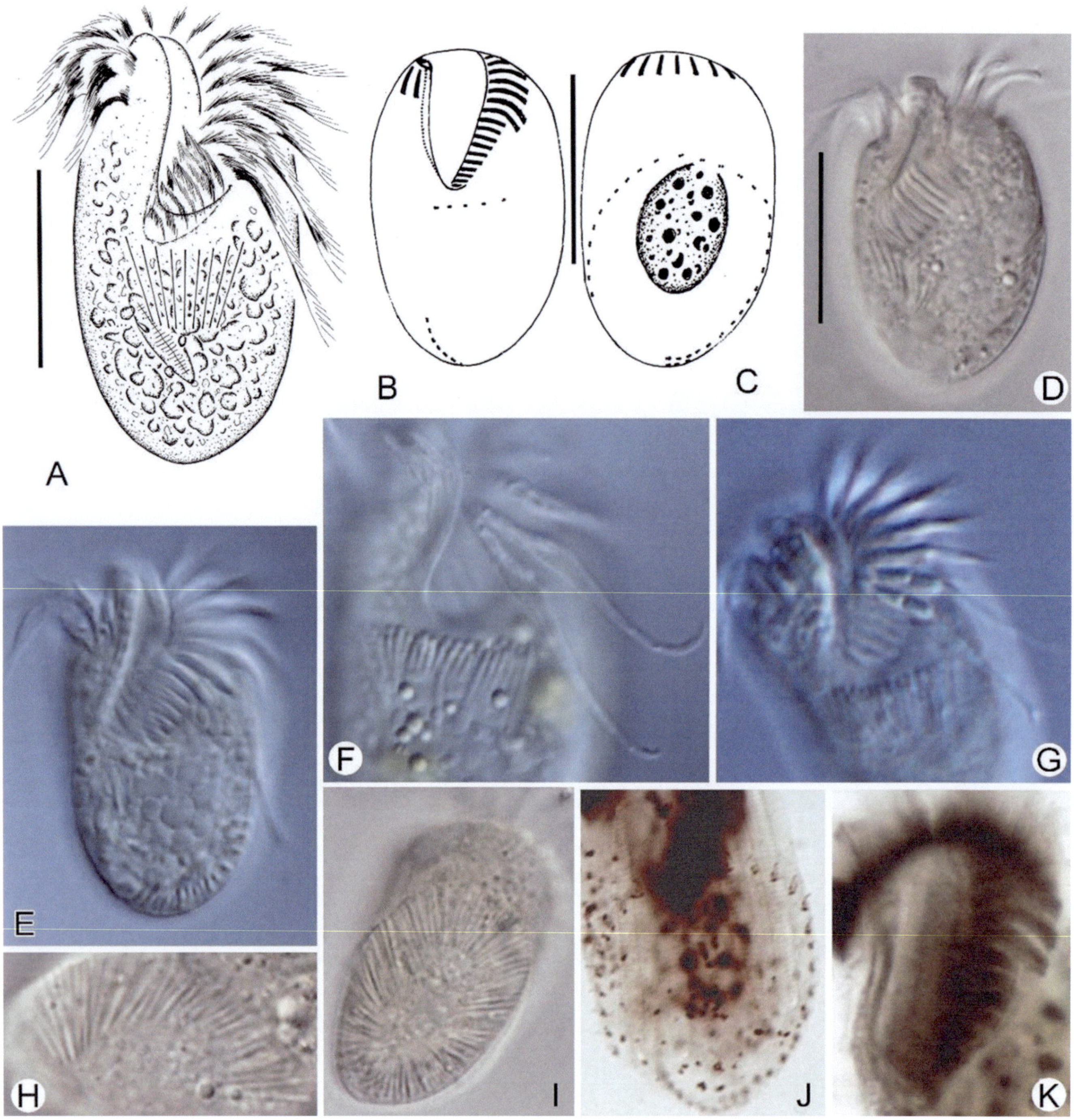


**Fig. 22B** *Omegastrombidium elegans*
(**a**, **d**, **e**) Ventral views of live cells. (**b**, **c**) Ventral (**b**) and dorsal (**c**) views of ciliature. (**f**, **g**) Ventral views of anterior portion of cell to show thigmotactic membranelles and extrusomes. (**h**) Extrusomes. (**i**) Dorsal view of live cell to show the arrangement of extrusomes. (**j**) Posterior portion of specimen to show girdle kinety. (**k**) Oral membranelles. Scale bars: 20 μm

## References

Song WB, Wang M, Warren A (2000) Redescriptions of three marine ciliates, *Strombidium elegans* Florentin, 1901, *Strombidium sulcatum* Claparède & Lachmann, 1859 and *Heterostrombidium paracalkinsi* Lei, Xu & Song, 1999 (Ciliophora, Oligotrichida). Eur J Protistol 36:327–342

Song WB, Warren A, Hu X (2009) Free-living Ciliates in the Bohai and Yellow Seas, China. Science Press, Beijing

Class Spirotrichea Bütschli, 1889
Order Strombidiida Petz & Foissner, 1992
Family Strombidiidae Fauré-Fremiet, 1970
*Parallelostrombidium conicum* Liu et al., 2013

**Ecological feature:** mangrove
**Water temperature:** 19.7 °C
**pH:** 7.8
**Salinity:** 23.9‰

**Fig. 23A** Sample site

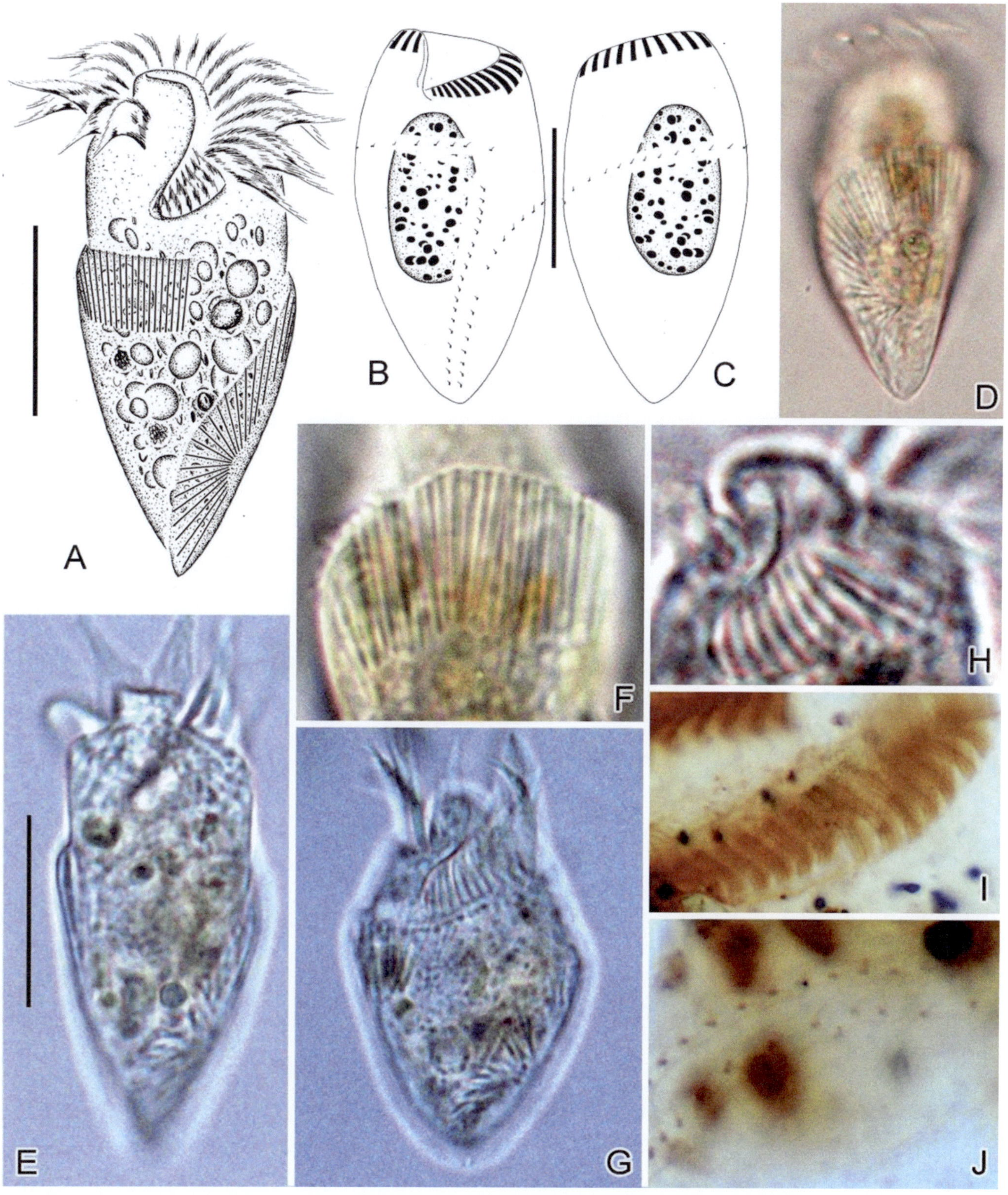


**Fig. 23B** *Parallelostrombidium conicum*
(**a**, **e**) Ventral views of live cells. (**b**, **c**) Ventral (**b**) and dorsal (**c**) views of ciliature. (**d**, **g**) Dorsal (**d**) and ventral (**g**) views in vivo showing the distribution of extrusomes. (**f**) Extrusomes. (**h**) Anterior portion of live cell to show ventral membranelles. (**i**) Bases of oral membranelles. (**j**) Part of somatic kineties. Scale bars: 20 μm

## Reference

Liu W, Yi Z, Li J, Warren A, Al-Farraj SA, Lin X (2013) Taxonomy, morphology and phylogeny of three new oligotrich ciliates (Protozoa, Ciliophora, Oligotrichia) from southern China. Int J Syst Evol Microbiol 63:4805–4817

Class Spirotrichea Bütschli, 1889
Order Strombidiida Petz & Foissner, 1992
Family Strombidiidae Fauré-Fremiet, 1970
*Parallelostrombidium dragescoi* Song et al., 2018

**Ecological feature:** estuary area
**Water temperature:** 29.9 °C
**pH:** 8.8
**Salinity:** 18.2‰

**Fig. 24A** Sample site

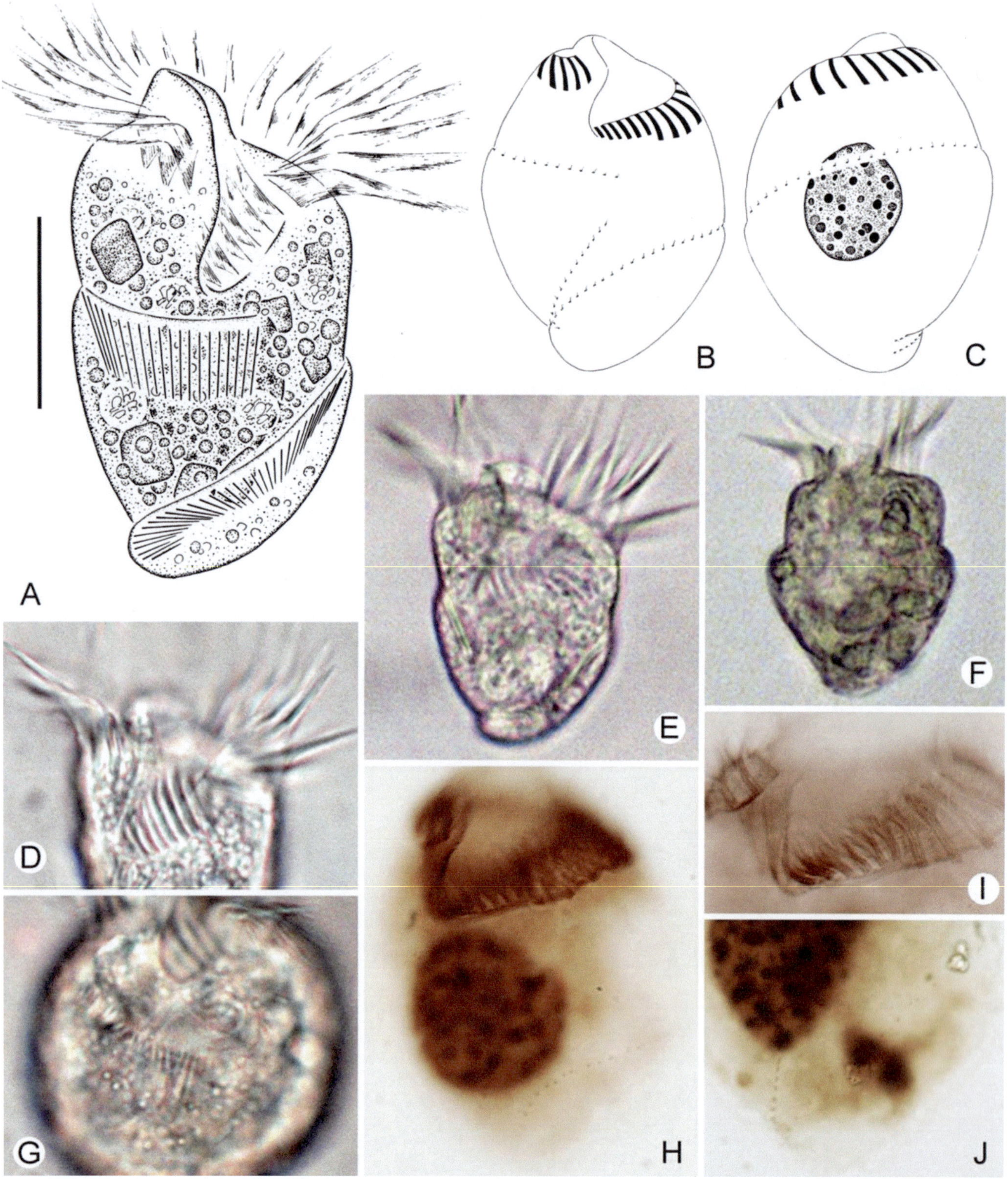


**Fig. 24B** *Parallelostrombidium dragescoi*
(**a**, **e**, **f**) Ventral views of live individuals. (**b**, **c**, **h**) Ventral (**b**, **h**) and dorsal (**c**) views of ciliature. (**d**) Ventral view of anterior portion of live individuals, showing oral membranelles. (**g**) Extrusomes on ventral side. (**i**) Adoral zone of membranelles. (**j**) Part of somatic kineties. Scale bar: 20 μm

## Reference

Song W, Wang L, Li LF, Al-Farraj SA, Aleidan A, Smith S, Hu X (2018) Morphological characterizations of four species of *Parallelostrombidium* (Ciliophora, Oligotrichia), with a note on the phylogeny of the genus. J Eukaryot Microbiol doi:10.1111/jeu.12513

Class Spirotrichea Bütschli, 1889
Order Strombidiida Petz & Foissner, 1992
Family Strombidiidae Fauré-Fremiet, 1970
*Parallelostrombidium ellipticum* Liu et al., 2015

**Ecological feature:** estuary area
**Water temperature:** 25.8 °C
**pH:** 8.9
**Salinity:** 1.7‰

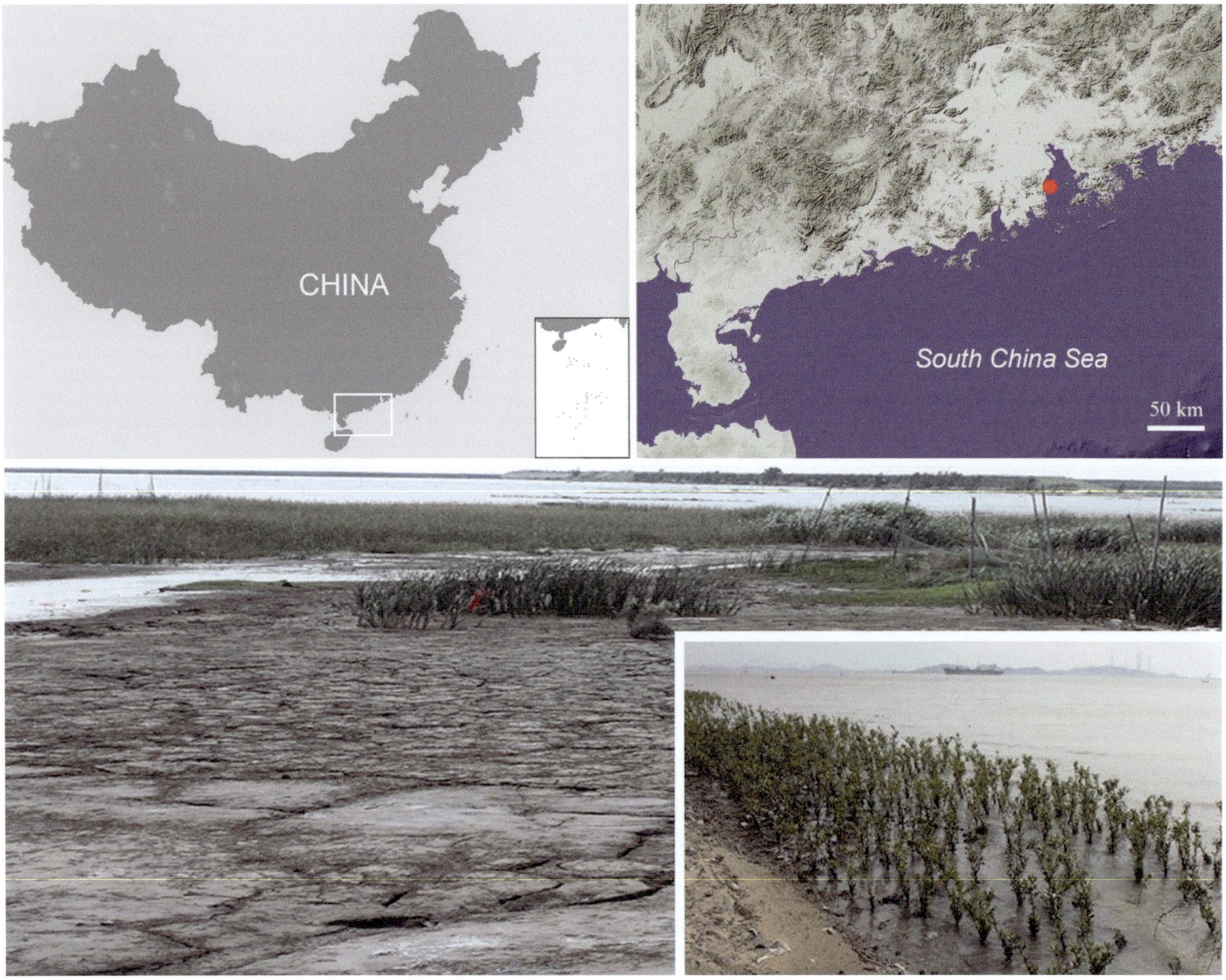


**Fig. 25A** Sample site

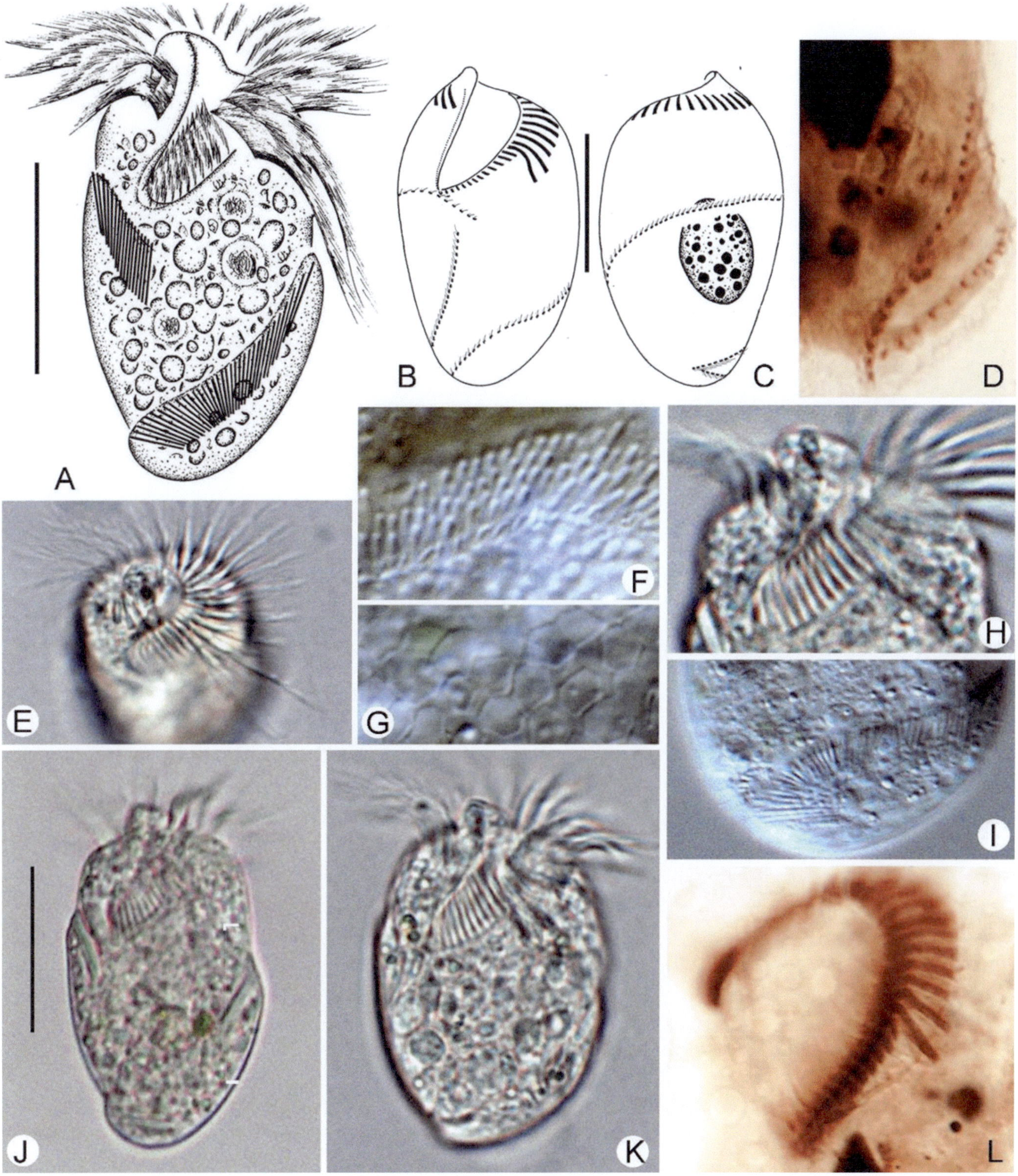


**Fig. 25B** *Parallelostrombidium ellipticum*
(**a**, **j**, **k**) Ventral views of live cells. (**b**, **c**) Ventral (**b**) and dorsal (**c**) views of ciliature. (**d**) Part of somatic kineties. (**e**) Apical view of oral membranelles. (**f**) Extrusomes. (**g**) Detail of polygonal platelets. (**h**) Ventral view of anterior portion of live cell showing oral membranelles. (**i**) Ventral view of posterior portion of live cell showing the distribution of extrusomes. (**l**) Adoral zone of membranelles. Scale bars: 30 μm

## Reference

Liu W, Yi Z, Lin X, Li J, Al-Farraj SA, Al-Rasheid KAS, Song WB (2015) Morphology and molecular phylogeny of three new oligotrich ciliates (Protozoa, Ciliophora) from the South China Sea. Zool J Linn Soc 174:653–665

Class Spirotrichea Bütschli, 1889
Order Strombidiida Petz & Foissner, 1992
Family Strombidiidae Fauré-Fremiet, 1970
*Parallelostrombidium jankowskii* (Xu et al., 2009) Song et al., 2018

**Ecological feature:** offshore water
**Water temperature:** 24.6 °C
**pH:** 7.6
**Salinity:** 22.6‰

**Fig. 26A** Sample site

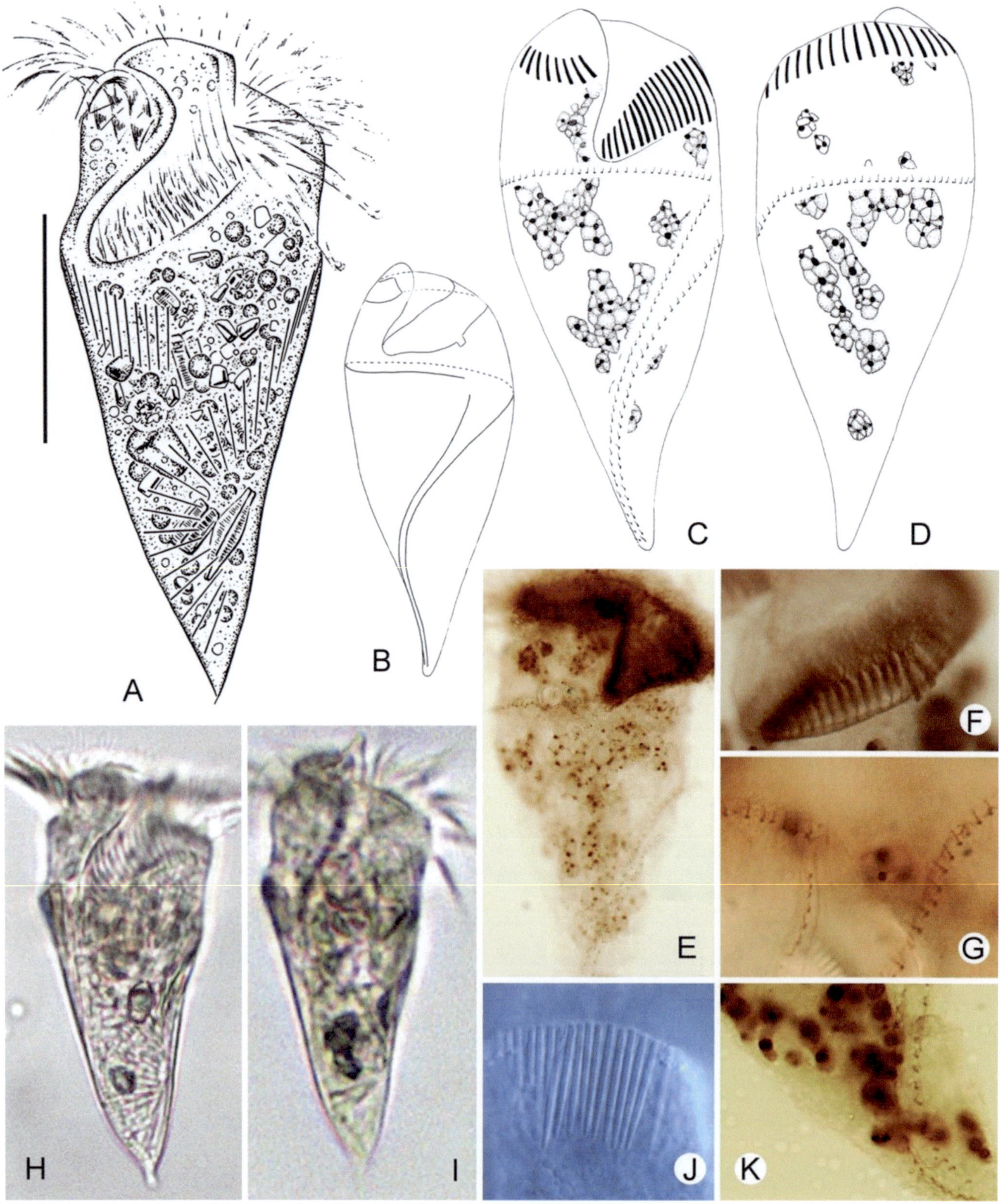


**Fig. 26B** *Parallelostrombidium jankowskii*
(**a**, **h**, **i**) Ventral views of live individuals. (**b**) Diagram of ciliature. (**c–e**) Ventral (**c**, **e**) and dorsal (**d**) views of ciliature. (**f**) Ventral membranelles. (**g**, **k**) Part of somatic kineties. (**j**) Extrusomes. Scale bar: 40 μm

## Reference

Song W, Wang L, Li LF, Al-Farraj SA, Aleidan A, Smith S, Hu X (2018) Morphological characterizations of four species of *Parallelostrombidium* (Ciliophora, Oligotrichia), with a note on the phylogeny of the genus. J Eukaryot Microbiozl doi:10.1111/jeu.12513

Class Spirotrichea Bütschli, 1889
Order Strombidiida Petz & Foissner, 1992
Family Strombidiidae Fauré-Fremiet, 1970
*Parallelostrombidium obesum* Liu et al., 2015

**Ecological feature:** mangrove
**Water temperature:** 24 °C; 27.0 °C
**pH:** 6.0; 8.2
**Salinity:** 5.0‰; 17.0‰

**Fig. 27A** Sample site

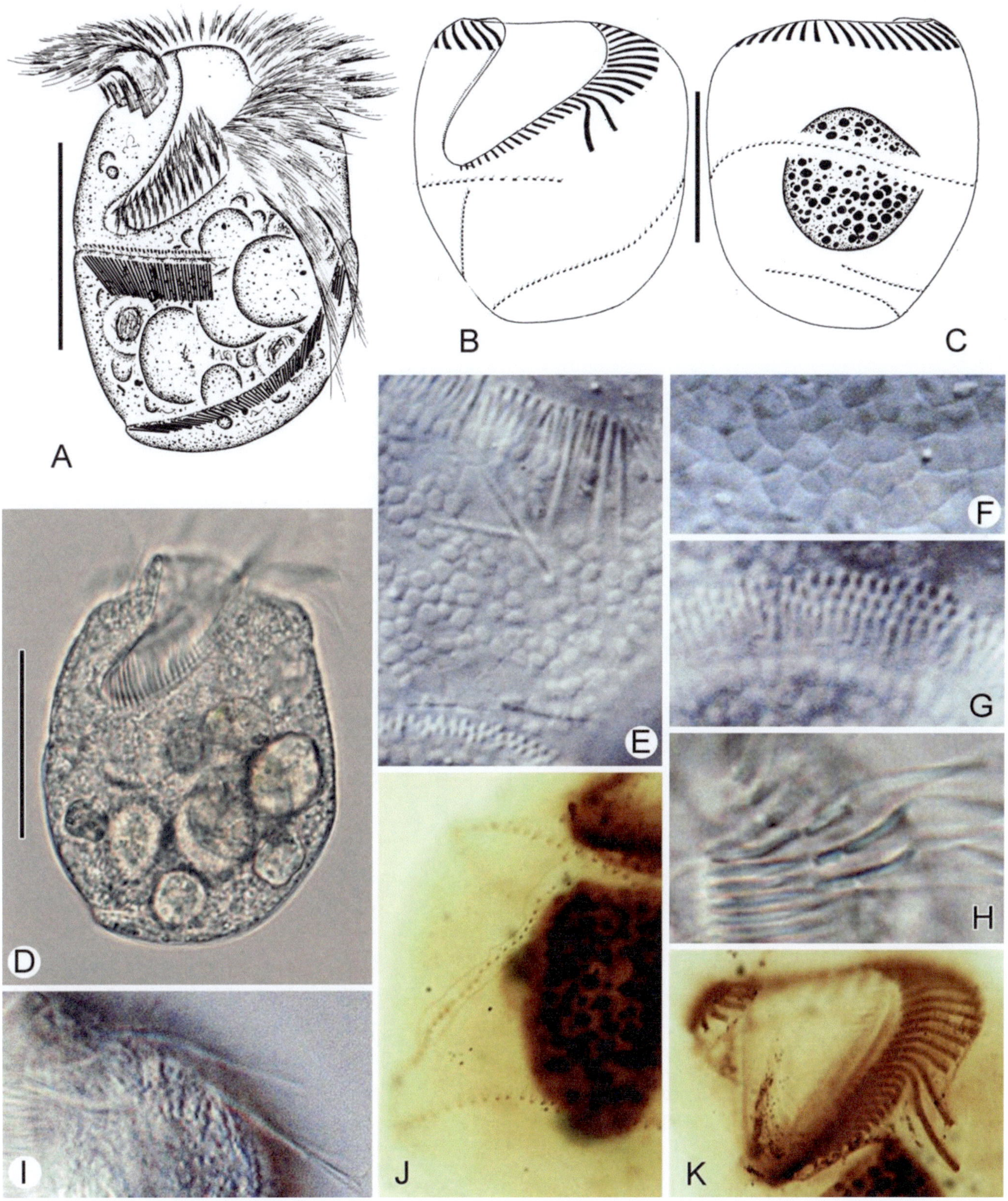


**Fig. 27B** *Parallelostrombidium obesum*
(**a**, **d**) Ventral views of live cells. (**b**, **c**) Ventral (**b**) and dorsal (**c**) views of ciliature. (**e**, **g**) Extrusomes. (**f**) Polygonal platelets. (**h**, **i**) Thigmotactic membranelles. (**j**) Somatic kineties. (**k**) Adoral zone of membranelles. Scale bars: 40 μm

## Reference

Liu W, Yi Z, Lin X, Li J, Al-Farraj SA, Al-Rasheid KAS, Song WB (2015) Morphology and molecular phylogeny of three new oligotrich ciliates (Protozoa, Ciliophora) from the South China Sea. Zool J Linn Soc 174:653–665

Class Spirotrichea Bütschli, 1889
Order Strombidiida Petz & Foissner, 1992
Family Strombidiidae Fauré-Fremiet, 1970
*Parallelostrombidium paraellipticum* Song et al., 2018

**Ecological feature:** offshore water
**Water temperature:** 29.2 °C
**pH:** 8.0
**Salinity:** 2.5‰

**Fig. 28A** Sample site

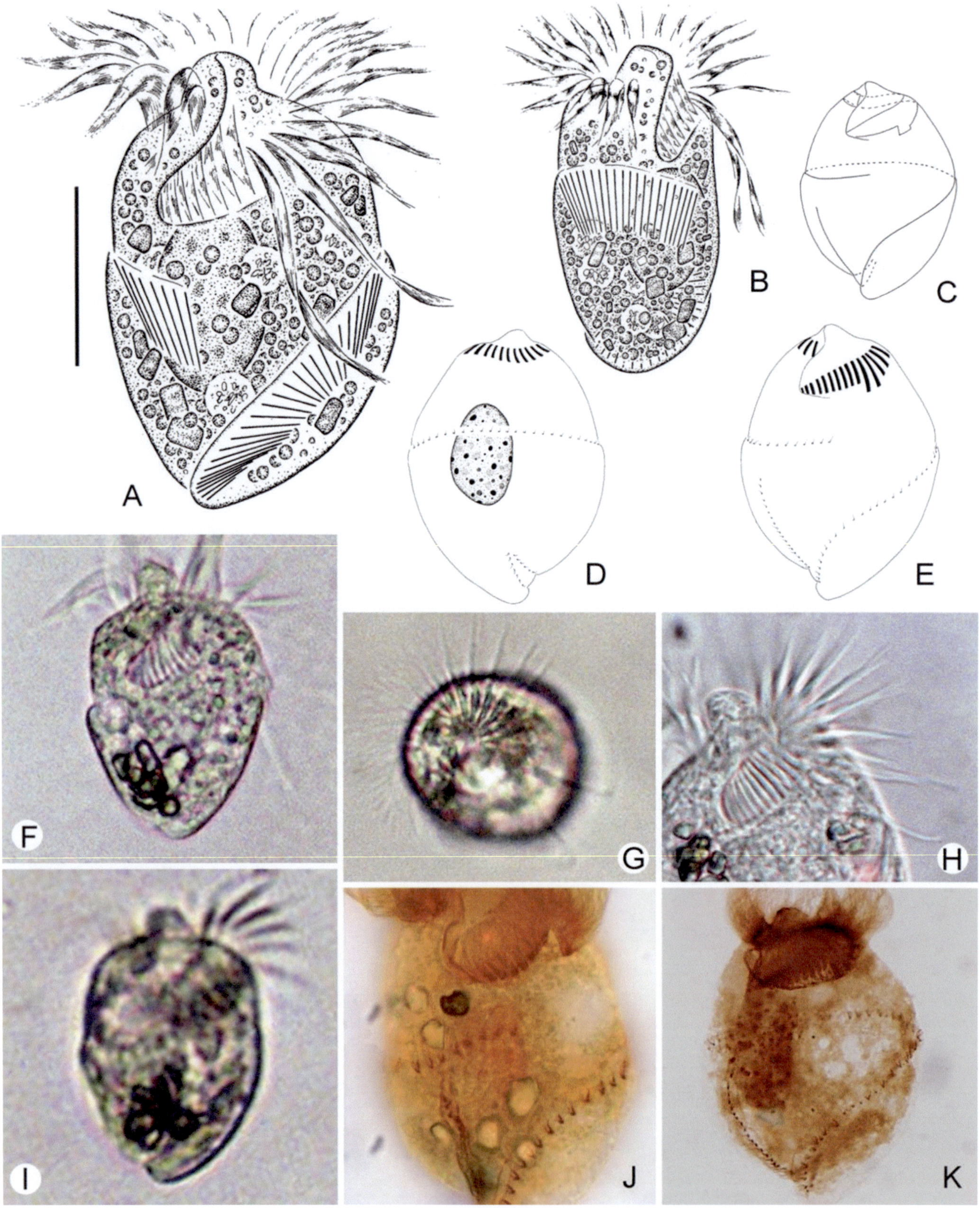


**Fig. 28B** *Parallelostrombidium paraellipticum*
(**a**, **b**, **f**, **i**) Ventral (A, F, I) and lateral (**b**) views of live individuals. (**c**) Ciliature. (**d**, **e**) Ventral (**d**) and dorsal (**e**) views of ciliature. (**g**) Apical view of membranelles. (**h**) Buccal field, showing thigmotactic membranelles. (**j**, **k**) Ventral views of ciliature. Scale bar: 20 μm

## Reference

Song W, Wang L, Li LF, Al-Farraj SA, Aleidan A, Smith S, Hu X (2018) Morphological characterizations of four species of *Parallelostrombidium* (Ciliophora, Oligotrichia), with a note on the phylogeny of the genus. J Eukaryot Microbiol doi:10.1111/jeu.12513

Class Spirotrichea Bütschli, 1889
Order Strombidiida Petz & Foissner, 1992
Family Strombidiidae Fauré-Fremiet, 1970
*Parallelostrombidium paralatum* Xu et al., 2006

**Ecological feature:** mangrove
**Water temperature:** 26.0 °C
**pH:** 8.9
**Salinity:** 25.9‰

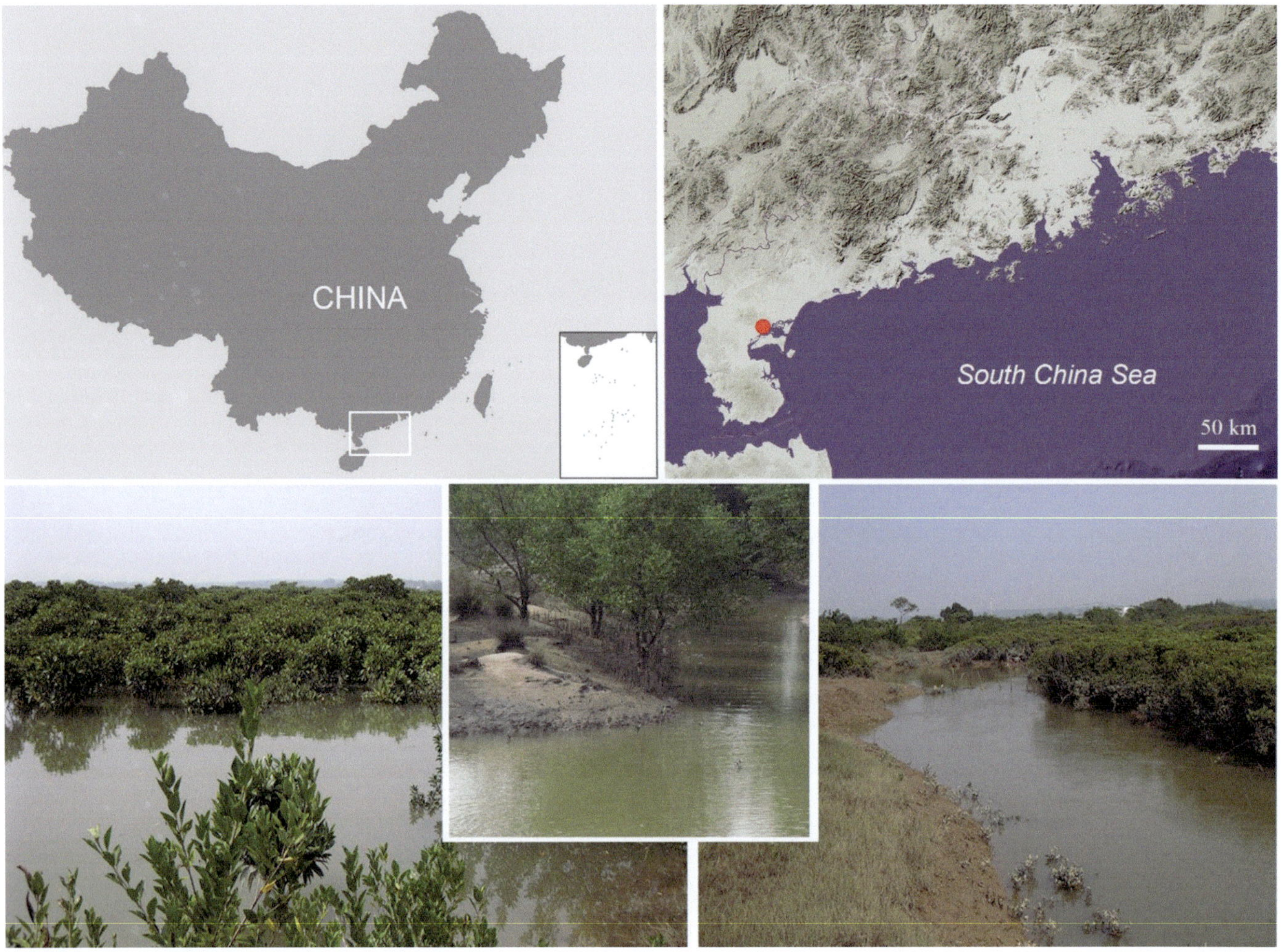


**Fig. 29A** Sample site

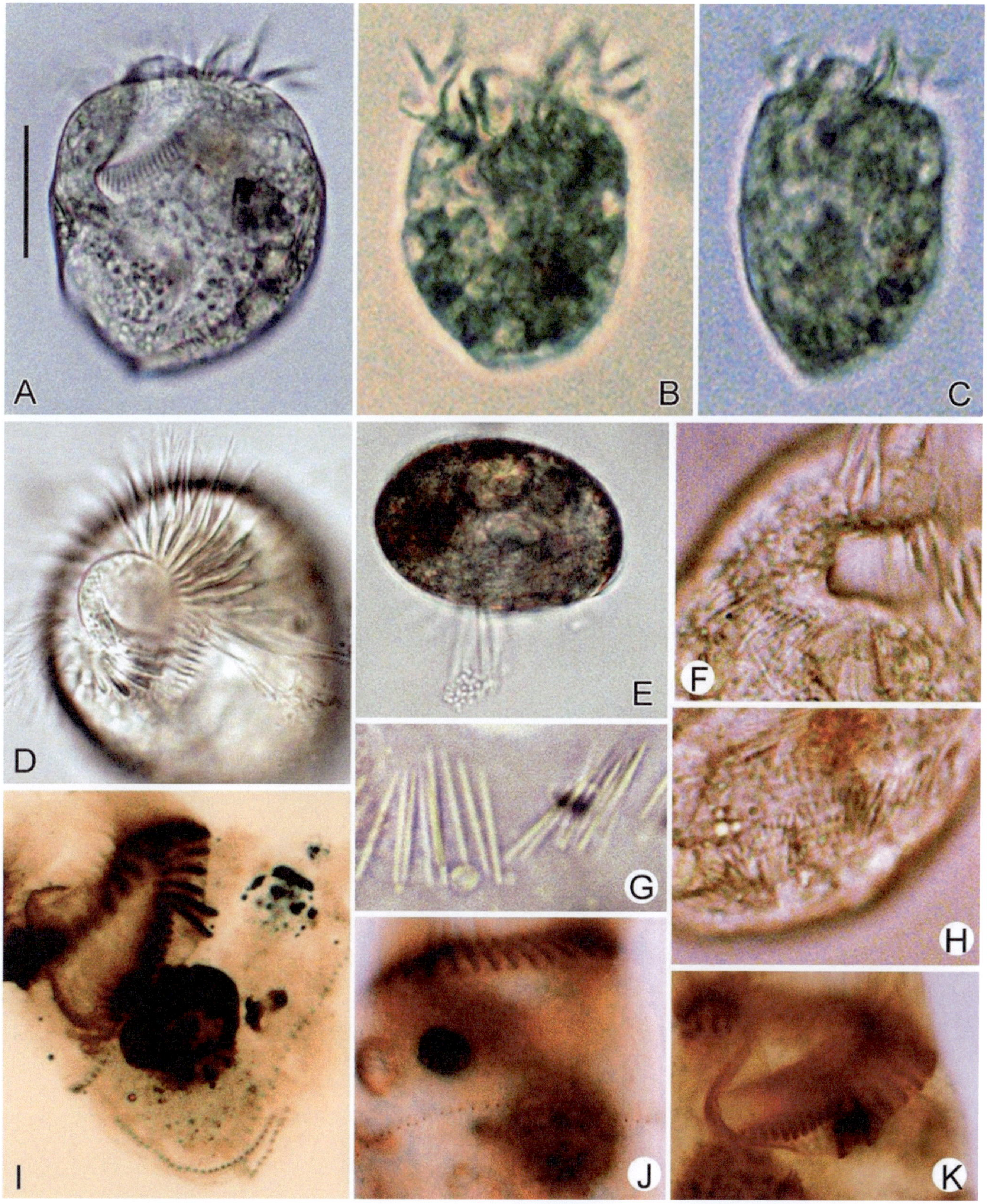


**Fig. 29B** *Parallelostrombidium paralatum*
(**a–c**) Ventral views of live cells, showing different body shapes. (**d**) Arrangement of adoral membranelles. (**e**) Lateral view to show thigmotactic membranelles. (**f–h**) Distribution of extrusomes. (**i**) Ventral view of ciliature. (**j**) Showing girdle kinety. (**k**) Adoral zone of membranelles. Scale bar: 30 μm

## Reference

Xu D, Song WB, Warren A (2006) Morphology and infraciliature of two new species of marine oligotrich ciliates (Ciliophora: Oligotrichida) from China. J Nat Hist 40:1287–1299

Class Spirotrichea Bütschli, 1889
Order Strombidiida Petz & Foissner, 1992
Family Strombidiidae Fauré-Fremiet, 1970
*Sinistrostrombidium cupiformum* Liu et al., 2015

**Ecological feature:** intertidal zone
**Water temperature:** 27.8 °C
**pH:** 8.6
**Salinity:** 31.0‰

**Fig. 30A** Sample site

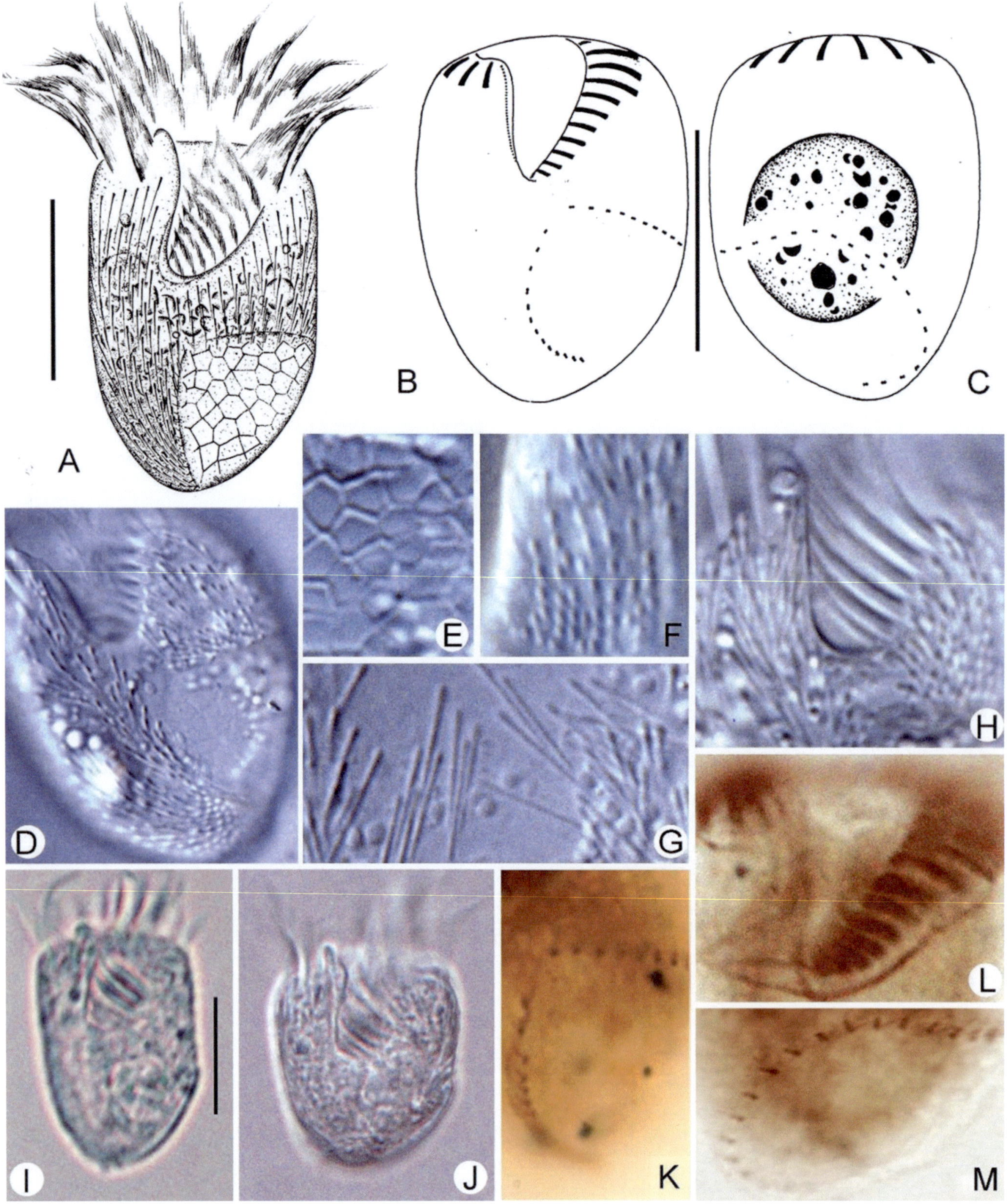


**Fig. 30B** *Sinistrostrombidium cupiformum*
(**a**, **i**, **j**) Ventral views of live cells. (**b**, **c**) Ventral (**b**) and dorsal (**c**) views of ciliature. (**d**, **g**, **h**) Extrusomes. (**e**) Polygonal platelets. (**k**, **m**) Part of somatic kineties. (**l**) Adora zone of membranelles. Scale bars: 20 μm

## Reference

Liu W, Yi Z, Xu D, Clamp J, Li J, Lin X, Song WB (2015) Two new genera of planktonic ciliates and insights into the evolution of the family Strombidiidae (Protista, Ciliophora, Oligotrichia). PLoS ONE 10(6):e0131726

Class Spirotrichea Bütschli, 1889
Order Strombidiida Petz & Foissner, 1992
Family Strombidiidae Fauré-Fremiet, 1970
*Spirostrombidium apourceolare* Liu et al., 2013

**Ecological feature:** intertidal zone
**Water temperature:** 21.7 °C
**pH:** 8.2
**Salinity:** 33.3‰

**Fig. 31A** Sample site

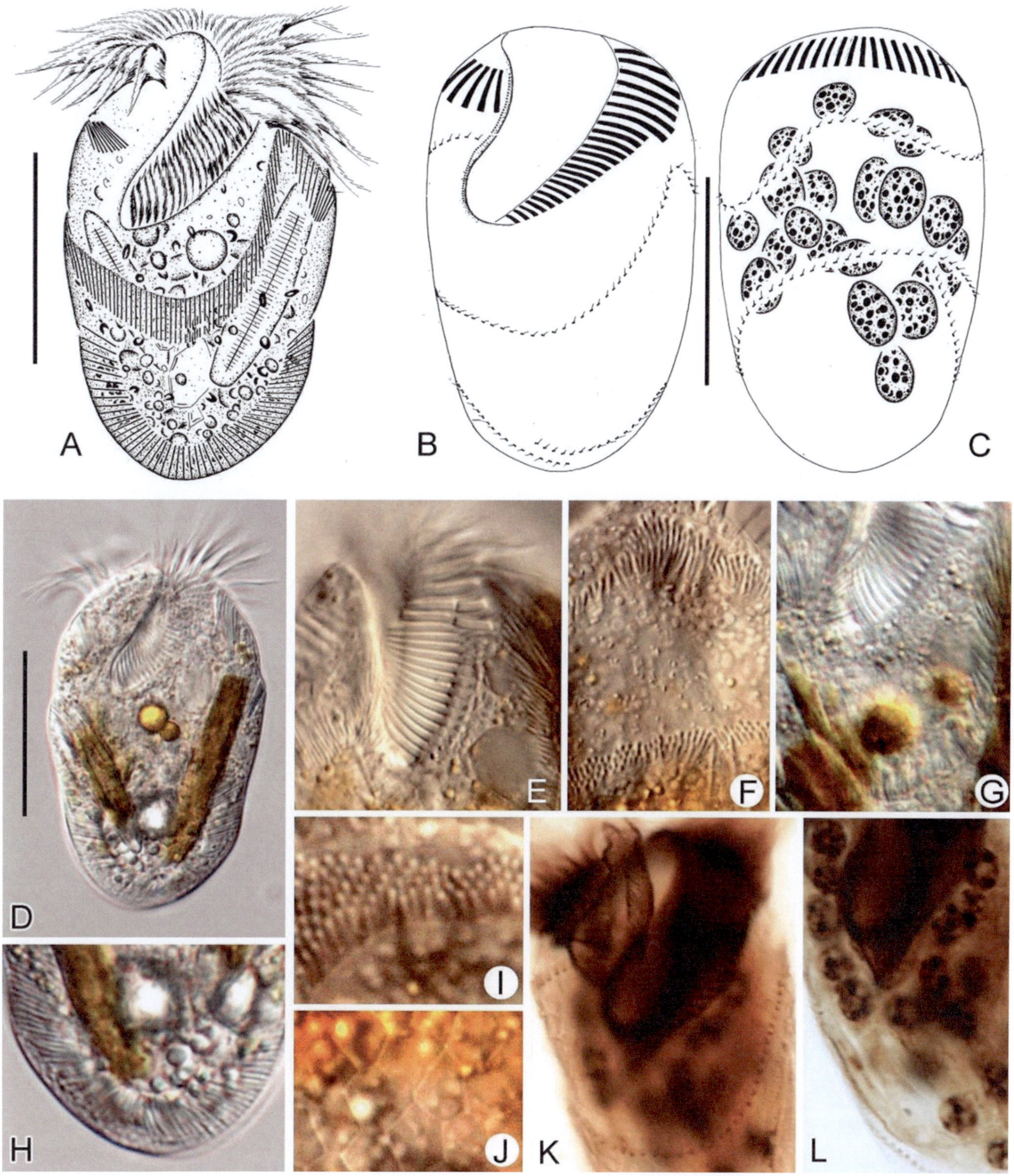


**Fig. 31B** *Spirostrombidium apourceolare*
(**a**, **d**) Ventral views in vivo. (**b**, **c**) Ventral (**b**) and dorsal (**c**) views of ciliature. (**e**) Ventral view of buccal field showing ventral membranelles. (**f–i**) Distribution of extrusomes. (**j**) Polygonal platelets. (**k**) Ventral view of ciliature. (**l**) Macronuclear nodules scattered in the body. Scale bars: 30 μm

## Reference

Liu W, Yi Z, Li J, Warren A, Al-Farraj SA, Lin X (2013) Taxonomy, morphology and phylogeny of three new oligotrich ciliates (Protozoa, Ciliophora, Oligotrichia) from southern China. Int J Syst Evol Microbiol 63:4805–4817

Class Spirotrichea Bütschli, 1889
Order Strombidiida Petz & Foissner, 1992
Family Strombidiidae Fauré-Fremiet, 1970
*Spirostrombidium cinctum* (Kahl, 1932) Petz et al., 1995

**Ecological feature:** mollusc-farming pond
**Water temperature:** 19.4 °C
**pH:** 8.2
**Salinity:** 32.6‰

**Fig. 32A** Sample site

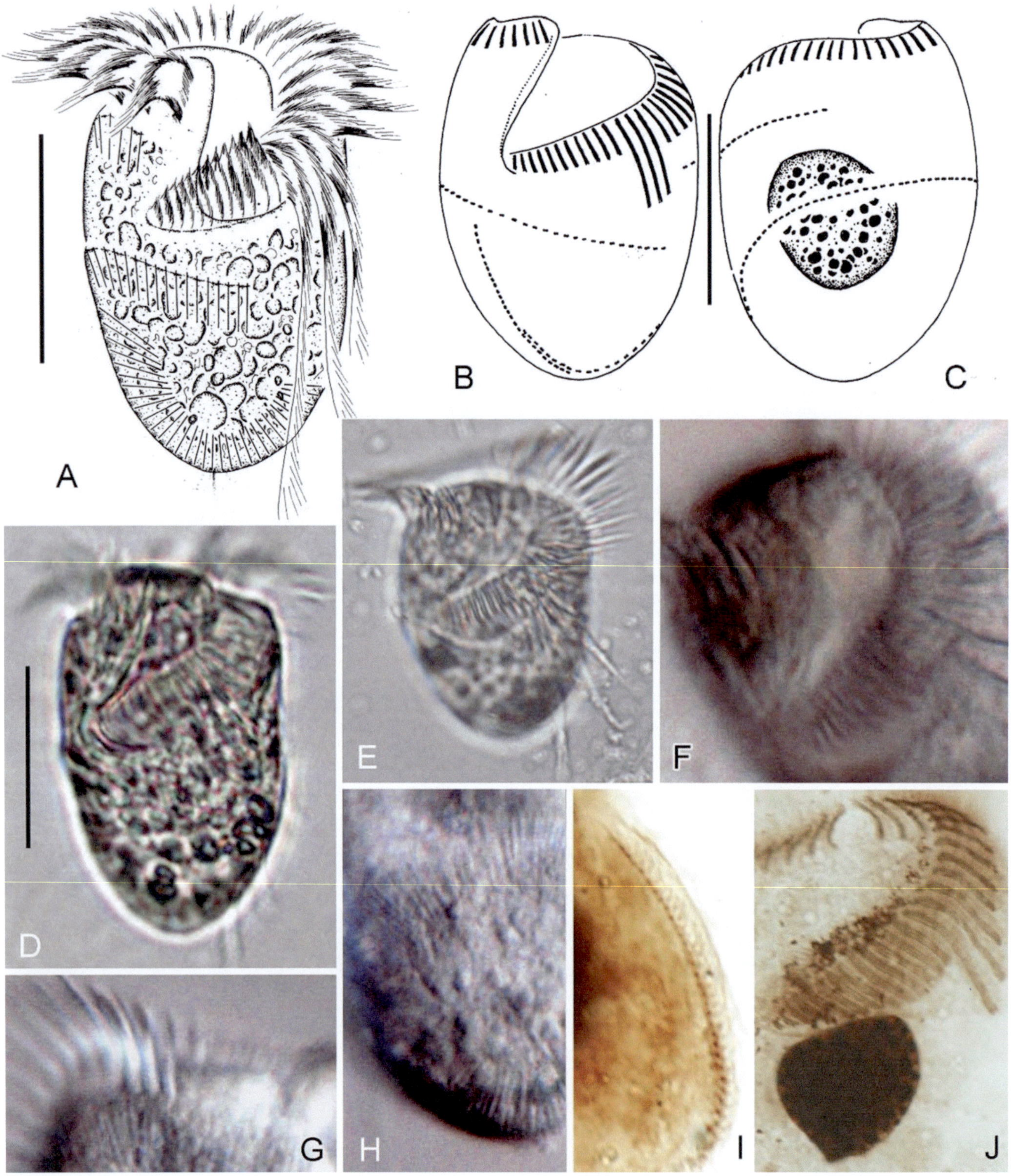


**Fig. 32B** *Spirostrombidium cinctum*
(**a**, **d**, **e**) Ventral views of live cells. (**b**, **c**) Ventral (**b**) and dorsal (**c**) views of ciliature. (**f**) Ventral view to show the oral membranelles. (**g**) Collar membranelles. (**h**) Extrusomes. (**i**) Ventral kinety. (**j**) Ventral view of anterior portion of cell to show adoral membranelles and macronucleus. Scale bars: 30 μm

## Reference

Xu D, Song WB (2006) Hapantotypification and morphological redescription of the marine planktonic ciliate, *Spirostrombidium cinctum* (Kahl, 1932) Petz et al., 1995 (Protozoa: Ciliophora: Oligotrichida). Acta Protozool 45:17–25

Class Spirotrichea Bütschli, 1889
Order Strombidiida Petz & Foissner, 1992
Family Strombidiidae Fauré-Fremiet, 1970
*Spirostrombidium paraurceolare* Wang et al., 2018

**Ecological feature:** mangrove
**Water temperature:** 24.5 °C
**pH:** 6.8
**Salinity:** 17.0‰

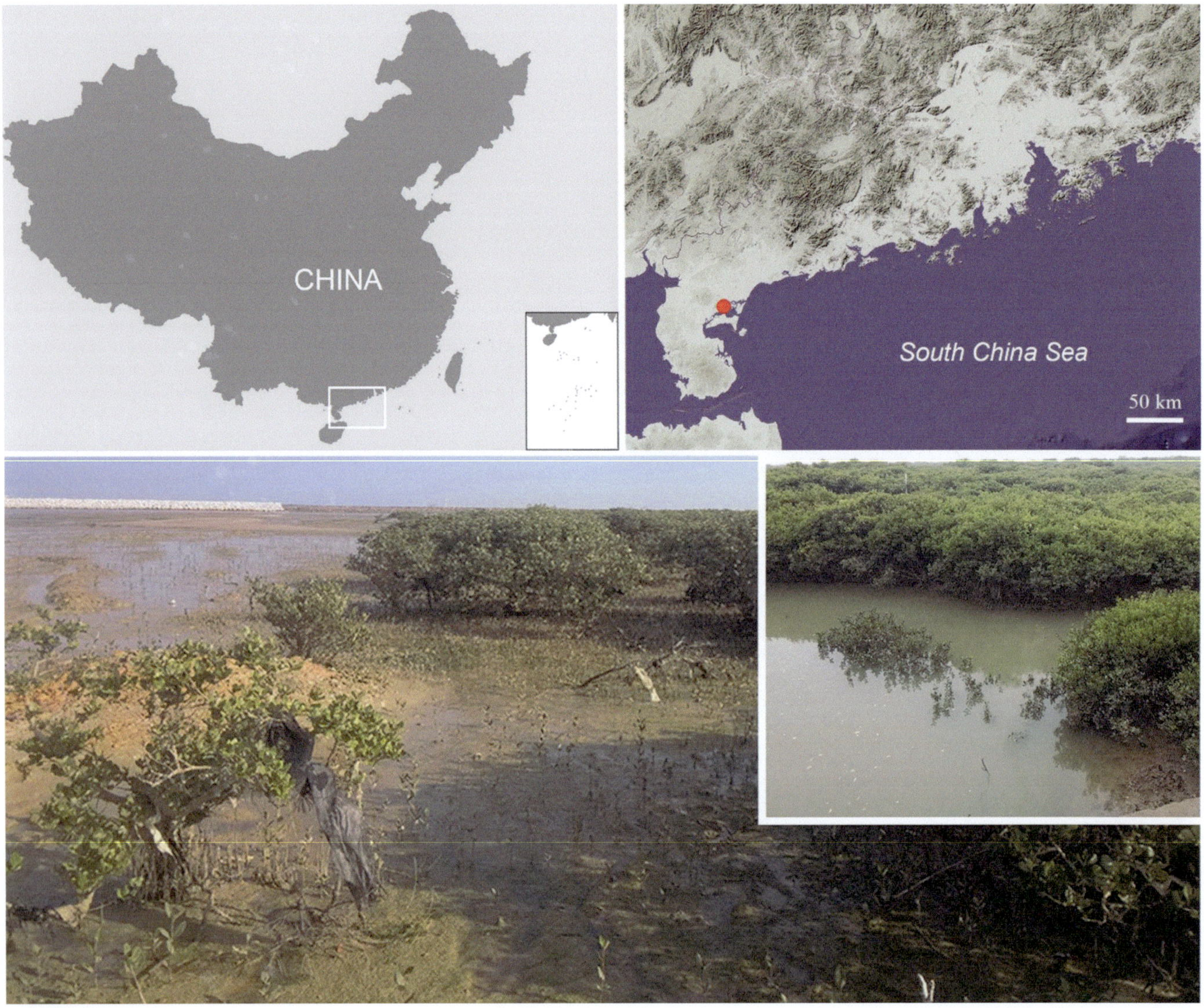


**Fig. 33A** Sample site

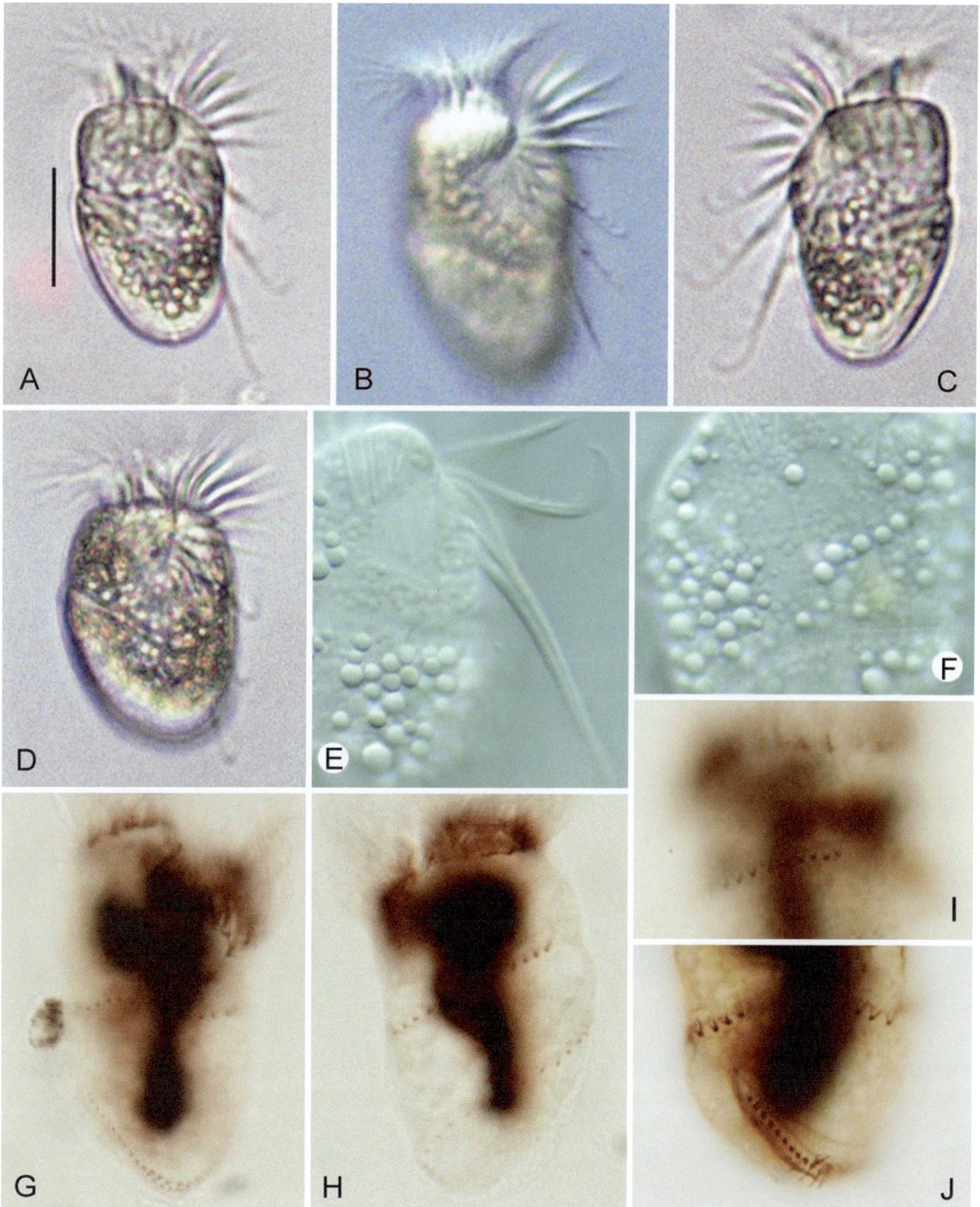


**Fig. 33B** *Spirostrombidium paraurceolare*
(**a**, **b**, **d**) Ventral views of live individuals. (**c**) Dorsal view of live individual. (**e**) Thigmotactic membranelles. (**f**) Extrusomes. (**g**, **h**) Ventral (**g**) and dorsal (**h**) views of ciliature. (**i**) Part of girdle kinety. (**j**) Part of somatic kineties. Scale bar: 20 μm

## Reference

Wang R, Song W, Yi Z, Al-Rasheid KAS, Hu X (2018) Morphology and molecular phylogeny of two new species of *Spirostrombidium* (Ciliophora, Oligotrichia), with a key to species in the genus. Syst Biodivers (accepted)

Class Spirotrichea Bütschli, 1889
Order Strombidiida Petz & Foissner, 1992
Family Strombidiidae Fauré-Fremiet, 1970
*Spirostrombidium schizostomum* (Kahl, 1932) Xu et al., 2006

**Ecological feature:** mangrove
**Water temperature:** 27.0 °C
**pH:** 8.2
**Salinity:** 17.0‰

**Fig. 34A** Sample site

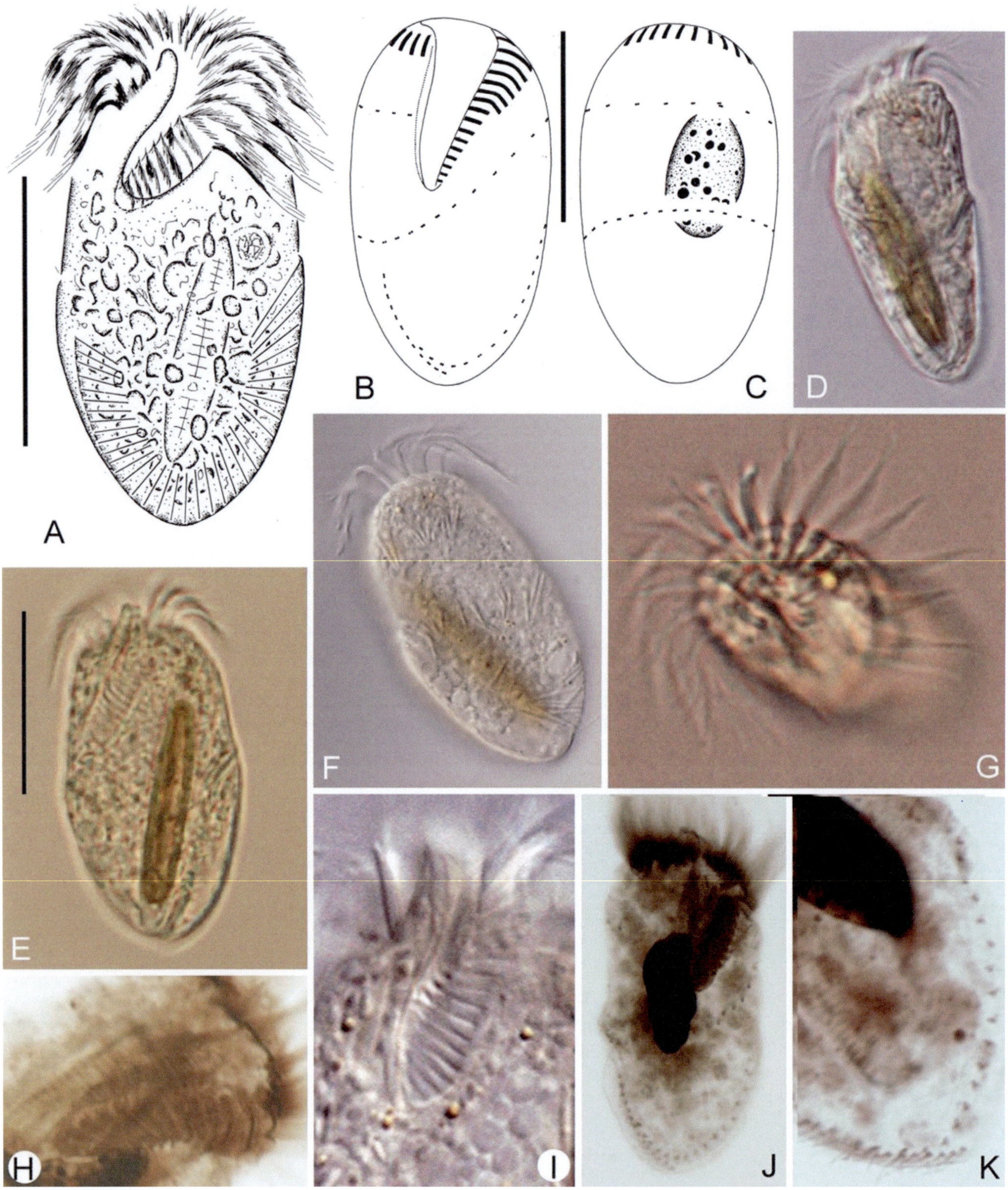


**Fig. 34B** *Spirostrombidium schizostomum*
(**a**, **e**) Ventral views of live cells. (**b**, **c**, **j**) Ventral (**b**, **j**) and dorsal (**c**) views of ciliature. (**d**) Lateral view of live cell. (**f**) Dorsal view of live cell, showing the distribution of extrusomes. (**g**) Apical view showing oral membranelles. (**h**, **i**) Ventral views of anterior portion of cell showing buccal field. (**k**) Part of somatic kineties. Scale bars: 30 μm

## Reference

Xu D, Song WB, Lin X (2006) On two marine oligotrich ciliates, *Spirostrombidium agathae* nov. spec. and *S. schizostomum* (Kahl, 1932) nov. comb. from China, with a key to the identification of seven well-characterized *Spirostrombidium* spp. (Ciliophora: Oligotrichida). Acta Protozool 45:433–442

Class Spirotrichea Bütschli, 1889
Order Strombidiida Petz & Foissner, 1992
Family Strombidiidae Fauré-Fremiet, 1970
*Spirostrombidium subtropicum* Liu et al., 2013

**Ecological feature:** mangrove
**Water temperature:** 20.8 °C
**pH:** 7.7
**Salinity:** 21.3‰

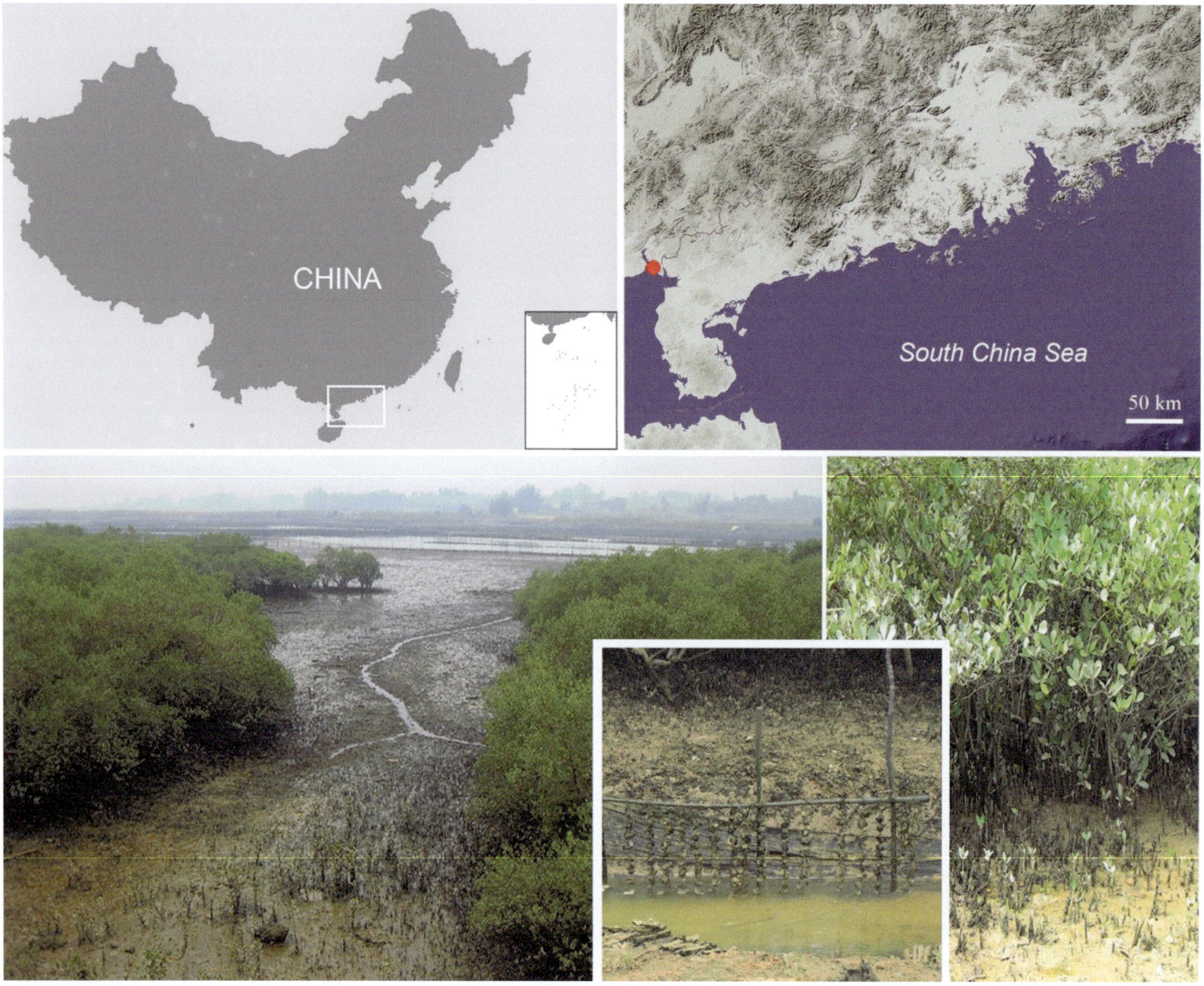


**Fig. 35A** Sample site

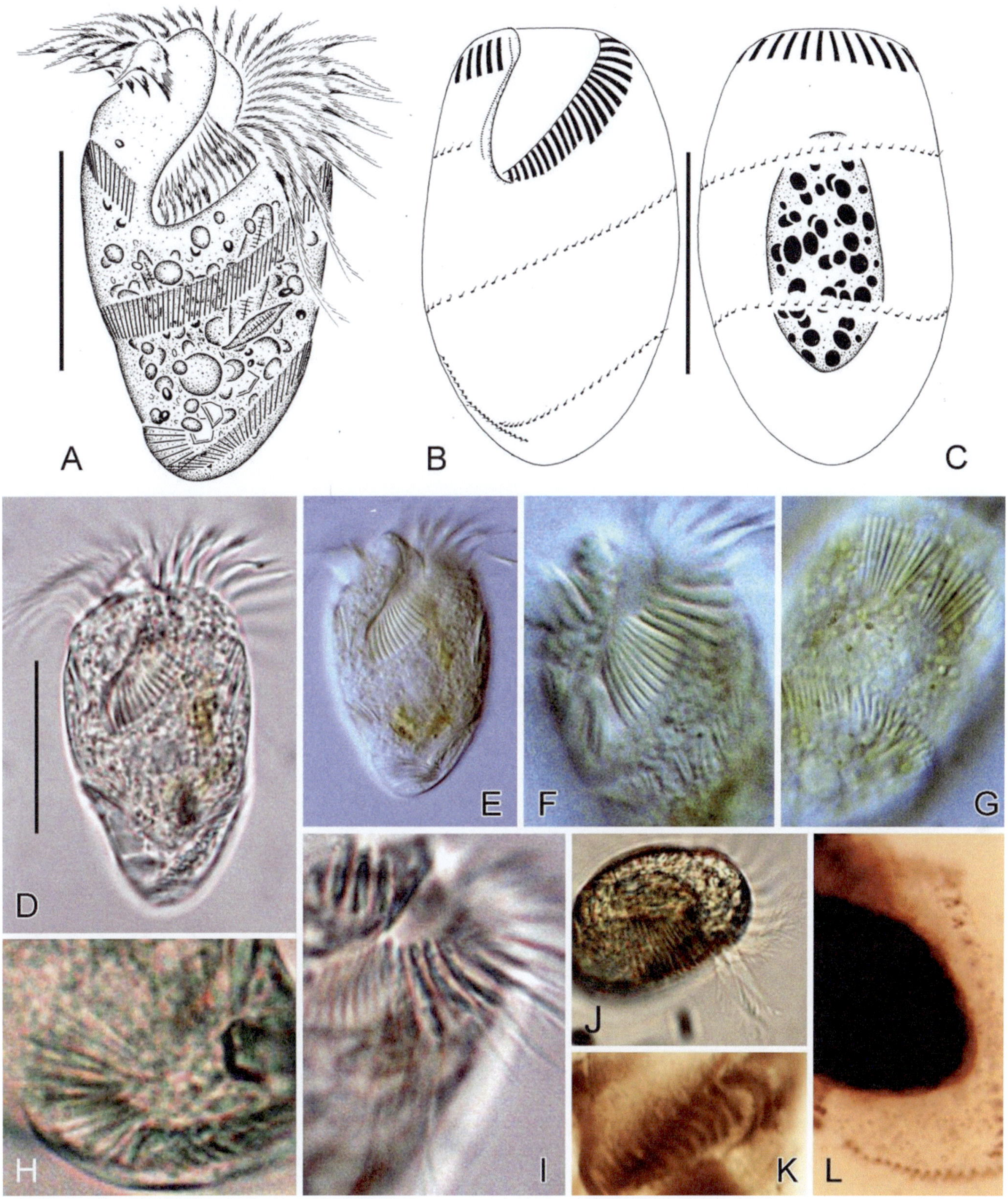


**Fig. 35B** *Spirostrombidium subtropicum*
(**a**, **d**, **e**) Ventral views of live cells. (**b**, **c**) Ventral (**b**) and dorsal (**c**) views of ciliature. (**f**) Ventral view of anterior portion to show buccal field and extrusomes. (**g**, **h**) Distribution of extrusomes. (**i**) Oral membranelles. (**j**) Apical view to show thigmotactic membranelles. (**k**) Bases of ventral membranelles. (**l**) Part of somatic kinety. Scale bars: 30 μm

## Reference

Liu W, Yi Z, Li J, Warren A, Al-Farraj SA, Lin X (2013) Taxonomy, morphology and phylogeny of three new oligotrich ciliates (Protozoa, Ciliophora, Oligotrichia) from southern China. Int J Syst Evol Microbiol 63:4805–4817

Class Spirotrichea Bütschli, 1889
Order Strombidiida Petz & Foissner, 1992
Family Strombidiidae Fauré-Fremiet, 1970
*Strombidium apolatum* Wilbert & Song, 2005

**Ecological feature:** offshore water
**Water temperature:** 22.7 °C
**pH:** 7.9
**Salinity:** 26.0‰

**Fig. 36A** Sample site

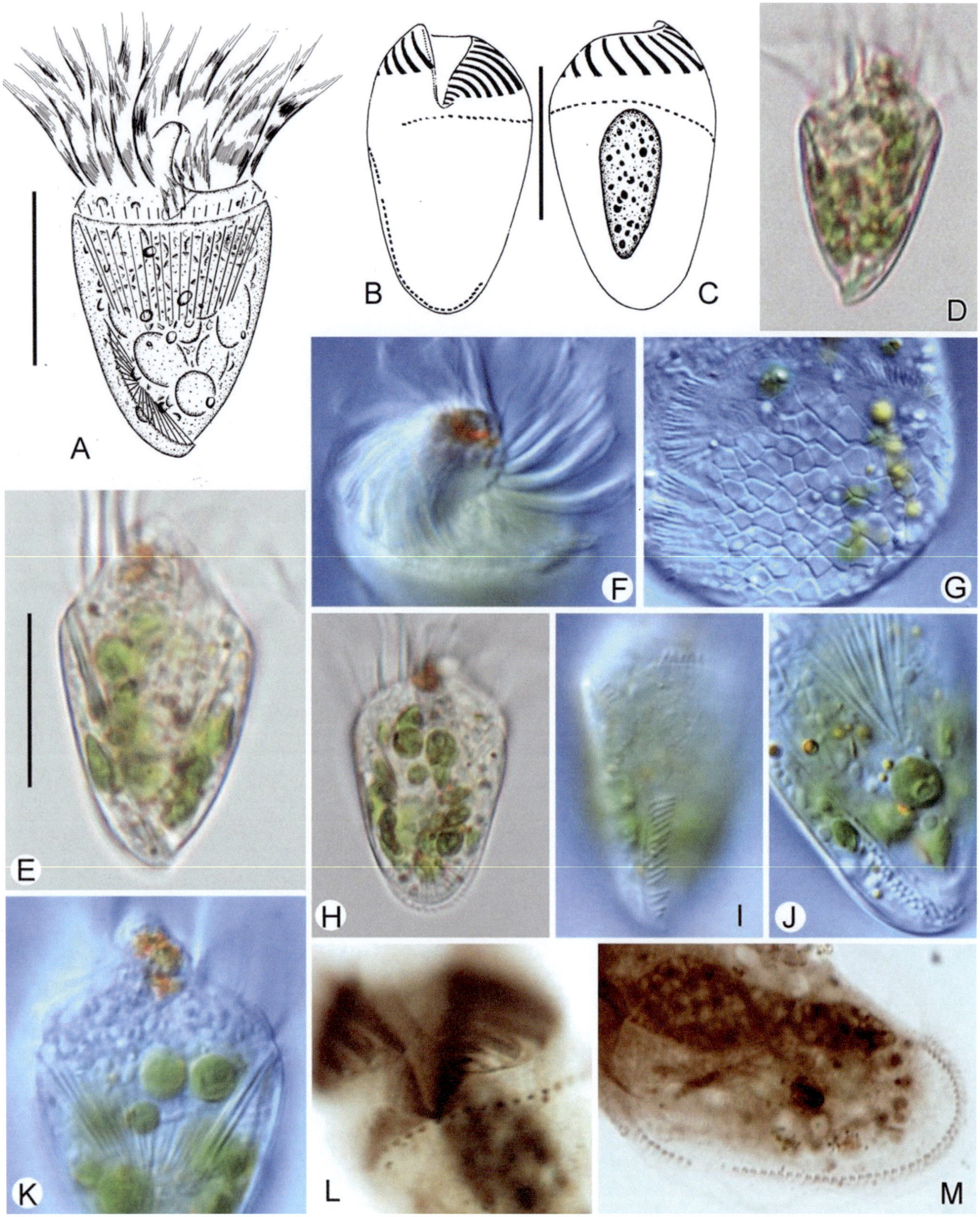


**Fig. 36B** *Strombidium apolatum*
(**a**, **d**, **e**, **h**) Ventral views of live cells. (**b**, **c**) Ventral (**b**) and dorsal (**c**) views of ciliature. (**f**) Pigment spot and collar membranelles. (**g**) Polygonal platelets and anterior ends of extrusomes. (**i**) Somatic cilia. (**j**, **k**) Arrangement of extrusomes. (**l**, **m**) Oral membranelles and somatic kineties. Scale bars: 20 μm

## References

Wilbert N, Song WB (2005) New contributions to the marine benthic ciliates from the Antarctic area, including description of seven new species (Protozoa, Ciliophora). J Nat Hist 39:935–973

Xu D, Song WB, Hu X (2005) Notes on two marine ciliates from the Yellow Sea, China: *Placus salinus* and *Strombidium apolatum* (Protozoa, Ciliophora). J Ocean Univ China 14:137–144

Class Spirotrichea Bütschli, 1889
Order Strombidiida Petz & Foissner, 1992
Family Strombidiidae Fauré-Fremiet, 1970
*Strombidium basimorphum* Martin & Montagnes, 1993

**Ecological feature:** mangrove
**Water temperature:** 18.0 °C
**pH:** 8.1
**Salinity:** 10.4‰

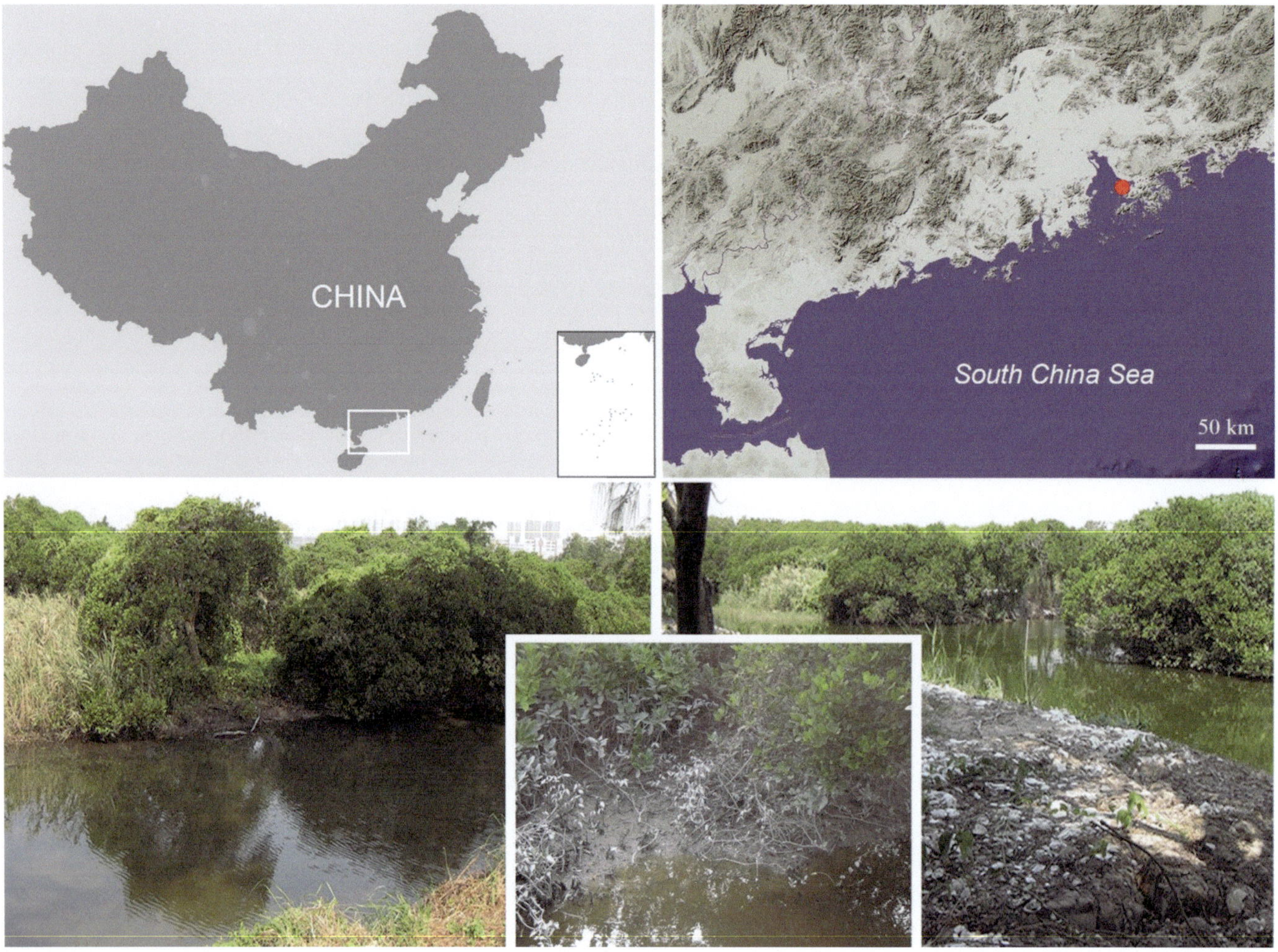


**Fig. 37A** Sample site

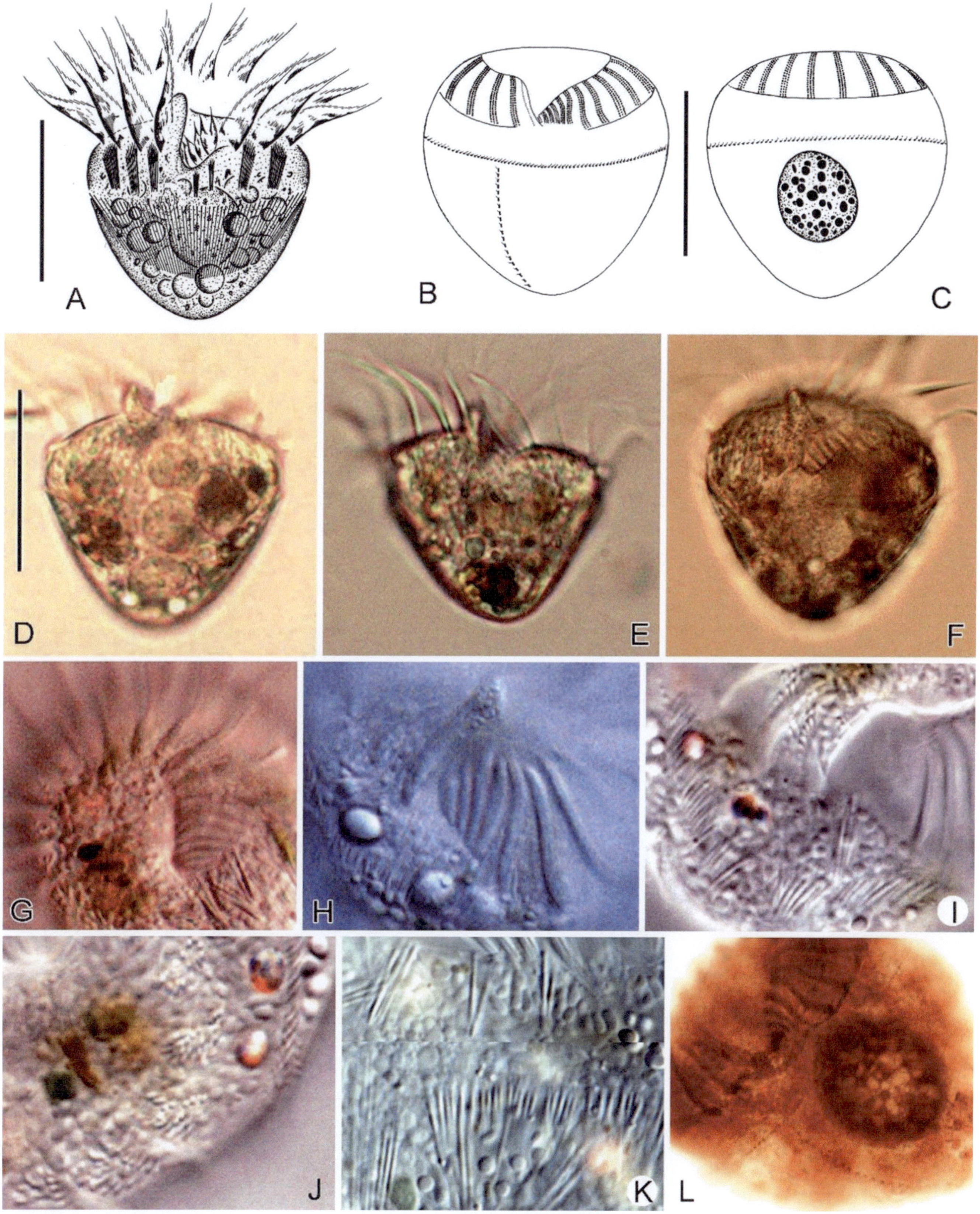


**Fig. 37B** *Strombidium basimorphum*
(**a**, **d–f**) Ventral views of live cells. (**b**, **c**) Ventral (**b**) and dorsal (**c**) views of ciliature. (**g**, **h**) Apical views of live cells to show oral membranelles. (**i–k**) Extrusomes. (**l**) Ventral view of ciliature. Scale bars: 40 μm

## References

Liu W, Yi Z, Warren A, Al-Rasheid KAS, Al-Farraj SA, Lin X, Song WB (2011) Taxonomy, morphology and molecular systematics of a new oligotrich ciliates, *Williophrya maedai* gen. nov., sp. nov., with redescriptions of *Strombidium basimorphum* and *Pseudotontonia simplicidens* (Protozoa, Ciliophora, Oligotrichia). Syst Biodiver 9:247–258

Martin A J, Montagnes D J S. 1993. Winter ciliates in a British Columbian Fjord: six new species and an analysis of ciliate putative prey. J Eukaryot Microbiol 40:535–549

Class Spirotrichea Bütschli, 1889
Order Strombidiida Petz & Foissner, 1992
Family Strombidiidae Fauré-Fremiet, 1970
*Strombidium capitatum* (Leegaard, 1915) Kahl, 1932

**Ecological feature:** offshore water
**Water temperature:** 23.1 °C
**pH:** 8.2
**Salinity:** 28.6‰

**Fig. 38A** Sample site

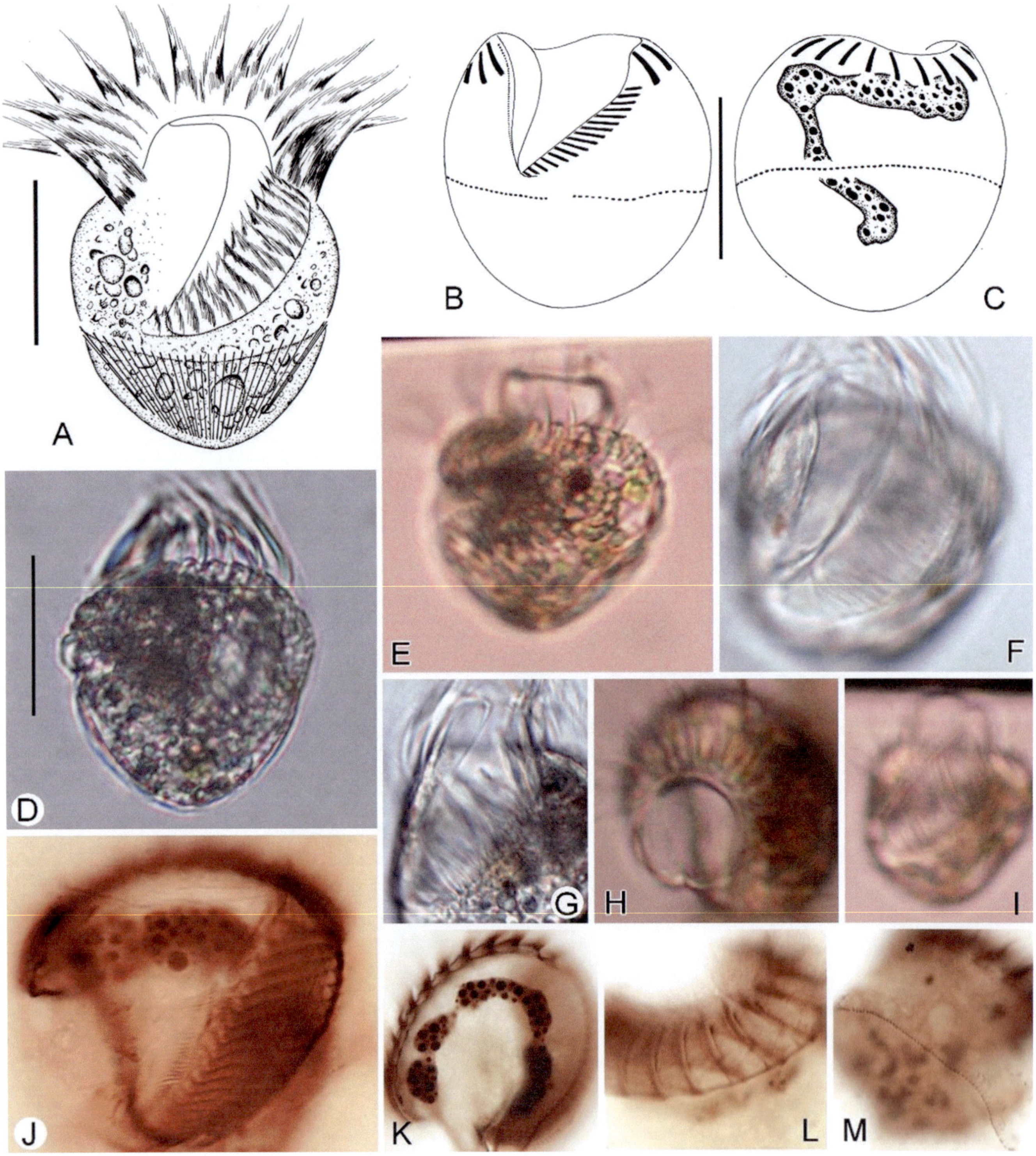


**Fig. 38B** *Strombidium capitatum*
(**a**, **d**, **e**) Ventral (**a**, **d**) and dorsal (**e**) views of live cells. (**b**, **c**) Ventral (**b**) and dorsal (**c**) views of ciliature. (**f**, **g**, **i**) Apical-ventral (F, G) and ventral (**i**) views to show wide buccal field. (**h**) Apical view to show collar membranelles. (**j**) Adoral zone of membranelles. (**k**) Macronucleus. (**l**) Collar membranelles. (**m**) Girdle kinety. Scale bars: 30 μm

## Reference

Agatha S, Riedel-Lorjé J (1997) Morphology, infraciliature, and ecology of halteriids and strombidiids (Ciliophora, Oligotrichea) from coastal brackish water basins. Arch Protistenk 148:445–459

Class Spirotrichea Bütschli, 1889
Order Strombidiida Petz & Foissner, 1992
Family Strombidiidae Fauré-Fremiet, 1970
*Strombidium caudispina* Song et al., 2015

**Ecological feature:** offshore water
**Water temperature:** 12.0 °C
**pH:** 8.1
**Salinity:** 33.4‰

**Fig. 39A** Sample site

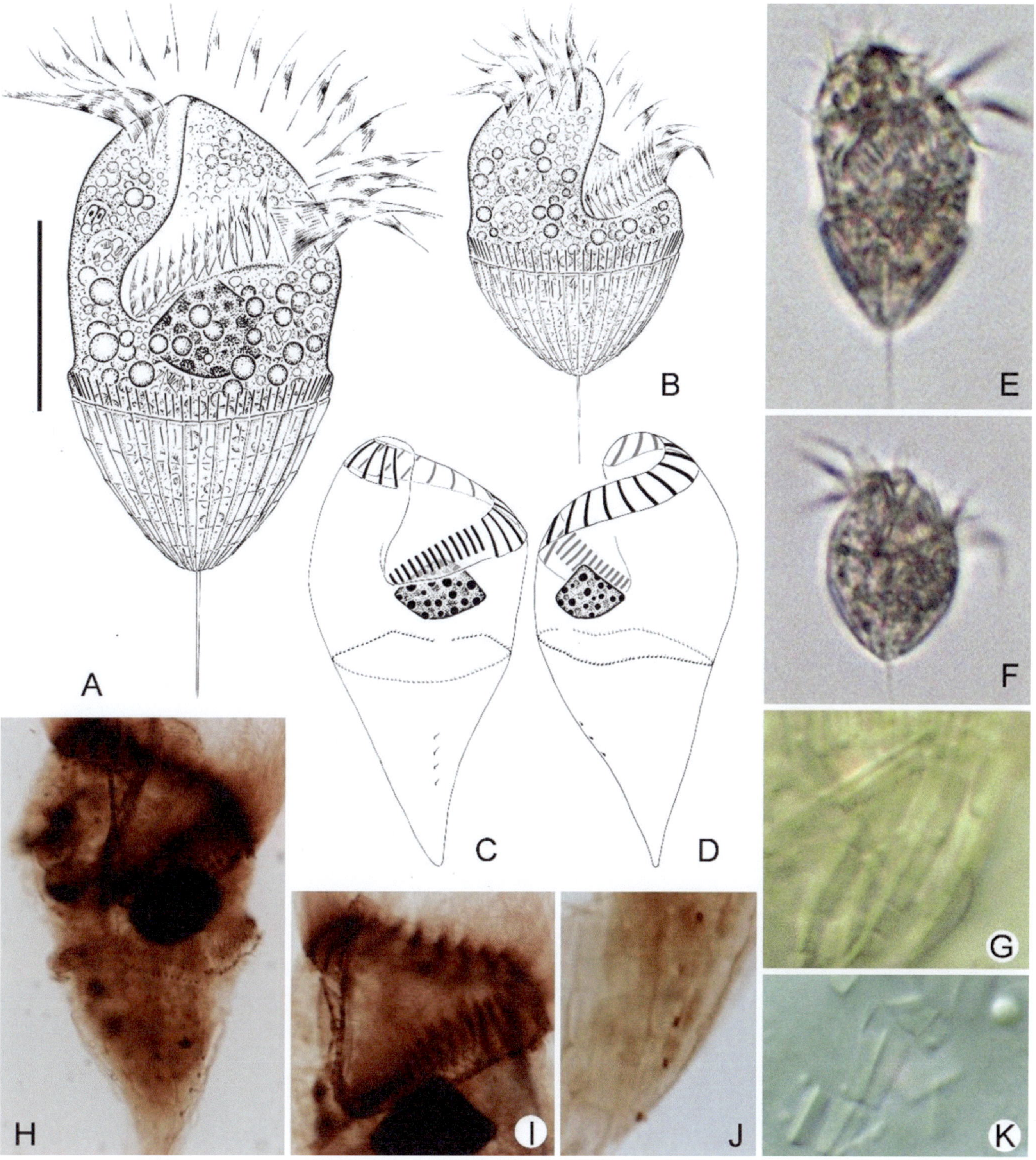


**Fig. 39B** *Strombidium caudispina*
(**a**, **b**, **e**, **f**) Ventral views of live individuals, showing posterior spine. (**c**, **d**) Ventral (**c**) and dorsal (**d**) views of ciliature and macronucleus. (**g**) Hemitheca. (**h**) Ventral view of ciliature. (**i**) Ventral view of buccal field, showing adoral zone of membranelles. (**j**) Ventral kinety. (**k**) Platelets of hemitheca. Scale bar: 20 μm

## Reference

Song W, Zhao X, Liu W, Hu X, Al-Farraj SA, Al-Rasheid KAS, Song WB, Warren A (2015) Biodiversity of oligotrich ciliates in the South China Sea: description of three new *Strombidium* species (Protozoa, Ciliophora, Oligotrichia) with phylogenetic analyses. Syst Biodivers 13:608–623

Class Spirotrichea Bütschli, 1889
Order Strombidiida Petz & Foissner, 1992
Family Strombidiidae Fauré-Fremiet, 1970
*Strombidium conicum* (Lohmann, 1908) Wulff, 1919

**Ecological feature:** intertidal zone
**Water temperature:** 21.7 °C
**pH:** 8.2
**Salinity:** 33.3‰

**Fig. 40A** Sample site

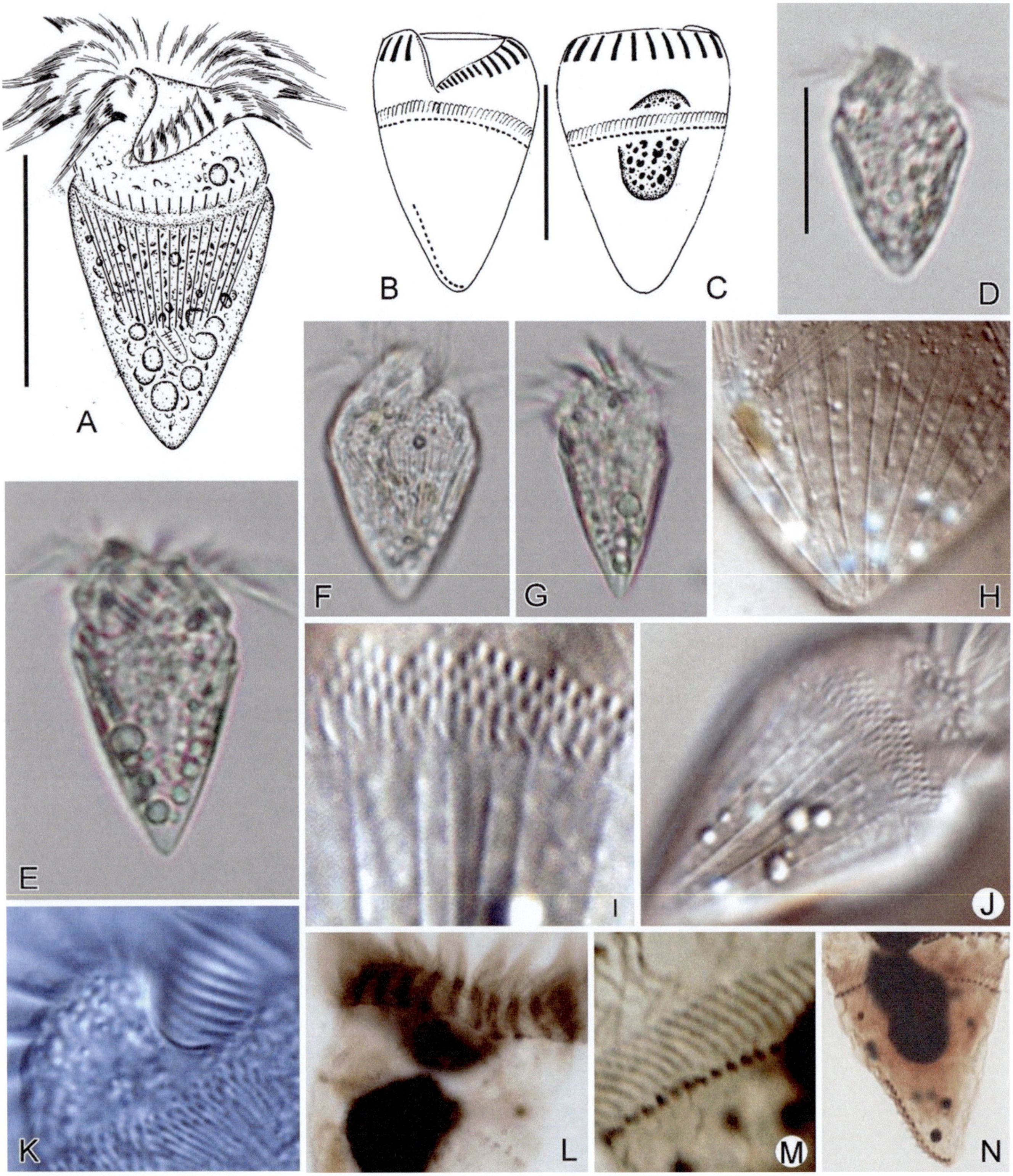


**Fig. 40B** *Strombidium conicum*
(**a**, **d–g**) Ventral views of live cells. (**b**, **c**) Ventral (**b**) and dorsal (**c**) views of ciliature. (**h**, **j**) Cell surface to show the perilemma with the longitudinal lines. (**i**) Anterior ends of extrusomes. (**k**) Ventral view of anterior portion in vivo to show the extrusomes and ventral membranelles. (**l**) Collar membranelles and girdle kinety. (**m**) Girdle kinety and extrusomes girdle. (**n**) Posterior portion of cell to show ventral and girdle kineties. Scale bars: 30 μm

## References

Agatha S, Riedel-Lorjé JC (1997) Morphology, infraciliature, and ecology of halteriids and strombidiids (Ciliophora, Oligotrichea) from coastal brackish water basins. Arch Protistenk 148:445–459

Montagnes D, Lynn D, Stoecker D, Small E (1988) Taxonomic description of one new species and redescription of four species in the family Strombidiidae (Ciliophora, Oligotrichida). J Protozool 35:189–197

Class Spirotrichea Bütschli, 1889
Order Strombidiida Petz & Foissner, 1992
Family Strombidiidae Fauré-Fremiet, 1970
*Strombidium cuneiforme* Song et al., 2015

**Ecological feature:** mangrove
**Water temperature:** 24.5 °C
**pH:** 6.8
**Salinity:** 17.0‰

**Fig. 41A** Sample site

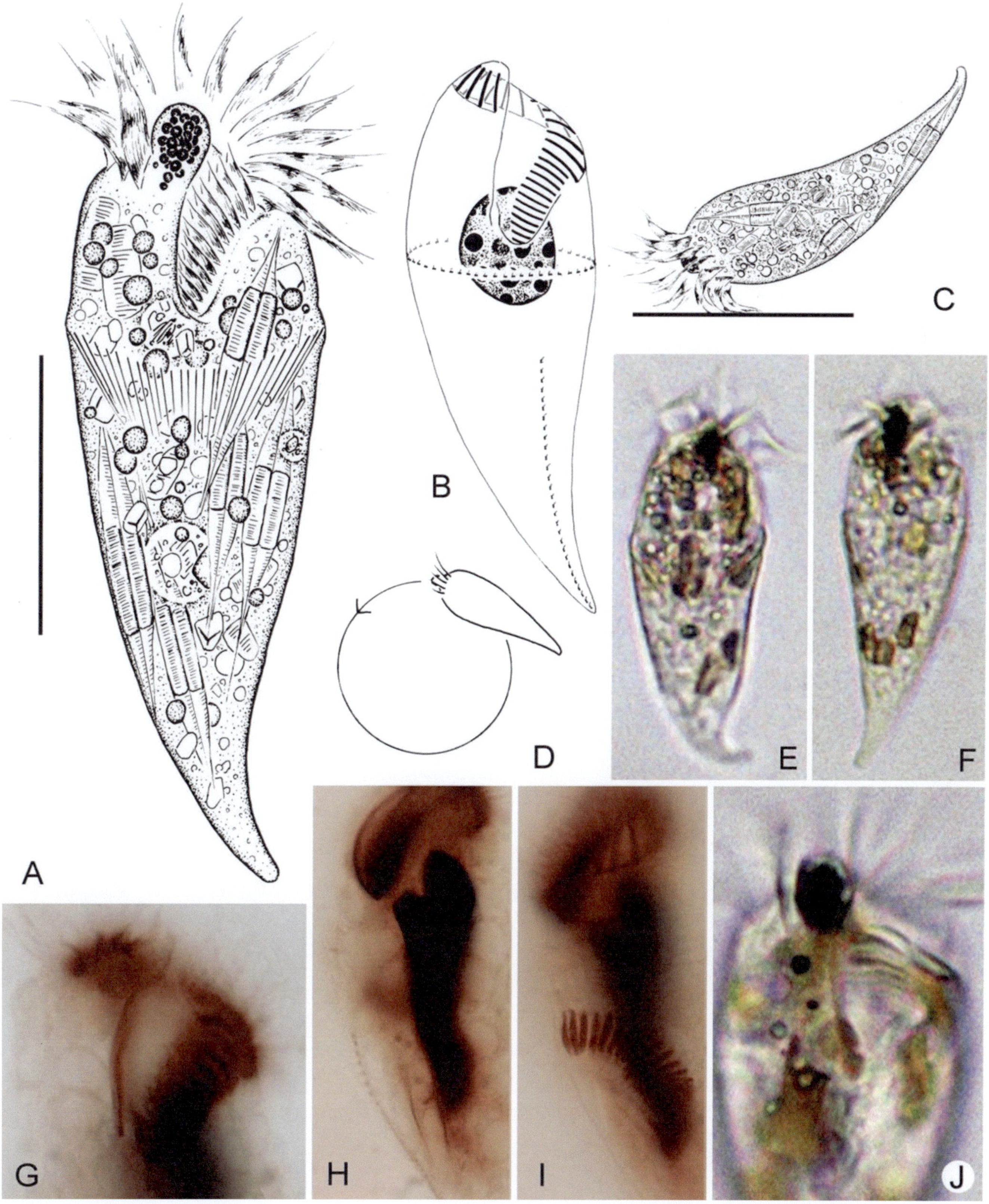


**Fig. 41B** *Strombidium cuneiforme*
(**a**, **c**, **e**, **f**) Ventral views of live cells. (**b**, **g**) Ventral view of ciliature. (**d**) Swimming path. (**h**) Ventral kinety and macronucleus. (**i**) A middle divider, showing new adoral membranelles. (**j**) Ventral view of anterior portion of live cell, showing pigment spot and buccal field. Scale bars: 30 μm

## Reference

Song W, Zhao X, Liu W, Hu X, Al-Farraj SA, Al-Rasheid KAS, Song WB, Warren A (2015) Biodiversity of oligotrich ciliates in the South China Sea: description of three new *Strombidium* species (Protozoa, Ciliophora, Oligotrichia) with phylogenetic analyses. Syst Biodivers 13:608–623

Class Spirotrichea Bütschli, 1889
Order Strombidiida Petz & Foissner, 1992
Family Strombidiidae Fauré-Fremiet, 1970
*Strombidium guangdongense* Liu et al., 2016

**Ecological feature:** mangrove
**Water temperature:** 24.7 °C
**pH:** 8.4
**Salinity:** 19.0‰

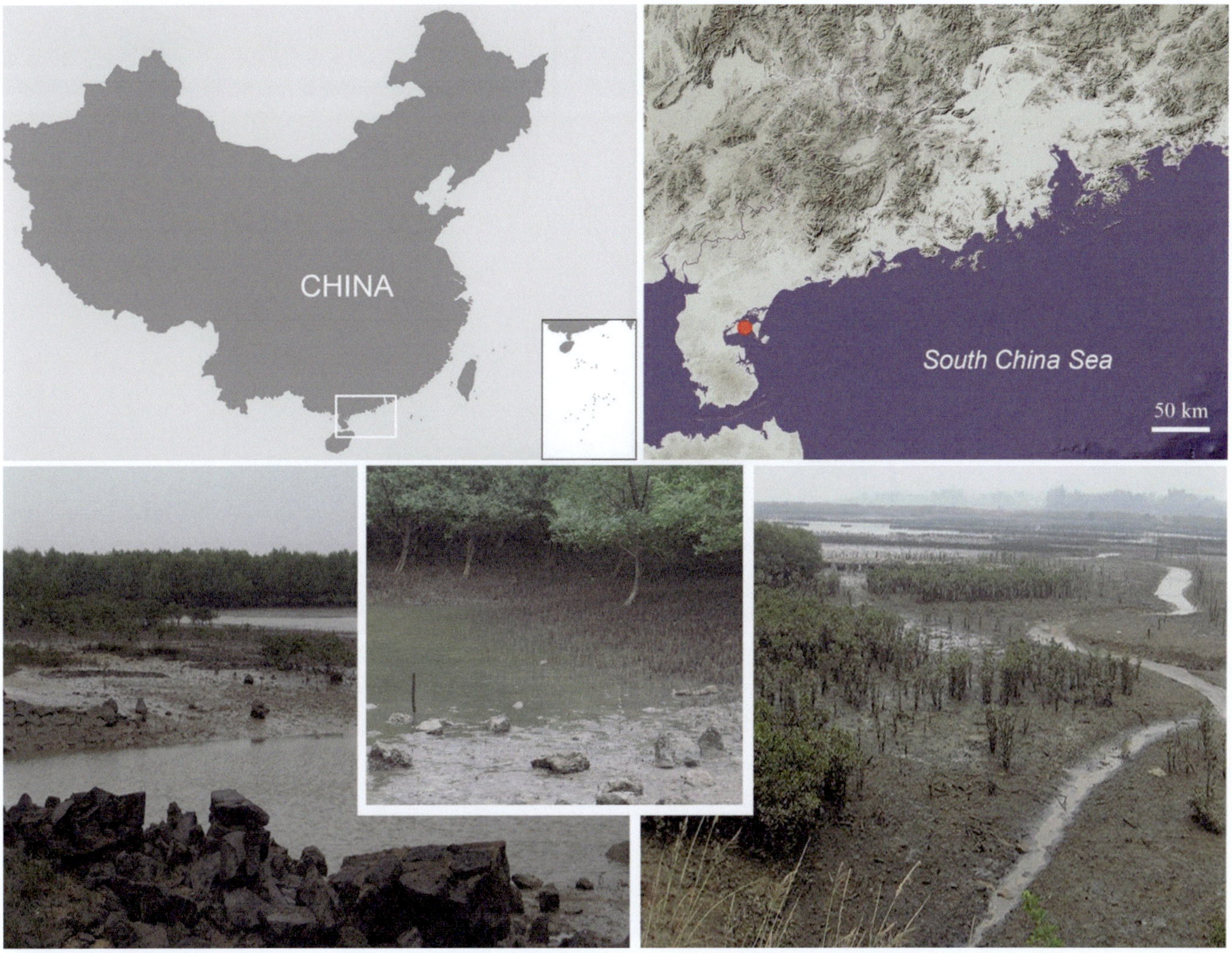


**Fig. 42A** Sample site

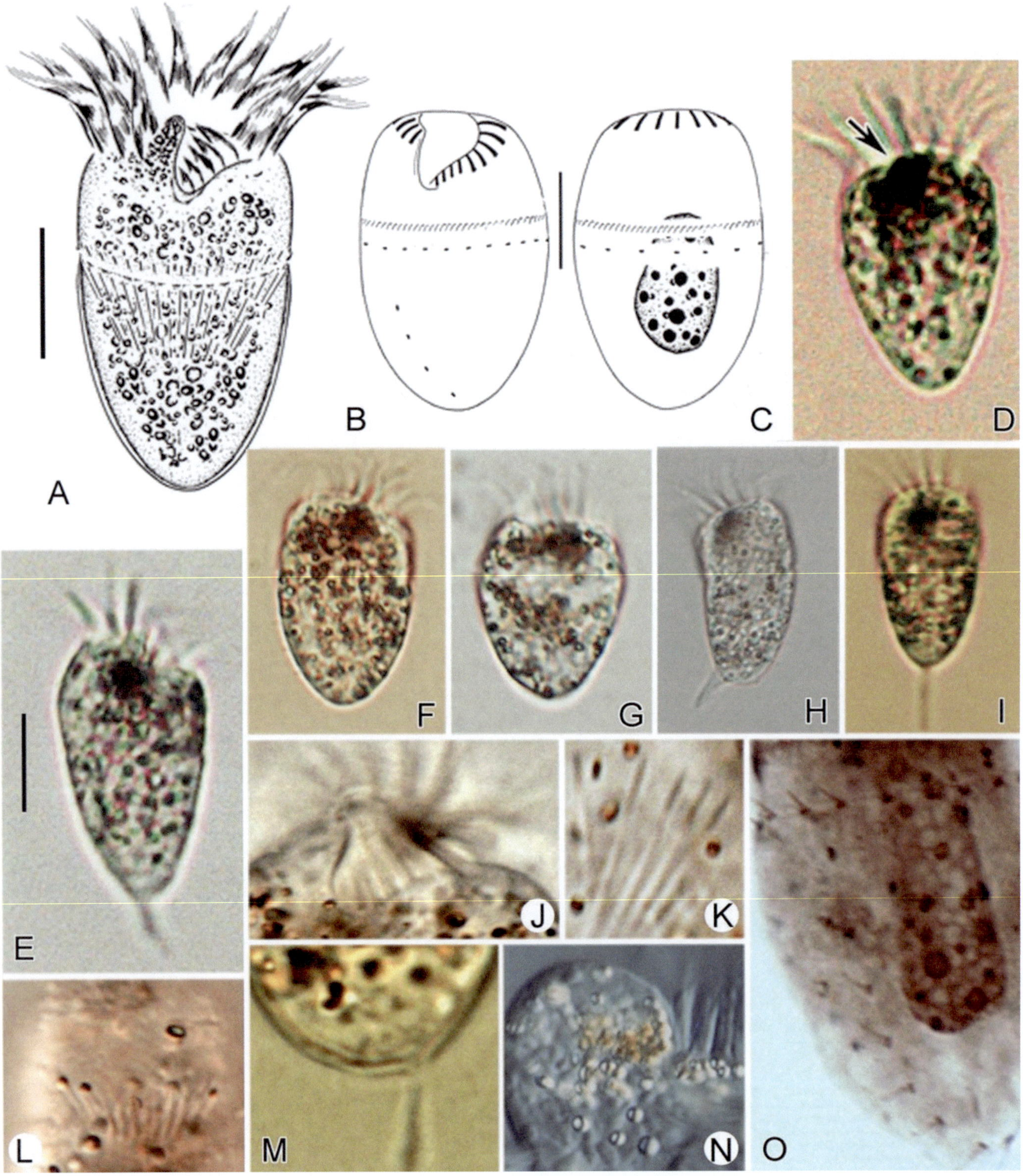


**Fig. 42B** *Strombidium guangdongense*
(**a**, **d–i**) Ventral views of live cells showing different body shapes; arrow marks the pigment spot. (**b**, **c**) Ventral (**b**) and dorsal (**c**) views of ciliature. (**j**) Ventral view of anterior portion showing buccal field. (**k**, **l**) Extrusomes. (**m**) Posterior portion of live cell showing hemitheca and tail. (**n**) Anterior end of live cell to show pigment spot on apical protrusion. (**o**) Ventral view of posterior portion to show somatic kinety. Scale bars: 10 μm

## Reference

Liu W, Xu D, Ma H, Al-Farraj SA, Warren A, Yi Z (2016) Taxonomy and molecular systematics of three oligotrich (s. l.) ciliates including descriptions of two new species, *Strombidium guangdongense* sp. nov. and *Strombidinopsis sinicum* sp. nov. (Protozoa, Ciliophora). Syst Biodivers 14:452–465

Class Spirotrichea Bütschli, 1889
Order Strombidiida Petz & Foissner, 1992
Family Strombidiidae Fauré-Fremiet, 1970
*Strombidium oculatum* Gruber, 1884

**Ecological feature:** mangrove
**Water temperature:** 22.5 °C
**pH:** 8.0
**Salinity:** 32.8‰

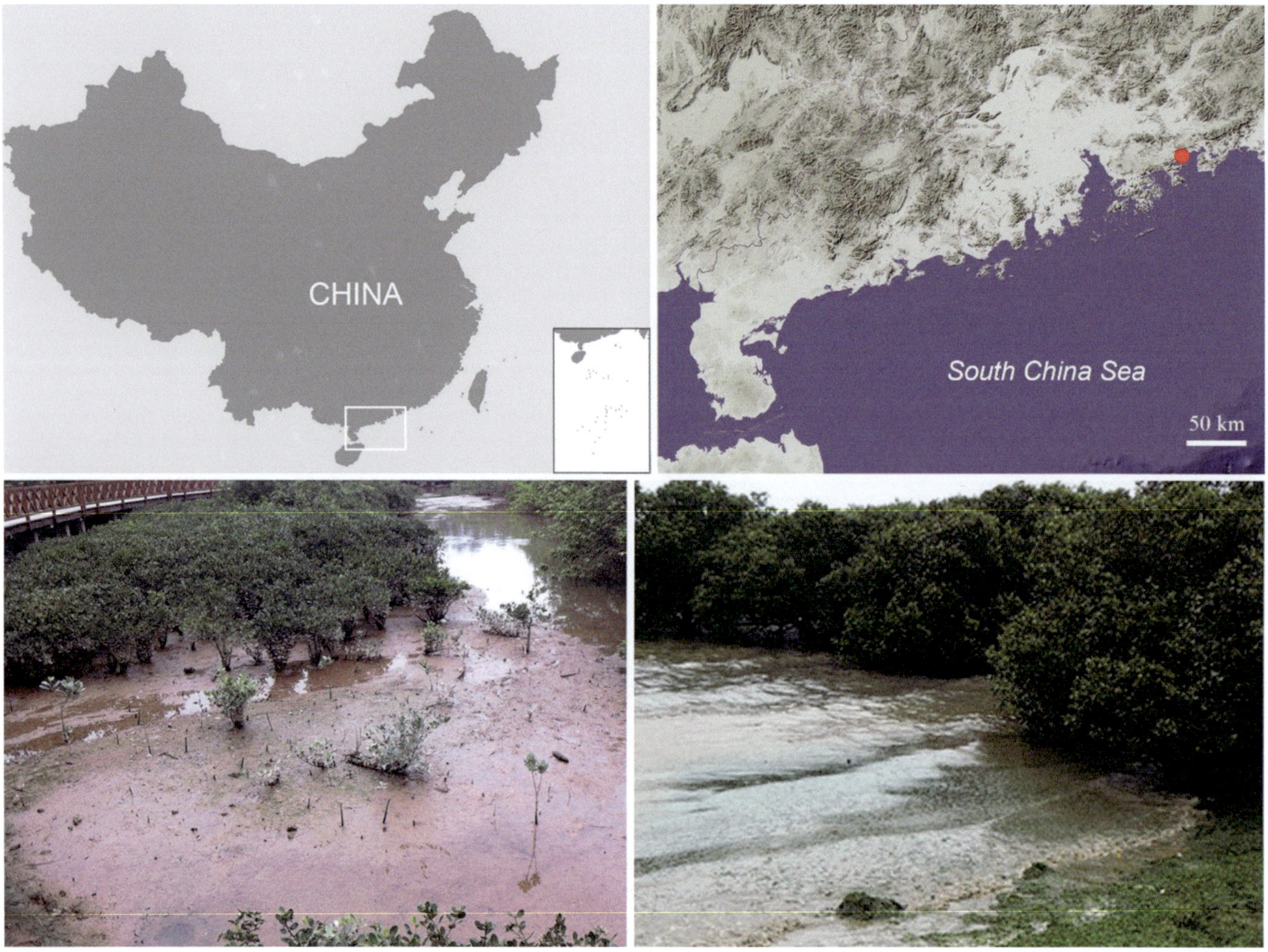


**Fig. 43A** Sample site

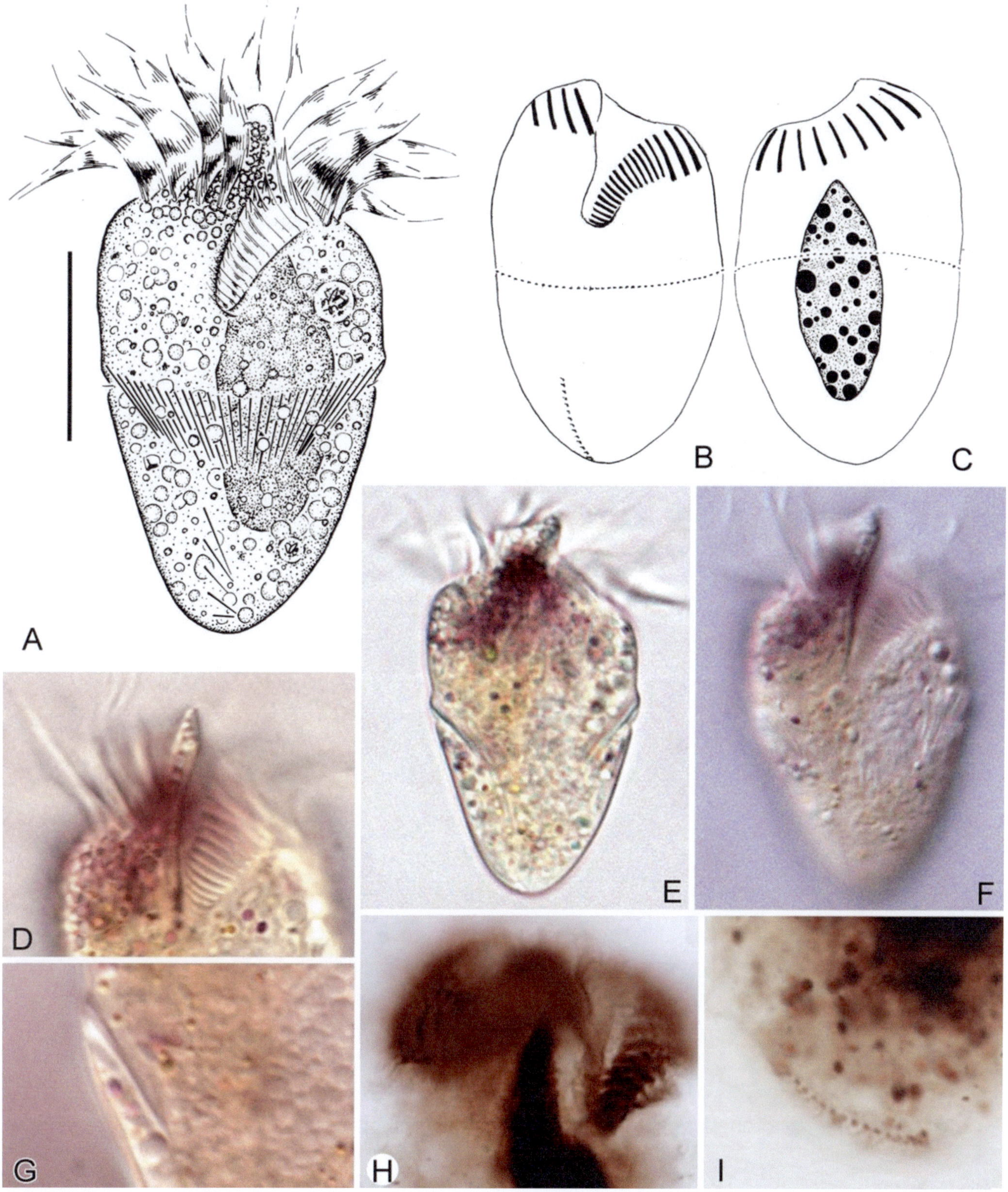


**Fig. 43B** *Strombidium oculatum*
(**a**, **e**, **f**) Ventral views of live individuals. (**b**, **c**) Ventral (**b**) and dorsal (**c**) views of ciliature. (**d**) Buccal field, showing oral membranelles. (**g**) Extrusomes. (**h**) Adoral zone of membranelles. (**i**) Ventral kinety. Scale bar: 20 μm

## Reference

Song W, Li J, Liu W, Al-Rasheid KAS, Hu X, Lin X (2015) Taxonomy and molecular phylogeny of four *Strombidium* species, including description of *S. pseudostylifer* sp. nov. (Ciliophora, Oligotrichia). Syst Biodivers 13:76–92

Class Spirotrichea Bütschli, 1889
Order Strombidiida Petz & Foissner, 1992
Family Strombidiidae Fauré-Fremiet, 1970
*Strombidium paracalkinsi* (Lei et al., 1999) Agatha, 2004

**Ecological feature:** offshore water
**Water temperature:** 20.1 °C
**pH:** 8.1
**Salinity:** 27.8‰

**Fig. 44A** Sample site

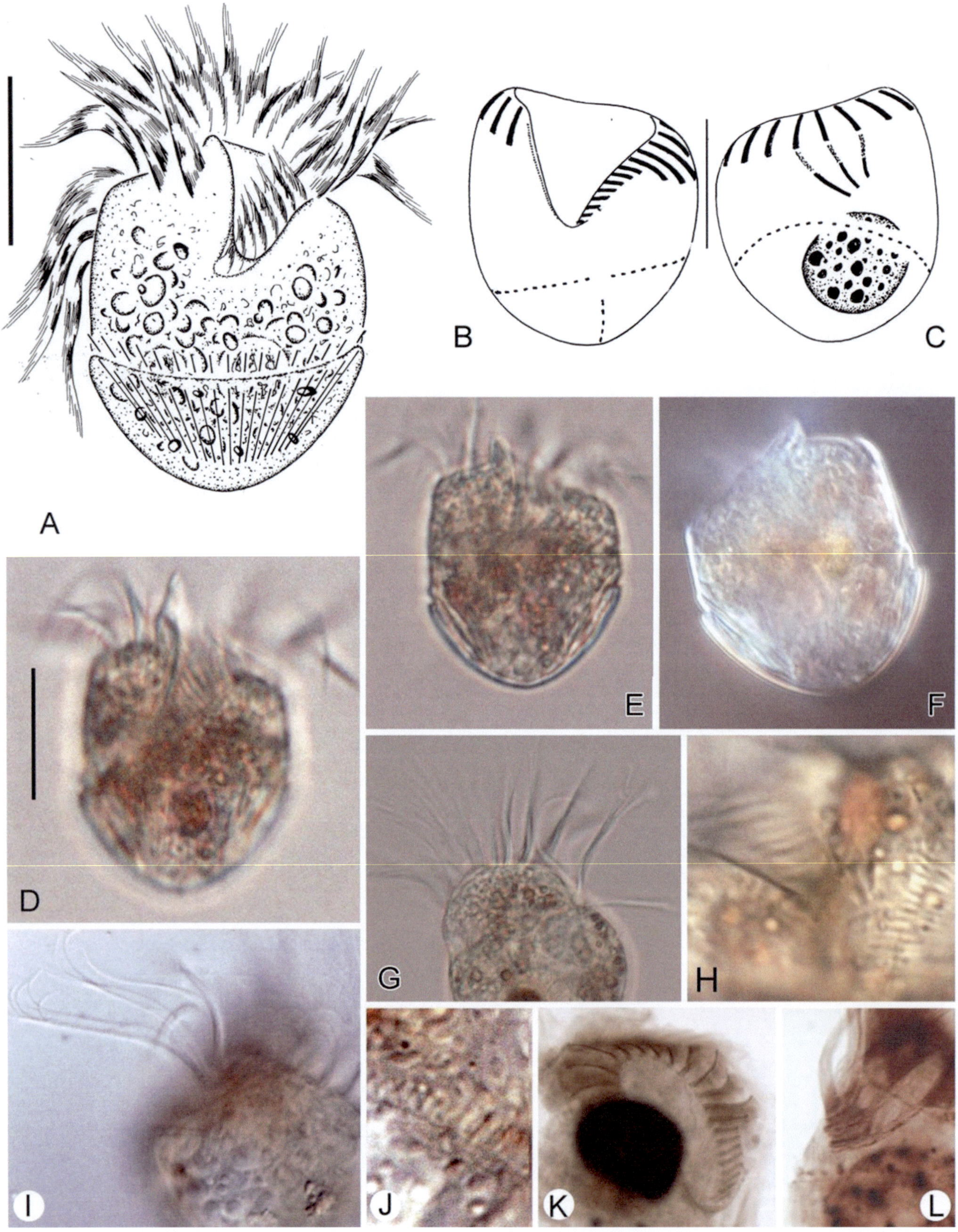


**Fig. 44B** *Strombidium paracalkinsi*
(**a**, **d**, **e**) Ventral views in vivo. (**b**, **c**) Ventral (**b**) and dorsal (**c**) view of ciliature. (**f**, **g**) Dorsal views of live cells. (**h**) Ventral membranelles and extrusomes. (**i**) Thigmotactic membranelles. (**j**) Extrusomes. (**k**) Adoral zone of membranelles. (**l**) Bases of thigmotactic membranelles. Scale bars: 20 μm

## References


Lee E S, Kim Y O, Agatha S, Jung J H, Xu D, Shin M. 2015. Revision of *Strombidium paracalkinsi* (Ciliophora: Oligotrichea: Oligotrichia), with comparison of strombidiids bearing thigmotactic membranelles. J Eukaryot Microbiol 62:400–409

Song WB, Wang M, Warren A (2000) Redescriptions of three marine ciliates, *Strombidium elegans* Florentin, 1901, *Strombidium sulcatum* Claparède & Lachmann, 1859 and *Heterostrombidium paracalkinsi* Lei, Xu & Song, 1999 (Ciliophora, Oligotrichida). Eur J Protistol 36:327–342

Class Spirotrichea Bütschli, 1889
Order Strombidiida Petz & Foissner, 1992
Family Strombidiidae Fauré-Fremiet, 1970
*Strombidium paracapitatum* Song et al., 2015

**Ecological feature:** estuary area
**Water temperature:** 28.0 °C
**pH:** no data available
**Salinity:** 5.0‰

**Fig. 45A** Sample site

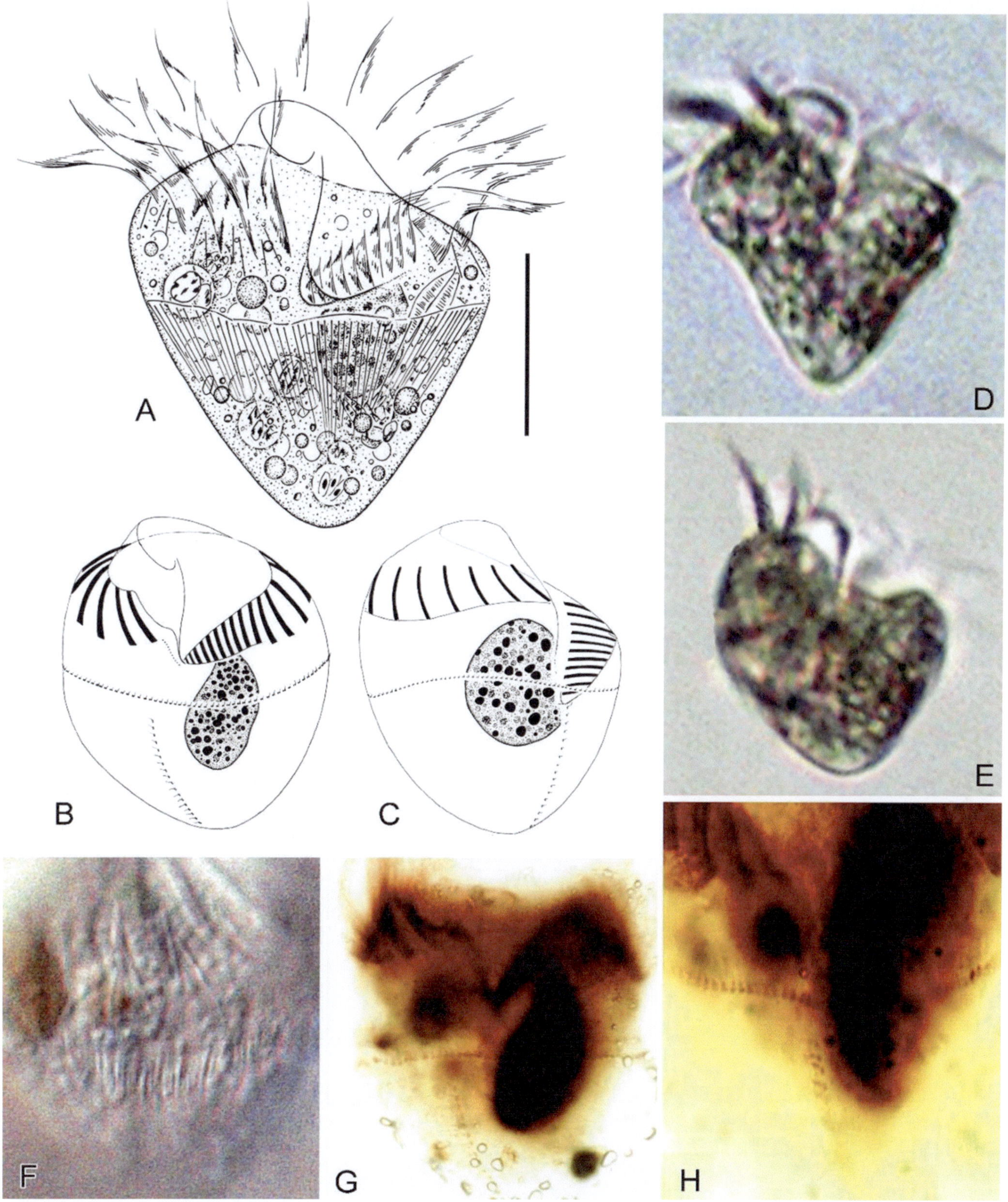


**Fig. 45B** *Strombidium paracapitatum*
(**a**, **d**, **e**) Ventral views of live individuals. (**b**, **c**) Ventral (**b**) and dorsal (**c**) views of ciliature and macronucleus. (**f**) Extrusomes. (**g**, **h**) Ventral views of ciliature. Scale bar: 20 μm

## Reference

Song W, Zhao X, Liu W, Hu X, Al-Farraj SA, Al-Rasheid KAS, Song WB, Warren A (2015) Biodiversity of oligotrich ciliates in the South China Sea: description of three new *Strombidium* species (Protozoa, Ciliophora, Oligotrichia) with phylogenetic analyses. Syst Biodivers 13:608–623

Class Spirotrichea Bütschli, 1889
Order Strombidiida Petz & Foissner, 1992
Family Strombidiidae Fauré-Fremiet, 1970
*Strombidium parastylifer* Song et al., 2009

**Ecological feature:** mangrove
**Water temperature:** no data available
**pH:** 7.5
**Salinity:** 19.0‰

**Fig. 46A** Sample site

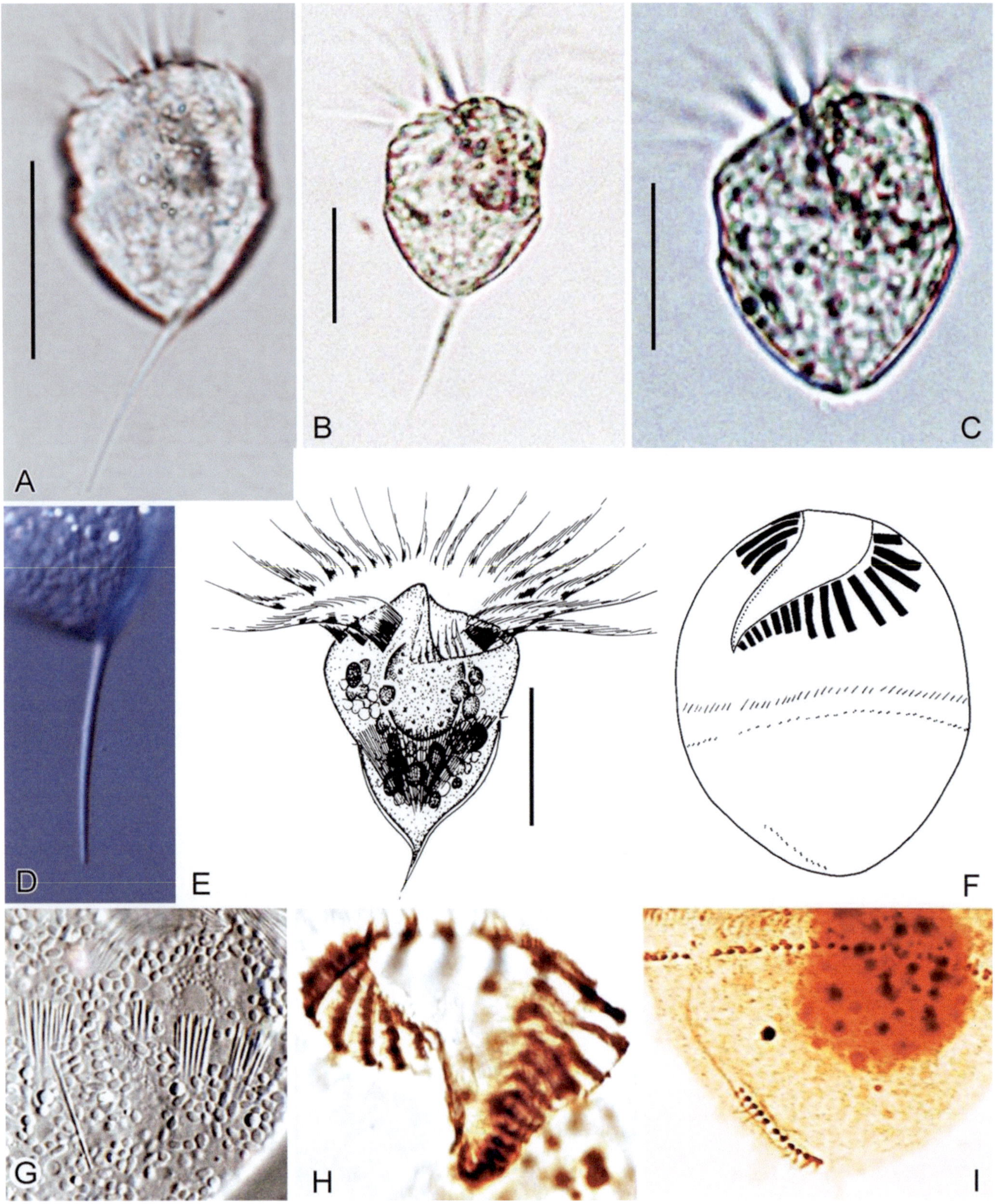


**Fig. 46B** *Strombidium parastylifer*
(**a**–**c**) Different shapes of live individuals. (**d**) Posterior spine. (**e**) Ventral view of live cell. (**f**) Ventral view of ciliature. (**g**) Extrusomes. (**h**) Adoral zone of membranelles. (**i**) Girdle kinety and ventral kinety. Scale bars: 30 μm

## Reference

Song WB, Warren A, Hu X (2009) Free-living Ciliates in the Bohai and Yellow Seas, China. Science Press, Beijing

Class Spirotrichea Bütschli, 1889
Order Strombidiida Petz & Foissner, 1992
Family Strombidiidae Fauré-Fremiet, 1970
*Strombidium pseudostylifer* Song et al., 2015

**Ecological feature:** mangrove
**Water temperature:** 28.0 °C
**pH:** 8.2
**Salinity:** 26.0‰

**Fig. 47A** Sample site

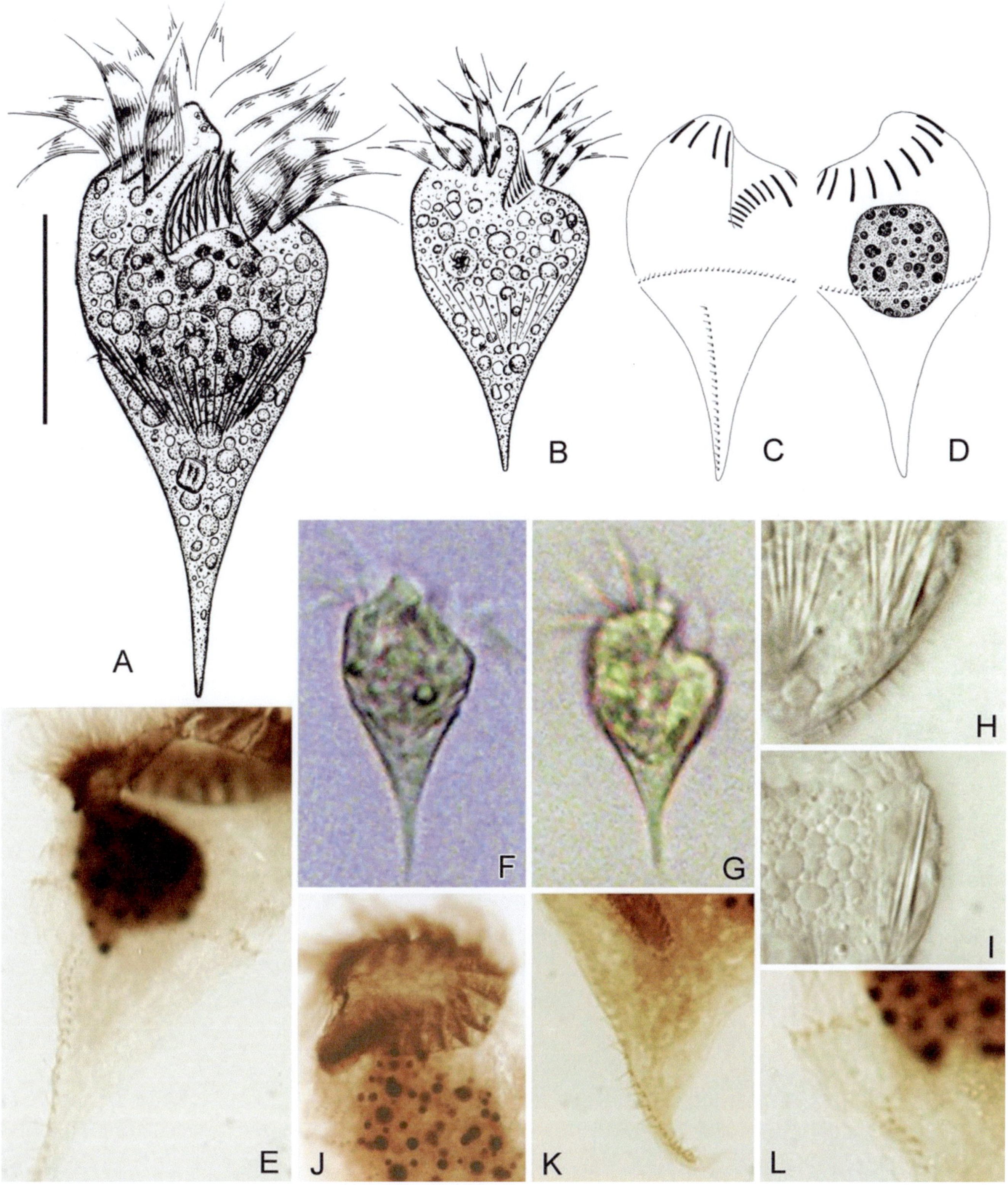


**Fig. 47B** *Strombidium pseudostylifer*
(**a**, **b**, **f**, **g**) Ventral views of live individuals. (**c**, **d**) Ventral (**c**) and dorsal (**d**) views of ciliature. (**e**) Ciliature and macronucleus. (**h**, **i**) Extrusomes. (**j**) Adora zone of membranelles. (**k**) Ventral kinety. (**l**) Somatic kineties. Scale bar: 20 μm

## Reference

Song W, Li J, Liu W, Al-Rasheid KAS, Hu X, Lin X (2015) Taxonomy and molecular phylogeny of four *Strombidium* species, including description of *S. pseudostylifer* sp. nov. (Ciliophora, Oligotrichia). Syst Biodivers 13:76–92

Class Spirotrichea Bütschli, 1889
Order Strombidiida Petz & Foissner, 1992
Family Strombidiidae Fauré-Fremiet, 1970
*Strombidium stylifer* Levander, 1894

**Ecological feature:** intertidal zone
**Water temperature:** 24.4 °C
**pH:** 8.5
**Salinity:** 29.5‰

**Fig. 48A** Sample site

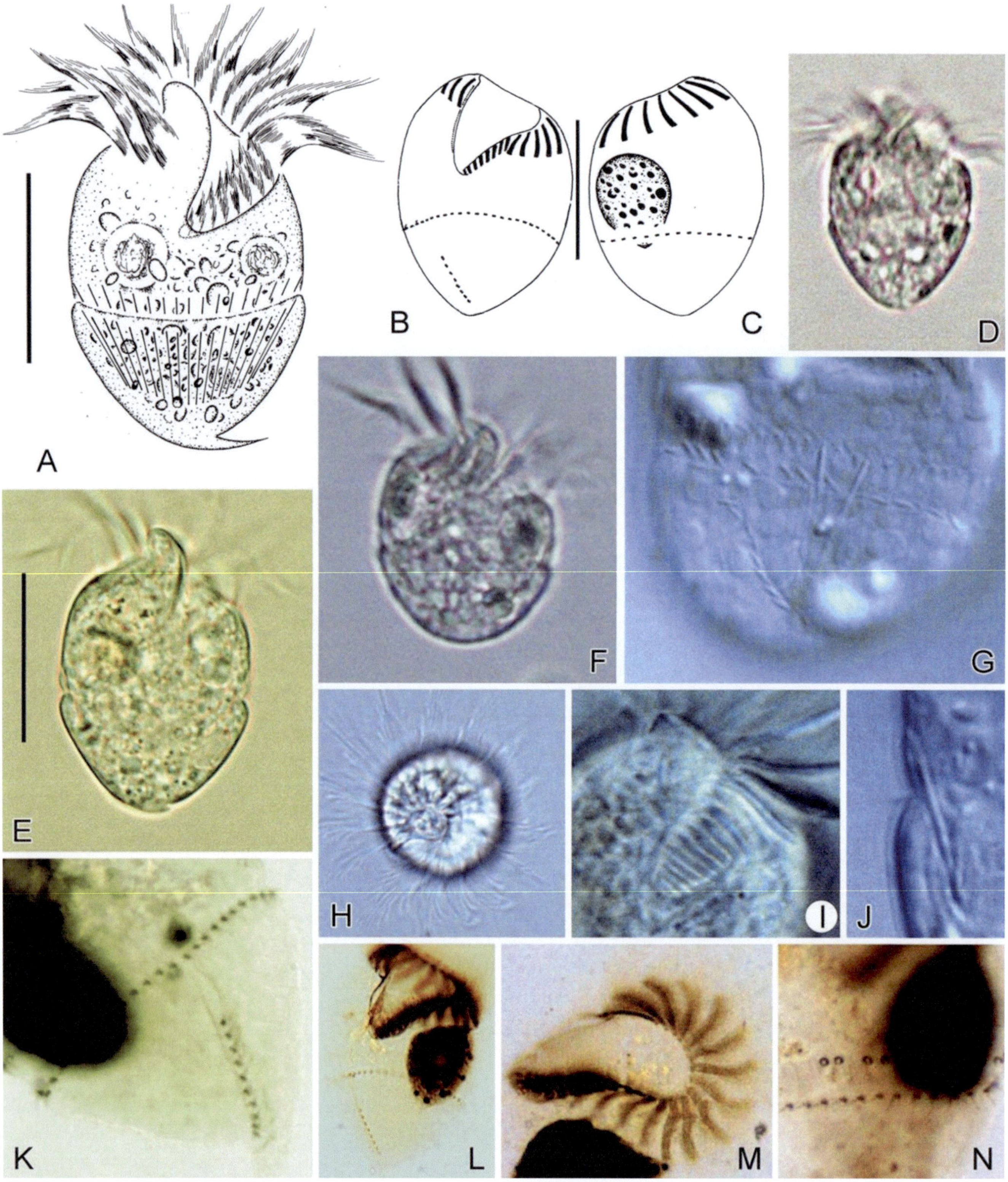


**Fig. 48B** *Strombidium stylifer*
(**a**, **d–f**) Ventral views of live cells; (**b**, **c**) ventral (**b**) and dorsal (**c**) views of ciliature. (**g**) Extrusomes and somatic cilia. (**h**) Apical view in vivo to show oral membranelles. (**i**) Ventral view of anterior portion of cell to show oral membranelles. (**j**) Extrusomes. (**k**) Girdle and ventral kineties. (**l**) Ventral view of ciliature. (**m**) Apical view to show oral membranelles. (**n**) Girdle kinety and extrusomes girdle. Scale bars: 30 μm

## References

Song WB, Packroff G (1997) Taxonomische Untersuchungen an marinen Ciliaten aus China mit Beschreibungen von neuen Arten, *Strombidium globosaneum* nov. spec. und *Strombidium platum* nov. spec. (Protozoa, Ciliophora). Arch Protistenk 147:331–360

Song WB, Xu K, Shi X, Hu X, Lei Y, Wei J, Chen Z, Shi X, Wang M (1999) Progress in Protozoology. Qingdao Ocean University Press, Qingdao

Class Spirotrichea Bütschli, 1889
Order Strombidiida Petz & Foissner, 1992
Family Strombidiidae Fauré-Fremiet, 1970
*Strombidium sulcatum* Claparède & Lachmann, 1859

**Ecological feature:** mangrove
**Water temperature:** 27.0 °C
**pH:** 6.5
**Salinity:** 10.0‰

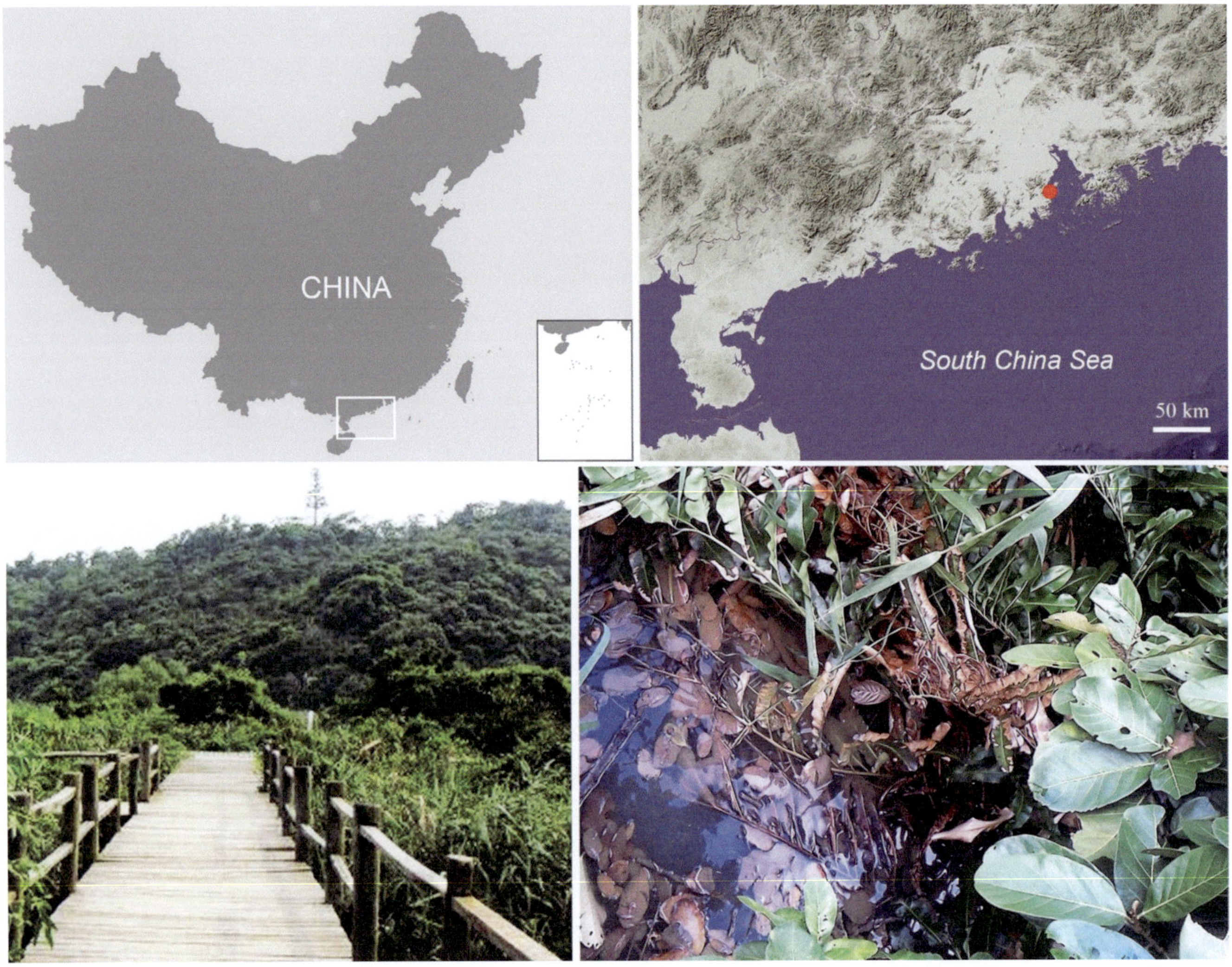


**Fig. 49A** Sample site

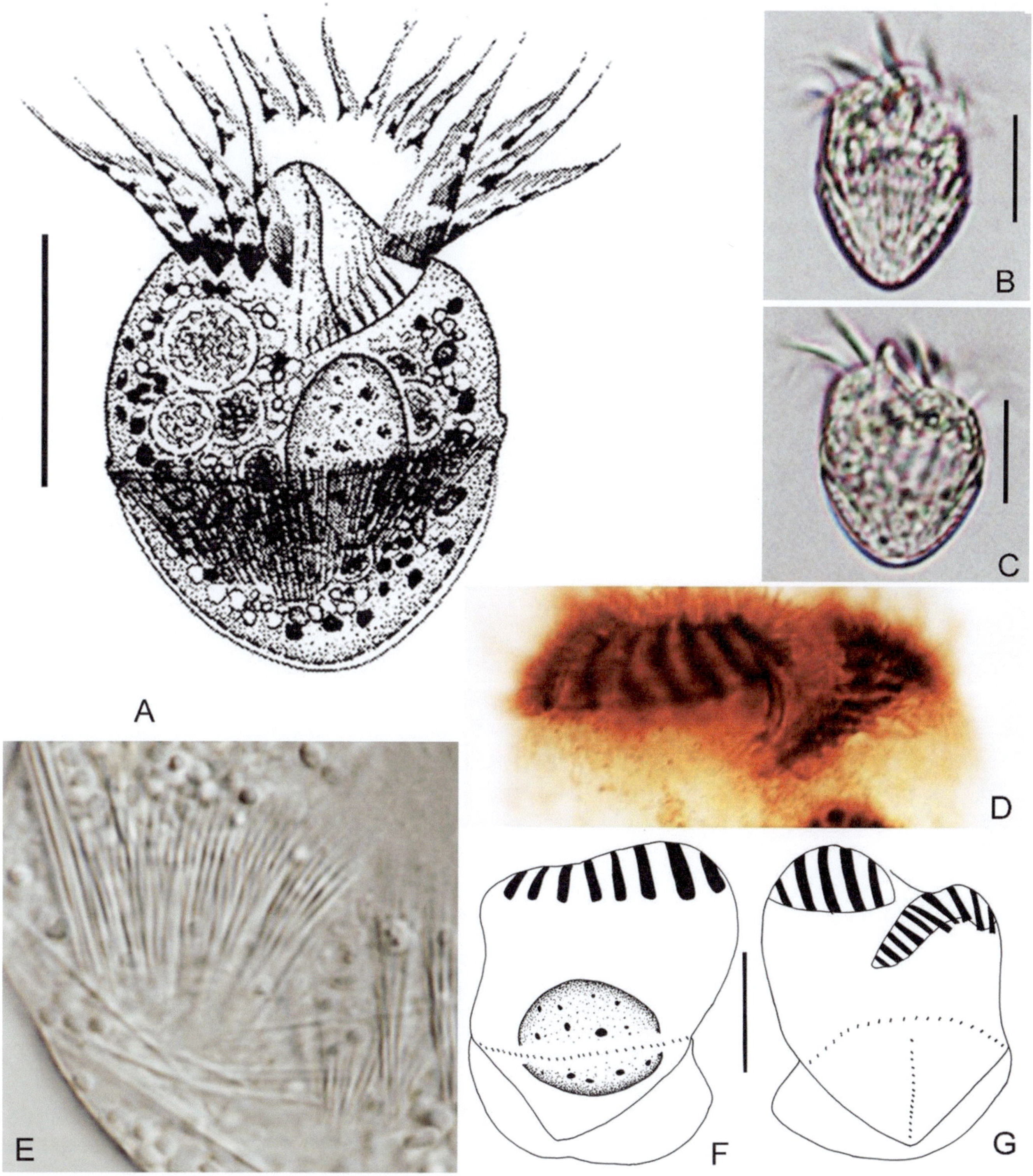


**Fig. 49B** *Strombidium sulcatum*
(**a–c**) View of live individuals. (**d**) Adoral membranelles. (**e**) Extrusomes. (**f**, **g**) Dorsal (**f**) and ventral (**g**) views of ciliature. Scale bars: 25 μm

## Reference

Song WB, Wang M, Warren A (2000) Redescriptions of three marine ciliates, *Strombidium elegans* Florentin, 1901, *Strombidium sulcatum* Claparède & Lachmann, 1859 and *Heterostrombidium paracalkinsi* Lei, Xu & Song, 1999 (Ciliophora, Oligotrichida). Eur J Protistol 36:327–342

Class Spirotrichea Bütschli, 1889
Order Strombidiida Petz & Foissner, 1992
Family Strombidiidae Fauré-Fremiet, 1970
*Strombidium tropicum* Liu et al., 2015

**Ecological feature:** intertidal zone
**Water temperature:** 27.8 °C
**pH:** 8.6
**Salinity:** 31.0‰

**Fig. 50A** Sample site

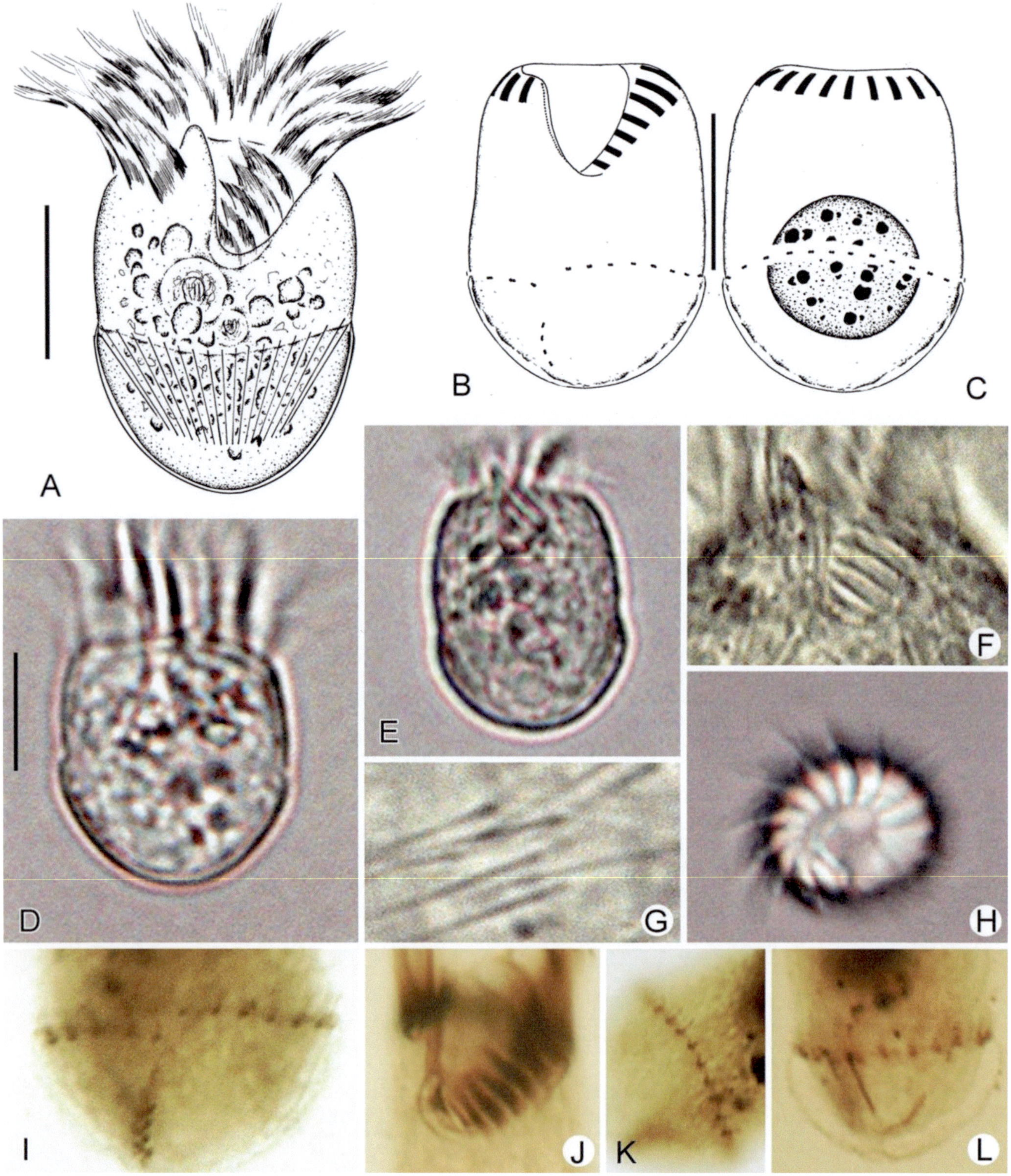


**Fig. 50B** *Strombidium tropicum*
(**a**, **d**, **e**) Ventral views of live cells. (**b**, **c**) Ventral (**b**) and dorsal (**c**) views of ciliature. (**f**, **j**) Ventral views of anterior portion of cell, showing buccal field. (**g**) Detail of ejected extrusomes. (**g**) Extrusomes. (**h**) Apical view showing oral membranelles. (**i**, **k**, **l**) Somatic kineties. Scale bars: 10 μm

## Reference

Liu W, Yi Z, Lin X, Li J, Al-Farraj SA, Al-Rasheid KAS, Song WB (2015) Morphology and molecular phylogeny of three new oligotrich ciliates (Protozoa, Ciliophora) from the South China Sea. Zool J Linn Soc 174:653–665

Class Spirotrichea Bütschli, 1889
Order Strombidiida Petz & Foissner, 1992
Family Strombidiidae Fauré-Fremiet, 1970
*Varistrombidium kielum* (Maeda & Carey, 1985) Xu et al., 2011

**Ecological feature:** mangrove
**Water temperature:** 24.7 °C
**pH:** 8.4
**Salinity:** 19.0‰

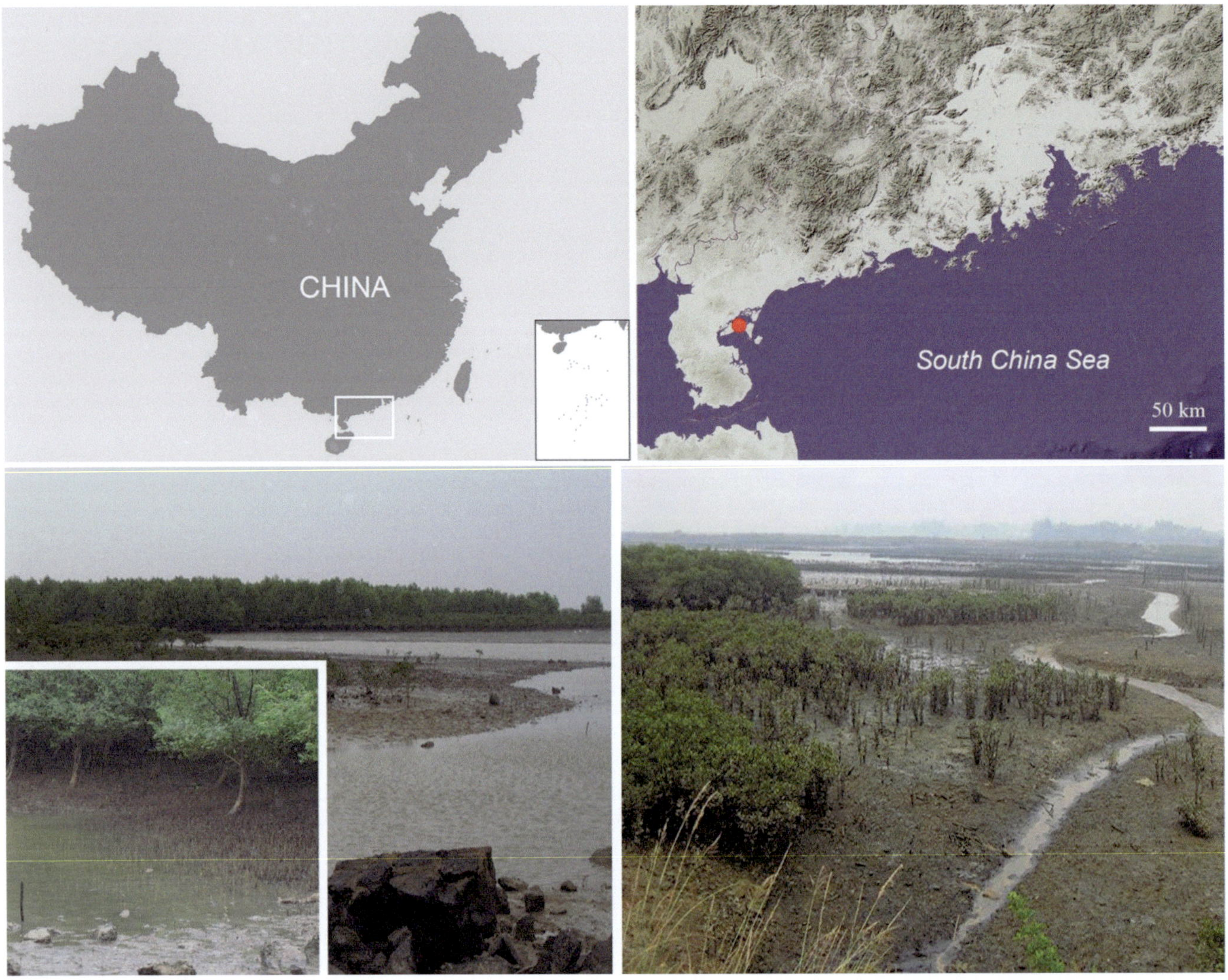


**Fig. 51A** Sample site

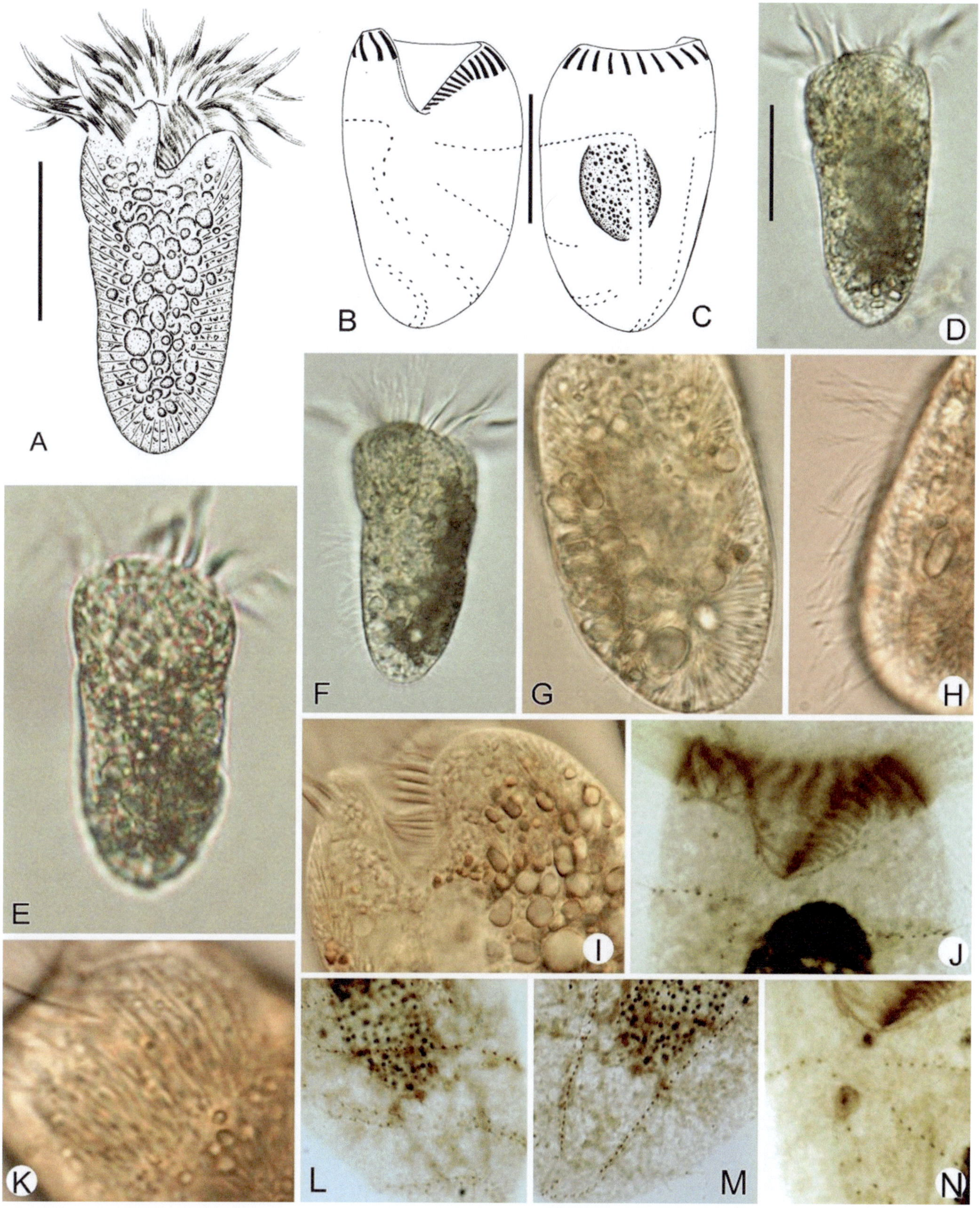


**Fig. 51B** *Varistrombidium kielum*
(**a**, **d**, **e**) Ventral views of live cells. (**b**, **c**) Ventral (**b**) and dorsal (**c**) views of ciliature. (**f**) Lateral view of live cell. (**g**, **k**) Extrusomes. (**h**) Lateral view of posterior portion showing somatic cilia. (**i**, **j**) Buccal membranelles. (**l**–**n**) Somatic kineties. Scale bars: 30 μm

## Reference

Xu D, Sun P, Clamp J, Ma H, Song WB (2011) The establishment of a new oligotrich genus *Varistrombidium* n. gen. and the morphology and phylogeny of a marine ciliate *V. kielum* (Maeda & Carey, 1985) n. comb. (Protista, Ciliophora). Acta Zootaxon Sin 36:502–511

Class Spirotrichea Bütschli, 1889
Order Strombidiida Petz & Foissner, 1992
Family Strombidiidae Fauré-Fremiet, 1970
*Williophrya maedai* Liu et al., 2011

**Ecological feature:** mangrove
**Water temperature:** 27.0 °C
**pH:** 8.2
**Salinity:** 17.0‰

**Fig. 52A** Sample site

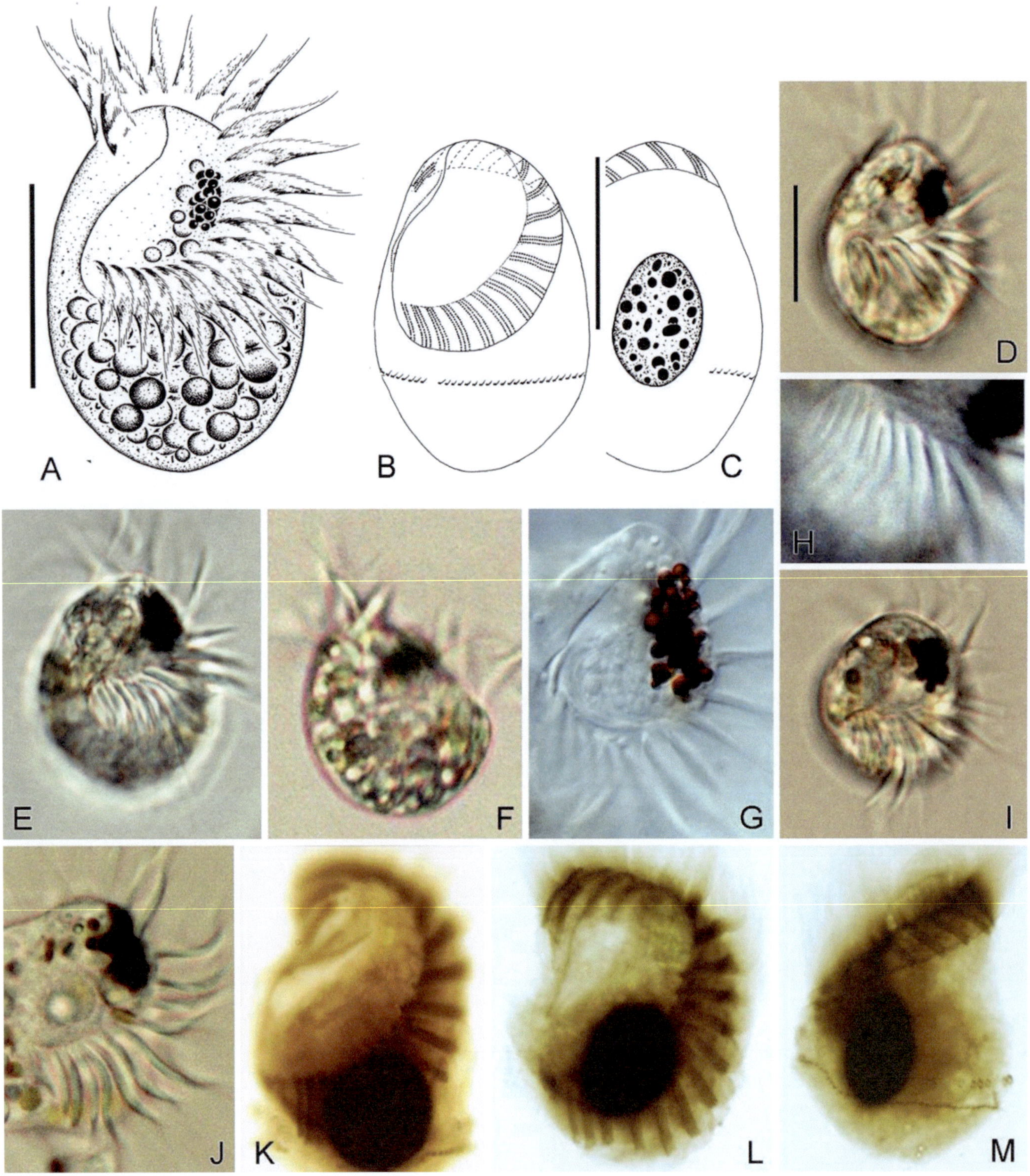


**Fig. 52B** *Williophrya maedai*
(**a**, **d**, **e**) Ventral views of live cells. (**b**, **c**) Ventral (**b**) and dorsal (**c**) views of ciliature. (**f**, **g**, **i**) Lateral (**f**) and apical (**g**, **i**) views of live cells. (**h**, **j**) Oral membranelles. (**k**, **l**) Adoral zone of membranelles. (**m**) Left view to show girdle kinety and collar membranelles. Scale bars: 25 μm

## Reference

Liu W, Yi Z, Warren A, Al-Rasheid KAS, Al-Farraj SA, Lin X, Song WB (2011) Taxonomy, morphology and molecular systematics of a new oligotrich ciliates, *Williophrya maedai* gen. nov., sp. nov., with redescriptions of *Strombidium basimorphum* and *Pseudotontonia simplicidens* (Protozoa, Ciliophora, Oligotrichia). Syst Biodiver 9:247–258

Class Spirotrichea Bütschli, 1889
Order Choreotrichida Small & Lynn, 1985
Family Tontoniidae Agatha, 2004
*Pseudotontonia simplicidens* (Lynn & Gilron, 1993) Agatha, 2004

**Ecological feature:** offshore water
**Water temperature:** 22.7 °C
**pH:** 7.9
**Salinity:** 26.0‰

**Fig. 53A** Sample site

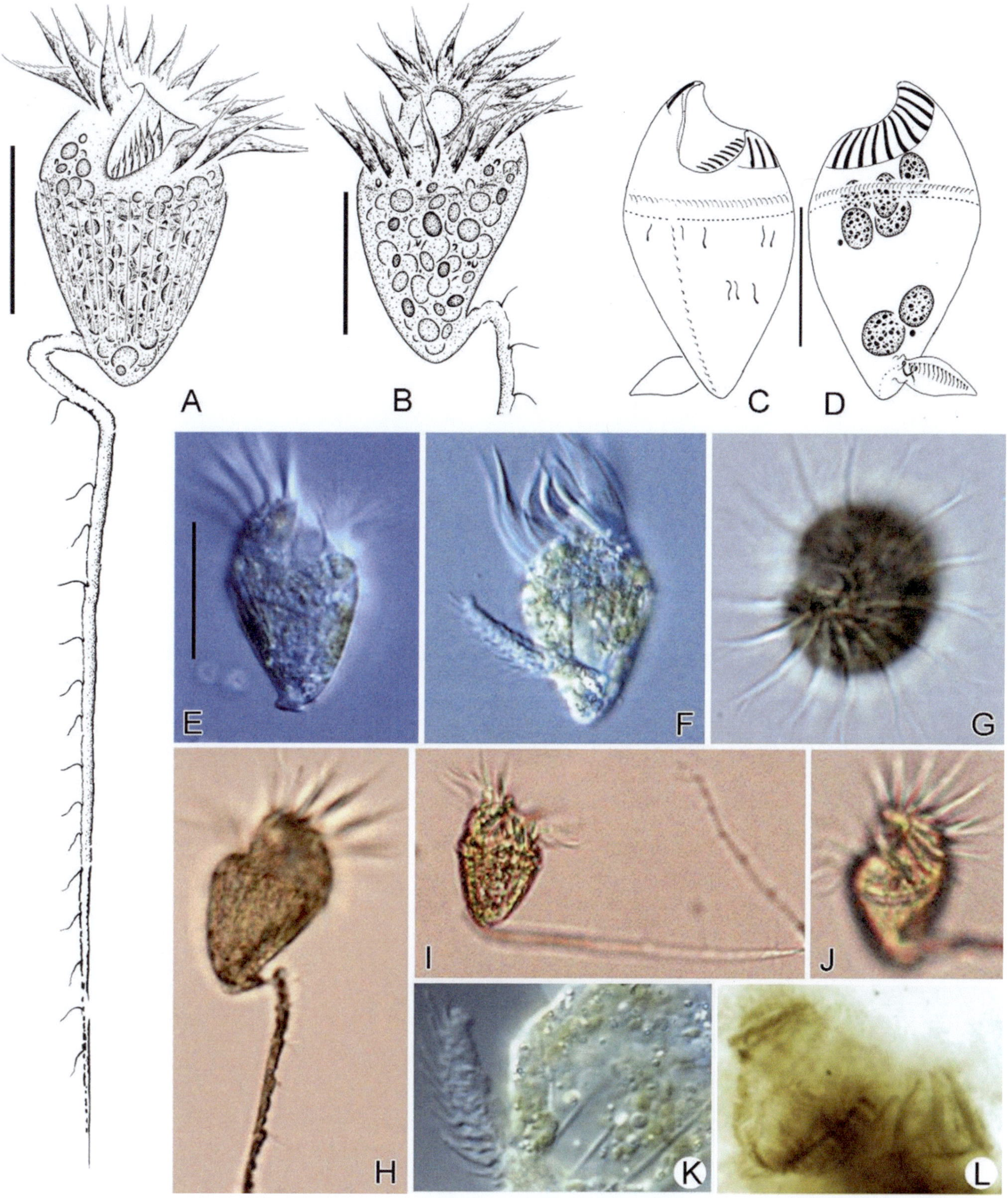


**Fig. 53B** *Pseudotontonia simplicidens*
(**a**, **b**, **e**, **h**, **i**) Ventral (**a**, **e**, **i**) and left-lateral (**b**, **h**) views of live cells. (**c**, **d**) Ventral (**c**) and dorsal (**d**) views of ciliature. (**f**) Dorsal view of live cell. (**g**, **j**) Apical (**g**) and apical-lateral (**j**) views to show collar membranelles. (**k**) Contracted tail and extrusomes. (**l**) Oral membranelles. Scale bars: 20 μm

## References

Lynn D, Gilron G (1993) Strombidiid ciliates from coastal waters near Kingston Harbour, Jamaica (Ciliophora, Oligotrichia, Strombidiidae). J Mar Boil Ass UK 73:47–65

Liu W, Yi Z, Warren A, Al-Rasheid KAS, Al-Farraj SA, Lin X, Song WB (2011) Taxonomy, morphology and molecular systematics of a new oligotrich ciliates, *Williophrya maedai* gen. nov., sp. nov., with redescriptions of *Strombidium basimorphum* and *Pseudotontonia simplicidens* (Protozoa, Ciliophora, Oligotrichia). Syst Biodivers 9:247–258

Class Spirotrichea Bütschli, 1889
Order Choreotrichida Small & Lynn, 1985
Family Tontoniidae Agatha, 2004
*Spirotontonia grandis* (Suzuki & Han, 2000) Agatha, 2004

**Ecological feature:** offshore water
**Water temperature:** 22.6 °C
**pH:** 7.6
**Salinity:** 22.6‰

**Fig. 54A** Sample site

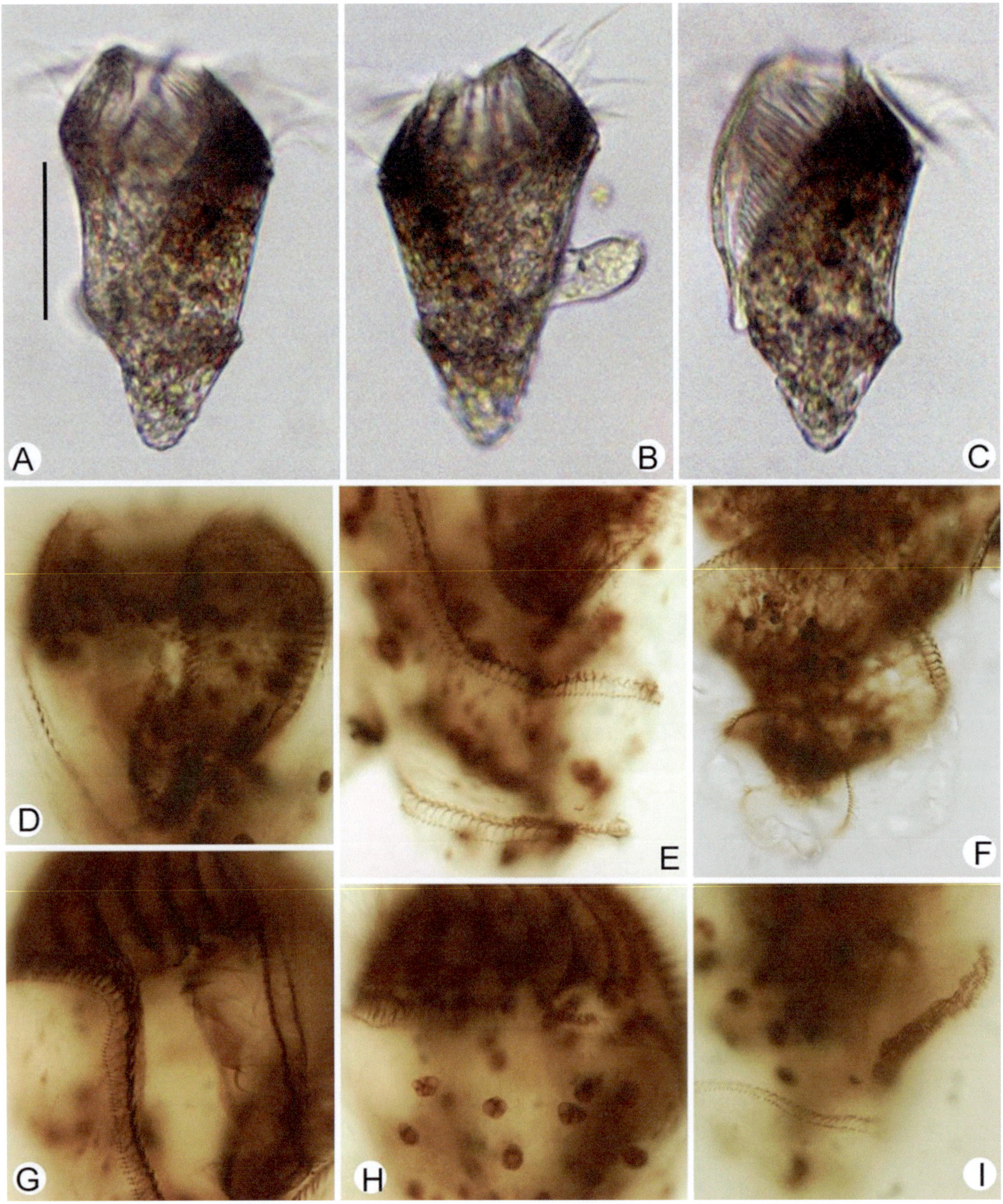


**Fig. 54B** *Spirotontonia grandis*
(**a**, **b**, **c**) Ventral, dorsal, lateral views of live individuals. (**d**) Buccal field. (**e–g**) Somatic kinety. (**h**) Macronuclei. (**i**) Tail kinety. Scale bars: 50 μm

## References

Agatha S (2004) Evolution of ciliary patterns in the Oligotrichida (Ciliophora, Spirotricha) and its taxonomic implications. Zoology 107:153–168

Suzuki T, Han M (2000) A study on a new species of *Tontonia* (Ciliophora: Oligotrichida) from the East China Sea and adjacent sea areas. J Mar Biol Ass UK 80:989–994

Class Spirotrichea Bütschli, 1889
Order Choreotrichida Small & Lynn, 1985
Family Tontoniidae Agatha, 2004
*Spirotontonia turbinata* (Song & Bradbury, 1998) Agatha, 2004

**Ecological feature:** offshore water
**Water temperature:** 19.4 °C
**pH:** 8.2
**Salinity:** 32.6‰

**Fig. 55A** Sample site

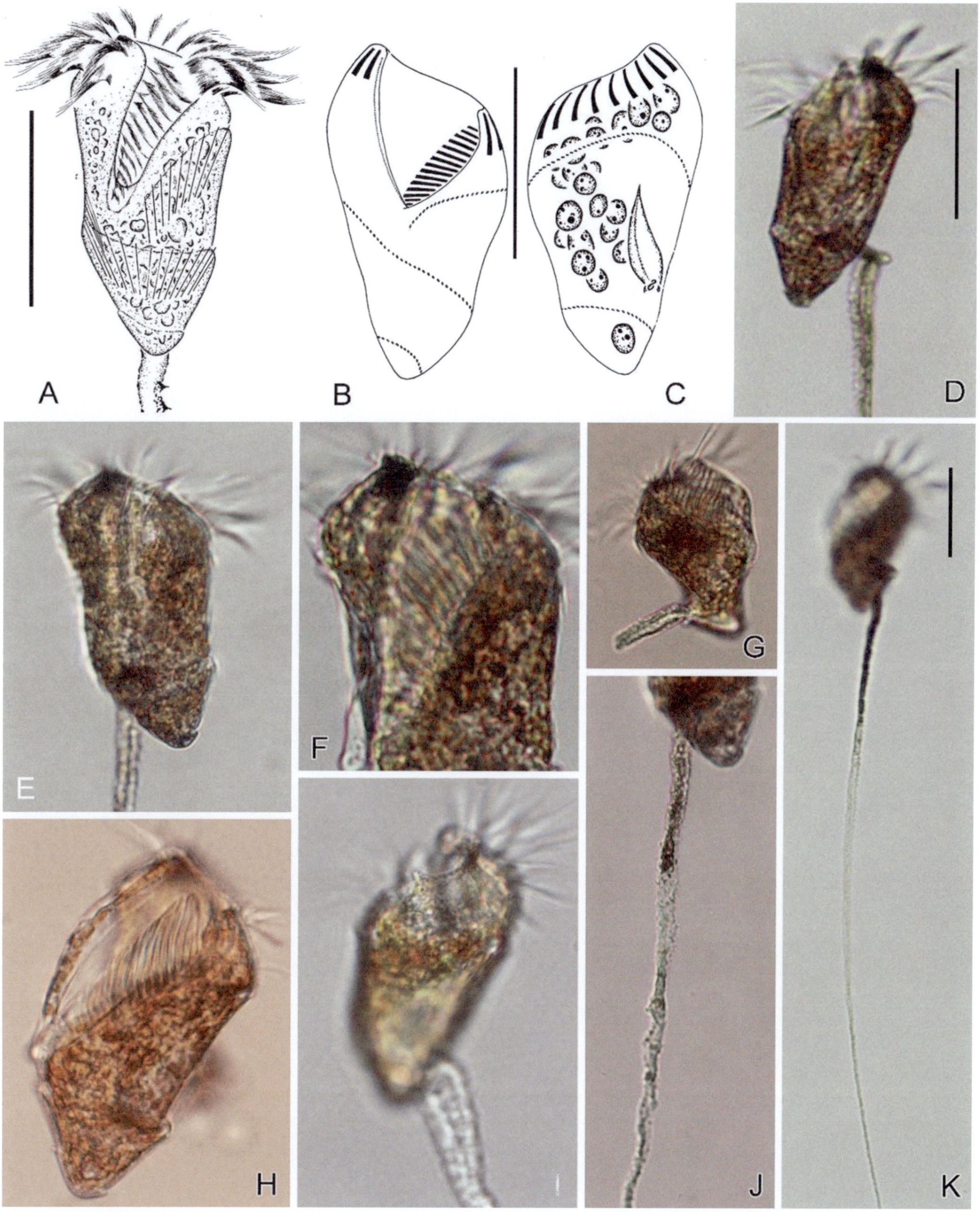


**Fig. 55B** *Spirotontonia turbinata*
(**a**, **d**, **e**, **h**, **i**) Ventral (**a**), dorsal (**e**, **i**), left-lateral (**d**, **h**) views of live cells. (**b**, **c**) Ventral (**b**) and dorsal (**c**) views of ciliature. (**f**) Anterior portion of live cell, showing ventral membranelles. (**g**) Contracted tail. (**j**, **k**) Elongated tails. Scale bars: 50 μm

## Reference

Song WB, Bradbury PC (1998) Studies on some new and rare reported marine planktonic ciliates (Ciliophora: Oligotricha) from coastal waters in North China. J Mar Biol Ass UK 78:767–794

Class Spirotrichea Bütschli, 1889
Order Tintinnida Kofoid & Campbell, 1929
Family Codonellidae Kent, 1881
*Tintinnopsis beroidea* Stein, 1867

**Ecological feature:** mangrove
**Water temperature:** 26.0 °C
**pH:** 7.0
**Salinity:** 5.0‰

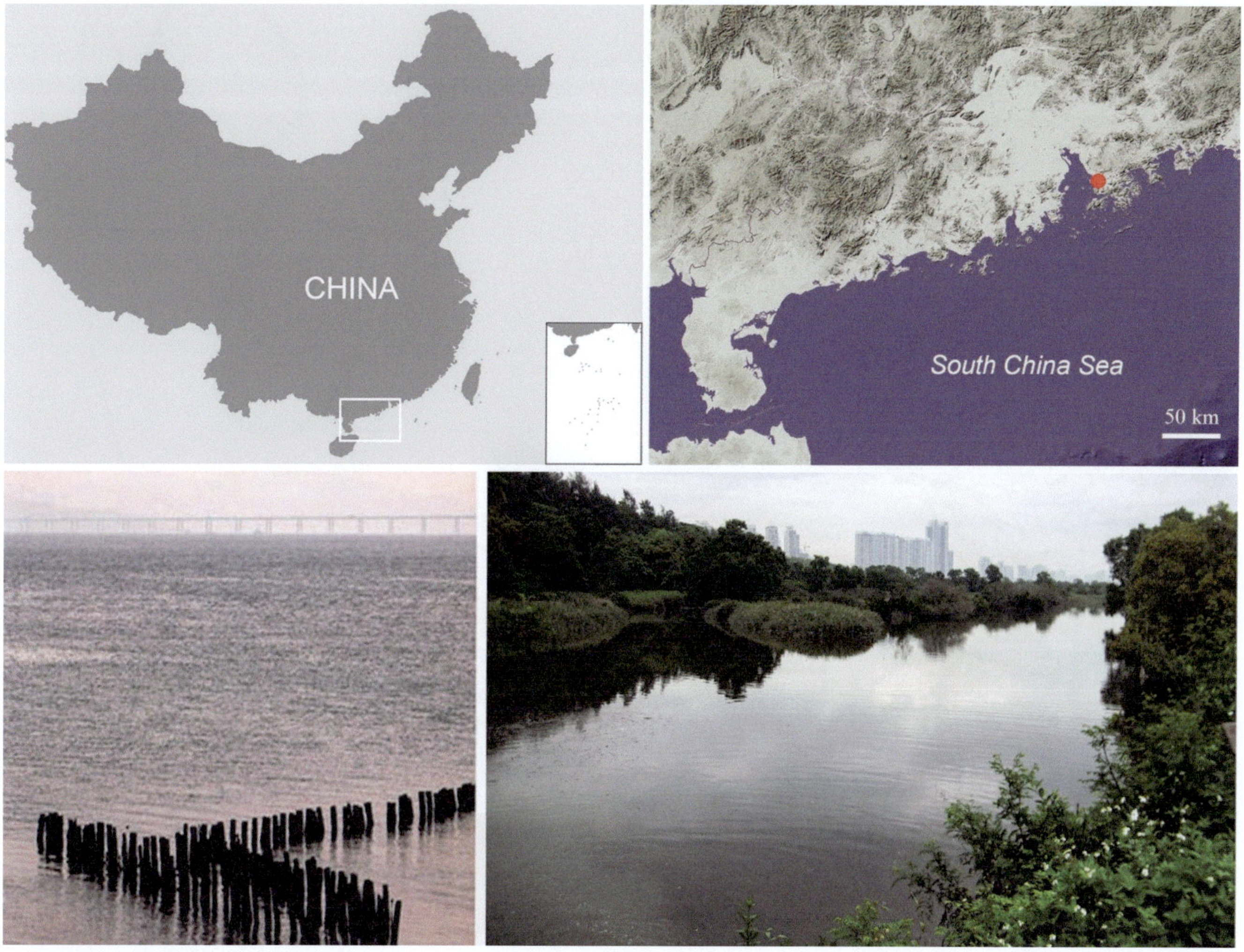


**Fig. 56A** Sample site

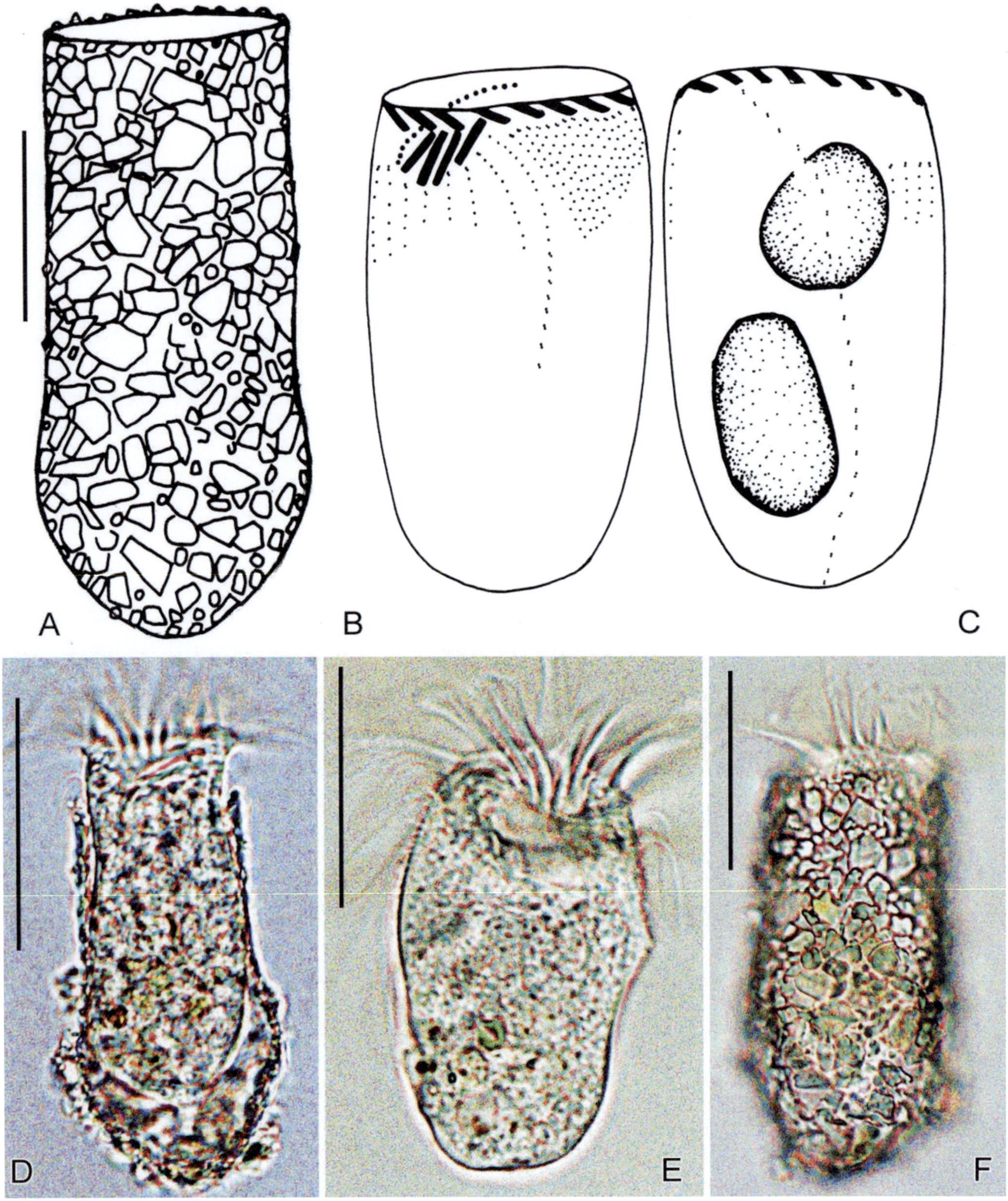


**Fig. 56B** *Tintinnopsis beroidea*
(**a**) Lateral view of lorica. (**e**, **f**) Lateral views of ciliature. (**d**, **f**) Lateral views of live cells within the lorica. (**e**) Lateral view of live cell broken away from lorica. Scale bars: 35 μm

## References

Dolan J, Jacquet S, Torreton J (2006) Comparing taxonomic and morphological biodiversity of tintinnids (planktonic ciliates) of New Caledonia. Limnol Oceanogr 51:950–958

Kofoid C, Campbell A (1929) A conspectus of the marine and fresh-water Ciliata belonging to the suborder Tintinnoinea, with descriptions of new species principally from the Agassiz Expedition to the eastern tropical Pacific 1904–1905. Univ Calif Publ Zool 34:1–403

Class Spirotrichea Bütschli, 1889
Order Tintinnida Kofoid & Campbell, 1929
Family Codonellidae Kent, 1881
*Tintinnopsis* cf. *beroidea*

**Ecological feature:** offshore water
**Water temperature:** 23.1 °C
**pH:** 8.2
**Salinity:** 28.6‰

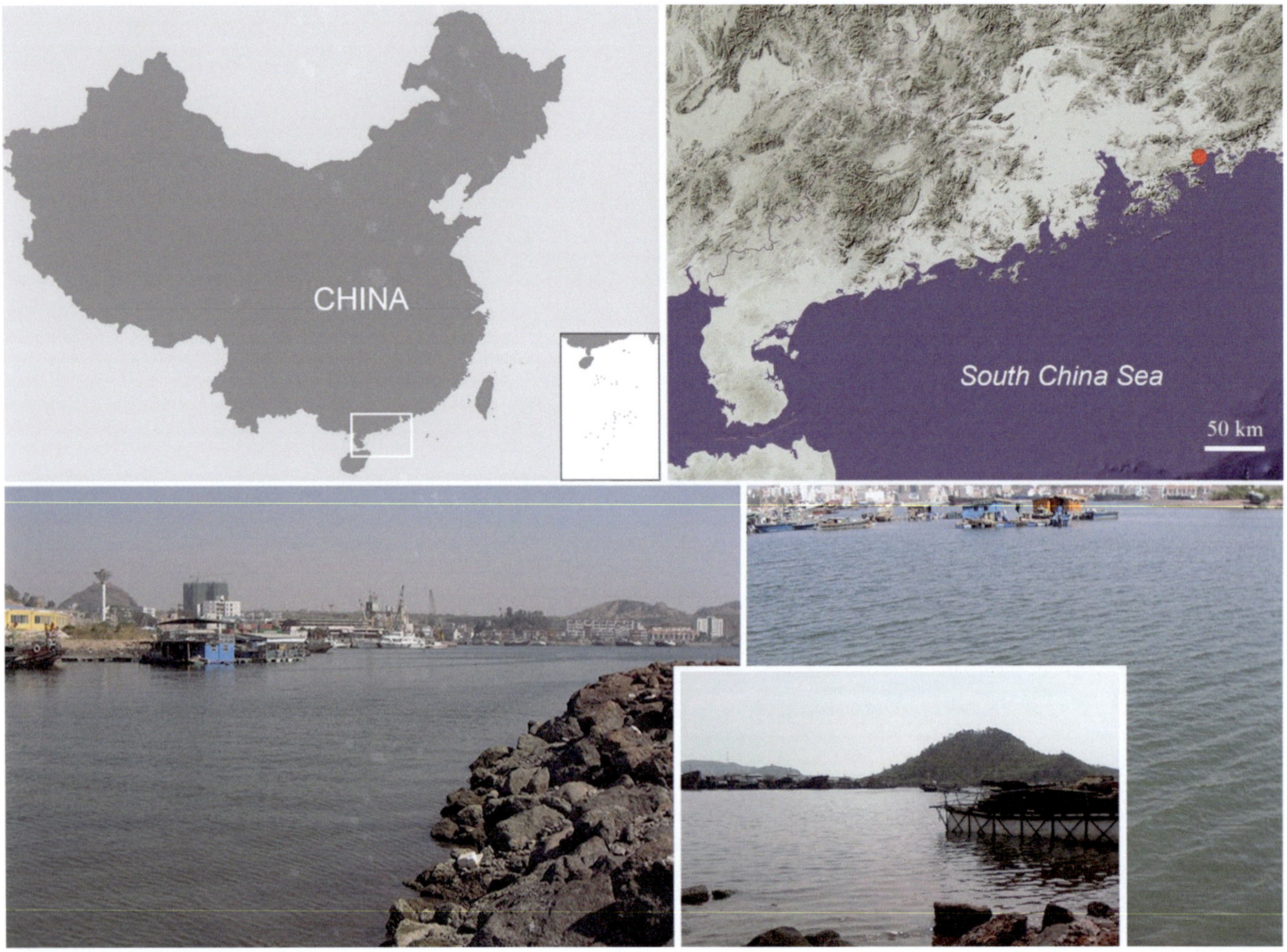


**Fig. 57A** Sample site

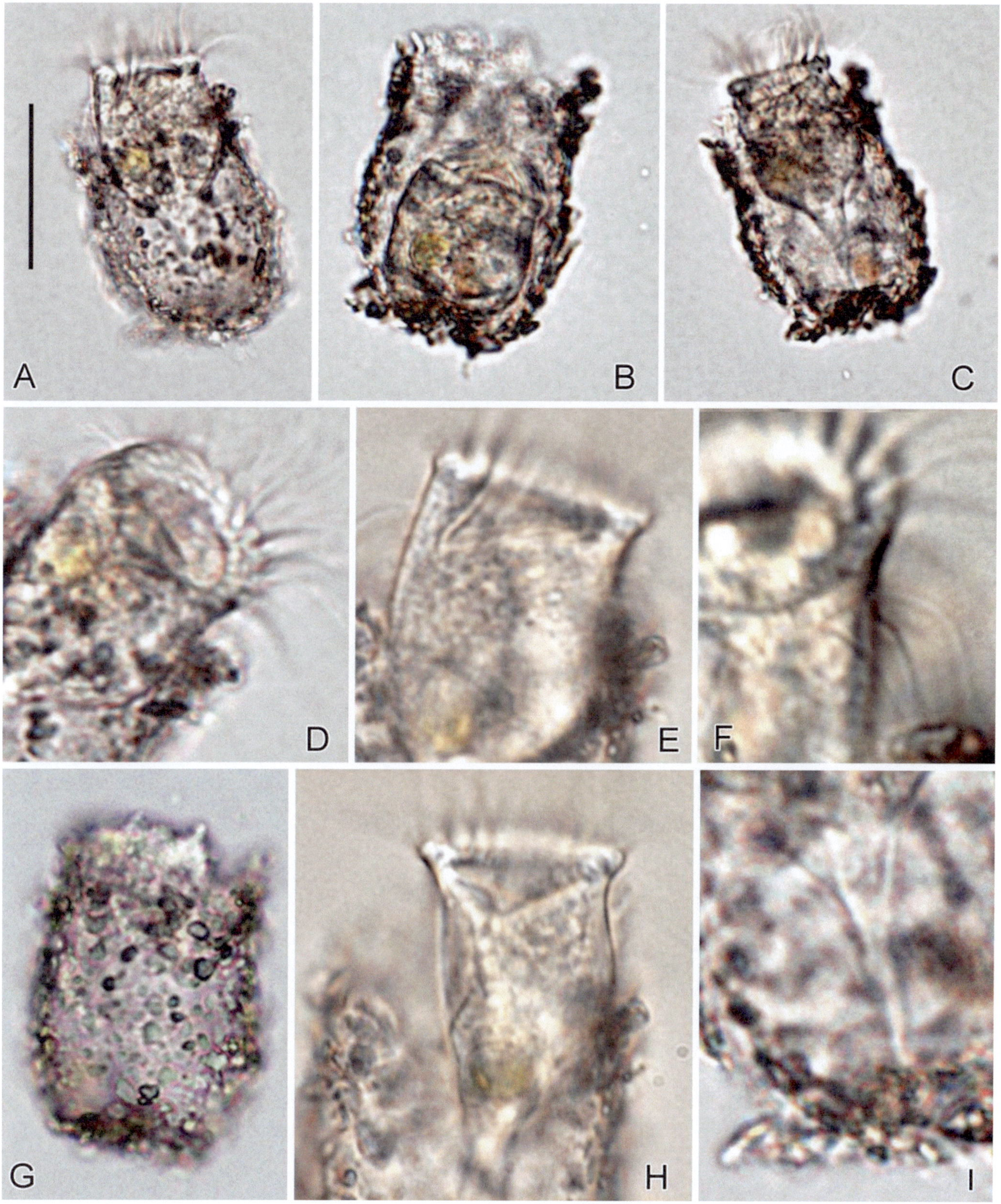


**Fig. 57B** *Tintinnopsis* cf. *beroidea*
(**a–c**) Lateral views of live cells within lorica. (**d**) Showing oral cilia extending through lorica aperture. (**e**, **h**) Anterior portion of cells to show the oral groove. (**f**) Detail of anterior portion showing the oral and somatic cilia. (**g**) Lorica wall with many silt particles. (**i**) To show the stalk. Scale bar: 30 μm

## References

Jiang Y, Yang J, Al-Farraj SA, Warren A, Lin X (2012) Redescriptions of three tintinnid ciliates, *Tintinnopsis tocantinensis*, *T. radix*, and *T. cylindrica* (Ciliophora, Spirotrichea), from coastal waters off China. Eur J Protistol 48:314–325

Kofoid C, Campbell A (1929) A conspectus of the marine and fresh-water Ciliata belonging to the suborder Tintinnoinea, with descriptions of new species principally from the Agassiz Expedition to the eastern tropical Pacific 1904–1905. Univ Calif Publ Zool 34:1–403

Class Spirotrichea Bütschli, 1889
Order Tintinnida Kofoid & Campbell, 1929
Family Codonellidae Kent, 1881
*Tintinnopsis cylindrica* Daday, 1887

**Ecological feature:** offshore water
**Water temperature:** 19.4 °C
**pH:** 8.2
**Salinity:** 32.6‰

**Fig. 58A** Sample site

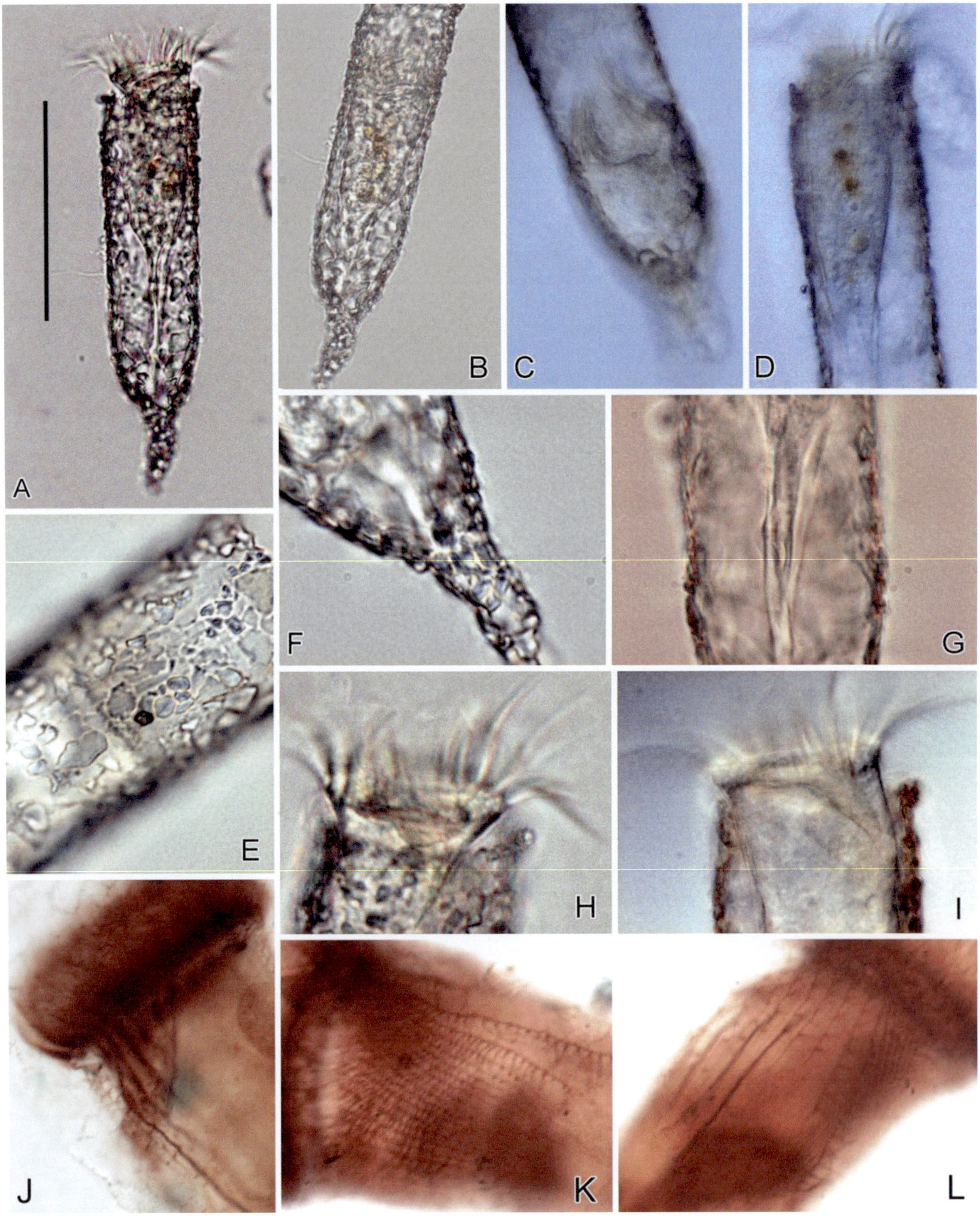


**Fig. 58B** *Tintinnopsis cylindrica*
(**a–d**, **h**, **i**) Lorica with live cells. (**e**, **f**) Lorica wall with many silt particles. (**g**) Cell stalk. (**j–l**) Ciliature. Scale bars: 80 μm

## Reference

Jiang Y, Yang J, Al-Farraj SA, Warren A, Lin X (2012) Redescriptions of three tintinnid ciliates, *Tintinnopsis tocantinensis*, *T. radix*, and *T. cylindrica* (Ciliophora, Spirotrichea), from coastal waters off China. Eur J Protistol 48:314–325

Class Spirotrichea Bütschli, 1889
Order Sporadotrichida Fauré-Fremiet, 1961
Family Oxytrichidae Ehrenberg, 1830
*Actinotricha saltans* Cohn, 1866

**Ecological feature:** mangrove (Zhanjiang); offshore water (Daya Bay)
**Water temperature:** 27.0 °C; 23.1 °C
**pH:** no data available; 8.2
**Salinity:** 25.0‰; 28.6‰

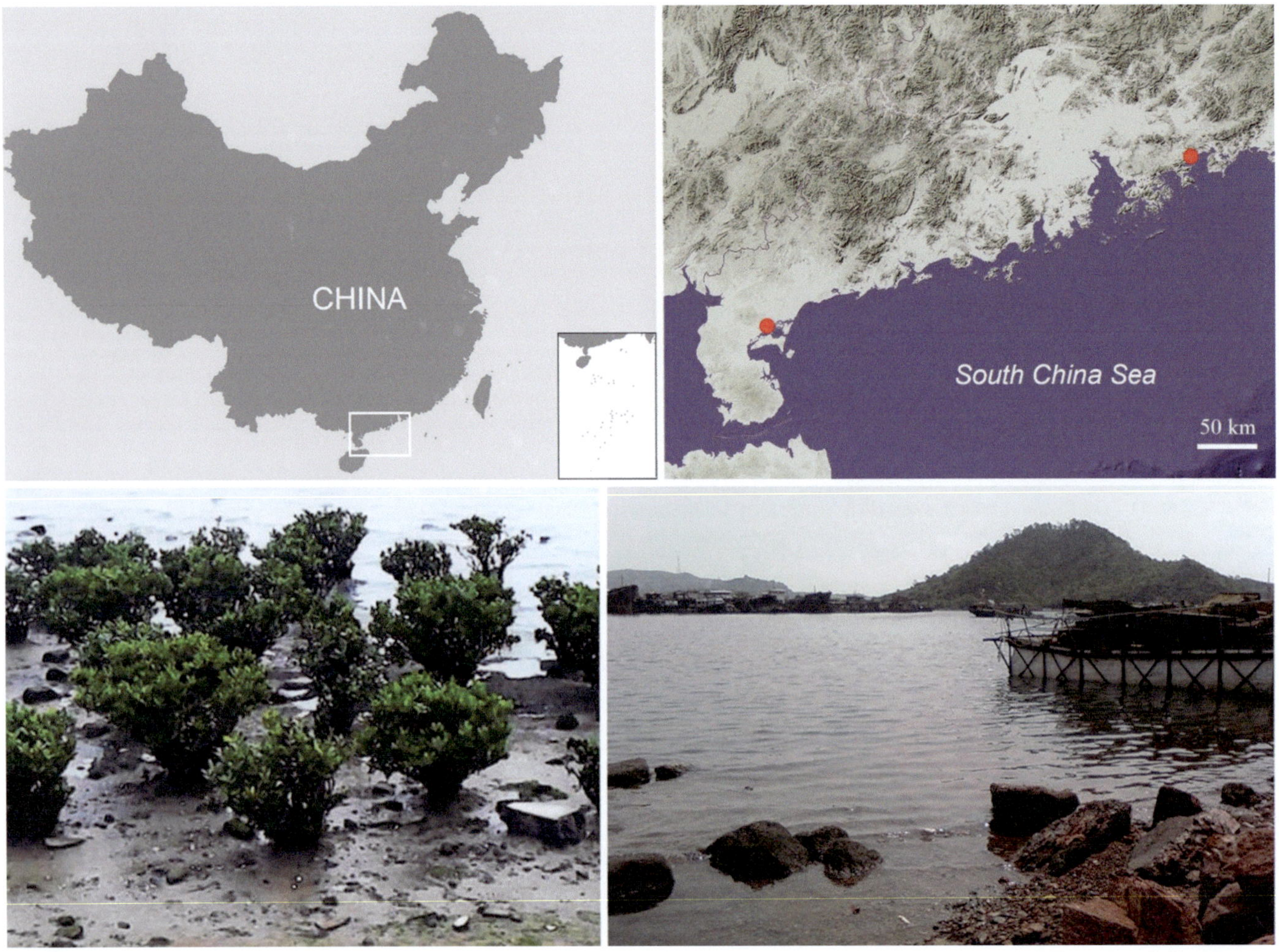


**Fig. 59A** Sample site

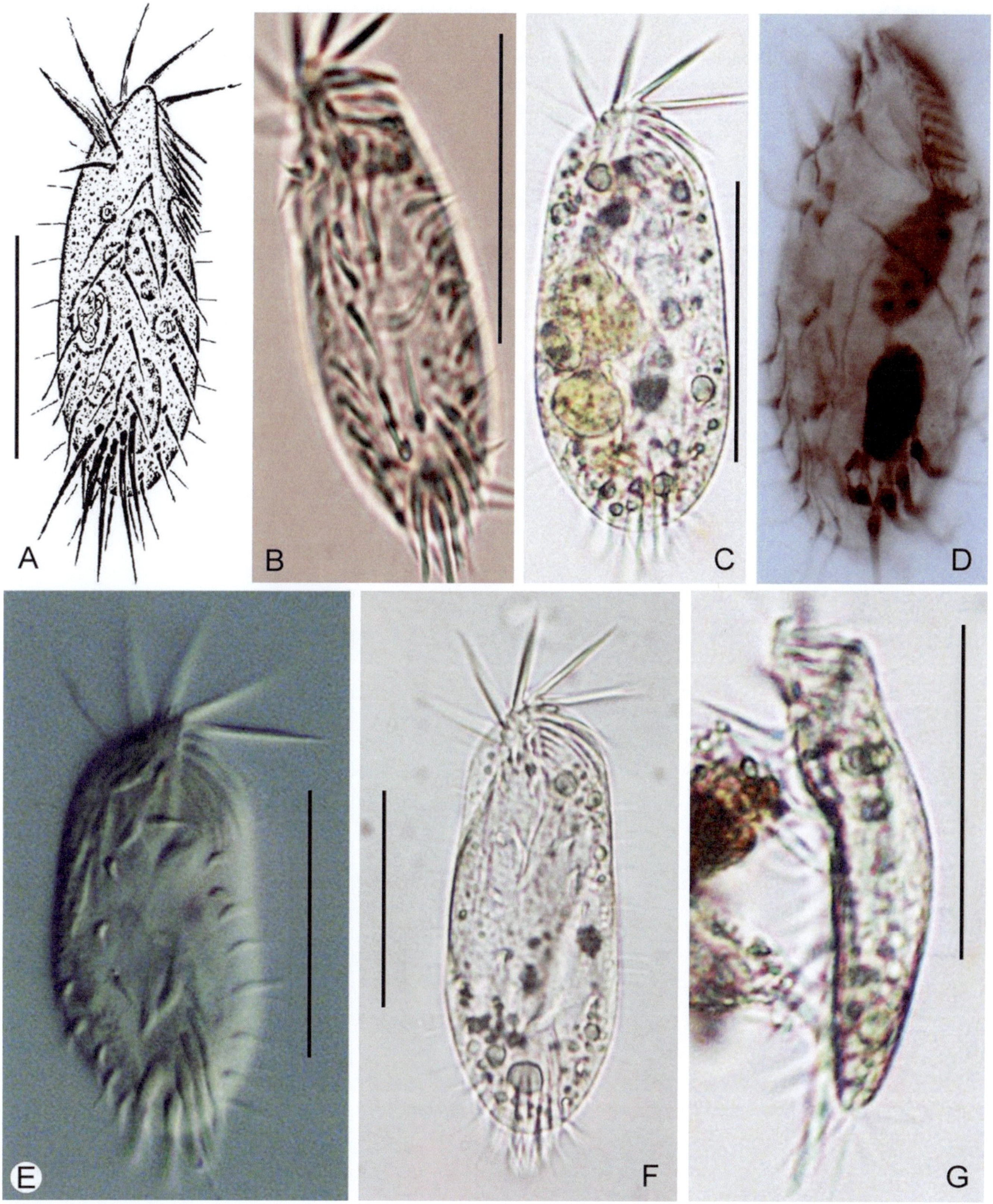


**Fig. 59B** *Actinotricha saltans*
(**a–c**, **e**, **f**) Ventral views of live cells. (**d**) Ventral view of ciliature. (**g**) Lateral view of live cell. Scale bars: 50 μm

## References

Song WB, Wilbert N (1997) Morphological investigations on some free living ciliates (Protozoa, Ciliophora) from China sea with description of a new hypotrichous genus, *Hemigastrostyla* nov. gen. Arch Protistenk 148:413–444

Song WB, Shin M, Kim W (1991) Morphology and infraciliature of the marine ciliate *Oxytricha saltans* (Cohn, 1886) Kahl, 1932 (Protozoa, Ciliophora, Potrichida). Korean J Syst Zool 7:233–240

Class Spirotrichea Bütschli, 1889
Order Sporadotrichida Fauré-Fremiet, 1961
Family Oxytrichidae Ehrenberg, 1830
*Hemigastrostyla paraenigmatica* (Dragesco & Dragesco-Kernéis, 1986) Shao et al., 2011

**Ecological feature:** mangrove
**Water temperature:** 24.0 °C; 25.0 °C
**pH:** no data available; 6.8
**Salinity:** 17.0‰; 18.0‰

**Fig. 60A** Sample site

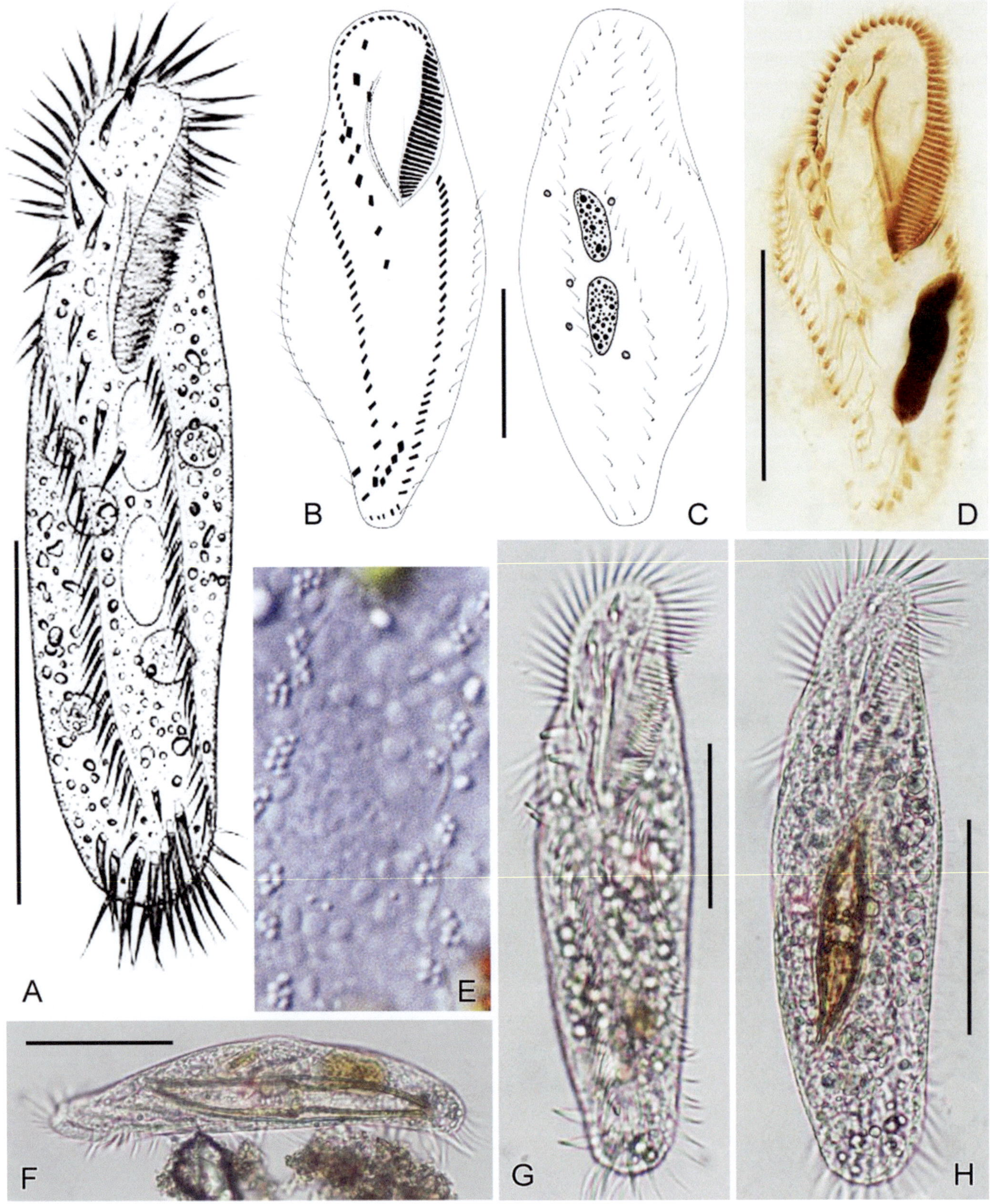


**Fig. 60B** *Hemigastrostyla paraenigmatica*
(**a**, **g**, **h**) Ventral views of live individuals. (**b–d**) Ventral (**b**, **d**) and dorsal (**c**) views of ciliature. (**e**) Cortical granules around dorsal cilia. (**f**) Lateral view of live cell. Scale bars: 60 μm

## Reference

Gong J, Kim S, Kim S, Min G, Roberts D, Warren A, Choi J (2007) Taxonomic redescriptions of two ciliates, *Protogastrostyla pulchra* n. g., n. comb. and *Hemigastrostyla enigmatica* (Ciliophora, Spirotrichea, Stichotrichia), with phylogenetic analyses based on 18S and 28S rRNA gene sequences. J Eukaryot Microbiol 54: 468–478

Class Spirotrichea Bütschli, 1889
Order Sporadotrichida Fauré-Fremiet, 1961
Family Oxytrichidae Ehrenberg, 1830
*Oxytricha lithofera* Foissner, 2016

**Ecological feature:** sandy sediment
**Water temperature:** 27.0 °C
**pH:** no data available
**Salinity:** 5.0‰

**Fig. 61A** Sample site

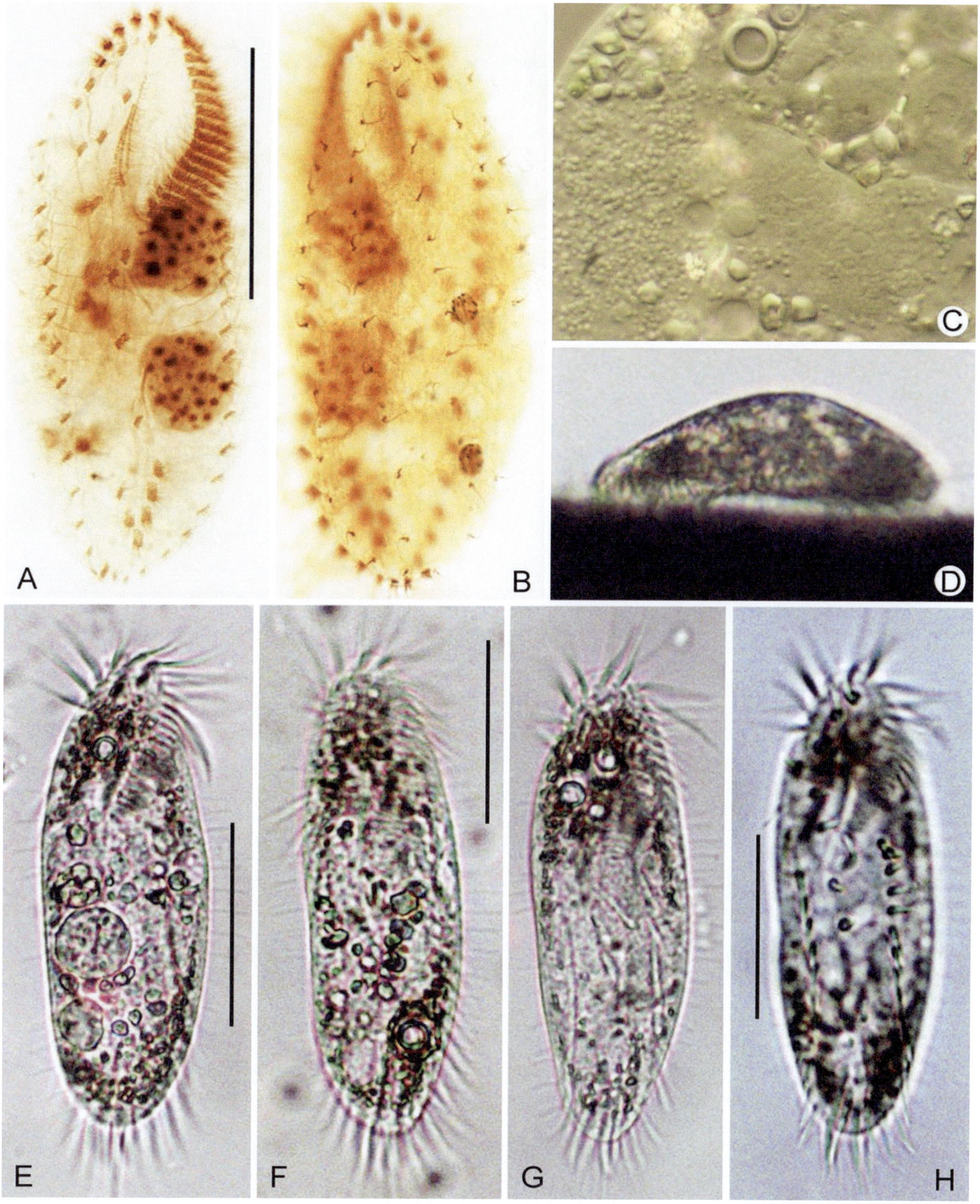


**Fig. 61B** *Oxytricha lithofera*
(**a**, **b**) Ventral (**a**) and dorsal (**b**) views of ciliature. (**c**) Cortical granules, macronuclear segment and ring-shaped structures. (**d**) Lateral view of live cell. (**e**–**h**)Ventral views of live individuals. Scale bars: 40 μm

## Reference

Foissner W (2016) Terrestrial and semiterrestrial ciliates (Protozoa, Ciliophora) from Venezuela and Galápagos. Denisia 35:1–912

Class Spirotrichea Bütschli, 1889
Order Sporadotrichida Fauré-Fremiet, 1961
Family Oxytrichidae Ehrenberg, 1830
*Polystichothrix monilata* Luo et al., 2017

**Ecological feature:** mangrove
**Water temperature:** 23.0 °C
**pH:** no data available
**Salinity:** 15.0‰

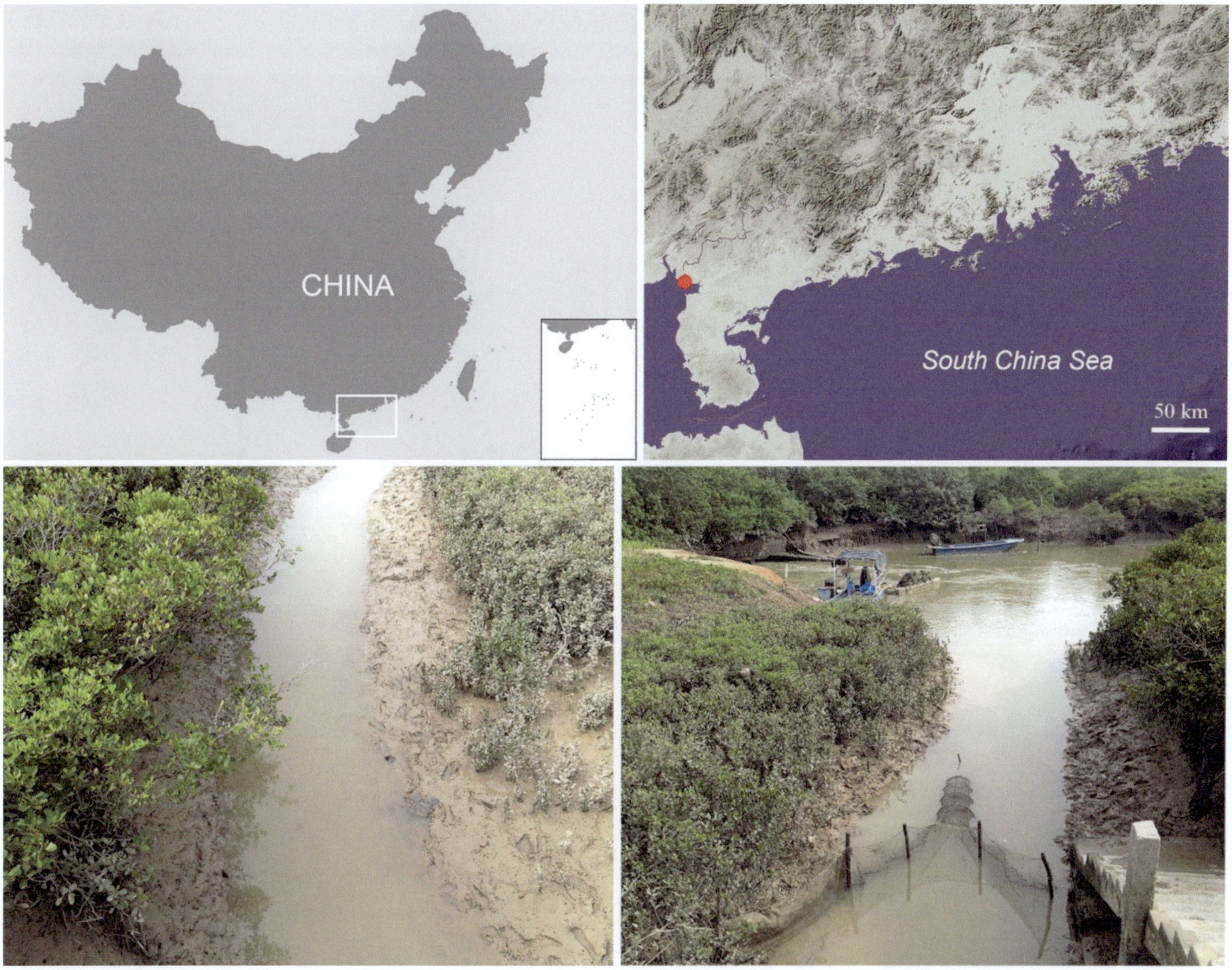


**Fig. 62A** Sample site

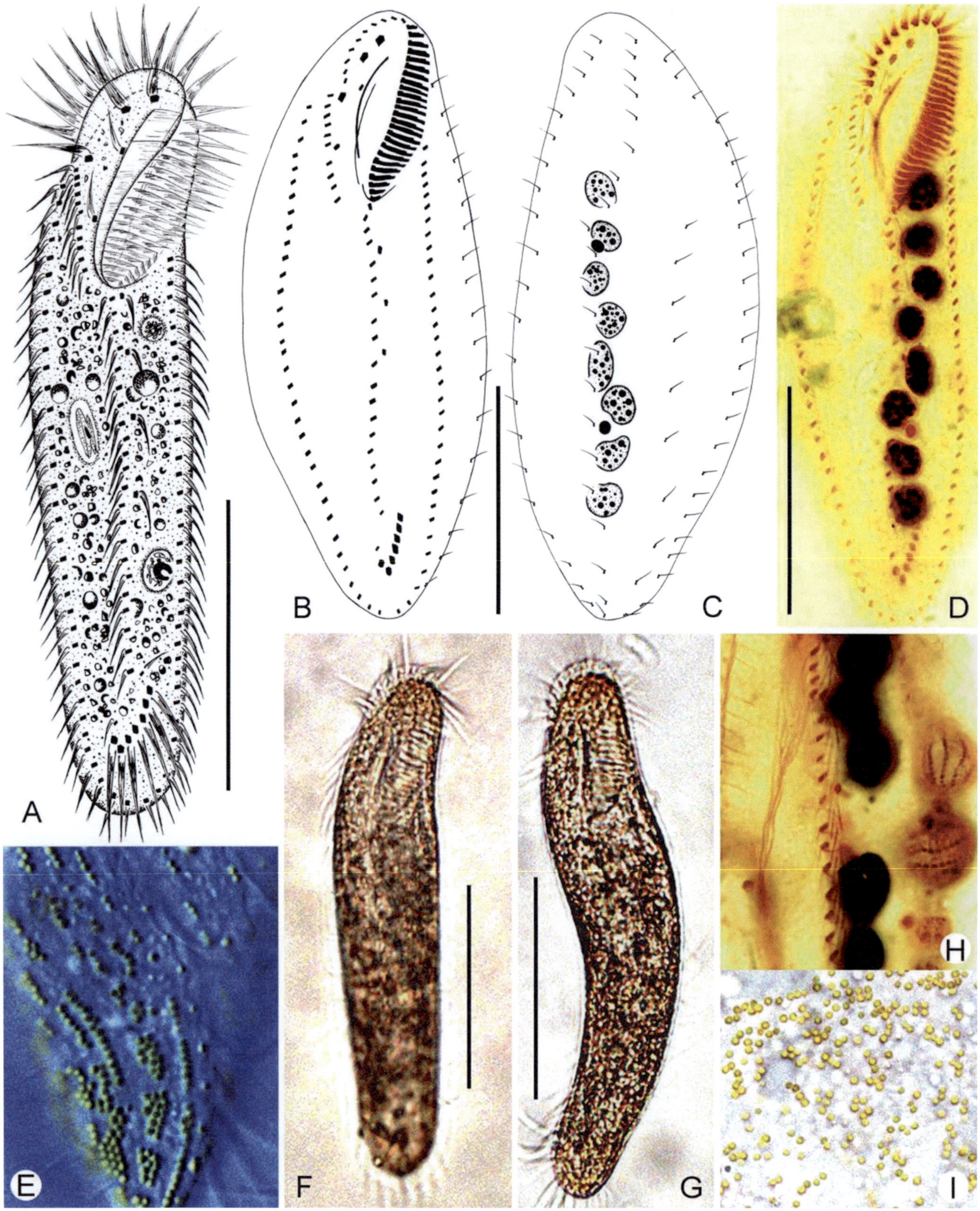


**Fig. 62B** *Polystichothrix monilata*
(**a**, **f**, **g**) Ventral views of live individuals. (**b**, **c**, **d**) Ventral (**b**, **d**) and dorsal (**c**) views of ciliature. (**e**, **i**) Brown-yellowish cortical granules. (**h**) Strong pharyngeal fibers and small ingested ciliates. Scale bars: 50 μm

## Reference

Luo X, Gao F, Yi Z, Pan Y, Al-Farraj SA, Warren A (2017) Taxonomy and molecular phylogeny of two new brackish hypotrichous ciliates, with the establishment of a new genus (Protozoa, Ciliophora). Zool J Linn Soc 179:475–491

Class Spirotrichea Bütschli, 1889
Order Sporadotrichida Fauré-Fremiet, 1961
Family Oxytrichidae Ehrenberg, 1830
*Ponturostyla enigmatica* (Dragesco & Dragesco-Kernéis, 1986) Jankowski, 1989

**Ecological feature:** mangrove
**Water temperature:** 28.0 °C; 20.0 °C; no data available
**pH:** 6.8; 7.5; 8.0
**Salinity:** 11.0‰; 10.0‰; 20.0‰

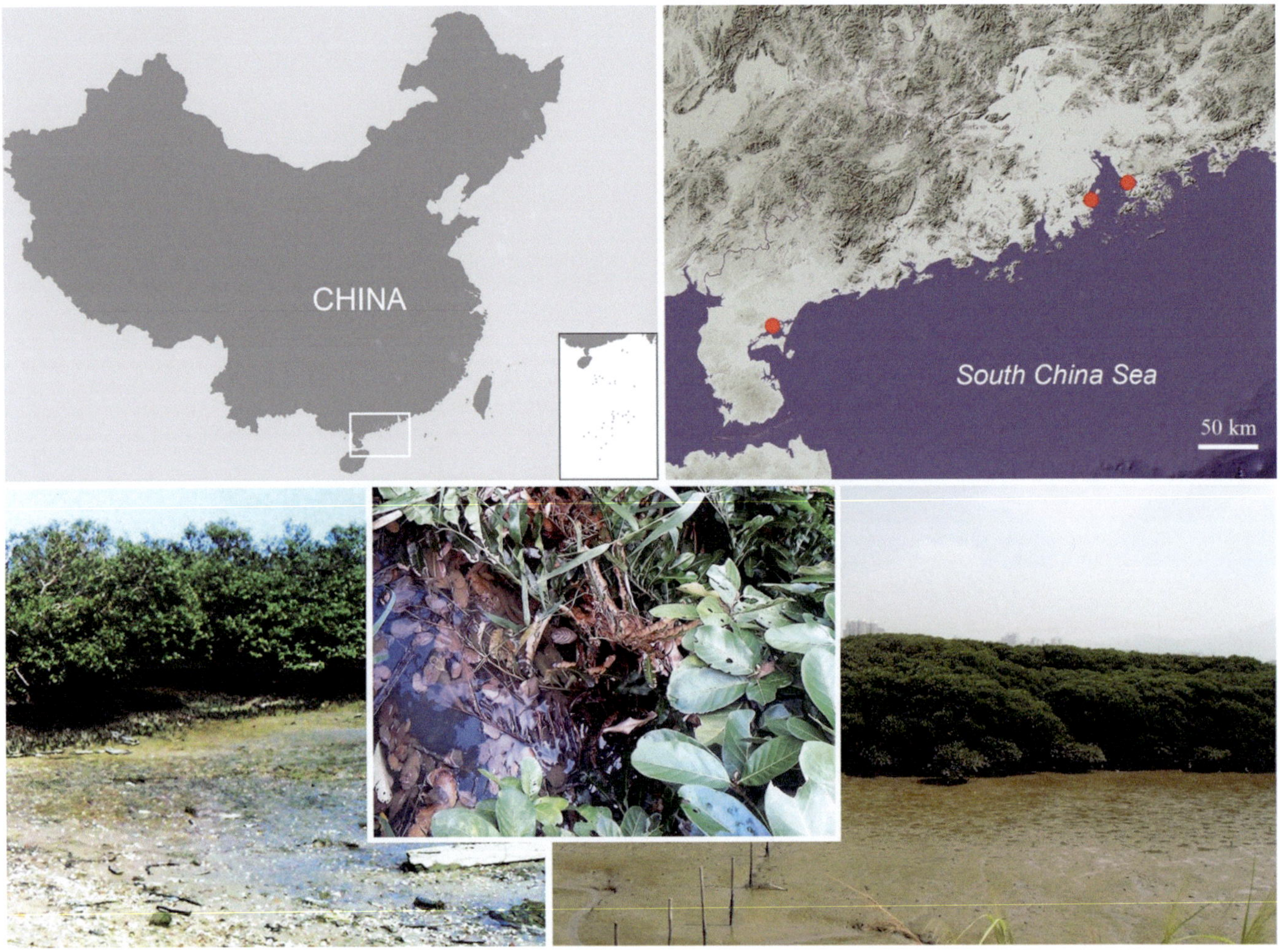


**Fig. 63A** Sample site

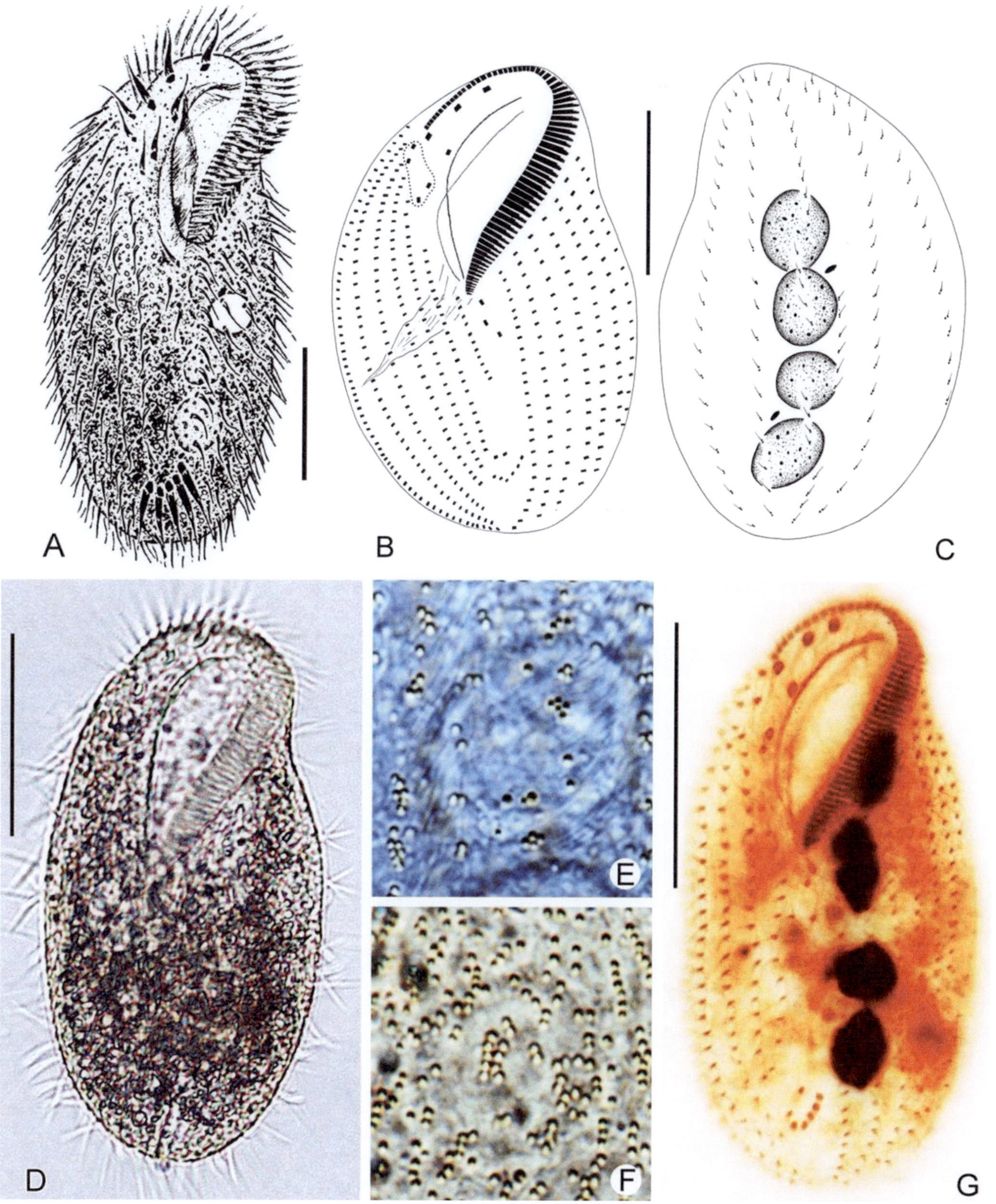


**Fig. 63B** *Ponturostyla enigmatica*
(**a**, **d**) Ventral views of live cells. (**b**, **c**, **g**) Ventral (**b**, **g**) and dorsal (**c**) views of ciliature. (**e**, **f**) Cortical granules on ventral (**e**) and dorsal (**f**) sides. Scale bars: 40 μm (**a**–**d**) and 70 μm (**g**)

## Reference

Song WB (2001) Morphology and morphogenesis of the marine ciliate *Ponturostyla enigmatica* (Dragesco & Dragesco-Kernéis, 1986) Jankowski, 1989 (Ciliophora, Hypotrichida, Oxytrichidae). Eur J Protistol 37:181–197

Class Spirotrichea Bütschli, 1889
Order Sporadotrichida Fauré-Fremiet, 1961
Family Oxytrichidae Ehrenberg, 1830
*Pseudogastrostyla flava* Fan et al., 2015

**Ecological feature:** mangrove
**Water temperature:** 21.3 °C
**pH:** 7.2
**Salinity:** 10.0‰

**Fig. 64A** Sample site

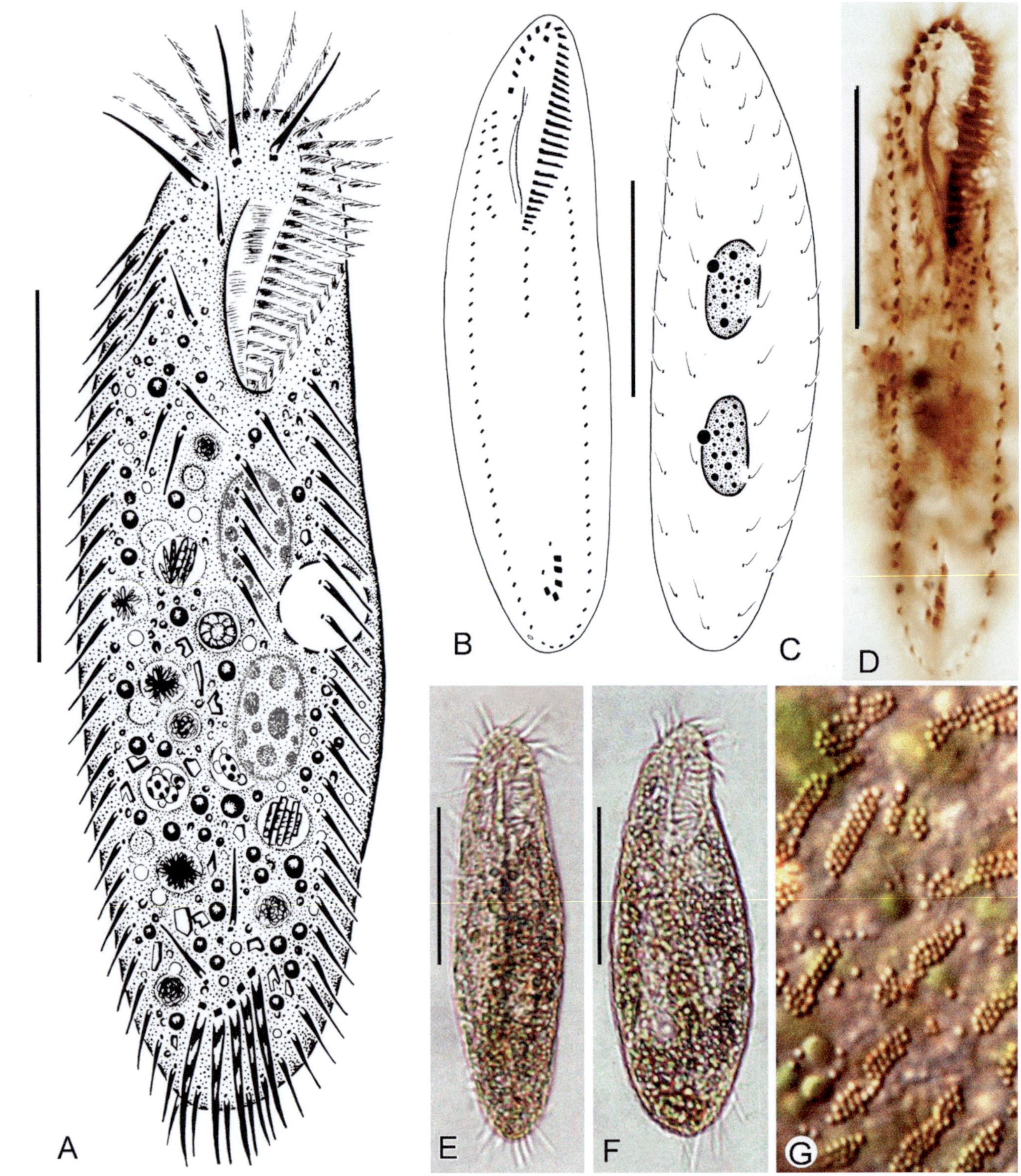


**Fig. 64B** *Pseudogastrostyla flava*
(**a**, **e**, **f**) Ventral views of live individuals. (**b–d**) Ventral (**b**, **d**) and dorsal (**c**) views of ciliature. (**g**) Arrangement of yellow cortical granules. Scale bars: 50 µm

## Reference

Fan Y, Zhao X, Hu X, Miao M, Warren A, Song WB (2015) Taxonomy and molecular phylogeny of two novel ciliates, with establishment of a new genus, *Pseudogastrostyla* n. g. (Ciliophora, Hypotrichia, Oxytrichidae). Eur J Protistol 51:374–385

Class Spirotrichea Bütschli, 1889
Order Sporadotrichida Fauré-Fremiet, 1961
Family Oxytrichidae Ehrenberg, 1830
*Rubrioxytricha haematoplasma* (Blatterer & Foissner, 1990) Berger, 1999

**Ecological feature:** brackish water
**Water temperature:** no data available
**pH:** no data available
**Salinity:** 8.6‰

**Fig. 65A** Sample site

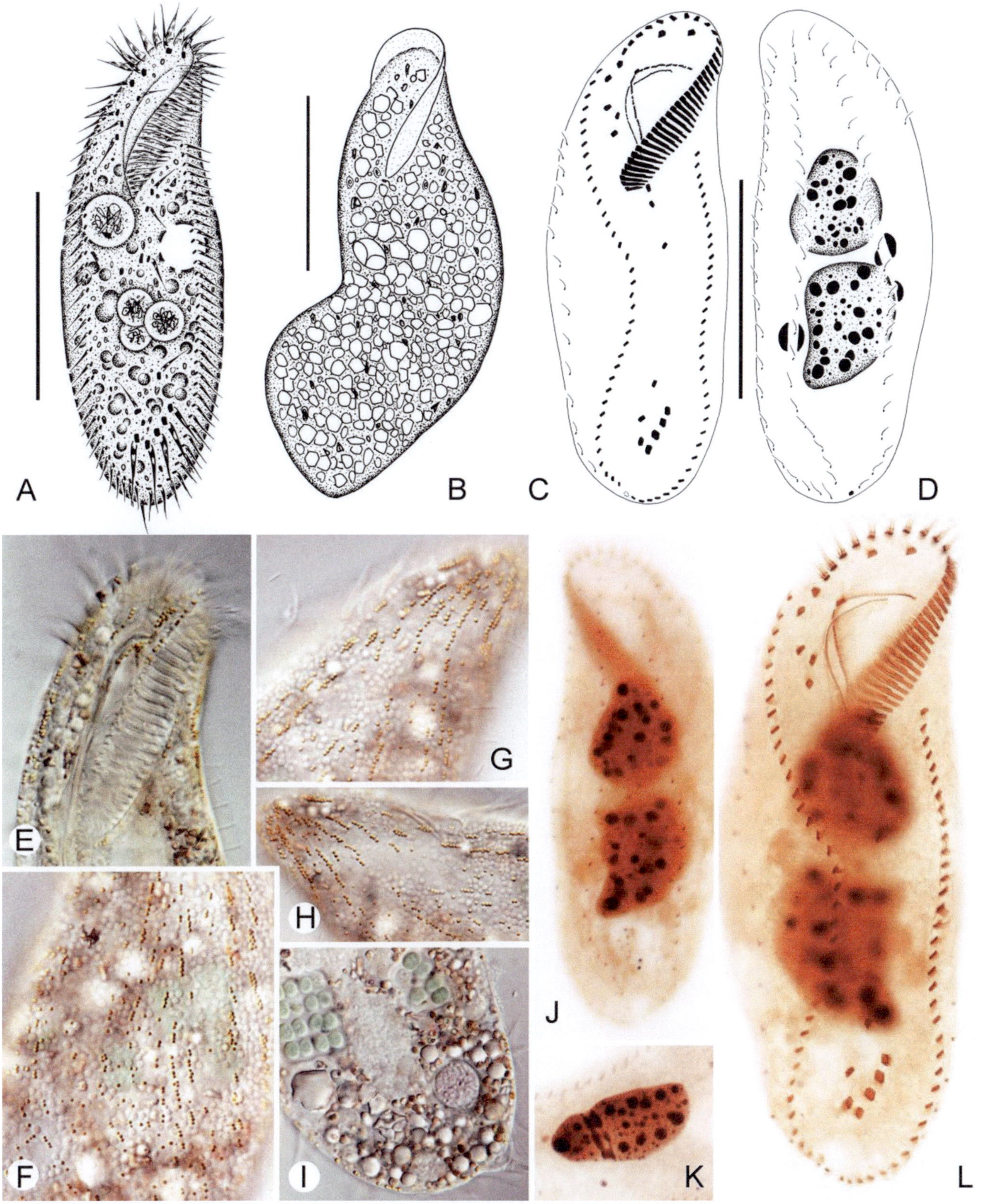


**Fig. 65B** *Rubrioxytricha haematoplasma*
(**a**) Ventral view of live cell. (**b**) Ventral view of live cell, to show the flexibility of the body. (**c**, **d**, **j**, **l**) Ventral (**c**, **l**) and dorsal (**d**, **j**) views of ciliature. (**e**–**i**) Cortical granules. (**k**) Macronuclear nodule. Scale bars: 60 μm

## Reference

Chen W, Chen X, Li L, Warren A, Lin X (2015) Morphology, morphogenesis and molecular phylogeny of an oxytrichid ciliate, *Rubrioxytricha haematoplasma* (Blatterer & Foissner, 1990) Berger, 1999 (Ciliophora, Hypotricha). Int J Syst Evol Microbiol 65:309–320

Class Spirotrichea Bütschli, 1889
Order Sporadotrichida Fauré-Fremiet, 1961
Family Oxytrichidae Ehrenberg, 1830
*Sterkiella cavicola* (Kahl, 1935) Foissner et al., 1991

**Ecological feature:** mangrove
**Water temperature:** 31.0 °C; no data available
**pH:** 7.0; 7.0
**Salinity:** 1.0‰; 5.0‰

**Fig. 66A** Sample site

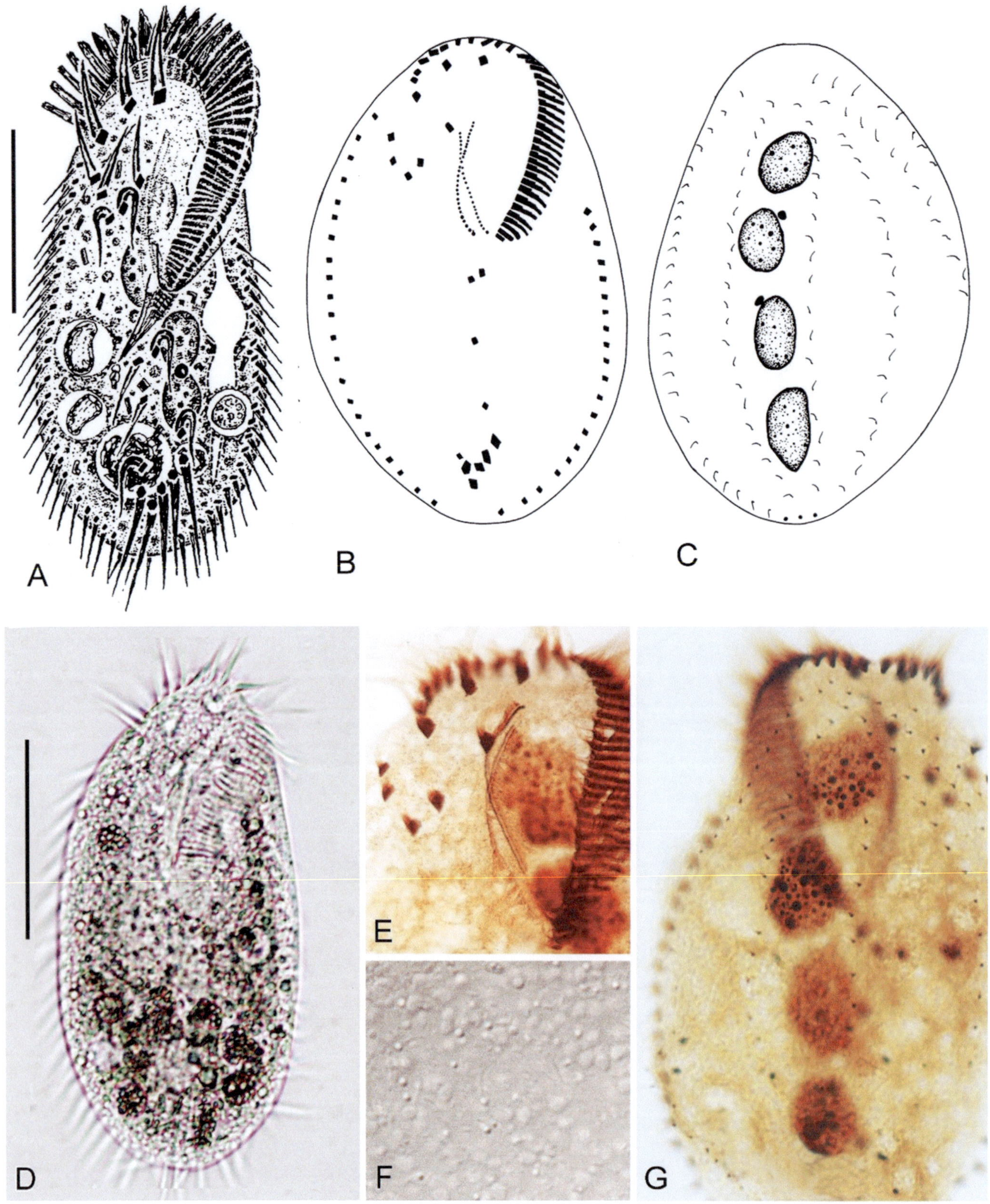


**Fig. 66B** *Sterkiella cavicola*
(**a**, **d**) Ventral views of live individuals, A from Berger and Foissner (1987). (**b**, **c**, **g**) Ventral (**b**) and dorsal (**c**, **g**) views of ciliature. (**e**) Ventral view of anterior portion of cell, showing frontal ciliature. (**f**) Cortical granules. Scale bars: 40 μm

## References

Berger H, Foissner W (1987) Morphology and biometry of some soil hypotirchs. Zool Jb Syst 114:193–239

Kahl A (1935) Urtiere oder Protozoa I: Wimpertiere oder Ciliata (Infusoria) 4. Peritrichaund Chonotricha. Tierwelt Dtl 30:651–886

Class Spirotrichea Bütschli, 1889
Order Sporadotrichida Fauré-Fremiet, 1961
Family Oxytrichidae Ehrenberg, 1830
*Sterkiella histriomuscorum* (Foissner et al., 1991) Foissner et al., 1991

**Ecological feature:** mangrove
**Water temperature:** no data available
**pH:** 7.0
**Salinity:** 9.0‰

**Fig. 67A** Sample site

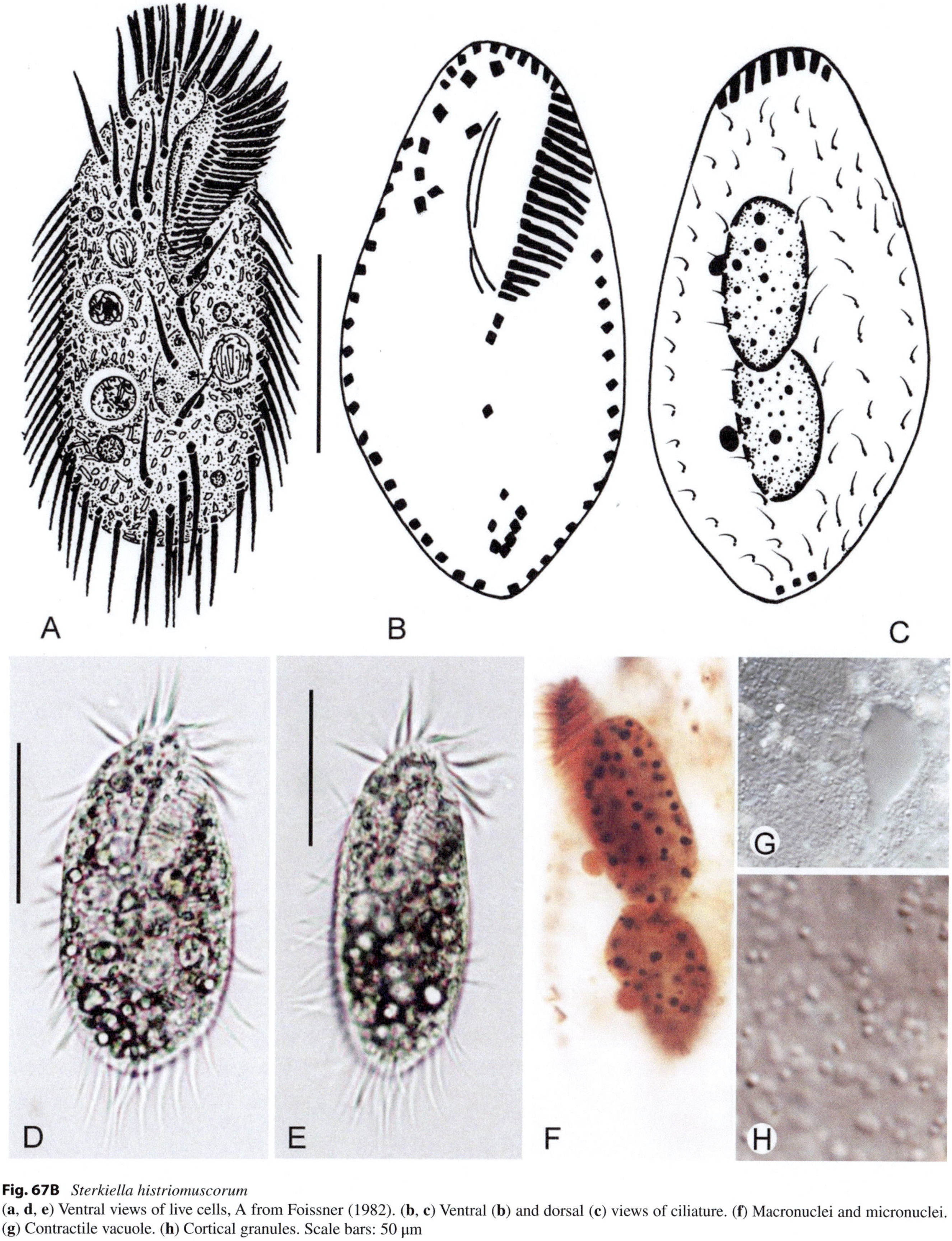


**Fig. 67B** *Sterkiella histriomuscorum*
(**a**, **d**, **e**) Ventral views of live cells, A from Foissner (1982). (**b**, **c**) Ventral (**b**) and dorsal (**c**) views of ciliature. (**f**) Macronuclei and micronuclei. (**g**) Contractile vacuole. (**h**) Cortical granules. Scale bars: 50 μm

## Reference

Foissner W (1982) Ökologie und Taxonomie der Hypotrichida (Protozoa: Ciliophora) einiger österreichischer Böden. Arch Protistenk 126:19–143

Class Spirotrichea Bütschli, 1889
Order Sporadotrichida Fauré-Fremiet, 1961
Family Oxytrichidae Ehrenberg, 1830
*Sterkiella subtropica* Chen et al., 2015

**Ecological feature:** mangrove
**Water temperature:** no data available
**pH:** no data available
**Salinity:** 29.0‰

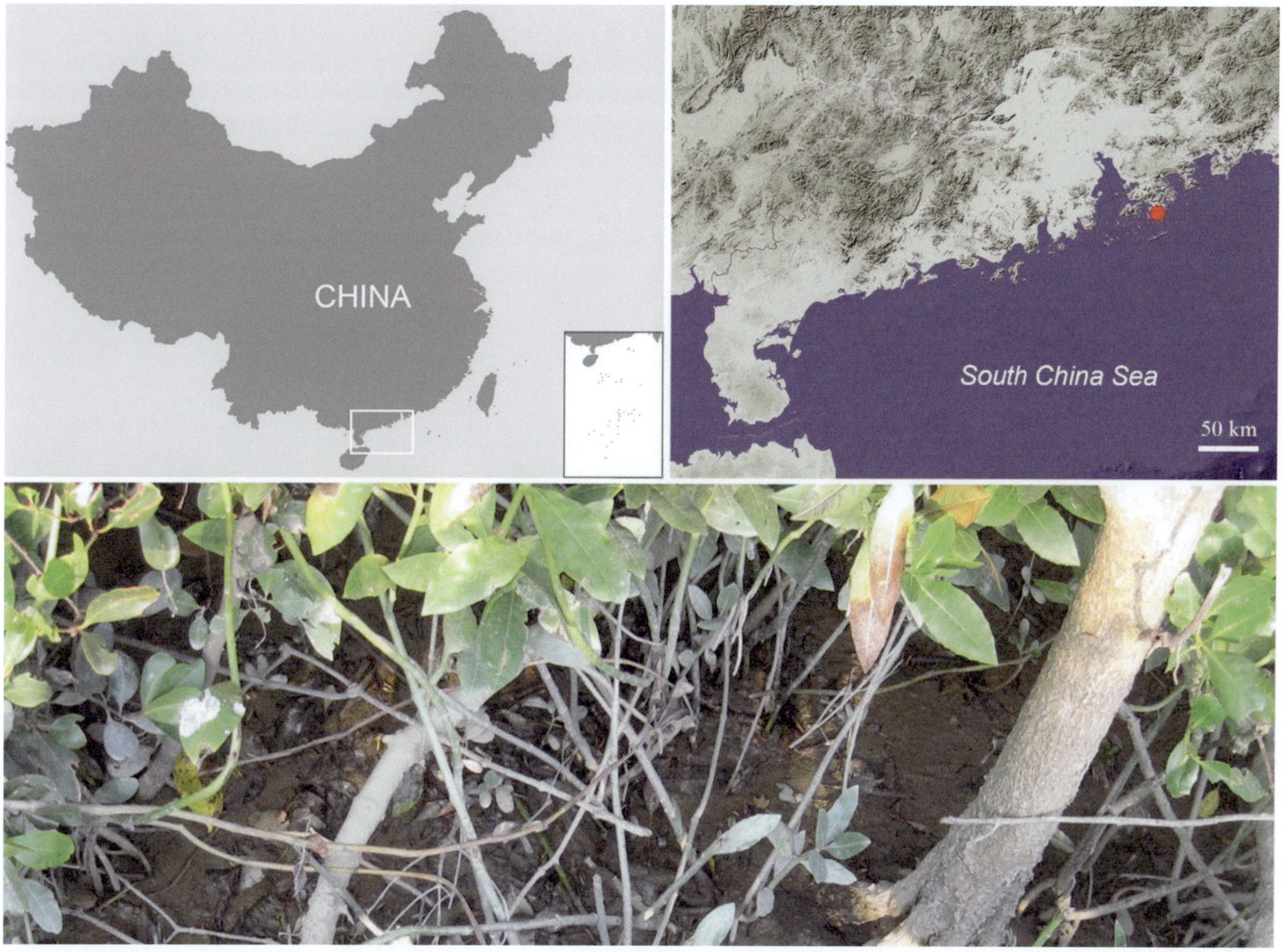


**Fig. 68A** Sample site

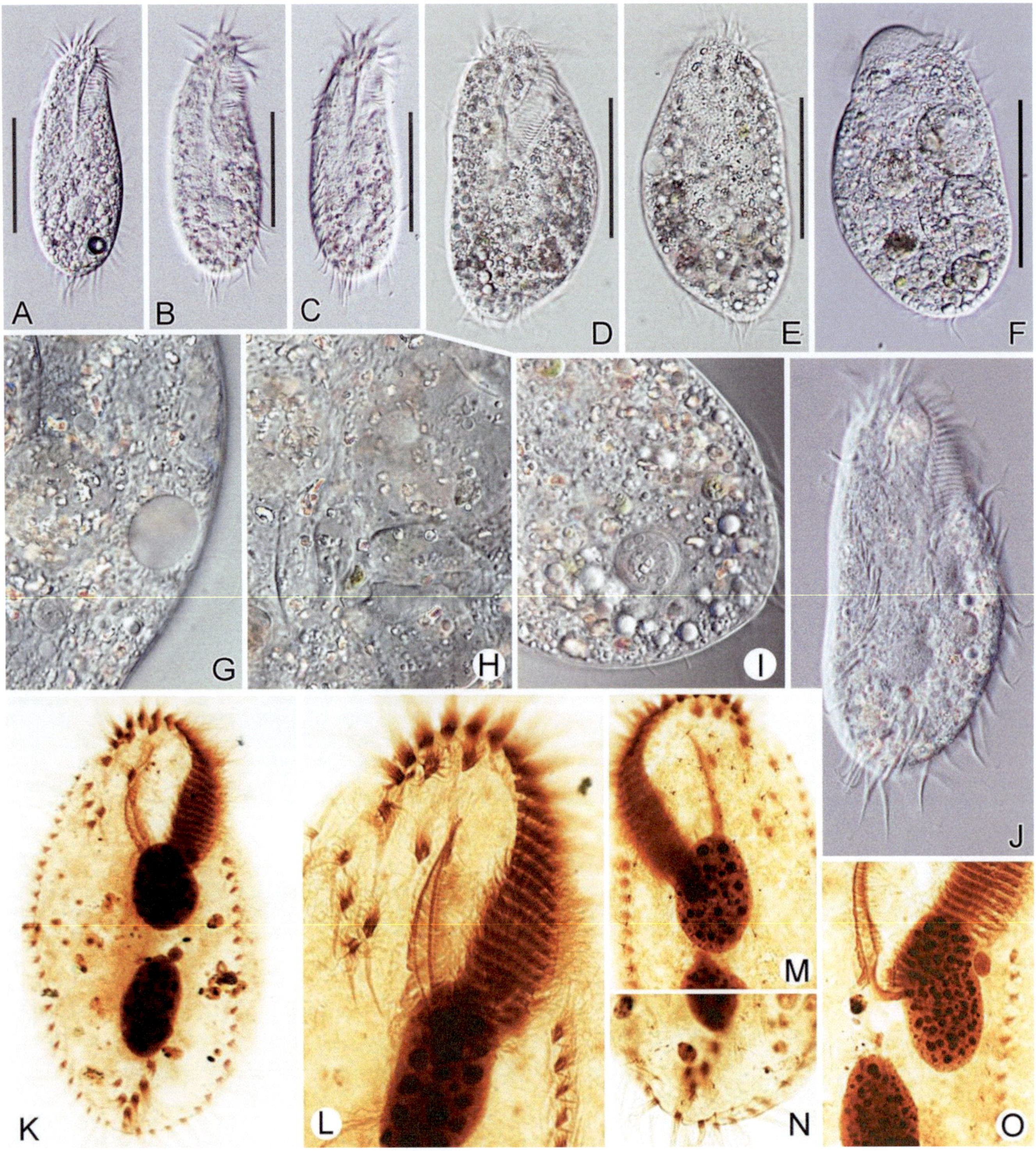


**Fig. 68B** *Sterkiella subtropica*
(**a–f**, **j**) Ventral (**a–d**, **j**) and dorsal (**e**, **f**) views of live individuals. (**g**) Contractile vacuole. (**h**, **i**) Inclusive granules. (**k–n**) Ventral (**k**, **l**) and dorsal (**m**, **n**) views of ciliature. (**o**) Macronuclei and micronucleus. Scale bars: 60 μm

## Reference

Chen XM, Gao F, Al-Farraj SA, Al-Rasheid KAS, Xu K, Song WB (2015) Morphology and morphogenesis of a novel mangrove ciliate, *Sterkiella subtropica* sp. nov. (Protozoa, Ciliophora, Hypotrichia), with phylogenetic analyses based on small-subunit rDNA sequence data. Int J Syst Evol Microbiol 65:2292–2303

Class Spirotrichea Bütschli, 1889
Order Sporadotrichida Fauré-Fremiet, 1961
Family Oxytrichidae Ehrenberg, 1830
*Tetmemena pustulata* (Müller, 1786) Eigner, 1999

**Ecological feature:** mangrove
**Water temperature:** 24.0 °C
**pH:** 6.5
**Salinity:** 10.0‰

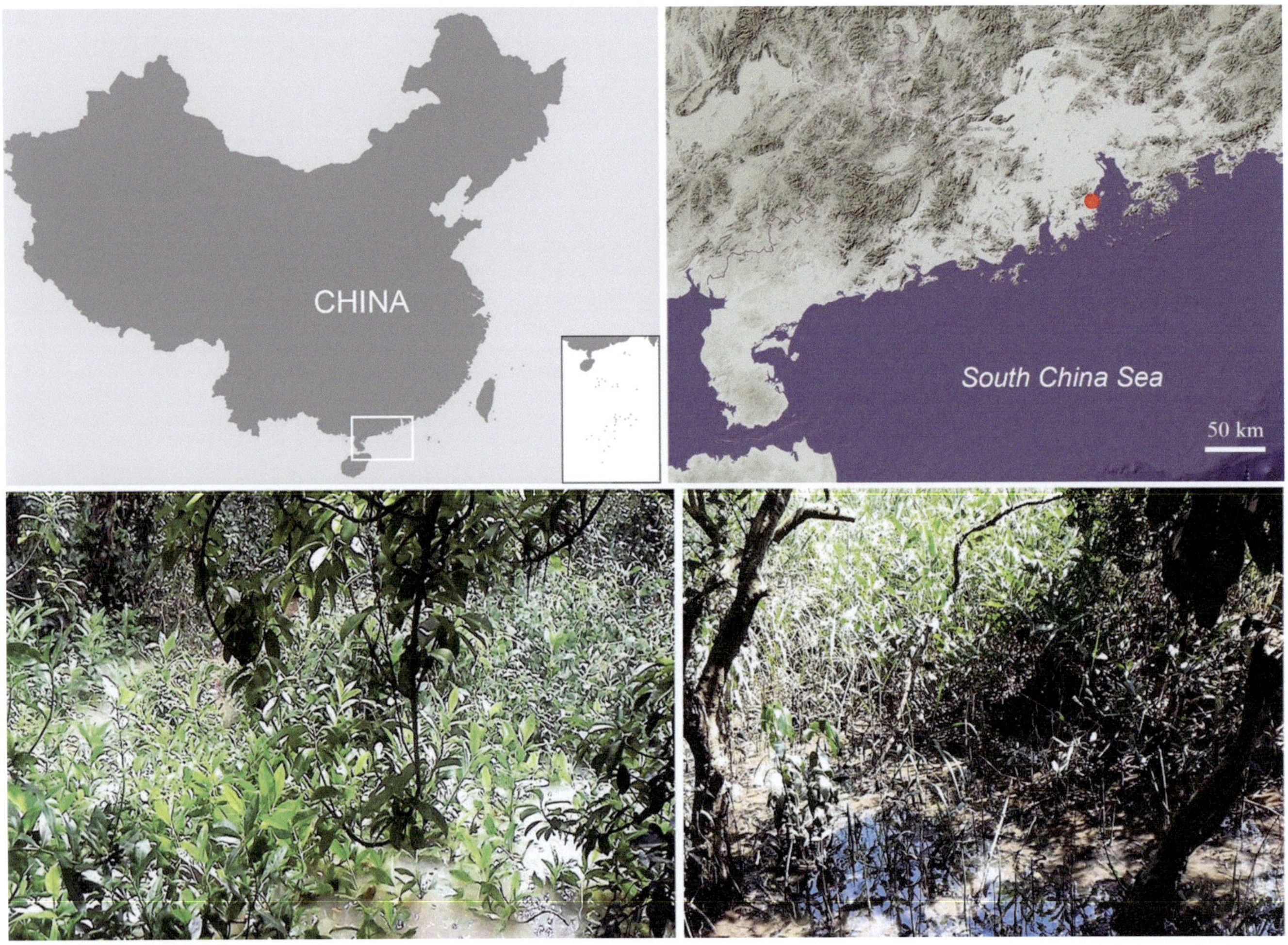


**Fig. 69A** Sample site

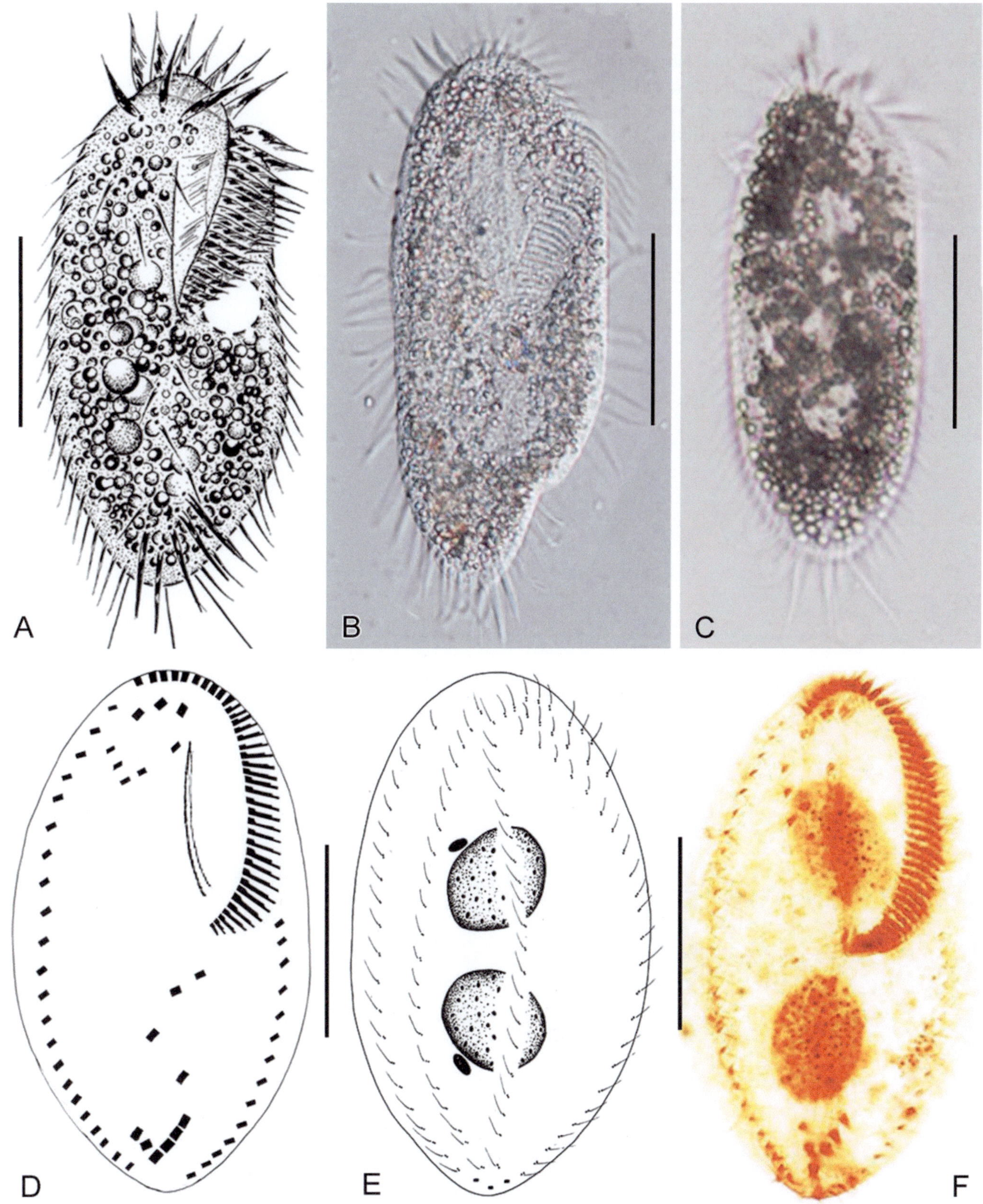


**Fig. 69B** *Tetmemena pustulata*
(**a–c**) Ventral views of live individuals. (**d–f**) Ventral (**d**, **f**) and dorsal (**e**) views of ciliature. Scale bars: 60 μm

## Reference

Shao C, Pan X, Jiang J, Ma H, Al-Rasheid KAS, Warren A, Lin X (2013) A redescription of the oxytrichid *Tetmemena pustulata* (Müller, 1786) Eigner, 1999 and notes on morphogenesis in the marine urostylid *Metaurostylopsis salina* Lei et al., 2005 (Ciliophora, Hypotrichia). Eur J Protistol 49:272–282

Class Spirotrichea Bütschli, 1889
Order Sporadotrichida Fauré-Fremiet, 1961
Family Oxytrichidae Ehrenberg, 1830
*Urosoma caudata* (Ehrenberg, 1833) Berger, 1999

**Ecological feature:** mangrove
**Water temperature:** no data available
**pH:** 6.8
**Salinity:** 8.0‰–10.0‰

**Fig. 70A** Sample site

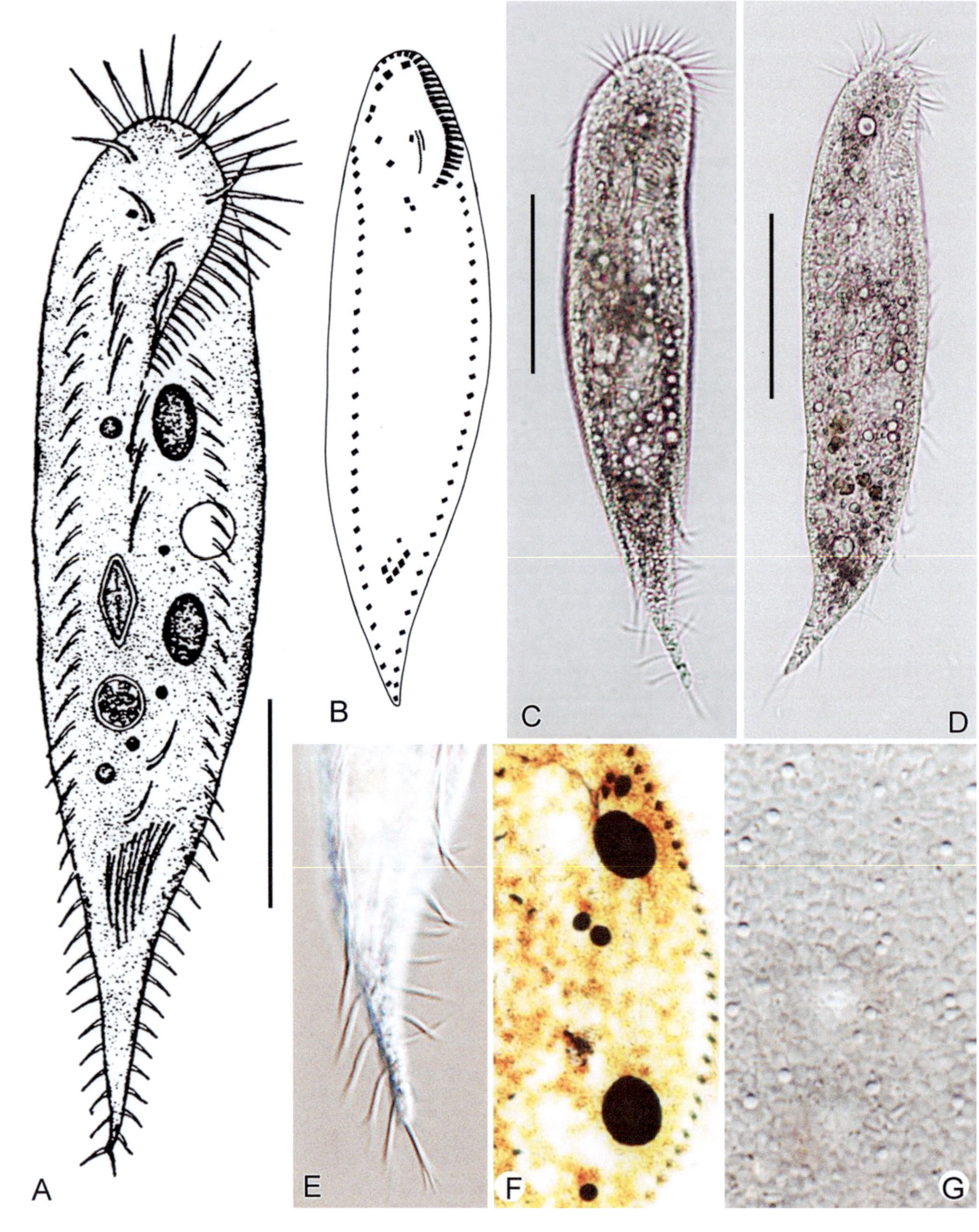


**Fig. 70B** *Urosoma caudata*

(**a**, **c**, **d**) Ventral views of live individuals, A from Wang (1930). (**b**) Ventral ciliature. (**e**) Pointed tail. (**f**) Macronuclear nodules and micronuclei. (**g**) Cortical granules. Scale bars: 50 μm

## References

Berger H (1999) Monograph of the Oxytrichidae (Ciliophora, Hypotrichia). Monogr Biol 78:1–1080

Wang CC (1930) Notes on some new and rare species of hypotrichous infusoria. Contr boil Lab Sci Soc China, Zool Ser 6:9–18

Class Spirotrichea Bütschli, 1889
Order Sporadotrichida Fauré-Fremiet, 1961
Family Oxytrichidae Ehrenberg, 1830
*Urosoma karini* Foissner, 1987

**Ecological feature:** mangrove
**Water temperature:** no data available
**pH:** 7.3
**Salinity:** 2.0‰–7.0‰

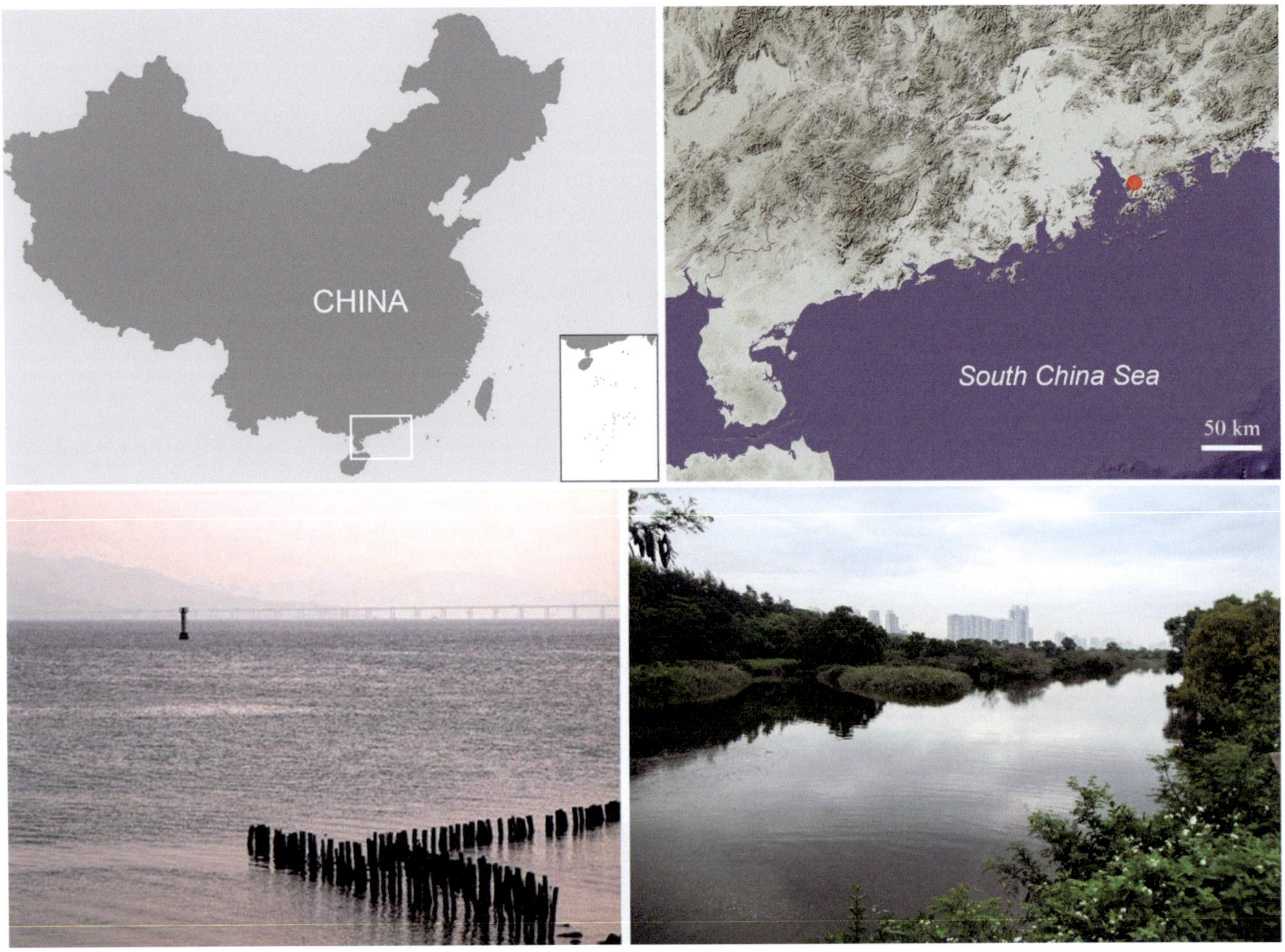


**Fig. 71A** Sample site

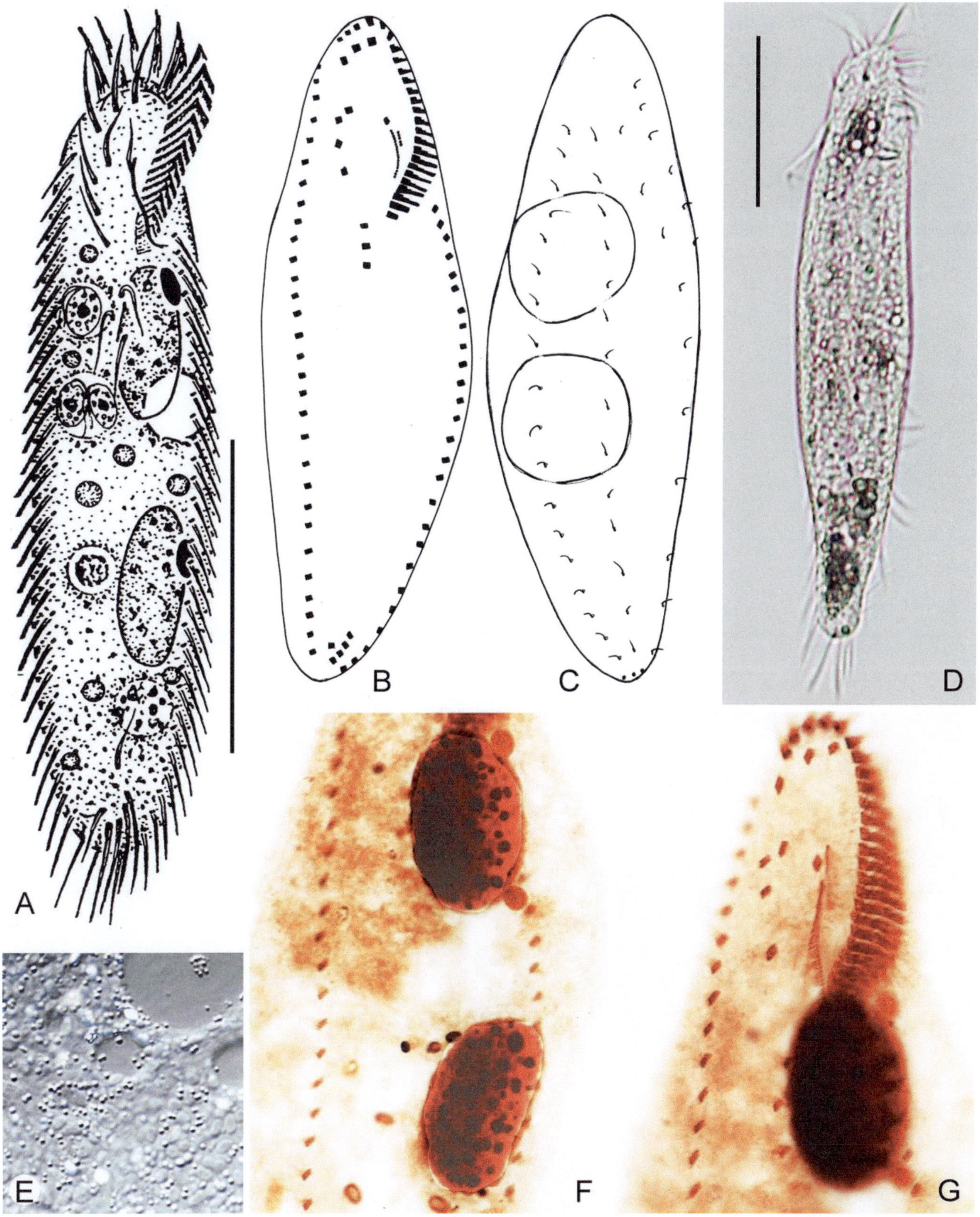


**Fig. 71B** *Urosoma karini*
(**a**) Ventral views of live individuals, A from Foissner (1987). (**b**, **c**) Ventral (**b**) and dorsal (**c**) views of ciliature. (**e**) Cortical granules. (**f**) Macronuclear nodules and micronuclei; (**g**) ventral view of anterior cell part, showing ciliature. Scale bars: 50 μm

## References

Berger H (1999) Monograph of the Oxytrichidae (Ciliophora, Hypotrichia). Monogr Biol 78:1–1080

Foissner W (1987) Faunistische und taxonomische Notizen über die Protozoen des Fuscher Tales (Salzburg, Österreich). Jber Haus Nat Salzburg 10:56–68

Class Spirotrichea Bütschli, 1889
Order Sporadotrichida Fauré-Fremiet, 1961
Family Trachelostylidae Small & Lynn, 1985
*Trachelostyla pediculiformis* (Cohn, 1866) Borror, 1972

**Ecological feature:** mangrove
**Water temperature:** 20.0 °C
**pH:** no data available
**Salinity:** 17.0‰

**Fig. 72A** Sample site

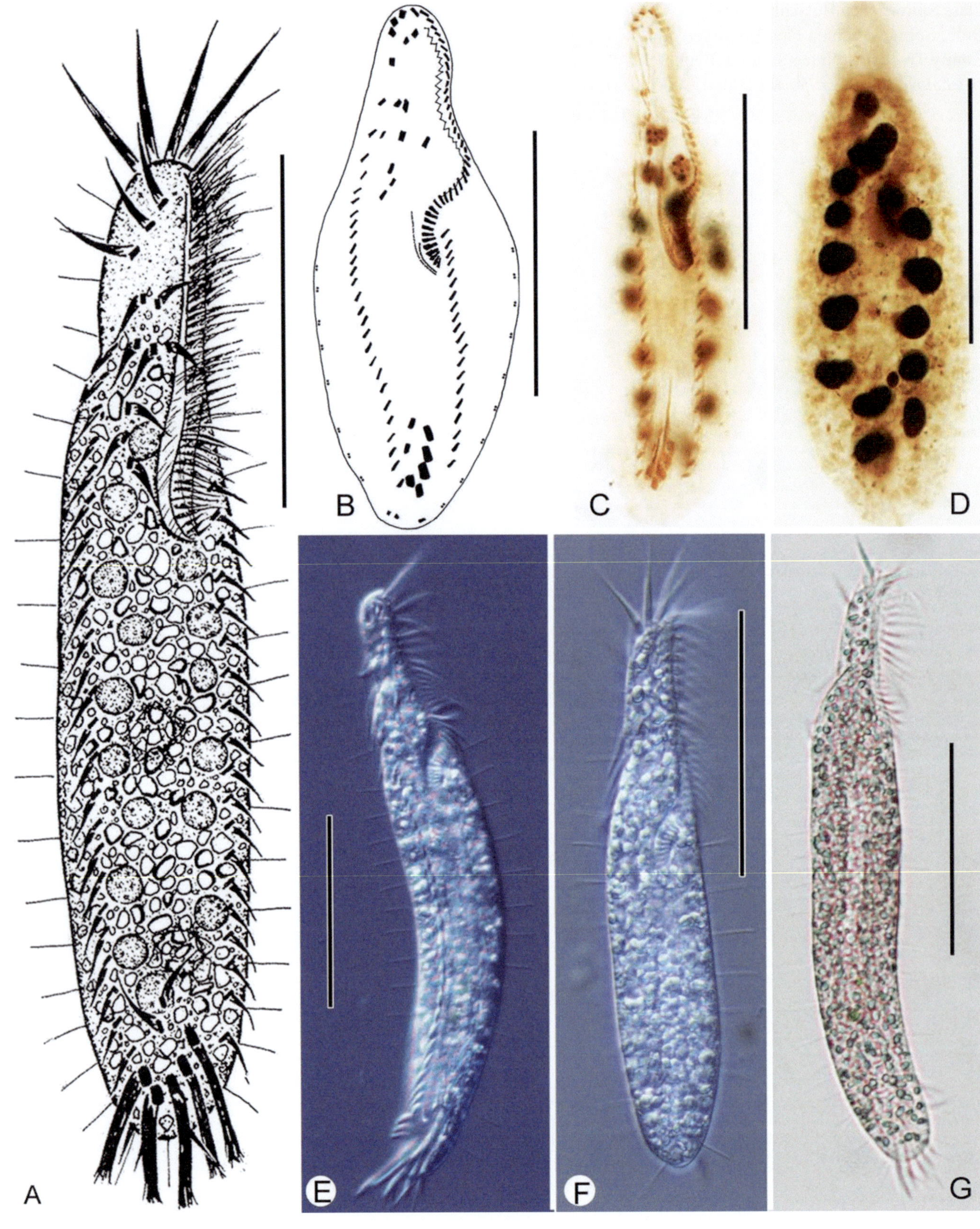


**Fig. 72B** *Trachelostyla pediculiformis*
(**a**, **e**–**g**) Ventral views of live individuals. (**b**–**d**) Ventral (**b**, **c**) and dorsal (**d**) views of ciliature. Scale bars: 40 μm

## Reference

Gong J, Song WB, Li L, Shao C, Chen Z (2006) A new investigation of the marine ciliate, *Trachelostyla pediculiformis* (Cohn, 1866) Borror, 1972 (Ciliophora, Hypotrichida), with establishment of a new genus, *Spirotrachelostyla* nov. gen. Eur J Protistol 42:63–73

Class Spirotrichea Bütschli, 1889
Order Sporadotrichida Fauré-Fremiet, 1961
Family Uroleptidae Foissner & Stoeck, 2008
*Parauroleptus musculus* Kahl, 1932

**Ecological feature:** mangrove
**Water temperature:** 20.0 °C
**pH:** 6.8
**Salinity:** 10.0‰

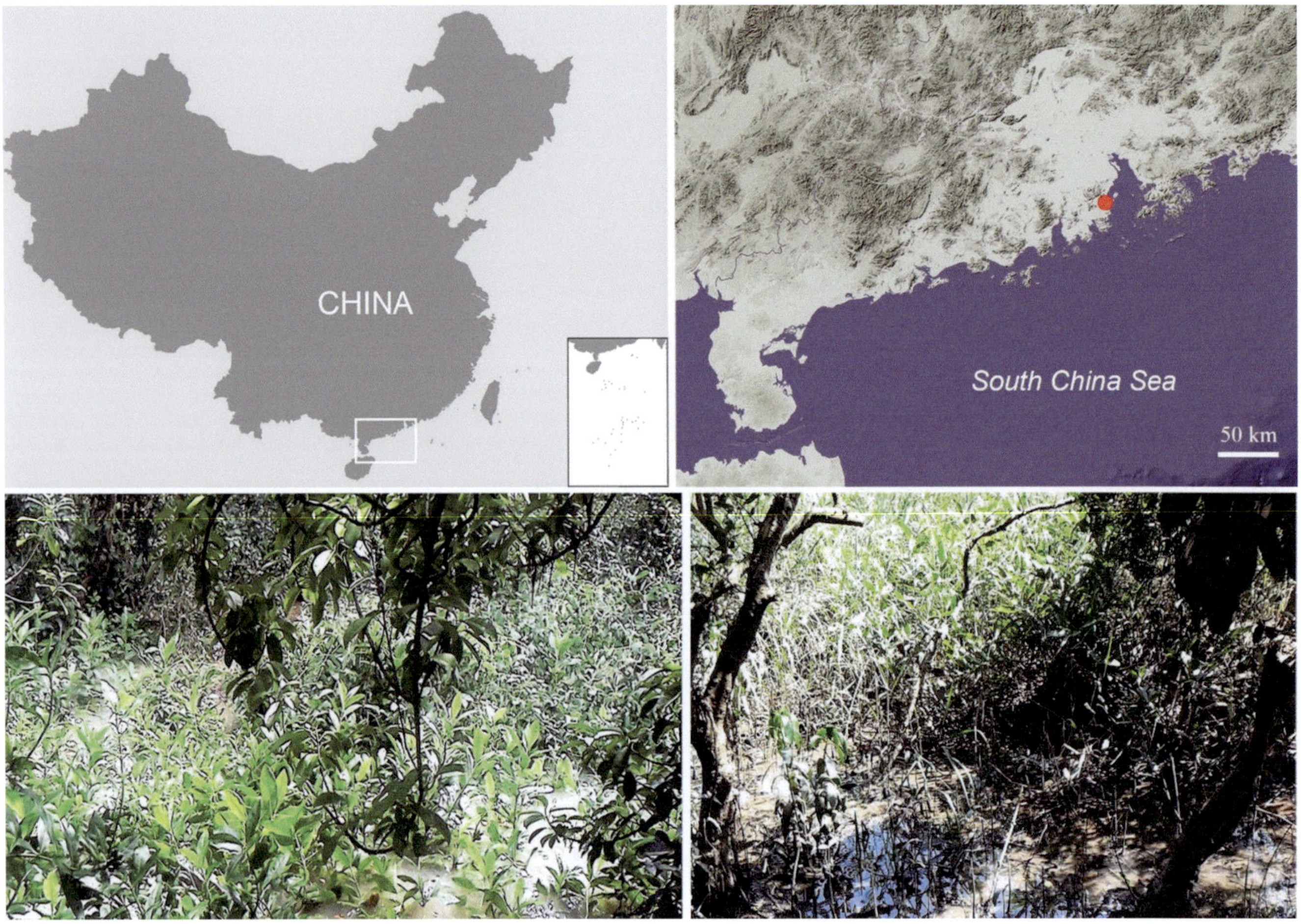


**Fig. 73A** Sample site

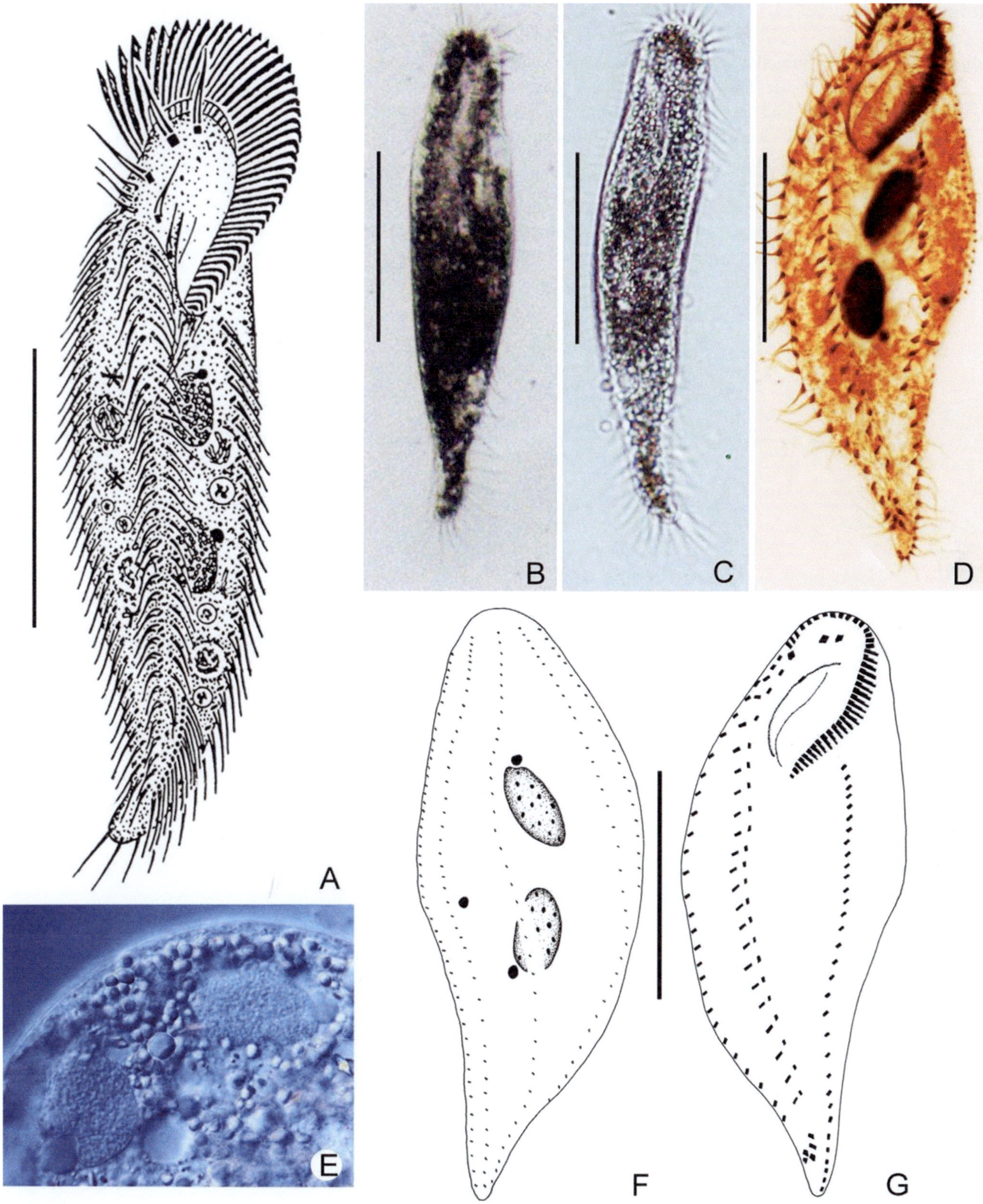


**Fig. 73B** *Paraurolepus musculus*
(**a–c**) Ventral views of live individuals, A from Foissner (1984). (**d**, **f**, **g**) Ventral (**d**, **g**) and dorsal (**f**) views of ciliature. (**e**) Nuclear apparatus. Scale bars: 60 μm

## References

Foissner W (1984) Infraciliatur, Silberliniensystem und Biometrie einiger neuer und wenig bekannter terrestrischer, limnischer und mariner Ciliaten (Protozoa: Ciliophora) aus den Klassen Kinetofragminophora, Colpodea und Polyhymenophora. Stapfia 12:1–165

Kahl A (1932) Urtiere oder Protozoa I: Wimpertiere oder Ciliata (Infusoria). 3. Spirotricha. Tierwelt Dtl 25:399–650

Class Spirotrichea Bütschli, 1889
Order Stichotrichida Fauré-Fremiet, 1961
Family Amphisiellidae Jankowski, 1979
*Amphisiella annulata* (Kahl, 1928) Borror, 1972

**Ecological feature:** sandy sediment (Zhanjiang); mangrove (Zhuhai)
**Water temperature:** 29.0 °C; 23.0 °C
**pH:** no data available; 6.5
**Salinity:** 9.0‰; 10.0‰

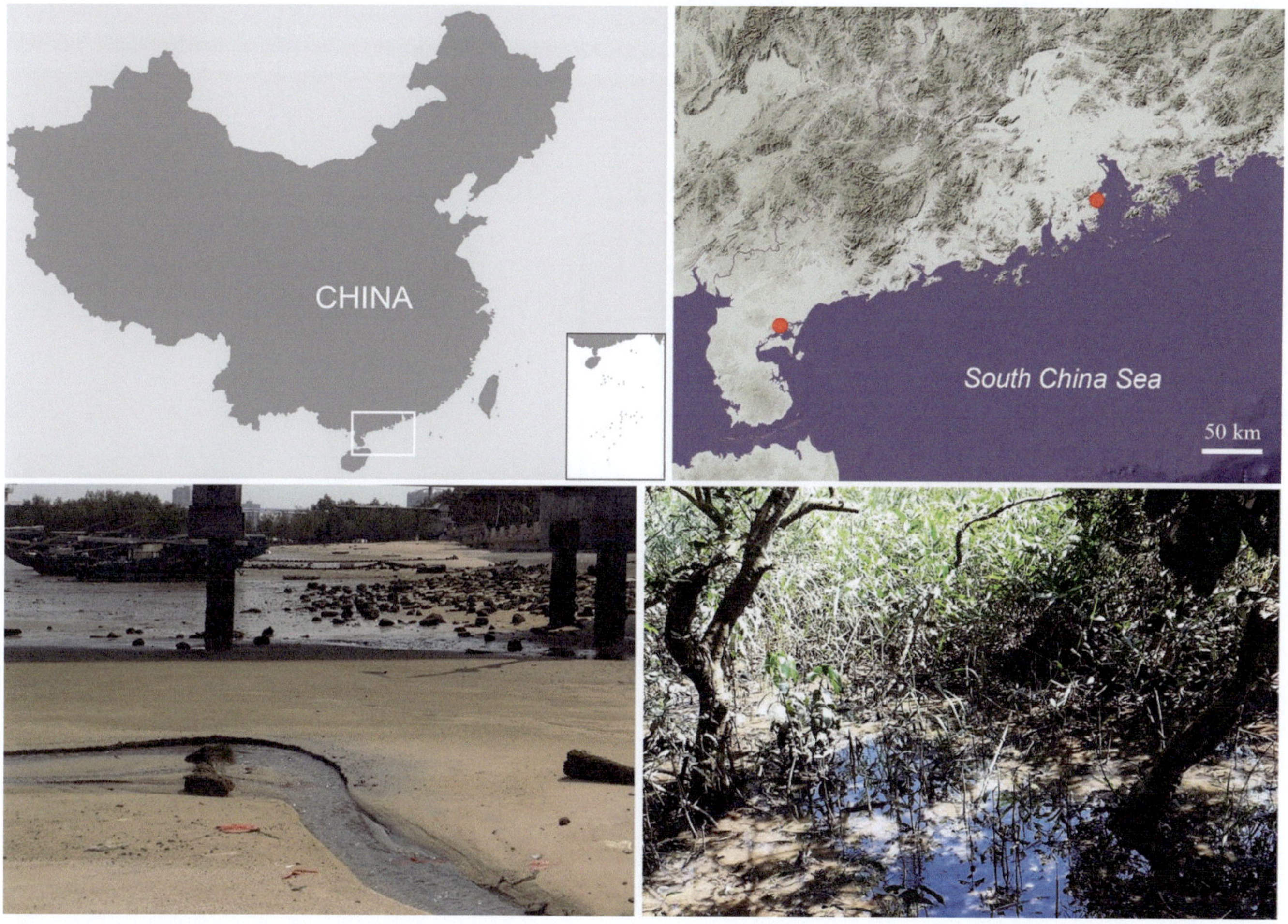


**Fig. 74A** Sample site

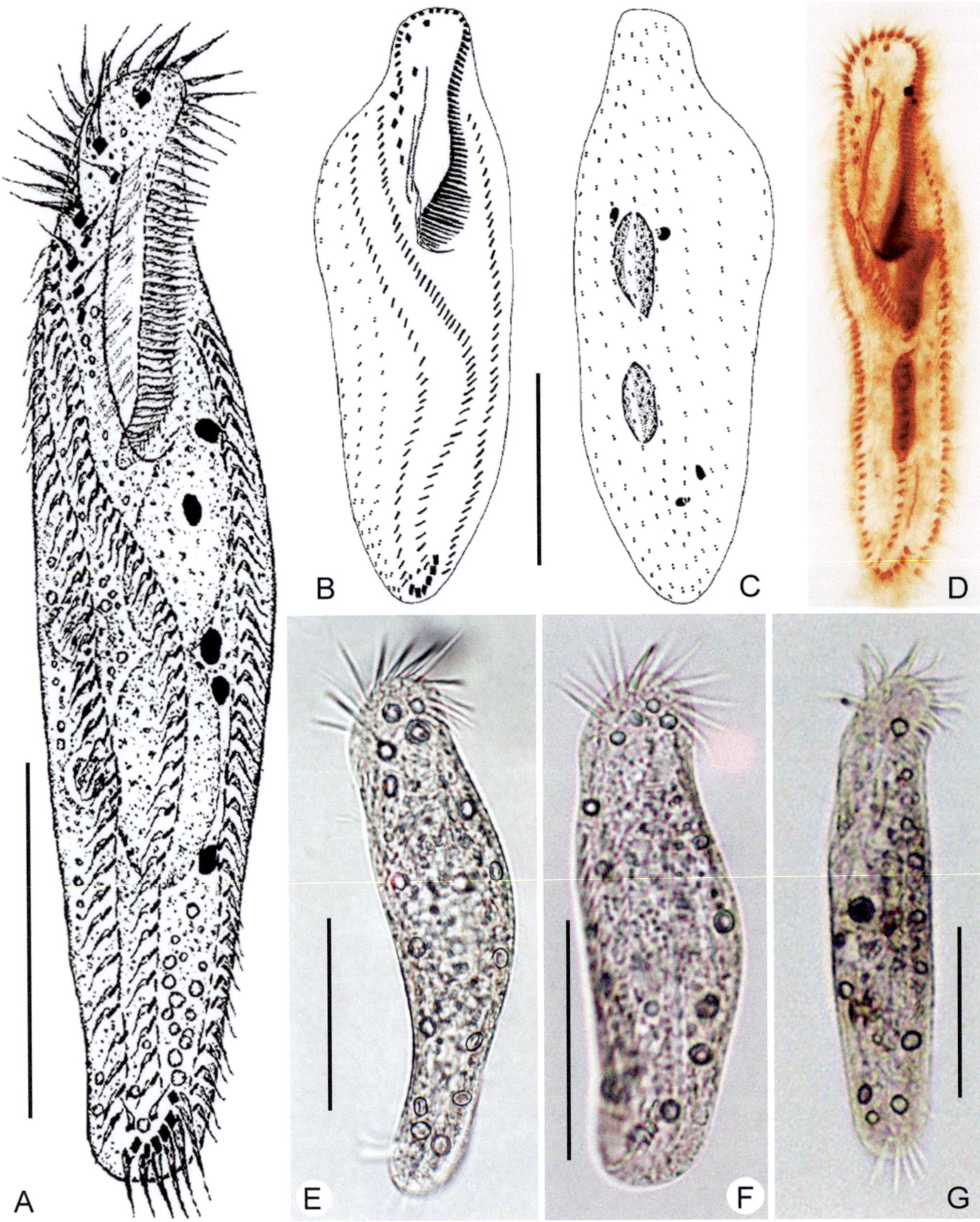


**Fig. 74B** *Amphisiella annulata*
(**a**, **e–g**) Ventral views of live individuals, showing variation of body shapes. (**b–d**) Ventral (**b**, **d**) and dorsal (**c**) views of ciliature. Scale bars: 50 μm

## References

Berger H (2004) *Amphisiella annulata* (Kahl, 1928) Borror, 1972 (Ciliophora: Hypotrichida): morphology, notes on morphogenesis, review of literature, and neotypification. Acta Protozool 43:1–14

Hu X, Warren A, Suzuki T. 2004. Morphology and morphogenesis of two marine ciliates, *Pseudokeronopsis pararubra* sp. n. and *Amphisiella annulata* from China and Japan (Protozoa: Ciliophora). Acta Protozool 43:351–368

Class Spirotrichea Bütschli, 1889
Order Stichotrichida Fauré-Fremiet, 1961
Family Amphisiellidae Jankowski, 1979
*Amphisiella candida* Chen et al., 2013

**Ecological feature:** brackish water
**Temperature:** 24.0 °C
**pH:** no data available
**Salinity:** 26.3‰

**Fig. 75A** Sample site

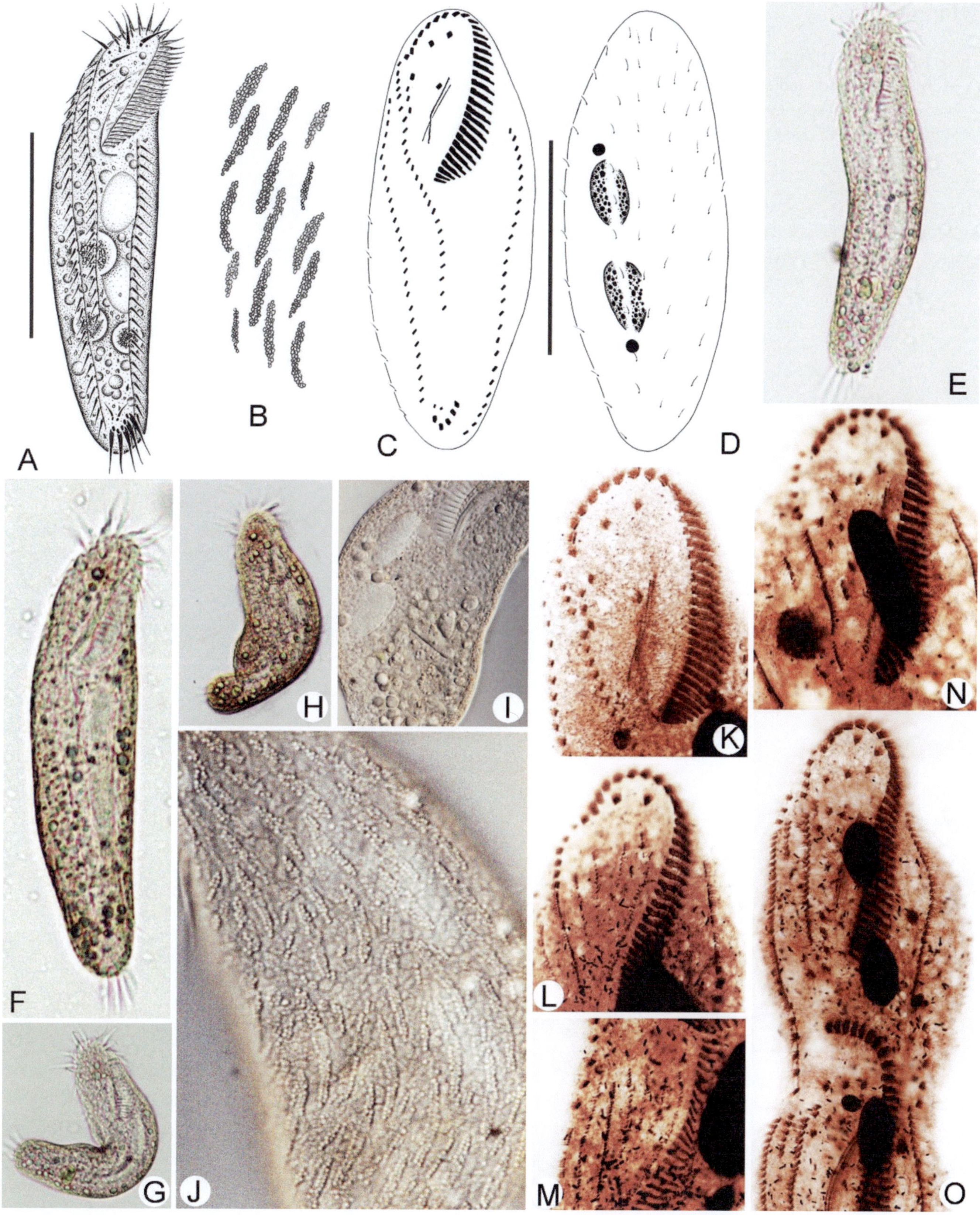


**Fig. 75B** *Amphisiella candida*
(**a**, **e–h**) Ventral views of live individuals. (**b**, **j**) Cortical granules. (**c**, **d**) Ventral (**c**) and dorsal (**d**) views of ciliature. (**k–o**) Ciliature in nondividing and dividing cells. Scale bars: 50 μm

## Reference

Chen XM, Shao C, Lin X, Clamp J, Song WB (2013) Morphology and molecular phylogeny of two new brackish-water species of *Amphisiella* (Ciliophora, Hypotricha), with notes on morphogenesis. Eur J Protistol 49:453–466

Class Spirotrichea Bütschli, 1889
Order Stichotrichida Fauré-Fremiet, 1961
Family Amphisiellidae Jankowski, 1979
*Amphisiella pulchra* Chen et al., 2013

**Ecological feature:** brackish water wetland
**Temperature:** 18.0 °C
**pH:** no data available
**Salinity:** 10.4‰

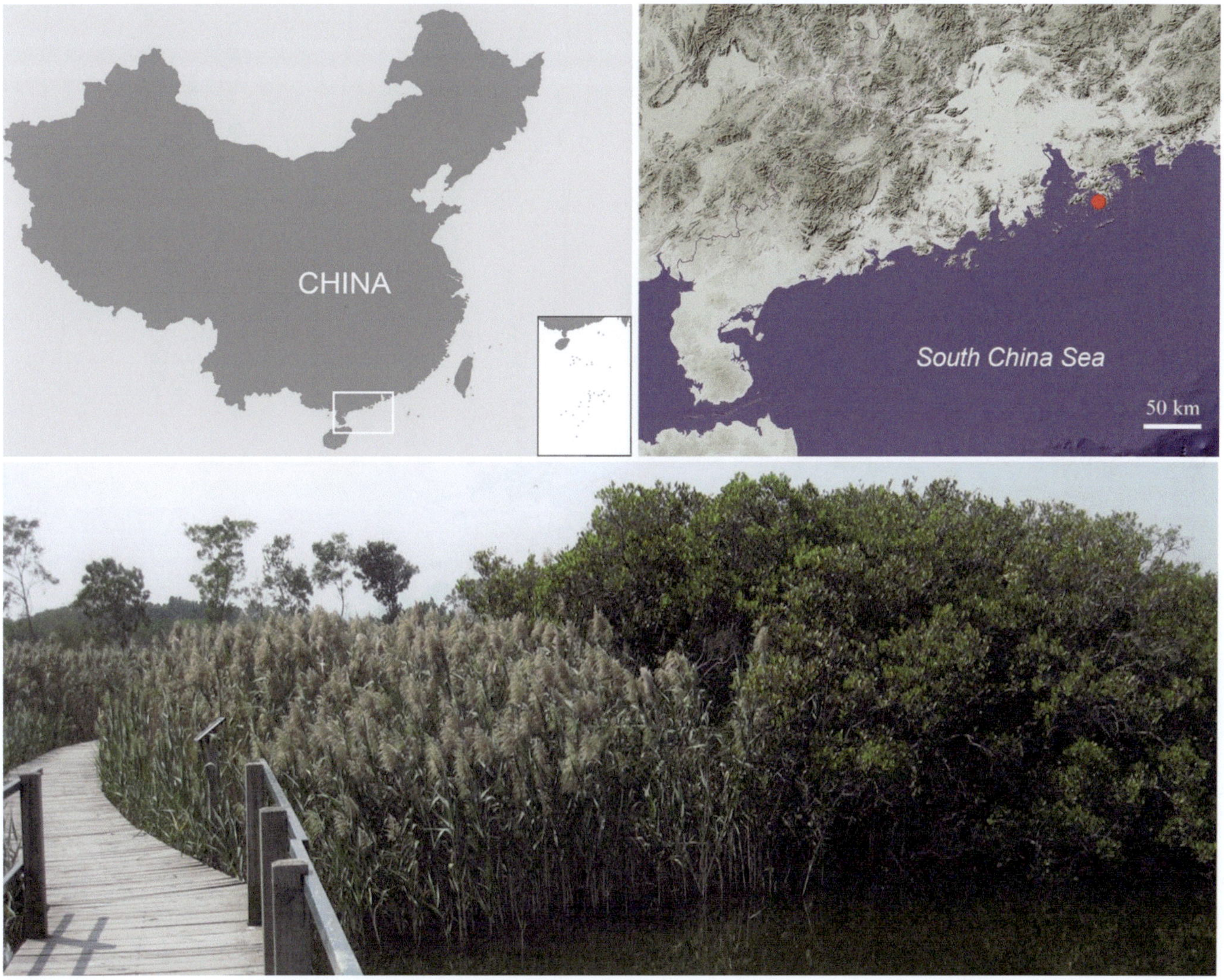


**Fig. 76A** Sample site

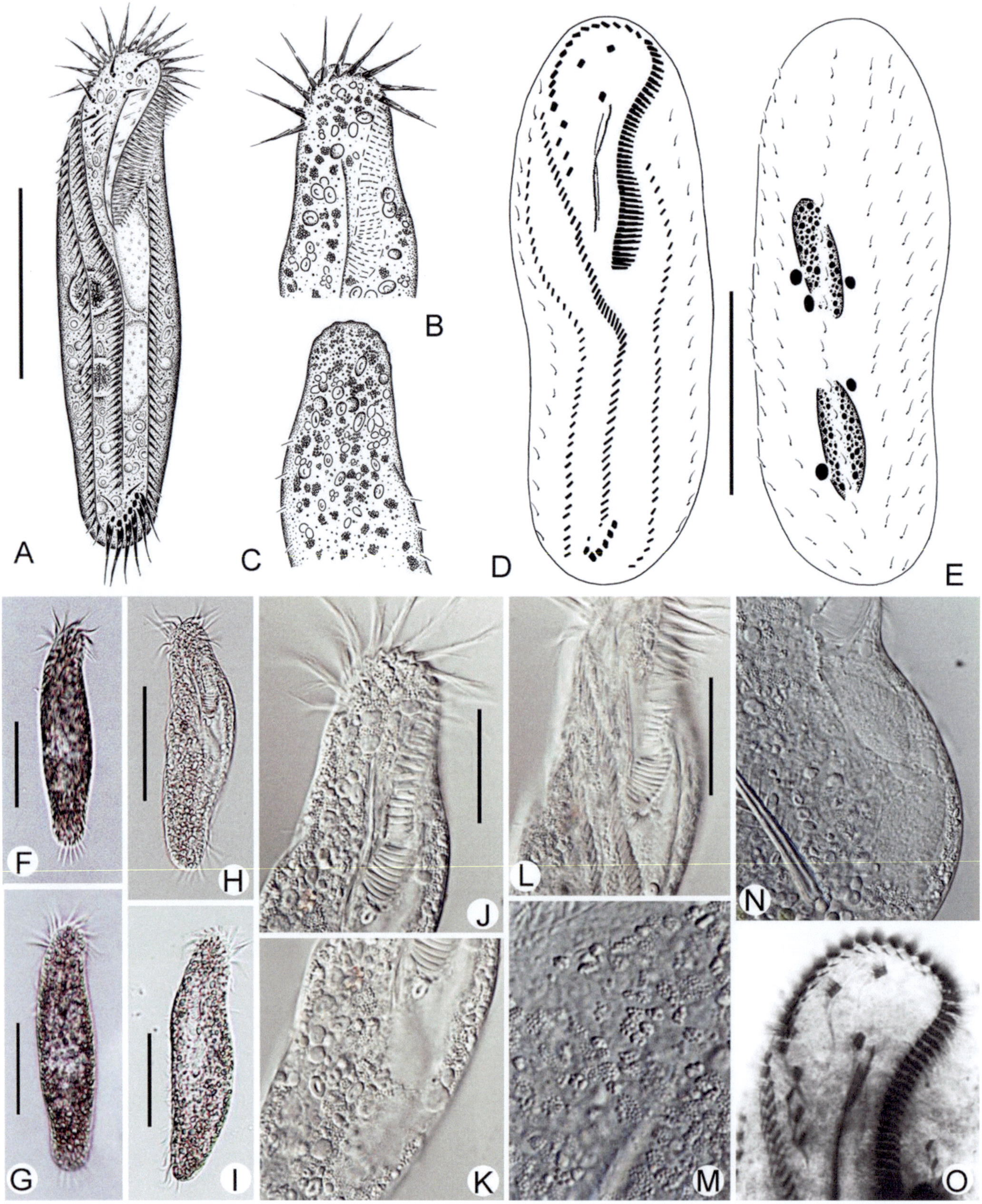


**Fig. 76B** *Amphisiella pulchra*
(**a**, **f–i**) Ventral views of live individuals. (**b**, **c**, **j**, **l**) Ventral (**b**, **j**, **l**) and dorsal (**c**) views of anterior parts. (**d**, **e**) Ventral (**d**) and dorsal (**e**) views of ciliature. (**k**, **m**) Cortical granules. (**n**) Macronuclei. (**o**) Ciliature of anterior portion. Scale bars: 50 μm (**a**), 60 μm (**d–i**), and 20 μm (**j**, **l**)

## Reference

Chen XM, Shao C, Lin X, Clamp JC, Song WB (2013) Morphology and molecular phylogeny of two new brackish-water species of *Amphisiella* (Ciliophora, Hypotricha), with notes on morphogenesis. Eur J Protistol 49:453–466

Class Spirotrichea Bütschli, 1889
Order Stichotrichida Fauré-Fremiet, 1961
Family Amphisiellidae Jankowski, 1979
*Amphisiella* sp.

**Ecological feature:** intertidal zone
**Water temperature:** 25.0 °C
**pH:** no data available
**Salinity:** 27.0‰

**Fig. 77A** Sample site

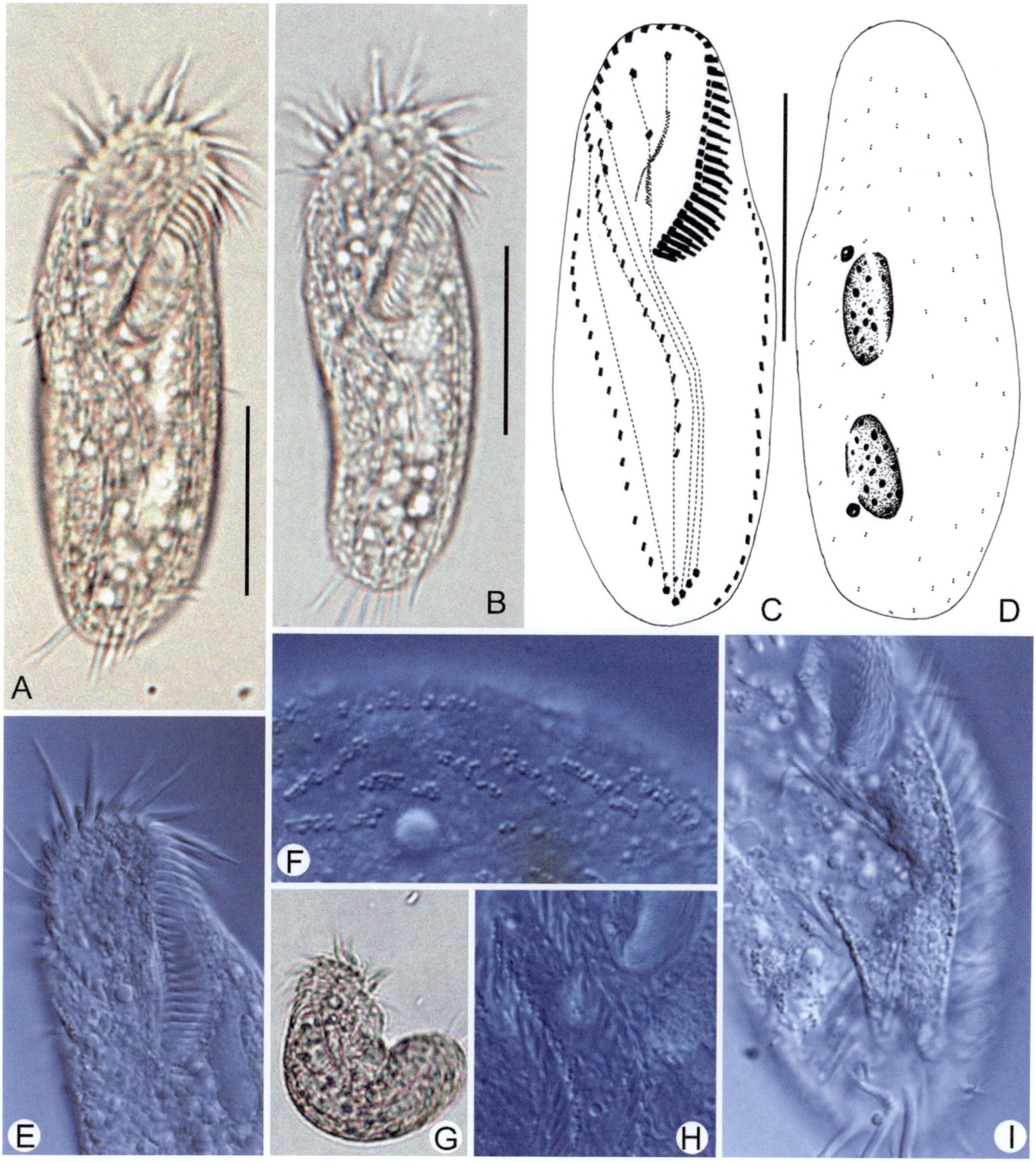


**Fig. 77B** *Amphisiella* sp.
(**a**, **b**) Ventral views of live cells. (**c**, **d**) Ventral (**c**) and dorsal (**d**) views of ciliature. (**e**, **i**) Ventral views of anterior (**e**) and posterior (**i**) portions. (**f**, **h**) Cortical granules on dorsal (**f**) and ventral (**h**) sides. (**g**) A bending individual. Scale bars: 40 μm

## Reference

Jiang J (2011) Taxonomic and morphogenetic studies on hypotrichs (Protozoa: Ciliophora). Ocean University of China, Ph D. thesis

Class Spirotrichea Bütschli, 1889
Order Stichotrichida Fauré-Fremiet, 1961
Family Amphisiellidae Jankowski, 1979
*Protogastrostyla pulchra* (Pereyaslawzewa, 1886) Gong et al., 2007

**Ecological feature:** mangrove
**Water temperature:** 20.0 °C; 12.0 °C
**pH:** 6.9; no data available
**Salinity:** 17.0‰; 6.0‰

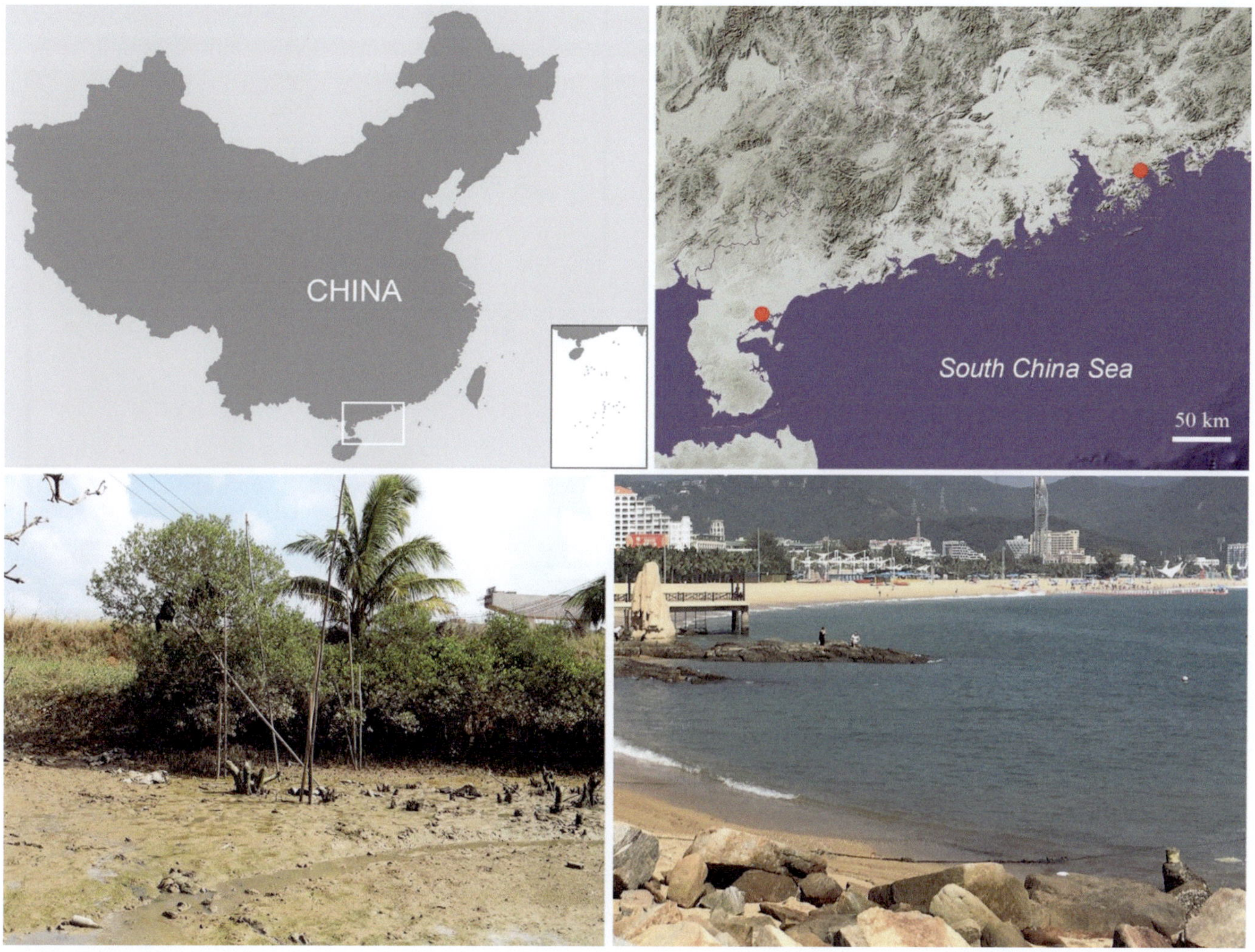


**Fig. 78A** Sample site

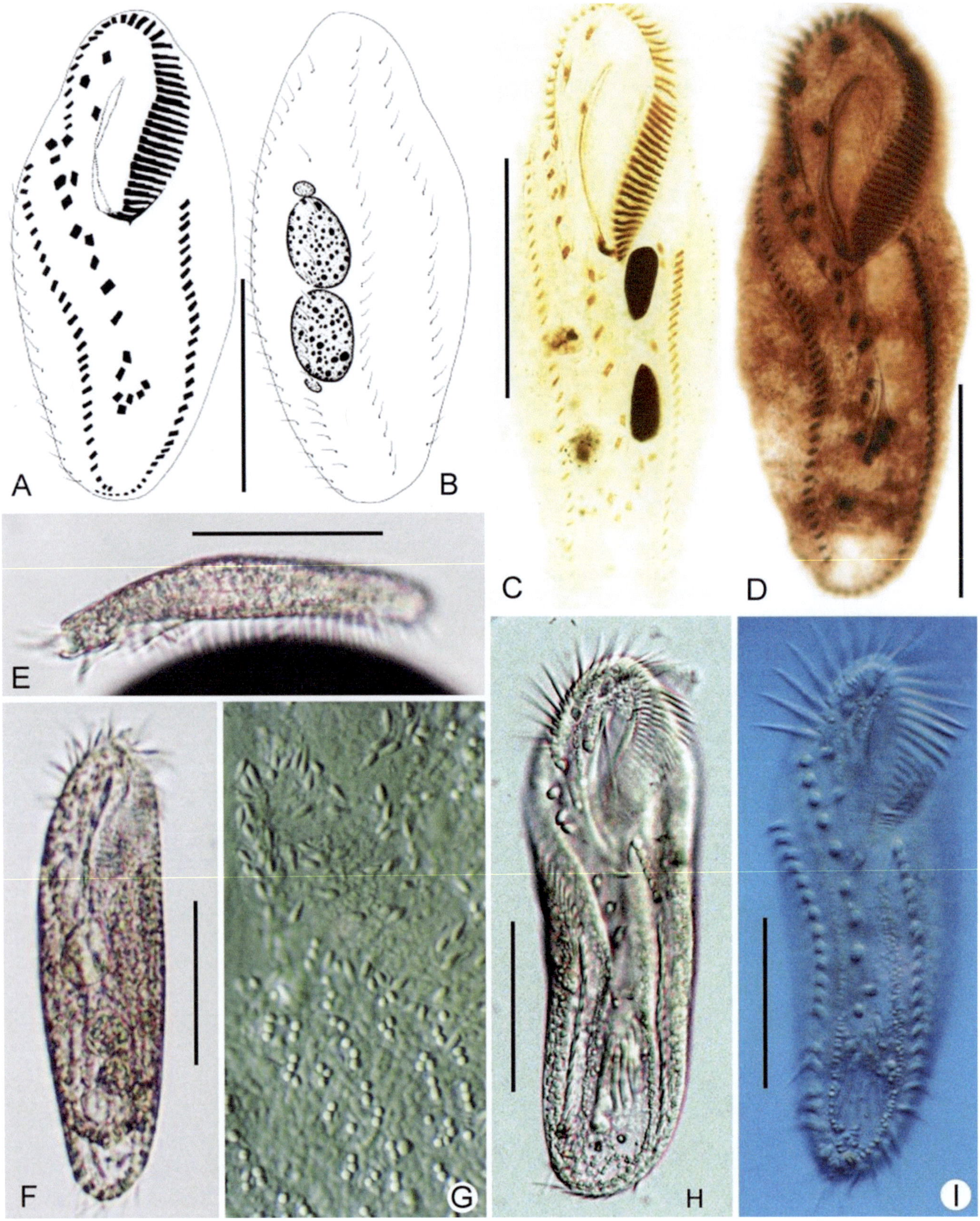


**Fig. 78B** *Protogastrostyla pulchra*
(**a**–**d**) Ventral (**a**, **c**, **d**) and dorsal (**b**) views of ciliature. (**e**) Lateral view of live cell. (**f**, **h**, **i**) Ventral views of live cells. (**g**) Cortical granules. Scale bars: 80 μm

## Reference

Gong J, Kim SJ, Kim SY, Min GS, Roberts D, Warren A, Choi JK (2007) Taxonomic redescriptions of two ciliates, *Protogastrostyla pulchra* n. comb. and *Hemigastrostyla enigmatica* (Ciliophora: Spirotrichea, Stichotrichia), with phylogenetic analyses based on 18S and 28S rRNA gene sequences. J Eukaryot Microbiol 54:468–478

Class Spirotrichea Bütschli, 1889
Order Stichotrichida Fauré-Fremiet, 1961
Family Amphisiellidae Jankowski, 1979
*Protogastrostyla sterkii* (Wallengren, 1900) Chen et al., 2013

**Ecological feature:** mangrove
**Water temperature:** 22 °C
**pH:** 8.2
**Salinity:** 16.0‰

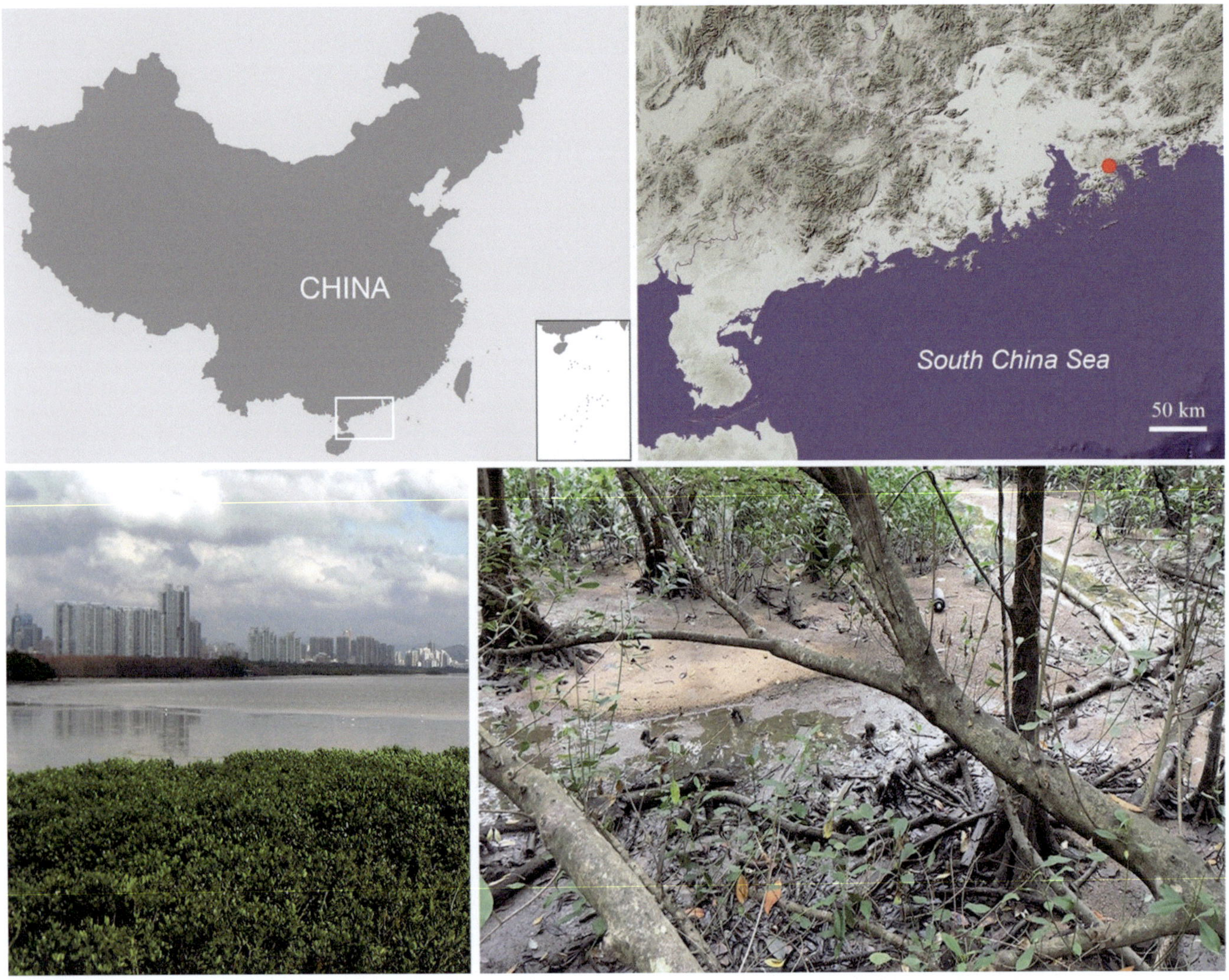


**Fig. 79A** Sample site

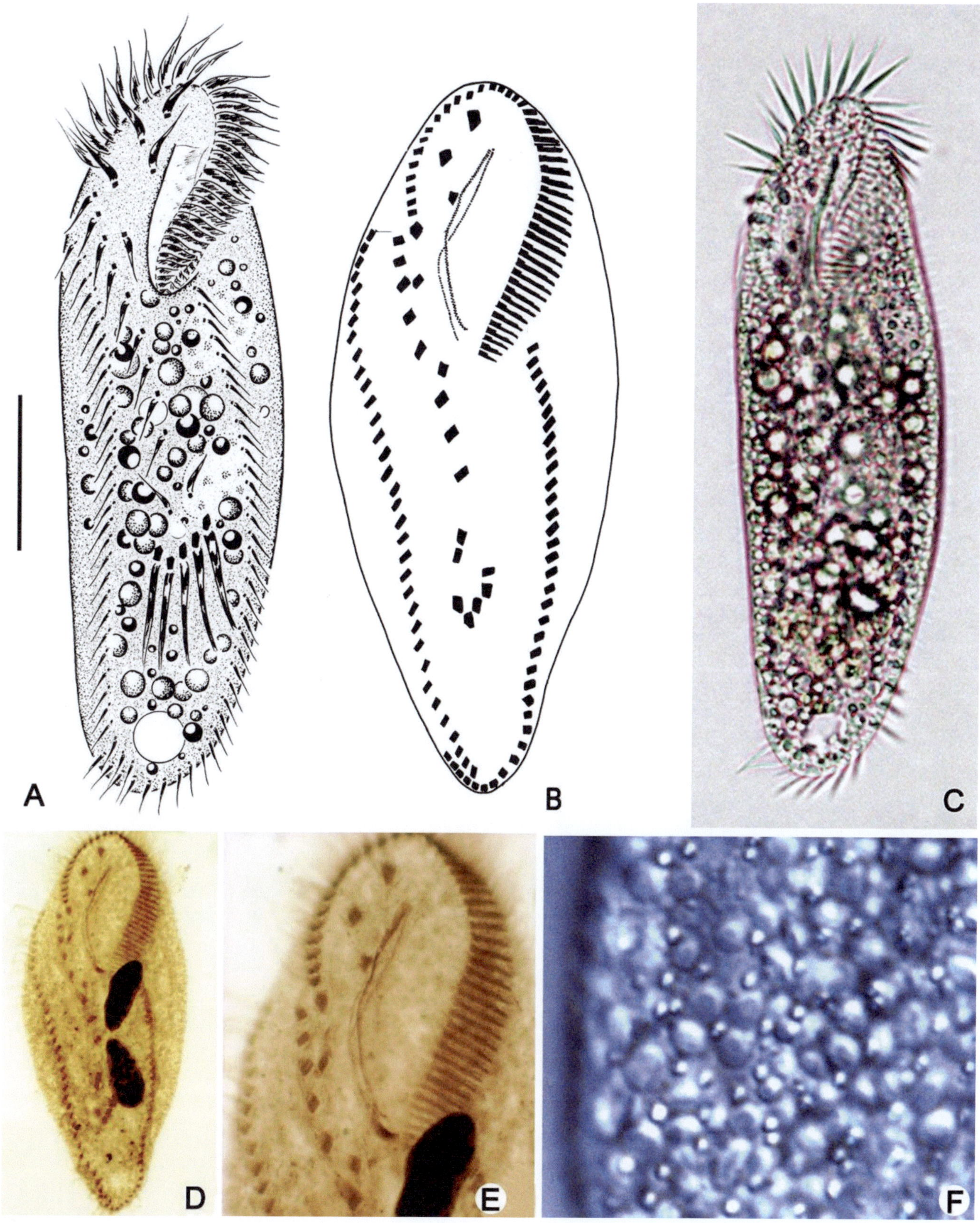


**Fig. 79B** *Protogastrostyla sterkii*
(**a**, **c**) Ventral views of live cells. (**b**, **d**) Ventral views of ciliature. (**e**) Anterior portion of ventral view to show frontal ciliature and oral apparatus. (**f**) Arrangement of cortical granules. Scale bar: 50 um

## Reference

Chen X, Shao C, Liu X, Huang J, Al-Rasheid KAS (2013) Morphology and phylogenies of two hypotrichous brackish ciliates, *Neourostylopsis orientalis* n. sp., and *Protogastrostyla sterkii* (Wallengren, 1900) n. comb., with establishment of a new genus *Neourostylopsis* (Protista, Ciliophora, Hypotrichia) from China. Int J Syst Evol Microbiol 63:1197–1209

Class Spirotrichea Bütschli, 1889
Order Stichotrichida Fauré-Fremiet, 1961
Family Kahliellidae Tuffrau, 1979
*Pseudokahliella marina* (Foissner et al., 1982) Berger et al., 1985

**Ecological feature:** sandy beach
**Temperature:** 17.0 °C
**pH:** no data available
**Salinity:** 30.0‰

**Fig. 80A** Sample site

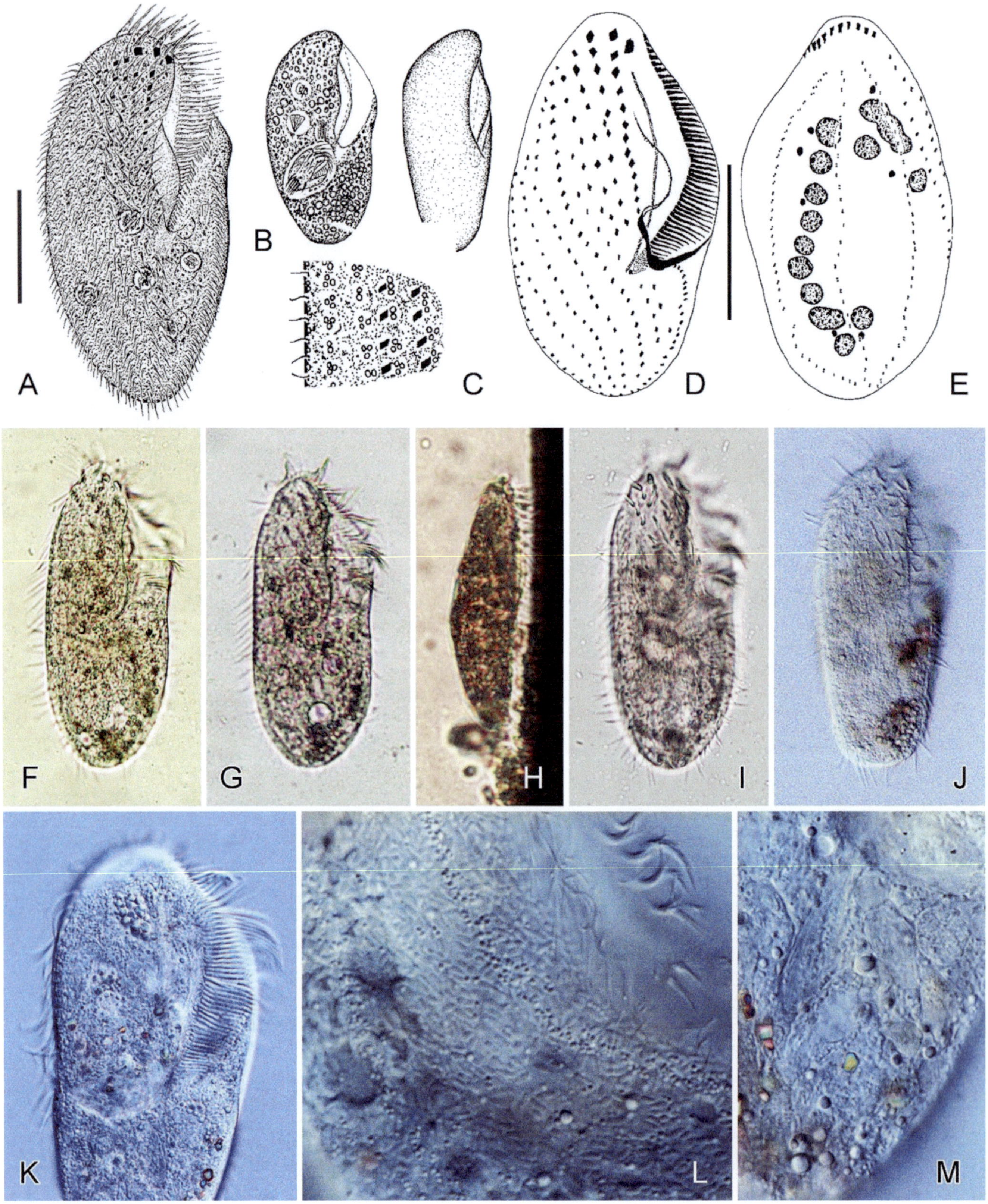


**Fig. 80B** *Pseudokahliella marina*
(**a**, **b**, **f–j**) Ventral and lateral views of live individuals. (**c**, **k–m**) Details of cortical granules. (**d**, **e**) Ventral (**d**) and dorsal (**e**) views of ciliature. Scale bars: 50 μm

## Reference

Hu X, Song WB (2003) Redescription of the morphology and divisional morphogenesis of the marine hypotrich *Pseudokahliella marina* (Foissner et al., 1982) from scallop-culture water of North China. J Nat Hist 37:2033–2043

Class Spirotrichea Bütschli, 1889
Order Stichotrichida Fauré-Fremiet, 1961
Family Spirofilidae von Gelei, 1929
*Hypotrichidium conicum* Ilowaisky, 1921

**Ecological feature:** mangrove
**Water temperature:** 18.0 °C
**pH:** 8.1
**Salinity:** 10.4‰

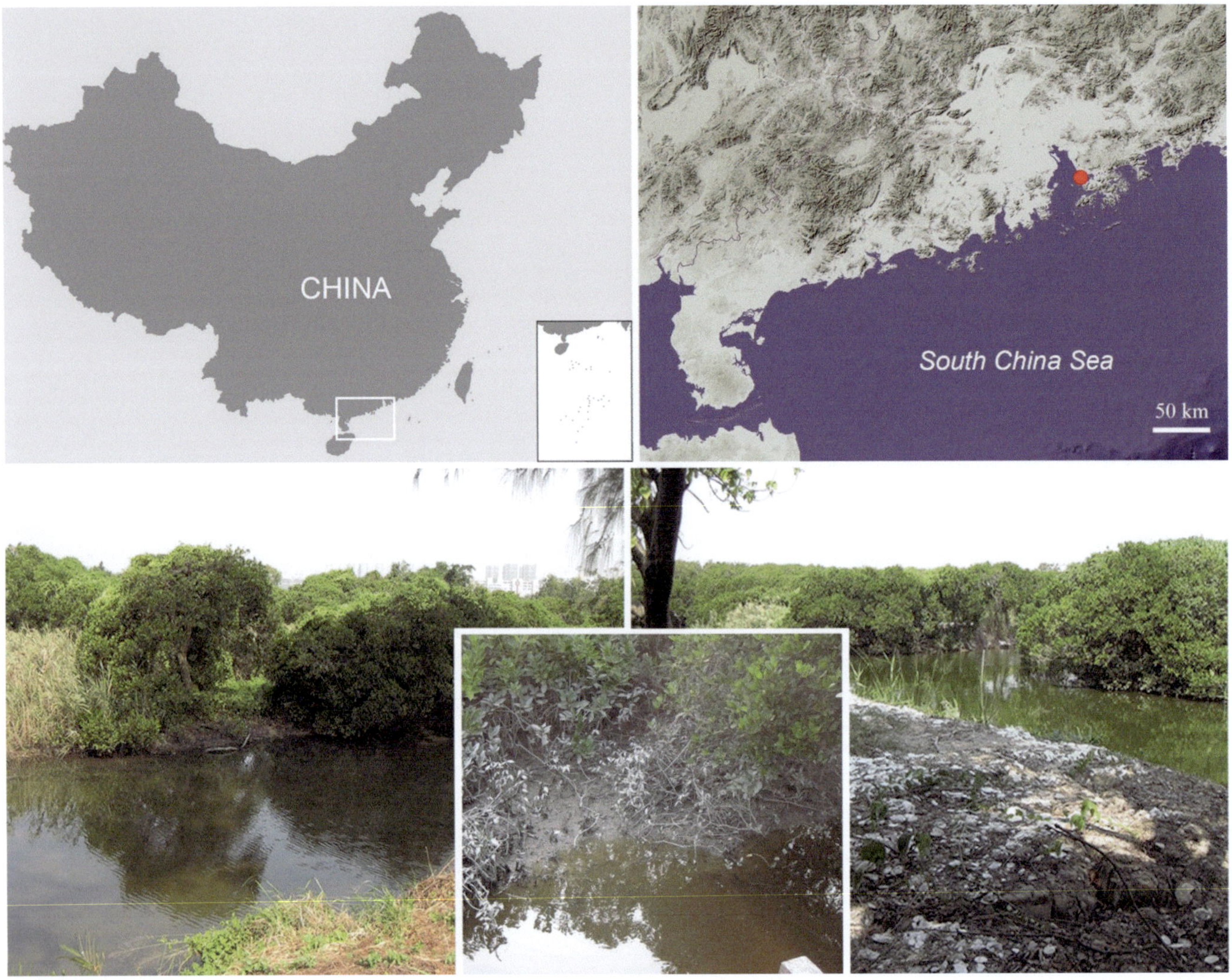


**Fig. 81A** Sample site

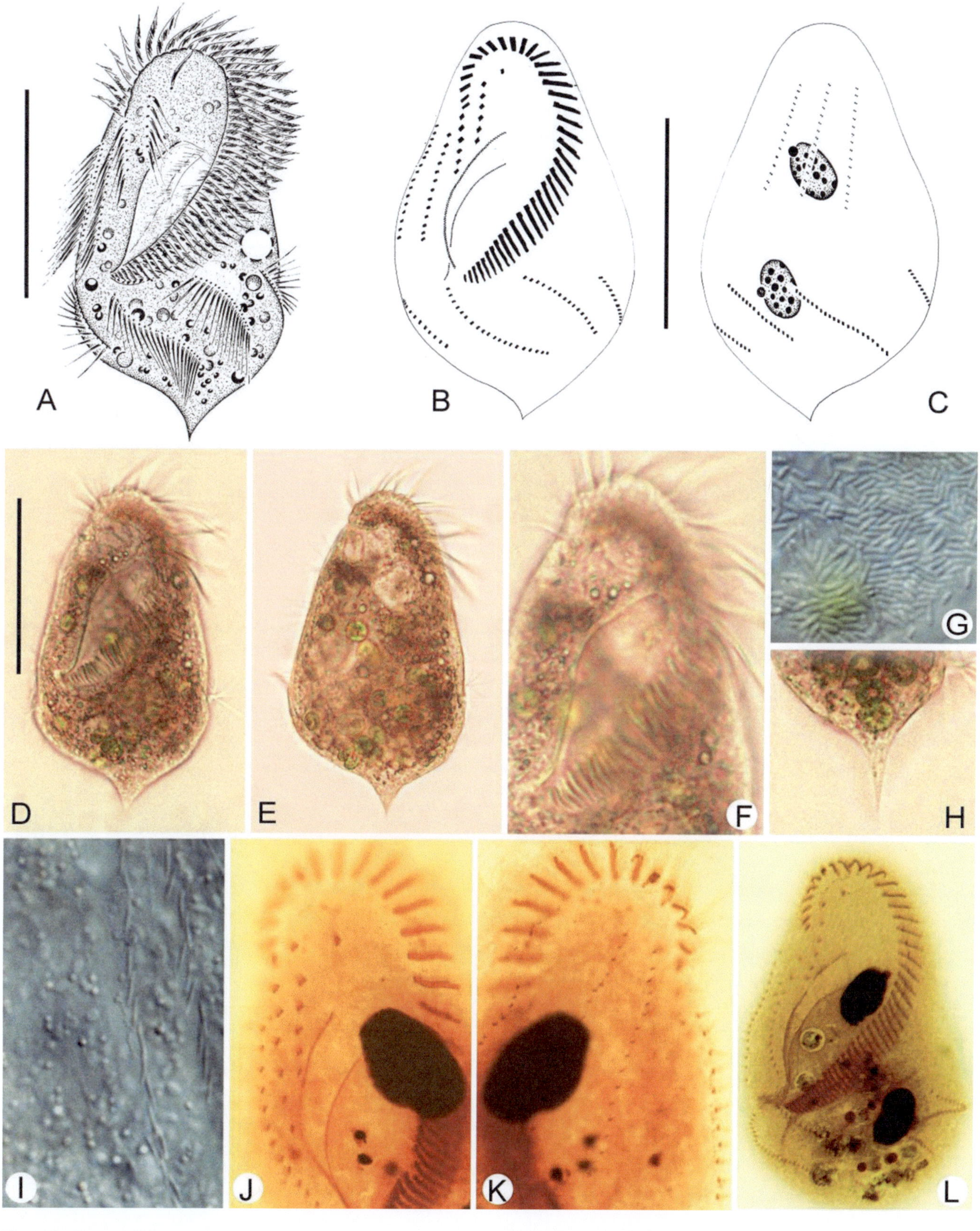


**Fig. 81B** *Hypotrichidium conicum*
(**a**, **d–f**) Ventral views of live cells. (**b**, **c**, **j–l**) Ventral (**b**, **j**, **l**) and dorsal (**c**, **k**) views of ciliature. (**g**) Cell surface, showing symbiotic bacteria. (**h**) Posterior tail. (**i**) Part of dorsal view, to show distribution of cortical granules and dorsal cilia. Scale bars: 75 μm

## Reference

Chen L, Liu W, Liu A, Al-Rasheid KAS, Shao C (2013) Morphology and molecular phylogeny of a new marine hypotrichous ciliate, *Hypotrichidium paraconicum* n. sp. (Ciliophora, Hypotrichia). J Eukaryot Microbiol 60:588–600

Class Spirotrichea Bütschli, 1889
Order Stichotrichida Fauré-Fremiet, 1961
Family Spirofilidae von Gelei, 1929
*Strongylidium guangdongense* Luo et al., 2018

**Ecological feature:** mangrove
**Water temperature:** 22.2 °C
**pH:** 7.2
**Salinity:** 7.0‰

**Fig. 82A** Sample site

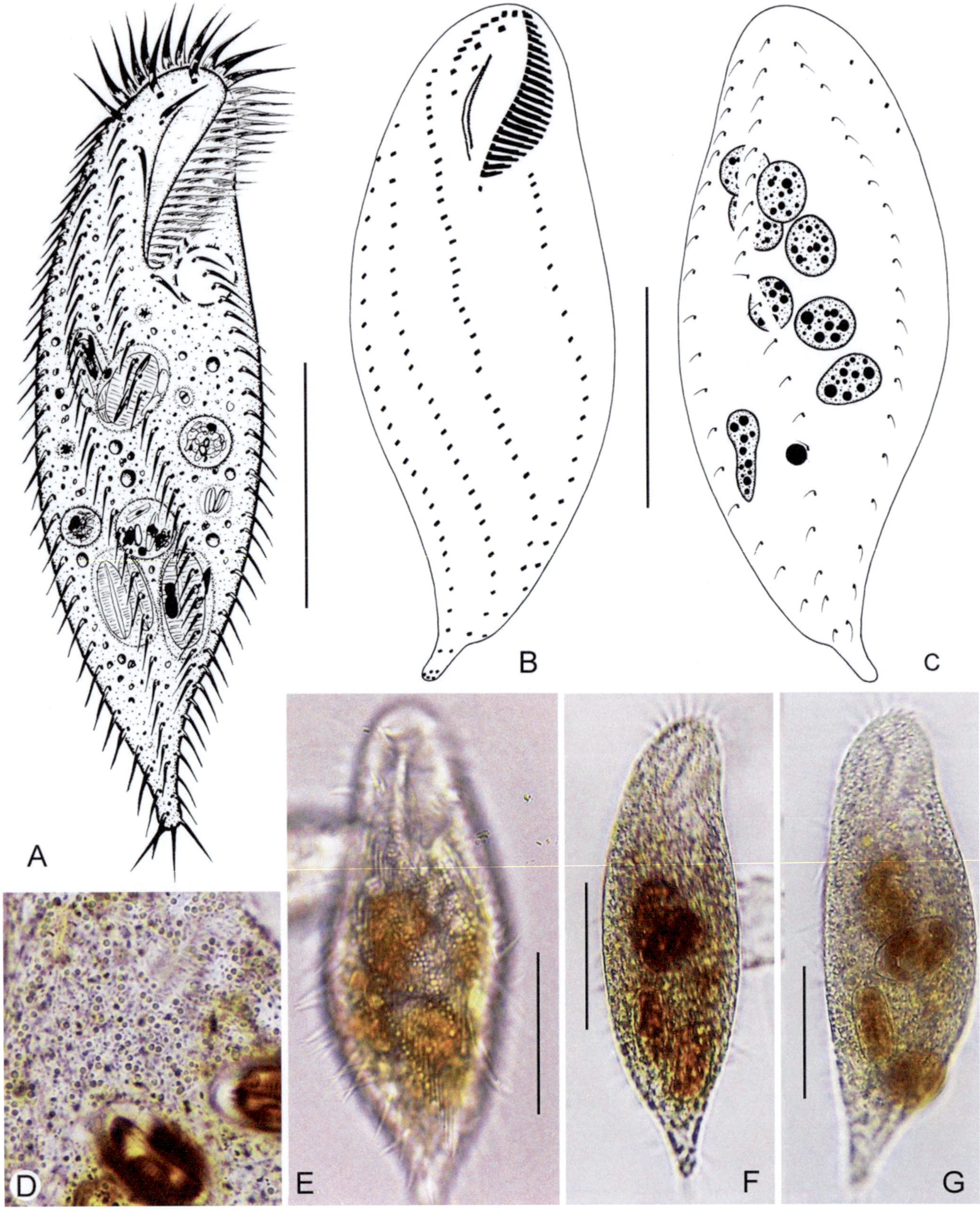


**Fig. 82B** *Strongylidium guangdongense*
(**a**, **f**, **g**) Ventral views of live individuals. (**b**, **c**) Ventral (**b**) and dorsal (**c**) views of ciliature. (**d**, **e**) Cortical granules on ventral (**e**) and dorsal (**d**) sides. Scale bars: 50 μm

## Reference

Luo X, Yan Y, Shao C, Al-Farraj SA, Bourland WA, Song WB (2018) Morphological, ontogenetic, and molecular data support strongylidiids as being closely related to Dorsomarginalia (Protozoa, Ciliophora) and reactivation of the family Strongylidiidae Fauré-Fremiet, 1961. Zool J Linn Soc doi/10.1093/zoolinnean/zly001/4962413

Class Spirotrichea Bütschli, 1889
Order Stichotrichida Fauré-Fremiet, 1961
Family Spirofilidae von Gelei, 1929
*Strongylidium orientale* Chen et al., 2013

**Ecological feature:** mangrove
**Temperature:** 32.0 °C; 20.0 °C
**pH:** 7.5; -
**Salinity:** 5.0‰; 15.7‰

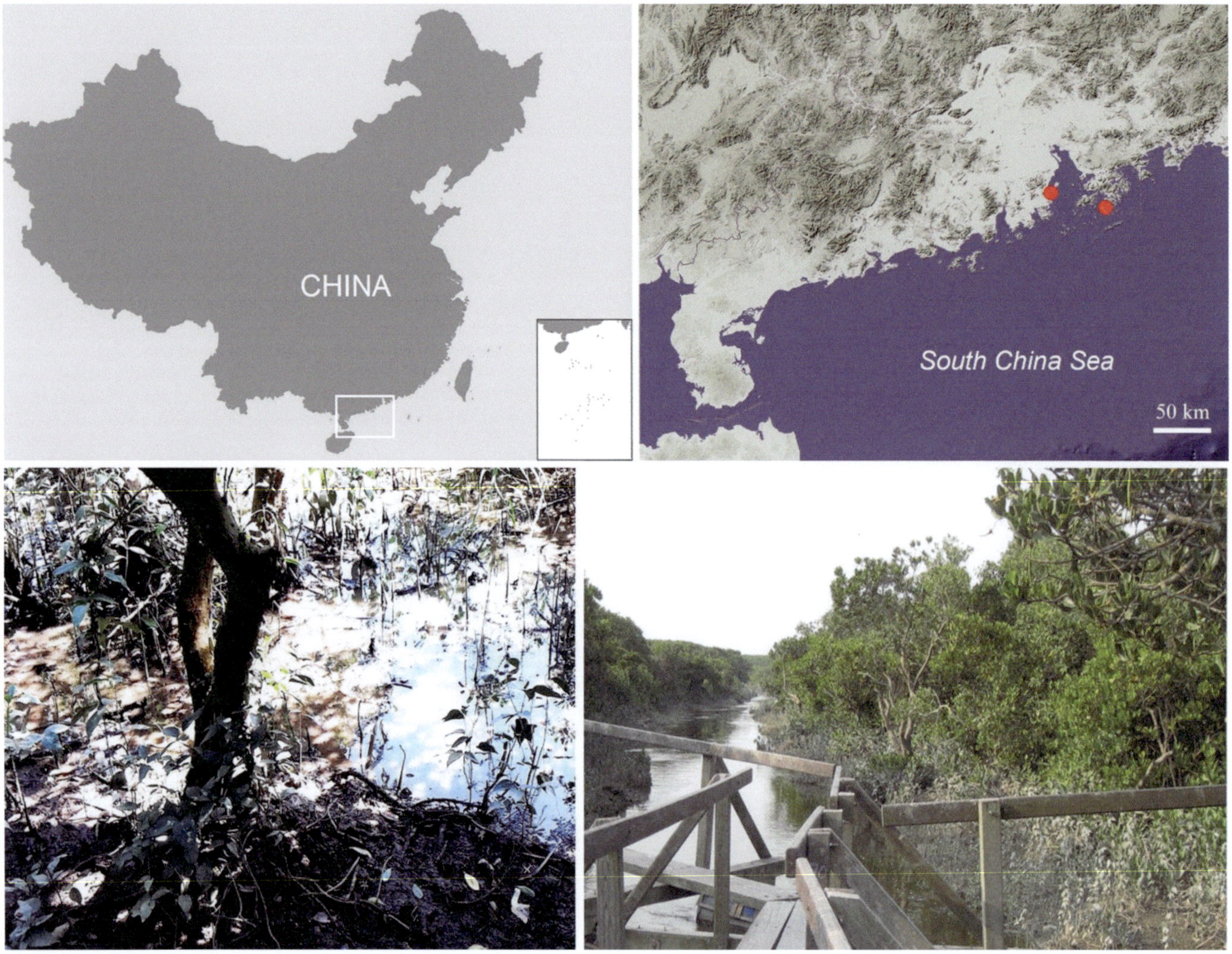


**Fig. 83A** Sample site

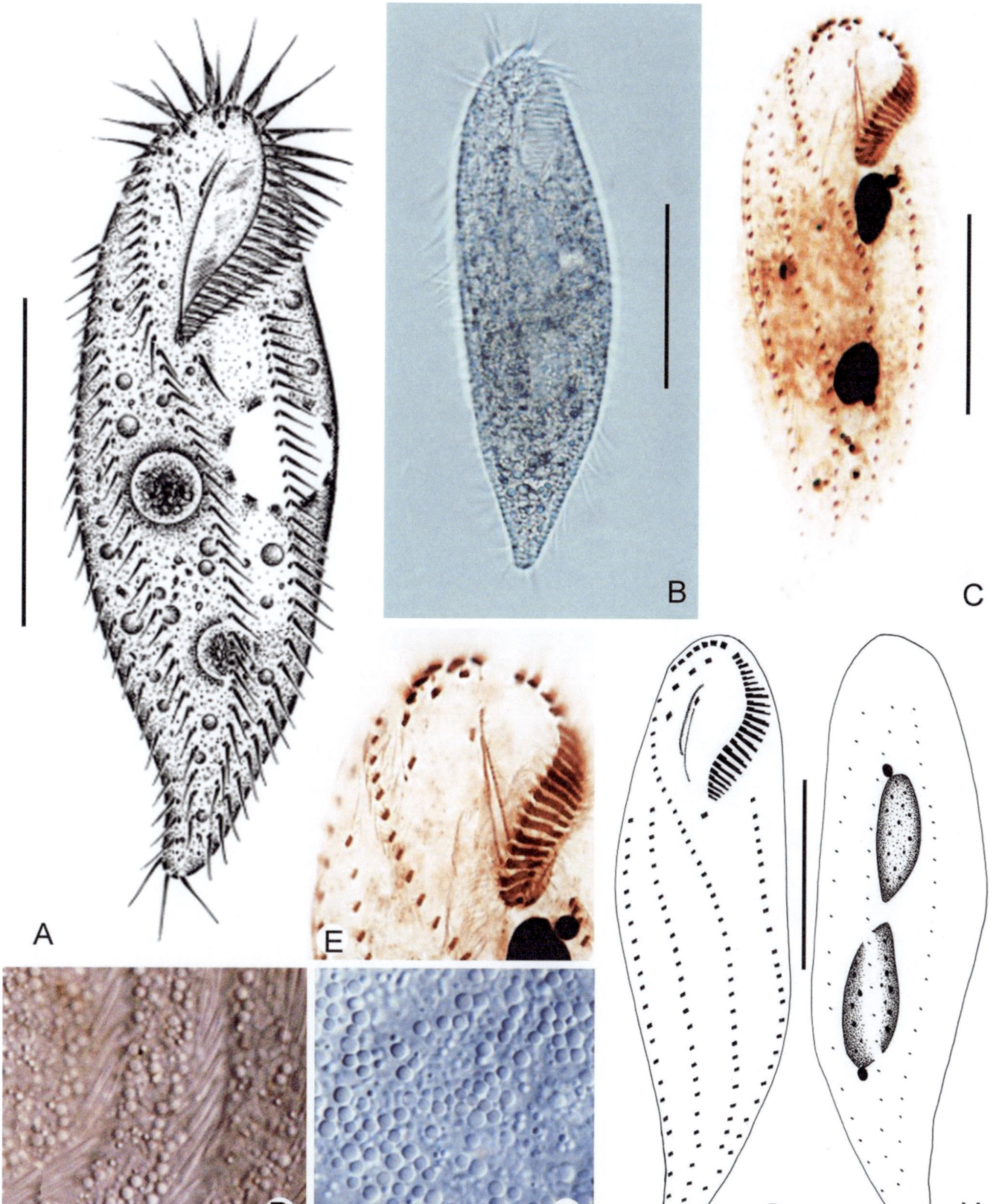


**Fig. 83B** *Strongylidium orientale*
(**a**, **b**) Ventral views of live individuals. (**c**, **g**, **h**) Ventral (**c**, **g**) and dorsal (**h**) views of ciliature. (**d**, **f**) Cortical granules. (**e**) Ventral view of anterior portion, showing cirri and oral apparatus. Scale bars: 50 μm

## Reference

Chen X, Miao M, Ma H, Shao C, Al-Rasheid KAS (2013) Morphology, morphogenesis and small-subunit rRNA gene sequence of the brackish-water ciliate *Strongylidium orientale* sp. nov. (Ciliophora, Hypotrichia). Int J Syst Evol Microbiol 63:1155–1164

Class Spirotrichea Bütschli, 1889
*Incertae sedis* in Order Stichotrichida
*Bistichella cystiformans* Fan et al., 2014

**Ecological feature:** mangrove
**Water temperature:** 27.0 °C
**pH:** 7.2
**Salinity:** 5.0‰

**Fig. 84A** Sample site

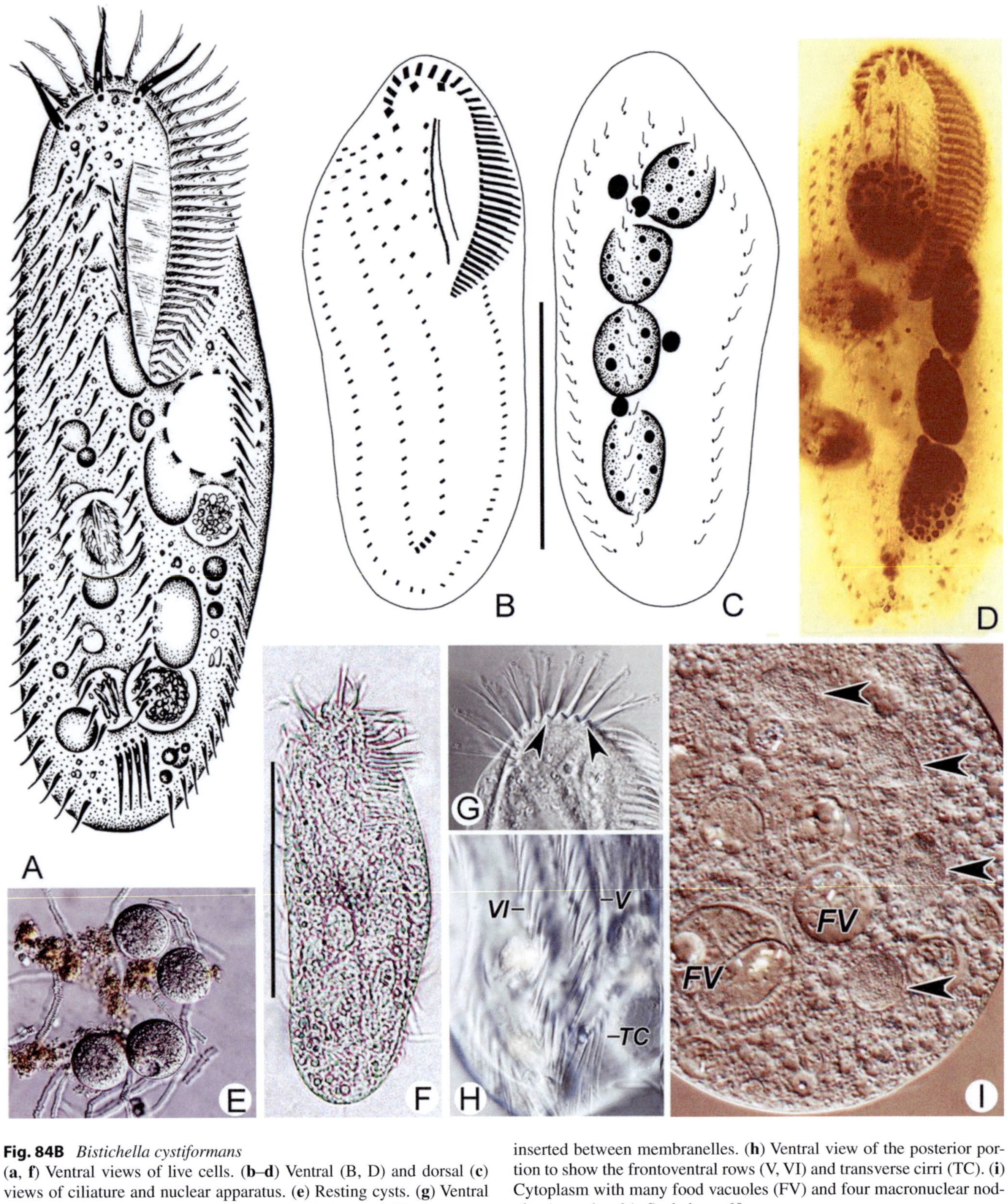


**Fig. 84B** *Bistichella cystiformans*
(**a**, **f**) Ventral views of live cells. (**b–d**) Ventral (B, D) and dorsal (**c**) views of ciliature and nuclear apparatus. (**e**) Resting cysts. (**g**) Ventral view of the anterior portion; arrowheads mark the tooth-like structures inserted between membranelles. (**h**) Ventral view of the posterior portion to show the frontoventral rows (V, VI) and transverse cirri (TC). (**i**) Cytoplasm with many food vacuoles (FV) and four macronuclear nodules (arrowheads). Scale bars: 60 μm

## Reference

Fan Y, Hu X, Gao F, Al-Farraj SA, Al-Rasheid KAS (2014) Morphology, ontogenetic features and SSU rRNA gene-based phylogeny of a soil ciliate, *Bistichella cystiformans* spec. nov. (Protista, Ciliophora, Stichotrichia). Int J Syst Evol Microbiol 64:4049–4060

Class Spirotrichea Bütschli, 1889
Order Urostylida Jankowski, 1979
Family Bergeriellidae Liu et al., 2010
*Bergeriella ovata* Liu et al., 2010

**Ecological feature:** intertidal zone
**Water temperature:** 21.7 °C
**pH:** 8.2
**Salinity:** 33.3‰

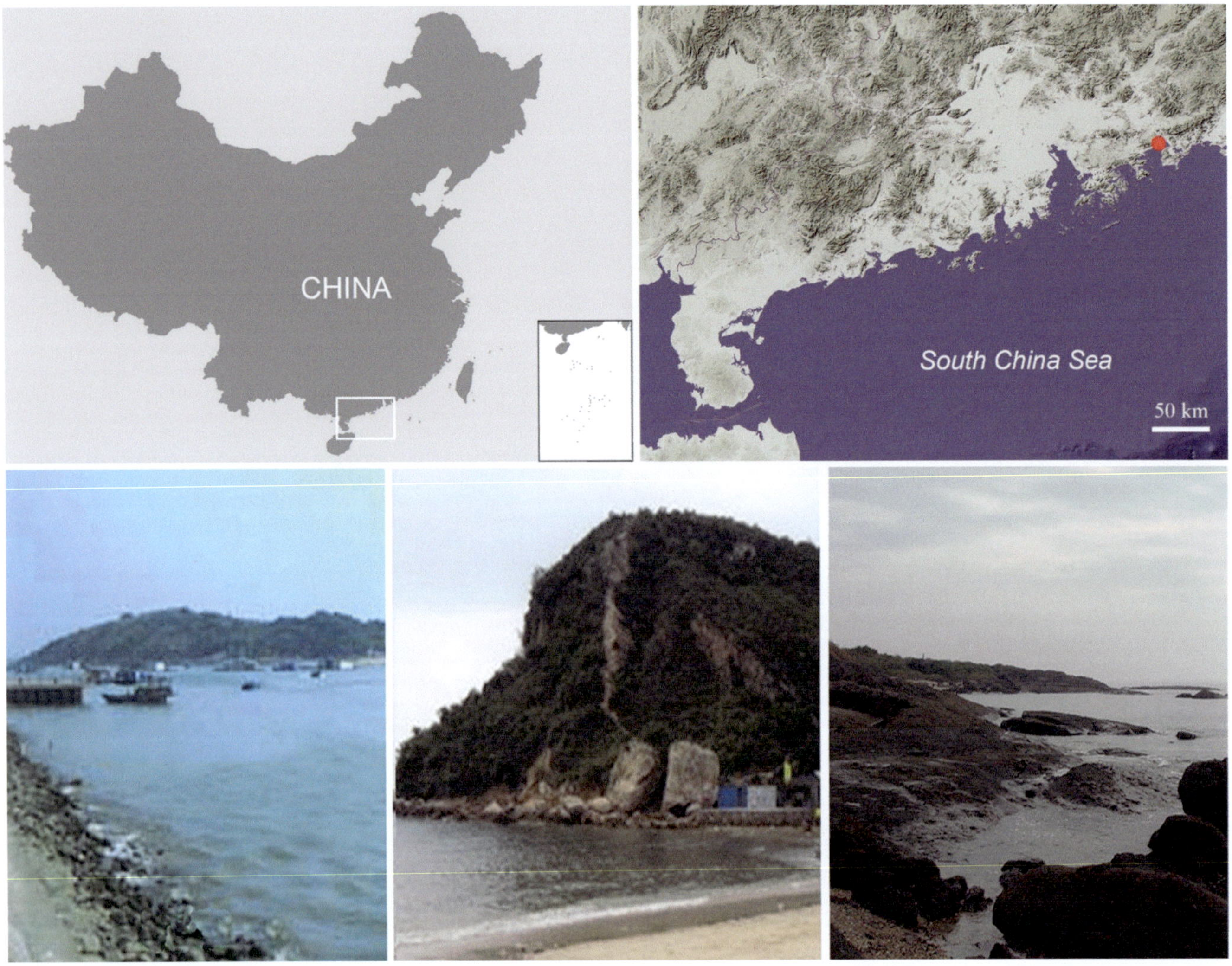


**Fig. 85A** Sample site

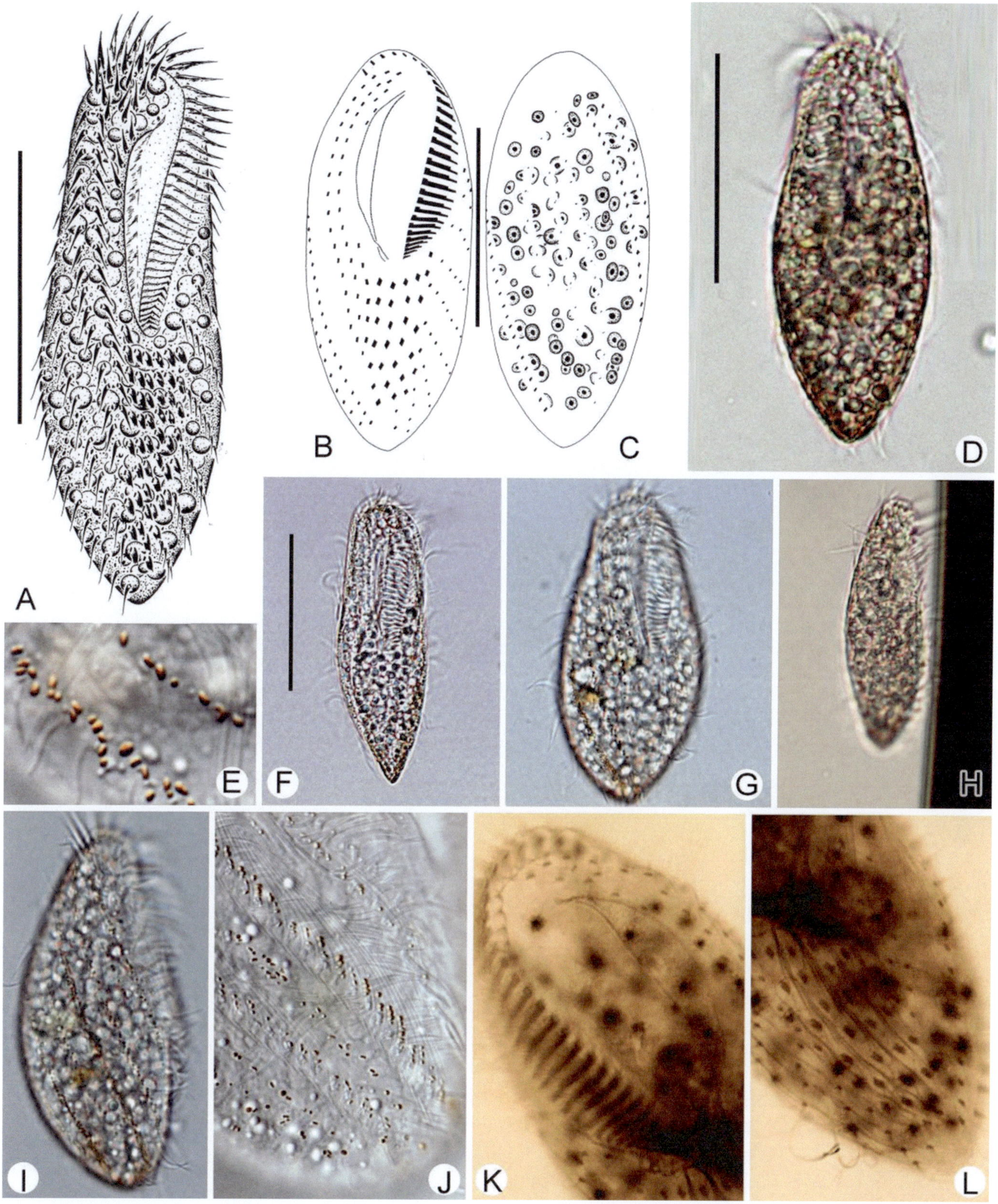


**Fig. 85B** *Bergeriella ovata*
(**a**, **f–h**) Ventral views of live individuals. (**b**, **c**) Ventral (**b**) and dorsal (**c**) views of ciliature. (**e**) Cortical granules. (**i**, **j**) Distribution of cortical granules. (**k**, **l**) Ciliature on ventral anterior (**k**) and posterior (**l**) portions. Scale bars: 50 μm

## Reference

Liu W, Shao C, Gong J, Li J, Lin X, Song WB (2010) Morphology, morphogenesis and molecular phylogeny of a new marine urostylid ciliate (Ciliophora, Stichotrichia) from the South China Sea, and an overview of the convergent evolution of midventral complex within the Spirotrichea. Zool J Linn Soc 158:697–710

Class Spirotrichea Bütschli, 1889
Order Urostylida Jankowski, 1979
Family Epiclintidae Wicklow & Borror, 1990
*Epiclintes auricularis* (Claparède & Lachmann, 1858) Stein, 1864

**Ecological feature:** sediment (mud); mangrove
**Water temperature:** 25.3 °C; 26.0 °C
**pH:** 7.9; 6.8
**Salinity:** 27.0‰; 22.0‰

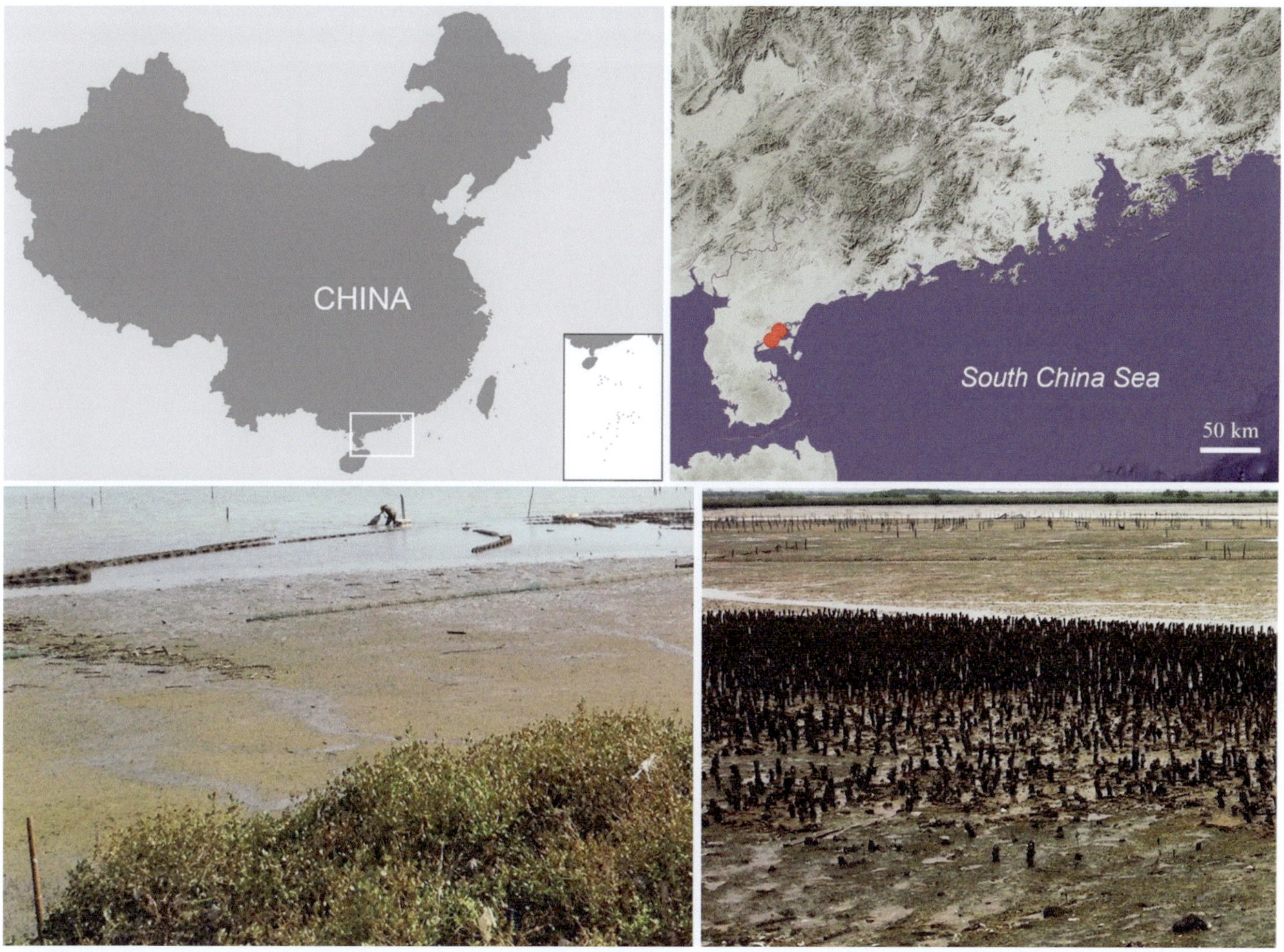


**Fig. 86A** Sample site

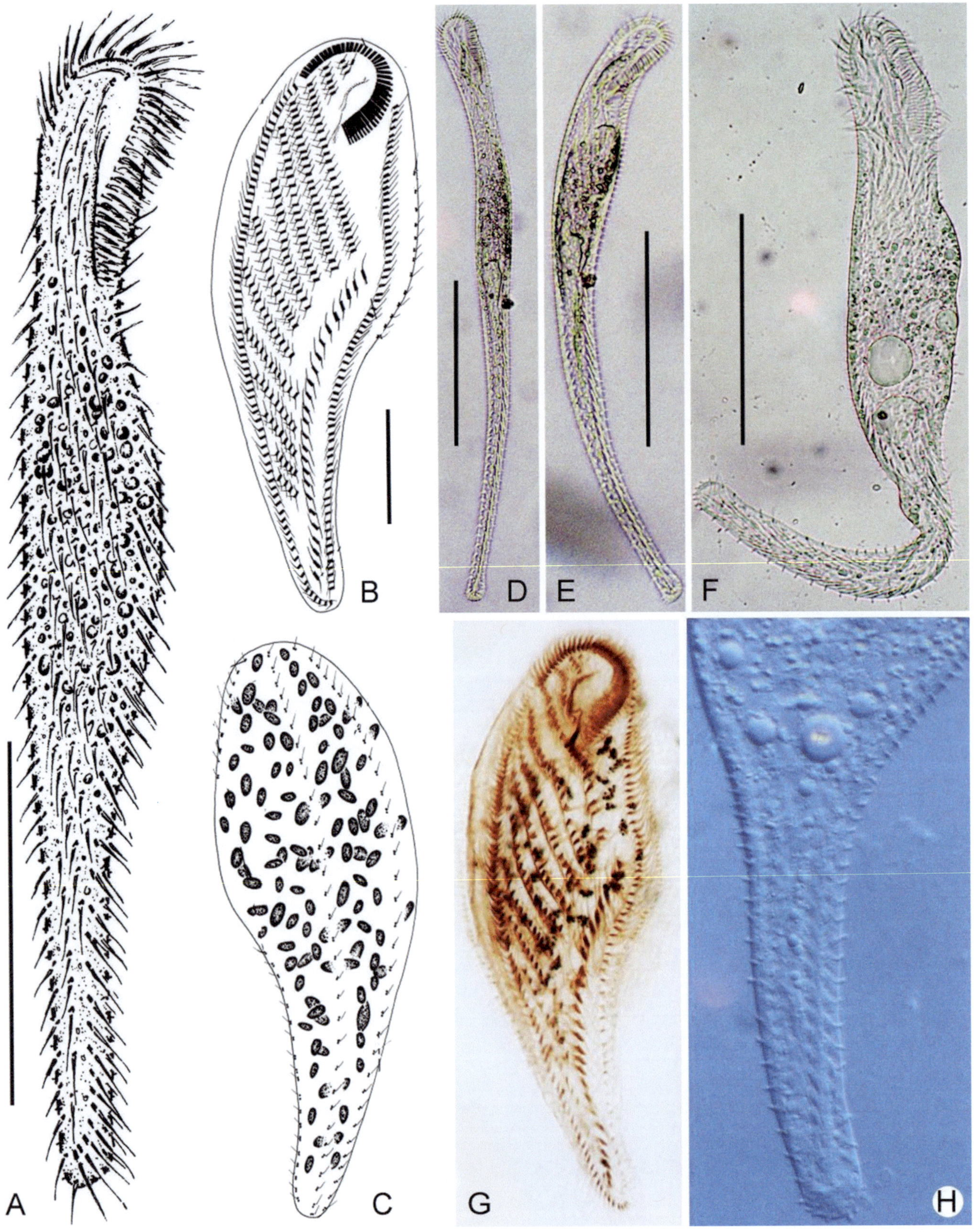


**Fig. 86B** *Epiclintes auricularis*
(**a**, **d**–**f**) Ventral views of live individuals. (**b**, **c**, **g**) Ventral (**b**, **g**) and dorsal (**c**) views of ciliature. (**h**) Posterior tail, showing dorsal bristles on papillae. Scale bars: 100 μm (**a**, **d**–**f**) and 70 μm (**b**)

## References

Hu X, Fan X, Lin X, Gong J, Song WB (2009) The morphology and morphogenesis of amarine ciliate, *Epiclintes auricularis rarisetus* nov. sspec. (Ciliophora, Epiclintidae), from the Yellow Sea. Eur J Protistol 45:281–291

Song WB, Warren A (1996) A redescription of the marine ciliates *Uroleptus retractilis* (Claparède and Lachmann, 1858) comb. n. and *Epiclintes ambiguus* (Müller, 1786) Bütschli, 1889 (Ciliophora, Hypotrichida). Acta Protozool 35:227–234

Class Spirotrichea Bütschli, 1889
Order Urostylida Jankowski, 1979
Family Parabirojimidae Dai & Xu, 2011
*Parabirojimia multinucleata* Chen et al., 2010

**Ecological feature:** culturing nets
**Water temperature:** 25 °C
**pH:** 7.8
**Salinity:** 28.0‰

**Fig. 87A** Sample site

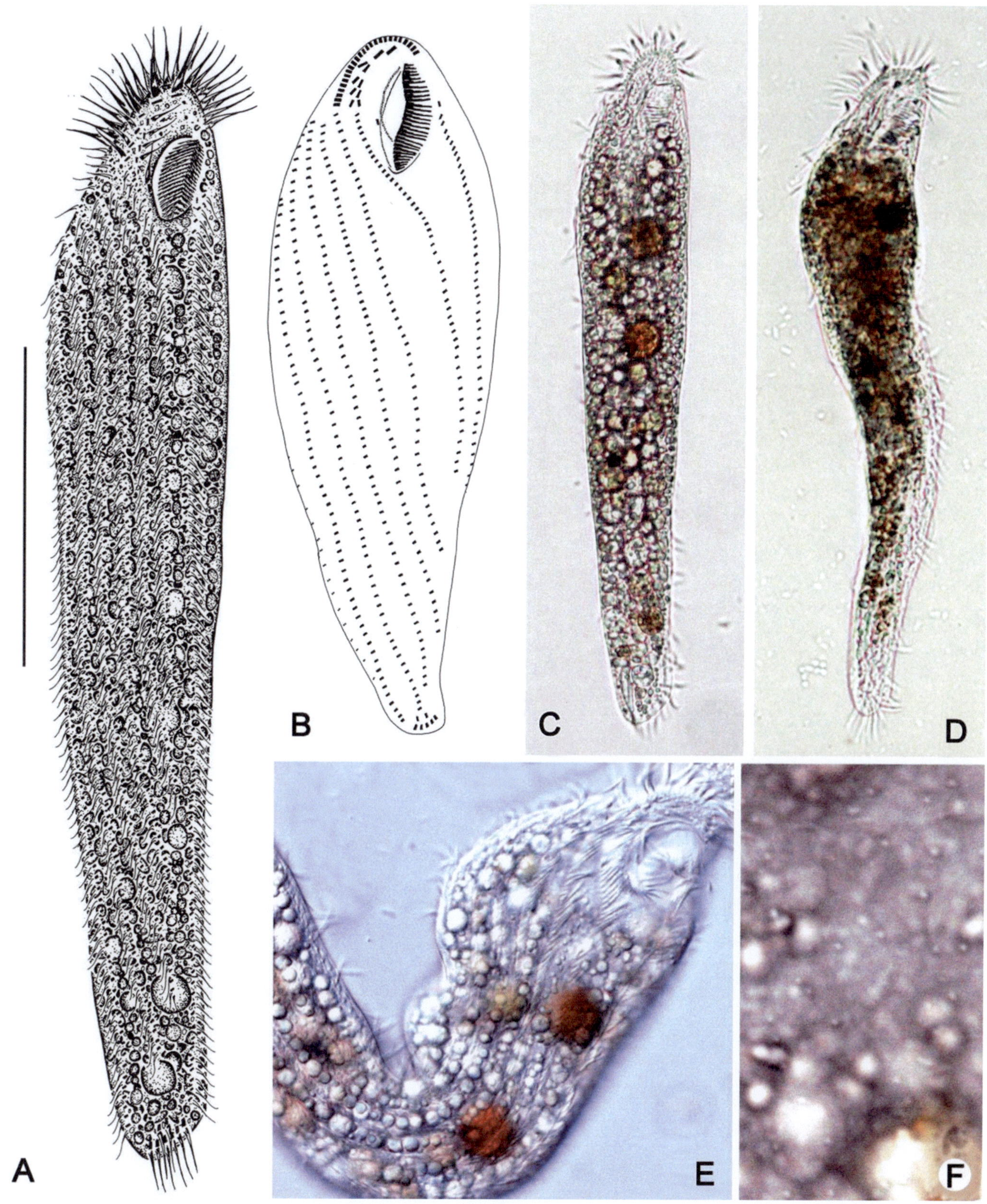


**Fig. 87B** *Parabirojimia multinucleata*
(**a**, **c**, **d**) Ventral views of live individuals. (**b**) Ventral view of ciliature. (**e**) Ventral view of anterior portion of cell, showing beak-like structure in buccal field. (**f**) Cortical granules. Scale bar: 100 μm

## Reference

Chen XR, Gao S, Song WB, Al-Rasheid KAS, Warren A, Gong J, Lin X (2010) *Parabirojimia multinucleata* nov. spec. and *Anteholosticha scutellum* (Cohn, 1866) Berger, 2003: two marine ciliates (Protozoa, Ciliophora, Hypotrichida) from tropical waters in south China, with report on SS rRNA gene sequences. Int J Syst Evol Microbiol 60:234–243

Class Spirotrichea Bütschli, 1889
Order Urostylida Jankowski, 1979
Family Parabirojimidae Dai & Xu, 2011
*Tunicothrix wilberti* (Lin & Song, 2004) Xu et al., 2006

**Ecological feature:** mangrove (Beihai, Guangzhou, Shenzhen); coastal water (Haikou)
**Water temperature:** 24.2 °C; 25.8 °C; –; 23.6 °C
**pH:** 7.2; 8.2; 8.0; no data available
**Salinity:** 9.7‰; 12.2‰; 15.0‰; 22.7‰

**Fig. 88A** Sample site

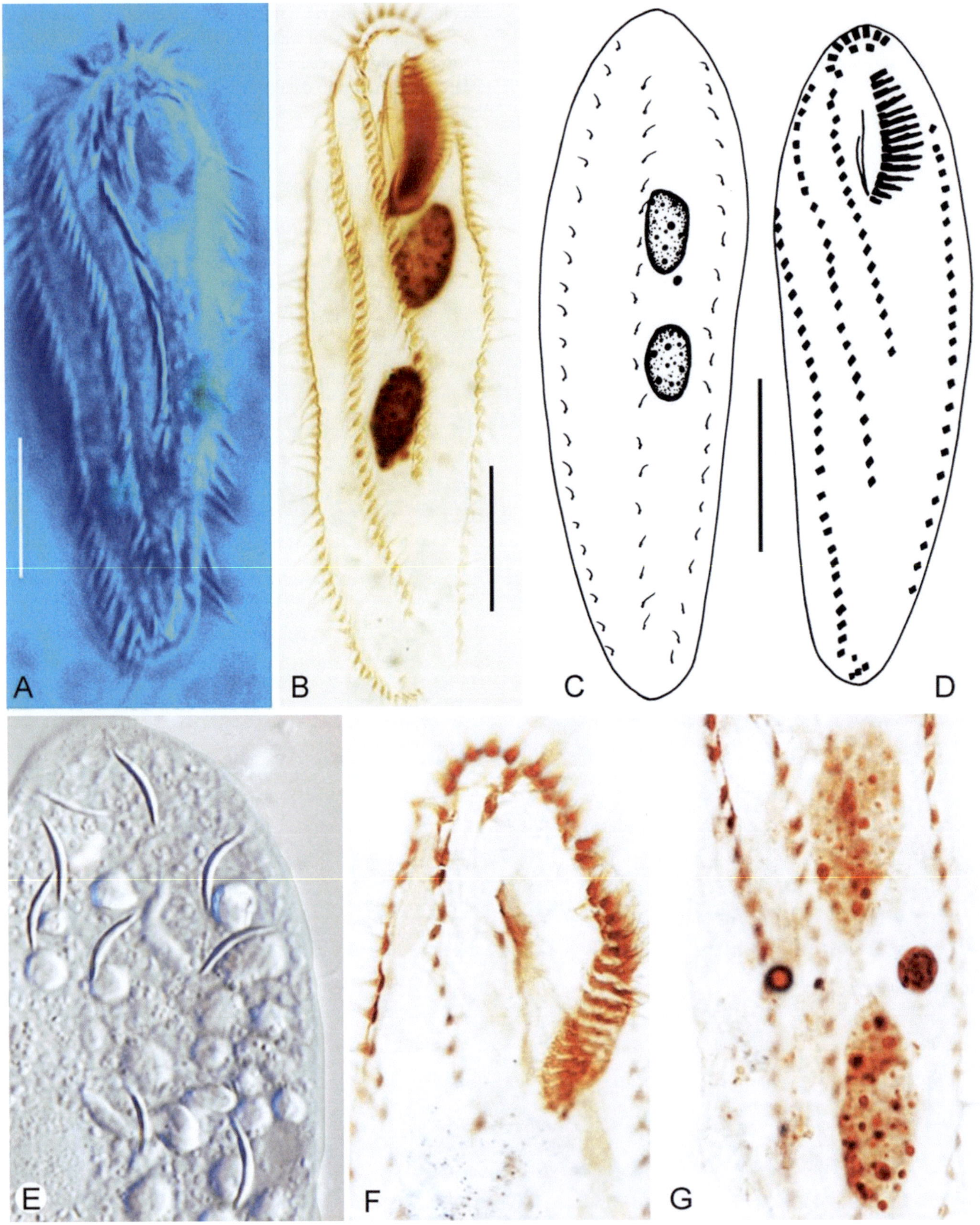


**Fig. 88B** *Tunicothrix wilberti*
(**a**) Ventral view of live cell. (**b–d**) Ventral (B, D) and dorsal (**c**) views of ciliature. (**e**) Ventral view of arc-shaped cortical structure. (**f**) Oral apparatus. (**g**) Nuclear apparatus. Scale bars: 40 μm

## References

Lin X, Song WB (2004) A new marine ciliate, *Erniella wilberti* sp. n. (Ciliophora: Hypotrichida), from shrimp culturing waters in North China. Acta Protozool 43:55–60

Xu K, Lei Y, Choi J (2006) *Tunicothrix rostrata* n. g., n. sp., a new urostylid ciliate (Ciliophora, Stichotrichia) from the Yellow Sea. J Eukaryot Microbiol 53:485–493

Class Spirotrichea Bütschli, 1889
Order Urostylida Jankowski, 1979
Family Pseudokeronopsidae Borror & Wicklow, 1983
*Heterokeronopsis pulchra* Pan et al., 2013

**Ecological feature:** mangrove
**Water temperature:** no data available; 26.0 °C
**pH:** 6.0; 7.3
**Salinity:** 1.0‰; 19.0‰

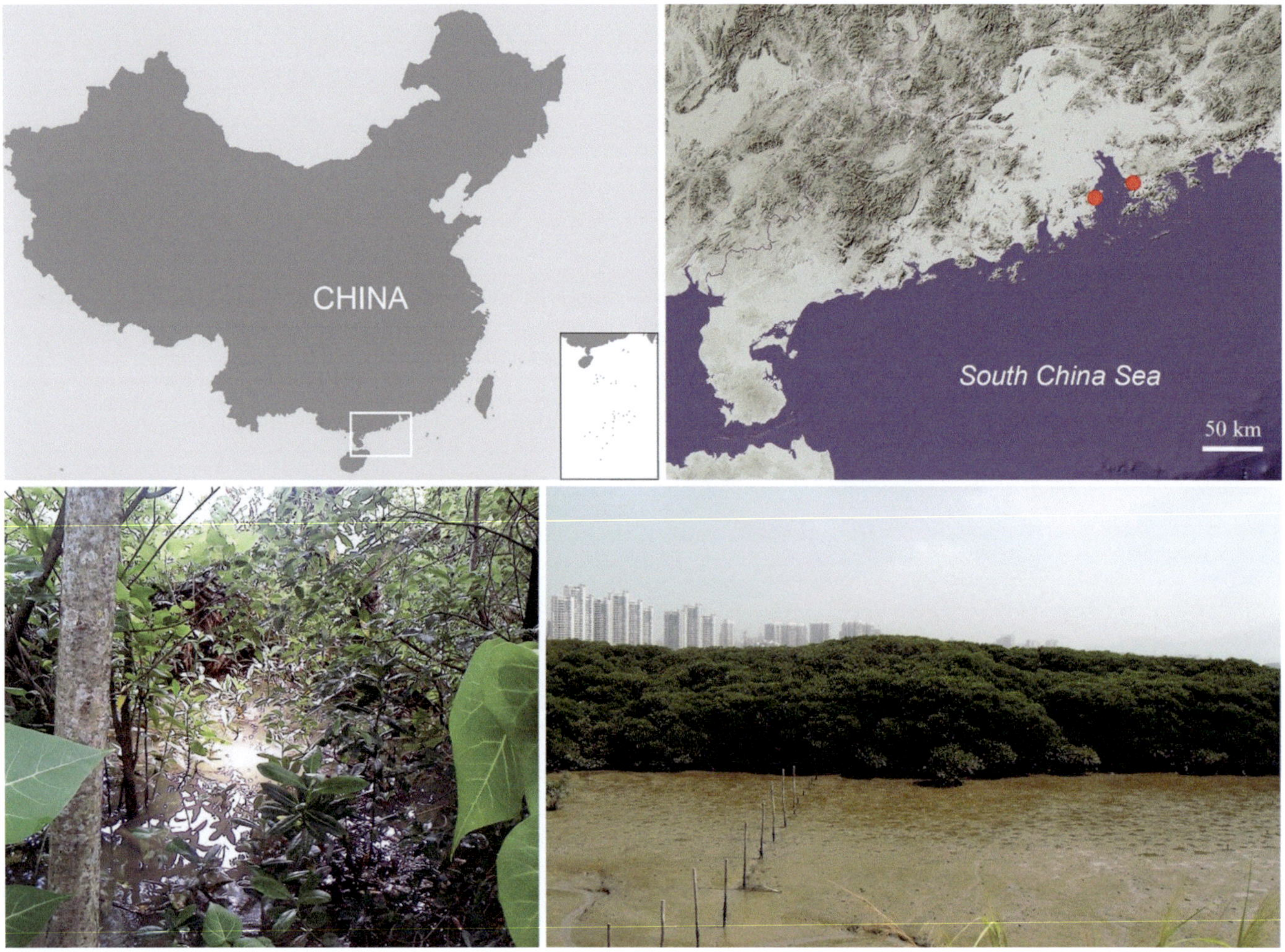


**Fig. 89A** Sample site

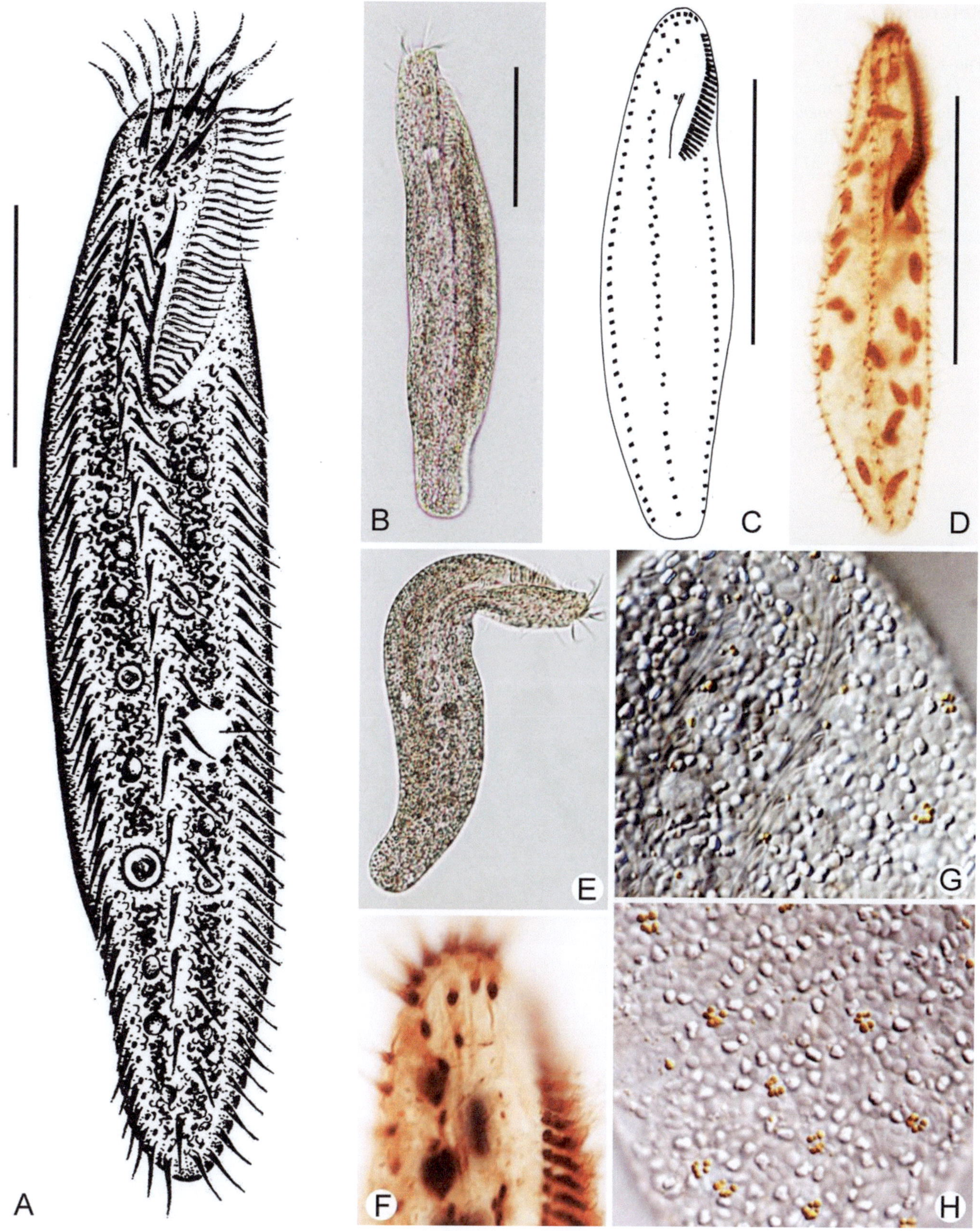


**Fig. 89B** *Heterokeronopsis pulchra*
(**a**, **b**) Ventral views of live individuals. (**c**, **d**, **f**) Ventral views of ciliature. (**e**) A bending cell. (**g**, **h**) Distribution of cortical granules. Scale bars: 90 μm

## Reference

Pan Y, Li J, Li L, Hu X, Al-Rasheid KAS, Warren A (2013) Ontogeny and molecular phylogeny of a new marine ciliate genus, *Heterokeronopsis* g. n. (Protozoa, Ciliophora, Hypotricha), with description of a new species. Eur J Protistol 49:298–311

Class Spirotrichea Bütschli, 1889
Order Urostylida Jankowski, 1979
Family Pseudokeronopsidae Borror & Wicklow, 1983
*Nothoholosticha* sp.

**Ecological feature:** mangrove
**Water temperature:** no data available
**pH:** 7.3
**Salinity:** 8.0‰

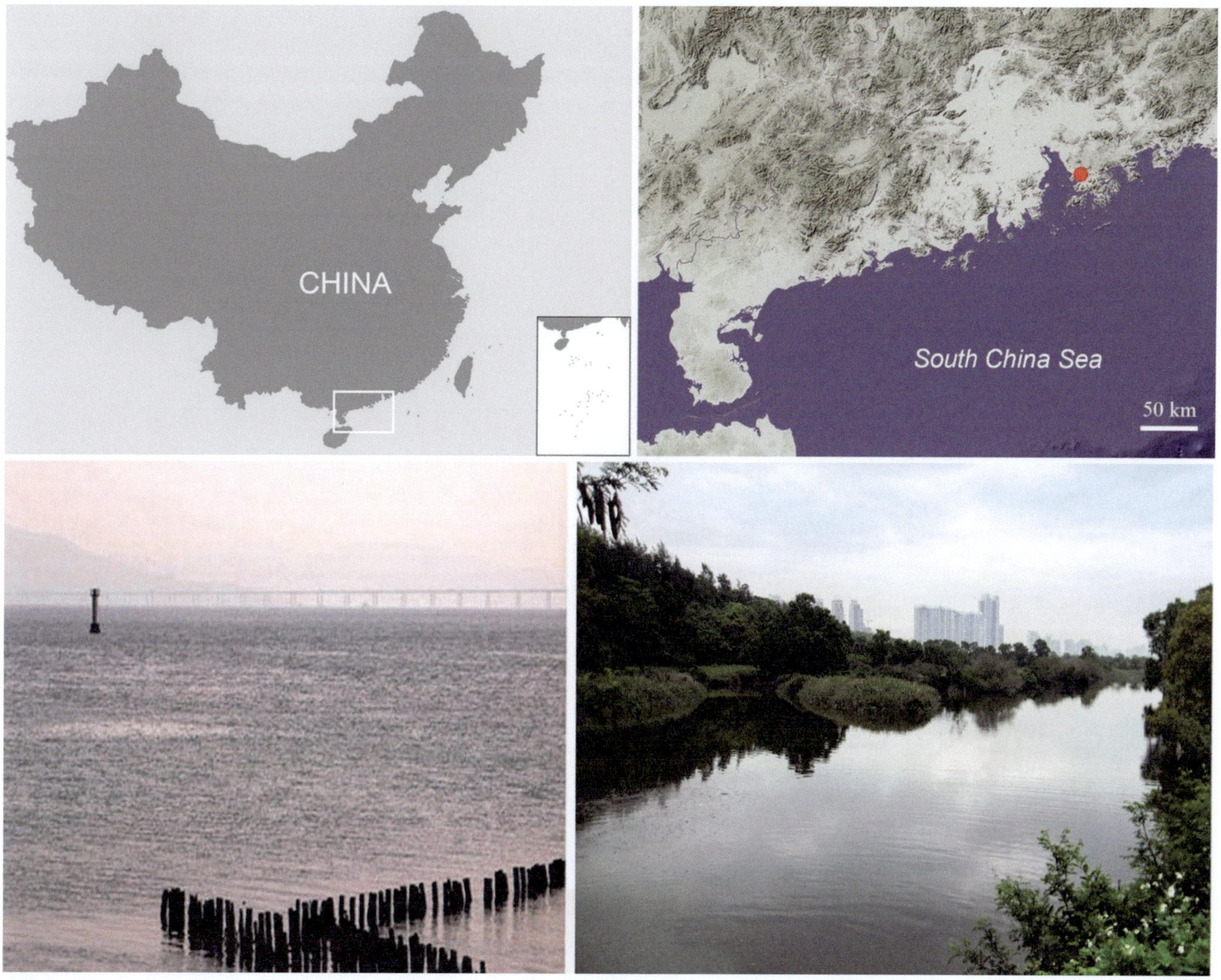


**Fig. 90A** Sample site

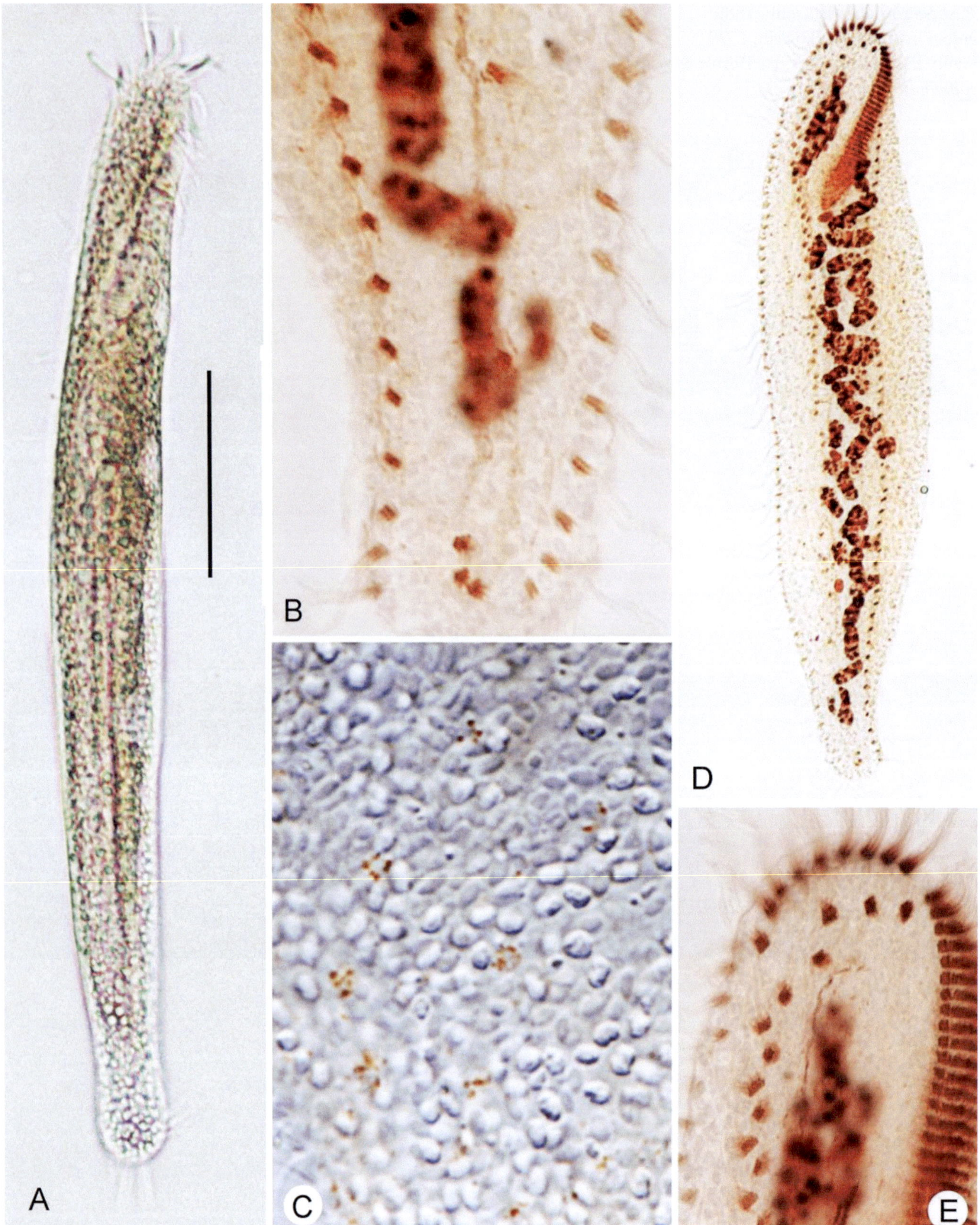


**Fig. 90B** *Nothoholosticha* sp.
(**a**) Ventral view of a live individual. (**b**) Ciliature of posterior (**b**) and anterior (**e**) portions of body. (**c**) Two types of cortical granules. (**d**) Ventral view of ciliature. Scale bar: 30 μm

## Reference

Berger H (2006) Monograph of the Urostyloidea (Ciliophora, Hypotricha). Monogr Biol 85:1–1303

Class Spirotrichea Bütschli, 1889
Order Urostylida Jankowski, 1979
Family Pseudokeronopsidae Borror & Wicklow, 1983
*Pseudokeronopsis carnea* (Cohn, 1866) Wirnsberger et al., 1987

**Ecological feature:** mangrove
**Water temperature:** 27.0 °C
**pH:** 6.8
**Salinity:** 11.0‰

**Fig. 91A** Sample site

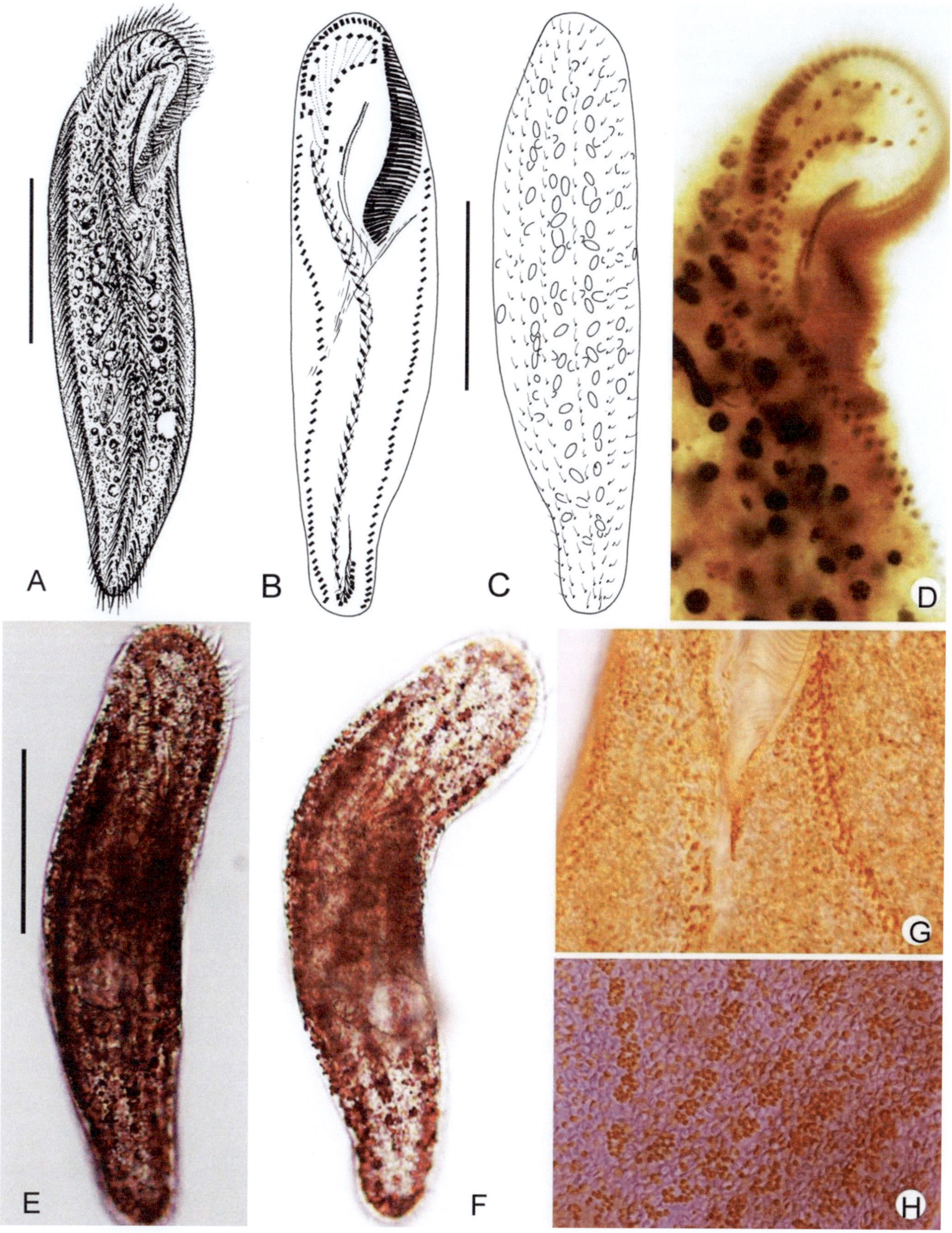


**Fig. 91B** *Pseudokeronopsis carnea*
(**a**, **e**, **f**) Views in vivo. (**b–d**) Ciliature. (**g**, **h**) Cortical granules on ventral and dorsal sides. Scale bars: 70 μm

## Reference

Song WB, Warren A, Roberts D, Wilbert N, Li L, Sun P, Hu X, Ma H (2006) Comparison and redefinition of four marine, coloured *Pseudokeronopsis* spp. (Ciliophora: Hypotrichida), with emphasis on their living morphology. Acta Protozool 45:271–287

Class Spirotrichea Bütschli, 1889
Order Urostylida Jankowski, 1979
Family Pseudokeronopsidae Borror & Wicklow, 1983
*Pseudokeronopsis* cf. *flava*

**Ecological feature:** mangrove
**Water temperature:** 21.0 °C
**pH:** no data available
**Salinity:** 14.0‰

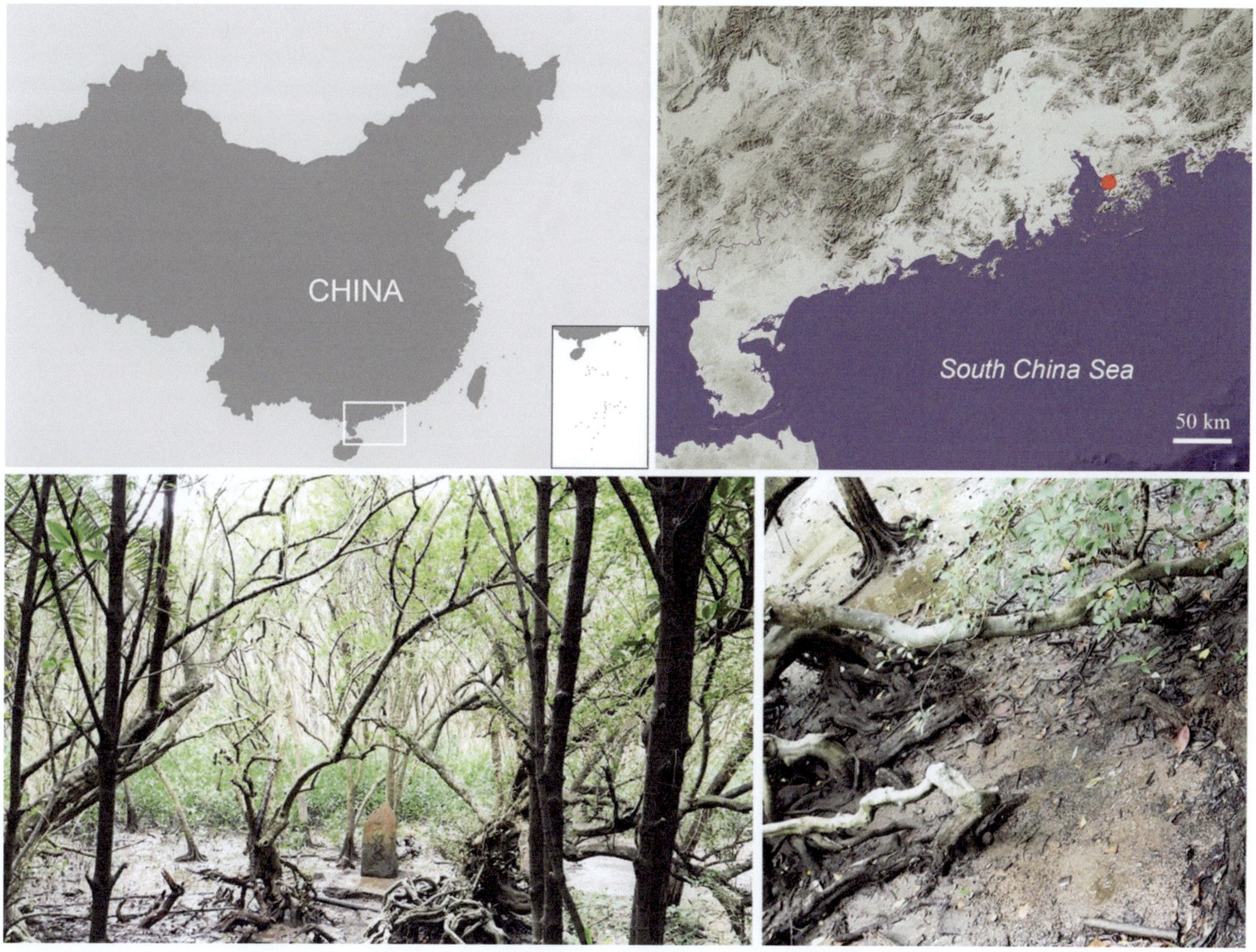


**Fig. 92A** Sample site

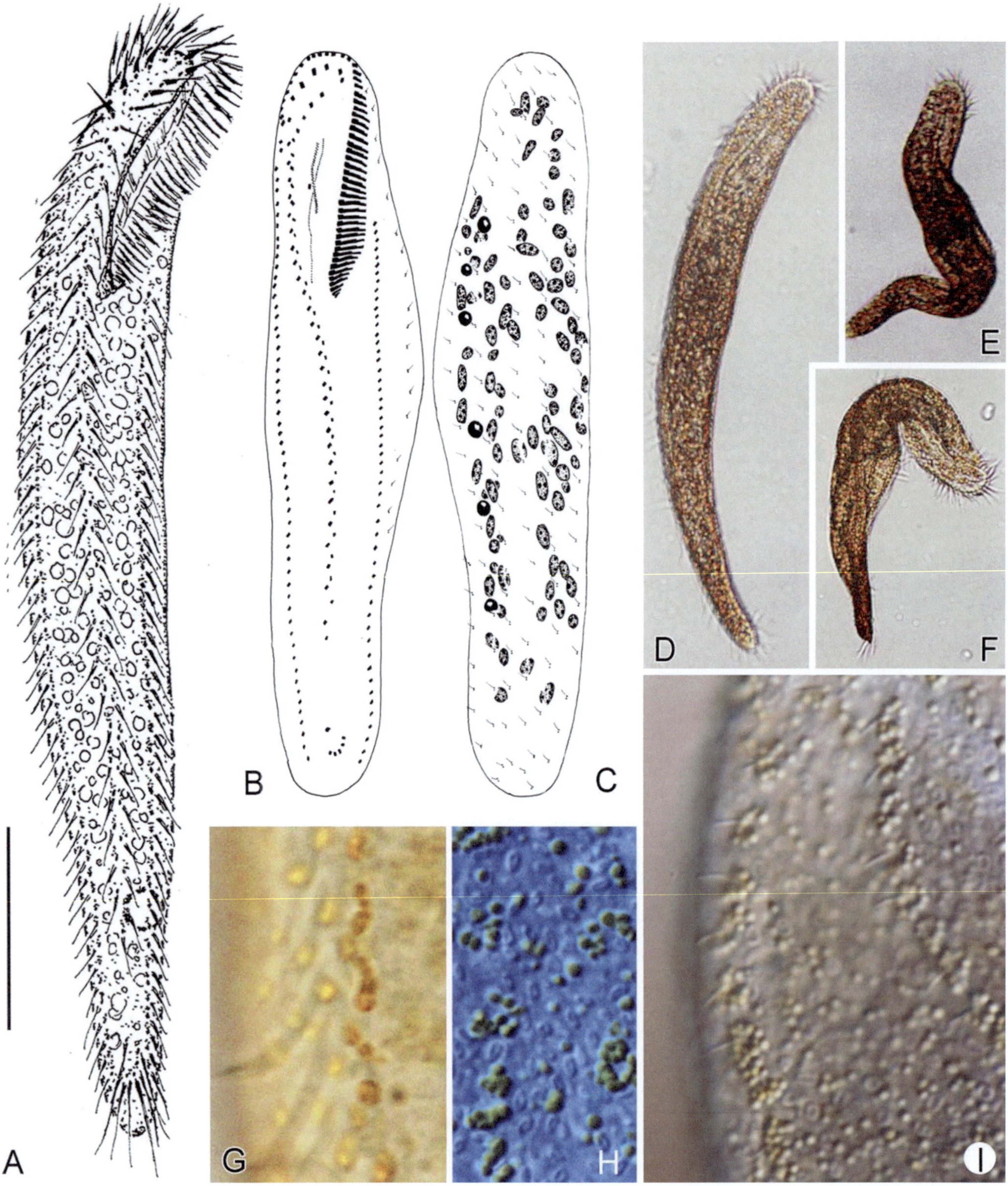


**Fig. 92B** *Pseudokeronopsis* cf. *flava*
(**a**, **d**) Ventral views of live cells. (**b**, **c**) Ventral (**b**) and dorsal (**c**) views of ciliature. (**e**, **f**) Flexible bodies. (**g–i**) Cortical granules along ventral cirral row (**g**) and on dorsal side (**h**, **i**). Scale bar: 50 μm

## Reference

Wirnsberger E, Larsen HF, Uhlig G (1987) Rediagnoses of closely related pigmented marine species of the genus *Pseudokeronopsis* (Ciliophora, Hypotrichida). Eur J Protistol 23:76–88

Class Spirotrichea Bütschli, 1889
Order Urostylida Jankowski, 1979
Family Pseudokeronopsidae Borror & Wicklow, 1983
*Pseudokeronopsis erythrina* Chen et al., 2011

**Ecological feature:** estuary
**Temperature:** 25.2 °C
**pH:** no data available
**Salinity:** 6.3‰

**Fig. 93A** Sample site

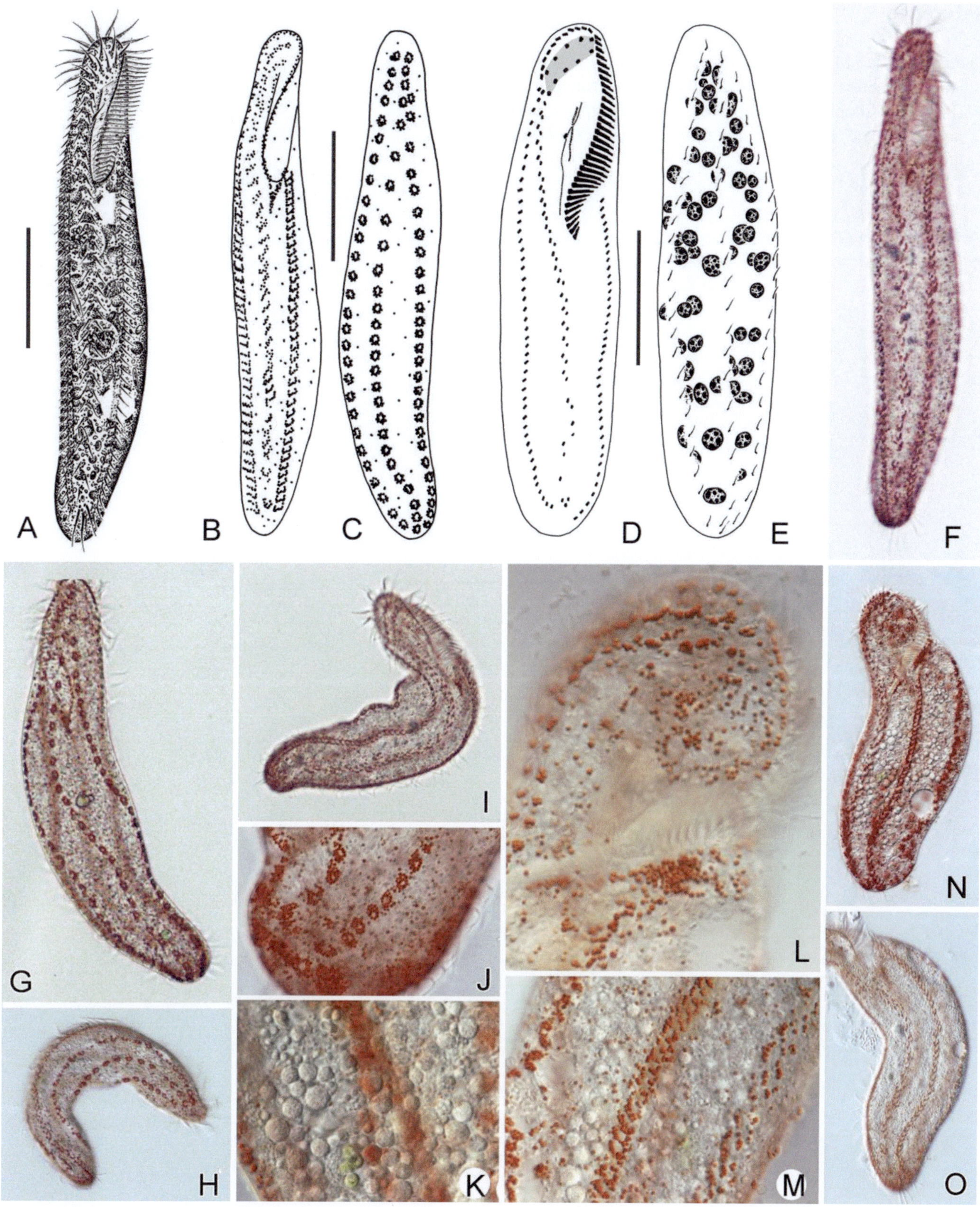


**Fig. 93B** *Pseudokeronopsis erythrina*
(**a**, **f**) Ventral views of live individuals. (**b**, **c**, **g–j**, **l–o**) Ventral (**b**, **i**, **l–o**) and dorsal (**c**, **g**, **h**) views of cortical granules. (**d**, **e**) Ventral (**d**) and dorsal (**e**) views of ciliature. Scale bars: 50 μm

## Reference

Chen XM, Miao M, Clamp JC, Song WB (2011) Phylogeny and systematic revision of the family Pseudokeronopsidae (Protista, Ciliophora, Hypotricha), with description of a new estuarine species of *Pseudokeronopsis*. Zool Scr 40:659–671

Class Spirotrichea Bütschli, 1889
Order Urostylida Jankowski, 1979
Family Pseudokeronopsidae Borror & Wicklow, 1983
*Pseudokeronopsis flava* (Cohn, 1866) Wirnsberger et al., 1987

**Ecological feature:** mangrove
**Water temperature:** 25.0 °C; 25.0 °C; 18.0 °C
**pH:** 6.7; 6.7; 6.0
**Salinity:** 17.0‰; 17.0‰; 2.0‰

**Fig. 94A** Sample site

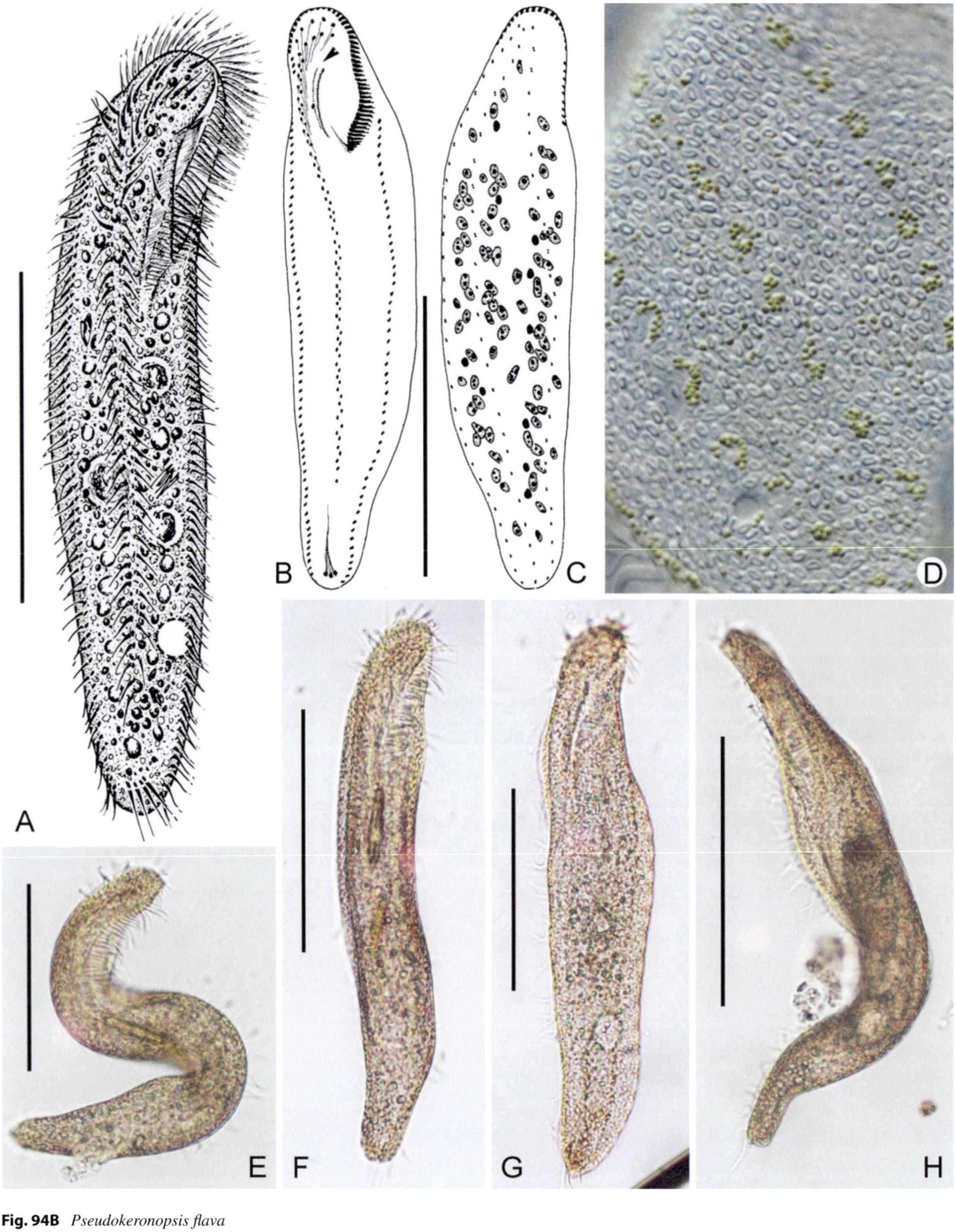


**Fig. 94B** *Pseudokeronopsis flava*
(**a**, **f**, **g**) Ventral views of live cells. (**b**, **c**) Ventral (**b**) and dorsal (**c**) views of ciliature. (**d**) Two types of cortical granules on dorsal side. (**e**) A twisted body. (**h**) Lateral view of live cell. Scale bars: 80 μm

## References

Song WB, Warren A, Roberts D, Wilbert N, Li L, Sun P, Hu X, Ma H (2006) Comparison and redefinition of four marine, coloured *Pseudokeronopsis* spp. (Ciliophora: Hypotrichida), with emphasis on their living morphology. Acta Protozool 45:271–287

Wirnsberger E, Larsen H, Uhlig G (1987) Rediagnoses of closely related pigmented marine species of the genus *Pseudokeronopsis* (Ciliophora, Hypotrichida). Eur J Protistol 23:76–88

Class Spirotrichea Bütschli, 1889
Order Urostylida Jankowski, 1979
Family Pseudokeronopsidae Borror & Wicklow, 1983
*Pseudokeronopsis flavicans* (Kahl, 1932) Borror & Wicklow, 1983

**Ecological feature:** mangrove
**Water temperature:** 24.7 °C
**pH:** no data available
**Salinity:** 17.0‰

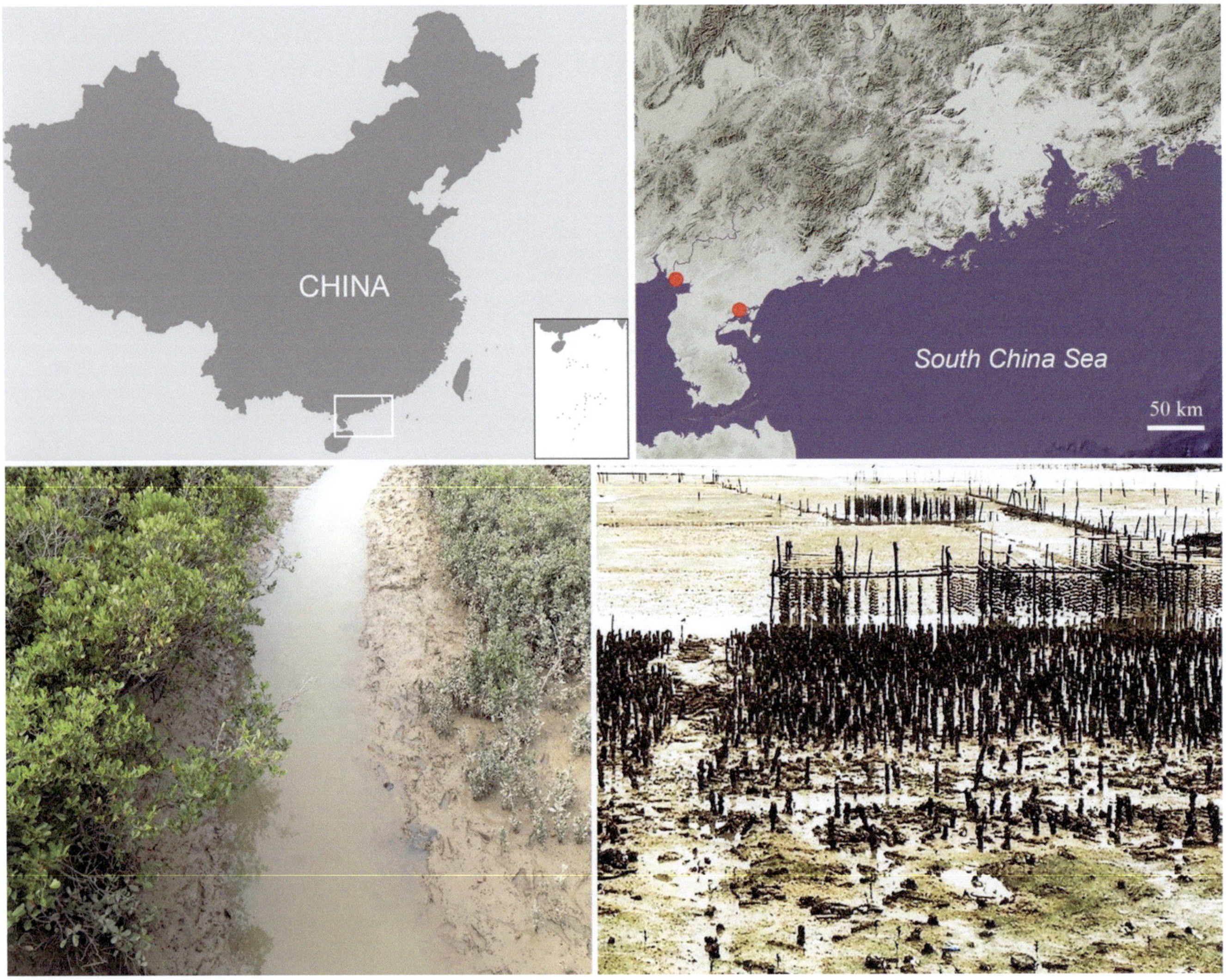


**Fig. 95A** Sample site

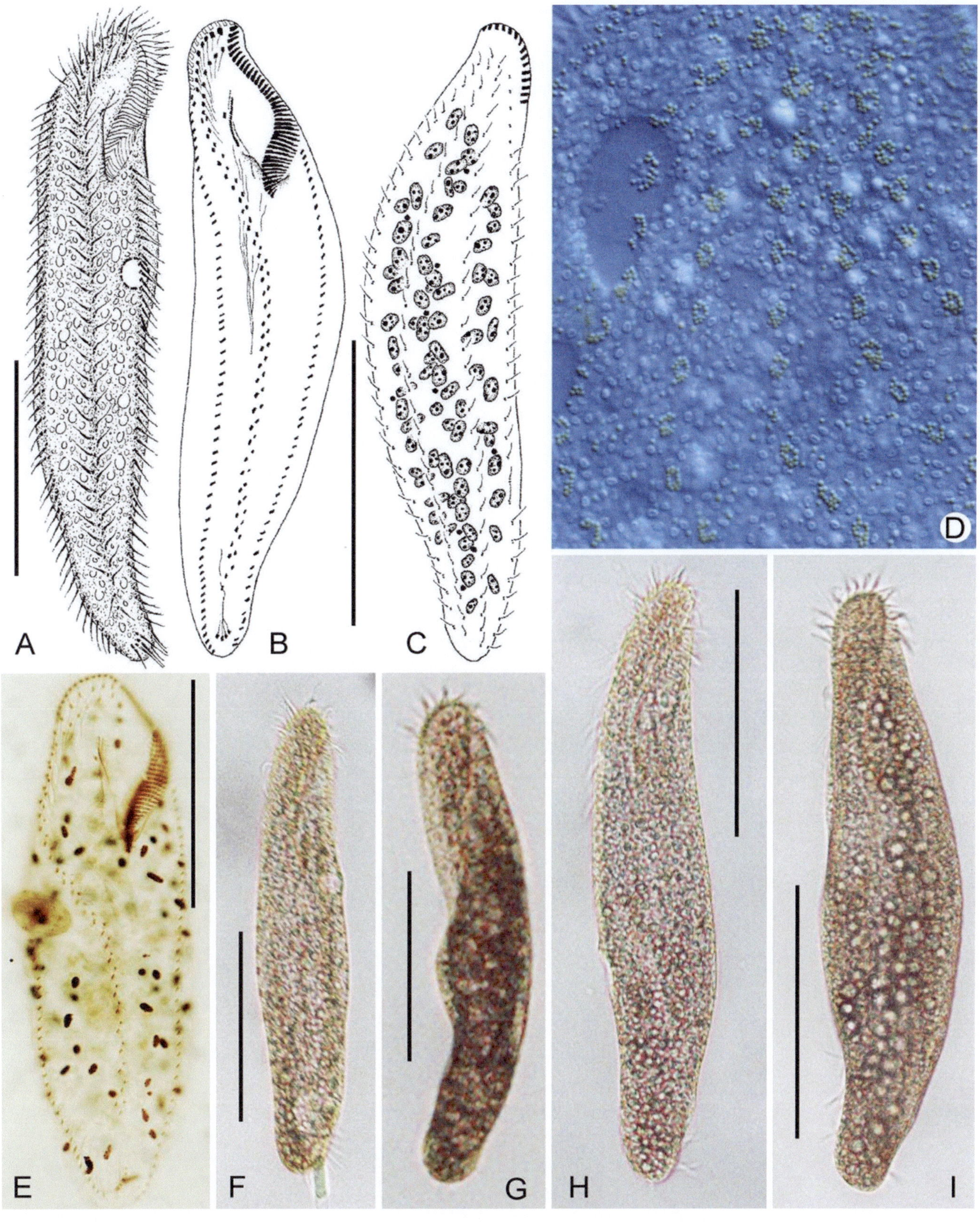


**Fig. 95B** *Pseudokeronopsis flavicans*
(**a**, **f–i**) Ventral views of live individuals. (**b**, **c**, **e**) Ventral (**b**, **e**) and dorsal (**c**) views of ciliature. (**d**) Two kinds of cortical granules on dorsal side. Scale bars: 100 μm

## Reference

Song WB, Wilbert N, Warren A (2002) New contribution to the morphology and taxonomy of four marine hypotrichous ciliates from Qingdao, China (Protozoa: Ciliophora). Acta Protozool 41:145–162

Class Spirotrichea Bütschli, 1889
Order Urostylida Jankowski, 1979
Family Pseudokeronopsidae Borror & Wicklow, 1983
*Pseudokeronopsis rubra* (Ehrenberg, 1838) Borror & Wicklow, 1983

**Ecological feature:** mangrove
**Water temperature:** 27.0 °C
**pH:** 6.0
**Salinity:** 13.0‰

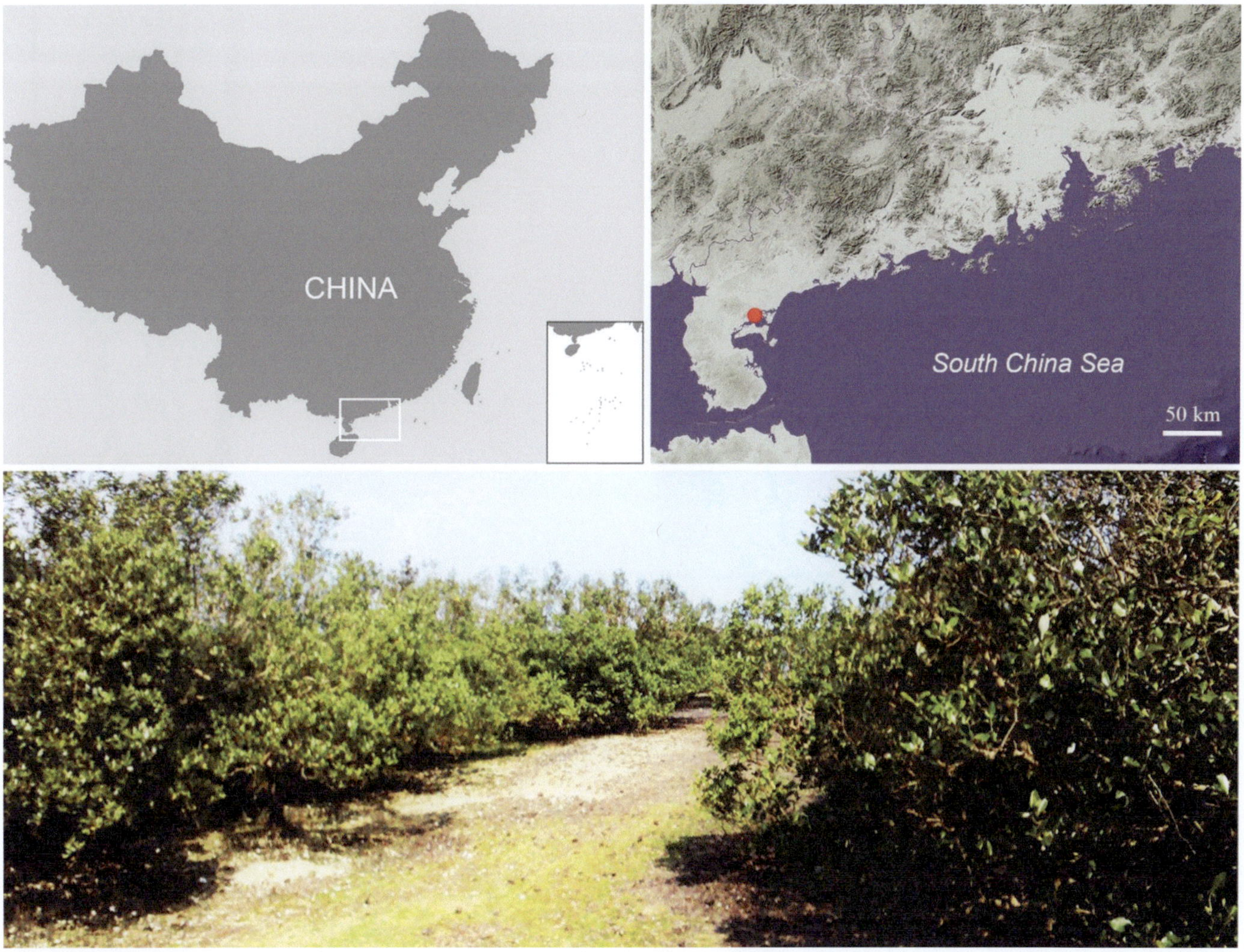


**Fig. 96A** Sample site

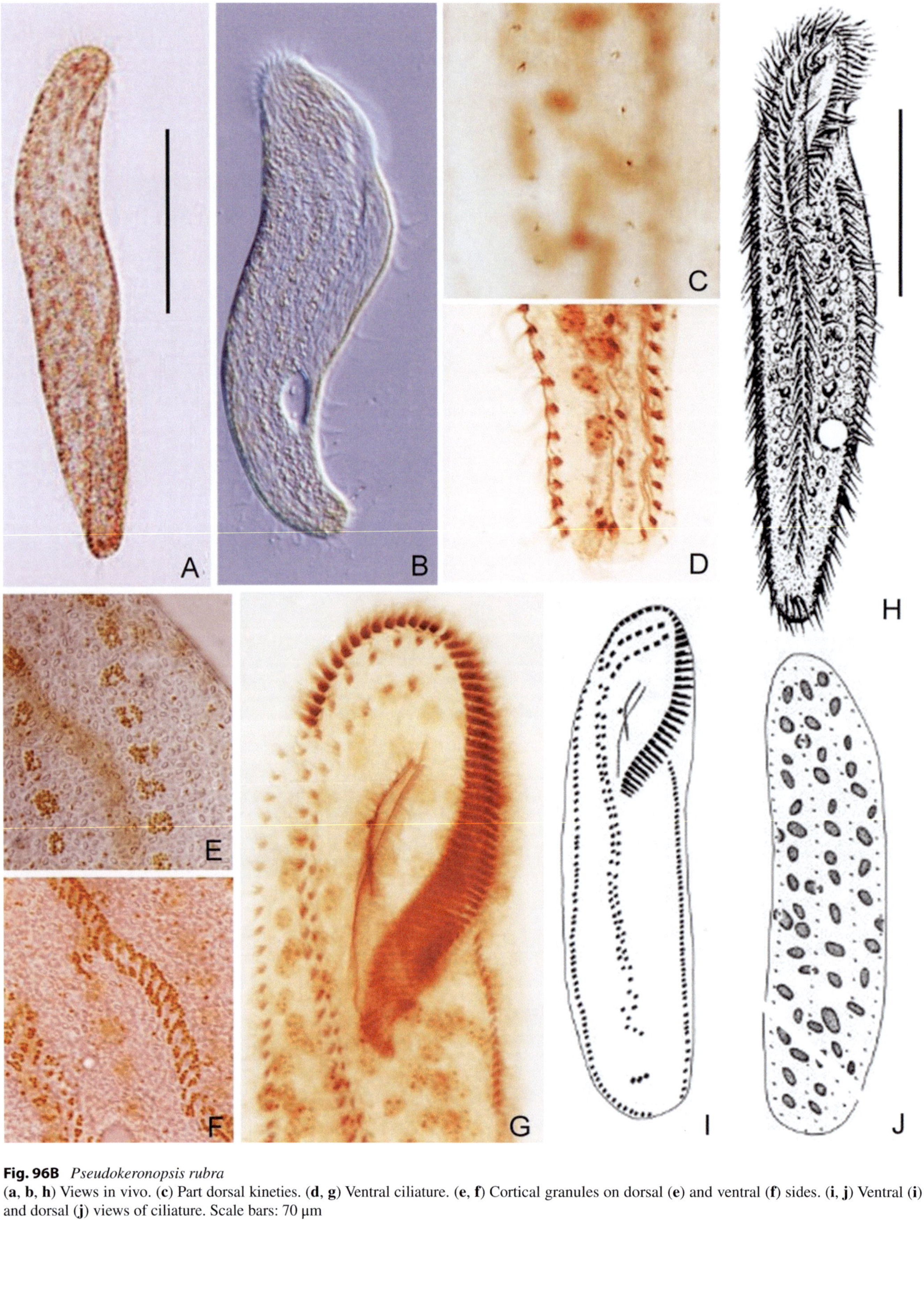


**Fig. 96B** *Pseudokeronopsis rubra*
(**a**, **b**, **h**) Views in vivo. (**c**) Part dorsal kineties. (**d**, **g**) Ventral ciliature. (**e**, **f**) Cortical granules on dorsal (**e**) and ventral (**f**) sides. (**i**, **j**) Ventral (**i**) and dorsal (**j**) views of ciliature. Scale bars: 70 μm

## Reference

Song WB, Warren A, Roberts D, Wilbert N, Li L, Sun P, Hu X, Ma H (2006) Comparison and redefinition of four marine, coloured *Pseudokeronopsis* spp. (Ciliophora: Hypotrichida), with emphasis on their living morphology. Acta Protozool 45:271–287

Class Spirotrichea Bütschli, 1889
Order Urostylida Jankowski, 1979
Family Pseudourostylidae Jankowski, 1979
*Pseudourostyla cristata* (Jerka-Dziadosz, 1964) Borror, 1972

**Ecological feature:** mangrove
**Water temperature:** 31.0 °C
**pH:** 6.8
**Salinity:** 1.0‰

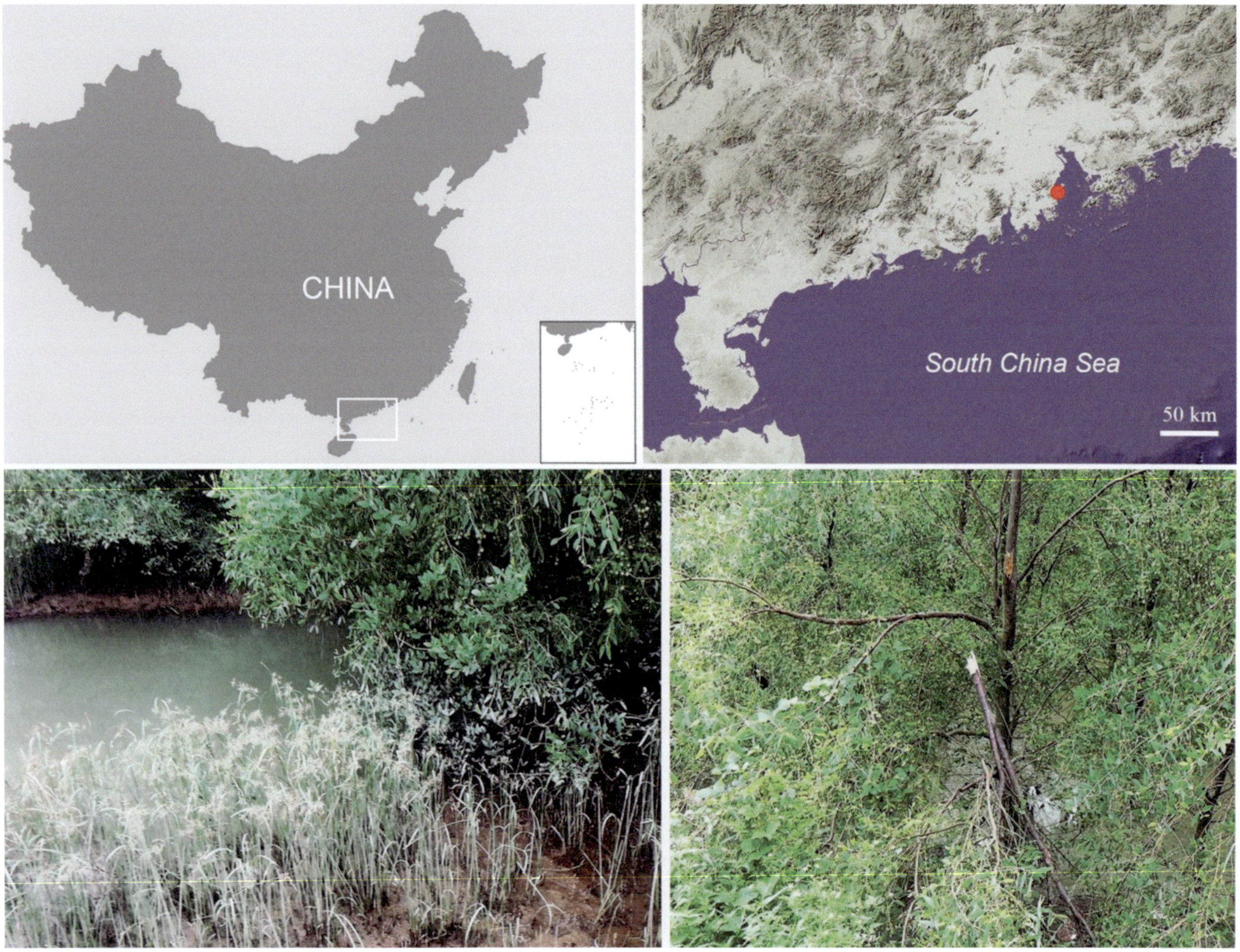


**Fig. 97A** Sample site

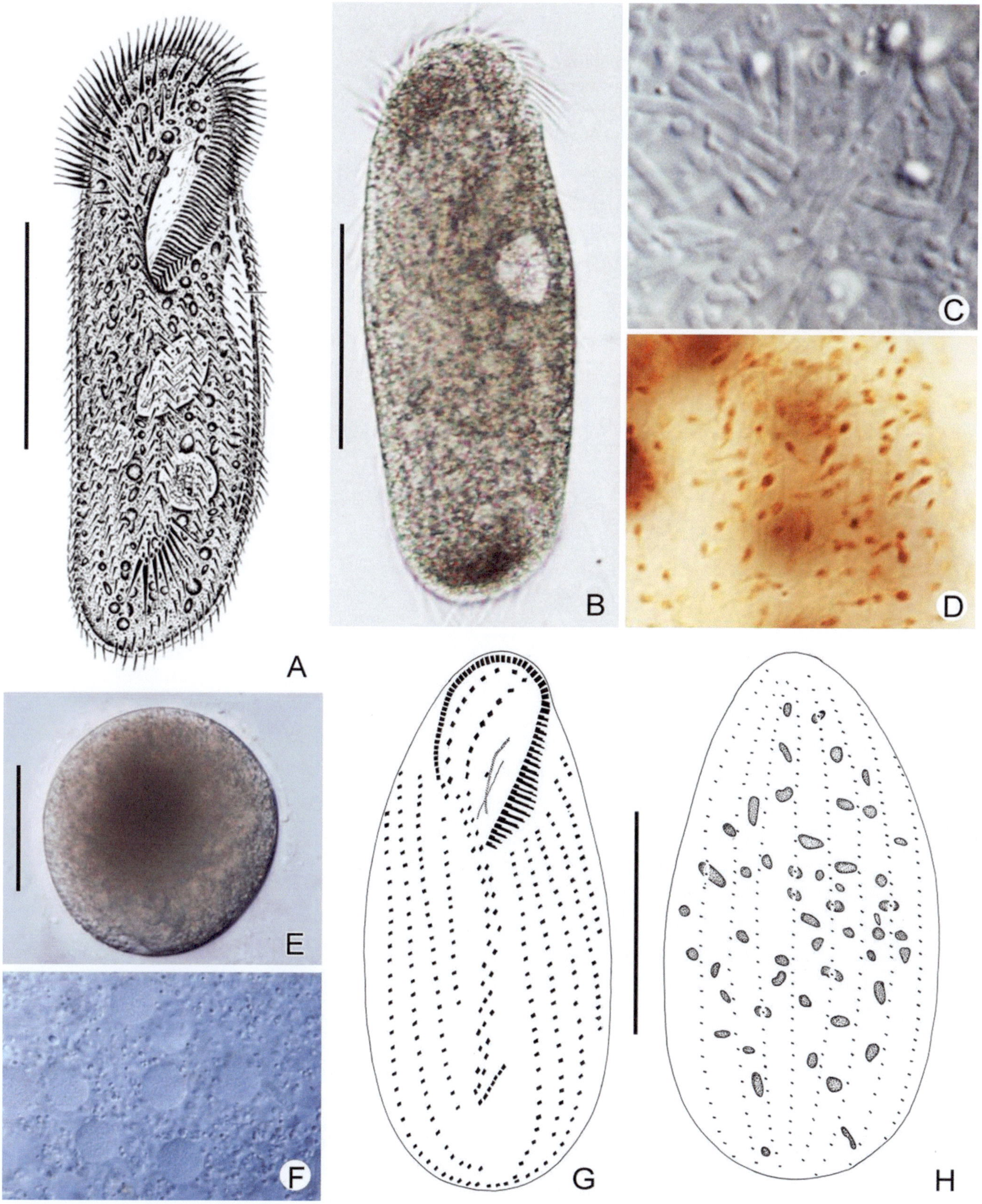


**Fig. 97B** *Pseudourostyla cristata*
(**a**, **b**) Ventral views of live cells. (**c**, **d**) Extrusomes. (**e**) Cyst. (**f**) Macronuclear nodules in vivo. (**g**, **h**) Ventral (**g**) and dorsal (**h**) views of ciliature. Scale bars: 100 μm (**a**, **b**, **g**, **h**) and 10 μm (**e**)

## Reference

Chen X, Li Z, Hu X, Kusuoka Y (2010) Morphology, morphogenesis and gene sequence of a freshwater ciliate, *Pseudourostyla cristata* (Ciliophora, Urostyloidea) from the ancient Lake Biwa, Japan. Eur J Protistol 46:43–60

Class Spirotrichea Bütschli, 1889
Order Urostylida Jankowski, 1979
Family Pseudourostylidae Jankowski, 1979
*Pseudourostyla nova* Wiackowski, 1988

**Ecological feature:** estuary
**Temperature:** 27.0 °C
**pH:** no data available
**Salinity:** 4.3‰

**Fig. 98A** Sample site

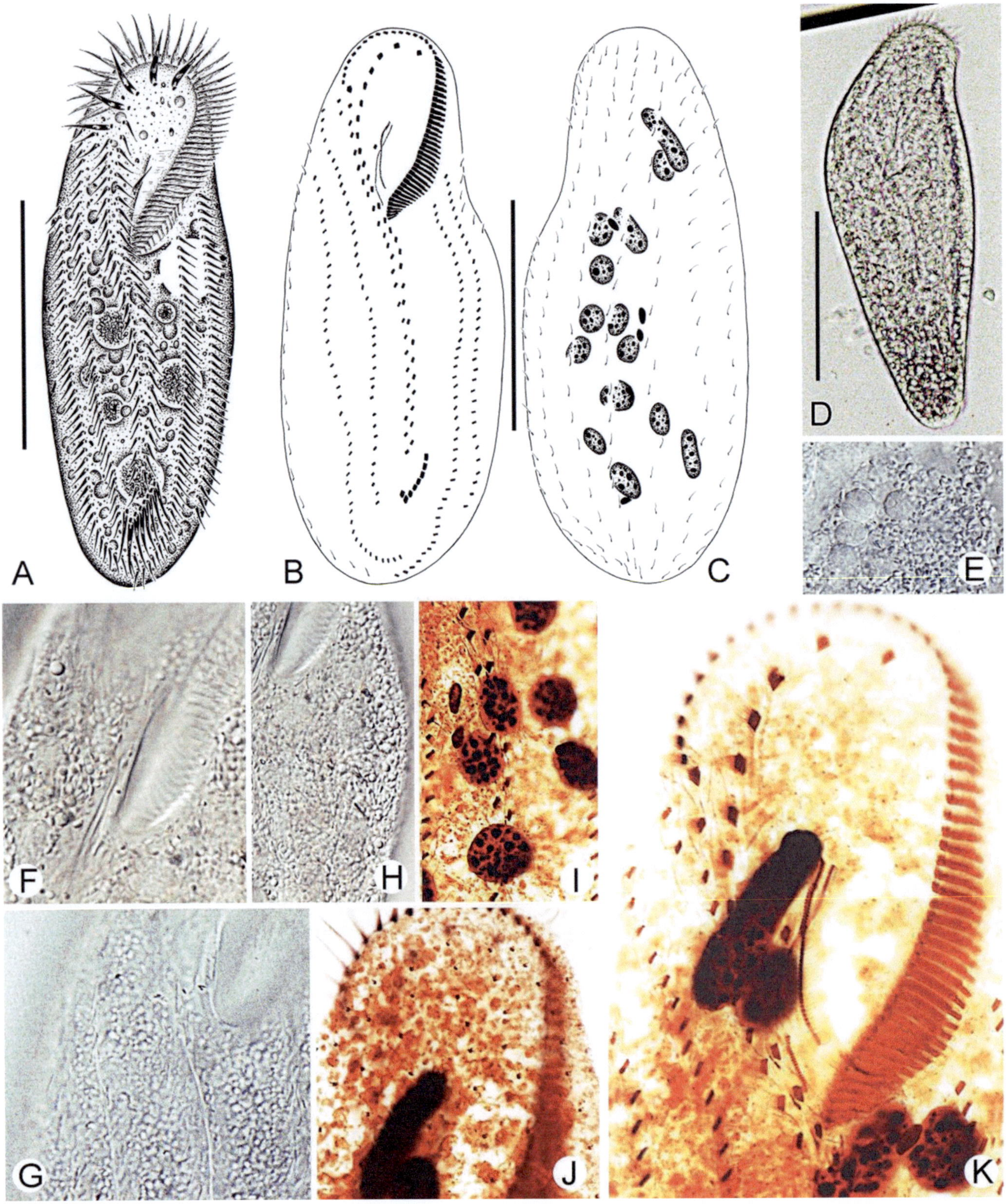


**Fig. 98B** *Pseudourostyla nova*
(**a**, **d**) Ventral views of live individuals. (**b**, **c**) Ventral (**b**) and dorsal (**c**) views of ciliature. (**e**) Macronuclear nodules. (**f**–**h**) Cortical granules. (**i**–**k**) Ventral (**i**, **k**) and dorsal (**j**) views of ciliature. Scale bars: 60 μm (**a**, **d**) and 100 μm (**b**, **c**)

## Reference

Chen X, Miao M, Ma H, Al-Rasheid KAS, Xu K, Lin X (2014) Morphology, ontogeny, and phylogeny of two brackish urostylid ciliates (Protist, Ciliophora, Hypotricha). J Eukaryot Microbiol 61:594–610

Class Spirotrichea Bütschli, 1889
Order Urostylida Jankowski, 1979
Family Pseudourostylidae Jankowski, 1979
*Pseudourostyla subtropica* Chen et al., 2014

**Ecological feature:** mangrove
**Temperature:** 16.5 °C
**pH:** no data available
**Salinity:** 4.5‰

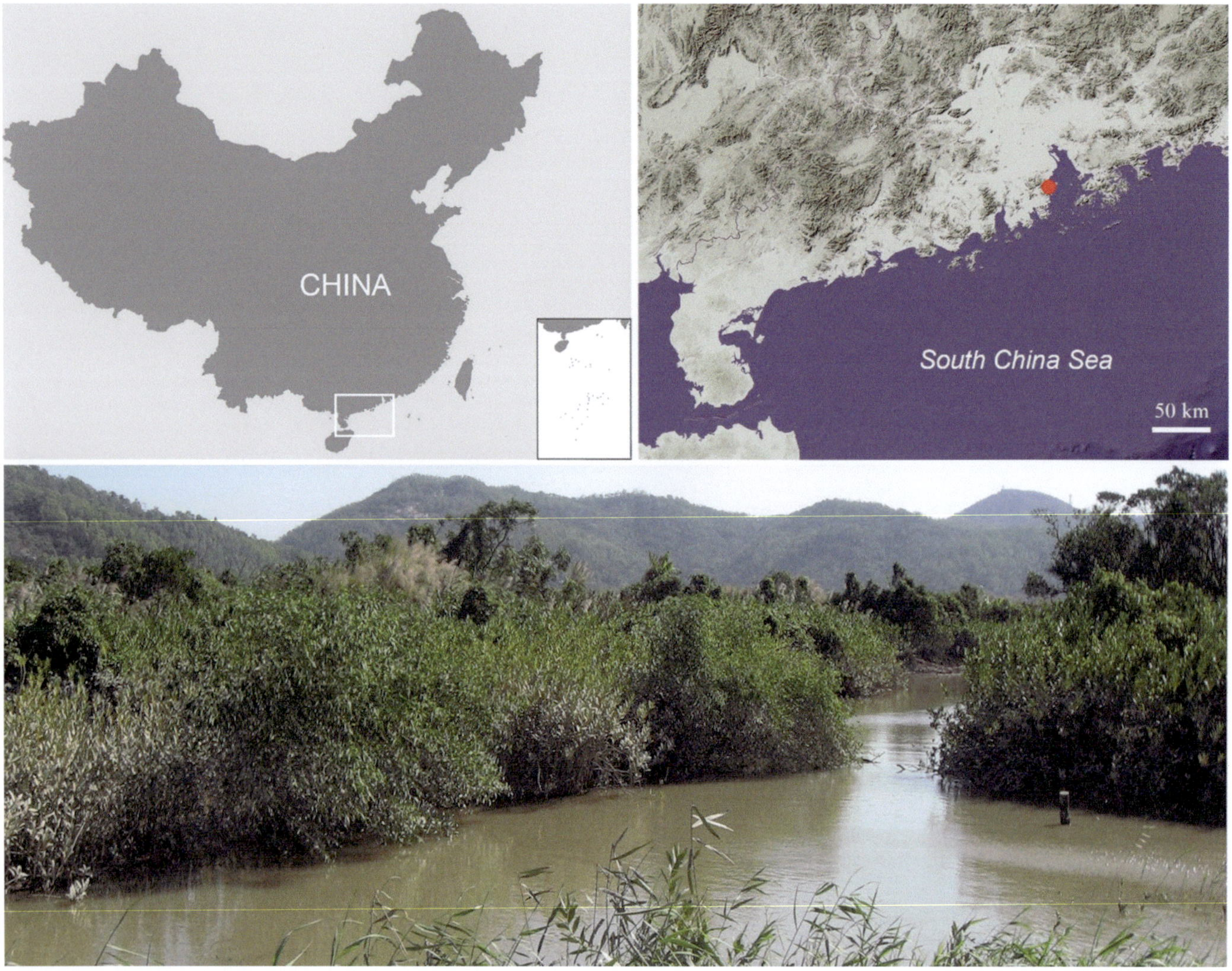


**Fig. 99A** Sample site

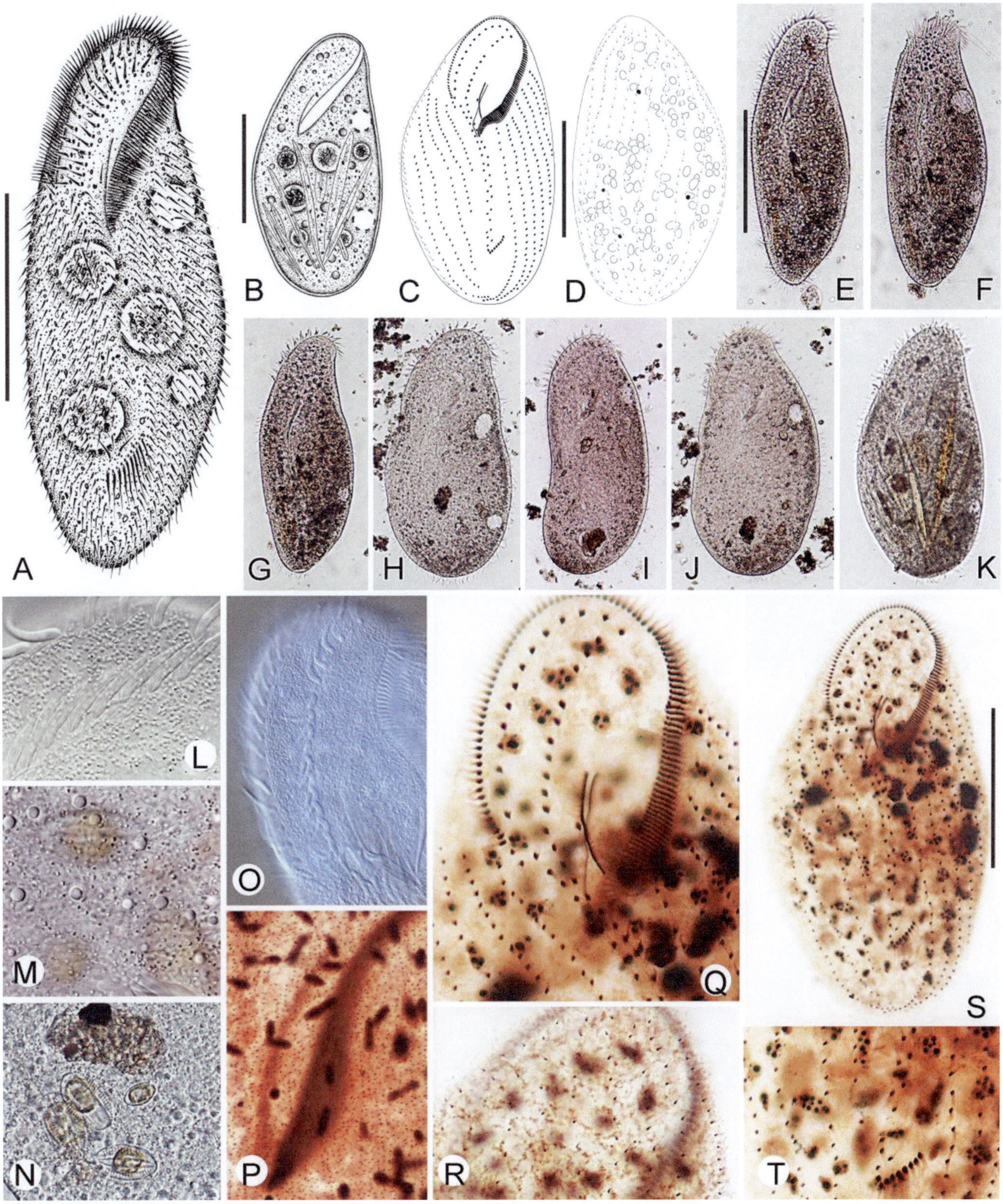


**Fig. 99B** *Pseudourostyla subtropica*
(**a**, **b**, **e**–**k**) Ventral view of live individuals. (**c**, **d**, **q**–**t**) Ventral (**c**, **q**, **s**, **t**) and dorsal (**d**, **r**) views of ciliature. (**l**, **m**, **p**) Arrangement of cortical granules. (**n**) Food vacuoles. (**o**) Arrangement of frontal cirri. Scale bars: 150 μm

## Reference

Chen X, Miao M, Ma H, Al-Rasheid KAS, Xu K, Lin X (2014) Morphology, ontogeny, and phylogeny of two brackish urostylid ciliates (Protist, Ciliophora, Hypotricha). J Eukaryot Microbiol 61:594–610

Class Spirotrichea Bütschli, 1889
Order Urostylida Jankowski, 1979
Family Urostylidae Bütschli, 1889
*Anteholosticha manca* (Kahl, 1932) Berger, 2003

**Ecological feature:** mangrove
**Water temperature:** 25.0 °C (Zhanjiang); 18.0 °C (Zhuhai)
**pH:** 6.6; 6.0
**Salinity:** 12.0‰; 4.6‰

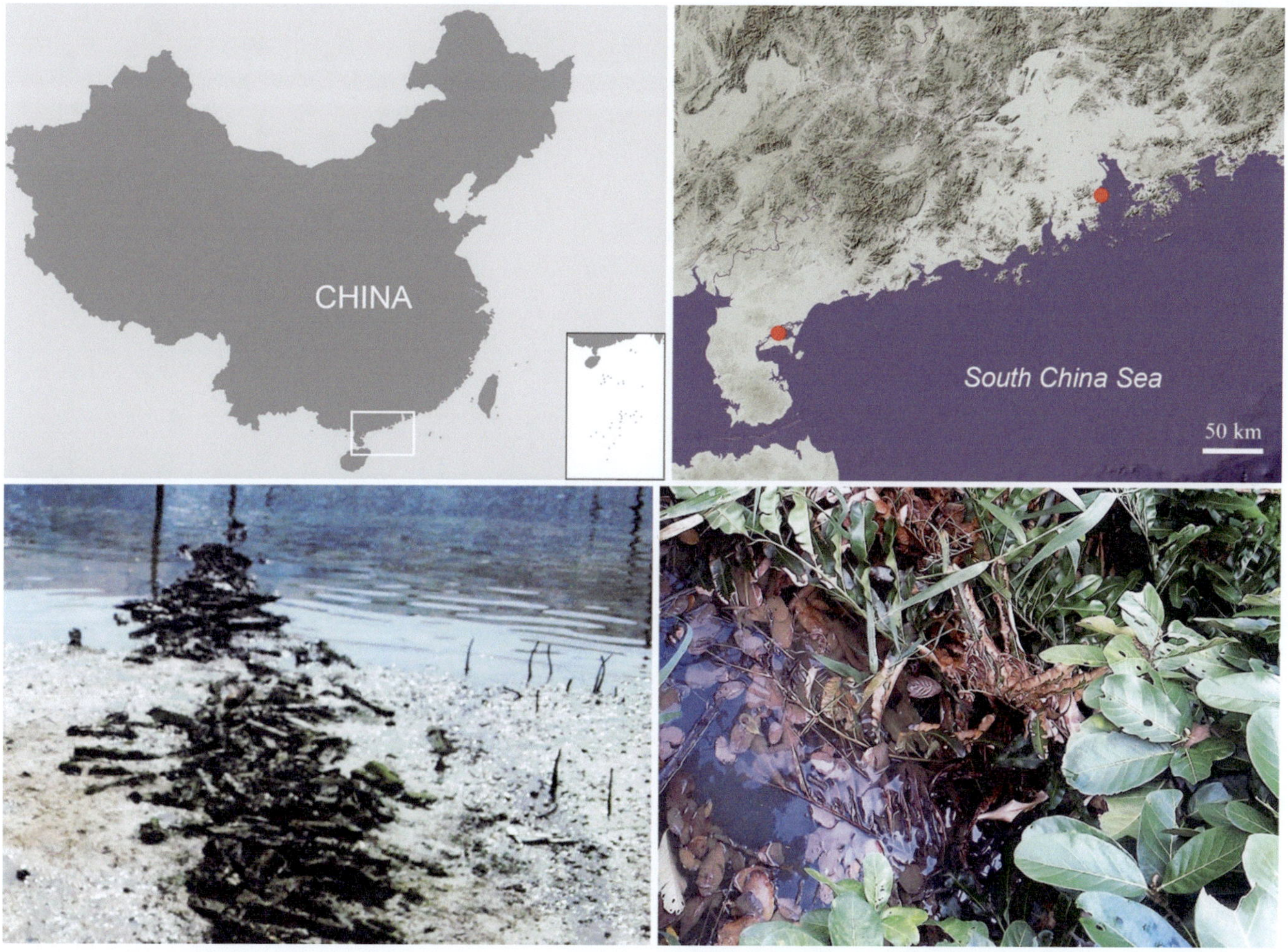


**Fig. 100A** Sample site

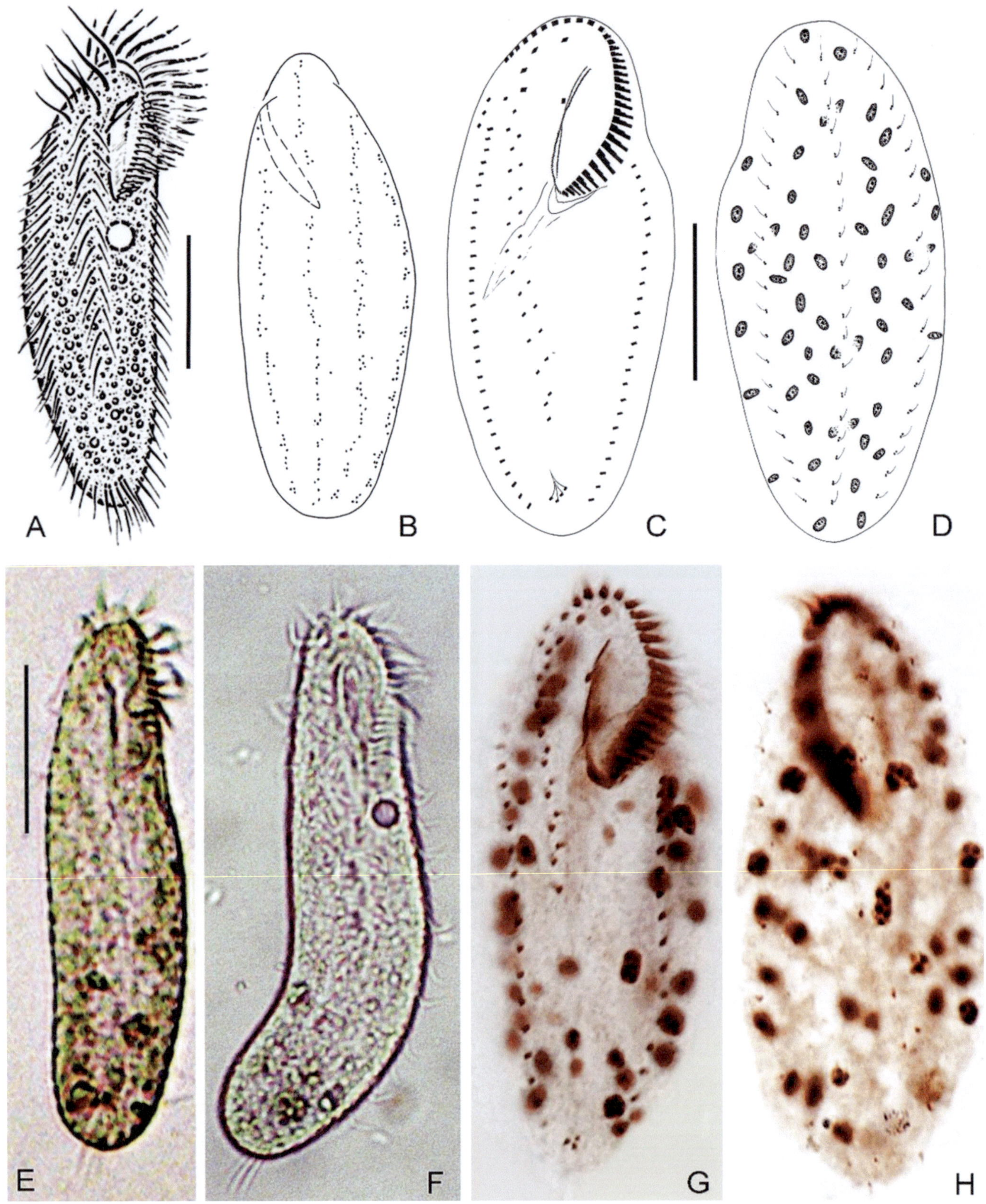


**Fig. 100B** *Anteholosticha manca*
(**a**, **e**, **f**) Ventral views of live cells. (**b**) Dorsal view showing arrangement of cortical granules. (**c**, **d**, **g**, **h**) Ventral (**c**, **g**) and dorsal (**d**, **h**) views of ciliature. Scale bars: 60 μm

## Reference

Song WB, Wilbert N (1997) Morphological investigations on some free living ciliates (Protozoa, Ciliophora) from China Sea with description of a new hypotrichous genus, *Hemigastrostyla* nov. gen. Arch Protistenk 148:413–444

Class Spirotrichea Bütschli, 1889
Order Urostylida Jankowski, 1979
Family Urostylidae Bütschli, 1889
*Anteholosticha paramanca* Fan et al., 2014

**Ecological feature:** mangrove
**Water temperature:** 24.0 °C (Zhanjiang); 26.0 °C (Shenzhen)
**pH:** 7.3; no data available
**Salinity:** 17.0‰; 26.0‰

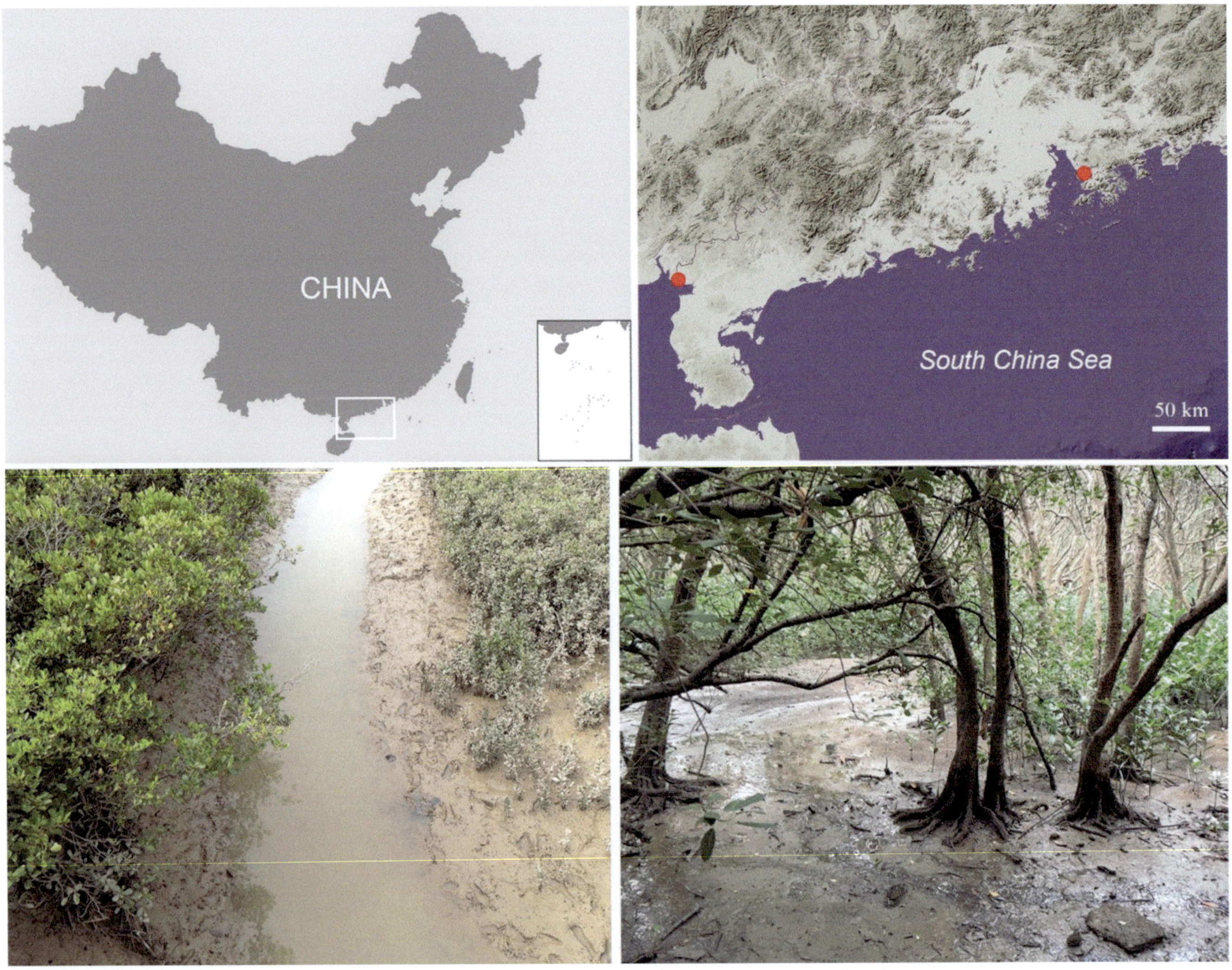


**Fig. 101A** Sample site

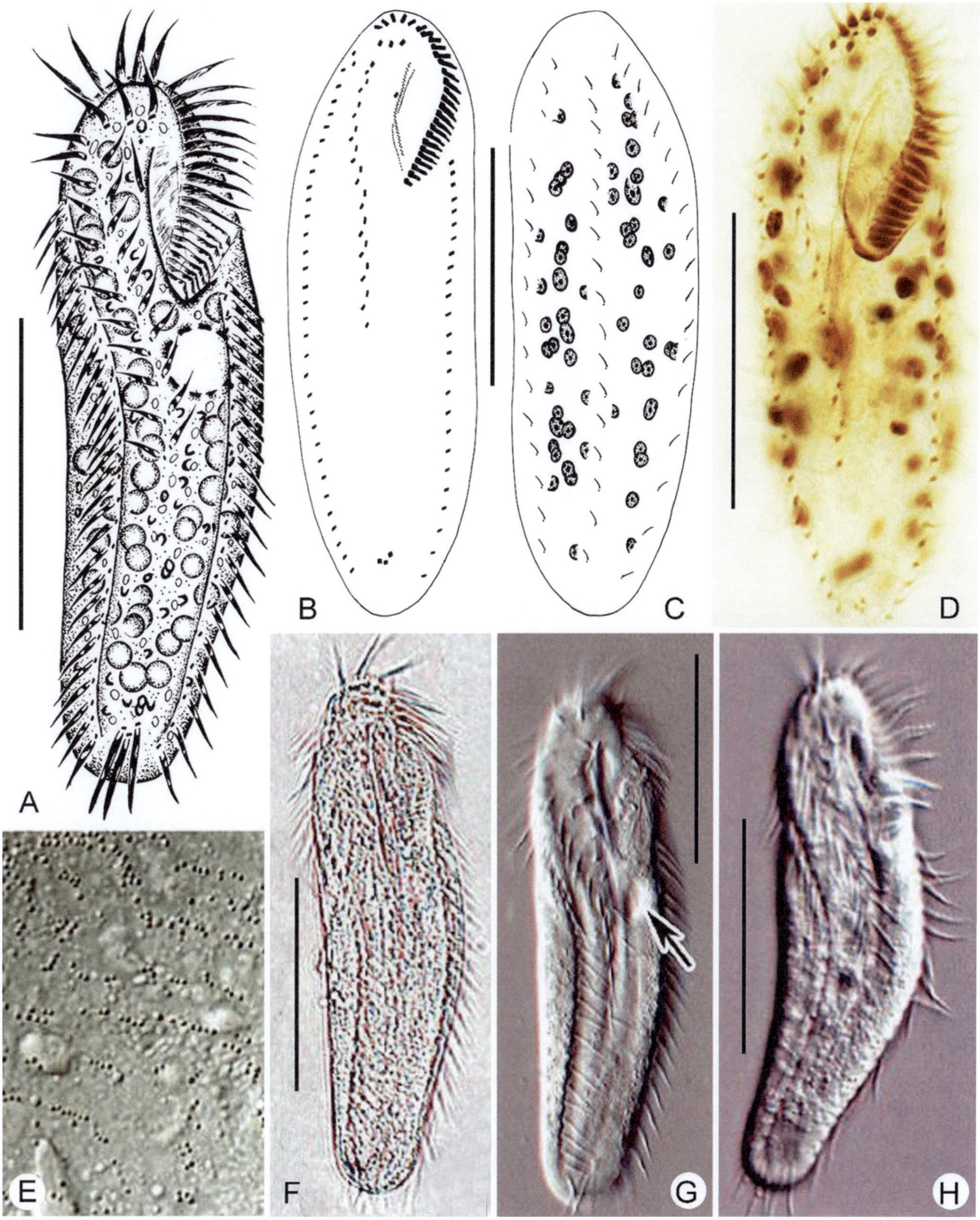


**Fig. 101B** *Anteholosticha paramanca*
(**a**, **f–h**) Ventral views of live individuals. (**b–d**) Ventral (**b**, **d**) and dorsal (**c**) views of ciliature. (**e**) Cortical granules. Scale bars: 50 μm

## Reference

Fan Y, Pan Y, Huang J, Lin X, Hu X, Warren A (2014) Molecular phylogeny and taxonomy of two novel brackish water hypotrich ciliates, with the establishment of a new genus, *Antiokeronopsis* gen. n. (Ciliophora, Hypotrichia). J Eukaryot Microbiol 61:449–462

Class Spirotrichea Bütschli, 1889
Order Urostylida Jankowski, 1979
Family Urostylidae Bütschli, 1889
*Apobakuella fusca* Jiang et al., 2013

**Ecological feature:** mangrove
**Water temperature:** 21.0 °C
**pH:** no data available
**Salinity:** 14.0‰

**Fig. 102A** Sample site

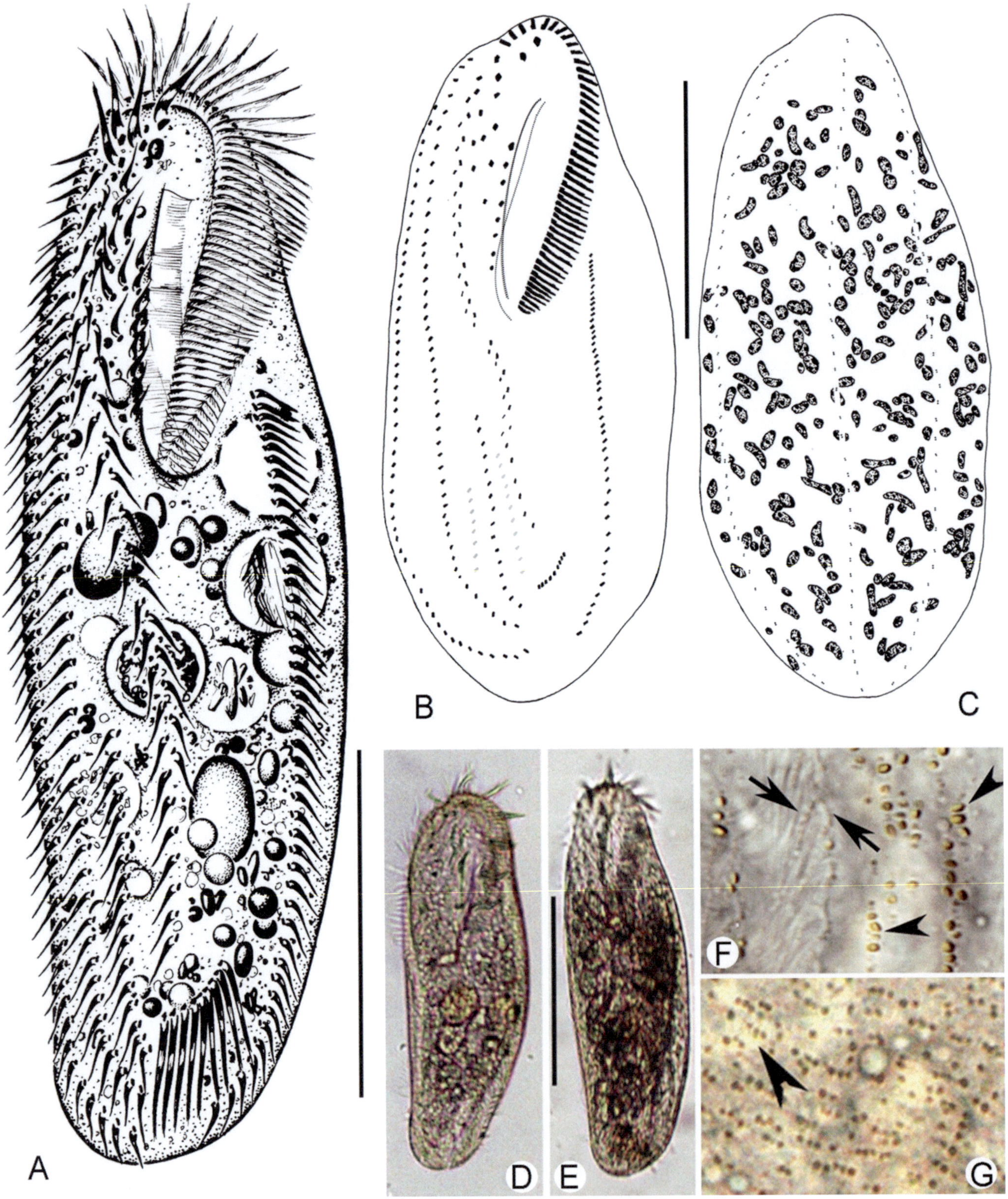


**Fig. 102B** *Apobakuella fusca*
(**a**, **d**, **e**) Ventral view of live cells. (**b**, **c**) Ventral and dorsal views of ciliature of the same specimen. (**f**, **g**) Cortical granules on ventral (**f**) and dorsal (**g**) sides; arrows show the smaller ones; arrowheads indicate the bigger ones. Scale bars: 60 μm (**a**), 80 μm (**b**, **c**), and 70 μm (**e**)

## Reference

Jiang J, Huang J, Li L, Shao C, Al-Farraj SA, Chen Z (2013) Morphology, ontogeny, and molecular phylogeny of two novel bakuellid-like hypotrichs (Ciliophora: Hypotricha), with establishment of two new genera. Eur J Protistol 49:78–92

Class Spirotrichea Bütschli, 1889
Order Urostylida Jankowski, 1979
Family Urostylidae Bütschli, 1889
*Apokeronopsis crassa* (Claparède & Lachmann, 1858) Shao et al., 2007

**Ecological feature:** mangrove
**Water temperature:** 25.0 °C
**pH:** 6.8
**Salinity:** 17.0‰

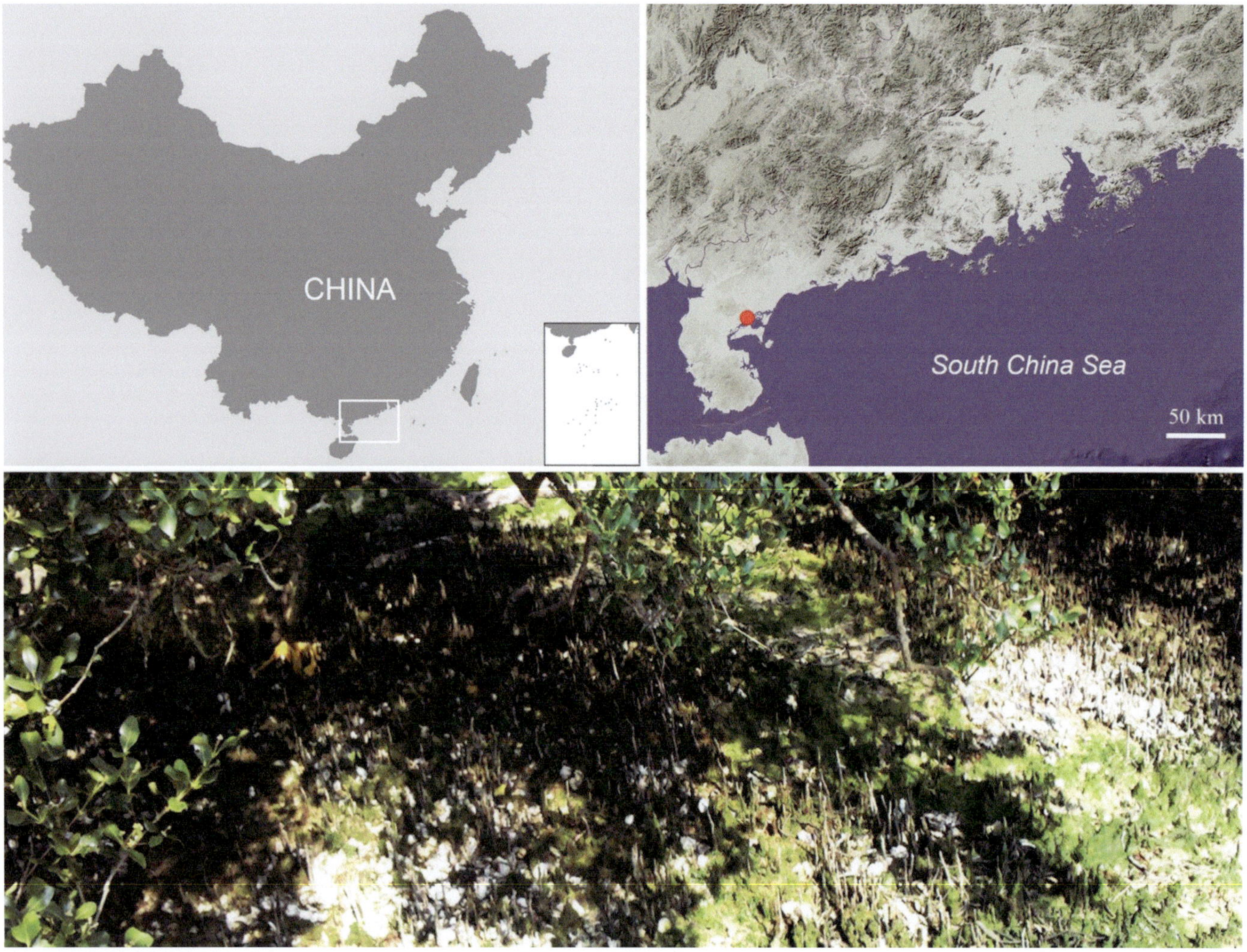


**Fig. 103A** Sample site

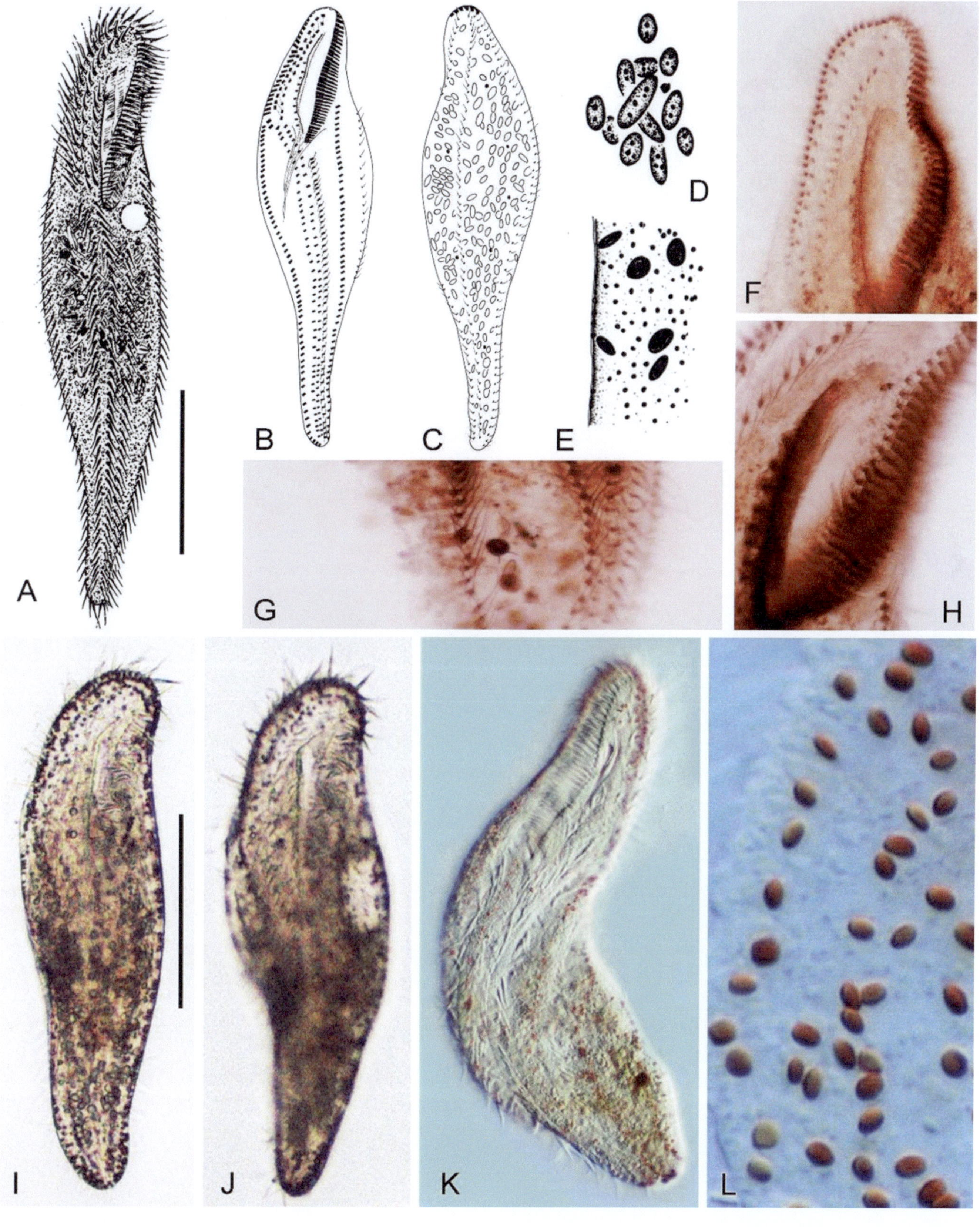


**Fig. 103B** *Apokeronopsis crassa*
(**a**, **i**, **j**) Ventral views of live cells. (**b**, **c**) Ventral and dorsal views of ciliature. (**d**) Macronuclear nodules. (**e**, **l**) Two types of cortical granules. (**f–h**) Ventral views of ciliature, showing oral apparatus and frontal cirri. (**k**) Twisted cell, showing arrangement of cortical granules. Scale bars: 50 μm

## Reference

Shao C, Hu X, Warren A, Al-Rasheid KAS, Al-Qurashy SA, Song WB (2007) Morphogenesis in the marine spirotrichous ciliate *Apokeronopsis crassa* (Claparède & Lachmann, 1858) nov. comb. (Ciliophora: Stichotrichia), with the establishment of a new genus, *Apokeronopsis* nov. gen. and redefinition of the genus *Thigmokeronopsis*. J Eukaryot Microbiol 54:392–401

Class Spirotrichea Bütschli, 1889
Order Urostylida Jankowski, 1979
Family Urostylidae Bütschli, 1889
*Apokeronopsis sinica* Liu et al., 2009

**Ecological feature:** intertidal zone
**Water temperature:** 24.0 °C
**pH:** 8.1
**Salinity:** 33.5‰

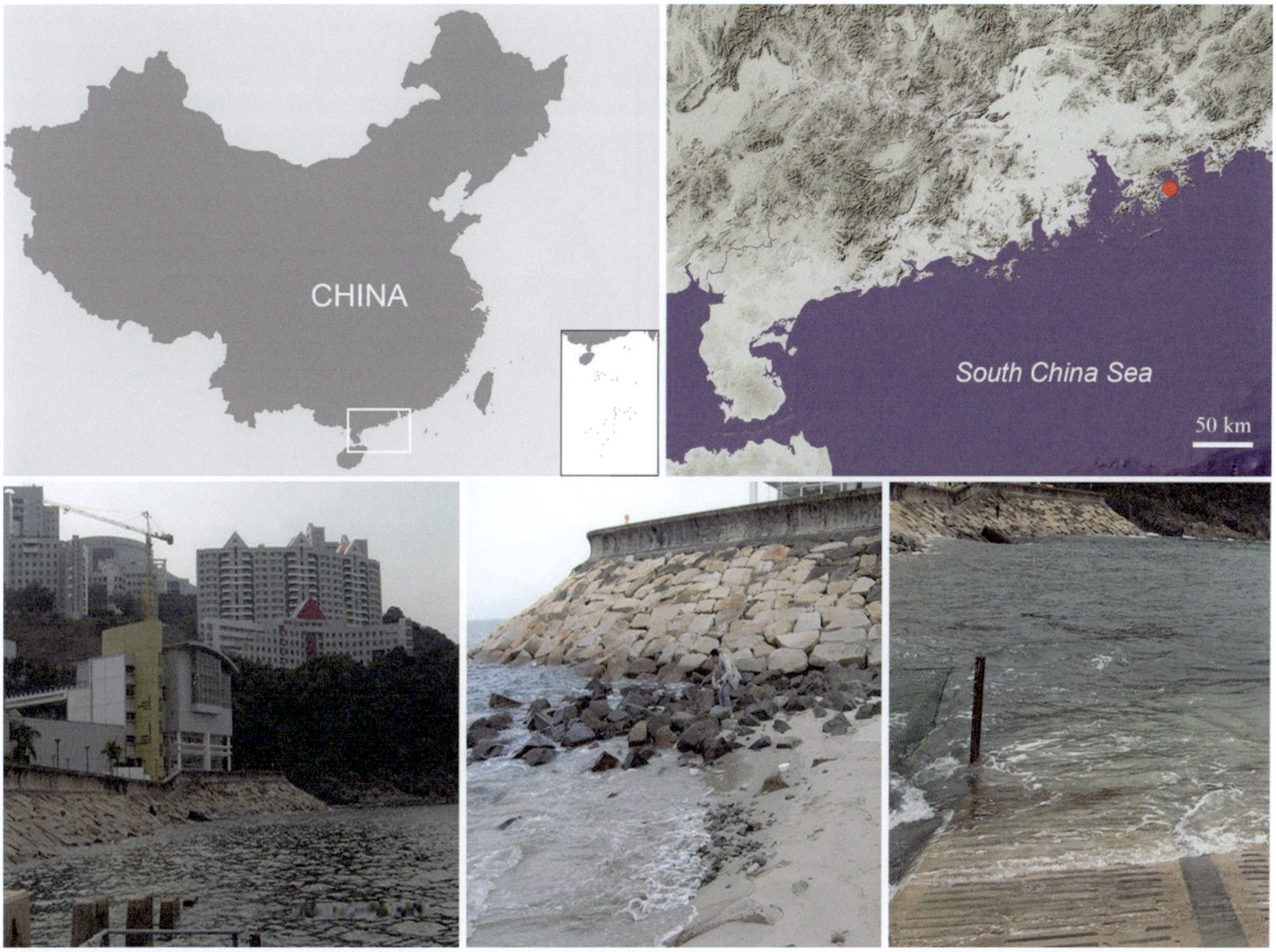


**Fig. 104A** Sample site

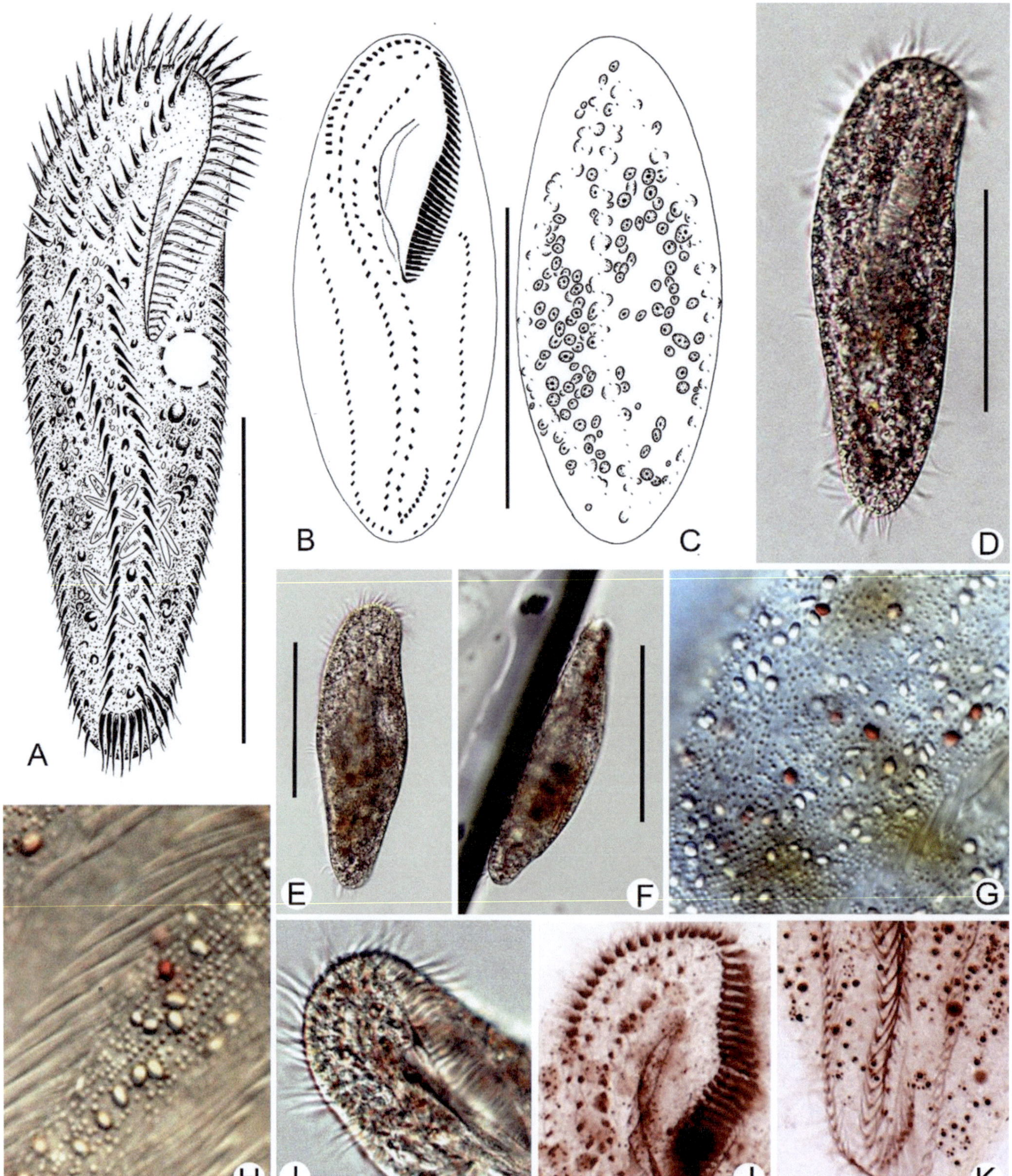


**Fig. 104B** *Apokeronopsis sinica*
(**a**, **d**, **e**) Ventral view of live individuals. (**b**, **c**) Ventral (**b**) and dorsal (**c**) views of iciliature. **f**. Lateral view of live cell. (**g**, **h**) Two types of cortical granules. (**i**) Ventral view of the anterior portion showing buccal field. (**j**, **k**) Ventral views of the anterior (**j**) and posterior (**k**) portions to show ciliature. Scale bars: 100 μm

## Reference

Liu W, Li J, Gao S, Shao C, Gong J, Lin X, Liu H, Song WB (2009) Morphological studies and molecular data on a new marine ciliate, *Apokeronopsis sinica* n. sp. (Ciliophora: Urostylida), from the South China Sea. Zootaxa 2005:57–66

Class Spirotrichea Bütschli, 1889
Order Urostylida Jankowski, 1979
Family Urostylidae Bütschli, 1889
*Apokeronopsis wrighti* Long et al., 2008

**Ecological feature:** coastal water
**Water temperature:** 24.0 °C; 24.0 °C;
**pH:** no data available; 8.1
**Salinity:** 34.0‰; 33.5‰

**Fig. 105A** Sample site

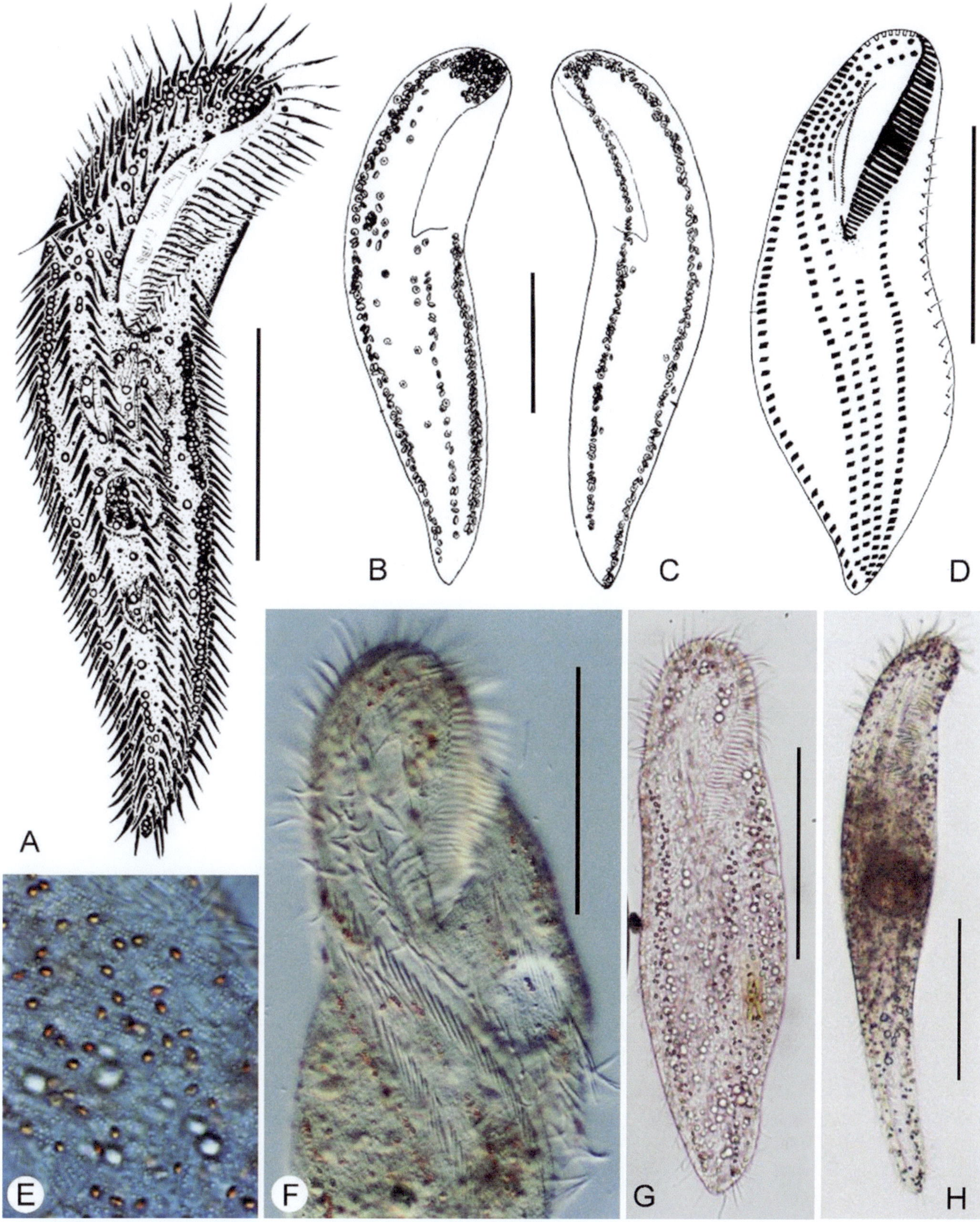


**Fig. 105B** *Apokeronopsis wrighti*
(**a**, **h**) Ventral views of live individuals. (**b**, **c**, **g**) Arrangement of blood-cell-shaped cortical granules on ventral (**b**, **g**) and dorsal (**c**) sides. (**d**) Ventral view of ciliature. (**e**) Two types of cortical granules. (**f**) Anterior part of ventral side, showing oral membranelles and cirri. Scale bars: 60 μm

## Reference

Long H, Liu H, Liu W, Miao M, Hu X, Lin X, Song WB (2008) Two new ciliates from Hong Kong coastal water: *Orthodonella sinica* n. sp. and *Apokeronopsis wrighti* n. sp. (Protozoa: Ciliophora). J Eukaryot Microbiol 55:321–330

Class Spirotrichea Bütschli, 1889
Order Urostylida Jankowski, 1979
Family Urostylidae Bütschli, 1889
*Arcuseries scutellum* (Cohn, 1866) Huang et al., 2014

**Ecological feature:** mangrove; sandy sediment
**Water temperature:** 24.0 °C; 25.0 °C
**pH:** 7.10; 7.80
**Salinity:** 30.0‰; 28.0‰

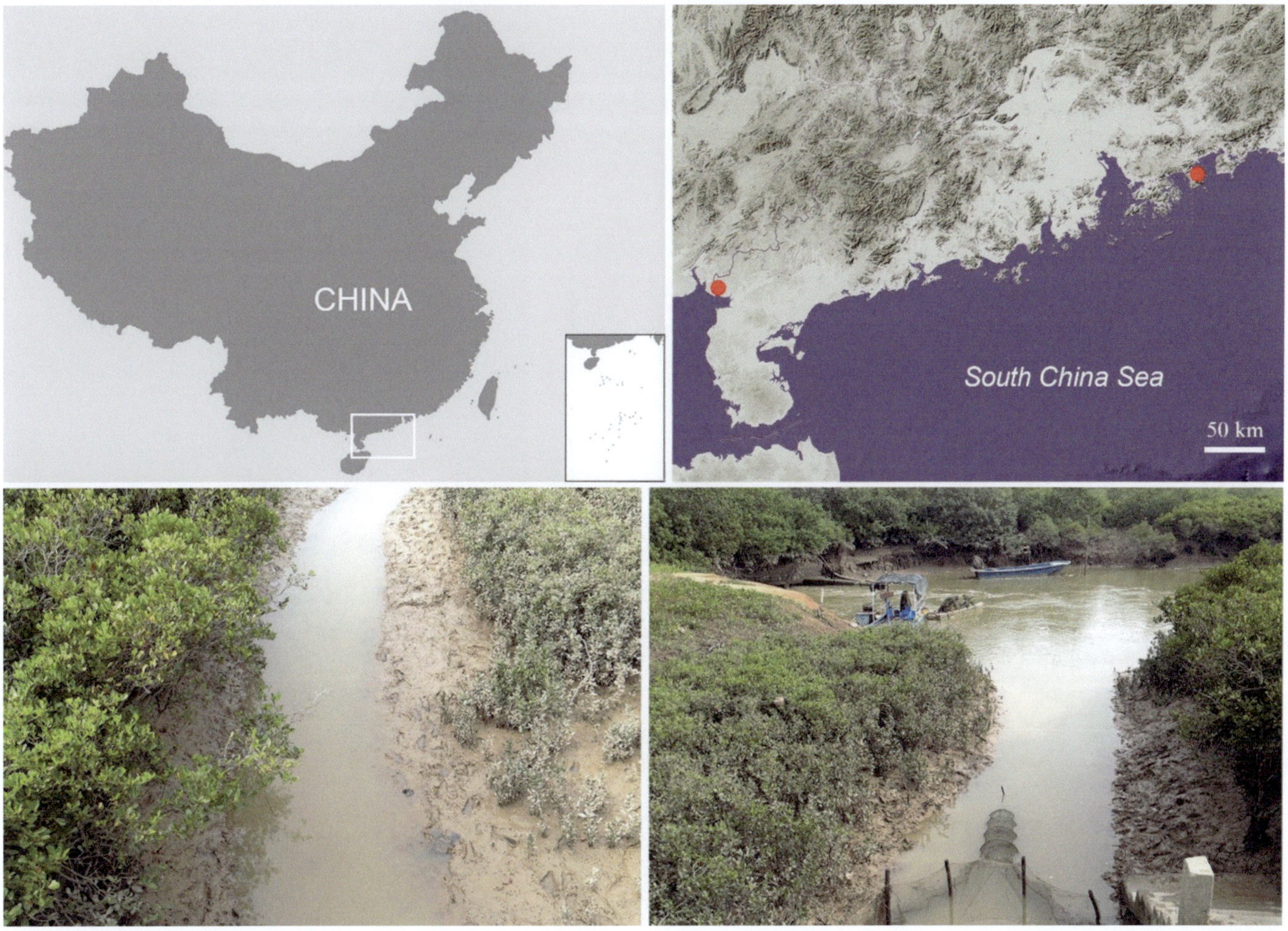


**Fig. 106A** Sample site

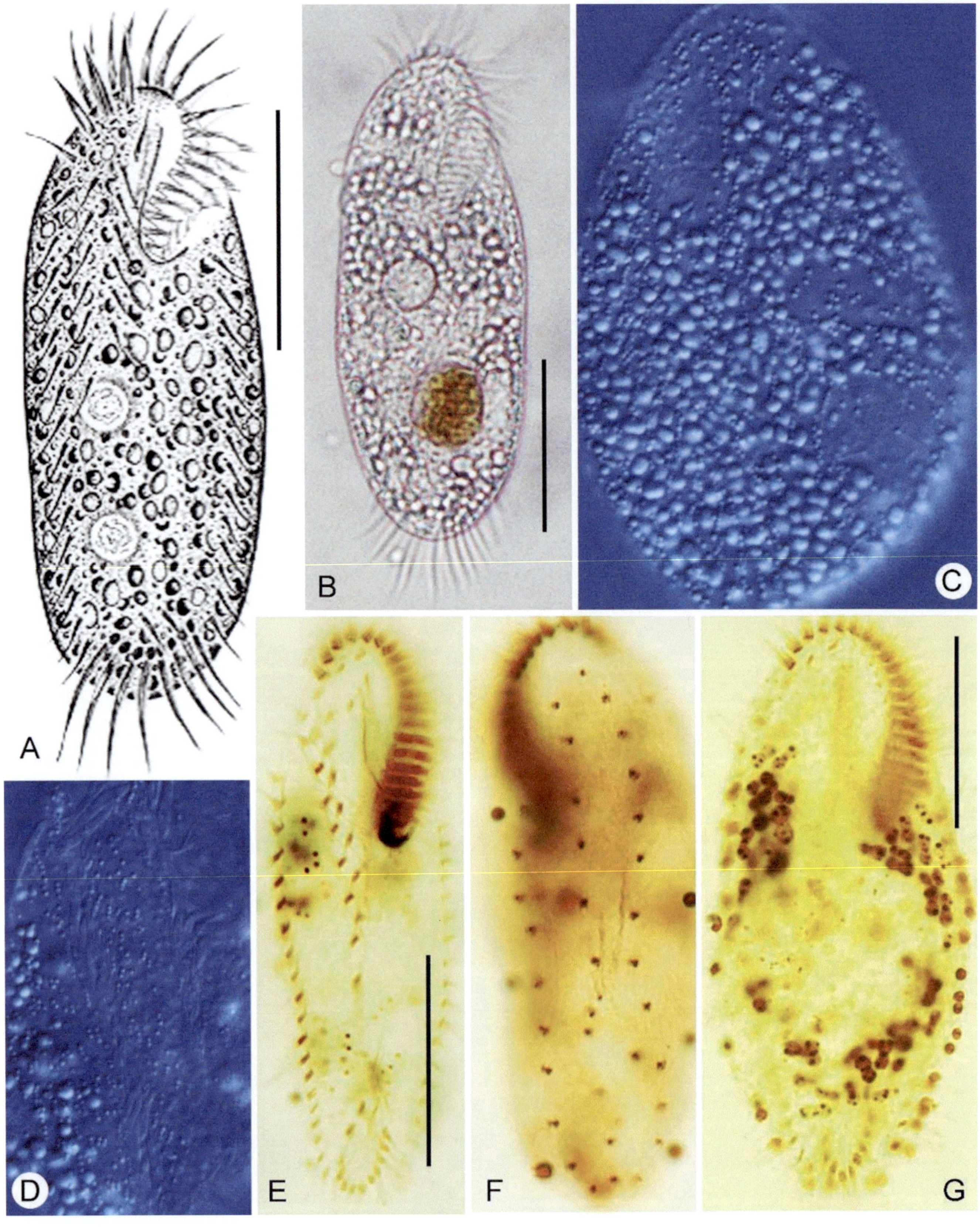


**Fig. 106B** *Arcuseries scutellum*
(**a**, **b**) Ventral views of live individuals. (**c**, **d**) Arrangement of cortical granules on ventral (**d**) and dorsal (**c**) sides. (**e**, **f**) Ventral (**e**) and dorsal (**f**) views of ciliature. (**g**) Macronuclear nodules and transverse cirri. Scale bars: 20 μm

## References

Chen XR, Gao S, Song WB, Al-Rasheid KAS, Warren A, Gong J, Lin X (2010) *Parabirojimia multinucleata* spec. nov. and *Anteholosticha scutellum* (Cohn, 1866) Berger, 2003, marine ciliates (Ciliophora, Hypotrichida) from tropical waters in southern China, with notes on their small-subunit rRNA gene sequences. Int J Syst Evol Microbiol 60:234–243

Huang J, Chen Z, Song WB, Berger H (2014) Three-gene based phylogeny of the Urostyloidea (Protista, Ciliophora, Hypotricha), with notes on classification of some core taxa. Mol Phylogenet Evol 70:337–347

Class Spirotrichea Bütschli, 1889
Order Urostylida Jankowski, 1979
Family Urostylidae Bütschli, 1889
*Arcuseries warreni* (Song & Wilbert, 1997) Huang et al., 2014

**Ecological feature:** beach
**Temperature:** 18.0 °C
**pH:** no data available
**Salinity:** 30.0‰

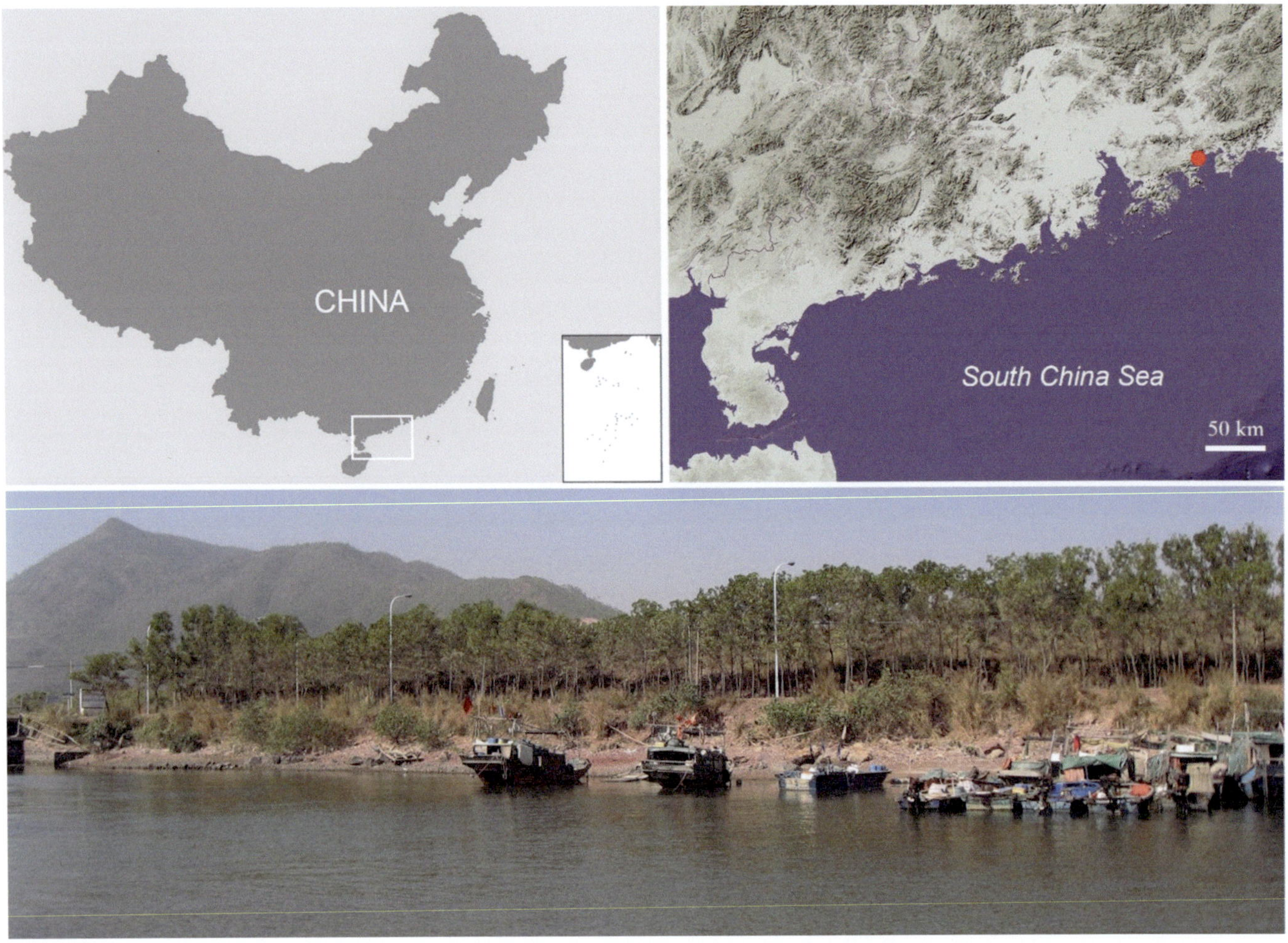


**Fig. 107A** Sample site

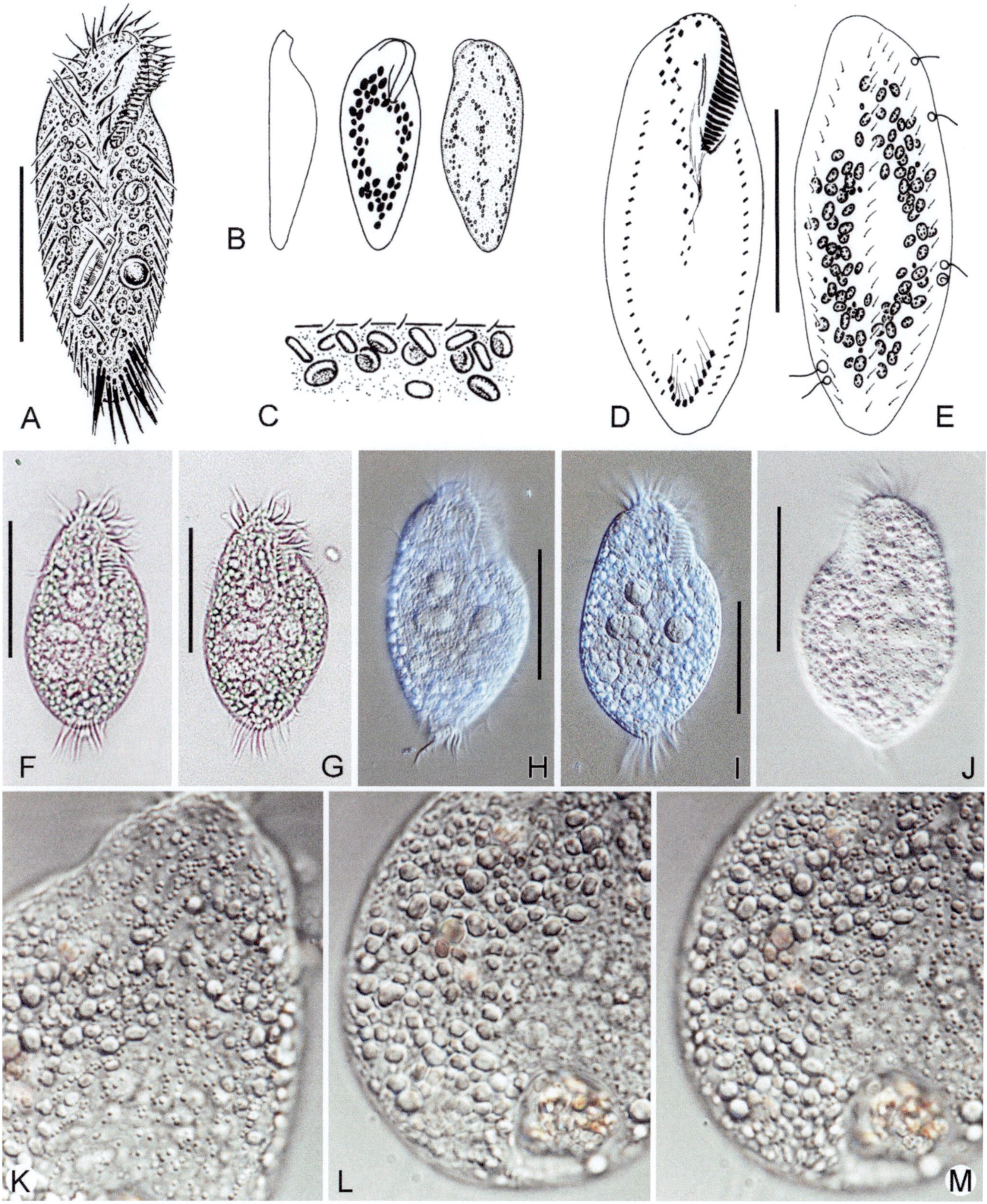


**Fig. 107B** *Arcuseries warreni*
(**a**, **f**–**j**) Views in vivo. (**b**) Lateral view of body, arrangement of macronuclei, and cortical granules on dorsal side. (**c**, **k**–**m**) Cortical granules. (**d**, **e**) Ventral and dorsal views of ciliature. Scale bars: 50 μm

## References

Huang J, Chen Z, Song WB, Berger H (2014) Three-gene based phylogeny of the Urostyloidea (Protista, Ciliophora, Hypotricha), with notes on classification of some core taxa. Mol Phylogenet Evol 70:337–347

Song WB, Wilbert N (1997) Morphological studies on some free living ciliates (Ciliophora: Heterotrichida, Hypotrichida) from marine biotopes in Qingdao, China, with descriptions of three new species: *Holosticha warreni* nov. spec., *Tachysoma ovata* nov. spec. and *T. dragescoi* nov. spec. Eur J Protistol 33:48–62

Class Spirotrichea Bütschli, 1889
Order Urostylida Jankowski, 1979
Family Urostylidae Bütschli, 1889
*Bakuella agamalievi* Borror & Wicklow, 1983

**Ecological feature:** mangrove
**Water temperature:** 24.0 °C
**pH:** no data available
**Salinity:** 13.0‰

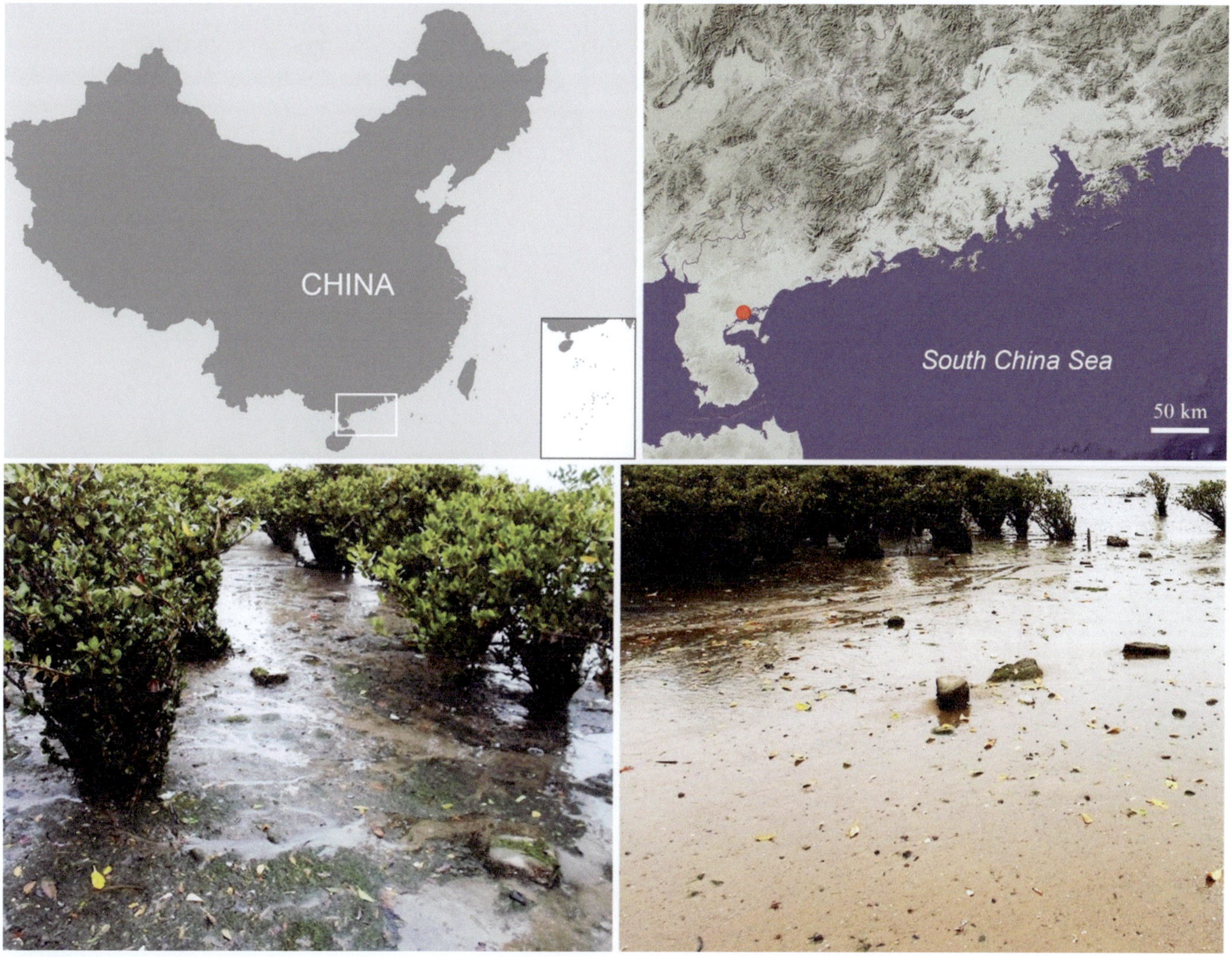


**Fig. 108A** Sample site

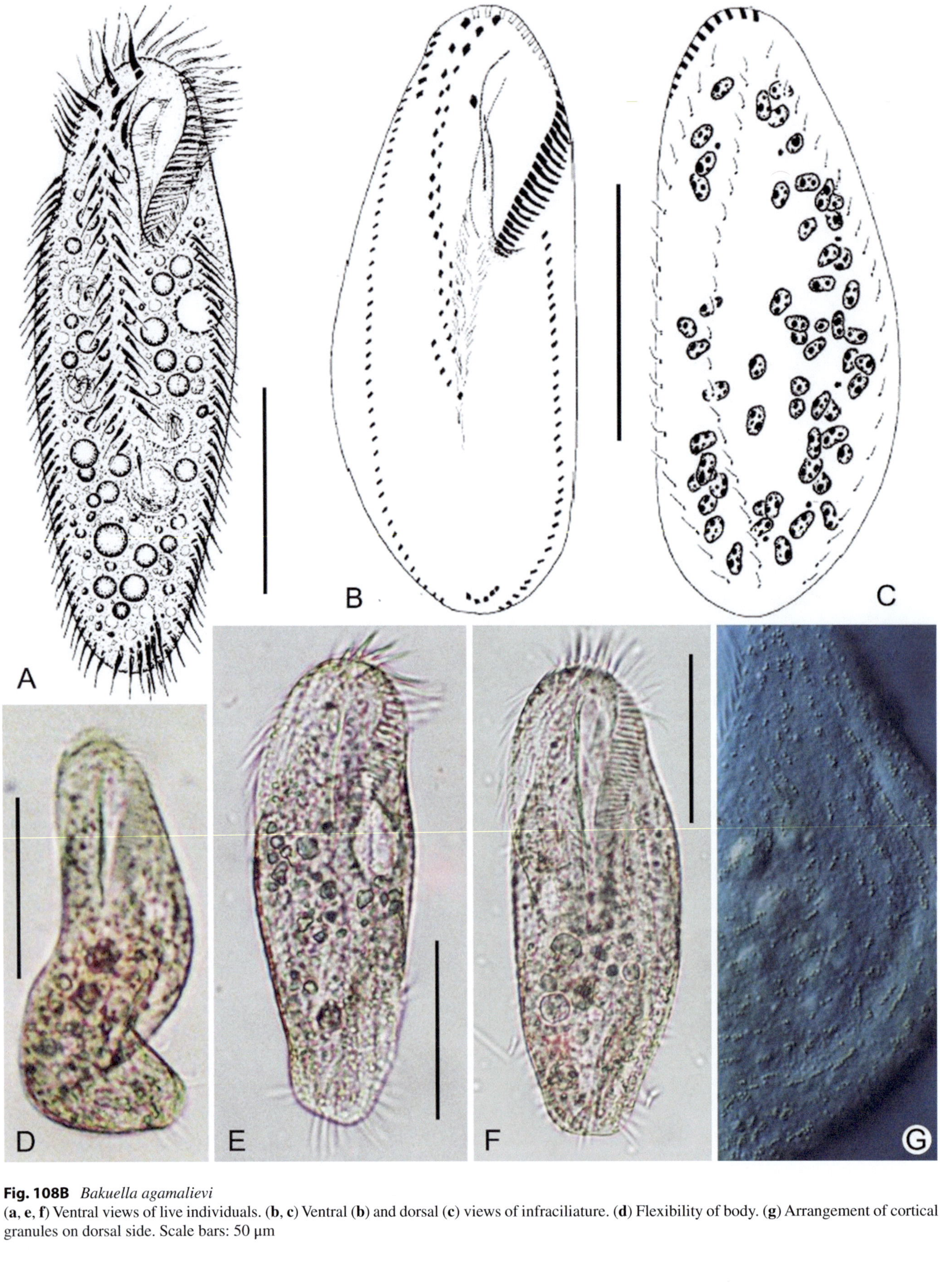


**Fig. 108B** *Bakuella agamalievi*
(**a**, **e**, **f**) Ventral views of live individuals. (**b**, **c**) Ventral (**b**) and dorsal (**c**) views of infraciliature. (**d**) Flexibility of body. (**g**) Arrangement of cortical granules on dorsal side. Scale bars: 50 μm

## Reference

Song WB, Wilbert N, Warren A (2002) New contribution to the morphology and taxonomy of four marine hypotrichous ciliates from Qingdao, China (protozoa: Ciliophora). Acta Protozool 41:145–162

Class Spirotrichea Bütschli, 1889
Order Urostylida Jankowski, 1979
Family Urostylidae Bütschli, 1889
*Bakuella subtropica* Chen et al., 2013

**Ecological feature:** estuary (Guangzhou); mangrove (Zhanjiang)
**Temperature:** 27.0 °C; 20.0 °C
**pH:** no data available; 6.8
**Salinity:** 4.3‰; 4.0‰

**Fig. 109A** Sample site

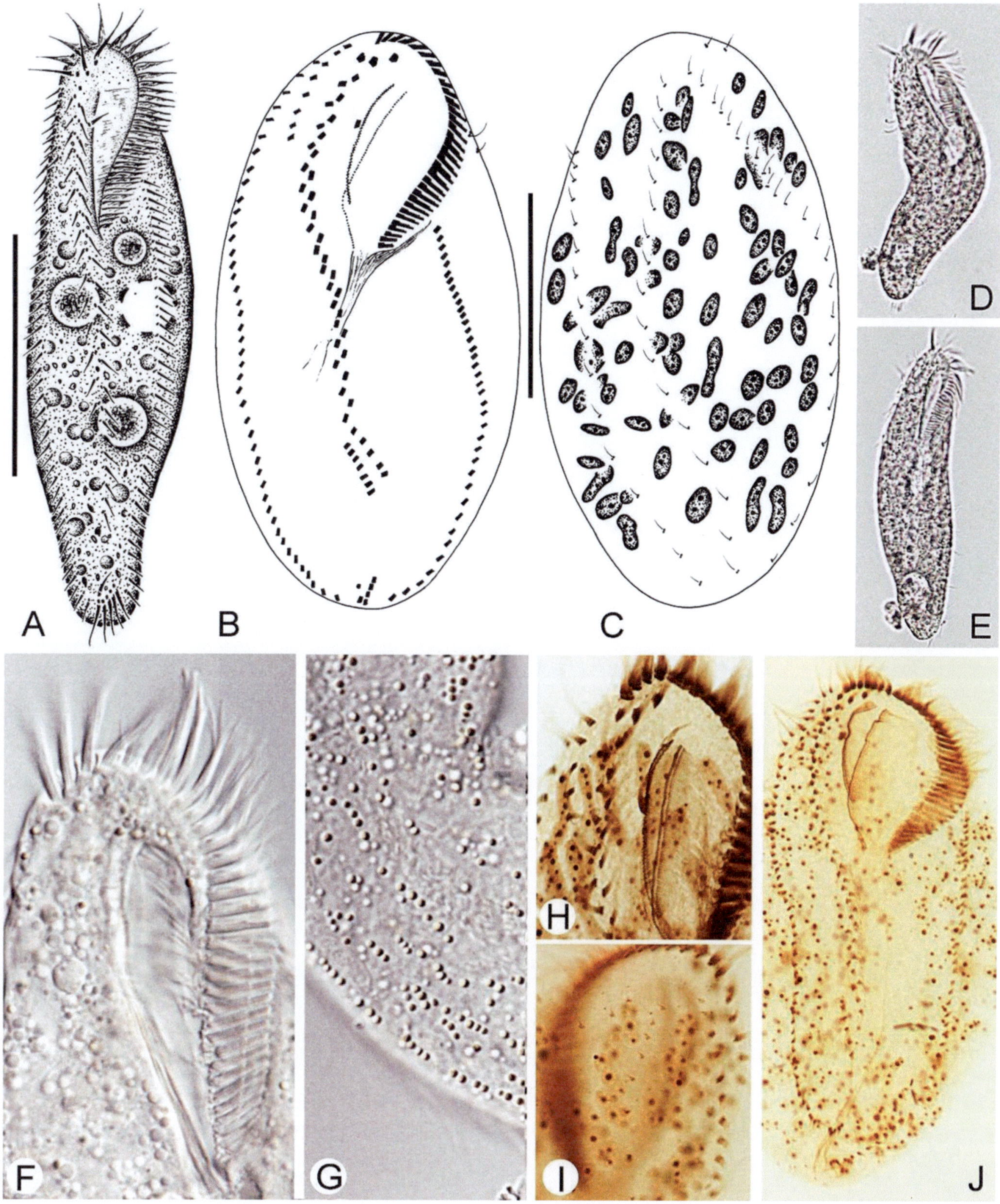


**Fig. 109B** *Bakuella subtropica*
(**a**, **d**, **e**) Ventral views of live individuals. (**b**, **c**) Ventral (**b**) and dorsal (**c**) views of ciliature. (**f**) Oral apparatus. (**g**) Cortical granules. (**h–j**) Ventral (**h**, **j**) and dorsal (**i**) views of ciliature. Scale bars: 60 μm (**a**) and 50 μm (**b**, **c**)

## Reference

Chen XM, Hu X, Lin X, Al-Rasheid KAS, Ma H, Miao M (2013) Morphology, ontogeny and molecular phylogeny of a new brackish water ciliate *Bakuella subtropica* sp. n. (Ciliophora, Hypotricha) from southern China. Eur J Protistol 49:611–622

Class Spirotrichea Bütschli, 1889
Order Urostylida Jankowski, 1979
Family Urostylidae Bütschli, 1889
*Holosticha* cf. *heterofoissneri*

**Ecological feature:** mangrove
**Water temperature:** 24.0 °C
**pH:** no data available
**Salinity:** 17.0‰

**Fig. 110A** Sample site

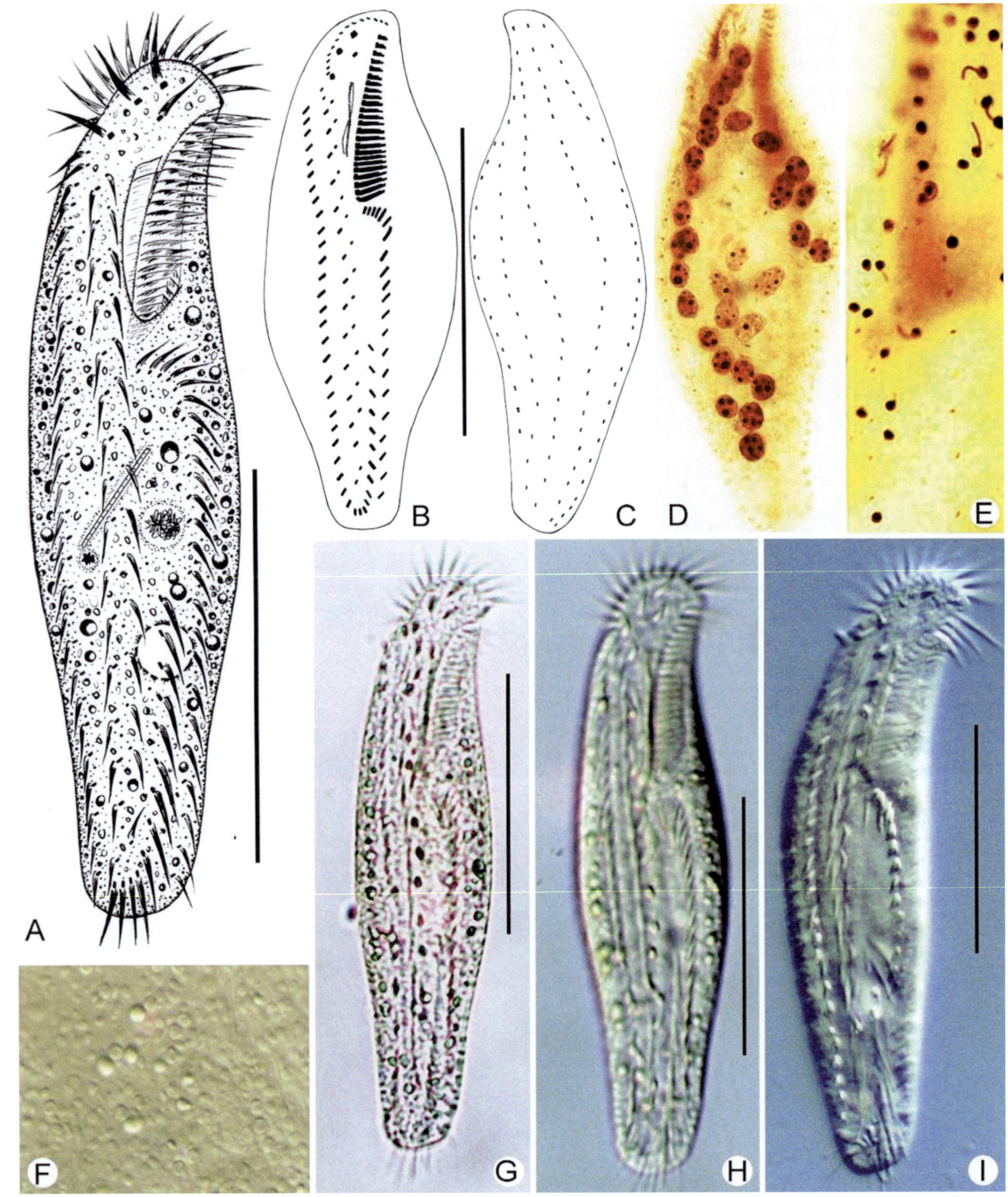


**Fig. 110B** *Holosticha* cf. *heterofoissneri*
(**a**, **g–i**) Ventral views of live individuals. (**b**, **c**) Ventral (**b**) and dorsal (**c**) views of ciliature. (**d**) Macronuclear nodules. (**e**, **f**) Ejected, hairlike cortical granule (**e**) and blood-cell-shaped cortical granules in vivo (**f**). Scale bars: 60 μm

## Reference

Luo X, Gao F, Al-Rasheid KAS, Warren A, Hu X, Song WB (2015) Redefinition of the hypotrichous ciliate *Uncinata*, with descriptions of the morphology and phylogeny of three urostylids (Protista, Ciliophora). Syst Biodivers 13:455–471

Class Spirotrichea Bütschli, 1889
Order Urostylida Jankowski, 1979
Family Urostylidae Bütschli, 1889
*Holosticha diademata* (Rees, 1883) Kahl, 1932

**Ecological feature:** mangrove
**Water temperature:** 24.0 °C; 25.0 °C; 22.0 °C; -
**pH:** no data available; no data available; 6.5; 7.0
**Salinity:** 25.0‰; 35.0‰; 28‰; 10‰

**Fig. 111A** Sample site

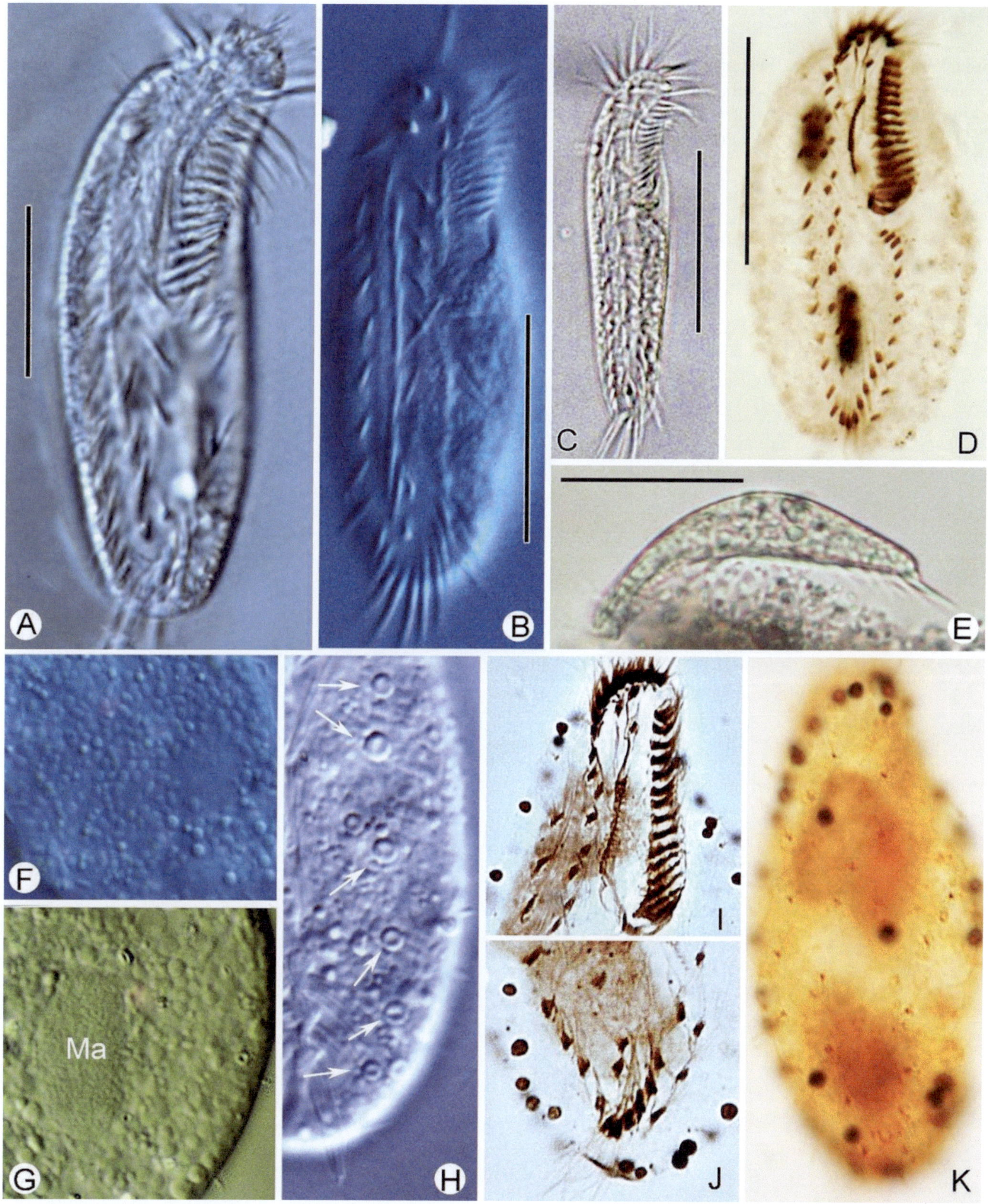


**Fig. 111B** *Holosticha diademata*
(**a–c**) Ventral views of live individuals. (**d**) Ventral (**d**) and dorsal (**k**) views of ciliature. (**e**) Lateral view of live cell. (**f**) Cortical granules. (**g**) Macronuclear nodule (Ma). (**h**) Blood-cell-shaped cortical granules (arrows). (**i**) Ventral view of anterior portion of cell showing oral apparatus. (**j**) Ventral view of posterior portion of cell showing transverse cirri. Scale bars: 40 μm (**a**) and 30 μm (**b–e**)

## Reference

Song WB, Wilbert N (2002) Faunistic studies on marine ciliates from the Antarctic benthic area, including descriptions of one epizoic form, 6 new species and 2 new genera (Protozoa: Ciliophora). Acta Protozool 41:23–61

Class Spirotrichea Bütschli, 1889
Order Urostylida Jankowski, 1979
Family Urostylidae Bütschli, 1889
*Holosticha heterofoissneri* Hu & Song, 2001

**Ecological feature:** Coastal water
**Water temperature:** 21.0 °C
**pH:** 7.5
**Salinity:** 30.0‰

**Fig. 112A** Sample site

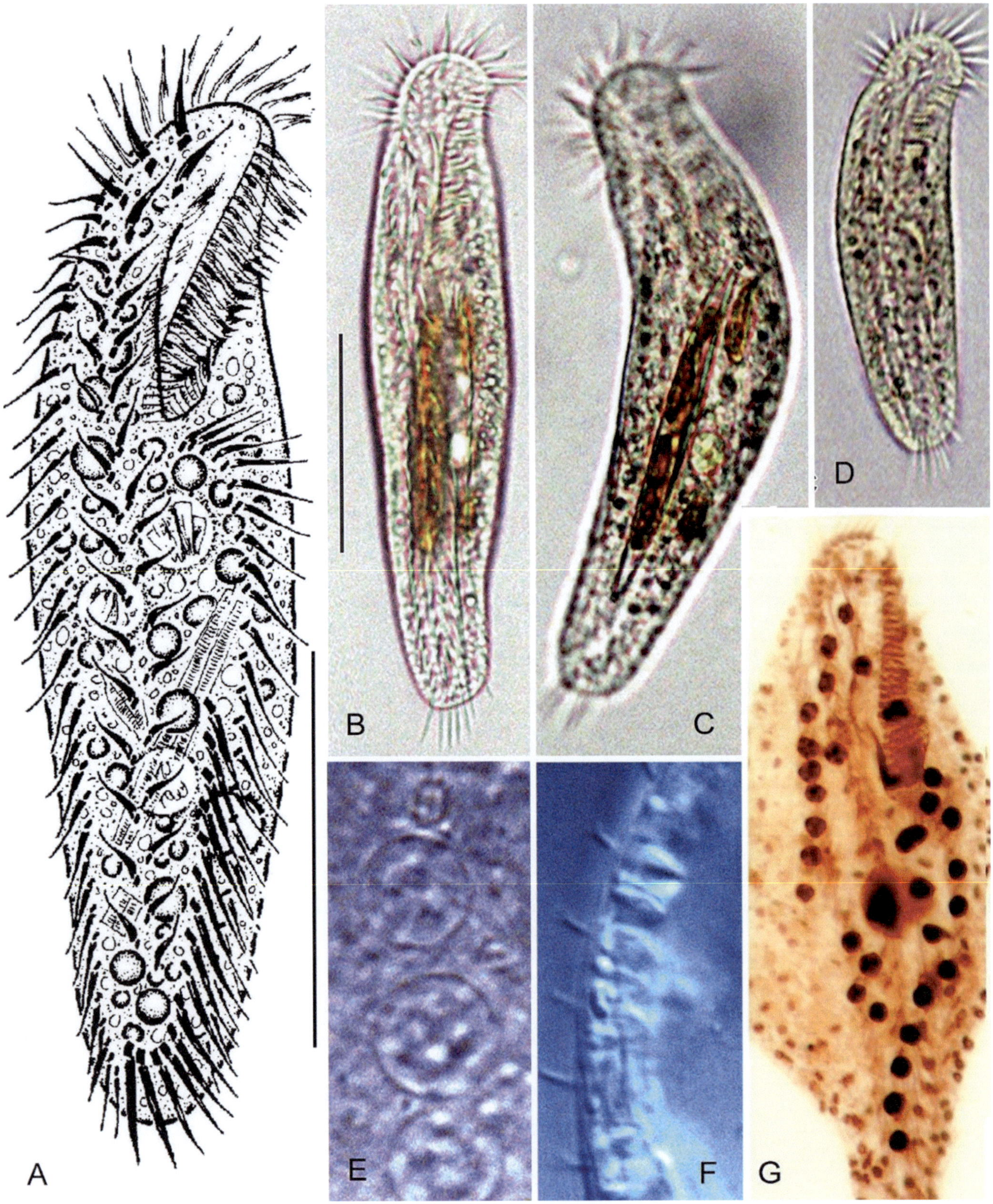


**Fig. 112B** *Holosticha heterofoissneri*
(**a–d**) Ventral views of live individuals. (**e**) The unidentified discoid structure in endoplasm. (**f**) Dorsal cilia. (**g**) Oral apparatus and macronuclear nodules. Scale bars: 50 μm

## References

Hu X, Song WB (2001) Morphology and morphogenesis of *Holosticha heterofoissneri* nov. spec. from the Yellow Sea, China (Ciliophora, Hypotrichida). Hydrobiologia 448:171–179

Song WB, Wilbert N, Warren A (2002) New contribution to the morphology and taxonomy of four marine hypotrichous ciliates from Qingdao, China (protozoa: Ciliophora). Acta Protozool 41:145–162

Class Spirotrichea Bütschli, 1889
Order Urostylida Jankowski, 1979
Family Urostylidae Bütschli, 1889
*Metaurostylopsis marina* (Kahl, 1932) Song et al., 2001

**Ecological feature:** muddy sediment
**Water temperature:** 24.0 °C
**pH:** no data available
**Salinity:** 35.0‰

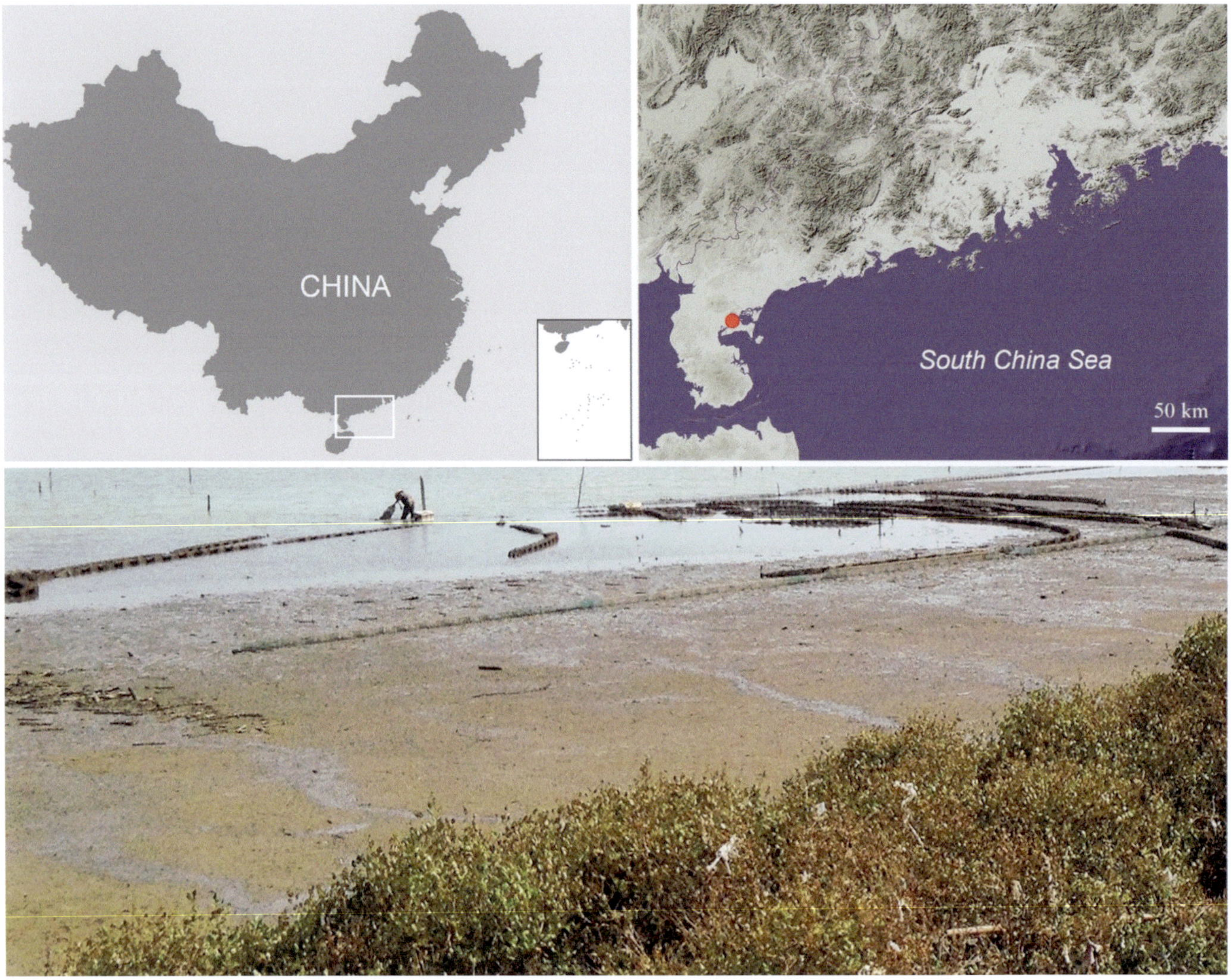


**Fig. 113A** Sample site

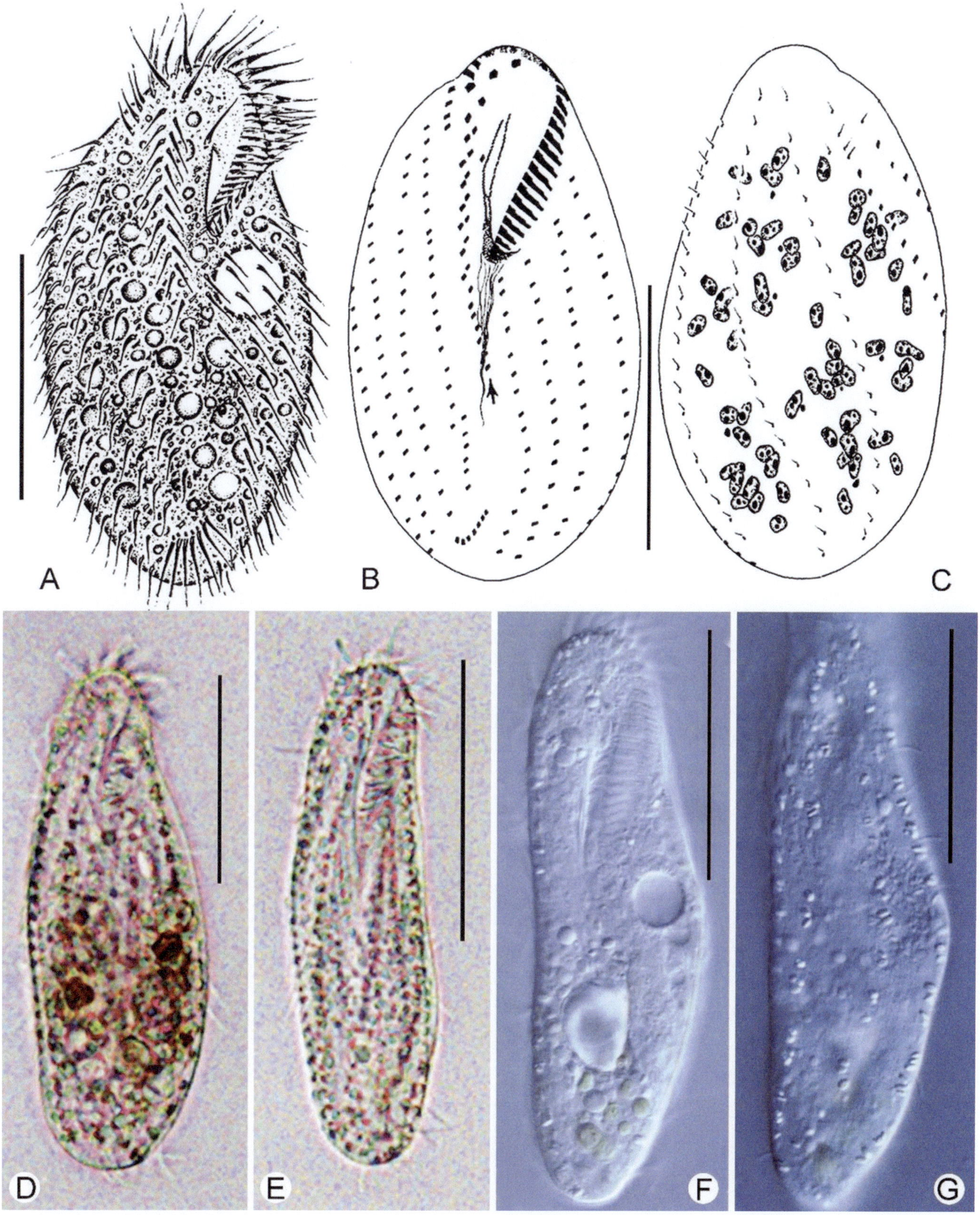


**Fig. 113B** *Metaurostylopsis marina*
(**a**, **d**, **e**) Ventral views of live individuals. (**f**) Ventral view of live cell showing contractile vacuole. (**b**, **c**) Ventral (**b**) and dorsal (**c**) views of ciliature. (**g**) Cortical granules on dorsal side. Scale bars: 50 μm

## Reference

Song WB, Petz W, Warren A (2001) Morphology and morphogenesis of the poorly-known marine urostylid ciliate, *Metaurostylopsis marina* (Kahl, 1932) nov. gen., nov. comb. (Protozoa, Ciliophora, Hypotrichida). Eur J Protistol 37:63–76

Class Spirotrichea Bütschli, 1889
Order Urostylida Jankowski, 1979
Family Urostylidae Bütschli, 1889
*Metaurostylopsis salina* Lei et al., 2005

**Ecological feature:** mangrove
**Water temperature:** 31.0 °C
**pH:** 6.8
**Salinity:** 18.0‰

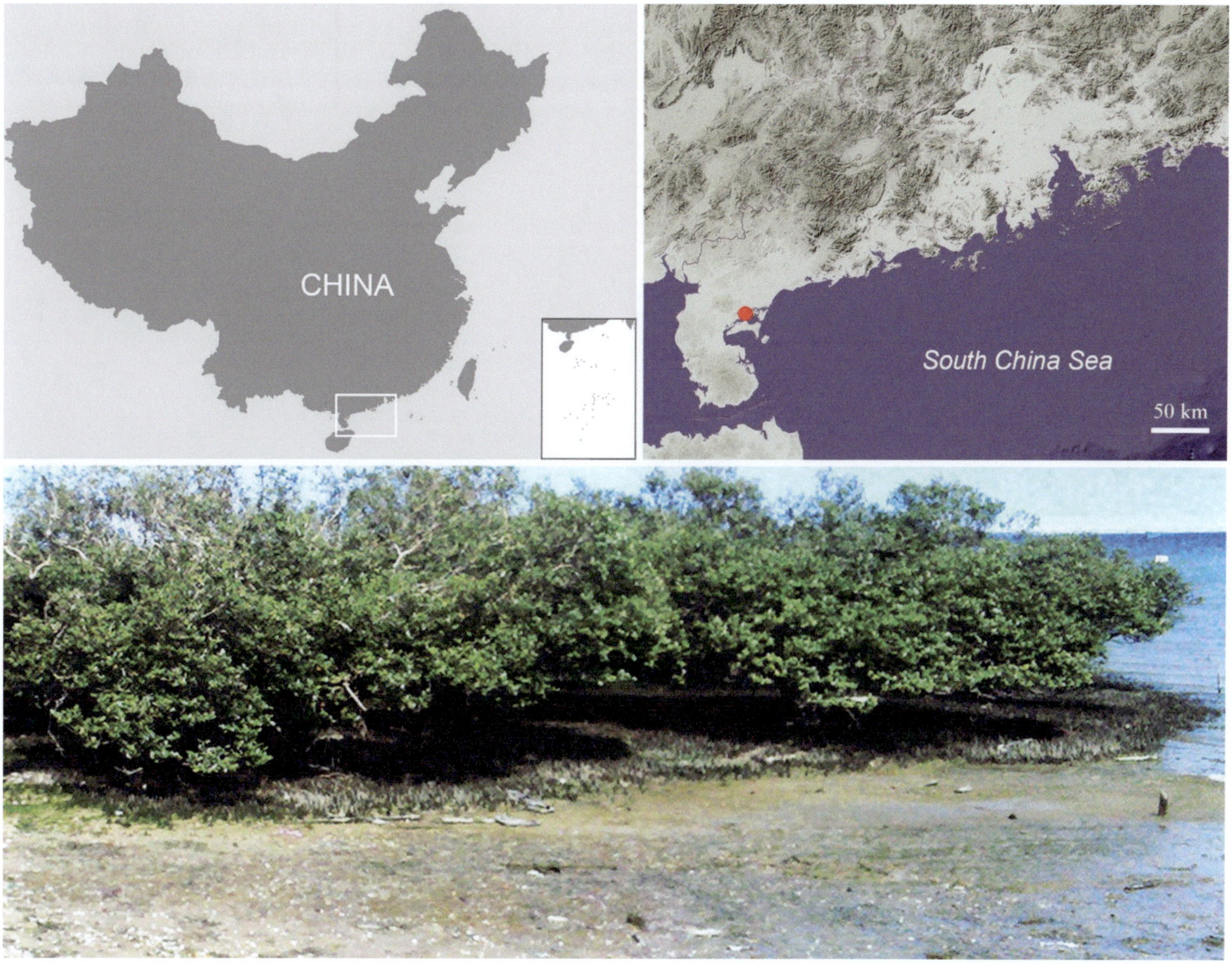


**Fig. 114A** Sample site

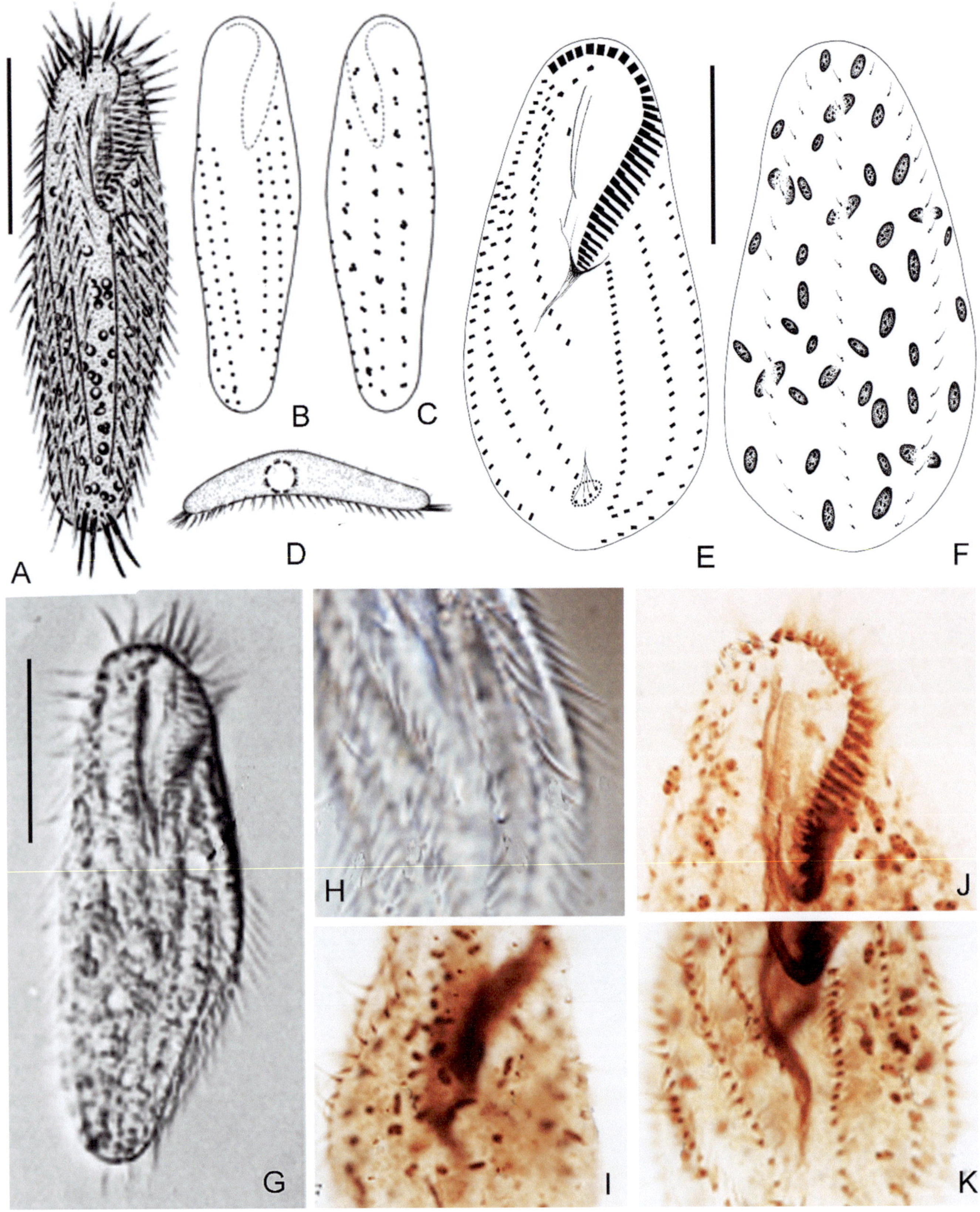


**Fig. 114B** *Metaurostylopsis salina*
(**a**, **g**) Ventral views of live individuals. (**b**, **c**) Arrangement of cortical granules on ventral (**b**) and dorsal (**c**) sides. (**d**) Lateral view of live cell. (**e**, **f**, **i–k**) Ventral (**e**, **j**, **k**) and dorsal (**f**, **i**) views of ciliature. (**h**) Arrangement of marginal cirri. Scale bars: 50 μm

## Reference

Shao C, Miao M, Song WB, Warren A, Al-Rasheid KAS, Al-Quaishy SA, Al-Farraj SA (2008) Studies on two marine *Metaurostylopsis* spp. from China with notes on morphogenesis in *M. sinica* nov. spec. (Ciliophora, Urostylida). Acta Protozool 47:95–112

Class Spirotrichea Bütschli, 1889
Order Urostylida Jankowski, 1979
Family Urostylidae Bütschli, 1889
*Metaurostylopsis struederkypkeae* Shao et al., 2008

**Ecological feature:** offshore water
**Water temperature:** 24.5 °C
**pH:** 8.4
**Salinity:** 28.2‰

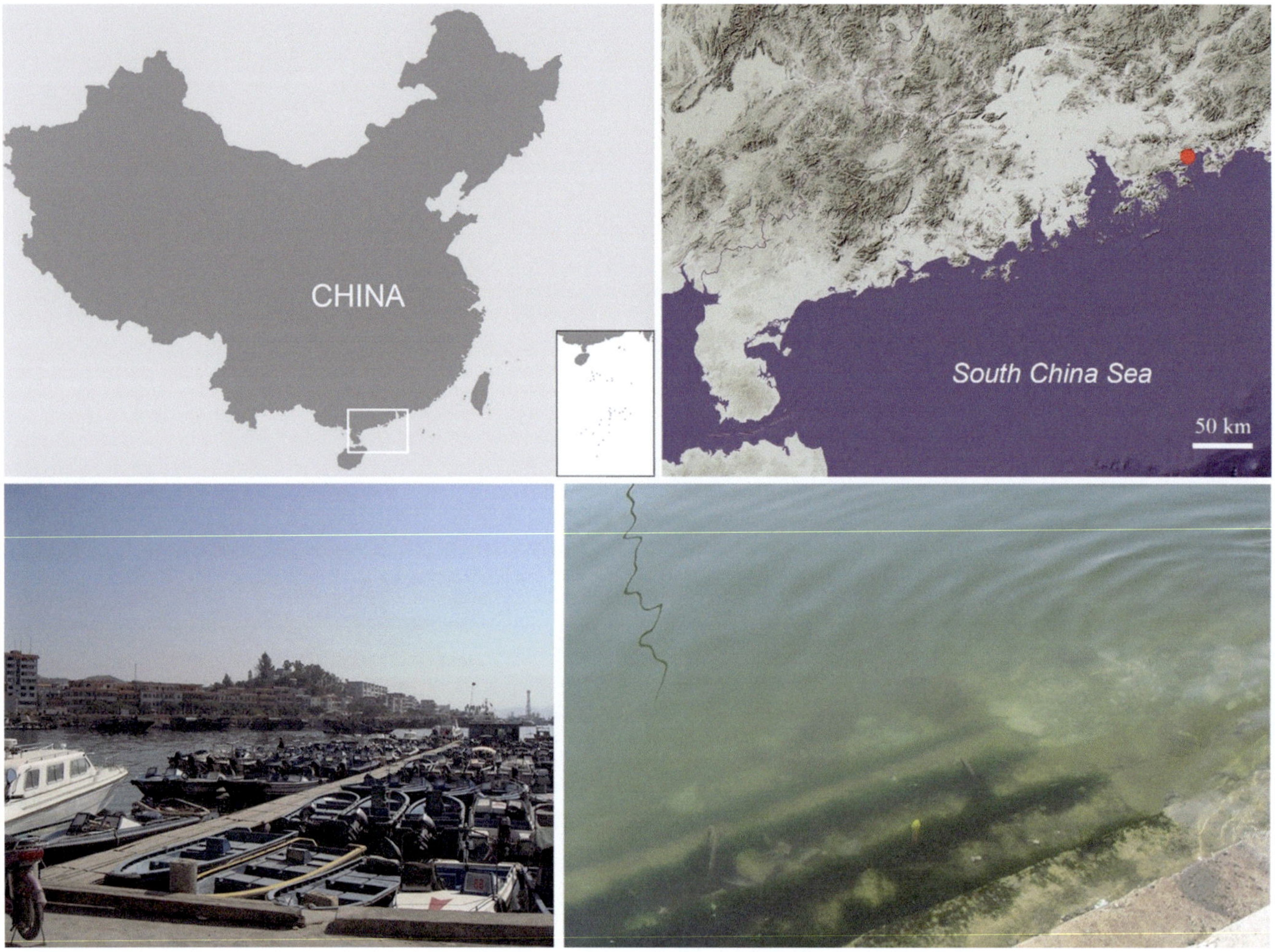


**Fig. 115A** Sample site

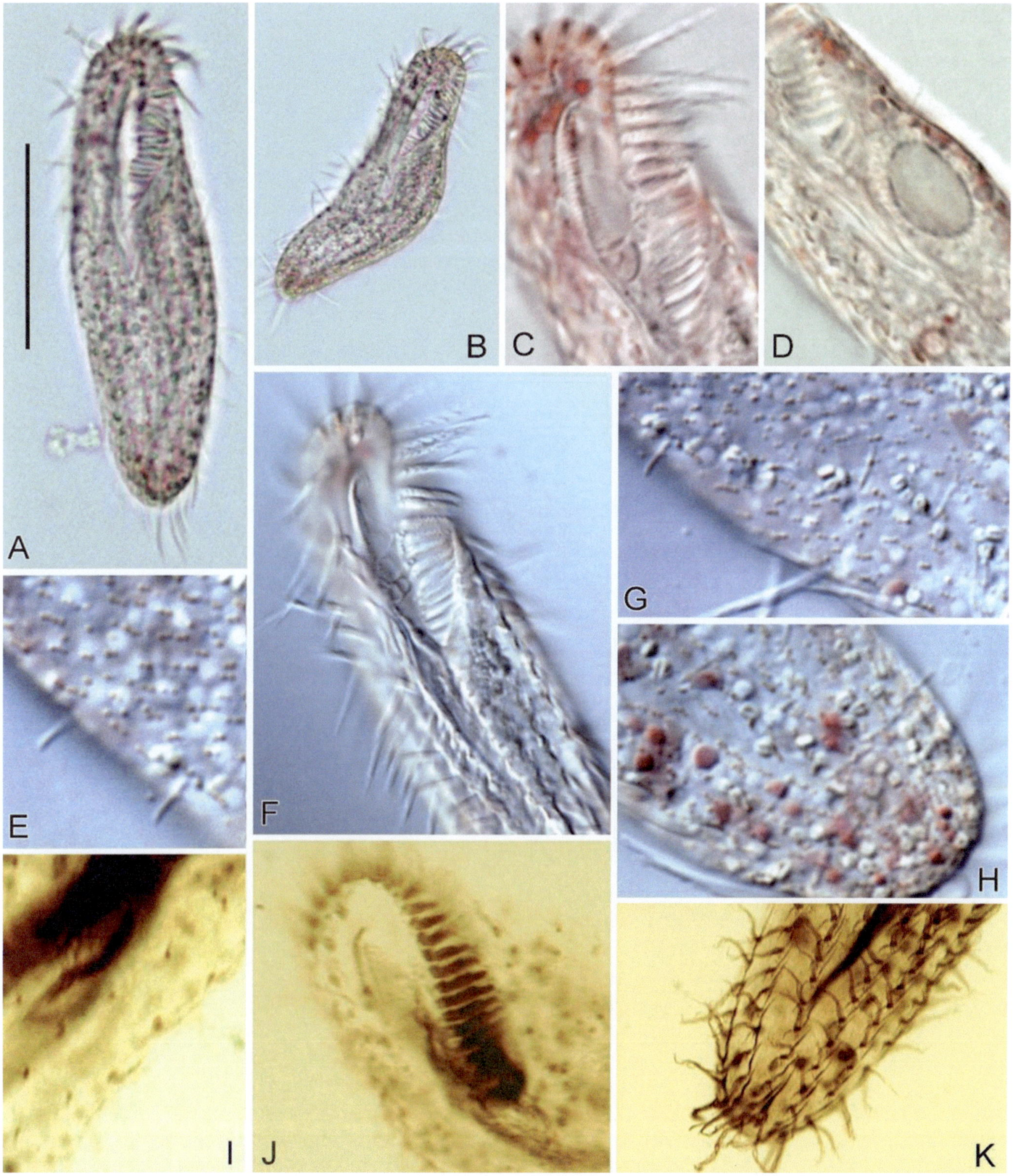


**Fig. 115B** *Metaurostylopsis struederkypkeae*
(**a**, **b**, **f**) Ventral views of live cells. (**c**) Ventral view of anterior portion showing adoral membranelles. (**d**) Contractile vacuole. (**e**, **g**, **h**) The larger and smaller cortical granules as well as dorsal cilia. (**i**) The dorsal kineties. (**j**) Adoral zone of membranelles. (**k**) Posterior portion of cell showing the marginal and transverse cirri. Scale bar: 40 μm

## Reference

Shao C, Song WB, Al-Rasheid KAS, Yi Z, Chen X, Al-Farraj SA, Al-Quraishy SA (2008) Morphology and infraciliature of two new marine urostylid ciliates: *Metaurostylopsis struederkypkeae* n. sp. and *Thigmokeronopsis stoecki* n. sp. (Ciliophora, Hypotrichida) from China. J Eukaryot Microbiol 55:286–296

Class Spirotrichea Bütschli, 1889
Order Urostylida Jankowski, 1979
Family Urostylidae Bütschli, 1889
*Monocoronella carnea* Chen et al., 2011

**Ecological feature:** coastal water
**Temperature:** 12.0 °C
**pH:** no data available
**Salinity:** 33.0‰

**Fig. 116A** Sample site

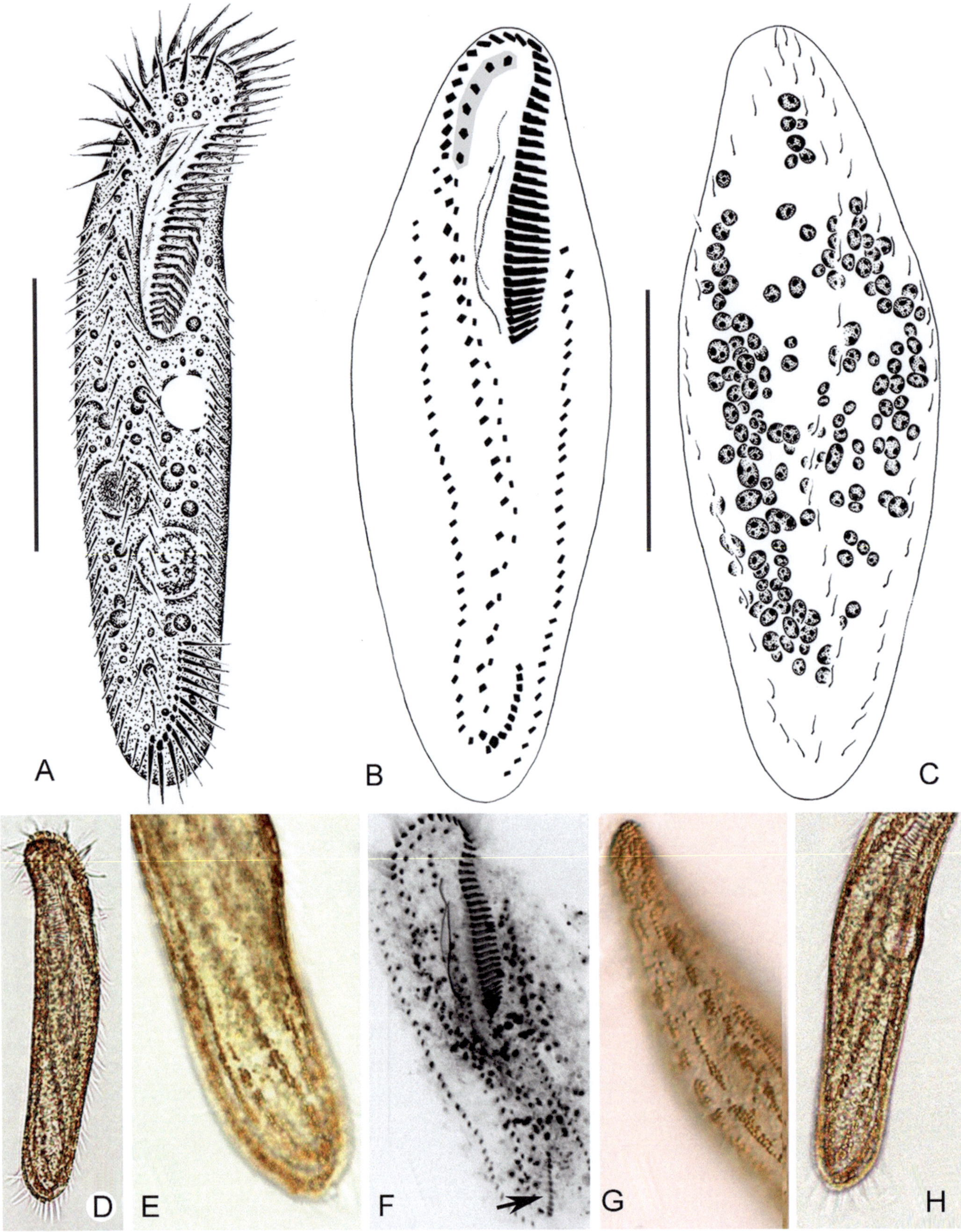


**Fig. 116B** *Monocoronella carnea*
(**a**) Ventral view of live cell. (**b**, **c**, **f**) Ventral (**b**, **f**) and dorsal (**c**) views of ciliature; arrow marks transverse cirri. (**d**, **e**, **g**, **h**) Arrangement of cortical granules. Scale bars: 50 μm

## Reference

Chen X, Dong J, Lin X, Al-Rasheid KAS (2011) Morphology and phylogeny of a new urostylid ciliate, *Monocoronella carnea* n. g., n. sp. (Ciliophora, Hypotricha) from Daya Bay, southern China. J Eukaryot Microbiol 58:497–503

Class Spirotrichea Bütschli, 1889
Order Urostylida Jankowski, 1979
Family Urostylidae Bütschli, 1889
*Neourostylopsis flavicana* (Wang et al., 2011) Chen et al., 2013

**Ecological feature:** mangrove (Zhanjiang, Shantou); aquaculture pond (Zhuhai)
**Water temperature:** 25.0 °C; 26.7 °C; 19.0 °C
**pH:** 7.9; 7.9; 9.0
**Salinity:** 10.0‰; 25.0‰; 12.9‰

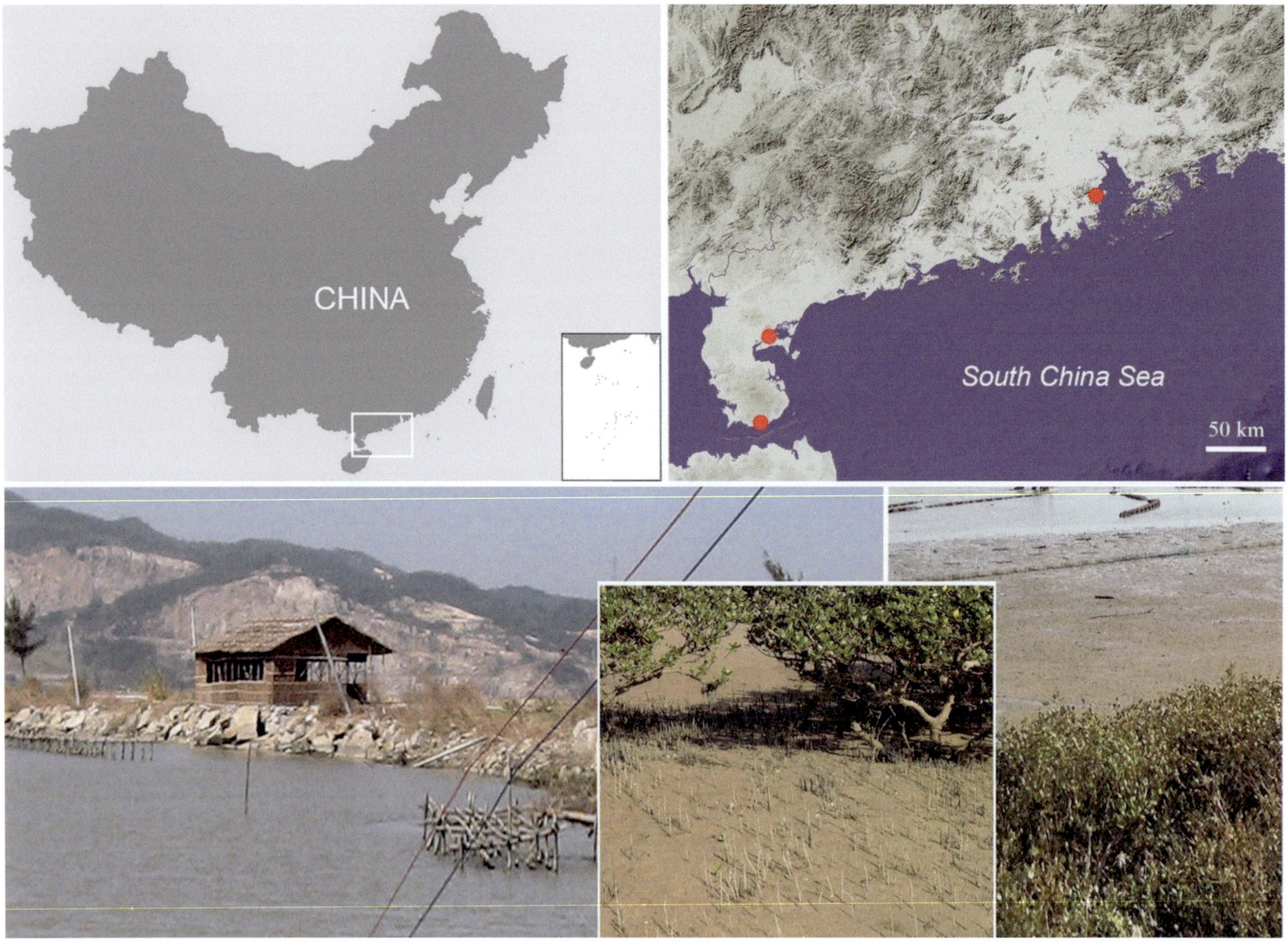


**Fig. 117A** Sample site

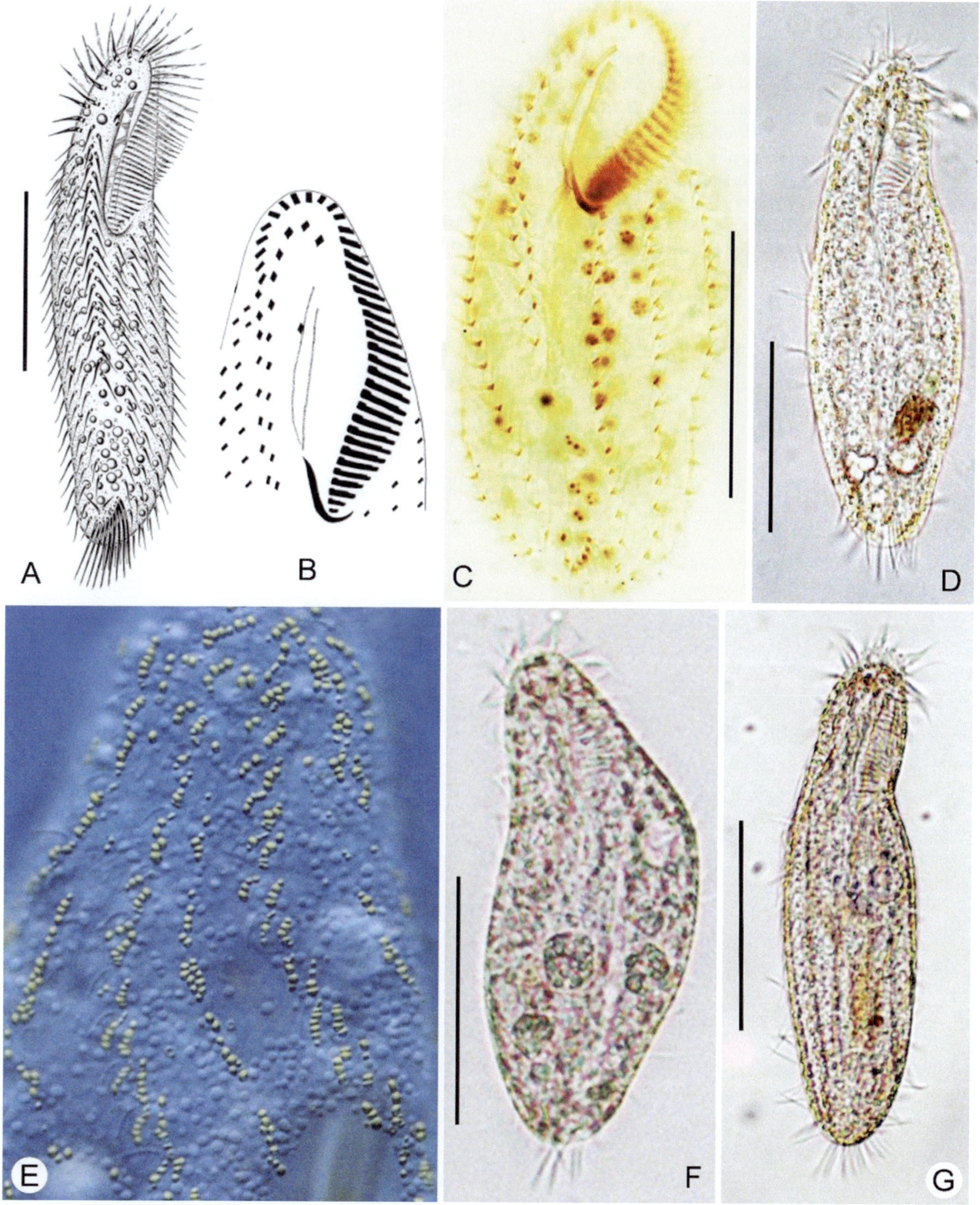


**Fig. 117B** *Neourostylopsis flavicana*
(**a**, **d**, **f**, **g**) Ventral views of live cells. (**b**) Oral apparatus. (**c**) Ventral view of ciliature. (**e**) Cortical granules on dorsal side and dorsal cilia. Scale bars: 50 μm

## References

Chen XR, Shao C, Liu X, Huang J, Al-Rasheid KAS (2013) Morphology and phylogenies of two hypotrichous brackish ciliates, *Neourostylopsis orientalis* n. sp., and *Protogastrostyla sterkii* (Wallengren, 1900) n. comb., with establishment of a new genus *Neourostylopsis* (Protista, Ciliophora, Hypotrichia) from China. Int J Syst Evol Microbiol 63:1197–1209

Wang Y, Hu X, Al-Rasheid KAS, Warren A (2011) Two urostylid ciliates, *Metaurostylopsis flavicana* spec. nov. and *Tunicothrix wilberti* (Lin & Song, 2004) Xu et al., 2006 (Ciliophora, Stichotrichia), from a mangrove nature protection area in China. Int J Syst Evol Microbiol 61:1740–1750

Class Spirotrichea Bütschli, 1889
Order Urostylida Jankowski, 1979
Family Urostylidae Bütschli, 1889
*Neourostylopsis orientalis* Chen et al., 2013

**Ecological feature:** wetland near estuary
**Water temperature:** 23.0 °C
**pH:** 8.5
**Salinity:** 9.0‰

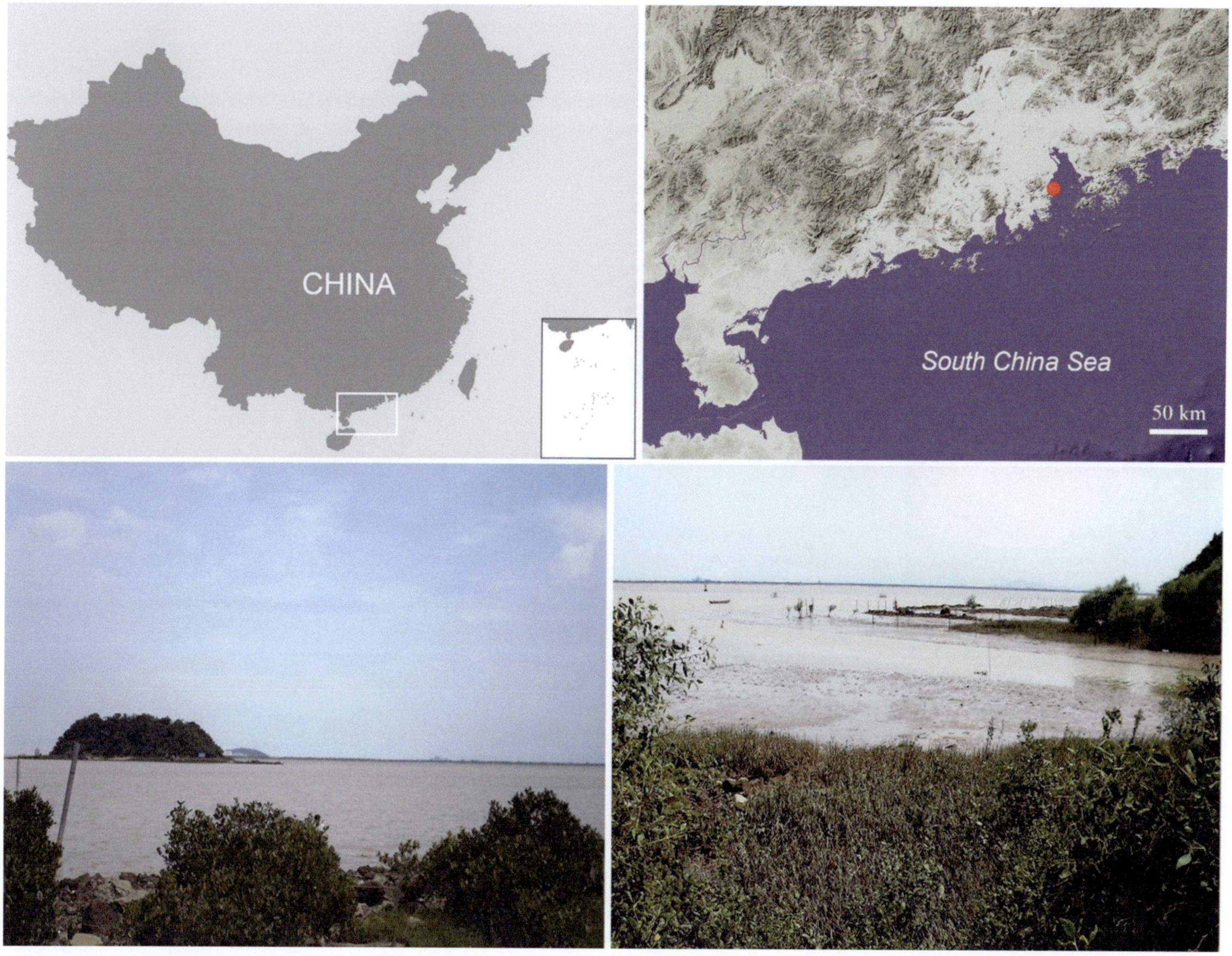


**Fig. 118A** Sample site

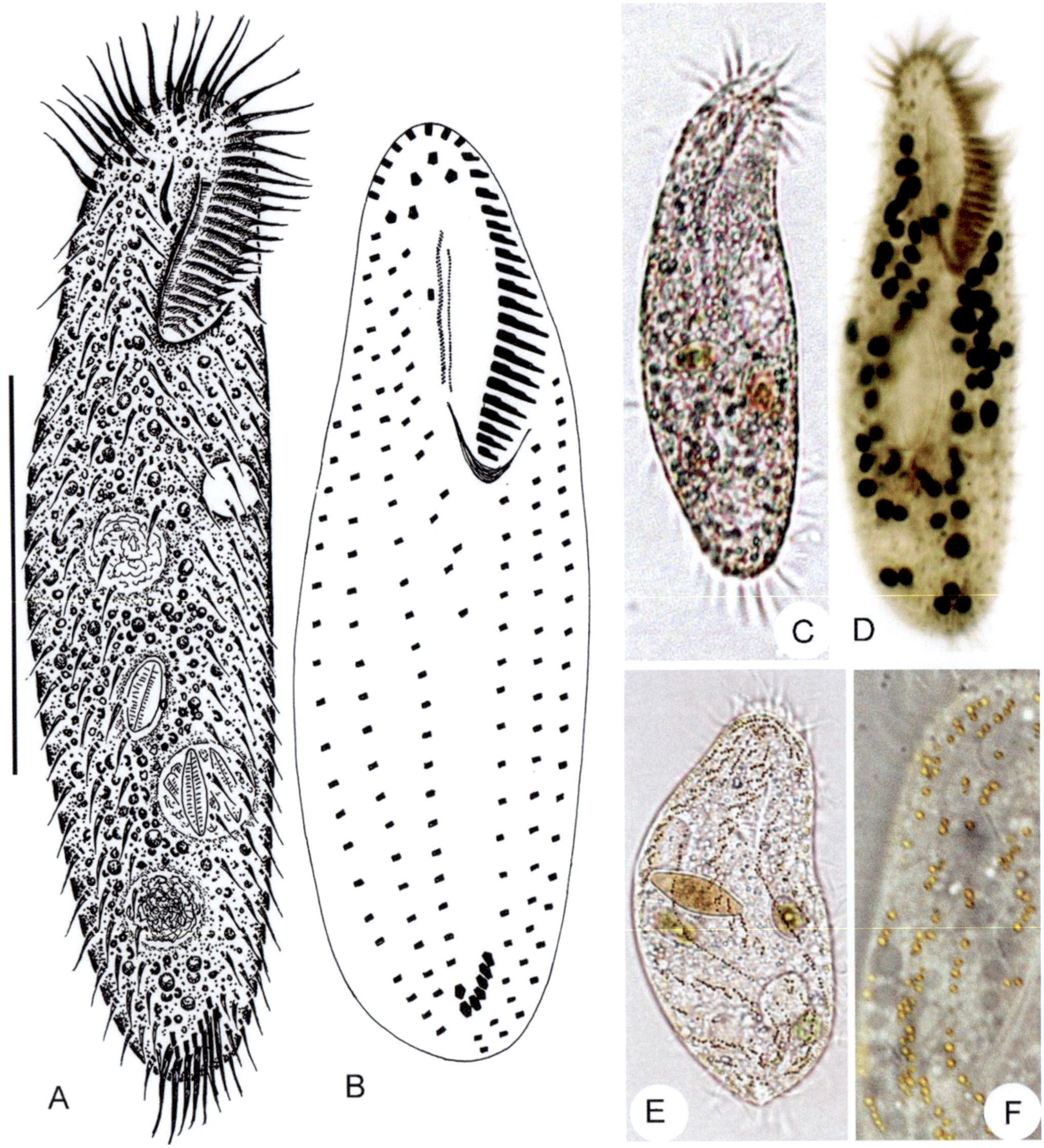


**Fig. 118B** *Neourostylopsis orientalis*
(**a**, **c**) Ventral views of live individuals. (**b**, **d**) Ventral view of ciliature. (**e**) Dorsal view of a compressed individual, to show the arrangement of cortical granules. (**f**) Dorsal view, to show cortical granules. Scale bar: 50 μm

## Reference

Chen XR, Shao C, Liu X, Huang J, Al-Rasheid KAS (2013) Morphology and phylogenies of two hypotrichous brackish ciliates, *Neourostylopsis orientalis* n. sp., and *Protogastrostyla sterkii* (Wallengren, 1900) n. comb., with establishment of a new genus *Neourostylopsis* (Protista, Ciliophora, Hypotrichia) from China. Int J Syst Evol Microbiol 63:1197–1209

Class Spirotrichea Bütschli, 1889
Order Urostylida Jankowski, 1979
Family Urostylidae Bütschli, 1889
*Urostyla grandis* Ehrenberg, 1830

**Ecological feature:** mangrove
**Water temperature:** 25.0 °C
**pH:** 6.8
**Salinity:** 10.0‰

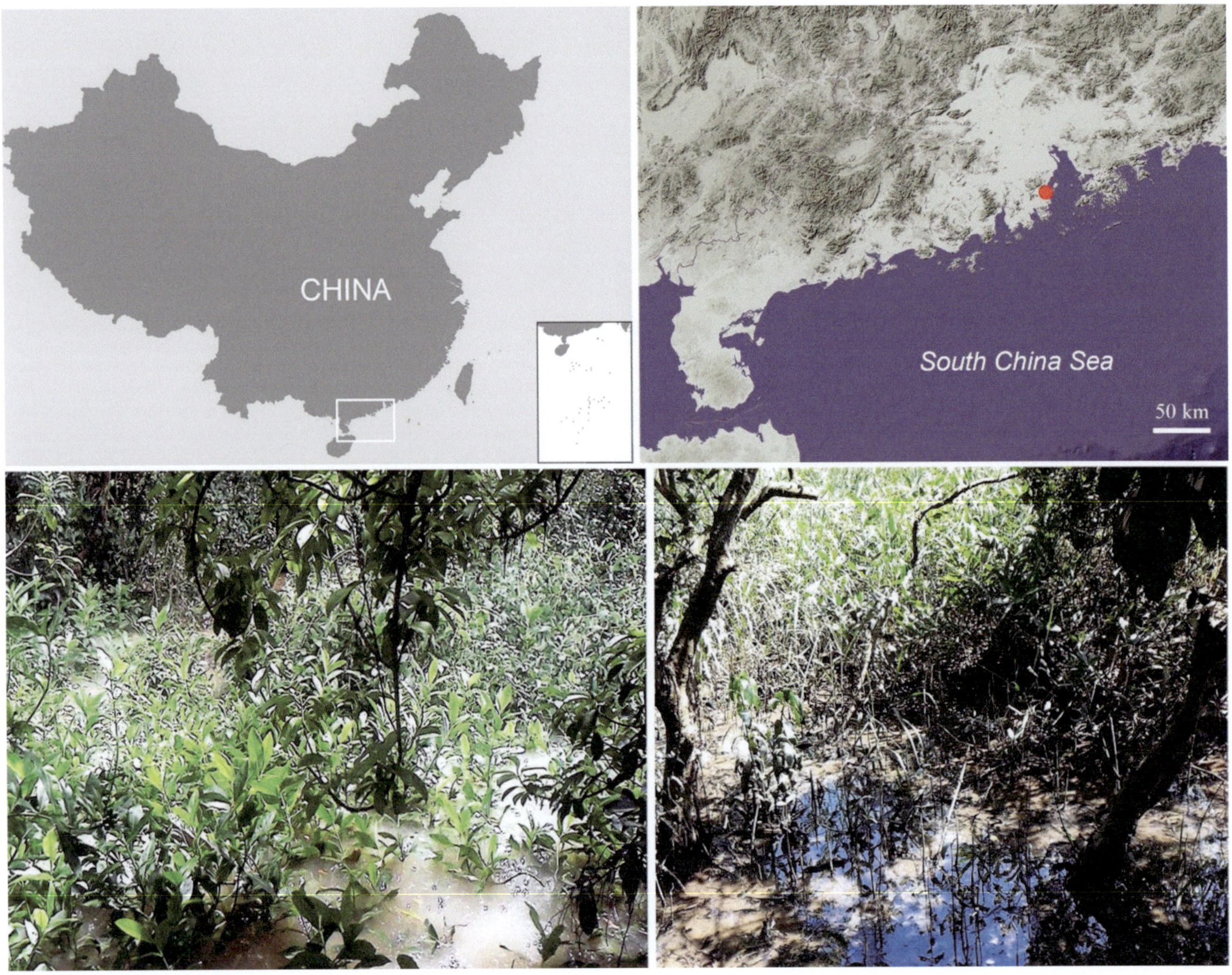


**Fig. 119A** Sample site

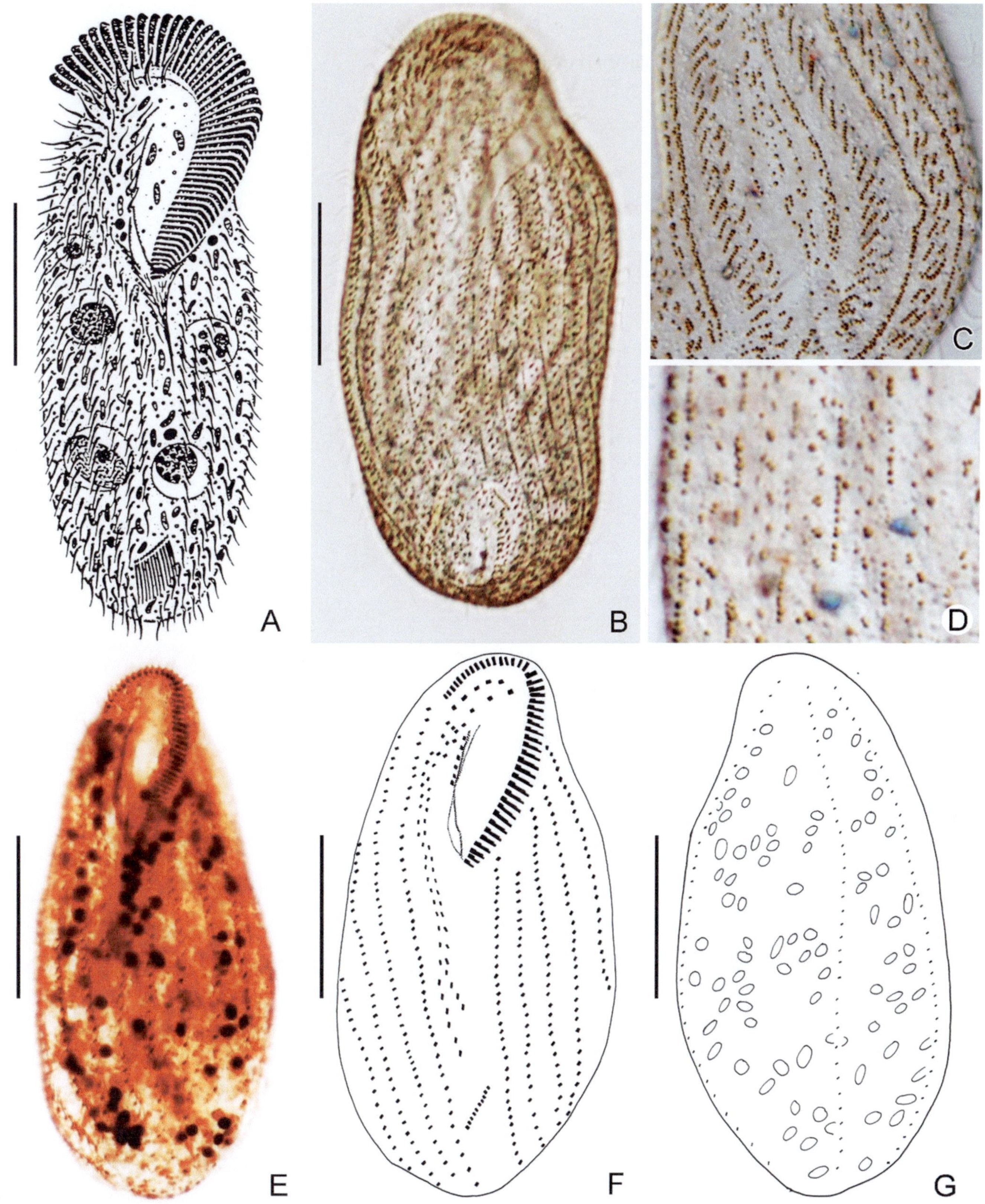


**Fig. 119B** *Urostyla grandis*
(**a**, **b**) Ventral views of live cells, A from Ganner (1991). (**c**, **d**) Cortical granules on ventral (**c**) and dorsal (**d**) sides. (**e–g**) Ventral (**e**, **f**) and dorsal (**g**) views of ciliature. Scale bars: 70 μm

## Reference

Ganner B (1991) Zur Taxonomie einiger Ciliaten (Protozoa) aus Fliessgewässern. Universität Salzburg, Ph D thesis

Class Spirotrichea Bütschli, 1889
Order Urostylida Jankowski, 1979
Incertae sedis in Order Urostylida
*Uncinata gigantea* Bullington, 1940

**Ecological feature:** muddy flat
**Water temperature:** 24.0 °C; 19.0 °C
**pH:** 7.3; 6.5
**Salinity:** 17.0‰; 15.0‰

**Fig. 120A** Sample site

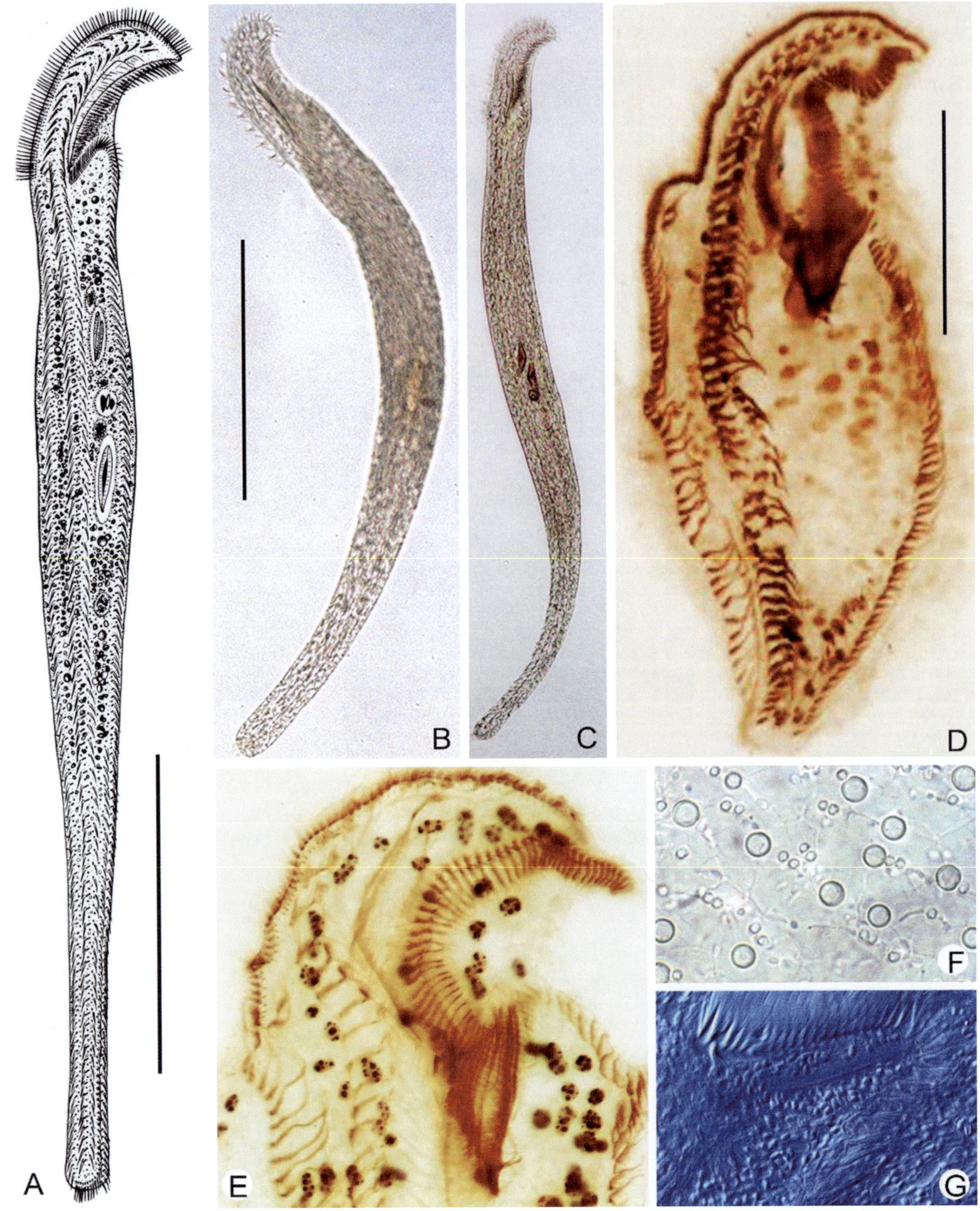


**Fig. 120B** *Uncinata gigantea*
(**a–c**) Ventral views of live individuals. (**d**) Ventral view of ciliature. (**e**) Frontal ciliature, showing the conspicuous gap in the adoral zone and the distinctly elongated proximal-most membranelles. (**f**) Cortical granules. (**g**) The transversely oriented left marginal cirri. Scale bars: 300 μm

## Reference

Luo X, Gao F, Al-Rasheid KAS, Warren A, Hu X, Song WB (2015) Redefinition of the hypotrichous ciliate *Uncinata*, with descriptions of the morphology and phylogeny of three urostylids (Protista, Ciliophora). Syst Biodivers 13:455–471

Class Spirotrichea Bütschli, 1889
Order Discocephalida Wicklow, 1982
Family Discocephalidae Jankowski, 1979
*Discocephalus pararotatorius* Jiang et al., 2013

**Ecological feature:** intertidal zone
**Water temperature:** 23.0 °C
**pH:** no data available
**Salinity:** 27.0‰

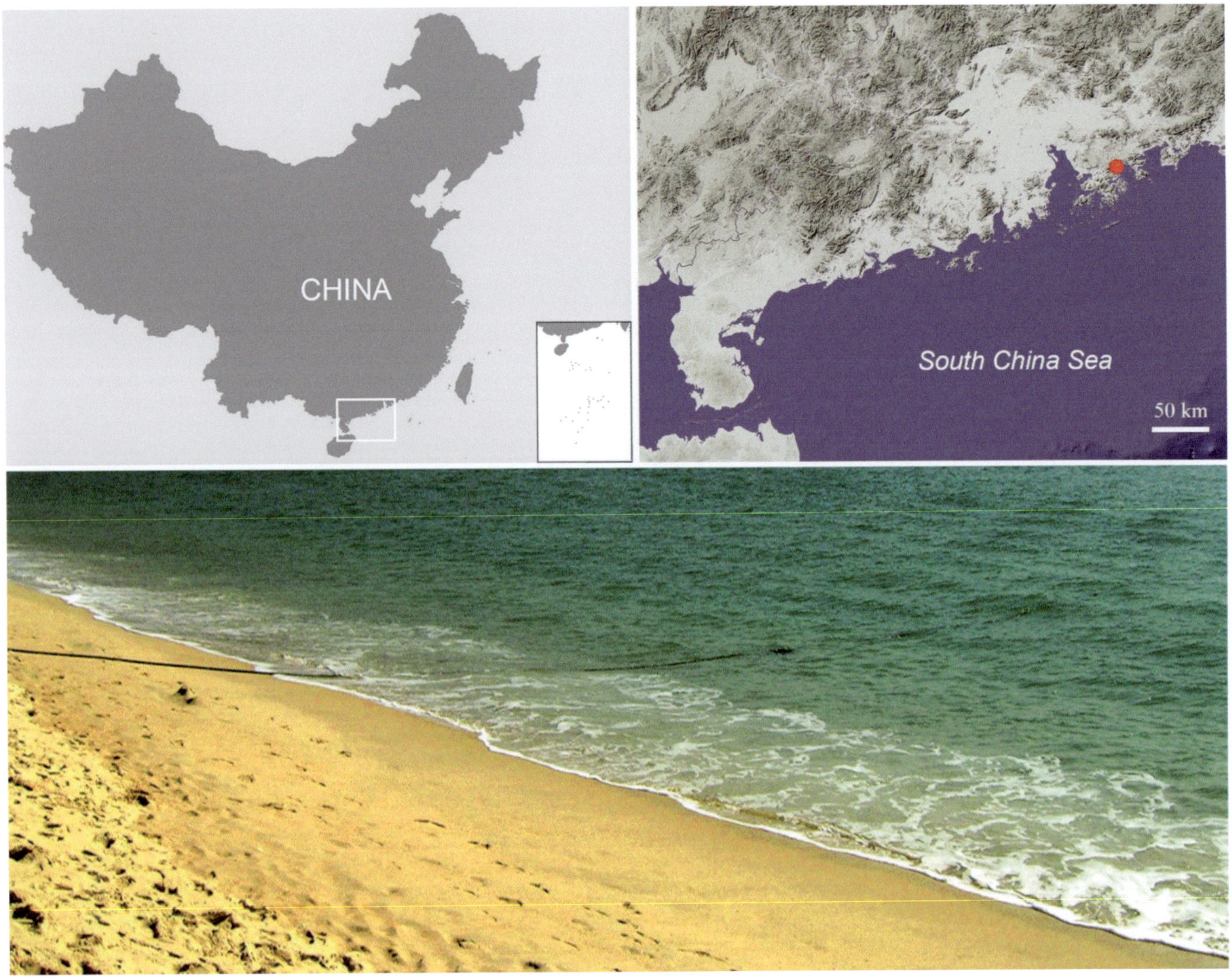


**Fig. 121A** Sample site

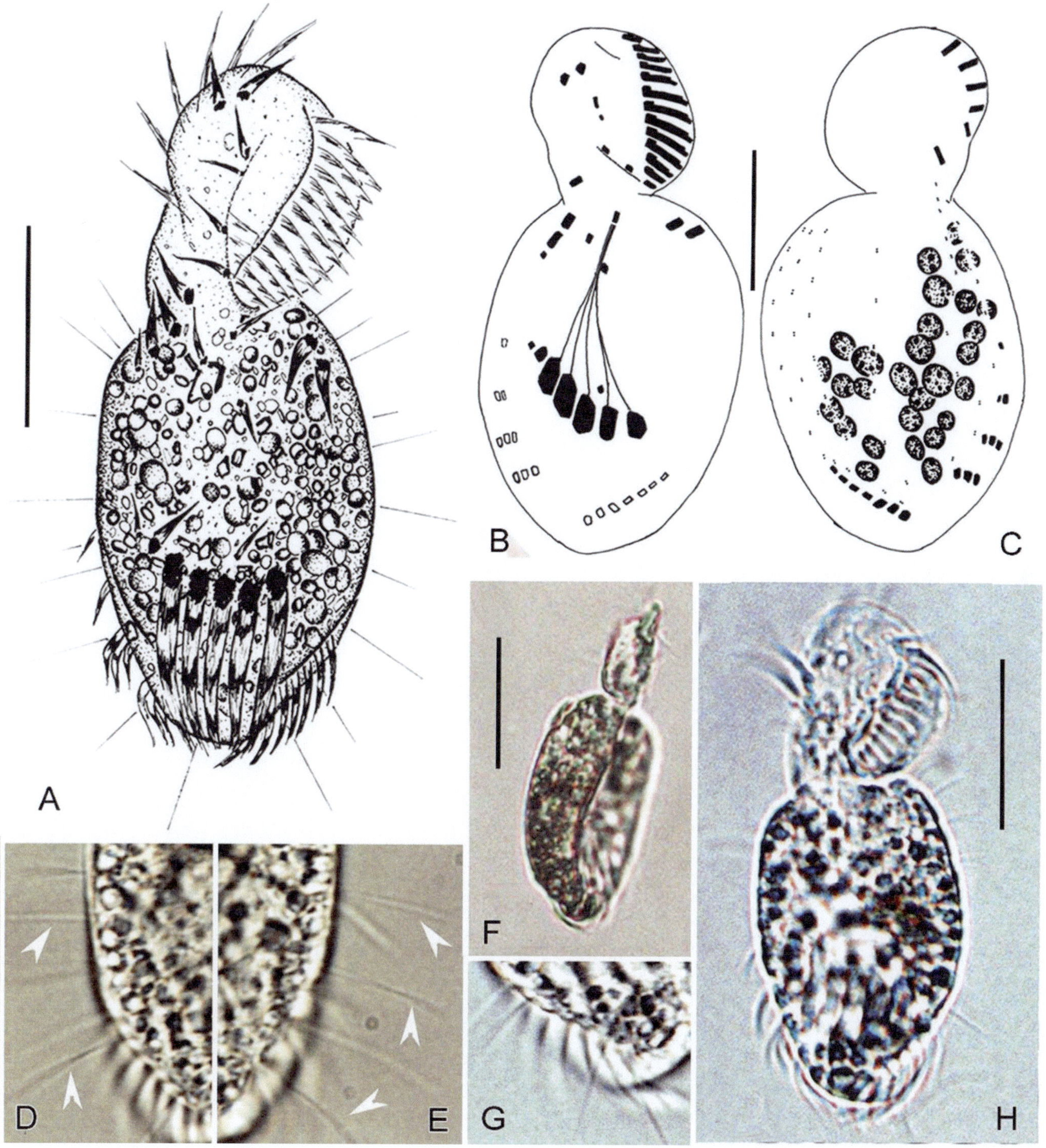


**Fig. 121B** *Discocephalus pararotatorius*
(**a**, **h**) Ventral views of live cells. (**b**, **c**) Ventral (**b**) and dorsal (**c**) views of ciliature. (**d**, **e**) Dorsal view of posterior portion of live cell; arrowheads mark the dorsal cilia. (**f**) Lateral view of live cell. (**g**) Show the posterolateral marginal cirri. Scale bars: 20 μm

## Reference

Jiang J, Xing Y, Miao M, Shao C, Warren A, Song WB (2013) Two new marine ciliates, *Caryotricha rariseta* n. sp. and *Discocephalus pararotatorius* n. sp. (Ciliophora, Spirotrichea), with phylogenetic analyses inferred from the small-subunit rRNA gene sequences. J Eukaryot Microbiol 60:388–398

Class Spirotrichea Bütschli, 1889
Order Discocephalida Wicklow, 1982
Family Pseudoamphisiellidae Song et al., 1996
*Pseudoamphisiella alveolata* (Kahl, 1932) Song & Warren, 2000

**Ecological feature:** mangrove (Zhanjiang); sandy sediment (Shenzhen)
**Water temperature:** 20.0 °C; 24.0 °C
**pH:** no data available; no data available
**Salinity:** 30.0‰; 30.0‰

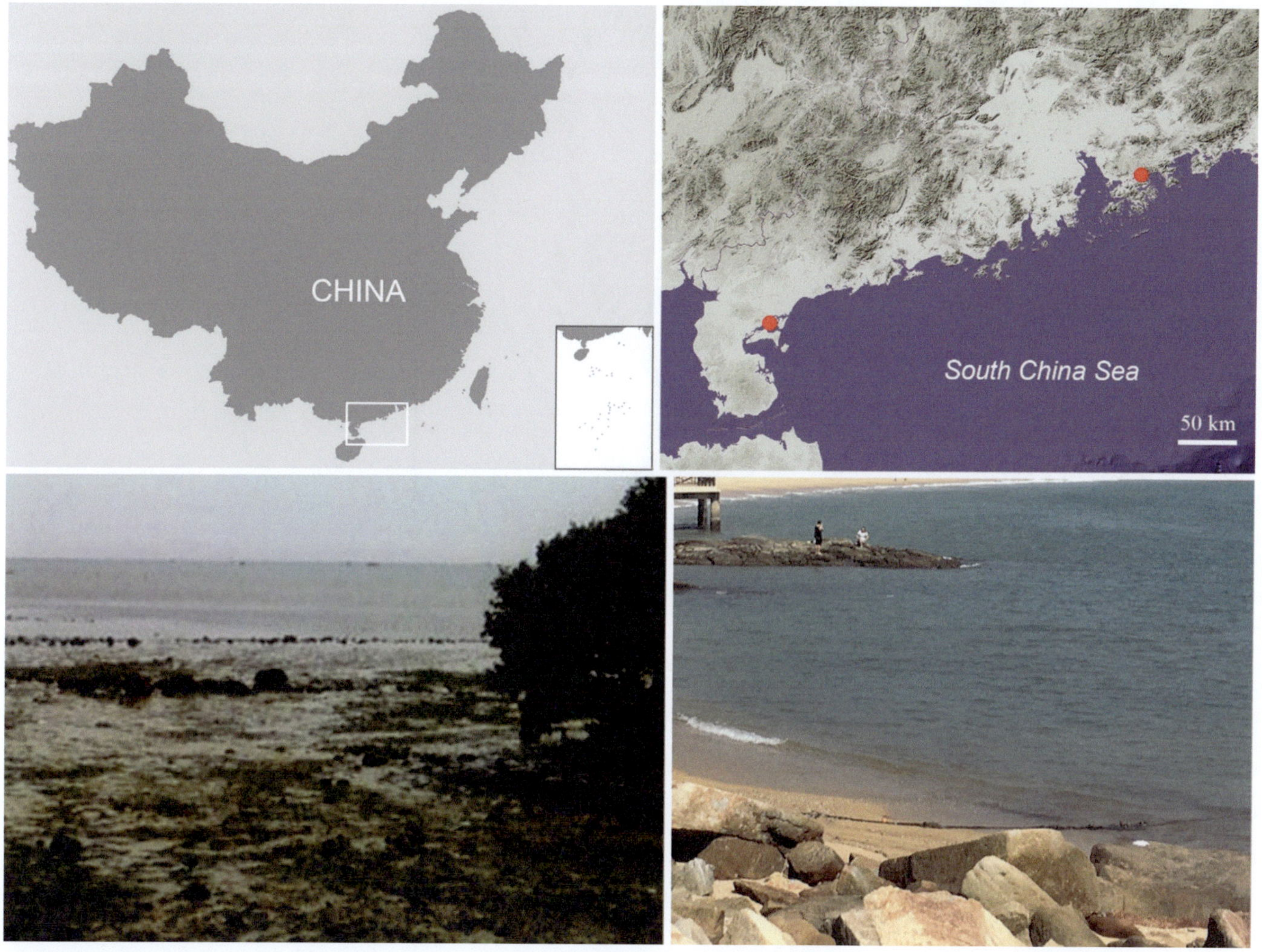


**Fig. 122A** Sample site

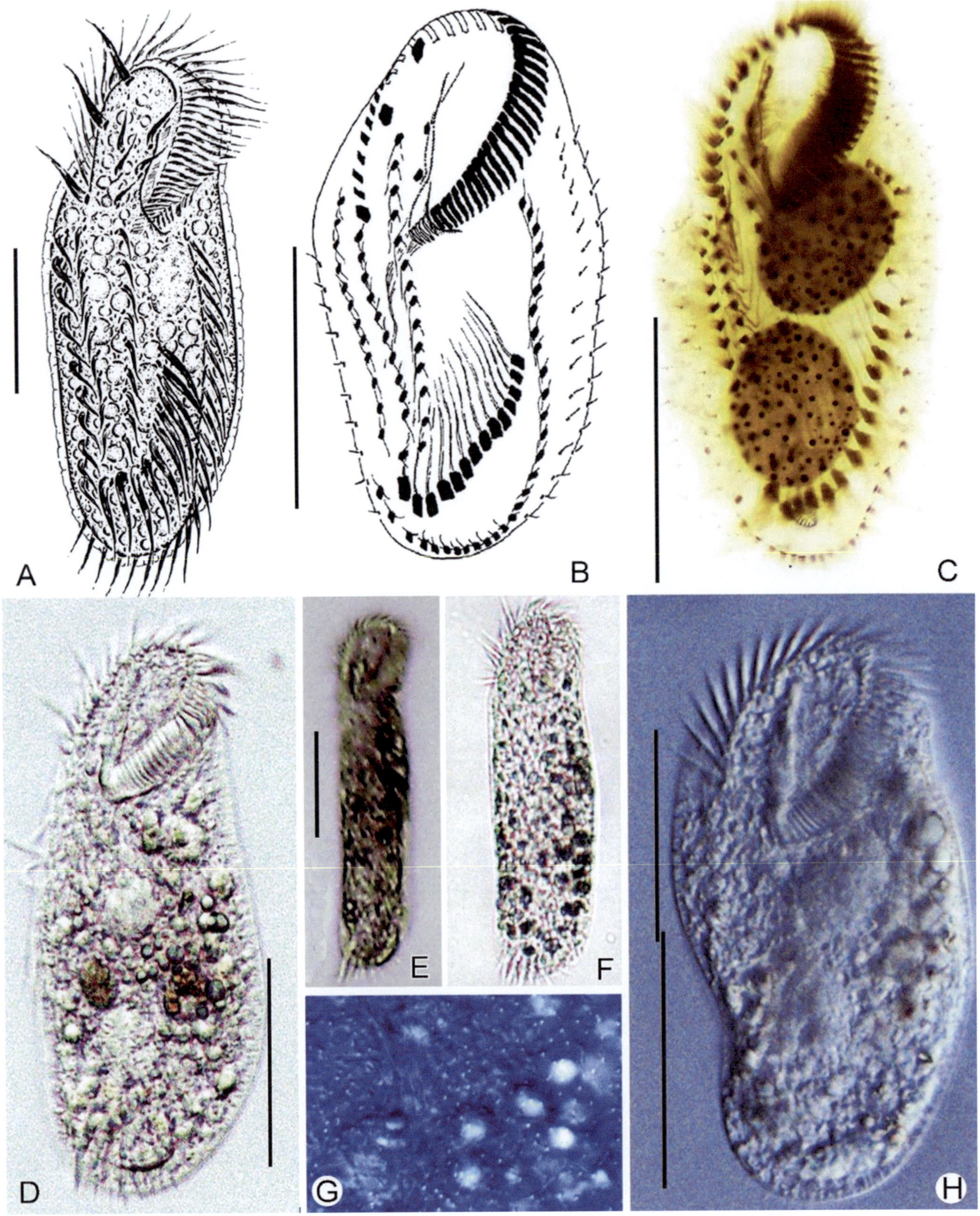


**Fig. 122B** *Pseudoamphisiella alveolata*
(**a**, **d–f**) Ventral views of live individuals. (**b**, **c**) Ventral views of ciliature. (**g**, **h**) Side (**h**) and top (**g**) views of polygonal-shaped alveoli and extrusomes. Scale bars: 50 μm

## Reference

Song WB, Warren W (2000) *Pseudoamphisiella alveolata* (Kahl, 1932) nov. comb., a large marine hypotrichous ciliate from China (Protozoa, Ciliophora, Hypotrichida). Eur J Protistol 36:451–457

Class Spirotrichea Bütschli, 1889
Order Discocephalida Wicklow, 1982
Family Pseudoamphisiellidae Song et al., 1996
*Pseudoamphisiella quadrinucleata* Shen et al., 2008

**Ecological feature:** fish-farming water (Daya Bay); offshore water (Hong Kong)
**Water temperature:** 23.0 °C; 24.0 °C
**pH:** 8.0; 7.9
**Salinity:** 19.0‰; 33.5‰

**Fig. 123A** Sample site

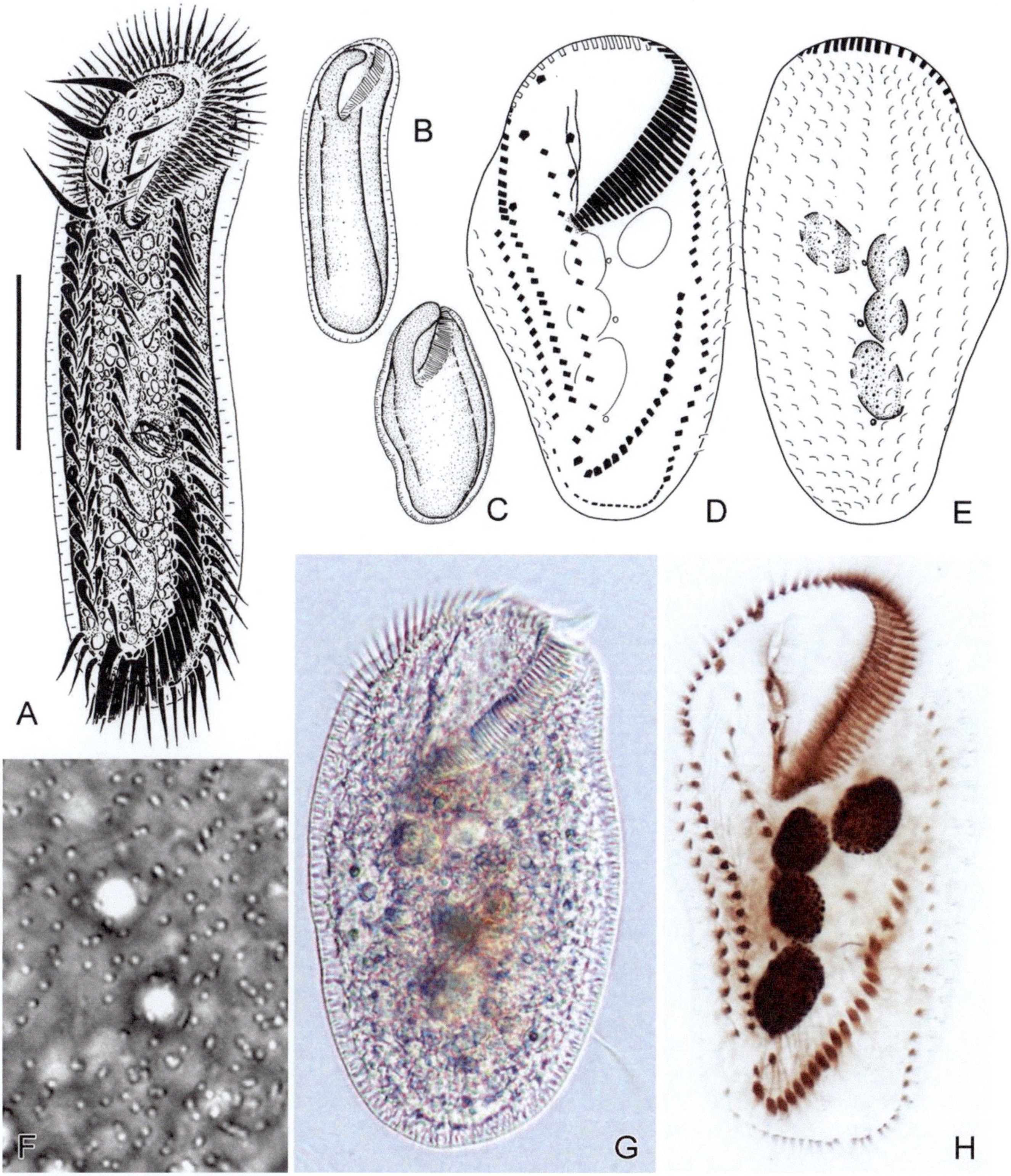


**Fig. 123B** *Pseudoamphisiella quadrinucleata*
(**a**, **g**) Ventral views of live individuals. (**b**, **c**) Shape variants of individuals. (**d**, **e**, **h**) Ventral (**d**, **h**) and dorsal (**e**) views of ciliature. (**f**) Dorsal side to show the distribution of extrusomes (as shining dots). Scale bar: 50 μm

## Reference

Shen Z, Lin X, Long H, Miao M, Liu H, Al-Rasheid KAS, Song WB (2008) Morphology and small subunit rDNA gene sequence of *Pseudoamphisiella quadrinucleata* n. sp. (Ciliophora, Urostylida) from the South China Sea. J Eukaryot Microbiol 55:510–514

Class Spirotrichea Bütschli, 1889
Order Euplotida Small & Lynn, 1985
Family Aspidiscidae Ehrenberg, 1830
*Aspidisca hexeris* Quennerstedt, 1869

**Ecological feature:** mangrove (Zhanjiang); intertidal zone (Shenzhen)
**Water temperature:** 20.0 °C; 23.0 °C
**pH:** no data available; no data available
**Salinity:** 26.0‰; 27.0‰

**Fig. 124A** Sample site

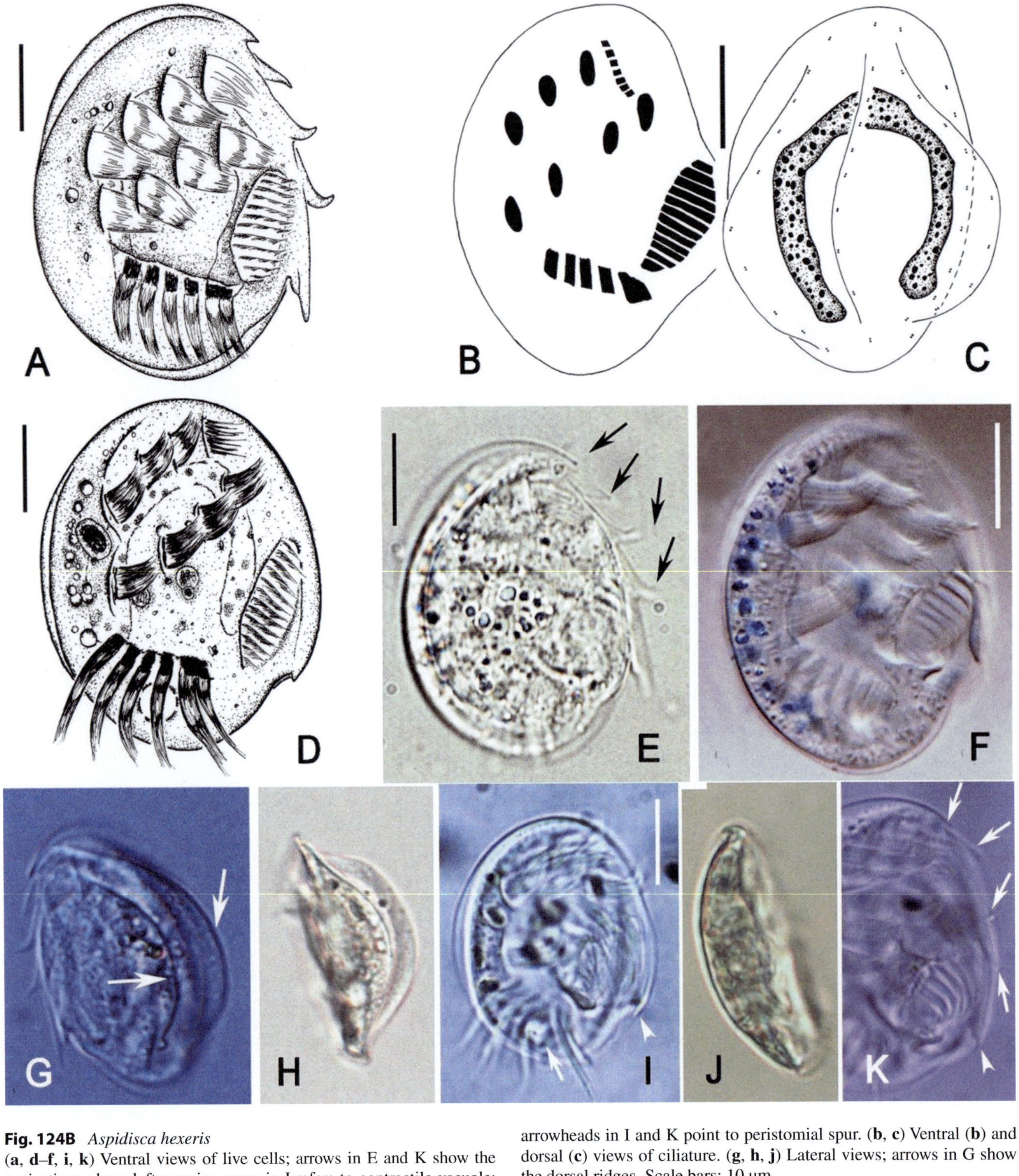


**Fig. 124B** *Aspidisca hexeris*
(**a**, **d–f**, **i**, **k**) Ventral views of live cells; arrows in E and K show the projections along left margin; arrow in I refers to contractile vacuole; arrowheads in I and K point to peristomial spur. (**b**, **c**) Ventral (**b**) and dorsal (**c**) views of ciliature. (**g**, **h**, **j**) Lateral views; arrows in G show the dorsal ridges. Scale bars: 10 μm

## Reference

Jiang J, Huang J, Li J, Al-Rasheid KAS, Al-Farraj SA, Lin X, Hu X (2013) Morphology of two marine euplotids (Ciliophora: Euplotida), *Aspidisca fusca* Kahl, 1928 and *A. hexeris* Quennerstedt, 1869, with notes on their small subunit rRNA gene sequences. Eur J Protistol 49:634–643

Class Spirotrichea Bütschli, 1889
Order Euplotida Small & Lynn, 1985
Family Aspidiscidae Ehrenberg, 1830
*Aspidisca hongkongensis* Shen et al., 2010

**Ecological feature:** offshore water
**Water temperature:** 22.0 °C
**pH:** 7.9
**Salinity:** 33.0‰

**Fig. 125A** Sample site

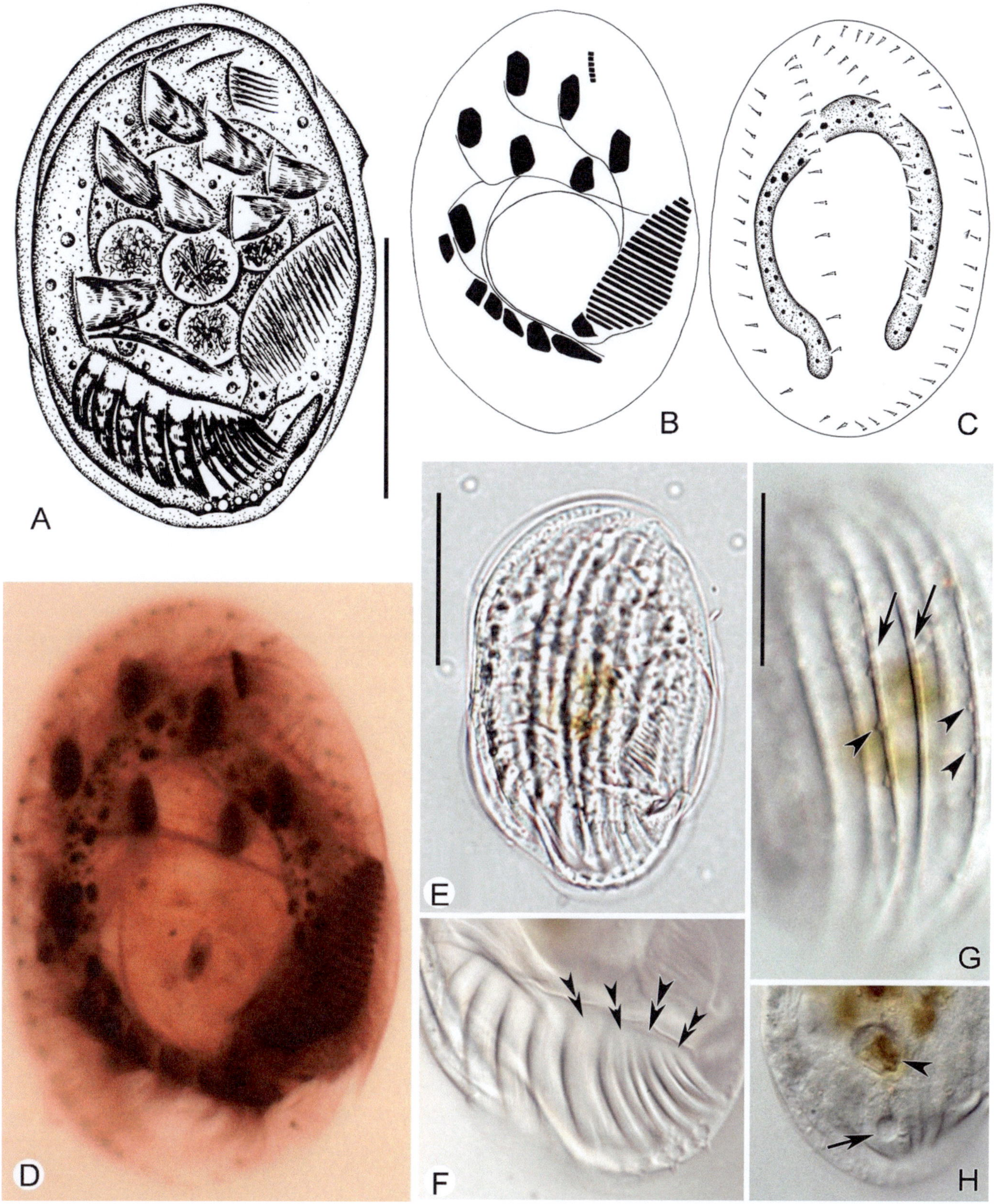


**Fig. 125B** *Aspidisca hongkongensis*
(**a**, **e**) Ventral views of live individuals. (**b–d**) Ventral (**b**, **d**) and dorsal (**c**) views of ciliature. (**f**) Transverse cirri (double arrowheads). (**g**) Dorsal ridges (arrows) and dorsal bristles (arrowheads). (**h**) Contractile vacuole (arrow) and food vacuole (arrowhead). Scale bars: 40 μm (**a**), 35μm (**e**), and 25μm (**g**)

## Reference

Shen Z, Huang J, Lin X F, Yi Z, Li J, Song WB (2010) Morphological and molecular characterization of *Aspidisca hongkongensis* spec. nov. (Ciliophora, Euplotida) from the South China Sea. Eur J Protistol 46:204–211

Class Spirotrichea Bütschli, 1889
Order Euplotida Small & Lynn, 1985
Family Aspidiscidae Ehrenberg, 1830
*Aspidisca magna* Kahl, 1932

**Ecological feature:** mangrove
**Water temperature:** 24.0 °C
**pH:** no data available
**Salinity:** 25.0‰

**Fig. 126A** Sample site

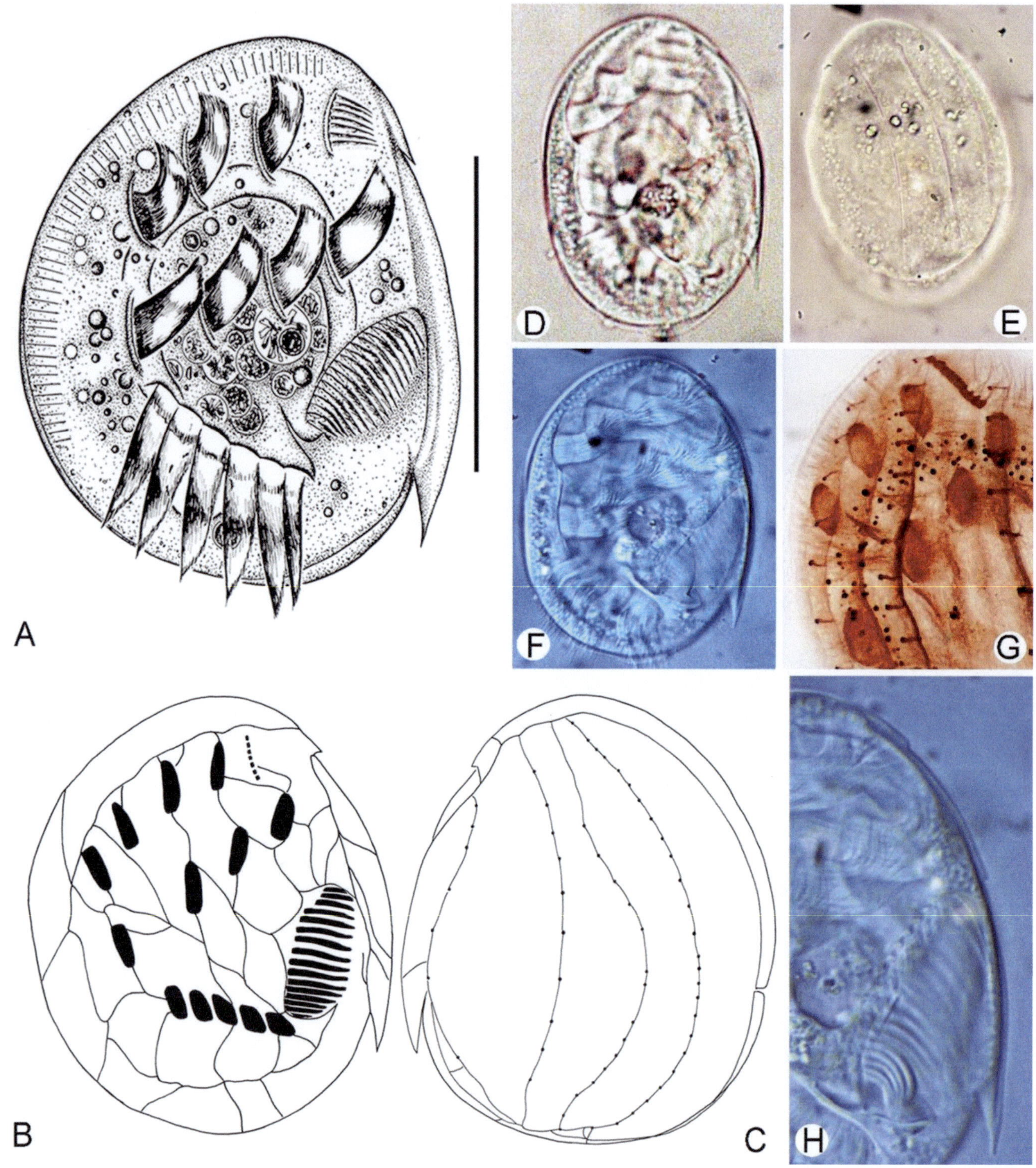


**Fig. 126B** *Aspidisca magna*
(**a**, **d**, **f**) Ventral views of live cells. (**b**, **c**, **g**) Ventral (**b**, **g**) and dorsal (**c**) views of ciliature. (**e**) Dorsal view of live cell. (**h**) Posterior membranelles and peristomial spur. Scale bar: 50 µm

## References

Kahl A (1932) Urtiere oder Protozoa I: Wimpertiere oder Ciliata (Infusoria) 3. Spirotricha. Tierwelt Dtl 25:399–650

Li L, Zhang Q, Al-Rasheid KAS, Kwon C, Shin M (2010) Morphological redescriptions of *Aspidisca magna*Kahl, 1932 and *A. leptaspis* Fresenius, 1865 (Ciliophora, Euplotida), with notes on morphogenetic process in *A. magna*. Acta Protozool 49:327–337

Tuffrau M (1964) La morphogenèse de bipartition et les structures neuromotrices dans le genre *Aspidisca* (ciliés hypotriches). Re-vue de quelques espèces. Cah Biol Mar 5:173–199

Class Spirotrichea Bütschli, 1889
Order Euplotida Small & Lynn, 1985
Family Euplotidae Ehrenberg, 1838
*Euplotes balteatus* Dujardin, 1841

**Ecological feature:** mangrove (Zhuhai); mollusc-farming pond (Daya Bay)
**Water temperature:** 25.0 °C; 19.4 °C
**pH:** 6.5; 8.2
**Salinity:** 13.0‰; 32.6‰

**Fig. 127A** Sample site

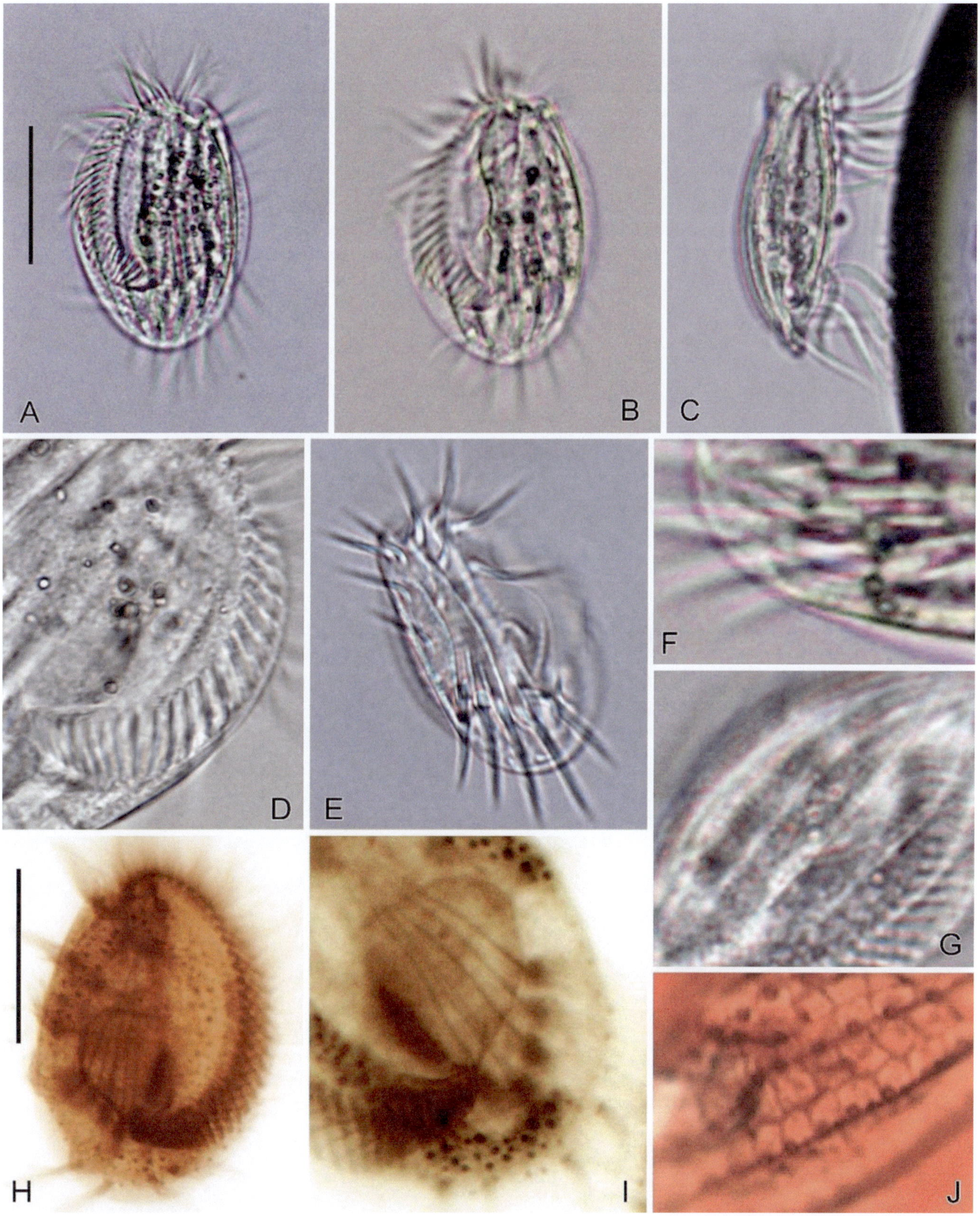


**Fig. 127B** *Euplotes balteatus*
(**a–c**, **e**) Views in vivo. (**d**)Adoral membranelles. (**f**) Contractile vacuole. (**g**) Dorsal ridges. (**h**, **i**) Ciliature. (**j**) Part silverline system on dorsal side. Scale bars: 30μm

## Reference

Song WB, Wilbert N (2002) Faunistic studies on marine ciliates from the Antarctic benthic area, including descriptions of one epizoic form, 6 new species and 2 new genera (Protozoa: Ciliophora). Acta Protozool 41:23–61

Class Spirotrichea Bütschli, 1889
Order Euplotida Small & Lynn, 1985
Family Euplotidae Ehrenberg, 1838
*Euplotes* cf. *parawoodruffi*

**Ecological feature:** mangrove; offshore water
**Water temperature:** no data available; no data available
**pH:** 7.0; 8.0
**Salinity:** 6.0‰; 29.0‰

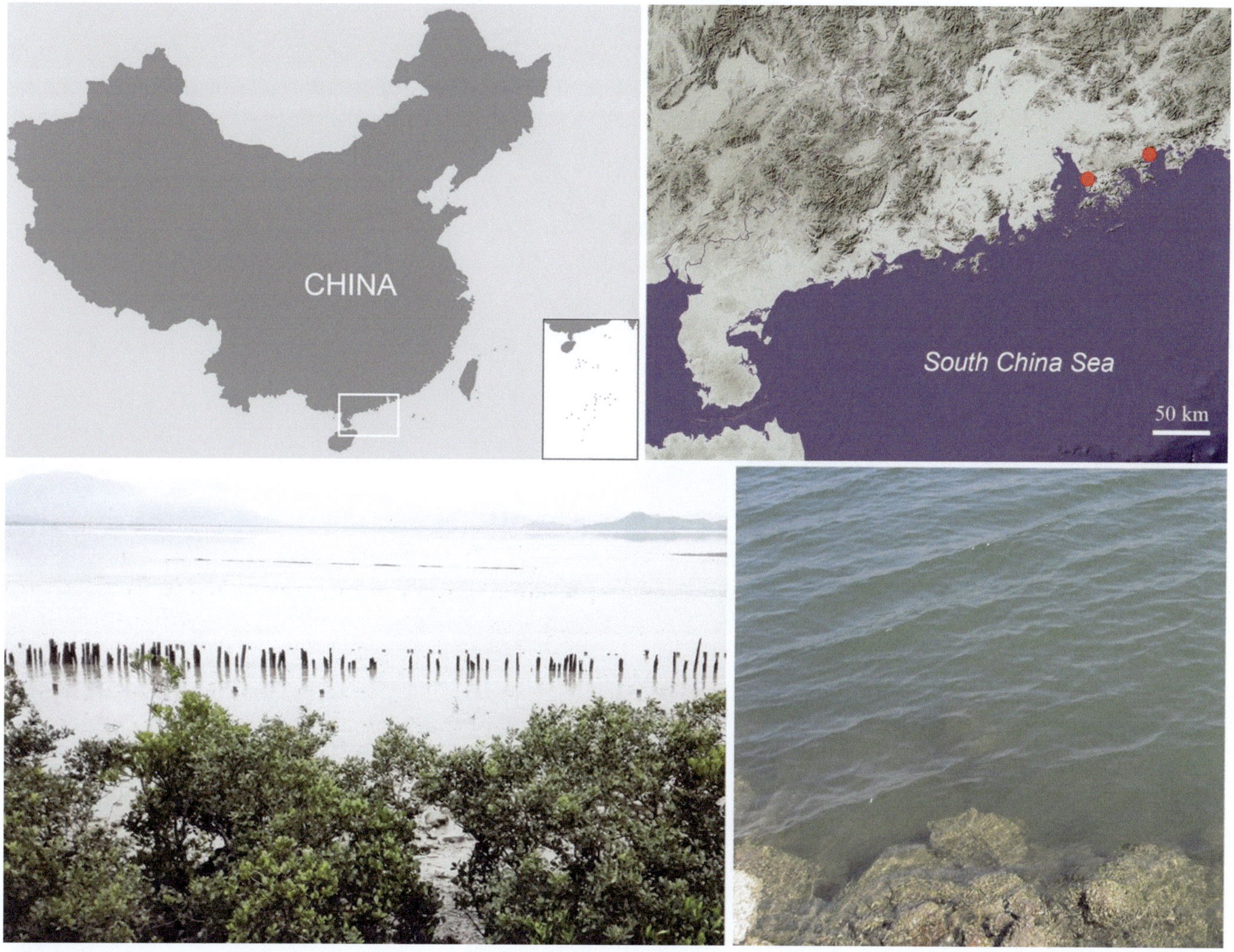


**Fig. 128A** Sample site

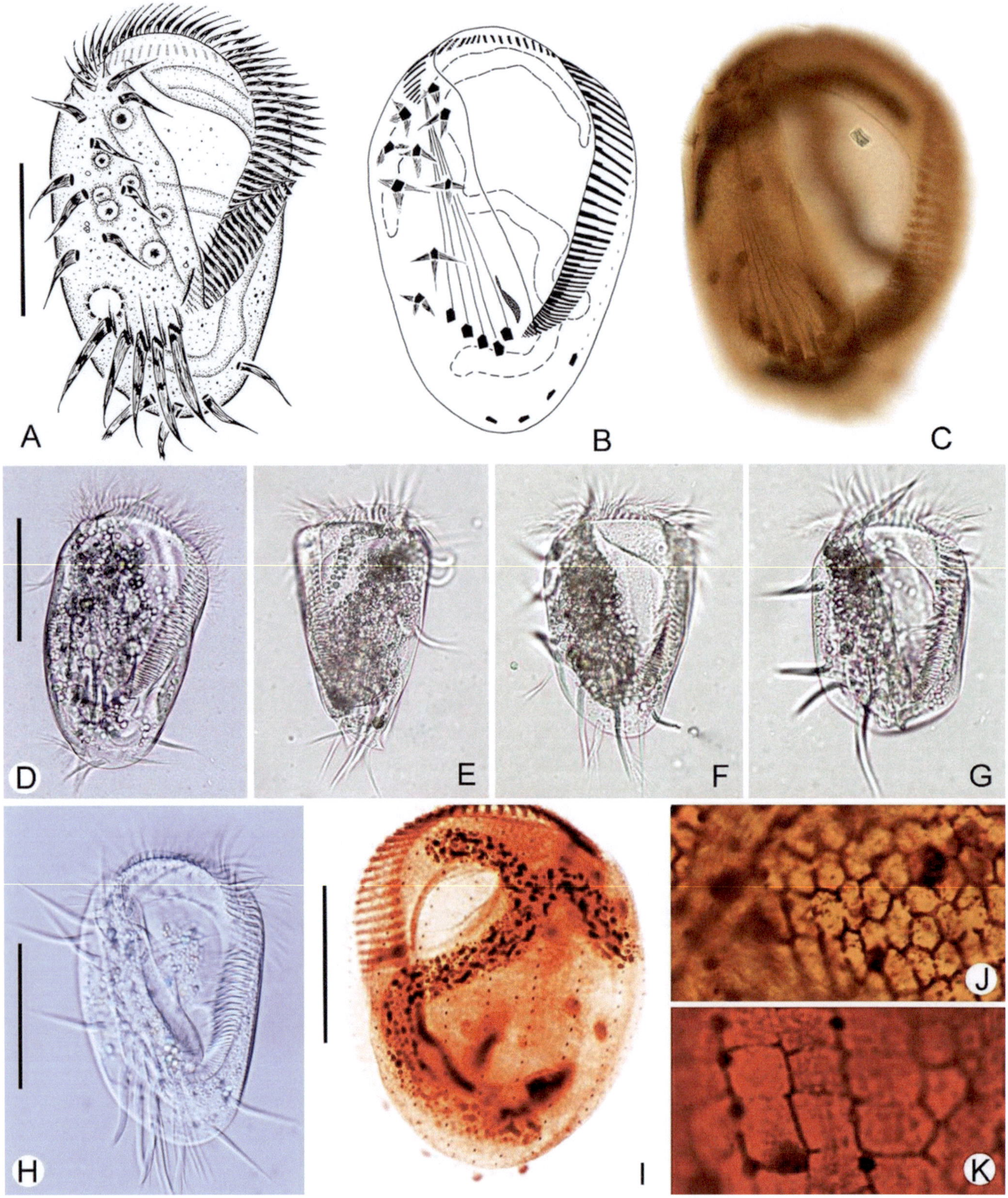


**Fig. 128B** *Euplotes* cf. *parawoodruffi*
(**a**, **d**–**h**) Views of live individuals. (**b**, **c**, **i**) Ventral (**b**, **c**) and dorsal (**i**) views of ciliature. (**j**, **k**) Silverlines on ventral (**j**) and dorsal (**k**) sides. Scale bars: 40 μm (A) and 50 μm (D, H, I)

## References

Shen Z, Lin X, Li J (2008) Morphological redescription of a rare marine euplotids, *Euplotes parawoodruffi* (Protozoa, Ciliophora, Euplotida). Acta Zootaxon Sin 33:335–339

Song WB, Bradbury P (1997) Comparative studies on a new brackish water *Euplotes*, *E. parawoodruffi* n. sp., and a redescription of *Euplotes woodruffi* Gaw, 1939 (Ciliophora: Hypotrichida). Arch Protistenk 148:399–412

Class Spirotrichea Bütschli, 1889
Order Euplotida Small & Lynn, 1985
Family Euplotidae Ehrenberg, 1838
*Euplotes* cf. *woodruffi*

**Ecological feature:** mangrove
**Water temperature:** 24.0 °C, 25.0 °C
**pH:** no data available; no data available
**Salinity:** 17.0‰, 15.0‰

**Fig. 129A** Sample site

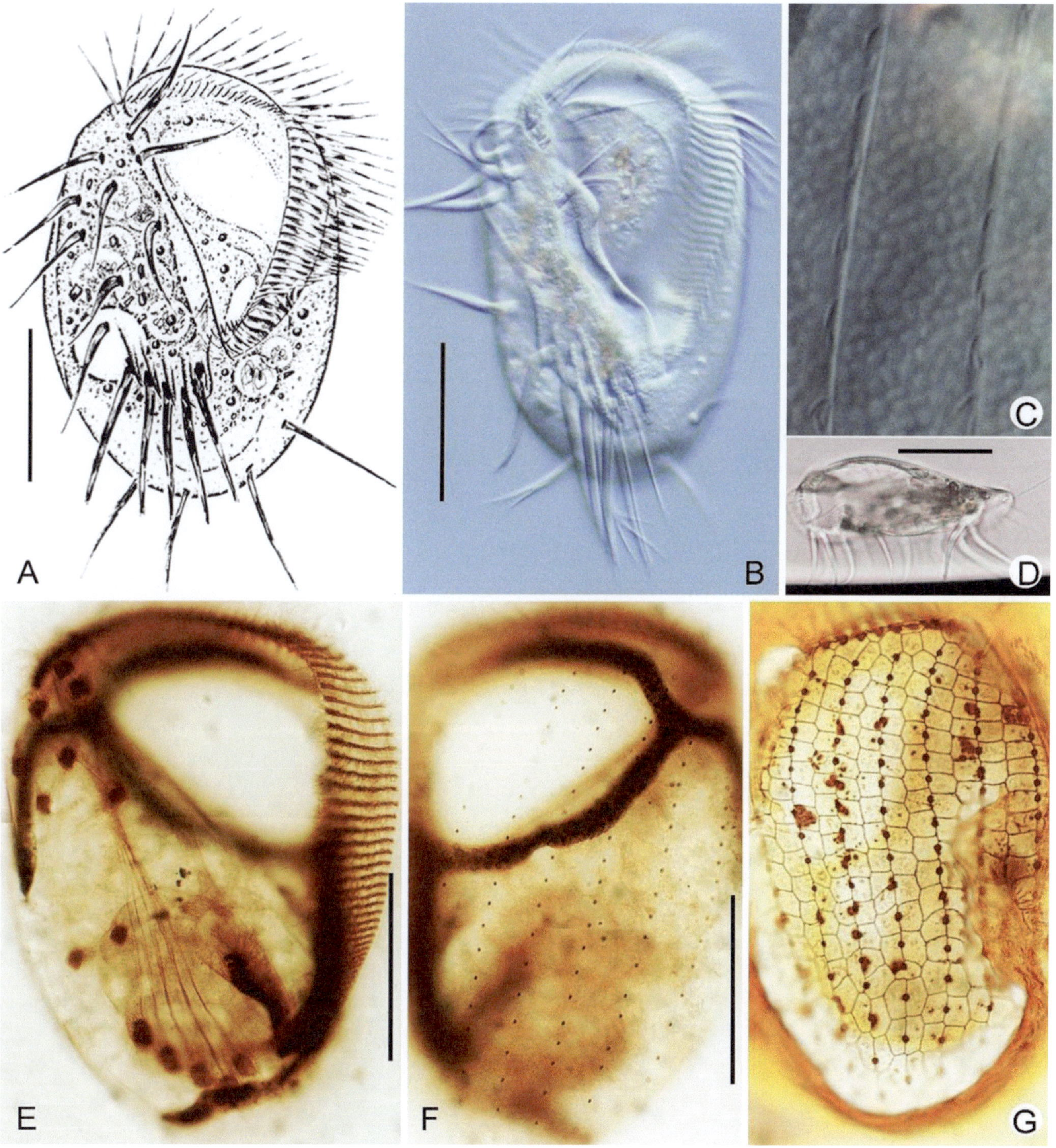


**Fig. 129B** *Euplotes* cf. *woodruffi*
(**a**, **b**) Ventral views of live individuals. (**c**) Cortical granules and dorsal cilia. (**d**) Lateral view of live cell. (**e**, **f**) Ventral (**e**) and dorsal (**f**) views of ciliature. (**g**) Dorsal view of silverline system. Scale bars: 50 μm

## Reference

Song WB, Bradbury P (1997) Comparative studies on a new brackish water *Euplotes*, *E. parawoodruffi* n. sp., and a redescription of *Euplotes woodruffi* Gaw, 1939 (Ciliophora; Hypotrichida). Arch Protistenk 148:399–412

Class Spirotrichea Bütschli, 1889
Order Euplotida Small & Lynn, 1985
Family Euplotidae Ehrenberg, 1838
*Euplotes charon* Müller, 1786

**Ecological feature:** mangrove (Zhanjiang, Zhuhai); mollusc-farming pond (Daya Bay)
**Water temperature:** 26.0 °C; 18.0 °C; 19.4 °C
**pH:** 6.7; 6.0; 8.2
**Salinity:** 6‰; 4.4‰; 32.6‰

**Fig. 130A** Sample site

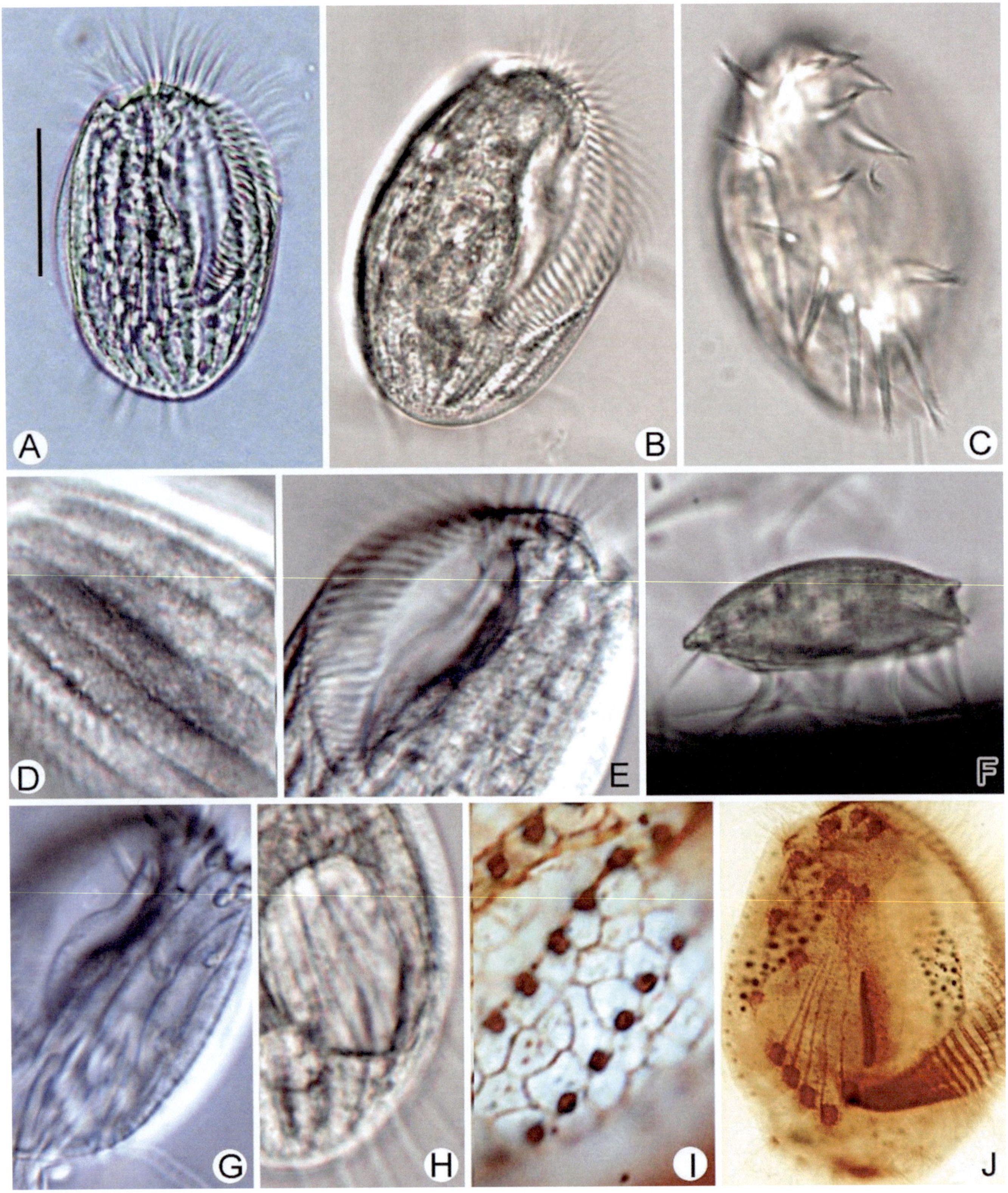


**Fig. 130B** *Euplotes charon*
(**a–c**) Ventral views of live cells. (**d**) Dorsal ridges. (**e**) Adoral membranelles. (**f**) Lateral view of live cell. (**g**) Three long ridges on ventral side. (**h**) Contractile vacuole. (**i**) Silverline system on dorsal side. (**j**) Ventral view of ciliature. Scale bar: 30 μm

## Reference

Song WB, Packroff G (1997) Taxonomische Untersuchungen an marinen Ciliaten aus China mit Beschreibungen von zwei neuen Arten, *Strombidium globosaneum* nov. spec. und *S. platum* nov. spec. (Protozoa, Ciliophora). Arch Protistenk 147:331–360

Class Spirotrichea Bütschli, 1889
Order Euplotida Small & Lynn, 1985
Family Euplotidae Ehrenberg, 1838
*Euplotes encysticus* Yonezawa, 1985

**Ecological feature:** sandy beach
**Water temperature:** 12.0 °C
**pH:** no data available
**Salinity:** 6.0‰

**Fig. 131A** Sample site

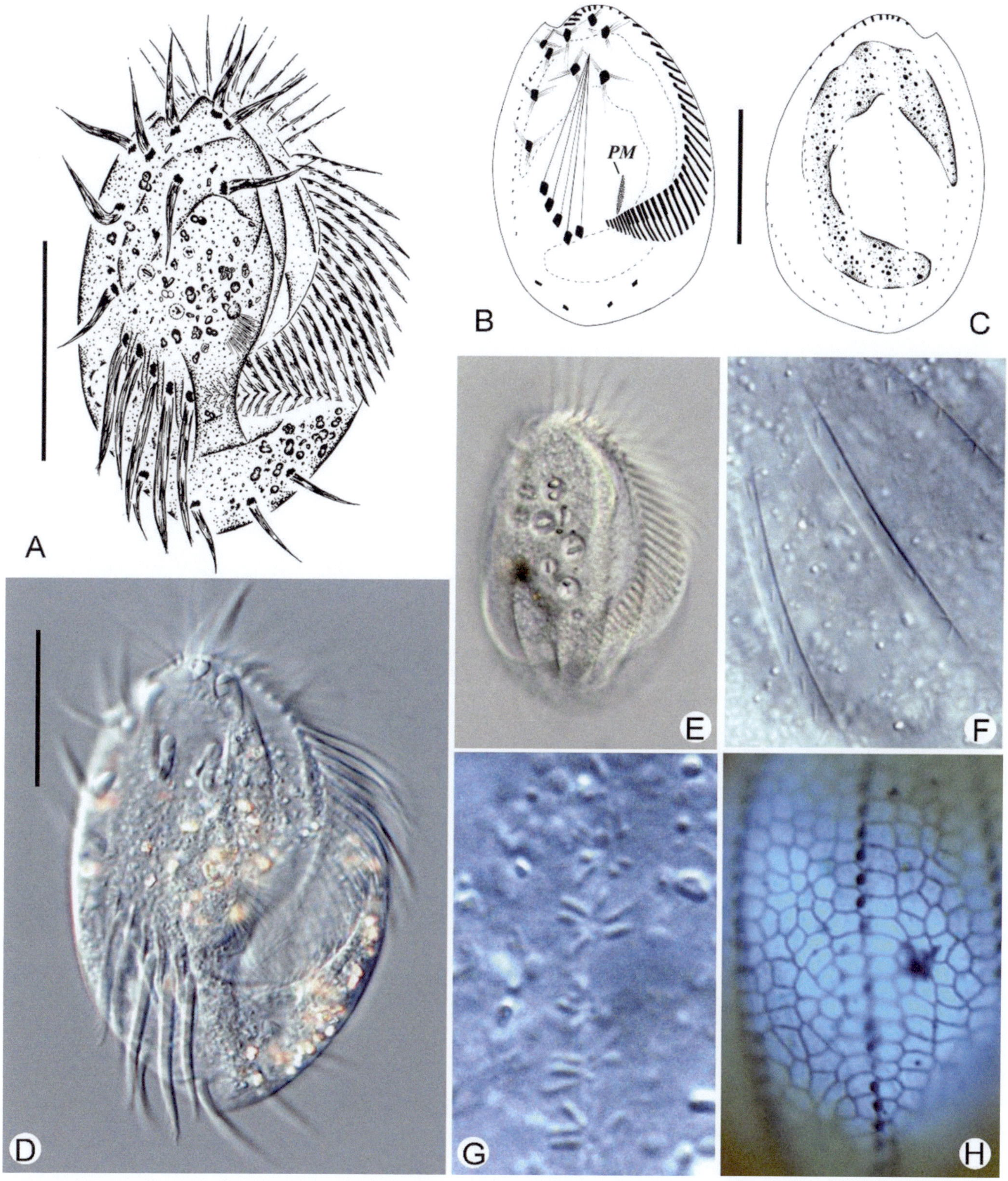


**Fig. 131B** *Euplotes encysticus*
(**a**, **d**) Ventral views of live cells. (**b**, **c**) Ventral (**b**) and dorsal (**c**) views of ciliature. (**e**) Dorsal ridges and food vacuoles. (**f**) Dorsal cilia near the ridges. (**g**) Granules around each dorsal cilia. (**h**) Silverlines on dorsal side. Scale bars: 40 μm

## References

Fan X, Huang J, Lin X, Li J, Al-Rasheid KAS, Hu X (2010) Morphological and molecular characterization of *Euplotes encysticus* (Protista: Ciliophora: Euplotida). J Mar Biol Ass UK 90:1411–1416

Yonezawa F (1985) New hypotrichous ciliate *Euplotes encysticus* sp. nov. J Sci Hiroshima Univ 32:35–45

Class Spirotrichea Bütschli, 1889
Order Euplotida Small & Lynn, 1985
Family Euplotidae Ehrenberg, 1838
*Euplotes minuta* Yocum, 1930

**Ecological feature:** offshore water
**Water temperature:** 23.1 °C
**pH:** 8.2
**Salinity:** 28.6‰

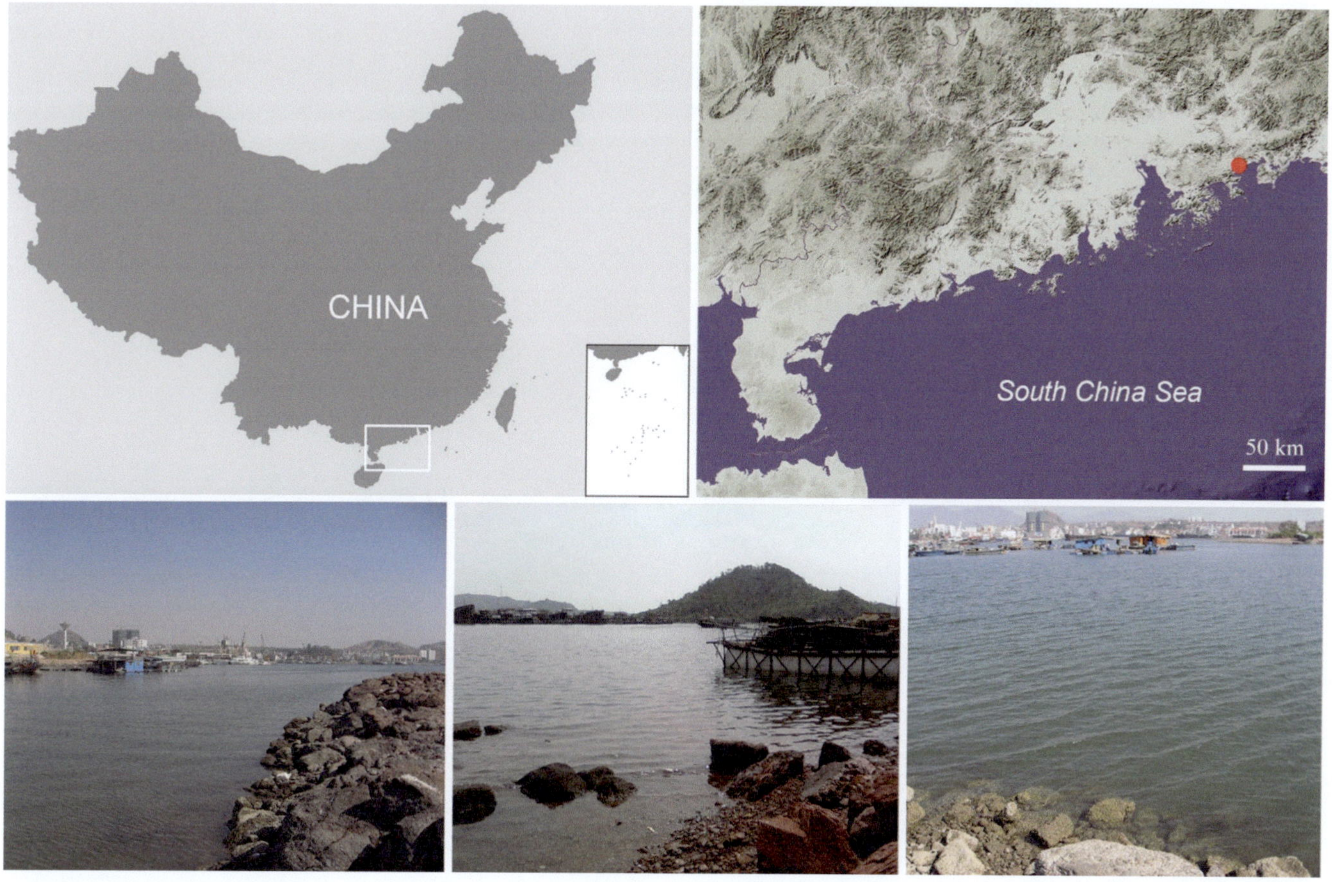


**Fig. 132A** Sample site

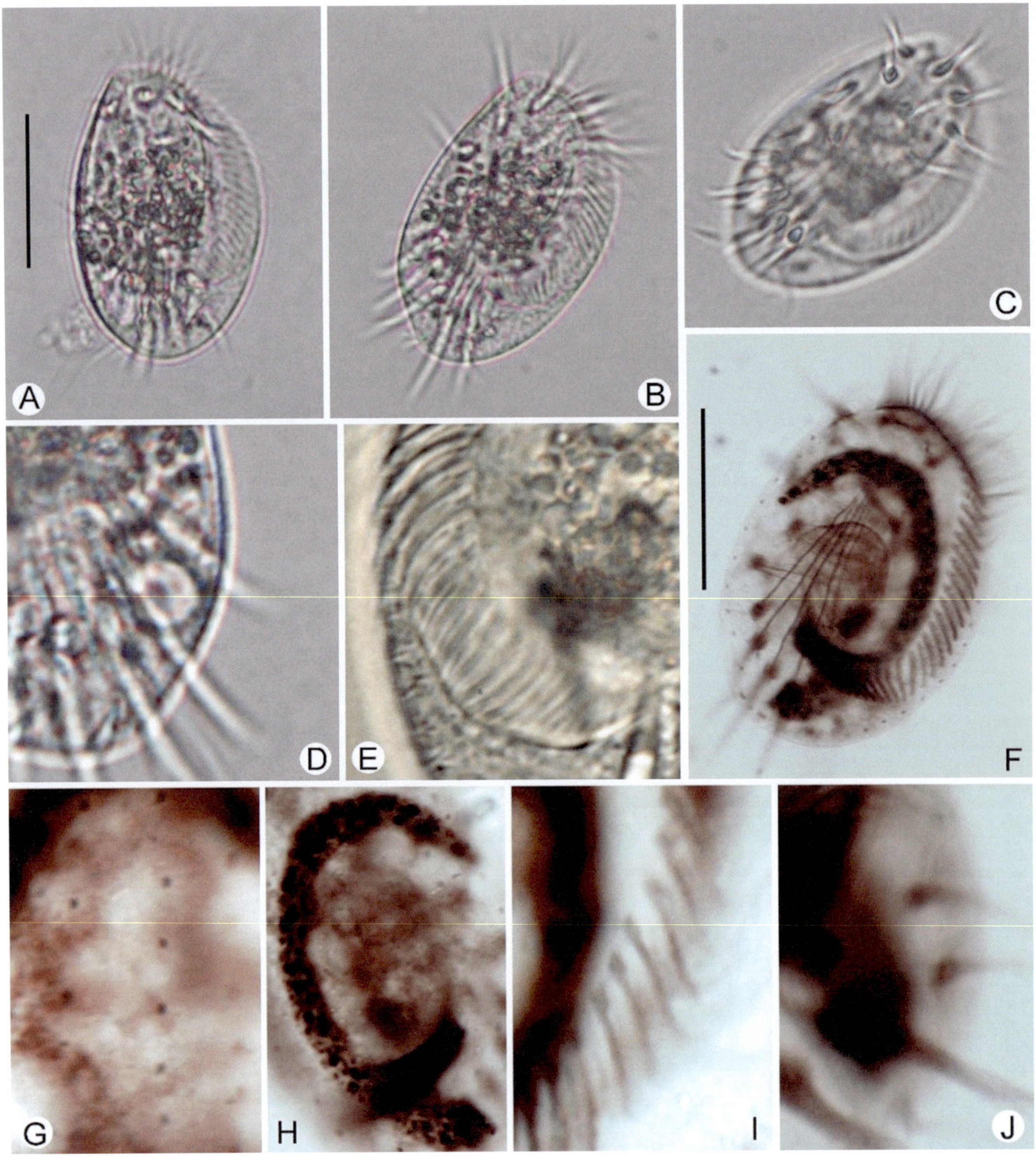


**Fig. 132B** *Euplotes minuta*
(**a**, **b**) Ventral views of live individuals. (**d**) Contractile vacuole. (**e**) Detail of buccal field showing adoral membranelles. (**f**) Ventral view of ciliature. (**g**) Part of dorsal kineties. (**h**) Macronucleus. (**i**) Adoral membranelles. (**j**) Posterior portion of cell, noting marginal cirri. Scale bars: 30 μm

## Reference

Song WB, Wilbert N (1997) Morphological investigation on some free living ciliates (Protozoa, Ciliophora) from China Sea with description of a new hypotrichous genus, *Hemigastrostyla* nov. gen. Arch Protistenk 148:413–444

Class Spirotrichea Bütschli, 1889
Order Euplotida Small & Lynn, 1985
Family Euplotidae Ehrenberg, 1838
*Euplotes moebiusi* Kahl, 1932

**Ecological feature:** offshore water
**Water temperature:** 23.1 °C
**pH:** 8.2
**Salinity:** 28.6‰

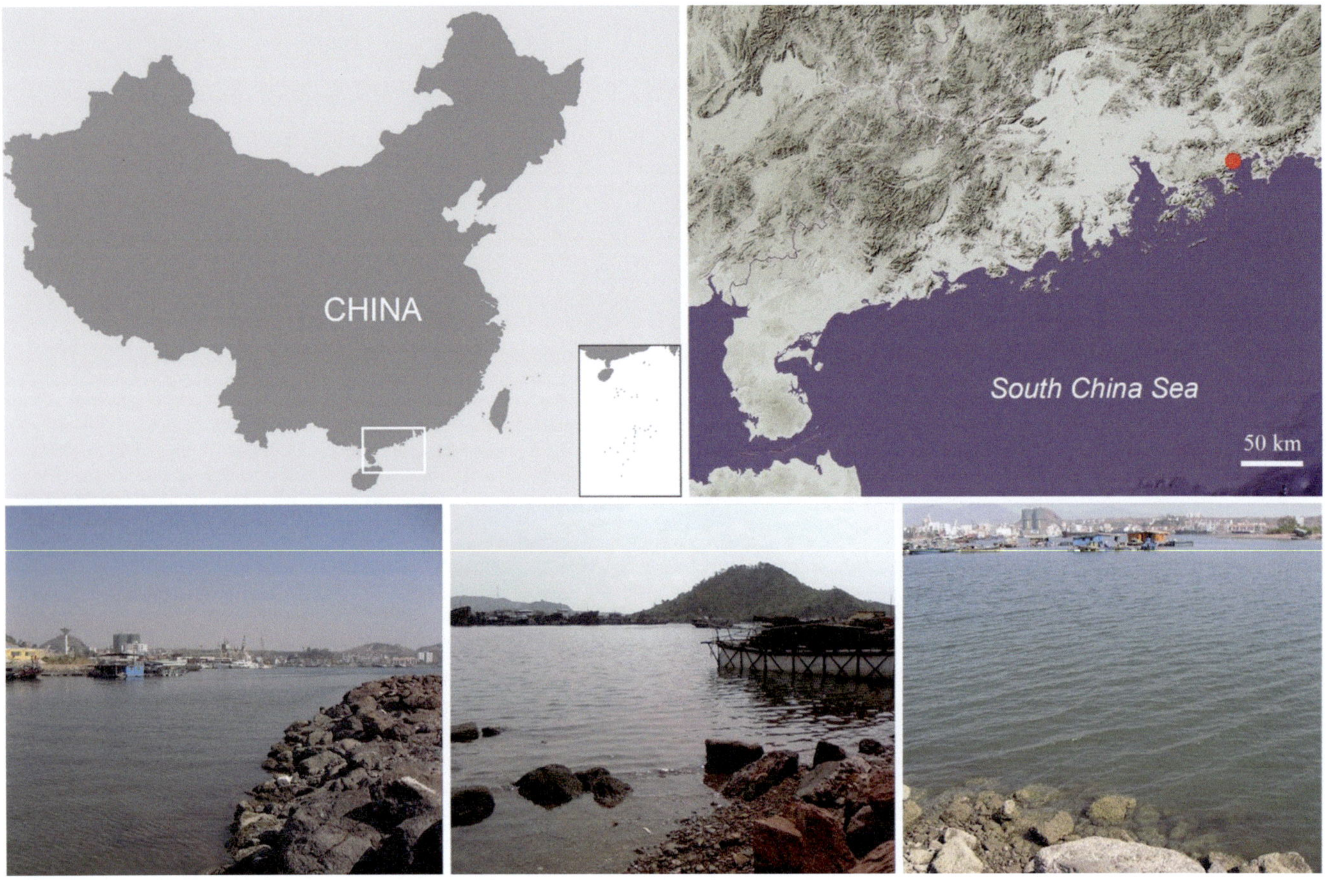


**Fig. 133A** Sample site

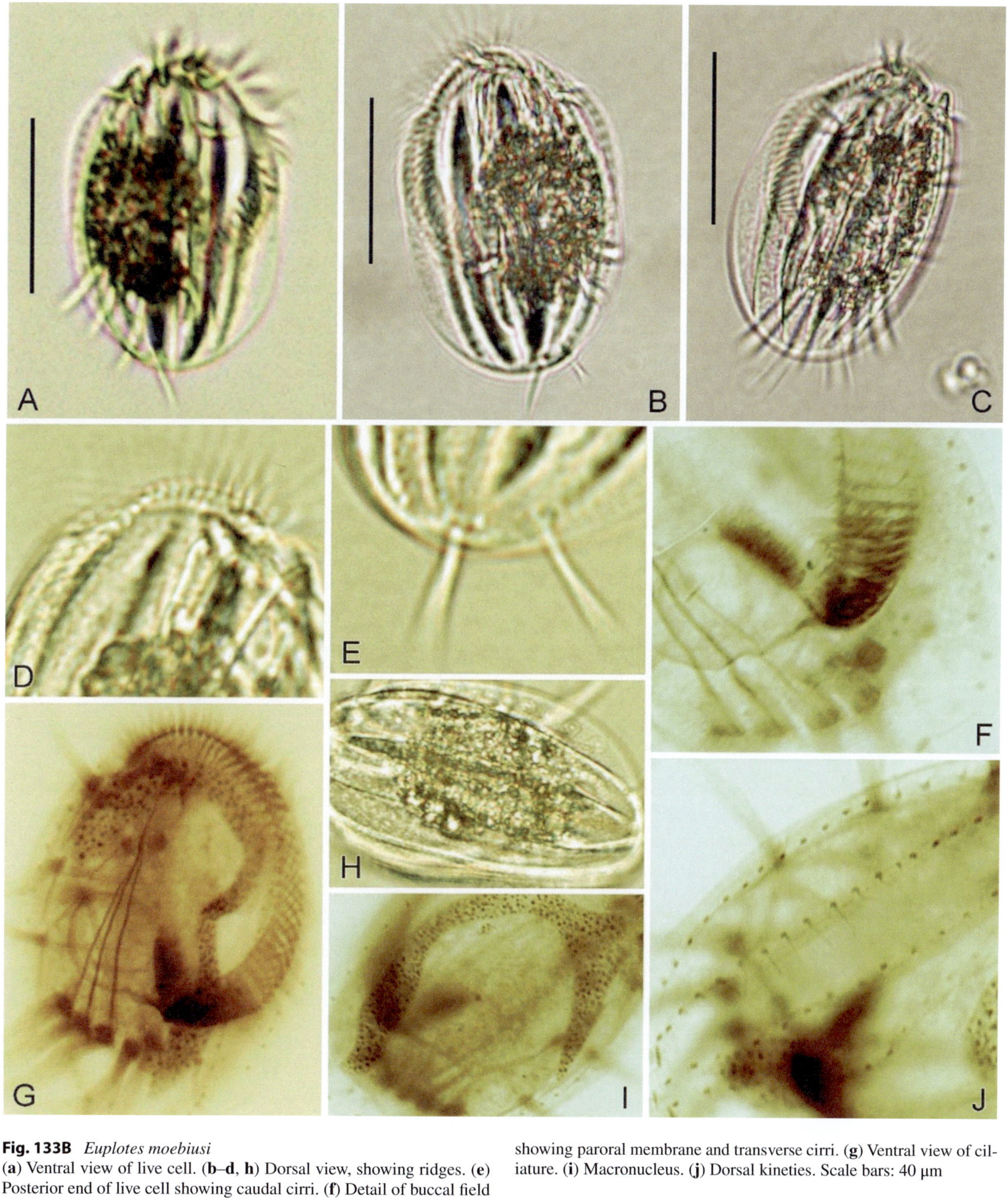


**Fig. 133B** *Euplotes moebiusi*
(**a**) Ventral view of live cell. (**b–d**, **h**) Dorsal view, showing ridges. (**e**) Posterior end of live cell showing caudal cirri. (**f**) Detail of buccal field showing paroral membrane and transverse cirri. (**g**) Ventral view of ciliature. (**i**) Macronucleus. (**j**) Dorsal kineties. Scale bars: 40 μm

## References

Foissner W (1978) *Euplotes moebiusi* f. *quadricirratus* (Ciliophora, Hypotrichida). I. Die Feinstruktur des Cortex und der argyrophilen Strukturen. Arch Protistenk 120:86–117

Kahl A (1932) Urtierre oder Protozoa I: Wimpertiere oder Ciliata (Infusoria) 3. Spirotricha. Tierwelt Dtl 25:399–650

Class Spirotrichea Bütschli, 1889
Order Euplotida Small & Lynn, 1985
Family Euplotidae Ehrenberg, 1838
*Euplotes orientalis* Jiang et al., 2010

**Ecological feature:** shrimp-farming pond
**Water temperature:** 25.0 °C
**pH:** no data available
**Salinity:** 24.0‰

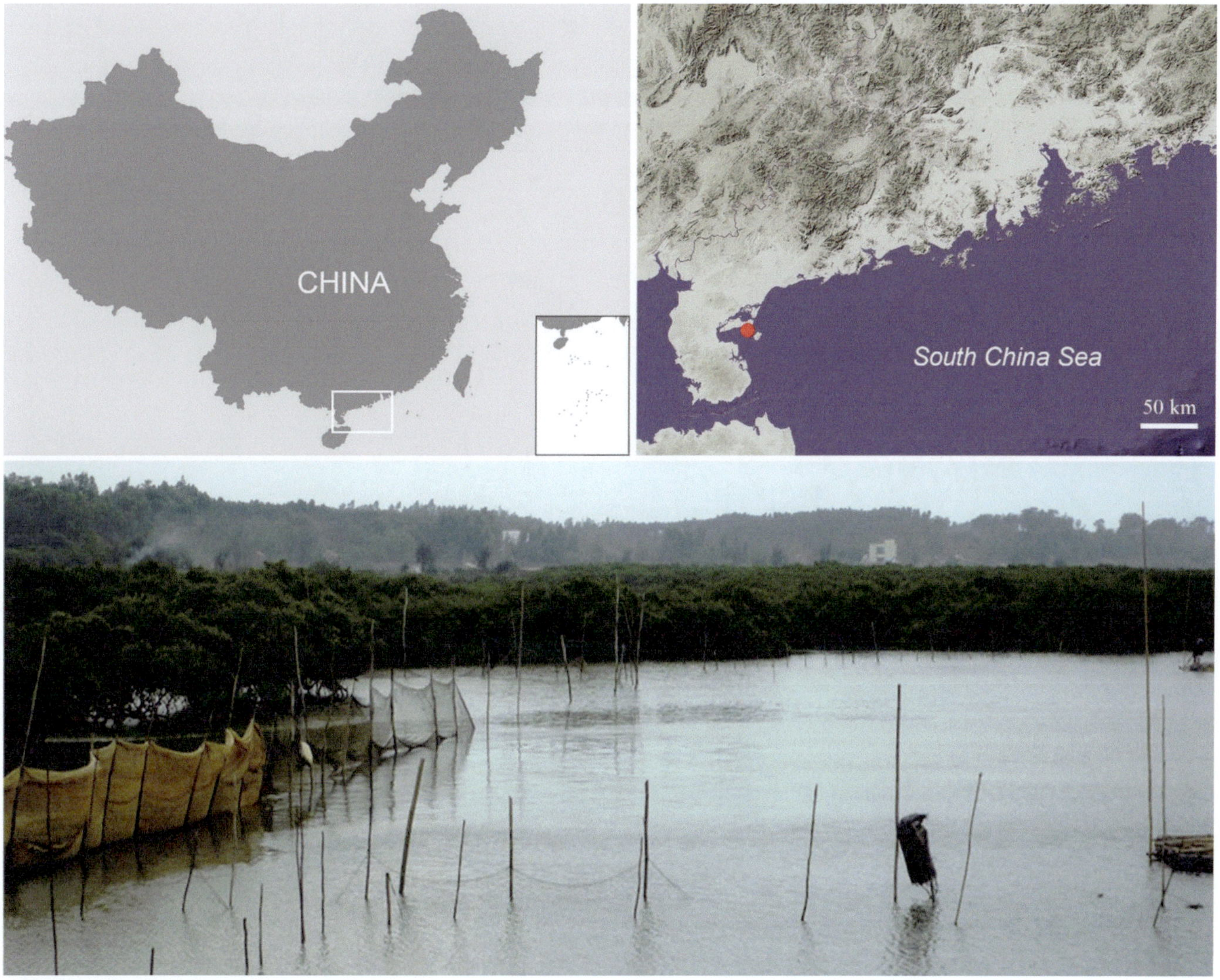


**Fig. 134A** Sample site

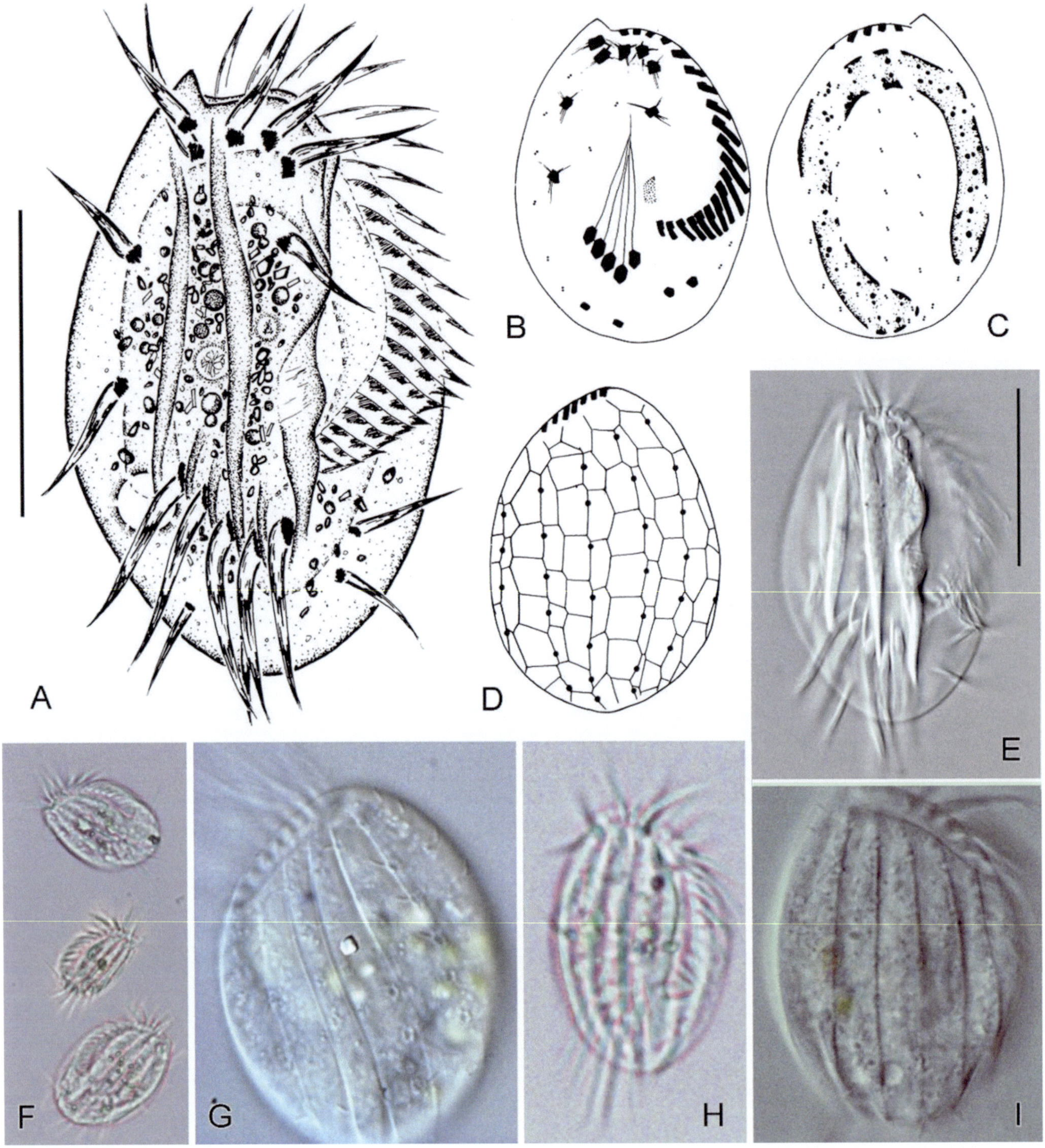


**Fig. 134B** *Euplotes orientalis*
(**a**, **e**, **h**) Ventral views of live cells. (**b**, **c**) Ventral (**b**) and dorsal (**c**) views of ciliature. (**d**) Silverline system on dorsal side. (**f**) Different individuals to show body shapes and sizes. (**g**, **i**) Dorsal ridges. Scale bars: 20 μm

## Reference

Jiang J, Zhang Q, Warren A, Al-Rasheid KAS, Song WB (2010) Morphology and SSU rRNA gene-based phylogeny of two marine *Euplotes* species, *E. orientalis* spec. nov. and *E. raikovi* Agamaliev, 1966 (Ciliophora, Euplotida). Eur J Protistol 46:121–132

Class Spirotrichea Bütschli, 1889
Order Euplotida Small & Lynn, 1985
Family Euplotidae Ehrenberg, 1838
*Euplotes rariseta* Curds et al., 1974

**Ecological feature:** mangrove (Zhanjiang); offshore water (Daya Bay)
**Water temperature:** 24.0 °C; 23.1 °C
**pH:** 6.8; 8.2;
**Salinity:** 23.0‰; 28.6‰

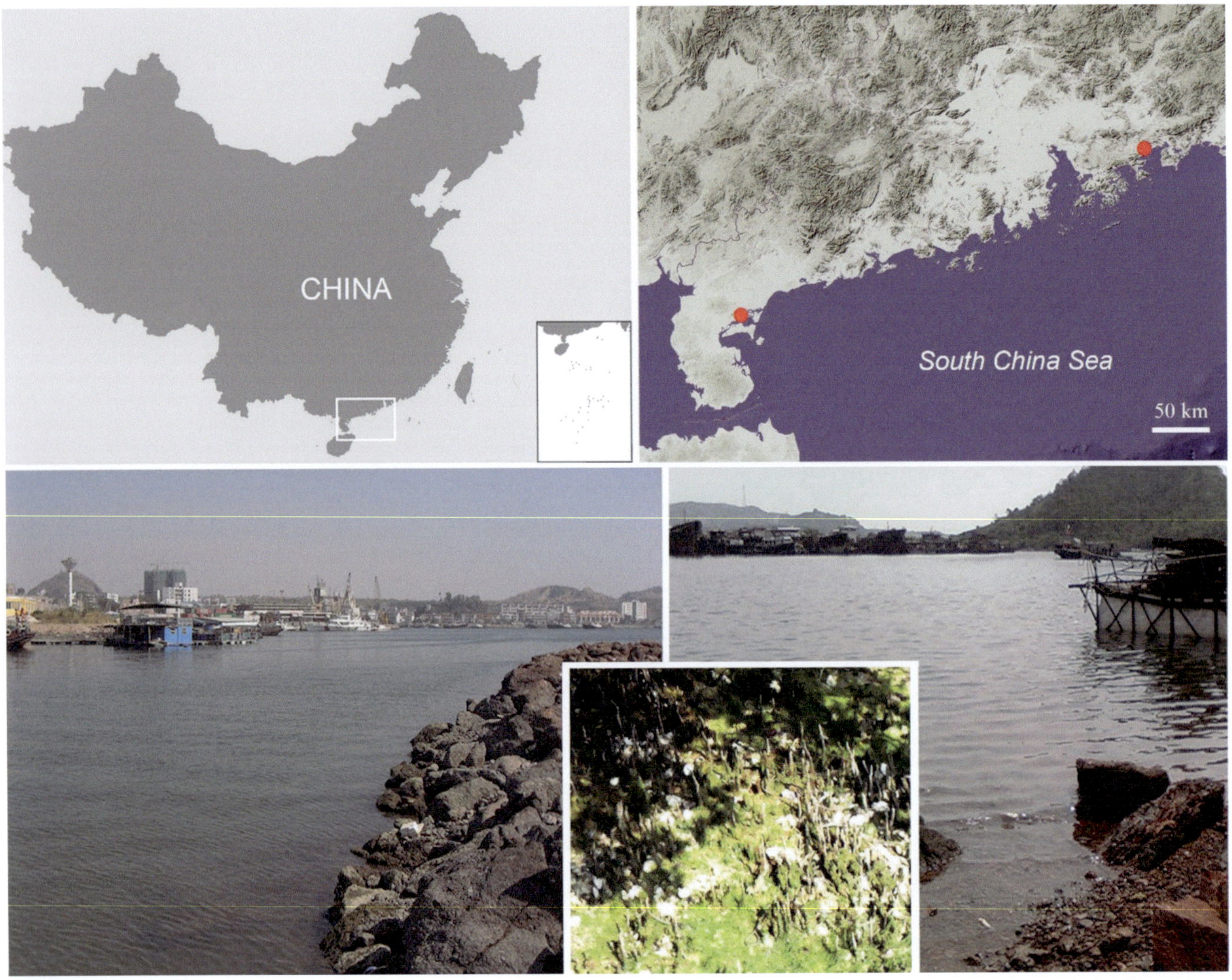


**Fig. 135A** Sample site

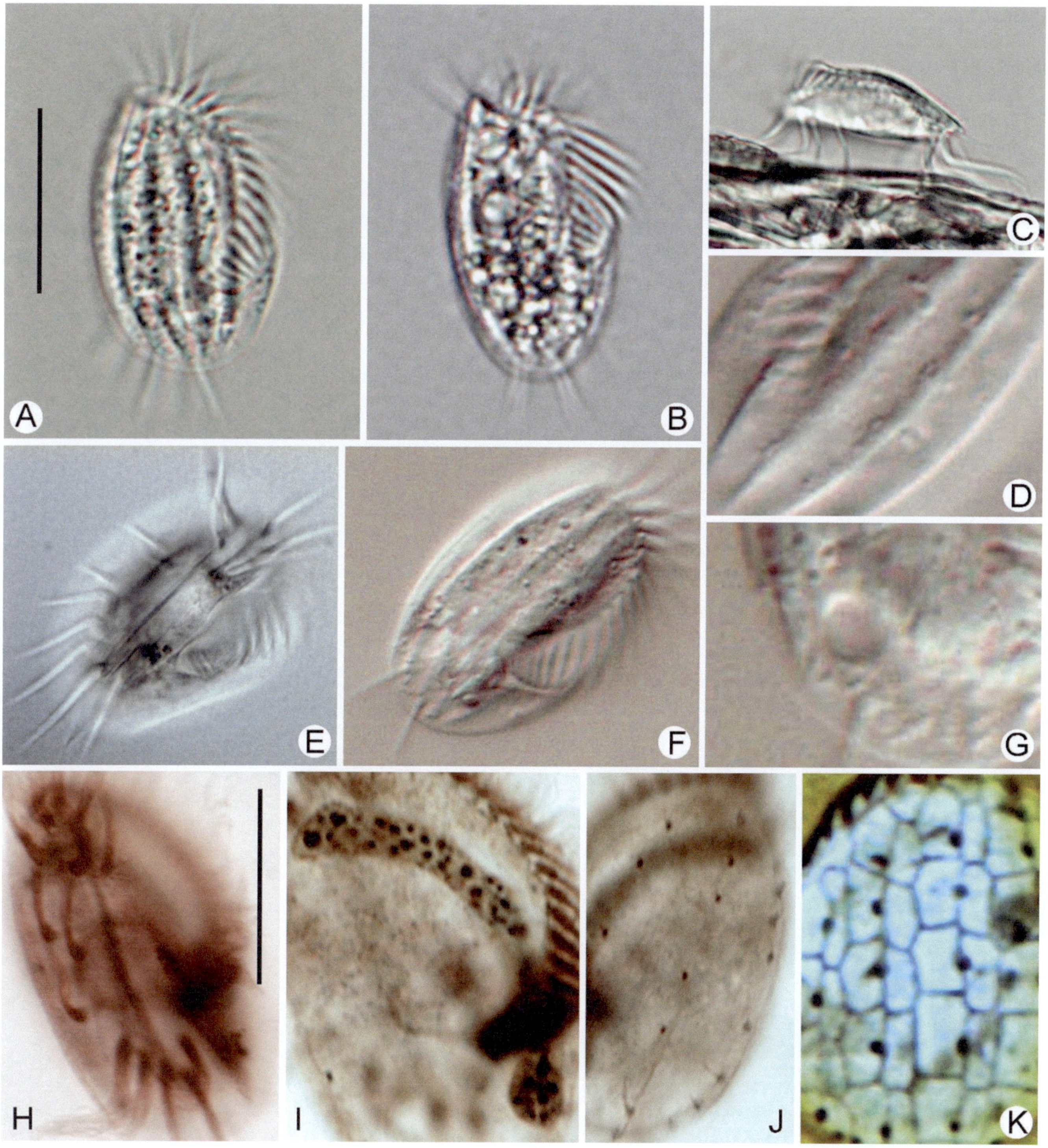


**Fig. 135B** *Euplotes rariseta*
(**a**, **b**, **e**) Ventral views of live cells. (**c**) Lateral view of live cell showing dorsal ridges. (**d**) Conspicuous dorsal ridges. (**f**) Dorsal ridges and narrow buccal field. (**g**) Contractile vacuole. (**h**) Ventral view of ciliature. (**i**) Macronucleus. (**j**) Part of dorsal kineties. (**k**) Silverline system on dorsal side. Scale bars: 20 μm

## Reference

Song WB, Packroff G (1997) Taxonomische Untersuchungen an marinen Ciliaten aus China mit Beschreibungen von zwei neuen Arten, *Strombidium globosaneum* nov. spec. und *S. platum* nov. spec. (Protozoa, Ciliophora). Arch Protistenk 147:331–360

Class Spirotrichea Bütschli, 1889
Order Euplotida Small & Lynn, 1985
Family Euplotidae Ehrenberg, 1838
*Euplotes vannus* (Müller, 1786) Minkjewicz, 1901

**Ecological feature:** mangrove
**Water temperature:** 24.0 °C; 23.1 °C
**pH:** no data available; 8.2
**Salinity:** 15.0‰; 28.6‰

**Fig. 136A** Sample site

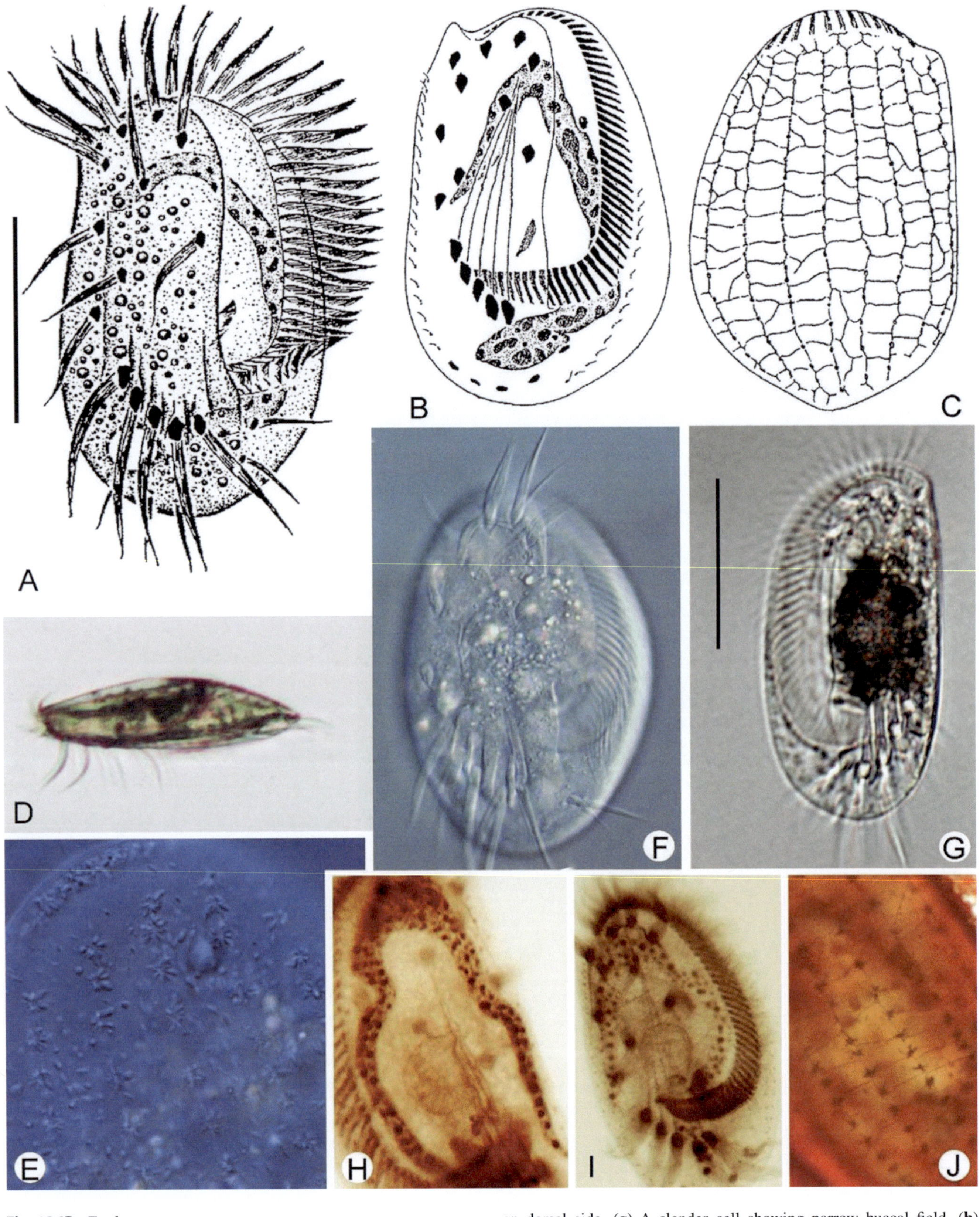


**Fig. 136B** *Euplotes vannus*
(**a**, **f**) Ventral views of live individuals. (**b**, **c**, **i**) Ventral (**b**, **i**) and dorsal (**c**) views of ciliature. (**d**) Lateral view of live cell. (**e**) Cortical granules on dorsal side. (**g**) A slender cell showing narrow buccal field. (**h**) Macronucleus. (**j**) Silverline system on dorsal side. Scale bars: 50 μm

## Reference

Song WB, Packroff G (1997) Taxonomy and morphology of marine ciliates from China with description of two new species, *Strombidium globosaneum* nov. spec. and *S. platum* nov. spec. (Protozoa, Ciliophora). Arch Protistenk 147:331–360

Class Spirotrichea Bütschli, 1889
Order Euplotida Small & Lynn, 1985
Family Gastrocirrhidae Fauré-Fremiet, 1961
*Gastrocirrhus monilifer* (Ozaki & Yagiu, 1942) Curds & Wu, 1983

**Ecological feature:** coastal water
**Water temperature:** 24.0 °C
**pH**: no data available
**Salinity:** 34.0‰

**Fig. 137A** Sample site

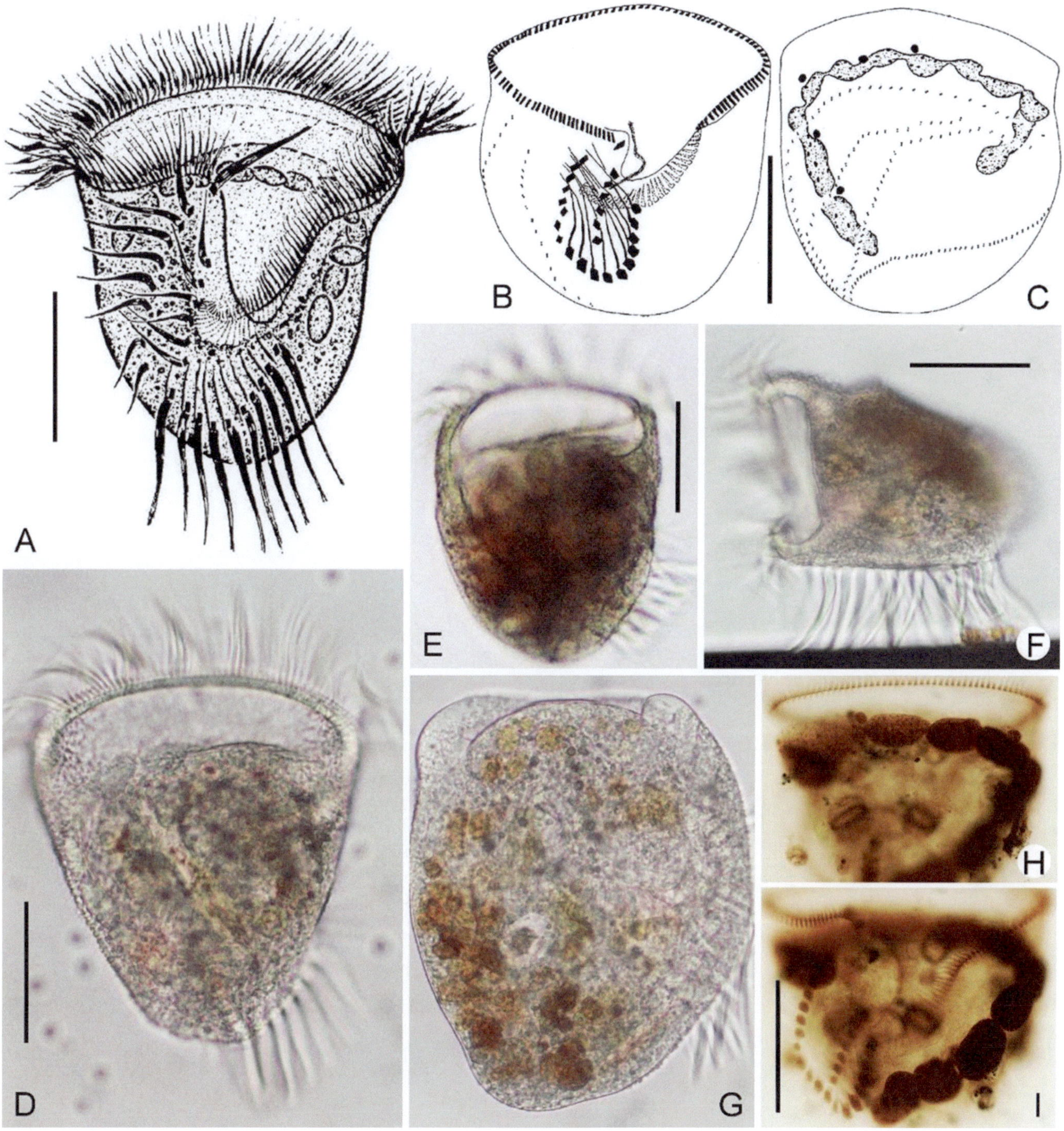


**Fig. 137B** *Gastrocirrhus monilifer*
(**a**, **d**, **e**) Ventral views of live individuals. (**b**, **c**, **i**) Ventral (**b**, **i**) and dorsal (**c**) views of ciliature. (**f**) Lateral view of live cell. (**g**) A cell full of algae. (**h**) Macronuclear nodules and micronuclei. Scale bars: 50 μm

## Reference

Hu X, Song WB (2003) Redescription of two known species, *Gastrocirrhus monilifer* (Ozaki et Yagiu, 1942) and *Gastrocirrhus stentoreus* Bullington, 1940, with reconsideration of the genera *Gastrocirrhus* and *Euplotidium*. Acta Protozool 42:345–355

Class Spirotrichea Bütschli, 1889
Order Euplotida Small & Lynn, 1985
Family Uronychiidae Jankowski, 1975
*Apodiophrys ovalis* Jiang et al., 2010

**Ecological feature:** intertidal zone
**Water temperature:** 22.0 °C
**pH:** no data available
**Salinity:** 28.0‰

**Fig. 138A** Sample site

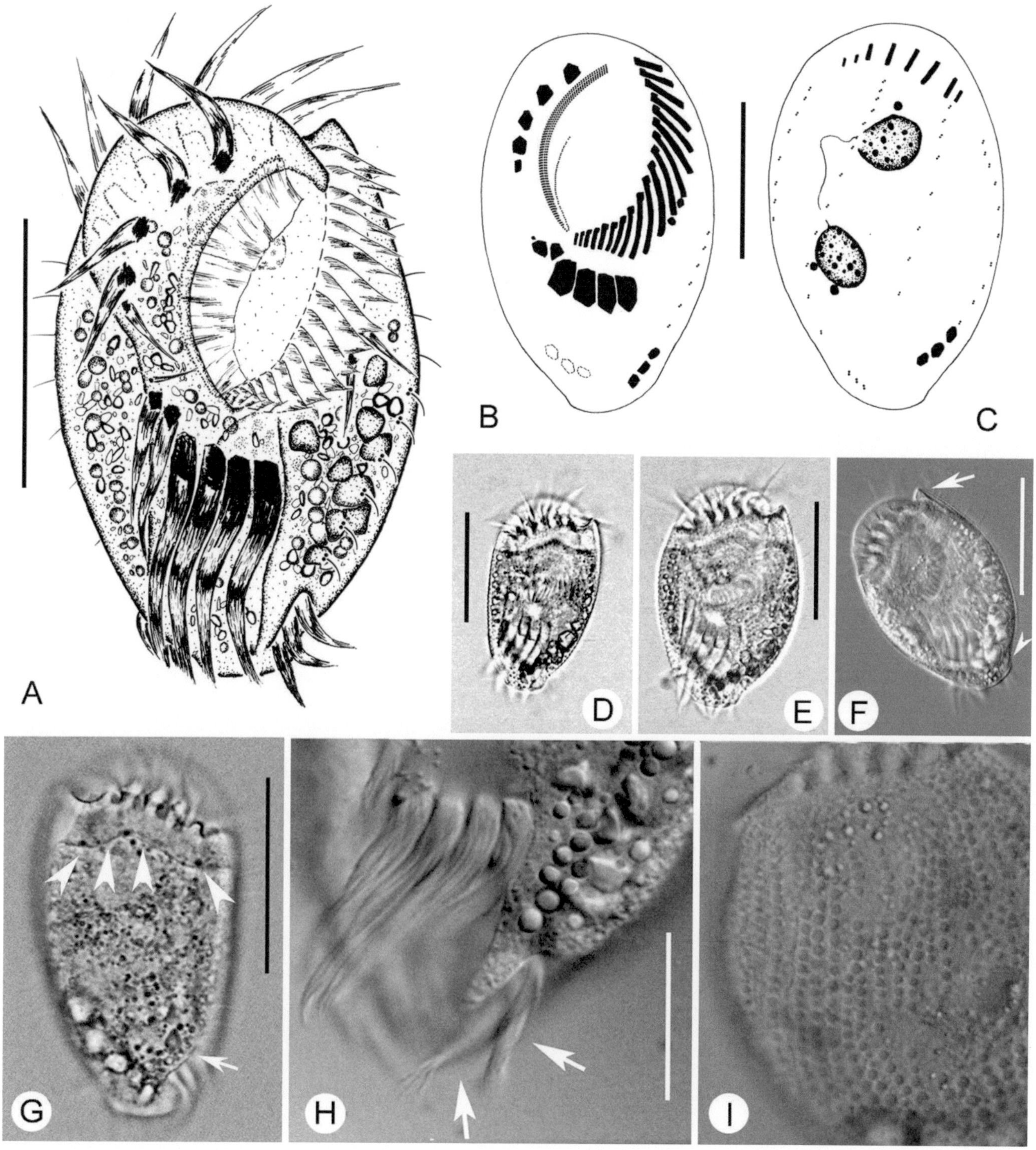


**Fig. 138B** *Apodiophrys ovalis*
(**a**, **d–f**) Ventral views of live cells; arrowhead (**f**) shows the concavity at the left margin, whereas arrow (**f**) marks the projection. (**b**, **c**) Ventral (**b**) and dorsal (**c**) views of ciliature. (**g**) Dorsal view to show the transverse dorsal groove (arrowheads) and the caudal cirri (arrow). (**h**) Ventral view of posterior region of cell; arrows indicate posterior marginal cirri. (**i**) Dorsal view, to show the cortical granules. Scale bars: 30 μm (**a–g**) and 20 μm (**h**)

## Reference

Jiang J, Song WB (2010) Two new *Diophrys*-like genera and their type species, *Apodiophrys ovalis* n. g., n. sp. and *Heterodiophrys zhui* n. g., n. sp. (Ciliophora: Euplotida), with notes on their molecular phylogeny. J Eukaryot Microbio 57:354–361

Class Spirotrichea Bütschli, 1889
Order Euplotida Small & Lynn, 1985
Family Uronychiidae Jankowski, 1975
*Diophrys oligothrix* Borror, 1965

**Ecological feature:** shrimp-farming pond; mangrove
**Water temperature:** 25.0 °C; 24.0 °C
**pH:** no data available; 7.0
**Salinity:** 29.0‰; 17.0‰

**Fig. 139A** Sample site

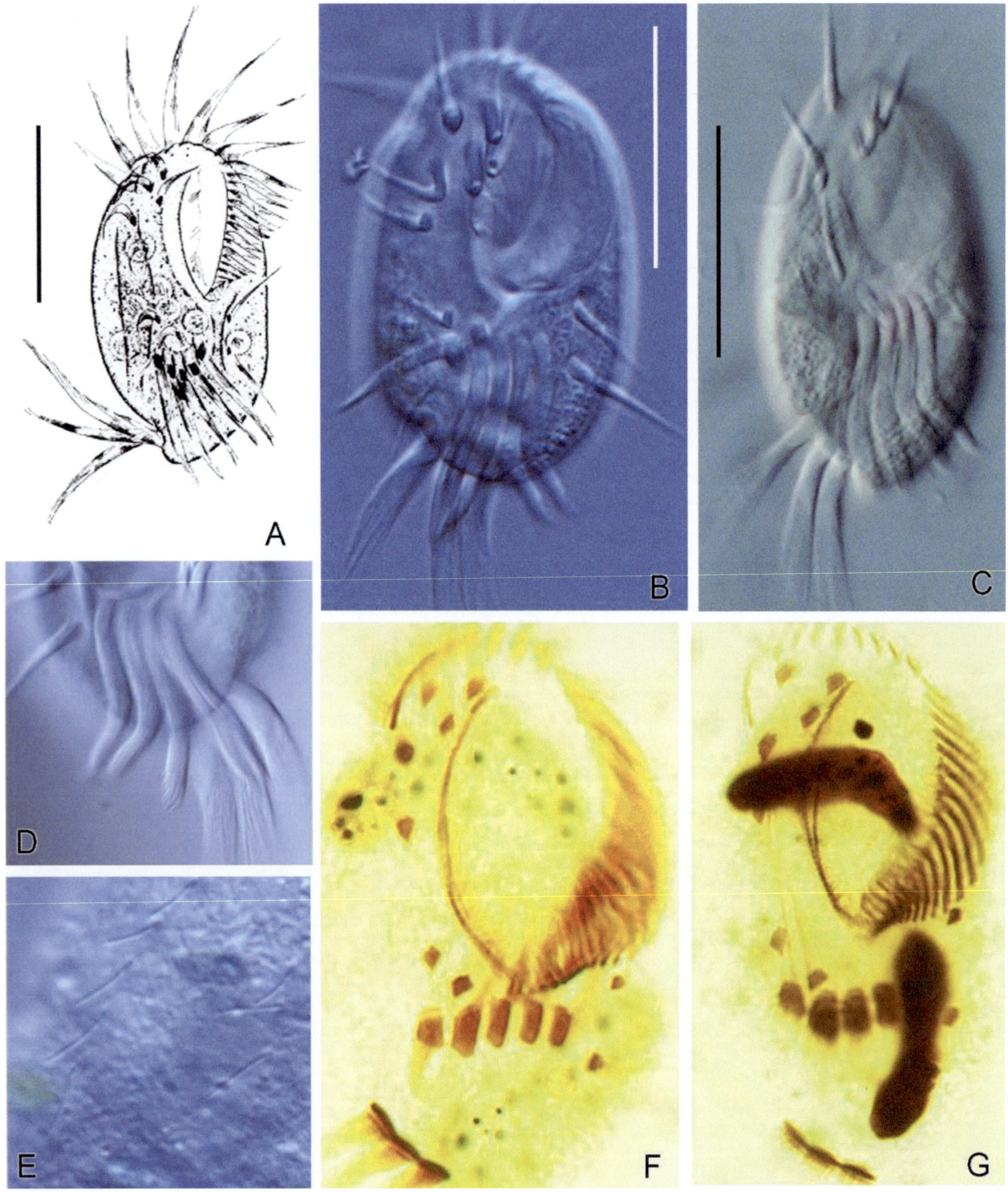


**Fig. 139B** *Diophrys oligothrix*
(**a–c**) Ventral views of live cells. (**d**) Ventral view of posterior portion of cell, to show transverse, left marginal and caudal cirri. (**e**) Dorsal cilia. (**f**, **g**) Ventral views of ciliature. Scale bars: 30 μm

## Reference

Song WB, Wilbert N (2002) Faunistic studies on marine ciliates from the Antarctic benthic area, including descriptions of one epizoic form, 6 new species and, 2 new genera (Protozoa: Ciliophora). Acta Protozool 41:23–61

Class Spirotrichea Bütschli, 1889
Order Euplotida Small & Lynn, 1985
Family Uronychiidae Jankowski, 1975
*Diophrys parappendiculata* Shen et al., 2011

**Ecological feature:** offshore water
**Water temperature:** 23.0 °C
**pH:** 8.0
**Salinity:** 30.0‰

**Fig. 140A** Sample site

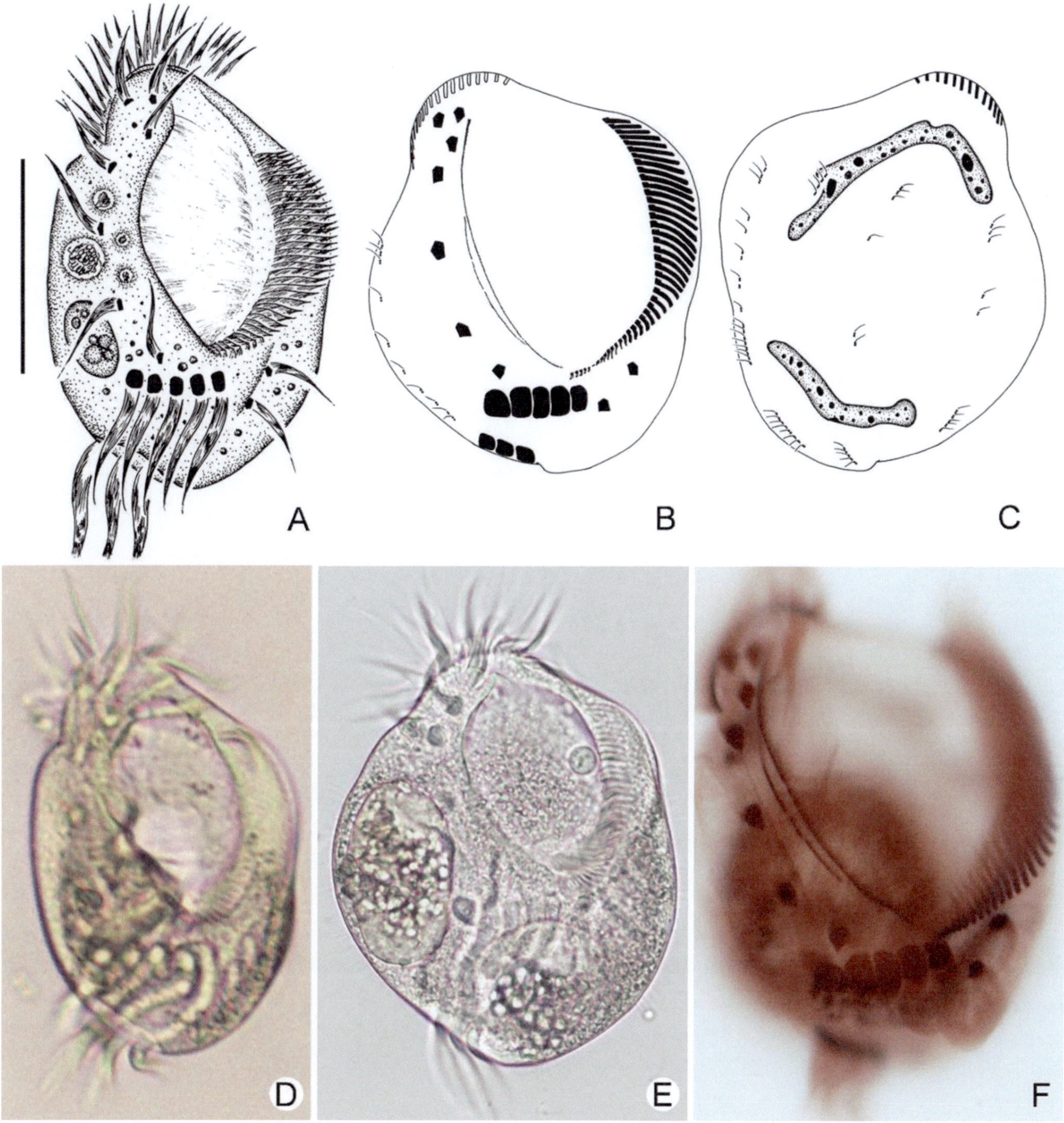


**Fig. 140B** *Diophrys parappendiculata*
(**a**, **d**) Ventral views of live individuals. (**b**, **c**) Ventral (**b**) and (**c**) dorsal views of ciliature. (**e**) A well-fed individual. (**f**) Ventral view of ciliature. Scale bar: 80 μm

## Reference

Shen Z, Yi Z, Warren A (2011) The morphology, ontogeny, and small subunit rRNA gene sequence analysis of *Diophrys parappendiculata* n. sp. (Protozoa, Ciliophora, Euplotida), a new marine ciliate from coastal waters of southern China. J Eukaryot Microbiol 58:242–248

Class Spirotrichea Bütschli, 1889
Order Euplotida Small & Lynn, 1985
Family Uronychiidae Jankowski, 1975
*Diophrys scutum* (Dujardin, 1841) Kahl, 1932

**Ecological feature:** mangrove
**Water temperature:** 25.0 °C
**pH:** no data available
**Salinity:** 27.0‰

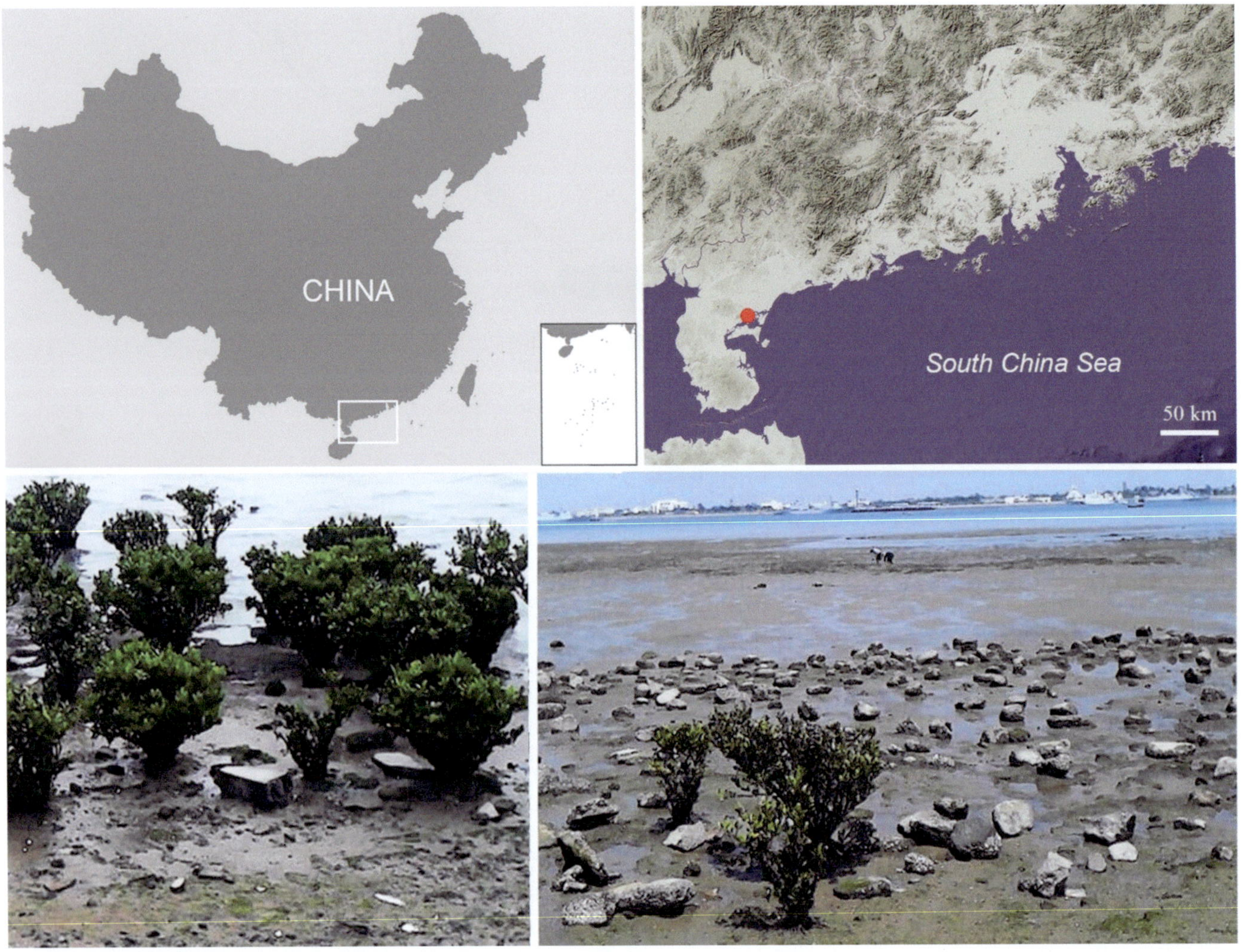


**Fig. 141A** Sample site

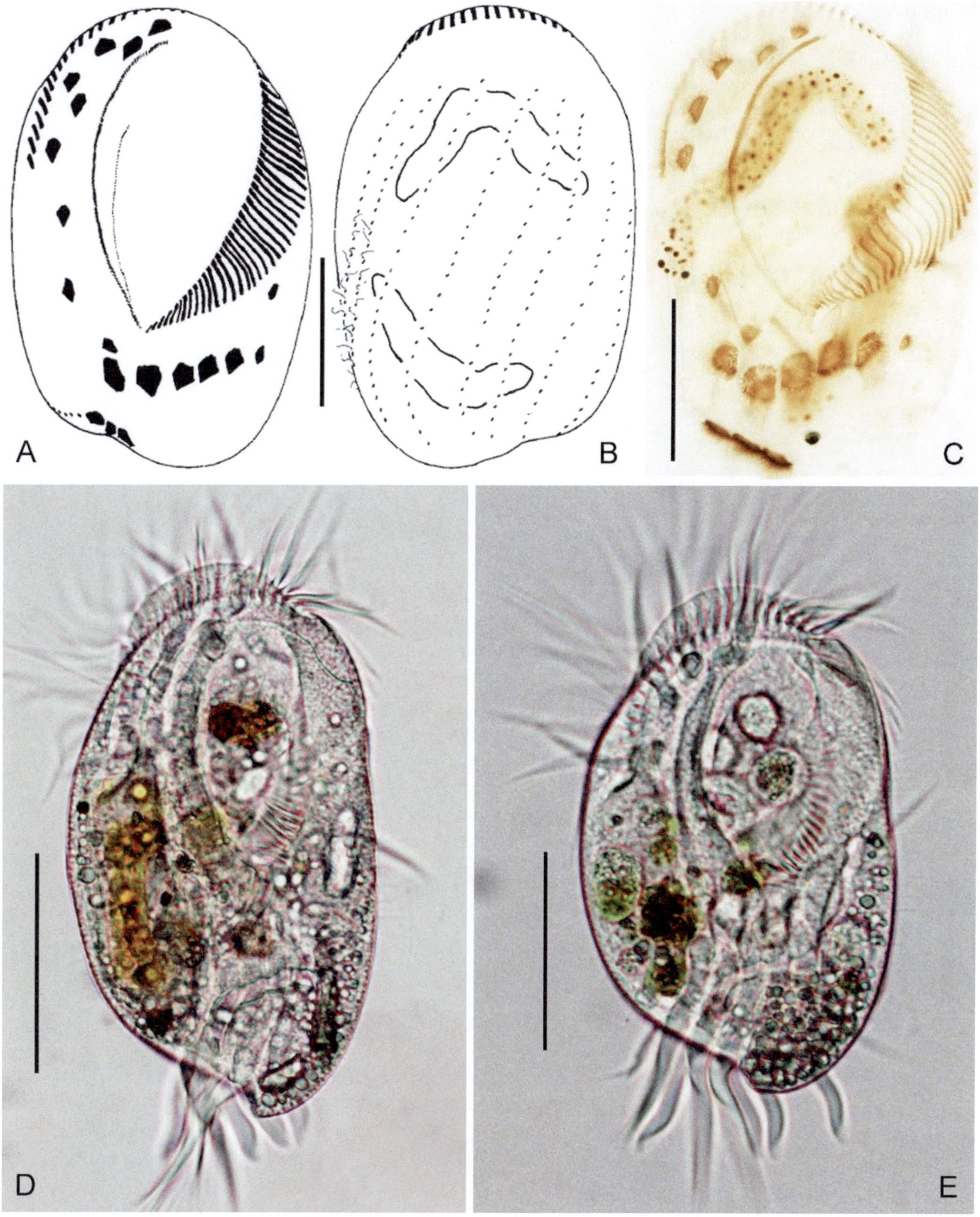


**Fig. 141B** *Diophrys scutum*
(**a–c**) Ventral (**a**, **c**) and dorsal (**b**) views of ciliature. (**d**, **e**) Ventral views of live individuals. Scale bars: 50 μm

## Reference

Song WB, Wilbert N (2002) Faunistic studies on marine ciliates from the Antarctic benthic area, including descriptions of one epizoic form, 6 new species and 2 new genera (Protozoa: Ciliophora). Acta Protozool 41:23–61

Class Spirotrichea Bütschli, 1889
Order Euplotida Small & Lynn, 1985
Family Uronychiidae Jankowski, 1975
*Heterodiophrys zhui* Jiang et al., 2010

**Ecological feature:** intertidal zone
**Water temperature:** 22.0 °C
**pH:** no data available
**Salinity:** 28.0‰

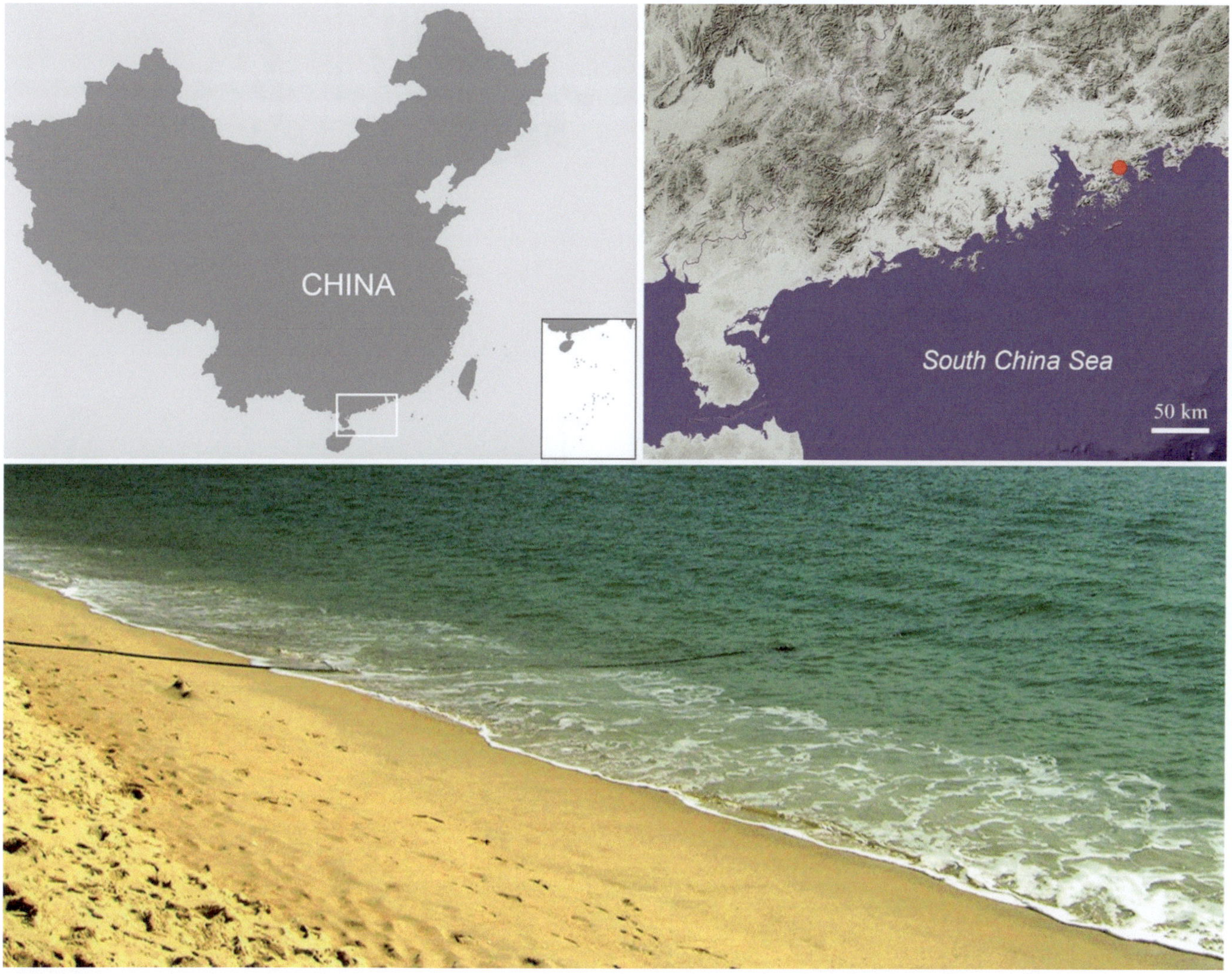


**Fig. 142A** Sample site

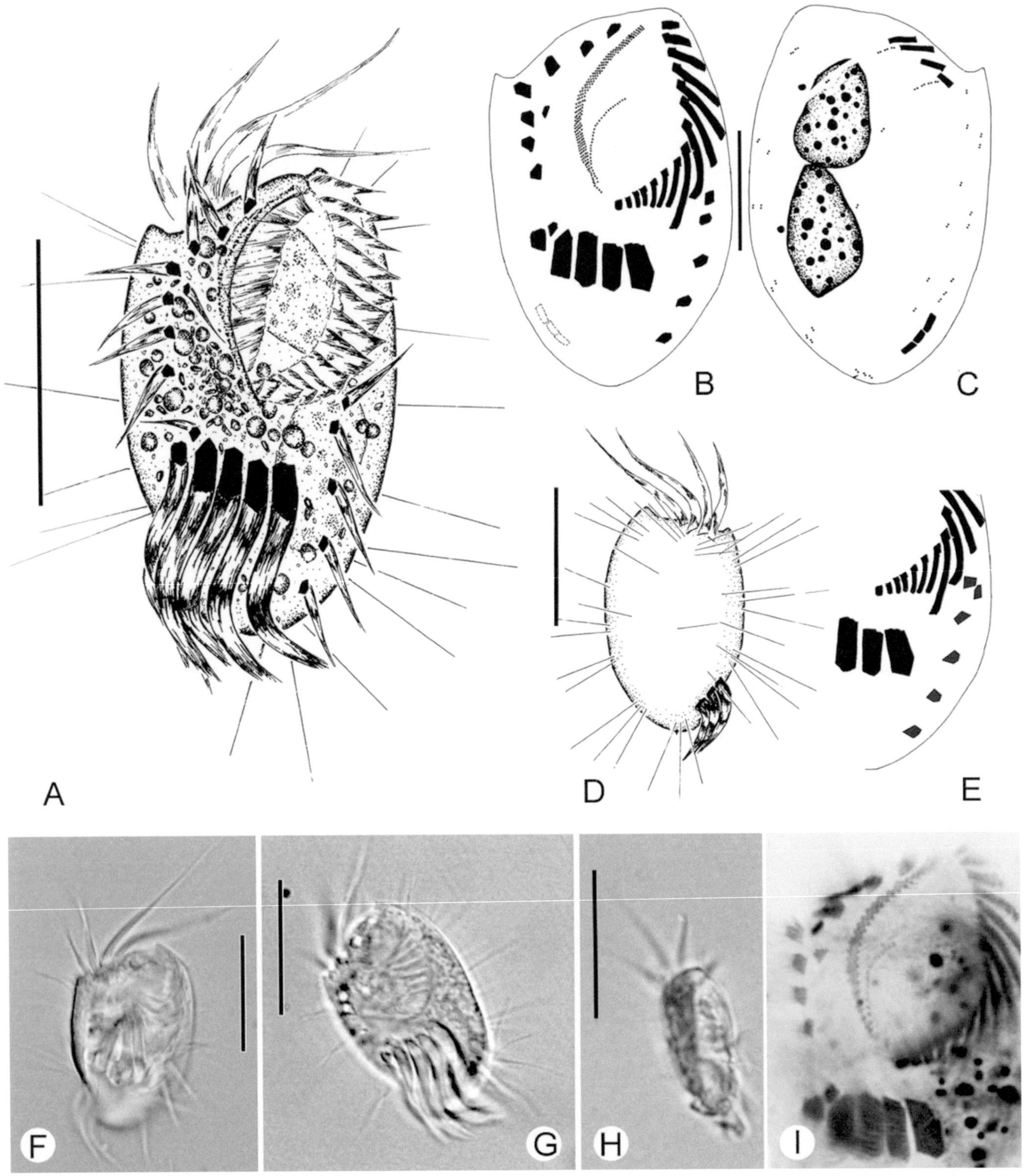


**Fig. 142B** *Heterodiophrys zhui*
(**a**, **f**, **g**) Ventral views of live cells. (**b**, **c**, **i**) Ventral (**b**, **i**) and dorsal (**c**) views of ciliature and macronuclear nodules. (**d**) Dorsal view to show conspicuous dorsal cilia. (**e**) Left portion of a specimen with six marginal cirri. (**h**) Lateral view of live cell. Scale bars: 20 μm (**a**, **d**, **f**–**h**) and 10 μm (**b**, **c**)

## Reference

Jiang J, Song WB (2010) Two new *Diophrys*-like genera and their type species, *Apodiophrys ovalis* n. g., n. sp. and *Heterodiophrys zhui* n. g., n. sp. (Ciliophora: Euplotida), with notes on their molecular phylogeny. J Eukaryot Microbiol 57:354–361

Class Spirotrichea Bütschli, 1889
Order Euplotida Small & Lynn, 1985
Family Uronychiidae Jankowski, 1975
*Uronychia binucleata* Young, 1922

**Ecological feature:** sandy sediment (Zhanjiang); mangrove (Zhuhai)
**Water temperature:** 30.0 °C; 27.0 °C
**pH:** no data available; 6.5
**Salinity:** 16.0‰; 10.0‰

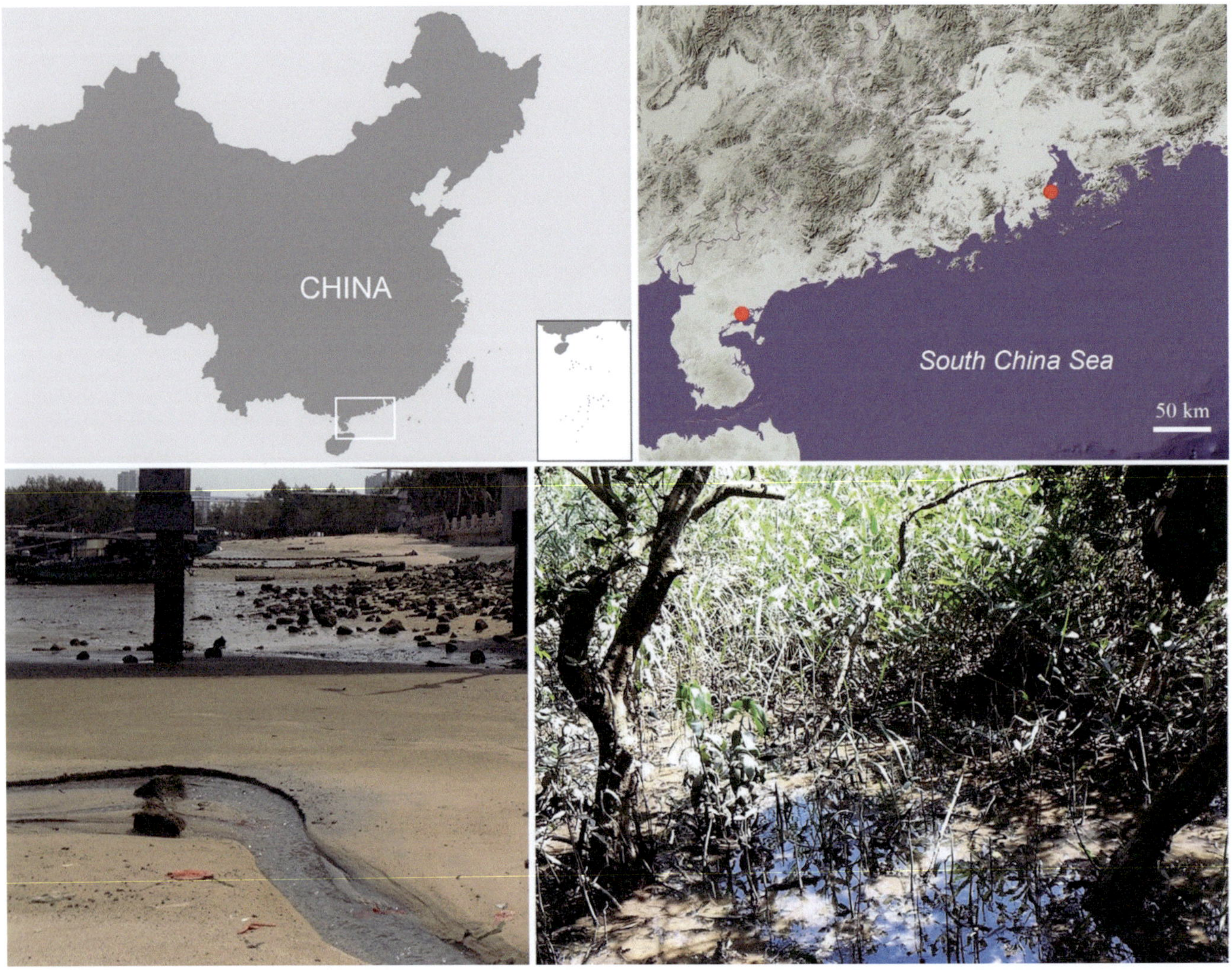


**Fig. 143A** Sample site

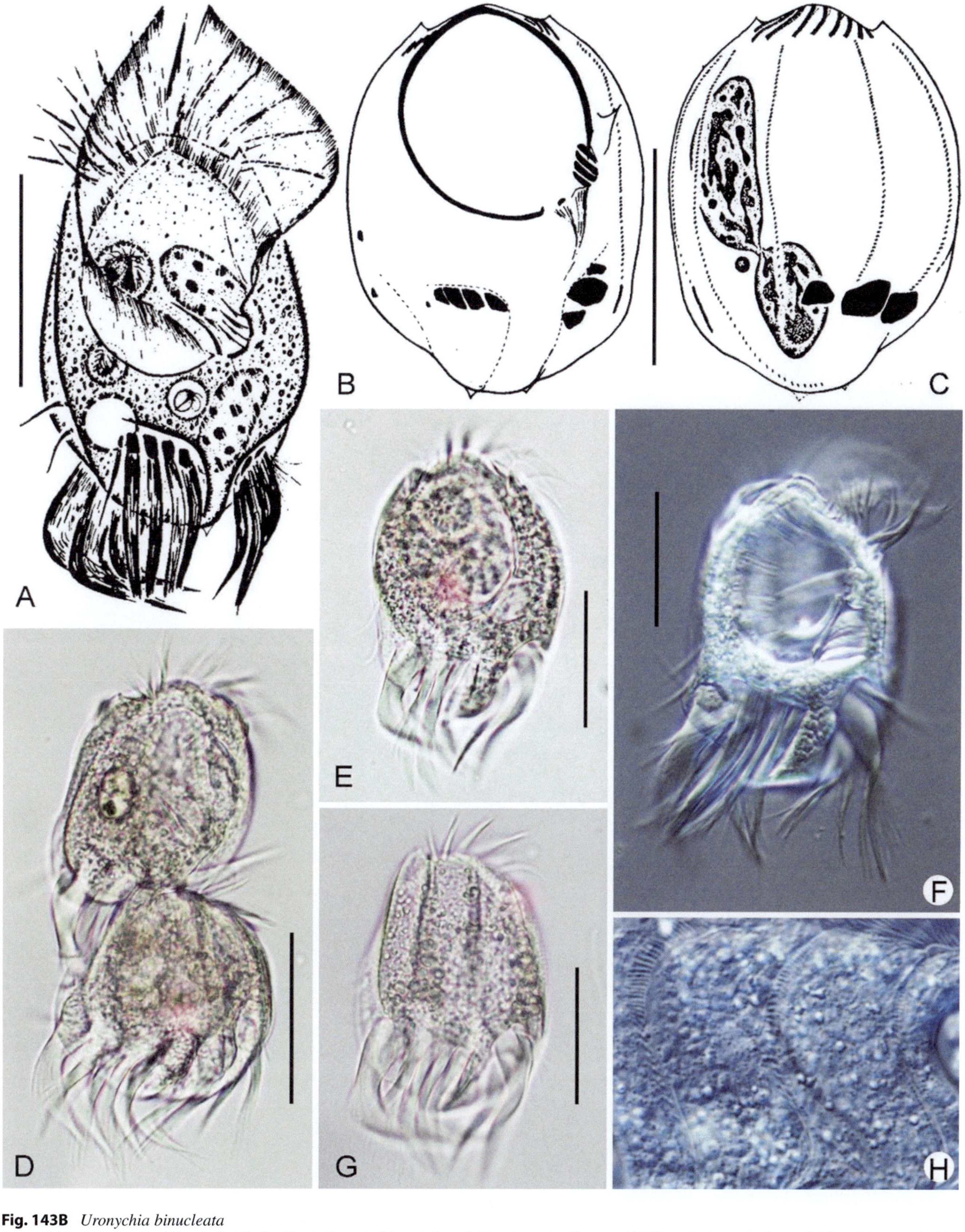


**Fig. 143B** *Uronychia binucleata*
(**a**, **e–g**) Ventral views of live individuals. (**b**, **c**) Ventral (**b**) and dorsal (**c**) views of ciliature. (**d**) Ventral view of a later divider. (**h**) Dorsal cilia. Scale bars: 50 μm

## Reference

Song WB, Wilbert N (1997) Morphological investigations on some free living ciliates (Protozoa, Ciliophora) from China sea with description of a new hypotrichous genus, *Hemigastrostyla* nov. gen. Arch Protistenk 148:413–444

Class Spirotrichea Bütschli, 1889
Order Euplotida Small & Lynn, 1985
Family Uronychiidae Jankowski, 1975
*Uronychia multicirrus* Song, 1997

**Ecological feature:** offshore water
**Water temperature:** 19.0 °C
**pH:** 8.0
**Salinity:** 29.0‰

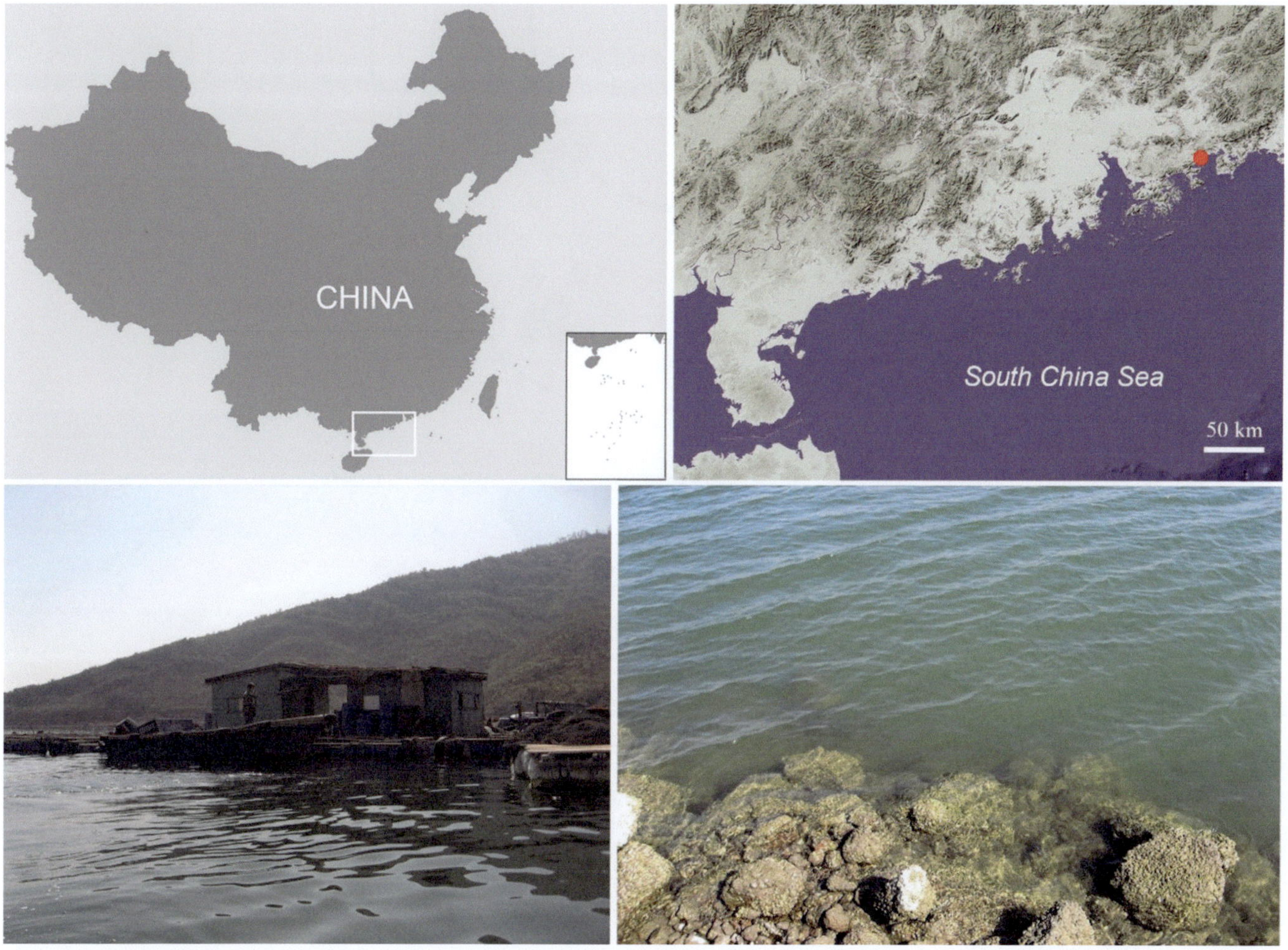


**Fig. 144A** Sample site

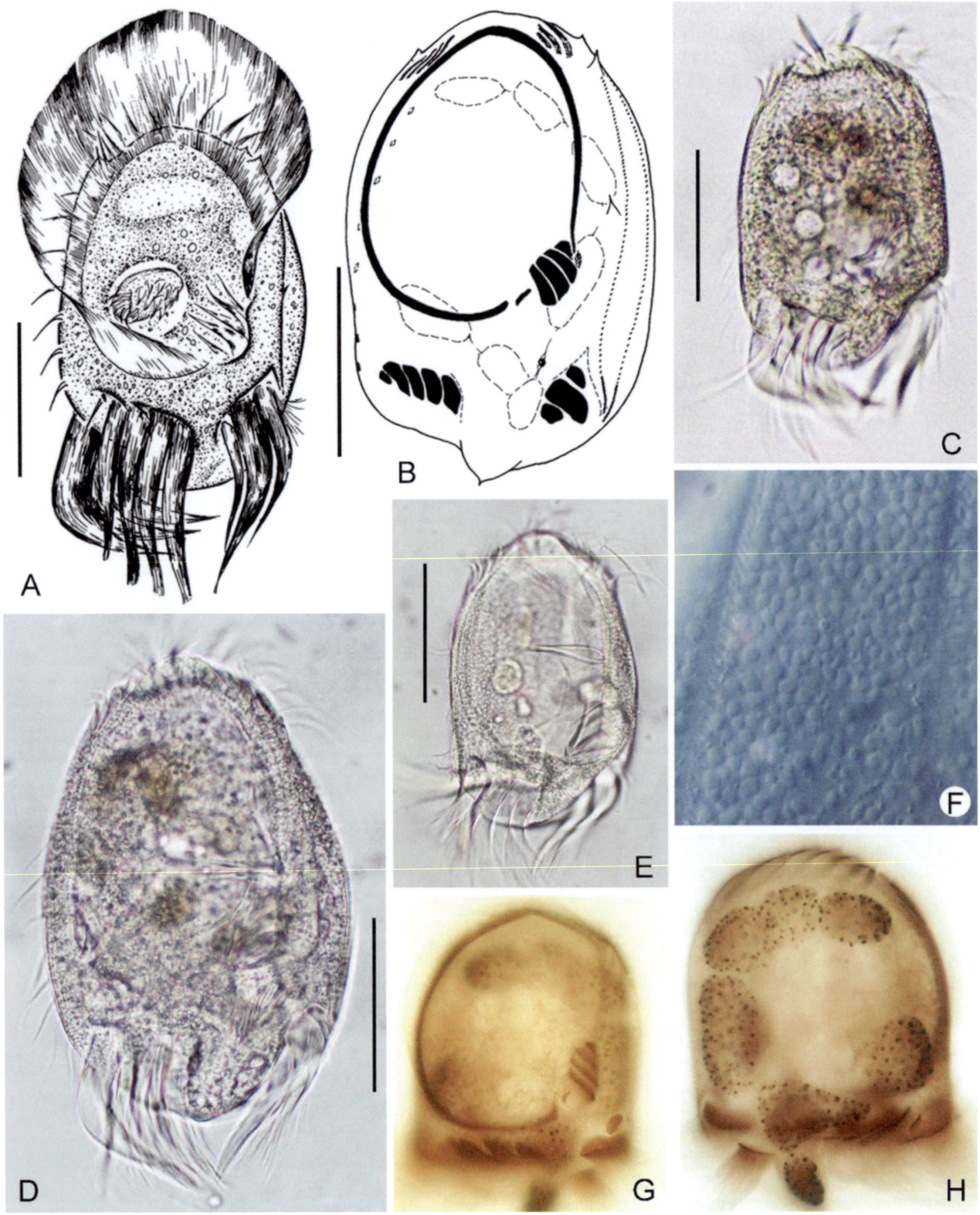


**Fig. 144B** *Uronychia multicirrus*
(**a**, **c**–**e**) Ventral views of live individuals. (**b**, **g**) Ventral views of ciliature. (**f**) Cortical granules and dorsal cilia. (**h**) Macronuclear nodules. Scale bars: 50 μm

## Reference

Song WB (1997) On the morphology and infraciliature of a new marine hypotrichous ciliate, *Uronychia multicirrus* sp. n. (Ciliophora: Hypotrichida). Acta Protozool 36:279–285

Class Spirotrichea Bütschli, 1889
Order Euplotida Small & Lynn, 1985
Family Uronychiidae Jankowski, 1975
*Uronychia setigera* Calkins, 1902

**Ecological feature:** muddy sediment (Zhanjiang); mangrove (Zhuhai)
**Water temperature:** 26.0 °C; 25.0 °C
**pH:** no data available; 6.5
**Salinity:** 22.0‰; 5.0‰

**Fig. 145A** Sample site

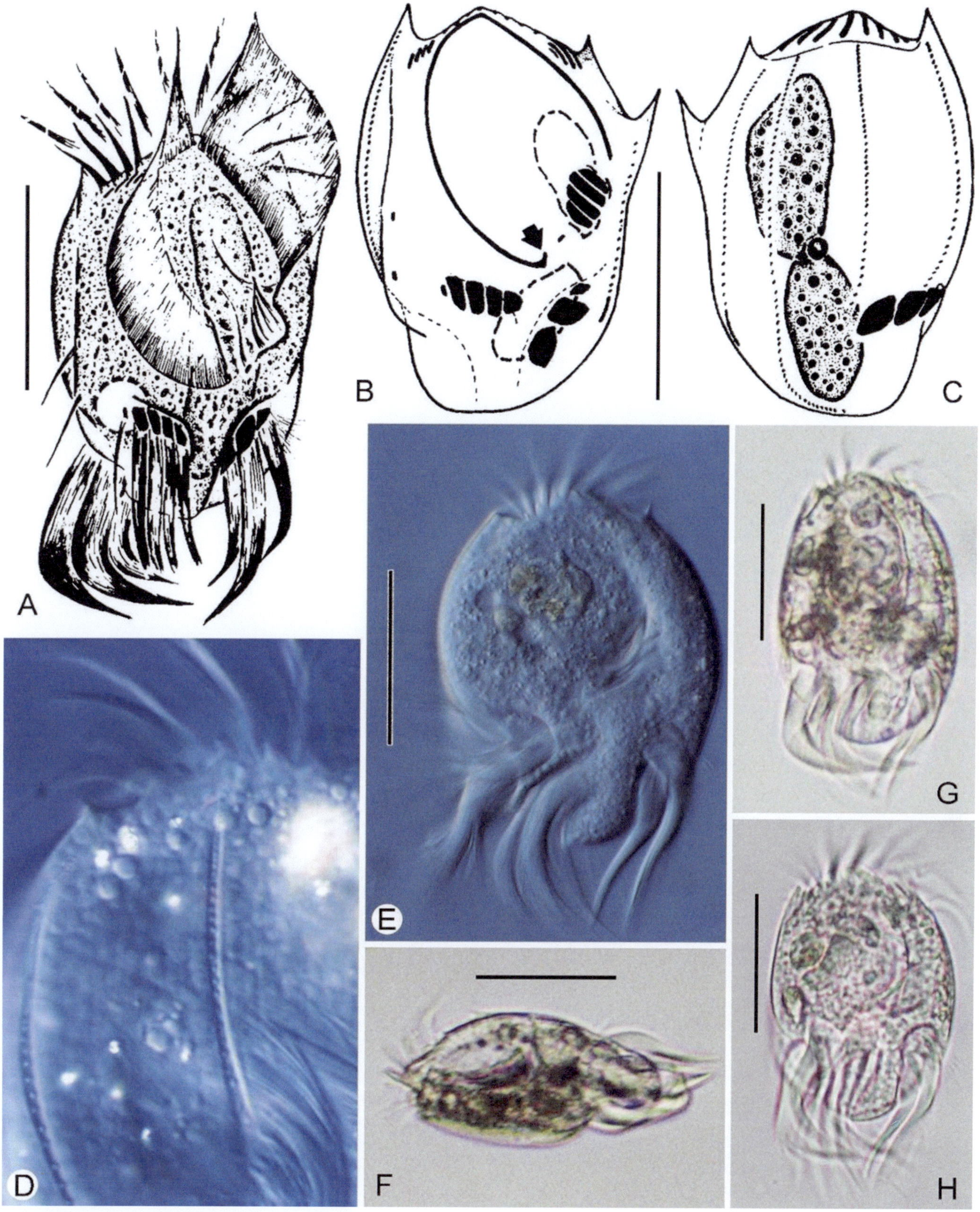


**Fig. 145B** *Uronychia setigera*
(**a**, **e**, **g**, **h**) Ventral views of live individuals. (**b**, **c**) Ventral (**b**) and dorsal (**c**) views of ciliature. (**d**) Dorsal cilia and cortical granules. (**f**) Lateral view of live cell. Scale bars: 30 μm

## Reference

Song WB, Wilbert N (1997) Morphological investigations on some free living ciliates (Protozoa, Ciliophora) from China sea with description of a new hypotrichous genus, *Hemigastrostyla* nov. gen. Arch Protistenk 148:413–444

# Index




X. Hu et al. (eds.), *Ciliate Atlas: Species Found in the South China Sea*,
https://doi.org/10.1007/978-981-13-5901-9

**Z**

MIX
Papier aus verantwortungsvollen Quellen
Paper from responsible sources
FSC® C105338

If you have any concerns about our products,
you can contact us on
**ProductSafety@springernature.com**

In case Publisher is established outside the EU,
the EU authorized representative is:
**Springer Nature Customer Service Center GmbH**
**Europaplatz 3, 69115 Heidelberg, Germany**

Printed by Libri Plureos GmbH
in Hamburg, Germany